Fundamentals of Continuum Mechanics of Soils

Yehuda Klausner

Fundamentals of Continuum Mechanics of Soils

With 210 Figures

Springer-Verlag
London Berlin Heidelberg New York
Paris Tokyo Hong Kong

Yehuda Klausner, PhD
Engineering Consultant, 19 Yarboa Lane, Beer-Sheva 84736, Israel

ISBN 3-540-19546-7 Springer-Verlag Berlin Heidelberg New York
ISBN 0-387-19546-7 Springer-Verlag New York Berlin Heidelberg

British Library Cataloguing in Publication Data

Klausner, Yehuda *1926–*
Fundamentals of continuum mechanics of soils.
1. Soils. Mechanics
I. Title
624.15136
ISBN 3-540-19546-7

Library of Congress Cataloging-in-Publication Data

Klausner, Yehuda, 1926–
Fundamentals of continuum mechanics of soils/Yehuda Klausner.
p. cm.
ISBN 0-387-19546-7 (New York). – ISBN 3-540-19546-7 (Berlin)
1. Soil mechanics. I. Title.
TA710.K549 1991
624.1′5136–dc20 90-20705
CIP

Printed in Great Britain

Typeset by KEYTEC, Bridport, Dorset
Printed by The Alden Press, Osney Mead, Oxford
69/3830-543210 Printed on acid-free paper

Preface

The world is a poor affair if it does not contain matter for investigation to everyone in every age. Nature does not reveal all her secrets at once. We imagine we are initiated in her mysteries: we are as yet, but hanging around her outer court.

Seneca

Between the Second International Conference on Soil Mechanics and Foundation Engineering (INCOSOMEFE) in 1948 and the 12th International Conference in 1989 the conferences were held at regular intervals of four years.* Over 2500 papers, not including discussions, were presented in the Proceedings. Around 10 000 additional articles were published during these 40 years, in the proceedings of regional and specialty conferences and symposia, as well as in scientific and professional journals. This is an enormous amount of information, and of this 75% or more was added after 1960. The question one must ask himself is: what has come of all these efforts?

In 1960 the ASCE Research Conference on Shear Strength of Cohesive Soils was held in Boulder, Colorado; the special significance of this conference was that it presented an up to date summary of the developments in soil mechanics. We can now say with certainty that, except for minor refinements, the body of knowledge in soil mechanics has not changed much since then, and the topics considered today had already been thoroughly studied by that time: consolidation; pore-water, pore-air, and negative pore pressures; effective pressures; critical void ratio; isotropic and anisotropic properties of soils; electrokinetic properties of soils; drained, consolidated-undrained and undrained tests; disturbed and undisturbed samples; phenomena of saturated, unsaturated and overconsolidated

* The interval between the second and third international conferences was five years. The first conference was held in 1936 in Cambridge, Massachusetts, with the establishment of the International Society of Soil Mechanics and Foundation Engineering.

soils, expansive and collapsible soils, cohesive and cohesionless soils; the critical state theory and the Coulomb–Mohr failure theory. All these topics were understood and discussed at the 1960 conference in great detail.

Nevertheless, several problems remained unresolved and continue to engage our interest today. Uniaxial versus triaxial consolidation, constitutive modeling, coaxiality of stresses and strains, conditions of anisotropy, elastic, viscous and plastic behavior, hysteretic behavior, the role of continuum mechanics, discontinuities, general failure conditions, workhardening and dilatancy – all are topics brought repeatedly to the surface but not resolved, apparently owing to shortcomings in the analytical and conceptual tools applied to the studies.

About the time of the Boulder conference, a small group of graduate students at Princeton University led by Professor W. E. Schmid, aware of these shortcomings, came to the conclusion that although the knowledge of the above topics was substantial, their formulation was deficient and resulted in inconsistencies and misinterpretations, thus obstructing further scientific advances. They began looking at things differently, and revised the formulation of the problems. Unfortunately they did not present their findings efficiently and appealingly and their work went unnoticed.

When Professor A. W. Skempton, in his presidential address at the 5th International Conference in 1961 in Paris, warned of two dangers: "the danger of what might be called *handbook* engineering", and "the second danger which can be foreseen and which we must strive to avoid can be expressed in the simple word *complacency*", he actually foresaw what was about to happen. The intellectual challenge posed by the discipline was replaced by computer techniques, numerical methods and data accumulating on job sites. The majority of publications mentioned earlier attest to that aspect of the discipline.

I had the good fortune to have some of the best teachers in my engineering studies. Among them the late M. Reiner of the Technion IIT, who introduced me to rheology, a discipline of which he was a cofounder; the late I. Haber-Schaim of the Technion IIT, an original thinker with a vast practical experience, my tutor and superior in my early engineering practice; the late G. P. Tschebotarioff of Princeton University, a man of integrity and a scientific thinker and leader; Professor J. G. Zeitlen, Professor Emeritus of the Technion IIT, who in 1954 brought the experience and tradition of the Corps of Engineers and the message of Classical Soil Mechanics to the Technion, and ushered me and a number of young faculty members into this field of knowledge, and Professor W. E. Schmid, my graduate study advisor at Princeton University, who led me into the realm of theoretical studies. This work is a product of their teaching effort.

An early version of the material presented here was given as a one-semester graduate course under the same name at Wayne State University, in 1961 and 1962 (Klausner 1962). Much scientific progress has been made in mechanics of materials and related disciplines since then and an abundance of experimental data on soils has appeared in the scientific and technical literature.

This work intends to review the present state of knowledge in mechanics of soils and place it abreast with the related disciplines. More specifically, it aims to close the gap between soil mechanics, a discipline based greatly on empirical impressions, and continuum mechanics and its many ramifications, and to present a long-needed general scheme, based on the laws of physics, for that important yet intricate material, soils.

The book is intended for soil scientists and scholars of soil engineering, engineering mechanics and material sciences, as well as for graduate students familar with the fundamentals of soil mechanics. It is not conceived as an undergraduate textbook, and is by no means intended to present practical applications of soil engineering.

The organization of the subjects is different from that of the existing textbooks. The phenomena constituting the behavior of soils are presented here not as separate topics like permeability, capillarity, consolidation, shear strength, etc. incidentally related, but as phenomena logically inferring from one another and depending upon one another, and governed by the laws of physics.

The book has two parts, although it is not formally so subdivided: the first part comprises Chapters 1–7 and the second Chapters 9–13, with Chapter 8 connecting the two. Each chapter is based on the previous one, however the experienced reader basically acquainted with the subject will have no problem in studying any individual chapter.

Chapters 1–7 contain the basic concepts of mechanics and serve as a foundation for the second part, concerned with the mechanical behavior of soils. Chapters 1–5 expound the concept of strains and stresses and the balance equations, Chapter 6 deals with the application of the balance equations to multiphase mixtures, a rather new branch of mechanics, and Chapter 7 discusses the constitutive equations, a topic generally not elaborated in textbooks.

Chapter 8, The Soil, is a self-contained chapter presenting the sub-structural approach to soils, and is not an absolutely required part of the book. It discusses the constituents of the soil, their properties and their interaction, and is instrumental in understanding the mechanical behavior of soils as a whole and the boundary conditions at the base of our treatment.

Chapters 9–13 deal with the wide range of phenomena concerning the mechanical behavior of soils, including flow, volumetric behavior, shear stress–strain behavior and failure. In Chapter 9 the balance equations of multiphase mixtures, discussed in Chapter 6 in a general manner, are applied to soils. Chapters 10–13 present the mechanical behavior of soils as derived from the dual constitutive equations, based also on the balance equations of multiphase mixtures. Chapters 10 and 11 are concerned with volumetric phenomena of soils, Chapter 10 looking at the motion of the phases within the voids, that is, the problem of flow in soils, and Chapter 11 at the motion of the solids phase and the free energies involved in these motions; this chapter relates to the topic of consolidation in its widest aspect. Chapter 12 discusses the effect of deviatoric stresses and strains and the respective free energies involved. Finally, Chapter 13 formulates a criterion for failure, based on the free energy of the soil.

Three appendices are found at the end of the book. Appendix A outlines the main rules of tensor calculus and can serve as a reference for the derivations in the main text. Appendix B discusses cylindrical coordinates and the transformation of several equations of mechanics into these coordinates. Appendix C is a discourse on rheological modeling. It is hoped that the material included in the appendices, which is beyond that required to master the subject, will serve further studies and help develop their application to soils and other materials.

Should the book raise interest, or controversy, by its stand, I would consider this my reward, as I believe it could stimulate scientific progress. I would like to think that the conceptual errors in the text are minimal. Much effort was invested, within my ability, in order to minimize mistakes, faulty mathematical expositions, and typing and printing errors. I would be grateful to readers who bring any remaining errors to my attention and welcome their constructive criticism.

As a practicing engineer I was fortunate to observe the behavior of matter very closely. On the other hand, I missed the day by day contact and exchange of views with fellow scientists, and could not benefit from their criticism and review of my work.

I have enjoyed a massive and unrepayable support, intellectually and practically, from my close family. My wife Yocheved did all the editing. Her composed judgment, sensible counsel and calm attitude more than once balanced and simmered down my own emotional and bold-tempered statements. My sons, David, Aviel, Meir and Moshe assisted me with their mathematical expertise and also helped solve many technical problems. To my family I would like to extend my love and deep thanks. May God bless them all.

I extend my thanks to the authors and publishers credited in the illustrations and tables of this book for releasing the publication rights of material in which their intelectual effort was invested.

I am indebted to Springer-Verlag and specifically to Springer-Verlag London Limited, who took the risk of publishing this book and who so patiently waited for the manuscript and encouraged me along the way. I hope their patience pays off.

Thanks to Him who created man in the image of His likeness and bestowed upon him His wisdom to understand, to learn and to teach.

תושלב״ע

Beer-Sheva
March, 1990

Y. Klausner

Contents

13 Failure

Symbols

Notation	*Name*	*Section of definition*
a_v	Coefficient of compressibility	11.4
a^i	Acceleration	3.3, B.7
a_i	Body force acceleration	10.3
a_n	Constant coefficients	C.1
a_r, a_θ, a_z	Acceleration components	B.7
A	Constant Velocity Reciprocal of time factor	C.1, C.13 3.19 C.14
$A(t)$	Retardation spectrum	C.11
A_i	ith coefficient of creep function	C.12
A	Dipole dependent function	8.21
A_i	Vector, tensor of order one	A.4
A_{rs}	Tensor of order two	A.4
A_{KM}	Metric tensor of the Euclidean space	2.9
b_n	Constant coefficients	C.1
B	Coefficient of transverse stretching Twist per unit length	11.12 12.11

Notation	Name	Section of definition
	Constant	C.1
	Relaxation spectrum	C.11
	Reciprocal of time factor	C.14
$B(t)$	Relaxation spectrum	C.11
B_i	ith coefficient of relaxation function	C.12
B_i	Body force	4.8
B_i^*	Body force acting on stress tetrahedron	4.5
$B_{\mathfrak{g}i}$	Body force on air	9.4
$B_{\mathfrak{n}i}$	Body force acting on the $\mathfrak{n}$th constituent	6.6
$B_{\mathfrak{s}i}$	Body force on solids	9.4
$B_{\mathfrak{w}i}$	Body force on water	9.4
B_{LM}	Transformation tensor	A.20
c_v	Coefficient of consolidation	11.4
c	Cohesive intercept	12.9
$c_{\mathfrak{s}}$	Effective cohesive intercept	12.9
c^{lm}, c_{lm}, $\mathbf{c}$	Cauchy's deformation tensor	2.1
C	Mean curvature of air–water interface	8.25
	Constant spherical pressure	11.26
	Closed contour of integration	A.18
	Constant	C.13
	Reciprocal of time factor	C.14
C_c	Compression index	11.7
C_1, C_2	Curvatures of air–water interface	8.25
CEC	Cation exchange capacity	8.12
C^{LM}, C_{LM}, $\mathbf{C}$	Green's deformation tensor	2.1

Notation	*Name*	*Section of definition*
$C_{\mathfrak{g}}$	Excess density supply rate of air	9.3
$C_{\mathfrak{n}}$	Excess density supply rate of the $\mathfrak{n}$th constituent	6.5
$C_{\mathfrak{s}}$	Excess density supply rate of solids	9.3
$C_{\mathfrak{w}}$	Excess density supply rate of water	9.3
c	Constant rate of vertical load	12.15
d	Dial reading Thickness of cylindrical tube	10.7 12.10
$\mathsf{d}(x)$	Linear extension in x^i	2.3
$\mathsf{d}(X)$	Linear extension in X^L	2.3
d_0	Initial dial reading	10.7
d	Half distance between particles	8.16
$\mathsf{D}_{\mathfrak{e}}$	Effective diameter of soil particles	10.8
D	Diffusivity Dielectric constant of water solute	6.5 8.14
$D_{\mathfrak{e}}$, $D'_{\mathfrak{e}}$, $D''_{\mathfrak{e}}$	Void ratio dependent diffusion coefficient	11.15
$D(w)$	Water content diffusivity	10.12
$D(\theta)_{ij}$	Moisture diffusivity	10.12
$D(\vartheta)$	Moisture ratio diffusivity	10.14
$\mathfrak{D}$	Dissipative energy	5.5
$\mathfrak{D}(t'')$	Lumped dispersed free energy	13.7
e	Electronic charge 4.803×10^{10} e.s.u./electron Void ratio 2.71828, constant base of natural logarithm	8.4 9.2 C.7
$e(t)$	Deviatoric strain factor	13.9

Notation	*Name*	*Section of definition*
e_r, e_θ, e_z	Base vectors	B.2
e_{ij}	Traceless strain tensor Distortion	2.3 7.5
e_{ijk}, e^{ijk}	Permutation symbol in x^i	A.14
e_{LMN}, e^{LMN}	Permutation symbol in X^L	A.14
e_c	Void ratio corresponding to pressure p_{czz}	11.7
E	Isotropic strain matrix Rheological response Response function	7.9 C.1 C.1
E_1, E_2, E_3	Rheological responses	C.7
E_∞	Response at infinite time	C.8
E_K	Response of the Kelvin model	C.14
E_S	Response of the St Venant element	C.14
$\dot{\mathsf{E}}$	Isotropic strain-rate matrix	7.9
EC	Exchangeable cations	8.12
E_n	Excess internal energy of the nth constituent	6.8
$\mathcal{E}$	Distortion Strain Constant rate of response Constant response	11.27 12.17 C.7 C.8
$\mathfrak{E}$	Energy efflux	5.4
$\mathfrak{E}_0$	Complex static elastic modulus	C.10
$\mathfrak{E}(\omega)$	Dynamic modulus	C.10
$\mathfrak{E}(\omega)_2$	Dynamic friction	C.10
$\mathfrak{E}(i\omega)$	Complex dynamic modulus	C.10

Notation	*Name*	*Section of definition*
$\mathbf{f}$	Force field	10.2
f_i	Force field, flux vector	10.2
$f_i^{\mathfrak{a}}$	Flux through cross-section $\mathfrak{a}$	10.2
$f(x)$	Unit force per volume	8.17
f_{ij}	Flow or stretching tensor	3.4
$f_{\mathfrak{g}ij}$	Flow tensor of air	9.5
$f_{\mathfrak{s}ij}$	Flow tensor of solids	9.5
$f_{\mathfrak{w}ij}$	Flow tensor of water	9.5
f	Counter of immiscible fluids	10.10
F	Rheological excitation	C.1
$\bar{\mathsf{F}}$	Excitation function	C.1
	Activating excitation	C.6
F_{K}	Excitation acting in the Kelvin model	C.14
F_{S}	Excitation acting in the St Venant element	C.14
$\mathsf{F}_1, \mathsf{F}_2, \mathsf{F}_3$	Rheological excitations	C.8
$\mathsf{F}(\mathscr{Z})$	Complex potential	10.9
F_i	Force vector	4.4
$F_{\mathfrak{n}}$	Unbalanced supply of the $\mathfrak{n}$th constituent	6.4
$F_{\mathfrak{n}i}$	Excess linear momentum supply of the $\mathfrak{n}$th constituent	6.6
$\mathscr{F}$	Function, functional relationship	3.9
	Viscoelastic differential operator	11.9
	Material coefficient function	11.26
	Constant excitation	C.7
	Constant rate of excitation	C.7

Notation	*Name*	*Section of definition*
$\mathfrak{F}$	Number of immiscible liquids	10.10
	Compliance	C.10
$\mathfrak{F}_0$	Proportionality constant of plastic restraint	C.5
$\mathfrak{F}(i\omega)$	Complex compliance	C.10
$\mathfrak{F}(i\omega)^*$	Complex elastic compliance	C.10
$g(\omega)$	Dynamic viscosity	C.10
g_i	Gravity acceleration	9.2
g^{lm}, g_{lm}, g^l_m	Metric tensor in x^i	2.1
$\mathfrak{g}_n$, $\mathfrak{g}_2$, $\mathfrak{g}_3$	General viscosity coefficients	C.3
g	Gravity acceleration	10.3
G	Elastic shear modulus	7.6
G_i	Elastic shear modulus of the ith element	7.6
$G_{\mathfrak{s}}$	Specific gravity of solids	9.2
$G_{\mathfrak{w}}$	Specific gravity of water	9.2
G_1, G_3	Elastic shear modulus of different elements	7.6
G^{LM}, G_{LM}, G^L_M	Metric tensor in X^L	2.1
$\mathcal{G}$	Function, functional relationship	7.2
	Viscoelastic differential operator	11.9
$\mathcal{G}(T)$	Measure of solubility of air in water	8.6
h_c	Head, capillary	8.26
h_i	Head, hydrostatic	8.26
h	Molar specific heat	8.5
h_i	Specific energy efflux	5.4

Notation	*Name*	*Section of definition*
$h_{\mathfrak{g}i}$	Specific energy efflux of air	9.5
$h_{\mathfrak{n}i}$	Specific energy flux of the $\mathfrak{n}$th-constituent	6.8
$h_{\mathfrak{n}i\mathfrak{d}}$	Deviatoric specific energy flux of the $\mathfrak{n}$th constituent	11.24
$h_{\mathfrak{s}i}$	Specific energy efflux of solids	9.5
$h_{\mathfrak{w}i}$	Specific energy efflux of water	9.5
H	Total entropy Piezometric head Length of consolidating path	5.9 10.8 11.3
$H_{\mathfrak{n}}$	Excess entropy supply of the $\mathfrak{n}$th constituent	6.9
i	Imaginary unit	C.10
$\mathfrak{i}$	$\mathfrak{i}$th viscoelastic element in the model	C.10
$\mathfrak{I}$	Internal energy Number of viscoelastic elements in the model	5.4 C.10
$\mathfrak{I}_{\mathfrak{s}}$	Internal energy of solids	C.10
$\mathfrak{I}_{\mathfrak{s}\mathfrak{d}}$	Disbursed internal energy of solids	11.24
$\mathfrak{I}_{\mathfrak{s}\mathfrak{i}}$	Stored internal energy of solids	11.24
$\mathfrak{I}_{\mathfrak{s}0}$	Residing stored internal energy of solids	11.24
$\Delta\mathfrak{I}_{\mathfrak{s}\mathfrak{i}}$	Excess stored internal energy of solids	11.24
$\mathcal{J}$	Jacobian	A.3, B.2
$\mathcal{J}'$	Inverse Jacobian	B.2
k	Bolzmann's constant 1.3805×10^{-16} erg/°K Isotropic coefficient of permeability Failure parameter Constant	8.14 10.5 13.2 C.1

Notation	*Name*	*Section of definition*
k_a	Restricted active coefficient of lateral earth pressure	13.16
k_p	Restricted passive coefficient of lateral earth pressure	13.16
k_{ij}	Coefficient of permeability	10.5
k_{11}, k_{22}, k_{33}	Coefficients of permeability in the principal directions	10.5
k_1, k_3	General elastic moduli	C.2
k	Coefficient of permeability	10.8
K	Coefficient of lateral earth pressure	11.6
K_0	Coefficient of lateral earth pressure at rest	11.9
K	Temperature in Kelvin	8.5
K_a	Active coefficient of lateral earth pressure	13.16
K_p	Passive coefficient of lateral earth pressure	13.16
K_{ij}	Intrinsic permeability	10.8
$\mathfrak{K}$	Kinetic energy	5.2
l_{ij}	Body moment	4.5
l_{nij}	Body moment of the nth constituent	6.7
L	Length of flow path Line, line element	10.8 A.18
L_0	Initial length	2.3
L_n	Length after straining	2.3
m	Mass element	4.1
m_g	Mass of air	9.2

Notation	*Name*	*Section of definition*
$m_{\mathfrak{n}}$	Mass of the $\mathfrak{n}$th constituent	6.2
$m_{\mathfrak{s}}$	Mass of solids	9.2
$m_{\mathfrak{w}}$	Mass of water	9.2
$m_{i\mathfrak{n}}$	Couple traction in direction n_i	4.5
$m_{ij\mathfrak{n}}$	Surface couple in direction n_i	4.5
m_{ijk}	Couple stress	4.5
$\dot{m}_{\mathfrak{n}ijk}$	Couple stress of the $\mathfrak{n}$th constituent	6.7
$\mathfrak{m}$	Summation counter of substates	5.7
M	Total mass	4.1
$M(s)$	Frequency function of creep	C.11
M_i	Torque, moment	4.4
$\mathfrak{M}$	Number of substates	5.7
n	Porosity	9.2
n_i	Unit vector in x^i normal to a surface	2.3, A.20
n^+, n^-	Ionic concentration	8.14
n_∞, n_∞^+, n_∞^-	Ionic concentration per unit volume in the bulk water	8.4
$\mathfrak{n}$	Number of constituents in a mixture	6.2
$N(s)$	Frequency function of relaxation	C.11
$\mathbf{N}$	Normal vector	A.19
N	Number of coordinates	A.3
N^L	Unit vector in X^L normal to a surface	2.3, A.20
$\mathfrak{p}_{\mathrm{b}}$	Spherical stress of the second order	

Notation	*Name*	*Section of definition*
$\mathfrak{N}$	Number of constituents in a mixture	6.2
p_c	Capillary pressure, suction Cell pressure	8.26 12.13
p_{czz}	Vertical pressure corresponding to void ratio e_c	11.7
p_e	Effective pressure	8.18
$p(d)_e$	Effective pressure at half-distance d between particles	8.19
$p_{\mathfrak{g}}$	Pressure of the air constituent or pore air pressure	8.7, 9.4
$p_{\mathfrak{g}c}$	Critical pressure of air	8.7
p_i	Air–water interfacial pressure	8.25
p_m	Mean spherical pressure	4.9
	Consolidation pressure	8.18
p_{m0}	Cohesion	13.5
p_0	Preconsolidation pressure Constant initial pressure Osmotic pressure	11.8 11.18 8.17
p_p	Pore pressure	11.9, 11.12
p_r	Repulsive pressure	8.17
p_u	Unconfined compression strength of undisturbed soil	8.23
p_{ur}	Unconfined compression strength of remolded soil	8.23
$p_{\mathfrak{w}v}$	Water vapor pressure	8.5
$p_{\mathfrak{w}}$	Pore water pressure	8.18

Notation	*Name*	*Section of definition*
$\bar{p}_{\mathfrak{w}}$	Average pore water pressure	11.4
$p(d)_{\mathfrak{w}}$	Pore water pressure at half-distance d between particles	8.19
pF	Measure of water tension	8.27
p_{ij}	Pressure tensor	4.8
p_{11}, p_{22}, p_{33}	Principal stresses	7.5
p', p'', p'''	Equivalent pressure corrections	12.14
$p_{\mathfrak{s}ij}$	Stress in solids or effective pressure	9.4
$p_{\alpha\alpha}$	Trace of pressure tensor	4.8
P	Isotropic stress matrix	7.9
$\dot{\mathsf{P}}$	Isotropic stress-rate matrix	7.9
$\mathsf{P}_{\mathfrak{d}}$	Traceless isotropic stress matrix	7.9
$\mathsf{P}_{\mathfrak{i}}$	Volumetric isotropic stress matrix	7.9
$\mathcal{P}$	Constant spherical pressure Differential operator	11.9 C.1
$\mathcal{P}_{\mathfrak{q}}$	$\mathfrak{q}$th step of constant deviatoric loading	13.7
$\mathfrak{P}$	Potential energy	5.3
q	Specific energy supply	5.4
$q_{\mathfrak{g}}$	Energy supply of air	9.5
$q_{\mathfrak{n}}$	Specific energy supply of the $\mathfrak{n}$th constituent	6.8
$q_{\mathfrak{n}\mathfrak{d}}$	Deviatoric specific energy supply of the $\mathfrak{n}$th constituent	11.24
$q_{\mathfrak{s}}$	Energy supply of solids	9.5
$q_{\mathfrak{w}}$	Energy supply of water	9.5

Notation	*Name*	*Section of definition*
q_i	Specific discharge vector	10.4
$q_{\mathfrak{g}i}$	Flux of air	10.4
$q_{\mathfrak{s}i}$	Flux of solids	10.4
$q_{\mathfrak{w}i}$	Flux of water	10.4
$\bar{q}_{\mathfrak{w}i}$	Diffusion velocity of water with respect to the solids	10.5
$\mathfrak{q}$	Counter of deviatoric step loadings	12.17
Q	Discharge	10.8
Q_i	Total discharge vector	10.4
$Q_{\mathfrak{g}i}$	Discharge of air	10.4
$Q_{\mathfrak{s}i}$	Discharge of solids	10.4
$Q_{\mathfrak{w}i}$	Discharge of water	10.4
$\mathcal{Q}$	Differential polynomial	C.1
$\mathfrak{Q}$	Number of deviatoric step loadings	12.17
r	Cylindrical coordinate	B.2
$\mathscr{r}$	Region in x^i	2.1
$\mathfrak{r}$	Counter of spherical stress increments	13.11
R	Linear stress–strain relationship function	7.3
	Gas constant, 8.314 J mole^{-1} deg^{-1}	8.7
	Radius of air–water interface	8.25
	Radius of cylindrical sample	12.11
R_1, R_2	Radii of air–water interface	8.25
R_{H}	Relative humidity	8.5
$\mathsf{R}_{\mathfrak{d}}$	Deviatoric linear stress–strain relationship function	7.3

Notation	Name	Section of definition
R_e	Reynolds number	10.8
R_i	Spherical linear stress–strain relationship function	7.5
$R_{\mathfrak{H}}$	Hydraulic radius	10.8
R	Radius in cylindrical coordinates	3.18
R^l_m, $\mathbf{R}(x)$	Rotation tensor in x^i	2.7
R^L_M, $\mathbf{R}(X)$	Rotation tensor in X^L	2.7
R_{KMPQ}	Riemann–Christoffel curvature tensor	2.9
$\mathcal{R}$	Region in X^L Differential polynomial	2.1 C.1
$\mathfrak{R}$	Number of spherical stress increments	13.11
s	Length of line element in x^i	A.10
s_{ij}	Traceless pressure tensor Deviatoric stress	4.8 7.5
$s_{\mathfrak{s}ij}$	Shear stress of soil	9.4
S	Surface area Length of line element in X^L	2.1 A.10
S	Degree of saturation	9.2
S_i	Projection vector of area	10.2
$S^{\mathfrak{a}}_i$	Projection of cross-section $\mathfrak{a}$ in direction i	10.2
$S_{\mathfrak{f}}$	Degree of saturation of the $\mathfrak{f}$th immiscible liquid	10.10
$S_{\mathfrak{t}}$	Sensitivity	8.23
$\mathcal{S}$	Constant shear stress Constant rate of deviatoric stress	12.27 12.17
$\mathcal{S}_{\mathfrak{r}}$	$\mathfrak{r}$th step of constant vertical load	12.17
$\mathfrak{S}$	Energy supply	5.4

Notation	*Name*	*Section of definition*
$\mathfrak{S}(t)_q$	Stored free energy of the qth step loading	13.7
$\mathfrak{S}(t')$	Lumped stored free energy	13.7
t	Time	3.1
$\mathbf{t}$	Traction vector	4.5
t_i	Traction vector	4.5
t_{in}	Traction in direction of vector n_i	4.5
t_{ij}, t^{ij}	Stress tensor, component of stress tensor	4.5
t^*_{in}	Stress vector acting on stress tetrahedron	4.5
t^*_{ij}	Stress tensor acting on stress tetrahedron	4.5
t_{nn}, $\mathbf{t}_n$	Normal traction	4.5
t_{nt}, $\mathbf{t}_t$	Tangential traction	4.5
t_{nij}	Stress tensor acting on the n-th constituent	6.6
T	Temperature in Kelvin degrees Tangential component in the octahedral plane Time factor	5.9 A.20 C.14
T_3	Volumetric retardation time	11.18
T_c	Critical temperature of gases	8.7
T_n	Temperature of the nth constituent	6.9
T_{rel}	Relaxation time	C.7
T_{reli}	Relaxation time of the ith element	C.11
T_{ret}	Retardation time	C.8
T_{reti}	Retardation time of the ith element	7.6
T'_{reti}	Shear retardation time of the ith element	7.6

Notation	Name	Section of definition
T_v	Consolidation time factor	11.4
T_{ni}	Excess linear momentum supply of the nth constituent	6.6
T_z	Torque	12.11
$\mathbf{T}$	Tangential vector	A.19
T	Surface tension	8.25
$\mathcal{T}$	Transmittance	C.1
$\mathfrak{T}$	Total energy	5.5
u_a	Pore air pressure (Soil Mechanics notation)	11.22
u_w	Pore water pressure (Soil Mechanics notation)	11.22
u^i, u_i	Displacement in x^i	2.1
u_θ	Tangential deformation	12.11
U	Average degree of consolidation	11.4
U_z	Degree of consolidation	11.4
U	Dimensionless mid-plane potential	8.21
U^L, U_L	Displacement in X^L	2.1
v	Speed	3.2
v^i	Velocity	3.2, B.7
v_i	Mean velocity or barycentric velocity	6.3
v_{gi}	Velocity of air	9.3
v_{mi}	Velocity of the reference constituent	6.3
v_{ni}	Velocity of the nth-constituent	6.3
v_{si}	Velocity of solids	9.3

Notation	*Name*	*Section of definition*
v_{wi}	Velocity of water	9.3
$\bar{v}_{gi}$	Diffusion velocity of air	9.4
$\bar{v}_{ni}$	Diffusion velocity or peculiar velocity of the nth constituent	6.3
$\bar{v}_{si}$	Diffusion velocity of solids	9.4
$\bar{v}_{wi}$	Diffusion velocity of water	9.4
v_{nmi}	Velocity of the nth constituent relative to the mth constituent	6.3
V	Volume	2.1
V_d	Volume of dry soil	8.27
V_f	Volume of the fth immiscible liquid	10.10
V_g	Volume of the air constituent	8.7
V_{gm}	Molal volume of air	8.7
V_n	Volume of the nth constituent	6.2
V_s	Volume of solids	9.2
V_w	Volume of water	9.2
V_v	Volume of voids	9.2
w	Water content	9.2
w_L	Liquid limit	8.2
w_P	Plastic limit	8.2
w_S	Shrinkage limit	8.27
w_i, $\mathbf{w}$	Vorticity vector	3.4
w_{ij}	Spin tensor	3.4

Notation	*Name*	*Section of definition*
$\mathfrak{w}_{\mathfrak{s}}$	Specific stress-work	5.2
$\mathfrak{w}_{\mathrm{m}}$	Specific mechanical work	5.5
$\mathfrak{w}_{\mathrm{n}}$	Specific non-mechanical work	5.5
W	Total weight of soil	6.2
$W_{\mathfrak{g}}$	Weight of air	9.2
$W_{\mathfrak{n}}$	Partial weight of the $\mathfrak{n}$th constituent	6.2
$W_{\mathfrak{s}}$	Weight of solids	9.2
$W_{\mathfrak{w}}$	Weight of water	9.2
$\mathfrak{W}_{\mathrm{m}}$	Mechanical power	5.5
$\mathfrak{W}_{\mathrm{n}}$	Non-mechanical power	5.5
$\mathfrak{W}_{\mathrm{r}}$	Work of repulsive force	8.17
$\mathfrak{W}_{\mathfrak{s}}$	Stress-power	5.2
x	Length of vector x^i	2.2
x^i, x_i	Deformed coordinates	2.1
x^i	Deformed position vector	A.3
$x^i_{,L}$	Deformation gradient in x^i	2.1
X^L, X_L	Undeformed coordinates	2.1
X^L	Undeformed position vector	A.3
$X^L_{,m}$	Deformation gradient in X^L	2.1
Y	Dimensionless potential at ξ	8.16
z	Cylindrical coordinate	B.2
z_{i}^+, z_i^-	Valency of the ith ion	8.4

Notation	*Name*	*Section of definition*
Z	Compressibility factor	8.7
Z	Dimensionless potential at particle surface	8.14
α	Angle of distortion Angle Normalization factor of creep	3.17 A.19 C.11
$\alpha_{\mathfrak{g}}$	Density concentration of air	9.5
$\alpha'_{\mathfrak{g}}$	Density concentration of the dissolved gas	8.6
$\alpha_{\mathfrak{n}}$	Density concentration of the $\mathfrak{n}$th constituent	6.2
$\alpha_{\mathfrak{s}}$	Density concentration of solids	9.5
$\alpha_{\mathfrak{w}}$	Density concentration of water	9.5
β	Angle Normalization factor of relaxation	A.20 C.11
γ	Angle, angle of rotation Angle between vectors	3.17, A.20 B.5
$\gamma_{\mathfrak{d}}$	Dry unit weight	9.2
$\gamma_{\mathfrak{g}}$	Unit weight of air	9.2
$\gamma_{\mathfrak{s}}$	Unit weight of solids	9.2
$\gamma_{\mathfrak{w}}$	Unit weight of water	8.26
$\gamma_{\mathfrak{w}0}$	Unit weight of water at 4 °C	8.26
$\Gamma_{k,ij}$	Christoffel symbol of the first kind in x^i	A.13
$\Gamma_{K,LM}$	Christoffel symbol of the first kind in X^L	A.13
Γ^k_{ij}	Christoffel symbol of the second kind in x^i	A.13
Γ^K_{LM}	Christoffel symbol of the second kind in X^L	A.13
$\delta(t)$	Dirac delta function	C.10

Notation	*Name*	*Section of definition*
δ^{ij}, δ_{ij}, δ^i_j	Kronecker unit tensor in x^i	A.5
δ^{LM}, δ_{LM}, δ^L_M	Kronecker unit tensor in X^L	A.5
Δ	Volumetric dilatation	2.4
	Vertical deformation	10.7
	Settlement	11.7
ε_{ij}	Strain tensor, general	2.3
ε_{11}, ε_{22}, ε_{33}	Principal strains	7.5
$\overset{\text{inf}}{\varepsilon}_{ij}$	Infinitesimal strain tensor	2.3
$\overset{A}{\varepsilon}_{ij}$	Almansi–Hamel strain tensor	2.3
$\overset{E}{\varepsilon}_{ij}$	Eulerian strain tensor	2.2
$\overset{G}{\varepsilon}_{ij}$	Green–St Venant strain tensor	2.3
$\overset{H}{\varepsilon}_{ij}$	Hencky strain tensor	2.3
$\overset{L}{\varepsilon}_{ij}$	Lagrangian strain tensor	2.2
$\overset{S}{\varepsilon}_{ij}$	Swainger strain tensor	2.3
ε_{m}	Volumetric strain	7.5
$\varepsilon_{\alpha\alpha}$	Trace of strain tensor	2.3
	Spherical strain	7.5
$\varepsilon_{\beta\beta}$	Dilatancy	11.27
ϵ_{i}	Specific internal energy	5.4
$\epsilon_{\mathrm{i}\mathfrak{g}}$	Specific internal energy of air	9.5
$\epsilon_{\mathrm{i}\mathfrak{n}}$	Specific internal energy of the $\mathfrak{n}$th constituent	6.8
$\epsilon_{\mathrm{i}\mathfrak{s}}$	Specific internal energy of solids	9.5
$\epsilon_{\mathrm{i}\mathfrak{w}}$	Specific internal energy of water	9.5
$\epsilon_{\mathfrak{K}}$	Specific kinetic energy	5.2

Notation	*Name*	*Section of definition*
$\epsilon_{\mathfrak{p}}$	Specific potential energy	5.3
$\epsilon_{\mathfrak{s}\mathfrak{d}}$	Disbursed specific internal energy	11.24
$\epsilon_{\mathfrak{s}\mathfrak{i}}$	Stored specific internal energy	11.24
$\Delta\epsilon_{\mathfrak{s}\mathfrak{i}}$	Excess stored specific internal energy	11.24
$\epsilon_{\mathfrak{s}\mathfrak{i}0}$	Residing stored specific internal energy	11.24
$\epsilon_{\mathfrak{t}}$	Specific total energy	5.5
ζ	Specific free enthalpy	5.10
ζ	Gravity potential	10.3
η	Specific entropy Shear viscosity coefficient Viscosity	5.9 7.6 10.8
η_2, η_3	Shear viscosity of elements	12.16
$\eta_{\mathfrak{g}}$	Specific entropy of air	9.5
$\eta_{\mathfrak{i}}$	Shear viscosity coefficient of the $\mathfrak{i}$th element	7.6
$\eta_{\mathfrak{n}}$	Specific entropy of the $\mathfrak{n}$th constituent	6.9
$\eta_{\mathfrak{s}}$	Specific entropy of solids	9.5
$\eta_{\mathfrak{w}}$	Specific entropy of water	9.5
ϑ	Moisture ratio	9.2, C.5
$\vartheta(t)$	Coefficient of volumetric plastic restraint Plastic restraint	11.5 C.5
θ	Angle of cylindrical coordinates Angle Failure angle	3.18, B.2 A.11 13.15
Θ	Angle of cylindrical coordinates	3.18
$\theta_{\mathfrak{cr}}$	Critical angle of failure	13.15

Notation	Name	Section of definition
$\theta_{\mathfrak{g}}$	Volumetric air fraction	9.2
$\theta_{\mathfrak{s}}$	Volumetric solids fraction	9.2
$\theta_{\mathfrak{w}}$	Volumetric moisture fraction	9.2
$\theta_{\mathfrak{w}i}$	Initial moisture content	10.13
$\theta_{\mathfrak{ws}}$	Saturated moisture content	10.13
$\theta_{\mathfrak{wr}}$	Moisture content between initial and saturated state	10.13
$\varkappa$	Elastic bulk modulus Reciprocal of double layer thickness	7.6 8.14
$\varkappa_1$, $\varkappa_3$	Elastic bulk moduli of elements	11.18
$\varkappa_{\mathfrak{m}}$	Elastic bulk modulus of the $\mathfrak{m}$th element	11.18
$\varkappa_{\mathfrak{s}}$	Bulk modulus of solids	11.4
$\varkappa_{\mathfrak{w}}$	Bulk modulus of water	11.4
λ	Scalar, proper number	A.15
λ_2	Vertical stretch	11.7
λ_T	Trouton's viscous traction	7.6
$\lambda_{\mathfrak{K}}$, $\lambda_{\mathfrak{M}}$	Latent roots of second order tensors	A.15
$\lambda_{\mathfrak{N}}$	Proper number	A.15
$\lambda(x)$	Linear stretch in x^i	2.3
$\lambda(X)$	Linear stretch in X^L	2.3
Λ	Volumetric stretch	2.4
μ	Coefficient of friction	13.5
μ_{i}	Volumetric viscosity coefficient of the ith element	7.6

Notation	Name	Section of definition
μ_2, μ_3	Bulk coefficients of viscosity of elements	11.18
ν	Kinematic viscosity	10.8
	Ratio of transverse dilatation	11.12
$\nu_{\mathfrak{nm}}$	The $\mathfrak{m}$th thermodynamic substate of the $\mathfrak{n}$th constituent	6.9
$\nu_{\mathfrak{nm}\mathfrak{d}}$	Thermodynamic deviatoric substates	11.24
$\xi_{\mathfrak{n}}$	Extensive quantity of the $\mathfrak{n}$th constituent	6.3
ξ	Extensive quantity per unit mass	4.2
	Specific total energy	5.5
	Dimensionless distance in the diffuse double layer	8.14
Ξ	Total extensive quantity of volume	4.2
Ξ_{LM}, Ξ_{ij}	Dimensionless deviatoric tensor	A.21
$\Pi_{\mathbf{A}}$	Octahedral invariant of a tensor A_{ij}	A.18
ρ	Density	4.1
	Charge density per unit volume of solution	8.10
$\rho_{\mathfrak{g}}$	Density of air	9.2
$\rho_{\mathfrak{m}}$	Density of the reference constituent	6.5
$\rho_{\mathfrak{n}}$	Density of the $\mathfrak{n}$th constituent	6.2
$\rho_{\mathfrak{s}}$	Density of solids	9.2
$\rho_{\mathfrak{w}}$	Density of water	9.2
σ	Supply of quantity per unit mass	4.2
	Surface charge density	8.14
	Total pressure (Soil Mechanics notation)	11.22
σ'	Effective pressure (Soil Mechanics notation)	11.22
σ_i	Body force	4.5
$\sigma_{\mathfrak{n}}$	Specific supply rate of the $\mathfrak{n}$-th constituent	6.4

Notation	*Name*	*Section of definition*
τ	Octahedral shear stress	13.5
$\tau_{\mathfrak{cu}}$	Shear stress of consolidated-undrained test	13.13
$\tau_{\mathfrak{d}}$	Shear stress of drained tests	13.12
$\tau_{\mathfrak{m}}$	$\mathfrak{m}$-th thermodynamic tension	5.7
$\tau_{\mathfrak{nm}}$	$\mathfrak{m}$-th thermodynamic tension of the $\mathfrak{n}$-th constituent	6.9
$\tau_{\mathfrak{nmd}}$	Thermodynamic deviatoric tension	11.24
$\tau_{\mathfrak{oct}}$	Octahedral shear stress	13.2
$\tau_{\mathfrak{u}}$	Shear stress of undrained test	13.14
ϕ	Angle of internal friction Potential Tensorial function	12.9 10.2 A.18
$\phi_{\mathfrak{s}}$	Effective angle of internal friction	12.9
$\phi_{\mathfrak{n}}$	Coefficients of polynomials	2.7
Φ	Mechanical and non-mechanical energy flux	5.5
$\Phi(x, y)$	Real part of complex potential	10.9
$\Phi_{\mathfrak{c}}$	Capillary potential	10.3
$\Phi_{\mathfrak{h}}$	Heat potential	10.3
$\Phi_{\mathfrak{e}}$	Electric potential Electrochemical potential	8.14 10.3
$\Phi_{\mathfrak{g}}$	Gravitational potential	10.3
$\Phi_{\mathfrak{n}}$	Specific efflux of the $\mathfrak{n}$-th constituent	6.4
$\Phi_{\mathfrak{p}}$	Pressure potential	10.3
$\Phi_{\mathfrak{t}}$	Total potential	10.3

Notation	*Name*	*Section of definition*
Φ_{ω}	Moisture potential	10.3
	Pore water potential	11.9
Φ_{γ}	Overburden potential	10.3
$\boldsymbol{\Phi}$	Total flow potential	10.3
φ	Scalar	A.16
χ	Specific enthalpy	5.10
	Bolzmann transformation function	10.13
	Pore pressure parameter	11.22
$\chi(t)$	Relaxation function	C.10
ψ	Specific free energy	5.10
	Imaginary part of flow line	10.9
$\psi(t)$	Creep function	C.10
$\psi_{\mathfrak{sd}}$	Disbursed specific free energy	11.25
$\psi_{\mathfrak{si}}$	Stored specific free energy	11.25
$\psi_{\mathfrak{si}0}$	Residing specific free energy	11.25
$\Delta\psi_{\mathfrak{si}}$	Excess specific free energy	11.25
$\boldsymbol{\Psi}$	Matric potential	10.3
ω	Frequency	C.10
Ω	Angle of twist	12.11
$\boldsymbol{\Omega}$	Overburden potential	10.3
$\aleph_{\mathfrak{n}}$	$\mathfrak{n}$th isotropic stress–strain relationship coefficients	7.9
$\beth_{\mathfrak{n}}$	$\mathfrak{n}$th volumetric isotropic stress–strain relationship coefficients	7.9, 12.17
$\daleth_{\mathfrak{n}}$	$\mathfrak{n}$th dilatancy coefficients of isotropic stress–strain relationship	7.9,

Notation	*Name*	*Section of definition*
$\mathfrak{w}_n$	nth shear isotropic stress–strain relationship coefficients	7.9, 12.17
$\mathrm{I}_{\mathbf{A}}, \mathrm{II}_{\mathbf{A}}, \mathrm{III}_{\mathbf{A}}$	Principal invariants of a tensor A_{ij}	A.18
$\bar{\mathrm{I}}_{\mathbf{A}}, \overline{\mathrm{II}}_{\mathbf{A}}, \overline{\mathrm{III}}_{\mathbf{A}}$	Moment invariants of a tensor A_{ij}	A.18
grad	Gradient of a scalar	A.16
div	Divergence of a vector	A.16
curl	Rotation of a vector	A.16
∇^2	Laplacian differential operator	10.2, A.14
$1_x, 1_y, 1_z$	Unit vector	B.2

1 Introduction

1.1 Scope

The appearance of Terzaghi's "Erdbaumechanik auf Boden physikalischer Grundlage" (1925) opened a new period in the study of soil engineering and gave a new impetus to the research of soils. The six decades that have passed since then may thus be considered the developing period of "classical soil mechanics". Governing principles such as effective pressure and pore water pressure, the spring-dashpot model of consolidation, pore air pressure in partially saturated soils, and others, were developed. Yet no one can deny that these concepts rely heavily on empirical notions, conceived through the ingenuity of a few and the hard work of many, while they remained unsupported theoretically. It is the purpose of this study to lay the theoretical foundation, and to show that these empirical concepts lean on a firm physical basis and are therefore correct theoretically as well. These concepts work in spite of the fact that their formulation was based mostly on experimental evidence and common sense rather than principles of physics.

The study presented here is based on theories that evolved from continuum mechanics, including the theories of multi-phase media and of the thermodynamics of irreversible processes.

It may well be that this kind of exposition was not possible earlier, when thermodynamics, and mechanics for that matter, were still bypassed by physicists in favor of other, more attractive, domains of physics. Only in the late fifties and early sixties did a group of physicists and mathematicians begin to re-examine this field of physics and formulate anew its concepts through a systematic approach, restoring its prominence. The works of Reiner (1958), Freudenthal and Geiringer (1958), Meixner and Reik (1959), Truesdell and Toupin (1960), Noll (1958), Ericksen (1960a), Coleman and Noll (1961a, b) and others, which mark the beginning of this new approach, have since become classics in their own right and serve as a basis for further studies in mechanics and thermodynamics of materials.

Much effort has been invested in retaining in this study the notation generally accepted in soil mechanics. However, in view of the fact that other disciplines are introduced and discussed, the notation had to be adjusted accordingly. The tensorial notation has been applied throughout the paper; the author assumes that the reader is familiar with this notation, which is a shorthand for lengthy mathematical equations and follows clearly defined rules. It is further assumed that all events, except for the general derivations, occur in an orthogonal coordinate system, and therefore no distinction is made between covariant and contravariant indices, and all indices are covariant. A brief review of tensorial mathematics is included in Appendix A. The symbols are defined as they appear.

1.2 Historical Notes

The history of the study of soil from the engineering point of view and its use to the advantage of man can roughly be divided into four major periods.

Archeological findings of earth dwellings, burial sites, roads, canals, bridges, fortifications and other structures mark the first period – prehistoric and ancient times – when soil was considered either as a structural material or as a supporting medium for structures (Verril 1942; Thomas 1889).

There are strong indications that the attention given to soils at that time was much more than superficial, as it involved dealing with problems of seepage, drainage, stability, distribution of stresses, apparently also physico-chemical stabilization of soils and other related problems, which seem quite sophisticated. Yet all this was accomplished without any basic knowledge of the processes occurring in the matter, processes which are partly hidden from us even today. It was mainly a trial and error method, applied with plenty of common sense, and only the successful attempts are apparent to us today; most of the unsuccessful attempts have naturally vanished in time. The master-builders of those times invested a lifetime in their artisanship, including early years of apprenticeship and later years as masters in their own right. Their skill lay in empiricism and long experience. They enjoyed status, power, and wealth as long as they succeeded in providing their sponsors, the kings and rulers of those times, with a becoming and safe product. Their fame traveled fast and far and commissions for work came from neighboring and from far away rulers. Consequently, they traveled worldwide and transported their trade from one place to another. "King Solomon sent for Hiram and brought him down from Tyre . . . And he cast two columns of bronze, one column was eighteen cubits high" (Kings I, 7.13; 7.15). There are indications that not all "imported" techniques were always necessary, or applied in the right place (Klausner 1970d), yet because they provided satisfactory results in one place they were assumed to be favorable in other instances as well.

On the other hand, heavy punishments lay in wait for the builders whenever public safety was breached (Harper 1904). Size and bulkiness were the results of fear of failure, which, combined with the abundance of cheap labor, resulted in heavy walls and monumental structures.

The fall of the Roman Empire, coinciding with many other important historical events, marked the end of the first period. The damages incurred by the Barbarian invasions thereafter were so great that remnants of civilization and indications of buildings during that time are scarce. Several centuries had to elapse before building on a large scale began anew, with renewed interest in soil problems. This second period is marked by the expansion of the feudal system of both the Church and the landlords, characterized on one hand by the construction of large cathedrals with domes and high bell-towers, and on the other hand by the construction of large castles and fortifications. In essence, the second period does not differ much from the first, except for more sophistication of the end product and the attention paid to artistic refinements and craftsmanship. Because of the delicacy of structural detail, more attention had to be paid to stability considerations (Painter 1953), mostly settlement problems. The bell-towers of Pisa, Venice, Bologna and others, which failed to meet the then non-existent stability requirements, serve today as tourist attractions for their out of plumb alignment (Terzaghi 1934). Excessive settlements, some of nearly 2 meters, became the fate of other structures as well, mostly cathedrals and bridges (Tiedemann 1932; Watson and Watson 1937).

The emergence of states, their continuous conflicts and warfare, and the rise of scientific thought in the fifteenth century, mark the beginning of the third period. The quest for knowledge and the attempts to frame that knowledge into a systematic order are considered to be the dawn of modern science, with Copernicus, Bruno, Galileo and Kepler as the forerunners. During this period the practical knowledge regarding the behavior of soils was advanced, and further developments ensued as studies in earth pressure and stability of slopes were carried out by engineers like Coulomb, Poncelet and Navier in France, Bruenings in Holland, Lambert, Mohr and Muller-Breslau in Germany and Rennie and R. Stephenson in England. These studies, valid to our day, culminated in the works of Atterberg (1911) and Fellenius (1918, 1927), who are considered the predecessors of the fourth period.

The industrial revolution and the changes introduced by the steam engine and later the internal combustion engine influenced the early years of the twentieth century. Coincidentally, the advancement of governments towards more enlightenment, the promotion of civil liberties, and the general emancipation of nations contributed to the progress of commercial and industrial development and the construction of housing, railroads, motor-roads and service facilities. However, it was not until the end of World War I that those developments picked up momentum. Terzaghi's "Erdbaumechanik" (1925) served as a turning point in soil engineering and established the beginning of "classical soil mechanics". Soon other engineers and investigators like A. Casagrande, H. P. Krey, D. W. Taylor, G. P. Tschebotarioff, R. B. Peck, A. Mayer and J. Feld helped to transform soil mechanics into a viable discipline.

1.3 Classical Soil Mechanics versus Mechanics of Soils

Today, when everything moves at a fast pace and each decade sets forth new basic developments and ideas, many disciplines have evolved and attained scientific stature in a comparatively short period. It could have been expected that soil mechanics, essential as a scientific basis of the ever increasing engineering undertakings, would develop steadily into a sound scientific discipline, according to the needs of engineering and in step with the scientific and technological developments.

However, soil mechanics has followed another course. The theoretical aspects have been neglected and the engineering approach, a legitimate approach in its own right, has been opted for. From a scientific point of view a reversal of ends and means has taken place. An abundance of experimental data was accumulated, and, rather than being used to illustrate and support theory, the experiments have become the main goal of the study, and scientific merits have been attributed to experiments as such. The "theories" that emerged from these experiments have not always conformed with the principles of physics, nor have the experiments been planned in accordance with those principles. Disregard of the fundamentals of physics, inconsistent discipline structure, use of mixed terminology, lack of interest in the study of sister disciplines such as soil physics, mechanics, theory of mixtures, etc., and inability of the "theories" to provide proper explanation of certain phenomena have often been tolerated in soil mechanics.

Discontent with the state of the art of soil mechanics has often been raised at symposia, at conferences and in published papers. The reason for this discontent was the slow rate of progress, which was not in proportion to the efforts invested in the laboratory and field investigations and with the numerous valuable data collected during years of study. The explanation was usually stated as lack of good correlation between experience (or experimental evidence) and theory and, as a remedy, further experiments were demanded. It seems, however, that the voices raised about the progress did not point to the real nature of the problem.

The main cause of this state of affairs seems to be the ambivalent attitude of the leading figures in soil mechanics, who, while voicing dissatisfaction with the state of the art and stating their conviction of the importance of basic research, nevertheless emphasized experimentation over basic theoretical studies.

Let us cite a few of the remarks on this subject made during the past fifty years.

In the opening address by Terzaghi (1948) at the Second International Conference of SOMEFE:

> Every science, pure and applied, is based on what is known as fundamental research.
> ... the laboratory tests disclose only one part of the real character of the material. The real character can only be discerned by carefully observing the performance of the clay in the field and comparing the

> results with predictions based on laboratory tests. Hence the further advance of our knowledge in soil mechanics depends to a large extent on the scope and the quality of our field observations.

Terzaghi (1953) in his invited lecture *Fifty Years of Subsoil Exploration* at the Third International Conference of SOMEFE:

> On account of the increase of the variety of procedures at our disposal, the selection of the techniques requires more and more experience and judgement. . . . Since programs of subsoil exploration have not always complied with these requirements, considerable amounts of time and money have been invested in rather unproductive investigations. However, the occasional misuse of procedures does not lessen the merits of the procedures as such, because the developments of the last thirty years vastly increased the amount of significant information.

It is also instructive to read the conclusions of Peltier (1953) at the same conference, which are not reproduced here.

Newmark (1960), in his opening address at the Boulder ASCE Shear Conference, mentions the paper of Dr Rutledge (1940) and adds his own consternation on the same topic while promoting certain theories of failure:

> Very little has changed in the past 20 years with regard to the knowledge of failure conditions; possibly some new theories have been developed, but they are still as inapplicable to actual conditions as those presented 20 years ago.

He also remarks on the confusion introduced in soil mechanics by erroneous terminology.

Skempton (1961), in the president's opening address at the Fifth International Conference of SOMEFE:

> Firstly, it is now proved beyond all doubt that the engineering behavior of soils in practice can be analysed in a rational manner.

Further he warns us of "complacency" and finally:

> In research, there is a very great deal of work still to be done in the laboratory, in the field and the derivation of new theoretical concepts . . . Unpromising they may be. But the problems are there and provide an inspiring challenge to the research mind.

Bjerum (1969), in the president's opening address of the Seventh International Conference of SOMEFE, admits:

> . . . we must nevertheless, admit that there is also available a relatively large number of carefully studied cases in which the classical approach of soil mechanics proved to be a failure. In these cases a comparison between a carefully performed prediction and the actual behavior has shown discrepancies so large that they could not be explained by inhomogeneities of the soil strata, nor by the scattering of the test results.

Ladd (1977), in the State of the Art Report on Soil Properties at the Ninth International Conference of SOMEFE still felt that "someday" there would be a major breakthrough in design capability. In his words:

> In order to improve the methods of design analysis it is necessary first to have better evaluation of the soil parameters. Actual behaviour is predicted by calculation with the soil parameters which are put into the model. This is a very complex topic but it will some day produce a major breakthrough in design capability.

Wroth and Houlsby (1985), in their presentation in *Soil Mechanics – Property Characterization and Analysis Procedure*, given at the Eleventh International Conference of SOMEFE, after describing the progress in soil mechanics in the last 30 years, said:

> As a consequence some of the basic properties and definitions used for characterising the behaviour of soil (and rock) need to be modified subtly or to be made more precise, e.g. the undrained shear strength, and the distinction between the undrained values of Young's modulus of elasticity.
> Soil behaviour is complex and depends on the complete geological history of the deposit as represented by the size, shape, mineral composition and packing of the particles, the stress history that has been experienced, the pore fluid and other factors. The response of a soil element to a particular test or engineering event will depend on the changes of effective stress that it undergoes; and further, this response will be inadequately represented by a few simplistic properties such as undrained shear strength, shear modulus, coefficient of consolidation, etc.

And in the summary of the presentation:

> It has been argued that the relationships between properties should be based on (i) physical insight, (ii) theoretical understanding and (iii) dimensionless forms of variables. Various examples have been discussed and, in some instances, it has been suggested that the variables need redefinition or need to be replaced.

Classical Soil Mechanics, as can be seen in the textbooks, is an agglomeration of several disciplines: geology and mineralogy, macromeritics, hydraulics, theory of elasticity, theory of plasticity, physical chemistry, mechanics, etc. Their interrelation is not always explicit, and their common basis not obvious, although it should be clear that it is physics.

Mechanics of Soils, with the accent on "mechanics", is not different from the known Soil Mechanics as far as final goals are concerned, but it is distinguished from it by the fact that it aims to restore the scientific stature and merit to soil science by applying warranted scientific methods of scrutiny, long accepted and employed in other disciplines. Mechanics of Soils should be considered the formalistic structure of all that has been gained in Soil Mechanics by empiricism.

1.4 Theory versus Experiment

When watching a chess game, even if we are not familiar with the rules of the game, we may learn the rules by which the pieces move around the board by observing for a long enough time. Eventually we might become acquainted with all the rules of the game.

The world around us is something like a great chess game, although more complex, and we are the observers. All we are allowed to do is to watch the playing. Of course if we watch long enough, we may eventually catch on to a few of the rules. The rules of the game are the *fundamental laws of physics*. Even if we knew every rule, however, we might not understand why a particular move is made in the game, merely because it is too complicated and our minds are limited. If you play chess you must know that it is easy to learn all the rules, and yet it is often very hard to select the best move, or to understand why a player moves as he does. With time and experience, however, the reasons for the moves become clear, and their selection more sophisticated. So it is in nature, only much more so; and, as in chess, we hope finally to find all the rules. This analogy of chessboard and nature is taken from Feynman, Leighton and Sands (1966), and our "world" is the world of soil, while the "rules" are, as above, the laws of physics.

The complex nature of the environmental circumstances in which soils appear, and the fact that we do not know all the laws that control the behavior of the constituents of soils, impede the blending of the laws into a sound theory – a model of their interaction.

The validity of a theory is probed by the following means:

First, it should check out experimentally with the phenomena described. Theory and experiment, therefore, go hand in hand: sometimes the theory develops and requires experimental support, other times experiments are advanced in order to seek out possible ramifications of the theory. Experiments are careful measurements, many times reproducible, that under identical conditions should yield identical data, irrespective of the person performing the measurements. Experimental data, under those constraints, are accepted factual evidence, to be reflected in or described by the theory. If experiments do not corroborate the theory, several possibilities arise:

1. The theory is incorrect and a new theory has to be devised to replace it.
2. Any of the laws on which the theory is based is incorrect, demanding a revision.
3. The theory requires some refinement in order to be confirmed by all experiments.

Second, the theory should satisfy accepted laws and formerly accepted theories. The laws on which the theories are built are two-fold. One set of laws are the laws that have passed all the rigid requirements and tests and are universally accepted as fundamental laws of physics. Such are the conservation laws, known also as balance equations, and several constitutive equations. The other set of laws are the ones still under scientific scrutiny, but nevertheless applied, awaiting further refinement or final corroboration.

While experiments are reproducible manifestations of physical processes, and invariant in that sense, theories may be devised by mere audacity and speculation with respect to the consequences of certain assumptions. Sometimes, the theory turns out to be irrelevant, or, at best, it might take a long time before it can be readily verified by experiments. Theories not based on the universally accepted laws of physics, or not proven to be compatible with them, are sheer speculations until proven.

In light of the above, science becomes a race between theory and experiment, while experience, thought, sharp instincts, common sense, and perhaps luck, serve as catalysts throughout the scientific process.

1.5 Levels of Investigation

Matter is the substance with which physics is concerned. When individualized through its characteristics, we talk about *materials*.

One way to analyze the composition of materials is to consider their *levels of aggregation*. Three levels of aggregation are considered here: the submolecular level, the structural level and the phenomenological level.

Molecules are the smallest parts of a given material structure that carry all its recognizable features and properties. Conversely, the physical behavior and properties of the molecules are characteristic of the specific material, and they determine its properties in the structural level. Since the behavior of the individual molecule can be satisfactorily predicted from the known physical laws, we may confidently say that we have an almost accurate knowledge of the physics and behavior of several structural aggregates and of molecular structures.

The *atoms* that form the molecules constitute a different level of aggregation. Their physico-chemical interaction is sometimes rather complex, but still analyzable by the methods of physics. The subatomic particles such as nuclei, electrons, protons, neutrons, photons and other elementary particles, whose properties are only partly known, represent other levels of aggregation. We shall refer to all these levels as the *submolecular level* of aggregation.

Investigation of materials takes place at all levels of aggregation. The lower,* submolecular level, is where most properties of the material originate. It would seem, therefore, most appropriate to prefer this level of investigation, the study of the interaction of corpuscles, in order to get acquainted with the physical and mechanical properties of materials. In principle this would be the most exact level, provided we were able to comprehend, expose and explore all intricacies and interactions occurring at this level, and project them to the higher levels. This appears, however, to be an almost endless task, (Freudenthal 1950; Truesdell and Toupin 1960), because of the following difficulties:

1. The wide ramification of the numerous particles, which prevents the

* The terms "lower" and "higher" used with regard to the levels of investigation are intended to indicate the order of magnitude of the unit investigated and have nothing to do with the relative importance of each level of investigation.

determination of all the laws and individual actions required for defining the properties of the material.

2. The mathematical problems that the complex nature of the interaction of the corpuscles might impose.
3. The limitations of measurement and observation, which at this level of investigation have to be very refined and are sometimes beyond our technical means.

Moreover, even if we succeeded in fully investigating certain properties, we would never be sure whether all properties and actions have been accounted for. Any discovery of a new and smaller entity could threaten the validity of the existing laws and theories and so set the investigation into a permanent "status nascendi".

The next higher level of investigation is the *structural level*. At this level the various structural components of the material, such as air, water, oil, mineralogical particles of all sorts, etc., are studied for their individual behavior and interaction. Some of the difficulties enumerated above are absent at the structural level of investigation. The structural components of the material, which are above the molecular size, contain most of the properties of the material, such as strength, friction, viscosity, elasticity, thixotropy; the number of the structural aggregates that compose the material is more limited than that of the corpuscles; and, finally, the phenomena at this level can be observed and measured with less sensitive instrumentation. It is, therefore, expected that such an investigation will lead to a much closer and less risky approximation of the quantitative data concerning the behavior of the material as a whole.

Even though we are dealing with a limited number of structural aggregates, in some cases there may be limitations, for example difficulties in mathematical formulation if a large number of properties has to be considered. Therefore, an investigation at the structural level will sometimes have to be aided by investigations of the molecular and submolecular interactions, and the latter will serve as a source of introspective knowledge, but in that case additional mathematical assumptions will have to be made.

The highest level of investigation of materials is the *phenomenological level*. Investigation at this level considers the material as an entity and deals with the phenomena as they appear and not with the large number of physical properties, the actions and interactions of the structural components. At this level the behavior pattern is the sum of the overall internal properties and interactions. It is assumed that the mechanical behavior of all materials follows the same phenomenological pattern and is governed by the same general equation, the difference between materials stemming from the quantitative values of the coefficients and the character of their combinations.

The mathematical formulation at this level of investigation becomes considerably simplified, and the boundary conditions are obtained from experiments performed with simpler and less sensitive instrumentation and from investigations at a lower level.

Comparing these three levels of investigation, we may summarize as follows:

1. At a lower level of investigation the factors are more numerous and their interaction is more complex.
2. At a lower level of investigation the measuring techniques become more refined, therefore more involved, difficult and sometimes technically impossible.
3. At a higher level of investigation the mathematical formulation becomes simpler.
4. At a higher level of investigation there are less assumptions concerning boundary conditions, and they are obtained from the lower levels.
5. In *principle*, all three levels of investigation should lead to the same quantitative results.
6. In *reality*, the higher the level the more of the overall behavior of the material it comprehends, therefore the resulting theory is a better quantitative approximation; however, the physical interactions in it become obscured.

Since today the mathematical tools are far more advanced than our measuring and testing techniques, we will logically conclude that the phenomenological and the structural levels, supported by information from the molecular and submolecular levels, are the most reasonable levels of investigation for good quantitative results in the behavior of the material.

1.6 The Continuum

Materials are composed of an array of molecules, and the constant motion, lattice vibration, collision and activation of their submolecular corpuscles determine their characteristic behavior. The physical processes in any aggregation emanate from the specific aggregants that contribute their characteristic behavior to the gross behavior of the aggregation. The discrete contributions of impulses and energy discharges do not occur all at the same time or location, nor do they all have the same intensity; however, if the number of contributions occurring and observed simultaneously is so large that the contribution of the individual aggregant is unidentifiable, being masked by the gross effect, we may consider the overall effect as a continuous phenomenon and we therefore introduce the concept of *continuity*. The physical behavior of such a system is called the *mechanics of the continuum* or simply the *continuum*. If the number of the individual contributions is small and the gross behavior of the system does not mask the individual contributions, the method of study of the system will be the *mechanics of particulate systems*, known also as *statistical mechanics*. Statistical mechanics is more comprehensive, in the sense that it can be applied both to continuous and discrete systems, whereas continuum mechanics is restricted by the limitations mentioned above. At the sub-molecular level, where particles do not always conform to Newtonian mechanics, any attempt to apply continuum mechanics would fail.

When observations indicate that certain phenomena are localized or sudden, in other words discontinuous in space or time, it could be a result of superficial observation or of our limitations in performing fine measurements of space and time. For example, the phenomenon of fracture seems to be such a discontinuous process, both in space and time, since it appears at once and at specific locations. The process, however, is a progressive separation of molecular bonds (Freudenthal 1950), starting at cavitation points and gradually expanding to adjacent points, all in a relatively short but finite time (Irwin 1958). The sudden appearance of the fracture is due to the release of accumulated energy in the corpuscles. This "chain reaction" in the corpuscles can certainly be described as a continuous process by that part of continuum mechanics which deals with discontinuities.

When we consider the material as a continuum, a great simplification is afforded, since instead of dealing with discrete phenomena of innumerable particles, we are concerned with the gross behavior of the substance.

1.7 Homogeneity and Isotropy

In the analysis of the mechanical behavior of materials at the phenomenological or at the structural level, *homogeneity* is very often postulated, as expressing the identity of properties in all fragments of the material under identical circumstances. Although the particles themselves vary somewhat in their properties, their distribution at random makes it statistically admissible to deal with the material as homogeneous when, as mentioned, the investigation is confined to the phenomenological or to the structural level. Homogeneity relates mainly to the physical properties of the material, like density, moisture, etc. Materials which are not homogeneous, are called *heterogeneous*.

The assumption of *isotropy* expresses the identity of properties in the material in all directions, in other words the fact that the properties are independent of directions in the body. When a property assumes different values in the different directions the material is *anisotropic* with respect to that property.

Most materials we are dealing with may be safely assumed homogeneous, or at least quasi-homogeneous (Reiner 1954), when the dimensions of the body investigated exceed by far the dimensions of the biggest particle. The assumption of isotropy, however, should be rated fair. Liquids and aggregates, composed of many small crystals, may or may not be isotropic with respect to a given property. Manufacturing processes such as rolling or extrusion of metals induce orientation in the crystals, thus being a source of anisotropy. Also many natural processes, like the structure of wood fibers or the formation of geological deposits such as varved clays, affect the material differently in different directions. Nevertheless, an assumption of isotropy, at least as an initial approach, is often stipulated.

It has become a common practice for engineers to claim lack of homogeneity or of isotropy whenever discrepancies between theory and experiments are encountered. This claim, however, should be critically viewed. In the majority of cases it can be repudiated and the explanation for the discrepancies should be sought rather in the "theories" themselves. In our present treatment of the subject we shall assume homogeneity as well as isotropy, unless otherwise indicated.

1.8 Soils as Multi-phase Mixtures

Soil, being a natural material, cannot be expected to have a single-phase structure. Unlike most materials encountered in our daily life or in engineering endeavors, which are manufactured and have constant and reproducible properties, the composition of soils is discriminate and varies randomly. Soils are multi-phase materials.

Soil is an aggregation of particles of different sizes and shapes, in an array with pore spaces between the particles. The pore spaces may be filled with a liquid such as water, liquid hydrocarbons, etc. containing various amounts of solubles, and/or a gas such as air, carbon monoxide, carbon dioxide, oxygen, hydrocarbons, etc. The particles representing the solid phase may be of organic or non-organic nature; of various mineralogical origins; of fractions varying from visible sizes to microscopic colloidal particles; inert or active.

This wide selection of soils is a result of a long process of decomposition and weathering of various parent rock materials by local climatic and geophysical effects, which makes soils vary from one location to another, as well as from one depth to another.

However, in spite of the multi-phase structure and variety of composition, the numerous components of the soil can be grouped, for practical purposes, into two or three distinct *phases* or *constituents*, each phase of distinctly observable and measurable physical properties. The multi-phase soil complex is considered *homogeneous* or *quasi-homogeneous*, when the constituents are distributed evenly, or *heterogeneous* when its phases are not evenly distributed. For instance, uneven distribution of density or moisture would render a soil heterogeneous.

In order to investigate a multi-phase material such as soil with the tools used for homogeneous materials, the following considerations have been adopted.

1. Discerning between primary or basic phenomena and secondary or marginal phenomena, and then considering the primary phenomena in the

first approach, and investigating the secondary phenomena only inasmuch as they may have an effect upon the behavior of the soil as a whole.

2. Applying the assumption, inherent in the physical approach, that macro-phenomenologically the mechanical behavior of all materials is qualitatively the same, and that the differences between materials stemming from their variety are quantitative and are reflected in the magnitude of the parameters of their constitutive equations.

3. Assuming that not only is each phase in itself considered homogeneous or quasi-homogeneous, but that the various phases are quasi-homogeneously distributed throughout the mixture. A mass point of the mixture of say $\mathfrak{N}$ constituents is considered to be occupied simultaneously by the $\mathfrak{N}$ different constituents. This state is known as the Fick–Stefan (Stefan 1871) *equipresence of constituents* or *superposition of constituents*.

The division of soils into two phases, solids and liquids when saturated, into three phases, solids, liquids and gas when partially saturated, or again into two phases, solids and gas when dry, is a fair assumption based on experience, with its origin well rooted in past studies. The division into a finite number of phases is only academic, as was already pointed out. The solid particles themselves are composed of several constituents, distinguished by their mineralogical nature, size and geometry. The liquid phase, the water in our case, is in liquid or vapor state, with electrolyte concentrations varying in nature and magnitude throughout the solution. The gas phase may contain one or several gaseous constituents of diverse properties. If any of the components exhibits a different and significant phenomenon not exhibited by the other components and worth considering, additional phases may be added to the study and/or a different distribution of phases may be assumed.

In order to apply the theory of multi-phase mixtures to soils, it is not necessary to present the entire study leading to the theory of multi-phase mixtures, which can be found in textbooks. Still, the author feels that a detailed treatment would enhance the understanding of the exposition, and at the same time it would allow further remarks relevant to the theory and its pertinence to soils, thereby making it more accessible to other investigators.

1.9 The Methodology of Continuum Mechanics

The solution of boundary-value problems in continuum mechanics entails the determination of the stress p_{ij} and the deformation field x_i within the material region. The solution has to satisfy four *continuity* or *balance equations*, namely density, moment, moment of momentum and internal energy; it has to comply with the *inequality of entropy production* known as the *postulate of irreversibility*; it has to satisfy the *dual constitutive equations* of the material in case; and it has to satisfy the *boundary conditions* on the envelope of the specific configuration. Our study concentrates on the balance equations of multi-phase materials, extended to soils. The balance equations express principles of mechanics and thermodynamics that are general enough to refer to any material, irrespective of its nature. The differences between materials appear when constitutive equations are introduced. It should be pointed out that while the balance equations enjoy universal acceptance when tested by the principles of physics, constitutive equations often do not satisfy certain physical or mathematical requirements.

As we shall see in Chap. 7, any material, soils included, is properly described by *two and only two* constitutive equations, subject to certain rules that qualify them as such. A mere correlation of engineering parameters, as is often done in soil mechanics, can hardly be considered a constitutive equation. Constitutive equations, required to give a unique account of the stress–strain relations of the soil constituents, are not supposed to vary from one engineering configuration to another, and must be confined to the two equations required for any material. In that respect, very little effort, if any, has been devoted in soil mechanics to determining the real constitutive equations of soils. This work is trying to improve this matter.

2 Deformation and Strain

2.1 Deformation and Displacement

The definition of space requires the use of a *measure* to relate individual points in this space. A system of axiomatics rectifies the existence of such a measure. Subsequently the measure determines the *geometry*, a formal concept of the properties of the space. In the first chapters of this study we shall adhere to the general coordinate system, in spite of the fact that it would be sufficient to use the simpler Cartesian coordinates, and occasionally the curvilinear orthogonal coordinates. As we progress and the derivations become more practical, we shall refer to the Cartesian coordinates. The tensorial notation will then be simplified to a great extent and will contain only covariant indices, i.e. subscripts.

Let a material point, described by a radius vector X^L, defined in a right-handed coordinate system X^K, $(K = K, L, M)$ of the continuous medium and contained in a volume S bounded by a surface A, be defined by a radius vector x^i in another coordinate system, Fig. 2.1.1. We say that the configuration of the region $\mathscr{R}$ including the point X^L is *deformed* into a configuration of a region $\mathscr{r}$ including the point x^l. The variables X^L are the *material* variables and form the *material coordinates*, while x^l are the *spatial* variables and form the *spatial coordinates*. Then x^l will be a function of X^L and vice versa

$$x^l = x(X^L)^l; \quad X^L = X(x^l)^L \tag{2.1.1}$$

The continuity axiom requires that the deformation throughout $\mathscr{R}$ and $\mathscr{r}$ be one-to-one valued, and consequently the deformation defines two *deformation gradients*, tensors of the second order, which, in a material field, are

$$x^l_{,M} = \frac{\partial x^l}{\partial X^M}; \quad X^L_{,m} = \frac{\partial X^L}{\partial x^m} \tag{2.1.2}$$

In a strained body, the motion of a point from an initial position X^L to a

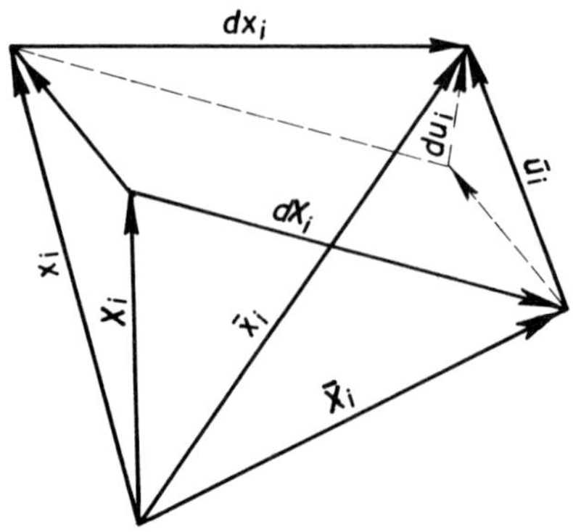

Fig. 2.1.1. Deformation of a line element.

position x^l is represented by the *displacement vector* $U(X)^L$, and defined by

$$U(X)^L = x(X)^l - X^L \tag{2.1.3}$$

Similarly, a point $\overline{X^L} = X^L + \mathrm{d}X^L$ in the neighborhood of X^L is displaced to a position $\bar{x}(X)^l = x(X)^l + \mathrm{d}x(X)^l$, Fig. 2.1.1, so that the displacement vector U^L representing the displacement of a point X^L becomes

$$\overline{U}(X)^L = \bar{x}(X)^l - \overline{X^L} \tag{2.1.4}$$

The magnitude and directions of the line element $\mathrm{d}X^L$ and $\mathrm{d}x^l$ before and after the deformation, respectively, can be determined in terms of the coordinate system in x and the coordinate system in X, according to Eqs. (2.1.1) and (2.1.2).

In symmetry with Eq. (2.1.3), the displacement vector u^l of an initial position x^l displaced to a position X^L may be defined

$$u(x)^l = X(x)^L - x^l \tag{2.1.5}$$

and so a position $\overline{x^l} = x^l + \mathrm{d}x^l$ in the neighborhood of x^l, if displaced to a position $\overline{X^L} = X^L + \mathrm{d}X^L$ in the neighborhood of X^L, may be represented by the displacement vector u^l

$$\bar{u}(x)^l = \overline{X}(x)^L - \overline{x^l} \tag{2.1.6}$$

This symmetry between the two coordinate systems X^L and x^l is referred to (Truesdell and Toupin 1960) as the *principle of duality* and is manifested in further symmetric relations, as we shall see.

The total differentials of X^L and of x^l from Eqs. (2.1.3) and (2.1.5), respectively, are

$$\mathrm{d}X(x)^L = X(x)^L{}_{,\alpha}\,\mathrm{d}x^\alpha = G^L_\alpha[G^\alpha_\beta - U(x)^\alpha_{,\beta}]\,\mathrm{d}x \tag{2.1.7}$$

$$\mathrm{d}x(x)^l = x(x)^l_{,\alpha}\,\mathrm{d}x^\alpha = g^l_\alpha[g^\alpha_\beta - u(x)^\alpha_{,\beta}]\,\mathrm{d}x \tag{2.1.8}$$

where G^{LM}, G_{LM} and g^{lm}, g_{lm} are the *metric tensors*, functions of X^L and x^l respectively.

Thus, a *displacement gradient* $u(x)^l_{,m}$ is defined in terms of the spatial or *Eulerian* coordinates, while another displacement gradient $U(X)^L_{,M}$ is defined in terms of the material or *Lagrangian* coordinates. It is evident from Eqs. (2.1.7) and (2.1.8) that the deformation gradients, $X^L_{,m}$ and $x^l_{,M}$, can be expressed in terms of the displacement gradients, $u^l_{,M}$ and $U^L_{,m}$, respectively, and vice versa.

The displacements

$$\mathrm{d}U^L = \overline{U^L} - U^L = \mathrm{d}x(X)^l - \mathrm{d}X^L \tag{2.1.9}$$

$$\mathrm{d}u^l = \overline{u^l} - u^l = \mathrm{d}X(x)^l - \mathrm{d}x^L \tag{2.1.10}$$

are the changes in the relative positions of the segment in the Lagrangian and Eulerian spaces, respectively, if those changes do not destroy the continuity of the region.

From Eqs. (2.1.9) and (2.1.10) the obvious relation follows

$$\mathrm{d}u^l \equiv -\mathrm{d}U^L \tag{2.1.11}$$

Since dX and dx are the distances between the displaced points before and after the displacement, then, applying Eqs. (2.1.7) and (2.1.8) to the equations defining those distances, we obtain

$$(\mathrm{d}X)^2 = G(X)_{\alpha\beta}\,\mathrm{d}X^\alpha\,\mathrm{d}X^\beta = C(x)_{\alpha\beta}\,\mathrm{d}x^\alpha\,\mathrm{d}x^\beta \tag{2.1.12}$$

$$(\mathrm{d}x)^2 = g(x)_{\alpha\beta}\,\mathrm{d}x^\alpha\,\mathrm{d}x^\beta = C(X)_{\alpha\beta}\,\mathrm{d}X^\alpha\,\mathrm{d}X^\beta \tag{2.1.13}$$

where c_{lm} and C_{LM} are the *Cauchy* and *Green deformation tensors*, respectively, defined

$$c_{lm} = g_{lm} - x_{l,m} - x_{m,l} + x_{\alpha,l}x_{\alpha,m} = g_{lm} - u_{l,m} - u_{m,l} + u_{\alpha,l}u_{\alpha,m} \tag{2.1.14}$$

$$\begin{aligned} C_{LM} &= G_{LM} + X_{L,M} + X_{M,L} + X_{\alpha,L}X_{\alpha,M} \\ &= G_{LM} + U_{L,M} + U_{M,L} + U_{\alpha,L}U_{\alpha,M} \end{aligned} \tag{2.1.15}$$

from which it can be seen that the deformation tensors c_{lm} and C_{LM} are symmetric second-order tensors. Another form for the deformation tensors may be given, considering Eq. (2.1.2)

$$c_{lm} = G_{\alpha\beta}X^\alpha_{,l}X^\beta_{,m} \tag{2.1.16}$$

$$C_{LM} = g_{\alpha\beta} x^{\alpha}_{,L} x^{\beta}_{,M} \tag{2.1.17}$$

It is obvious that if $C_{LM} = c_{lm}$, we have a *rigid body* deformation.

2.2 Strain

If a length $\mathrm{d}X$ in a region $\mathcal{R}$ of a body, defined in the X^L coordinate system, changes during the displacement, one says that the body is *strained*, thus the difference $\mathrm{d}x^2 - \mathrm{d}X^2$ or $\mathrm{d}X^2 - \mathrm{d}x^2$, where $\mathrm{d}X$ is the initial length and $\mathrm{d}x$ is the deformed length, can be taken as a measure of *strain*. Considering Eqs. (2.1.12) and (2.1.13) we obtain

$$\begin{aligned}
(\mathrm{d}x)^2 - (\mathrm{d}X)^2 &= (g_{\alpha\beta} - c_{\alpha\beta})\,\mathrm{d}x^{\alpha}\,\mathrm{d}x^{\beta} \\
&= (\delta_{\alpha\gamma} u^{\gamma}_{,\beta} + \delta_{\alpha\rho} u^{\rho}_{,\beta} - \delta_{\gamma\rho} u^{\gamma}_{,\alpha} u^{\rho}_{,\beta})\,\mathrm{d}x^{\alpha}\,\mathrm{d}x^{\beta} \\
&= 2\overset{E}{\varepsilon}_{\alpha\beta}\,\mathrm{d}x^{\alpha}\,\mathrm{d}x^{\beta} \\
&= (C_{\alpha\beta} - G_{\alpha\beta})\,\mathrm{d}X^{\alpha}\,\mathrm{d}X^{\beta} \\
&= (\delta_{\alpha\gamma} U^{\gamma}_{,\beta} + \delta_{\alpha\rho} U^{\rho}_{,\beta} + \delta_{\gamma\rho} U^{\gamma}_{,\alpha} U^{\rho}_{,\beta})\,\mathrm{d}X^{\alpha}\,\mathrm{d}X^{\beta} \\
&= 2\overset{L}{\varepsilon}_{\alpha\beta}\,\mathrm{d}X^{\alpha}\,\mathrm{d}X^{\beta}
\end{aligned} \tag{2.2.1}$$

where

$$\overset{E}{\varepsilon}_{lm} = \tfrac{1}{2}(g_{lm} - c_{lm})$$

is the *Eulerian strain tensor* and

$$\overset{L}{\varepsilon}_{LM} = \tfrac{1}{2}(G_{LM} - C_{LM})$$

is the *Lagrangian strain tensor* (Hanin and Reiner 1956).

Obviously, if the differences $(\mathrm{d}x)^2 - (\mathrm{d}X)^2$ or $(\mathrm{d}X)^2 - (\mathrm{d}x)^2$ vanish, i.e. $\mathbf{c} = \mathbf{C} = \mathbf{1}$, the displacement is a *rigid body motion*, and since no strain is involved the changes in the position of the lines $\mathrm{d}X^L$ and $\mathrm{d}x^l$, respectively, are either translational or rotational motions.

Replacing the values of $\mathrm{d}x^i$ from Eq. (2.1.8) and of $\mathrm{d}X^i$ from Eq. (2.1.9) into Eq. (2.2.1) we obtain

$$\overset{E}{\varepsilon}_{ij} = \tfrac{1}{2}[\delta_{i\gamma} u^{\gamma}_{,j} + \delta_{j\rho} u^{\rho}_{,i} - \delta_{\gamma\rho} u^{\gamma}_{,i} u^{\rho}_{,j}] \tag{2.2.2}$$

$$\overset{L}{\varepsilon}_{ij} = \tfrac{1}{2}[\delta_{i\gamma} U^{\gamma}_{,j} + \delta_{j\rho} U^{\rho}_{,i} + \delta_{\gamma\rho} U^{\gamma}_{,i} U^{\rho}_{,j}] \tag{2.2.3}$$

which represent the components of the Eulerian and Lagrangian strain tensors, respectively.

2.3 Strain Measures

The Eulerian and the Lagrangian strain tensors are but two measures of strain. According to a *theorem of equivalence* formulated by Rivlin and Ericksen (1955): *Any uniquely invertible isotropic second-order tensor functions of c_{lm} or C_{LM} are strain measures in the x^l and X^L coordinates, respectively*. Among them there is a class of strain measures of the form $\frac{1}{2}\mathfrak{n}^{-1}(\mathbf{1}-\mathbf{c}^{\mathfrak{n}})$ and $\frac{1}{2}\mathfrak{n}^{-1}(\mathbf{C}^{\mathfrak{n}}-\mathbf{1})$, some quite well known. Since no restrictions have been imposed on the magnitude of the deformations u^l and U^L or of the deformation gradients $u^l_{,M}$ and $U^L_{,m}$ all strain measures are finite measures, and as such are formally suitable to represent a straining phenomenon. The choice of any particular measure is mostly a matter of convenience, either to simplify the mathematical structure of the problem or to render a specific physical meaning to the phenomena or expressions.

Consequently, many strain measures have been studied and proposed, mostly from a practical point of view. Strains are never measured. What is measured, in the laboratory or in the field, are the changes in lengths.

One defines *linear extension* $d(X)$, $d(x)$ as the ratio of the change in length of the line element considered, in the undeformed and deformed reference coordinates, respectively

$$d(X) = \frac{dx - dX}{dX} = \frac{dx}{dX} - 1 = \lambda(X) - 1 \tag{2.3.1}$$

$$d(x) = \frac{dX - dx}{dx} = \frac{dX}{dx} - 1 = \lambda(x) - 1 \tag{2.3.2}$$

where $\lambda(X)$ and $\lambda(x)$ are the *stretches* with respect to the undeformed and deformed states. Hence, if $\lambda(X) > 1 > \lambda(x)$, the stretch is an *increase* in length, while if $\lambda(X) < 1 < \lambda(x)$ it is a decrease in length; $\lambda = 0$ gives a rigid body motion. The range of possible values of λ and d, are

$$0 < \lambda < \infty; \quad -1 < d < \infty \tag{2.3.3}$$

Substituting Eqs. (2.1.12) and (2.1.13) into Eqs. (2.3.1) and (2.3.2) we obtain

$$\lambda(X)^2 = \frac{(dx)^2}{(dX)^2} = \frac{C_{\alpha\beta}\, dX^\alpha\, dX^\beta}{G_{\gamma\rho}\, dX^\gamma\, dX^\rho} = C_{\alpha\beta}N(X)^\alpha N(X)^\beta = \frac{g_{\alpha\beta}\, dx^\alpha\, dx^\beta}{c_{\gamma\rho}\, dx^\gamma\, dx^\rho}$$
$$= \frac{1}{c_{\alpha\beta}n(x)^\alpha n(x)^\beta} \tag{2.3.4}$$

$$\lambda(x)^2 = \frac{(dX)^2}{(dx)^2} = \frac{c_{\alpha\beta}\, dx^\alpha\, dx^\beta}{g_{\gamma\rho}\, dx^\gamma\, dx^\rho} = c_{\alpha\beta}n(x)^\alpha n(x)^\beta = \frac{G_{\alpha\beta}\, dX^\alpha\, dX^\beta}{C_{\gamma\rho}\, dX^\gamma\, dX^\rho}$$
$$= \frac{1}{C_{\alpha\beta}N(X)^\alpha N(X)^\beta} \tag{2.3.5}$$

where $N(X)^L$ and $n(x)^l$ are unit vectors in the directions of the coordinates, in the respective frame of reference of the undeformed and deformed coordinates.

While we may consider $\lambda(X)^2$ and $\lambda(x)^2$ as the squares of the lengths of the stretch vectors $\lambda(X)^L$ and $\lambda(x)^l$, it has been shown (Ericksen 1960a) that these components are in the directions $N(X)^L$ and $n(x)^l$ of the principal components of C_{LM} and c_{lm}.

In a simple extension, in the Cartesian coordinates for example, where the body is strained only in the i direction, the only existing component of the strain tensor ε_{ij} is $\varepsilon_{ii} = \varepsilon_{11}$. All other components vanish identically. A length L_0 before straining becomes L_1 after straining and the measured change in length is

$$\Delta L = L_1 - L_0 \tag{2.3.6}$$

and the matrix of the strain tensor ε_{ij} has the following form

$$|\varepsilon_{ij}| = \begin{vmatrix} \varepsilon_{ij} & 0 & 0 \\ 0 & 0 & 0 \\ 0 & 0 & 0 \end{vmatrix}$$

Any function of $\lambda(X)$ or $\lambda(x)$ may be considered as a measure of strain. For that particular straining the measures studied look as follows.

1. The *Cauchy strain measure*

$$\overset{C}{\varepsilon}_{ii} = \lambda(X)_i - 1; \quad n = \tfrac{1}{2}; \quad \overset{C}{\varepsilon}_{ij} = \sqrt{C_{ij}} - \delta_i \tag{2.3.7}$$

2. The *Green–St Venant strain measure*, a simpler version of the Lagrangian strain measure

$$\overset{G}{\varepsilon}_{ii} = \tfrac{1}{2}(\lambda(X)^2{}_i - 1); \quad n = 1; \quad \overset{G}{\varepsilon}_{ij} = \tfrac{1}{2}(C_{ij} - \delta_{ij}) \tag{2.3.8}$$

3. The *Almansi–Hamel strain measure*, a simpler version of the Eulerian strain measure

$$\overset{A}{\varepsilon}_{ii} = \tfrac{1}{2}\left(1 - \frac{1}{\lambda(X)^2{}_i}\right); \quad n = 1; \quad \overset{A}{\varepsilon}_{ij} = \tfrac{1}{2}(c_{ij} - \delta_{ij}) \tag{2.3.9}$$

4. The *Swainger strain measure*

$$\overset{S}{\varepsilon}_{ii} = 1 - \frac{1}{\lambda(X)_i}; \quad n = \tfrac{1}{2}; \quad \overset{S}{\varepsilon}_{ij} = \delta_{ij} - \sqrt{c_{ij}} \tag{2.3.10}$$

5. The *Hencky strain measure*

$$\overset{H}{\varepsilon}_{ii} = \ln \lambda(X)_i = -\ln \lambda(x)_i; \tag{2.3.11}$$

All these measures are *finite* strain measures, and Reiner (1948) stated the following requirements to be satisfied by any strain measure.

1. For $dx^l = dX^L$ there is no change in length, and the strain vanishes.
2. For infinitesimal deformations, the strain reduces to

$$\overset{\text{inf}}{\varepsilon}_{lm} = \tfrac{1}{2}(u_{l,M} + u_{m,L}) \equiv \tfrac{1}{2}(U_{L,m} + U_{M,l}).$$

3. The strain is dimensionless.

The Cauchy measure is the most commonly known strain measure applied in engineering. Another commonly used strain measure is the Swainger strain measure; it is particularly attractive, since the proper numbers of $\sqrt{\mathbf{C}}$ and $\sqrt{\mathbf{c}}$ are the stretches $\lambda(X)$ and $\lambda(x)$, as may be seen from Eqs. (2.3.4) and (2.3.5), meaning that by rotation of the coordinate system (see Sect. A.15 of Appendix A) the stretches $\lambda(X)$ and $\lambda(x)$ become the principal components of tensor $\sqrt{C_{LM}}$ and $\sqrt{c_{lm}}$, respectively, in their principal axes. Each of the other measures has its own merit, being derived for specific purposes and conditions. Further in our study we will emphasize the Hencky measure and its advantages when studying compressible materials such as soils.

To appreciate the differences between the strain measures, let us consider a bar extending to twice its length. The different measures will yield*

$$\overset{C}{\varepsilon}_{ll} = -1, \quad \overset{G}{\varepsilon}_{ll} = -1.5, \quad \overset{A}{\varepsilon}_{ll} = 0.375, \quad \overset{S}{\varepsilon}_{ll} = 0.5, \quad \overset{H}{\varepsilon}_{ll} = -0.66,$$

while if the bar is compressed to half its length, we get

$$\overset{C}{\varepsilon}_{ll} = 0.5, \quad \overset{G}{\varepsilon}_{ll} = 0.375, \quad \overset{A}{\varepsilon}_{ll} = 1.5, \quad \overset{S}{\varepsilon}_{ll} = 1, \quad \overset{H}{\varepsilon}_{ll} = 0.66.$$

Since any function of C_{LM} or c_{lm} is a strain measure, many strain measures have been proposed (Biot 1939; Mooney 1948; Murnaghan 1941; Oldroyd 1950; Signorini 1930) to suit specific materials or test conditions. They are, however, of no particular interest to us at present. Some of the strain measures suggested were not expressed to form a tensorial quantity and to transform as such, among them the Hencky strain measure (Klausner 1969a), which has only recently (Fitzgerald 1980) received its proper treatment.

Irrespective of its definition, the strain ε_{ij} is a symmetric tensor of order two. Its three diagonal components ε_{ii}, ε_{jj} and ε_{kk} are the *normal* strain components resulting in extension when negative, or contraction when positive, of the infinitesimal volume element in the respective direction. The six remaining strain components of ε_{ij}, where $i \neq j$, of which each two are equal, are the three *shear* strain components, representing distortions that may occur in the three planes of the respective coordinate system.

The strain tensor ε_{ij}, as a symmetric second-order tensor, can be resolved into two tensors, a *spherical* or *isotropic* strain tensor $\frac{1}{3}\varepsilon_{\alpha\alpha}\delta_{ij}$, and a *traceless*

* Here the notation used in soil mechanics has been adopted. In mechanics an extension is denoted positive and a contraction negative.

or *deviatoric* strain tensor e_{ij}, as follows from Eqs. (A.20.1)–(A.20.3), Appendix A

$$\varepsilon_{ij} = \tfrac{1}{3}\varepsilon_{\alpha\alpha}\delta_{ij} + e_{ij} + \tfrac{1}{3}\mathrm{I}_{\varepsilon}\delta_{ij} + e_{ij} \tag{2.3.12}$$

where

$$\varepsilon_{\alpha\alpha} = \varepsilon_{ii} + \varepsilon_{jj} + \varepsilon_{kk} = \varepsilon_{11} + \varepsilon_{22} + \varepsilon_{33} = \mathrm{I}_{\varepsilon} = \overline{\mathrm{I}_{\varepsilon}} \tag{2.3.13}$$

$$e_{ij} = \varepsilon_{ij} - \tfrac{1}{3}\varepsilon_{\alpha\alpha}\delta_{ij}$$

$$\equiv \frac{1}{3}\begin{vmatrix} 2\varepsilon_{ii} - \varepsilon_{jj} - \varepsilon_{kk} & \varepsilon_{ij} & \varepsilon_{ik} \\ \varepsilon_{ij} & 2\varepsilon_{jj} - \varepsilon_{ii} - \varepsilon_{kk} & \varepsilon_{jk} \\ \varepsilon_{ik} & \varepsilon_{jk} & 2\varepsilon_{kk} - \varepsilon_{ii} - \varepsilon_{jj} \end{vmatrix}$$

$$= \frac{1}{3}\begin{vmatrix} 2\varepsilon_{11} - \varepsilon_{22} - \varepsilon_{33} & 0 & 0 \\ 0 & 2\varepsilon_{22} - \varepsilon_{11} - \varepsilon_{33} & 0 \\ 0 & 0 & 2\varepsilon_{33} - \varepsilon_{11} - \varepsilon_{22} \end{vmatrix} \tag{2.3.14}$$

The resolution of the strain tensor, as shown here, can be performed on any of the strain measures, whether infinitesimal or finite. The physical sense of the resolved strain components, however, is questionable, except for the Hencky strain measure, as will be shown in Sect. 2.8 and, of course, for the infinitesimal strain. In the investigation of soils the volume changes play an important role, and one of the outstanding properties that only the Hencky and the infinitesimal strain possess is that when resolved, according to Eq. (2.3.12), the spherical component $\varepsilon_{\alpha\alpha}$ represents the volumetric strain and the traceless component e_{ij} represents the distortion. The implications of this property will be explored further in the succeeding chapters.

2.4 Invariants of the Deformation Tensor

From Eqs. (2.3.4) and (2.3.5) we may infer that the principal axes of the deformation tensors C_{LM} and c_{lm} are in the direction of the principal axes of the strains, in X and x, corresponding to the squares of stretches $\lambda(X)_L$ and the reciprocal squares of stretches $\lambda(x)_l$, respectively, since $\lambda(X) = 1/\lambda(x)$. We shall therefore use C_{LM} denoting $C(X)_{LM}$ and consequently c_{lm} denoting $c(x)_{lm}$

$$C_{LL} = \lambda_L^2; \quad c_{ll} = \frac{1}{\lambda_l^2} \tag{2.4.1}$$

and in matricial form

$$|C_{LM}| = \begin{vmatrix} \lambda_{11}^2 & 0 & 0 \\ 0 & \lambda_{22}^2 & 0 \\ 0 & 0 & \lambda_{33}^2 \end{vmatrix} \tag{2.4.2}$$

$$|c_{lm}| = \begin{vmatrix} \lambda_{11}^{-2} & 0 & 0 \\ 0 & \lambda_{22}^{-2} & 0 \\ 0 & 0 & \lambda_{33}^{-2} \end{vmatrix} \tag{2.4.3}$$

The invariants of the deformation tensor C_{LM} may now be expressed as follows, according to Sect. A17, Appendix A:
the *principal invariants*

$$\begin{aligned} \mathrm{I}_C &= \lambda_{11}^2 + \lambda_{22}^2 + \lambda_{33}^2 \\ \mathrm{II}_C &= \lambda_{11}^2\lambda_{22}^2 + \lambda_{22}^2\lambda_{33}^2 + \lambda_{33}^2\lambda_{11}^2 \\ \mathrm{III}_C &= \lambda_{11}^2\lambda_{22}^2\lambda_{33}^2 \end{aligned} \tag{2.4.4}$$

the *moment invariants*

$$\begin{aligned} \bar{\mathrm{I}}_C &= \mathrm{I}_C = \lambda_{11}^2 + \lambda_{22}^2 + \lambda_{33}^2 \\ \overline{\mathrm{II}}_C &= \lambda_{11}^4 + \lambda_{22}^4 + \lambda_{33}^4 \\ \overline{\mathrm{III}}_C &= \lambda_{11}^6 + \lambda_{22}^6 + \lambda_{33}^6 \end{aligned} \tag{2.4.5}$$

and the *octahedral invariant*

$$\Pi_C = \tfrac{1}{4}[(\lambda_{11}^2 - \lambda_{22}^2)^2 + (\lambda_{22}^2 - \lambda_{33}^2)^2 + (\lambda_{33}^2 - \lambda_{11}^2)^2] \tag{2.4.6}$$

The corresponding principal invariants of c_{ij} are

$$\begin{aligned} \mathrm{I}_c &= \frac{1}{\lambda_{11}^2} + \frac{1}{\lambda_{22}^2} + \frac{1}{\lambda_{33}^2} \\ \mathrm{II}_c &= \frac{1}{\lambda_{11}^2\lambda_{22}^2} + \frac{1}{\lambda_{22}^2\lambda_{33}^2} + \frac{1}{\lambda_{33}^2\lambda_{11}^2} \\ \mathrm{III}_c &= \frac{1}{\lambda_{11}^2\lambda_{22}^2\lambda_{33}^2} \end{aligned} \tag{2.4.7}$$

the corresponding moment invariants of c_{ij} are

$$\bar{\mathrm{I}}_c = \mathrm{I}_c = \frac{1}{\lambda_{11}^2} + \frac{1}{\lambda_{22}^2} + \frac{1}{\lambda_{33}^2}$$

$$\overline{\mathrm{II}}_c = \frac{1}{\lambda_{11}^4} + \frac{1}{\lambda_{22}^4} + \frac{1}{\lambda_{33}^4} \tag{2.4.8}$$

$$\overline{\mathrm{III}}_c = \frac{1}{\lambda_{11}^6} + \frac{1}{\lambda_{22}^6} + \frac{1}{\lambda_{33}^6}$$

and the corresponding octahedral invariant of c_{ij} is

$$\Pi_c = \tfrac{1}{4}\left[\left(\frac{1}{\lambda_{11}^2 - \lambda_{22}^2}\right)^2 + \left(\frac{1}{\lambda_{22}^2 - \lambda_{33}^2}\right)^2 + \left(\frac{1}{\lambda_{33}^2 - \lambda_{11}^2}\right)^2\right] \tag{2.4.9}$$

Eq. $(2.4.4)_3$ represents the square of the *volumetric stretch*, Λ, defined

$$\Lambda = \frac{\mathrm{d}v}{\mathrm{d}V} = \frac{\mathrm{d}x^1\,\mathrm{d}x^2\,\mathrm{d}x^3}{\mathrm{d}X^1\,\mathrm{d}X^2\,\mathrm{d}X^3}\,|\mathcal{J}| = \lambda_{11}\lambda_{22}\lambda_{33} = \sqrt{\mathrm{III}_C} = \frac{1}{\sqrt{\overline{\mathrm{III}}_c}} \tag{2.4.10}$$

The *volumetric dilatation* Δ may also be defined

$$\Delta = \Lambda - 1 = \frac{\mathrm{d}x^1\,\mathrm{d}x^2\,\mathrm{d}x^3}{\mathrm{d}X^1\,\mathrm{d}X^2\,\mathrm{d}X^3} - 1 \tag{2.4.11}$$

It should be remarked that a negative value of the dilatation is an increase in volume and a positive value a decrease in volume, as is conventional in geotechnical engineering, unlike the convention in mechanics.

If $\mathrm{d}x^1$, $\mathrm{d}x^2$, $\mathrm{d}x^3$ are infinitesimally small the last part of Eq. (2.4.11) can be written

$$\begin{aligned}\Lambda_{\text{inf}} &= \frac{\mathrm{d}x^1\,\mathrm{d}x^2\,\mathrm{d}x^3}{\mathrm{d}X^1\,\mathrm{d}X^2\,\mathrm{d}X^3} = \left(\frac{\mathrm{d}x^1}{\mathrm{d}X^1} - 1\right)\left(\frac{\mathrm{d}x^2}{\mathrm{d}X^2} - 1\right)\left(\frac{\mathrm{d}x^3}{\mathrm{d}X^3} - 1\right)\\ &= \frac{\mathrm{d}x^1}{\mathrm{d}X^1} + \frac{\mathrm{d}x^2}{\mathrm{d}X^2} + \frac{\mathrm{d}x^3}{\mathrm{d}X^3} - 1\\ &= \lambda_1 + \lambda_2 + \lambda_3 - 1\end{aligned} \tag{2.4.12}$$

where the second and third powers of $\mathrm{d}x^i$ are omitted, since they are very small.

From Eqs. (2.4.11) and (2.4.12) it is seen that for infinitesimally small strains the sum of the linear stretches is equal to the volumetric stretch.

2.5 Small Deformations and Infinitesimal Strains

We have seen, Eq. (2.2.1), that the Lagrangian and Eulerian strain tensors are related to the respective displacement vectors. The expressions for the components of the strain tensor simplify a great deal when expressed in Cartesian coordinates, since in these coordinates there is no distinction between covariant, contravariant or mixed indices, as has already been explained. All indices reduce to subscripts. This simplifies somewhat the components of the respective strains. The Lagrangian strain components, for instance, read as follows

$$\overset{L}{\varepsilon}_{XX} = \tfrac{1}{2}(C_{XX} - 1) = U_{X,X} + \tfrac{1}{2}(U^2_{X,X} + U^2_{Y,X} + U^2_{Z,X}) \tag{2.5.1}$$

$$\overset{L}{\varepsilon}_{XY} = \tfrac{1}{2}C_{XY} = \tfrac{1}{2}(U_{X,Y} + U_{Y,X} + U_{X,X}U_{X,Y} + U_{Y,X}U_{Y,Y} + U_{Z,X}U_{Z,Y})$$

$$\overset{L}{\varepsilon}_{XZ} = \ldots \tag{2.5.2}$$

The components of the Eulerian strain become, similarly

$$\overset{E}{\varepsilon}_{xx} = \tfrac{1}{2}(1 - c_{xx}) = u_{x,x} - \tfrac{1}{2}(u^2_{x,x} + u^2_{y,x} + u^2_{z,x}) \tag{2.5.3}$$

$$\overset{E}{\varepsilon}_{xy} = -\tfrac{1}{2}c_{xy} = \tfrac{1}{2}(u_{x,y} + u_{y,x} - u_{x,x}u_{x,y} - u_{y,x}u_{y,y} - u_{z,x}u_{z,y})$$

$$\overset{E}{\varepsilon}_{xz} = \ldots \tag{2.5.4}$$

The displacements U^L in X^L and u^l in x^l have the dimensions of length. The deformation tensors C_{LM} and c_{lm}, therefore, are dimensionless.

If the deformations and displacements are small so are their gradients. The squared terms and the multiplied terms of Eqs. (2.5.1)–(2.5.4) might become even smaller, so that in many approximations they are neglected. The strain tensors are reduced then to the *infinitesimal strain tensor*, $\overset{\text{inf}}{\varepsilon}_{ij}$ and their components retain only their first-order terms. In that case there will be no distinction between the different strain measures, as they all reduce to the same infinitesimal strain. Moreover, the distinction between the Lagrangian and Eulerian strains disappears as well. The infinitesimal strain components become then

$$\overset{\text{inf}}{\varepsilon}_{ii} = U_{i,i} = u_{i,i}; \quad \overset{\text{inf}}{\varepsilon}_{ij} = \tfrac{1}{2}(U_{i,j} + U_{j,i}) = \tfrac{1}{2}(u_{i,j} + u_{j,i}) \tag{2.5.5}$$

2.6 The Strain Invariants

As a second-order symmetric tensor, and as discussed in Appendix A, the strain tensor ε_{ij} allows the definition of its invariant quantities.

The *principal strain invariants* are

$$\mathrm{I}_\varepsilon = \delta_{\alpha\beta}\varepsilon_{\alpha\beta} = \varepsilon_{ii} + \varepsilon_{jj} + \varepsilon_{kk} = \varepsilon_{11} + \varepsilon_{22} + \varepsilon_{33} \tag{2.6.1}$$

$$\mathrm{II}_\varepsilon = \frac{1}{2!}\,\delta_{\alpha\beta}\delta_{\gamma\rho}\varepsilon_{\alpha\beta}\varepsilon_{\gamma\rho} = |E_{ij}|$$

$$= \varepsilon_{ii}\varepsilon_{jj} + \varepsilon_{jj}\varepsilon_{kk} + \varepsilon_{kk}\varepsilon_{ii} - \varepsilon_{ij}^2 - \varepsilon_{jk}^2 - \varepsilon_{ki}^2$$

$$= \varepsilon_{11}\varepsilon_{22} + \varepsilon_{22}\varepsilon_{33} + \varepsilon_{11}\varepsilon_{33} \tag{2.6.2}$$

$$\mathrm{III}_\varepsilon = \frac{1}{3!}\,e_{\alpha\beta\gamma}e_{\rho\varphi\nu}\varepsilon_{\alpha\beta}\varepsilon_{\gamma\rho}\varepsilon_{\varphi\nu} = |\varepsilon_{ij}|$$

$$= \varepsilon_{ii}\varepsilon_{jj}\varepsilon_{kk} - \varepsilon_{ii}\varepsilon_{jk} - \varepsilon_{jj}\varepsilon_{ki} - \varepsilon_{kk}\varepsilon_{ij} + 2\varepsilon_{ij}\varepsilon_{jk}\varepsilon_{ki}$$

$$= \varepsilon_{11}\varepsilon_{22}\varepsilon_{33} \tag{2.6.3}$$

where $|E_{ij}|$ is the minor of the ε_{ij} components of the matrix formed by the strain tensor.

The *moment invariants*, defined as the sum of the powers of the proper numbers of the strain tensor ε_{ij} and functions of the principal invariants, are

$$\bar{\mathrm{I}}_\varepsilon = \varepsilon_{\alpha\alpha} = \mathrm{I}_\varepsilon = \varepsilon_{ii} + \varepsilon_{jj} + \varepsilon_{kk} = \varepsilon_{11} + \varepsilon_{22} + \varepsilon_{33} \tag{2.6.4}$$

$$\overline{\mathrm{II}}_\varepsilon = \varepsilon_{\alpha\beta}\varepsilon_{\alpha\beta} = \mathrm{I}_\varepsilon^2 - 2\mathrm{II}_\varepsilon = \varepsilon_{ii}^2 + \varepsilon_{jj}^2 + \varepsilon_{kk}^2 + 2\varepsilon_{ij}^2 + 2\varepsilon_{jk}^2 + 2\varepsilon_{ki}^2$$

$$= \varepsilon_{11}^2 + \varepsilon_{22}^2 + \varepsilon_{33}^2 \tag{2.6.5}$$

$$\overline{\mathrm{III}}_\varepsilon = \varepsilon_{\alpha\beta}\varepsilon_{\beta\gamma}\varepsilon_{\gamma\alpha} = \mathrm{I}_\varepsilon^3 - 3\mathrm{I}_\varepsilon\mathrm{II}_\varepsilon + 3\mathrm{III}_\varepsilon$$

$$= \varepsilon_{ii}^3 + \varepsilon_{jj}^3 + \varepsilon_{kk}^3 + 3\varepsilon_{ij}^2(\varepsilon_{ii} + \varepsilon_{jj}) + 3\varepsilon_{jk}^2(\varepsilon_{jj} + \varepsilon_{kk}) + 3\varepsilon_{ki}^2(\varepsilon_{kk} + \varepsilon_{ii})$$

$$+ 6\varepsilon_{ij}\varepsilon_{jk}\varepsilon_{ki} = \varepsilon_{11}^3 + \varepsilon_{22}^3 + \varepsilon_{33}^3 \tag{2.6.6}$$

The *octahedral strain invariant* may be derived in several forms

$$\Pi_\varepsilon = \tfrac{1}{2}(\overline{\mathrm{II}}_\varepsilon - \mathrm{II}_\varepsilon) = \tfrac{1}{2}(\mathrm{I}_\varepsilon^2 - 3\mathrm{II}_\varepsilon) = \tfrac{1}{4}(3\overline{\mathrm{II}}_\varepsilon + \mathrm{I}_\varepsilon^2)$$

$$= \tfrac{1}{4}[(\varepsilon_{ii} - \varepsilon_{jj})^2 + (\varepsilon_{jj} - \varepsilon_{kk})^2 + (\varepsilon_{kk} - \varepsilon_{ii})^2] + 6(\varepsilon_{ij}^2 + \varepsilon_{jk}^2 + \varepsilon_{ki}^2)$$

$$= \tfrac{1}{2}(\varepsilon_{11}^2 + \varepsilon_{22}^2 + \varepsilon_{33}^2 - \varepsilon_{11}\varepsilon_{22} - \varepsilon_{22}\varepsilon_{33} - \varepsilon_{11}\varepsilon_{33}) \tag{2.6.7}$$

Considering Eq. (2.3.12) and resolving the strain tensor ε_{ij} into a spherical and a traceless tensor, $\varepsilon_{\alpha\alpha} = 0$, the following invariant relations can be obtained from Eqs. (A.17.9) and (A.17.10), Appendix A, for the traceless strain tensor

$$\bar{\mathrm{I}}_e = \mathrm{I}_e = 0 \tag{2.6.8}$$

$$\overline{\mathrm{II}}_e = -\mathrm{II}_e = e_{ii}^2 + e_{jj}^2 + e_{kk}^2 + 2e_{ij}^2 + 2e_{jk}^2 + 2e_{ik}^2$$

$$= e_{11} + e_{22} + e_{33} \tag{2.6.9}$$

$$\overline{\mathrm{III}}_e = \mathrm{III}_e = e_{ii}^3 + e_{jj}^3 + e_{kk}^3 + 3e_{ij}^2(e_{ii} + e_{jj}) + 3e_{jk}^2(e_{jj} + e_{kk})$$

$$+ 3e_{ik}^2(e_{ii} + e_{kk}) + 6e_{ij}e_{jk}e_{ik} = \tfrac{1}{3}(e_{11}^3 + e_{22}^3 + e_{33}^3) \tag{2.6.10}$$

We may also derive the octahedral strain invariant for the traceless strain tensor

$$\Pi_e = \tfrac{3}{2}\mathrm{II}_e = \tfrac{3}{4}\overline{\mathrm{II}}_e = \Pi_\varepsilon = \tfrac{3}{4}(e_{ii}^2 + e_{jj}^2 + e_{kk}^2 + 2e_{ij}^2 + 2e_{jk}^2 + 2e_{ik}^2)$$

$$= \tfrac{3}{4}(e_{11}^2 + e_{22}^2 + e_{33}^2) \tag{2.6.11}$$

The significance of the octahedral strain invariant as a "*measure of the intensity of the distortion suffered by the material*" was realized by Nadai (1950), and will be examined in Chap. 13.

2.7 The Hencky Measure of Strain

As mentioned in Sect. 2.3, the formal theorems concerning the formulation of the various strain measures have been laid down by Rivlin and Ericksen (1955). The introduction of the Hencky measure, known also as the *natural strain measure*, has been a process of finding a strain measure to describe certain phenomena properly (Hencky 1929; Murnaghan 1941; Richter 1949), rather than a possible consequence of the theorem of equivalence, (Sect. 2.3). We shall nevertheless define the Hencky measure on the basis of this theorem, and subsequently discuss some of its qualities and applications.

Many investigators have suggested several forms for the natural strain measure. What all those forms share is that they all present a logarithmic function, common or natural. Yoshimura (1953) presented the form $\log\tan\frac{1}{2}[(\pi/2) - \gamma]$ as having some merits in shear problems. Truesdell and Toupin (1960) presented a common logarithmic function of the Green and Cauchy deformation tensors, $\frac{1}{2}\log \mathbf{C}$ and $\frac{1}{2}\log \mathbf{c}$. Reiner (1948) proposed the natural logarithm of the Cauchy deformation tensor, $\ln \sqrt{(1 - \mathbf{c})}$. The importance of this measure to highly compressible materials was pointed out by Klausner (1969a). However, they failed to show that the natural strain measure forms a tensor and transforms as such. Fitzgerald (1980) presented this formulation and derived some of its applications for kinematics and energetics, showing at the same time its limitations. Here Fitzgerald's presentation is adopted, whereby two strain measures may be defined.

According to a theorem by Finger (1892) generalized by Toupin (1956), the nth powers of C^L_M and $1/\sqrt{c^l_m}$ are related through the *rotation* tensors R^L_M and R^l_m by the following equation

$$x^l_{,m} = R^l_\alpha \sqrt{C^\alpha_M} = \frac{R^\alpha_m}{\sqrt{c^\alpha_M}} \tag{2.7.1}$$

where the rotation tensors R^L_M and R^l_m are in the X^L and x^l coordinates, respectively, $x^l_{,m}$ is the deformation gradient, and $\mathbf{C}$ and $\mathbf{c}$ are the deformation tensors in X and x.

Since the deformation tensors serve as measures of length, the stretches can be expressed in terms of them, and they also serve as proper numbers for the matrices that the deformation tensors form in the direction of the unit vectors $N(X)$ and $n(x)$

$$\lambda(X) = \sqrt{(C_{\alpha\beta}N_\alpha N_\beta)}; \quad \lambda(x) = \sqrt{(c_{\alpha\beta}n_\alpha n_\beta)} \tag{2.7.2}$$

and in the direction of the principal triad the stretches become

$$\lambda(N)_1 = \sqrt{C_{11}}; \quad \lambda(N)_2 = \sqrt{C_{22}}; \quad \lambda(N)_3 = \sqrt{C_{33}}; \tag{2.7.3}$$

$$\lambda(n)_1 = \sqrt{c_{11}}; \quad \lambda(n)_2 = \sqrt{c_{22}}; \quad \lambda(n)_3 = \sqrt{c_{33}}; \tag{2.7.4}$$

Furthermore, an additively symmetric measure of strain $\mathcal{F}(\,)$ must satisfy (Truesdell and Toupin 1960)

$$\mathcal{F}\left(\frac{1}{\lambda}\right) = -\mathcal{F}(\lambda) \tag{2.7.5}$$

and one of the many smooth functions to conform to Eq. (2.6.5) is

$$\mathcal{F}(\lambda) = \ln \lambda \quad -\infty < \mathcal{F}(\lambda) < +\infty \tag{2.7.6}$$

The problem is to generalize Eq. (2.6.6) to form a tensor representing a finite strain measure. If we show that $\ln \lambda$ is a tensorial function of the diagonal tensor λ_{ij} we have shown that $\ln \sqrt{\mathbf{C}} = \frac{1}{2}\ln \mathbf{C}$ forms also a tensorial quantity. Our derivation will be given for the Green deformation tensor; for the Cauchy deformation tensor the approach is similar.

According to the Cayley–Hamilton theorem, a function of a symmetric matrix may be expressed as a polynomial of that matrix up to the second order. The coefficients of the polynomial are scalar functions of the invariants of the matrix. Since $\sqrt{\mathbf{C}}$ is analytic except for a singularity at $\sqrt{\mathbf{C}} = 0$, the Cayley–Hamilton equation implies

$$\ln \sqrt{C_{LM}} = \phi_0\delta_{LM} + \phi_1\sqrt{C_{LM}} + \phi_2 C_{LM} \tag{2.7.7}$$

with the restriction $\sqrt{\mathbf{C}} > 0$.

If tensor $\sqrt{C_{LM}}$ is diagonalized according to Eq. (2.7.2), it transforms into a tensor λ_{LM} with diagonals λ_{11}, λ_{22} and λ_{33} and all non-diagonal components vanish, $\lambda_{LM} = 0$ for $L \neq M$. This permits the exchange of matrix $|\sqrt{C_{LM}}|$ with matrix $|\lambda_{LM}|$, and Eq. (2.7.7) yields

$$\ln \lambda_{LL} = \phi_0 \delta_{LL} + \phi_1 \lambda_{LL} + \phi_2 \lambda_{LL} \quad (L = 1, 2, 3) \tag{2.7.8}$$

which explicitly is

$$\begin{aligned} \ln \lambda_{11} &= \phi_0 + \phi_1 \lambda_{11} + \phi_2 \lambda_{11}^2 \\ \ln \lambda_{22} &= \phi_0 + \phi_1 \lambda_{22} + \phi_2 \lambda_{22}^2 \\ \ln \lambda_{33} &= \phi_0 + \phi_1 \lambda_{33} + \phi_2 \lambda_{33}^2 \end{aligned} \tag{2.7.9}$$

Eq. (2.7.9) may be solved for the scalar coefficients ϕ_0, ϕ_1 and ϕ_2

$$\begin{aligned} \phi_0 &= \frac{\lambda_2 \lambda_3 \ln \lambda_1}{(\lambda_1 - \lambda_2)(\lambda_1 - \lambda_3)} + \frac{\lambda_1 \lambda_3 \ln \lambda_2}{(\lambda_2 - \lambda_1)(\lambda_2 - \lambda_3)} + \frac{\lambda_1 \lambda_2 \ln \lambda_3}{(\lambda_3 - \lambda_1)(\lambda_3 - \lambda_2)} \\ \phi_1 &= \frac{(\lambda_2 + \lambda_3)(\ln \lambda_1)^{-1}}{(\lambda_1 - \lambda_2)(\lambda_1 - \lambda_3)} + \frac{(\lambda_1 + \lambda_3)(\ln \lambda_2)^{-1}}{(\lambda_2 - \lambda_1)(\lambda_2 - \lambda_3)} + \frac{(\lambda_1 + \lambda_2)(\ln \lambda_3)^{-1}}{(\lambda_3 - \lambda_1)(\lambda_3 - \lambda_2)} \\ \phi_2 &= \frac{\ln \lambda_1}{(\lambda_1 - \lambda_2)(\lambda_1 - \lambda_3)} + \frac{\ln \lambda_2}{(\lambda_2 - \lambda_1)(\lambda_2 - \lambda_3)} + \frac{\ln \lambda_3}{(\lambda_3 - \lambda_1)(\lambda_3 - \lambda_2)} \end{aligned} \tag{2.7.10}$$

If Eqs. (2.7.10) are substituted into Eq. (2.7.7), the symmetric matrix $\ln \sqrt{C_{LM}}$ is obtained

$$|\ln \sqrt{C_{LM}}| = \begin{vmatrix} \phi_0 + \phi_1 \sqrt{C_{11}} + \phi_2 C_{11} & \phi_1 \sqrt{C_{12}} + \phi_2 C_{12} & \phi_1 \sqrt{C_{13}} + \phi_2 C_{13} \\ \phi_1 \sqrt{C_{12}} + \phi_2 C_{12} & \phi_0 + \phi_1 \sqrt{C_{22}} + \phi_2 C_{22} & \phi_1 \sqrt{C_{23}} + \phi_2 C_{23} \\ \phi_1 \sqrt{C_{13}} + \phi_2 C_{13} & \phi_1 \sqrt{C_{23}} + \phi_2 C_{23} & \phi_0 + \phi_1 \sqrt{C_{33}} + \phi_2 C_{33} \end{vmatrix} \tag{2.7.11}$$

The principal invariants of the Hencky strain measure presented here are

$$\overset{\mathbf{H}}{\mathrm{I}} = \ln \sqrt{\mathbf{C}} = \ln \lambda_1 + \ln \lambda_2 + \ln \lambda_3 = \ln \lambda_1 \lambda_2 \lambda_3 = \ln \Lambda \tag{2.7.12}$$

$$\overset{\mathbf{H}}{\mathrm{II}} = \tfrac{1}{2}[(\ln \sqrt{\mathbf{C}})_{\alpha\alpha}]^2 - [(\ln \sqrt{\mathbf{C}})^2]_{\alpha\alpha} \tag{2.7.13}$$

$$\overset{\mathbf{H}}{\mathrm{III}} = (\ln \sqrt{\mathbf{C}})^3 \tag{2.7.14}$$

the moment invariants are

$$\overset{\mathbf{H}}{\bar{\mathrm{I}}} = \overset{\mathbf{H}}{\mathrm{I}} = \ln \lambda_1 + \ln \lambda_2 + \ln \lambda_3 = \ln \lambda_1 \lambda_2 \lambda_3 = \ln \Lambda \tag{2.7.15}$$

$$\overset{H}{\overline{\text{II}}} = [(\ln \sqrt{\mathbf{C}})^2]_{\alpha\alpha} \tag{2.7.16}$$

$$\overset{H}{\overline{\text{III}}} = [(\ln \sqrt{\mathbf{C}})^3]_{\alpha\alpha} \tag{2.7.17}$$

and the octahedral invariant is

$$\overset{H}{\Pi} = \tfrac{1}{2}[3((\ln \sqrt{\mathbf{C}})^2)_{\alpha\alpha} - ((\ln \sqrt{\mathbf{C}})_{\alpha\alpha})^2] \tag{2.7.18}$$

As a finite measure of strain, the Hencky measure is non-linear in the displacement gradients $U(X)_{L,M}$ or $u(x)_{l,m}$, respectively, as may be easily seen.

Several of the properties of the Hencky measure are discussed in the following section.

2.8 The Properties of the Hencky Measure

The Hencky strain displays many favorable properties, particularly with regard to highly compressible materials such as soils. Its properties can be listed as follows:

1. Like any of the other finite strain measures, the Hencky measure is non-linear in the displacement gradients $u(x)_{l,m}$ and $U(X)_{L,M}$. Fig. 2.8.1 shows the stretch of a one-dimensional compression–extension in the Hencky measure as compared with that of the infinitesimal strain measure. It is seen in Fig. 2.8.1 that at $\lambda_3 \neq 1$, where there is no deformation, the Hencky and the infinitesimal Lagrangian measures have the same value. As the stretch λ_3 departs from the value $\lambda_3 = 0$, either by extension or by contraction, a discrepancy between the two measures is noticed, the difference increasing as the deformation increases. For $\lambda_3 = 2$ the discrepancy amounts to over 30% and for $\lambda_3 = \frac{1}{2}$ it reaches some 40%.

2. The second important property of the Hencky measure has been pointed out by Reiner (1948) and is named the *progressive association*. A progressive straining expressed in the Hencky measure forms an additive group, which makes it suitable for describing viscous and plastic deformations.

If, for instance, L_n is the measured length after the nth deformation, then the total strain is

$$\overset{H}{\varepsilon} = \ln\frac{L_n}{L_0} = \ln\frac{L_n L_{n-1} L_{n-2} \ldots L_2 L_1}{L_{n-1} L_{n-2} L_{n-3} \ldots L_1 L_0} = \ln\lambda_n + \ln\lambda_{n-1} + \ldots + \ln\lambda_1$$

$$= \overset{H}{\varepsilon}_n + \overset{H}{\varepsilon}_{n-1} + \overset{H}{\varepsilon}_{n-2} + \ldots \overset{H}{\varepsilon}_2 + \overset{H}{\varepsilon}_1 \tag{2.8.1}$$

Fitzgerald (1980) has shown that Eq. (2.8.1) does not hold if the strain is non-diagonal or if the directions change with time. This, however, very

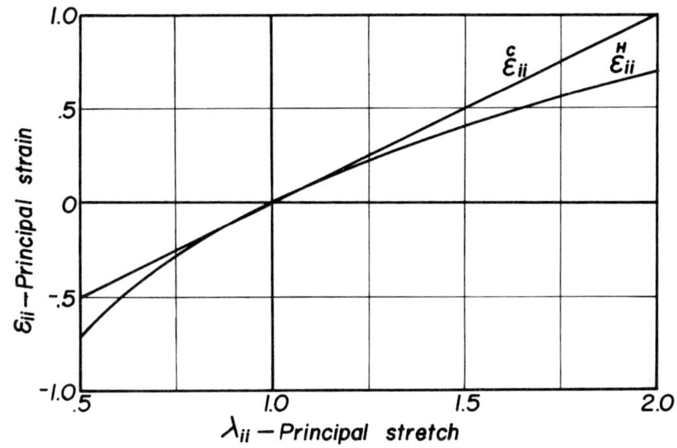

Fig. 2.8.1. The Hencky measure of strain compared to the Cauchy measure for one-dimensional stretching.

seldom happens in a deformable medium. The error introduced by using an alternative infinitesimal strain, or the physical meaning that the Hencky measure provides compared to any of the other finite strain measures, makes the Hencky strain become indispensable for highly compressible visco-plastic materials.

3. The material derivative of the Hencky strain measure in its principal directions is equal to the material derivatives of the logarithm of the principal stretches, and is equal to the flow tensor $f(d)_{ij}$ in its principal directions, as shown in Sect. 3.4.

$$\dot{\overset{\mathrm{H}}{\varepsilon}}_{\mathrm{d}} = \overline{\ln \dot{\sqrt{C_{\mathrm{d}}}}} = \frac{\overline{\dot{\sqrt{C_{\mathrm{d}}}}}}{\sqrt{C_{\mathrm{d}}}} = \frac{\dot{\lambda}_{\mathrm{d}}}{\lambda} = d_{\mathrm{d}} \tag{2.8.2}$$

This derivation has been given by Eringen (1962).

4. If the stretches in the principal directions are λ_i forming the diagonalized stretch tensor $\lambda(d)_{ij}$ the *volumetric stretch* becomes

$$\Lambda = \lambda_1 \lambda_2 \lambda_3 \tag{2.8.3}$$

and the Lagrangian *volumetric strain* or *dilatation* Δ becomes

$$\Delta = \Lambda - 1 = \lambda_1 \lambda_2 \lambda_3 - 1 \tag{2.8.4}$$

It may be shown that the third invariant of the diagonalized deformation tensor C_{LM} is equal to its determinant $|C_{LM}|$, to the square of the Jacobian, and to the square of the volumetric stretch

$$\mathrm{III}_C = |C_{LM}(d)| = J^2 = \left(\frac{\mathrm{d}v}{\mathrm{d}V}\right)^2 = \lambda_1^2 \lambda_2^2 \lambda_3^2 \tag{2.8.5}$$

The Hencky measure is not as widely used as would be expected judging from the advantages it offers. It is hoped that in the future the benefits of the versatile Hencky measure will be common knowledge, and it will be used in place of the infinitesimal strain.

2.9 Compatibility Equations

The deformation tensors C_{LM} and c_{lm} may be easily calculated from Eqs. (2.1.16) and (2.1.17). All we have to consider are the deformation gradients $X^L_{,m}$ and $x^l_{,M}$. If, however, C_{LM} is known and we want to calculate the displacements U_K we get the six equations (2.1.15)

$$C_{LM} = g_{\alpha\beta} X^{\alpha}_{,L} X^{\beta}_{,M} = F(X_1, X_2, X_3)_{LM} \tag{2.9.1}$$

with three variables, that is, we have an overdetermined system. The same holds for c_{lm}. In order to solve that problem certain restrictions have to be imposed on C_{LM}. According to a well known theorem by Riemann (1876), the necessary and sufficient condition (Lodge 1951) that any second-order symmetric tensor A_{KM} be a metric tensor of the Euclidean space is that in addition to A_{KM} being non-singular positive-definite, the Riemann–Christoffel fourth-order curvature tensor $R(A)_{KMPQ}$ formed from it has to vanish identically. In the general curvilinear coordinates we have

$$\begin{aligned} R(A)_{KMPQ} &= \tfrac{1}{2}(A_{KQ,MP} - A_{KP,MQ} + A_{MP,KQ} - A_{MQ,KP}) \\ &\quad + A_{\alpha\beta}(\Gamma^{\alpha}_{KQ}\Gamma^{\beta}_{MP} - \Gamma^{\alpha}_{KP}\Gamma^{\beta}_{MQ}) = 0 \end{aligned} \tag{2.9.2}$$

where Γ^K_{LM} are the Christoffel symbols of the second kind, (see Sects. A.13 and A.14, Appendix A).

Applying Eq. (2.9.2) to the deformation tensors, we obtain

$$R(C)_{KMPQ} = 0; \quad R(c)_{kmpq} = 0 \tag{2.9.3}$$

while applying it to any of the strain tensors we obtain explicitly

$$\begin{aligned} R(\varepsilon)_{ijkl} &= \varepsilon_{il,jk} - \varepsilon_{ik,jl} + \varepsilon_{jk,il} - \varepsilon_{jl,ik} \\ &\quad + \varepsilon_{\alpha\beta}[(\varepsilon_{l\alpha,i} + \varepsilon_{i\alpha,l} - \varepsilon_{il,\alpha})(\varepsilon_{\psi\beta,k} + \varepsilon_{\varkappa\beta,j} - \varepsilon_{\psi\varkappa,\beta}) \\ &\quad - (\varepsilon_{l\alpha,k} + \varepsilon_{k\alpha,l} - \varepsilon_{kl,\alpha})(\varepsilon_{j\beta,i} + \varepsilon_{i\beta,j} - \varepsilon_{ij,\beta})] = 0 \end{aligned} \tag{2.9.4}$$

Eqs. (2.9.3) and (2.9.4) are *compatibility equations* for the deformation tensors and the strain tensor, respectively. Only 6 out of 81 components are distinct for the three-dimensional space. The others, because of symmetry in the indices, are either identically satisfied or repetitions. Eqs. (2.9.4) are valid for any of the strain measures, as well as the displacement vectors in which they are expressed. For Cartesian coordinates the following simple form of the compatibility equation (2.9.4) is the most familiar

$$\varepsilon_{il,jk} + \varepsilon_{jk,il} - \varepsilon_{ik,jl} - \varepsilon_{jl,ik} = 0 \tag{2.9.5}$$

If the strains are infinitesimal, we can neglect the second-order terms in Eq. (2.9.4) and the same equation, (2.9.5), is obtained. The simplicity of Eq. (2.9.5) in comparison with Eq. (2.9.4) explains why investigators prefer the use of the infinitesimal strain measure.

3 Kinematics

3.1 Material Derivatives

Let $A(x, t)_i$ be a vector, function of space and time. We define $\dot{A}(x, t)_i$ as the *material derivative* of $A(x, t)_i$ as follows

$$\dot{A}_i = \frac{\mathrm{d}A(x, t)_i}{\mathrm{d}t} \equiv \frac{\partial \mathbf{A}}{\partial t} + A_{i,\alpha}\,\dot{x}^{\alpha} = \frac{\partial \mathbf{A}}{\partial t} + A_{i,\alpha}\,v^{\alpha} \tag{3.1.1}$$

where the partial time derivative $\partial\mathbf{A}/\partial t$ is taken with x held constant, and the *convection* $A_{i,\alpha}\dot{x}^{\alpha}$ with t held constant. $\dot{x}^i$ is the velocity, defined in Eq. (3.2.2). Similarly, the material derivative of a tensor of any order, say A_{ijkl}, function of space x and time t, may be defined

$$\dot{A}_{ijkl} = \frac{\mathrm{d}A_{ijkl}}{\mathrm{d}t} \equiv \frac{\partial A_{ijkl}}{\partial t} + A_{ijkl,\alpha}\dot{x}^{\alpha} \tag{3.1.2}$$

Analogous to Eq. (3.1.1), we can obtain a different material derivative $\hat{A}_i$ designated by a circumflex over vector A_i

$$\hat{A}_i = \frac{\partial A_i}{\partial t} + A_{i,\alpha}\hat{x}^{\alpha} \tag{3.1.3}$$

Subtracting Eq. (3.1.1) from Eq. (3.1.3) we obtain

$$\hat{A}_i - \dot{A}_i = A_{i,\alpha}\,(\hat{x}^{\alpha} - \dot{x}^{\alpha}) \tag{3.1.4}$$

3.2 Velocity and Speed

Let a point defined by the position vector X^L, of the deforming continuous medium, be in its undeformed state at time t_0 and be carried into a point x^i

at time t. The *axiom of continuity* requires x^i to be single valued in X^L and vice versa, and continuously differentiable to any order required, except for some singular points. We may write then

$$x^i = x(X^L, t)^i \quad \text{or} \quad X^L = X(x^i, t)^L \tag{3.2.1}$$

The *velocity* v^i is then the rate of change of the position vector x^i that occurs during the time $t - t_0$

$$v^i = \frac{\mathrm{d}x^i}{\mathrm{d}t} = \dot{x}^i = \dot{u}^i \tag{3.2.2}$$

We shall assume that x^i is not only a continuously differentiable function of t, but of the material coordinate system X^L as well. Thus the explicit form of (3.2.2) will be

$$v^i = \frac{\partial x^i}{\partial t} + x^i_{,\alpha} \frac{\mathrm{d}X^\alpha}{\mathrm{d}t} = \frac{\partial x^i}{\partial t} + x^i_{,\alpha}\dot{X}^\alpha \tag{3.2.3}$$

where the partial derivative $\partial x^i/\partial t$ is the rate of change as it appears to the observer stationed on the moving point, and $x^i_{,\alpha}\dot{X}^\alpha$ is the rate of change of the moving coordinate system.

In Chapter Two we discussed at length the two frames of reference: the undeformed coordinates and the deformed coordinates. The velocity can be determined in either of these coordinates, but its magnitude should be independent of the choice.

If we desire to express the velocity in the undeformed coordinate system X, namely, that the position vector $x^i = x(X^\mathrm{L}, t)^i$ be a function of X^L, the intrinsic derivative of the contravariant tensor x^i will have to be introduced, according to Eq. (A.14.1), Appendix A

$$v^i = \frac{\partial x^i}{\partial t} + \left(\frac{\partial x^i}{\partial X^\alpha} + \Gamma^i_{\alpha\beta}\, X^\beta\right)\frac{\mathrm{d}X^\alpha}{\mathrm{d}t} \tag{3.2.4}$$

If, however, the velocity should be expressed in the deformed coordinates $u^i = u(x^i, t)^i$, the expression would be much simpler

$$v^i = \frac{\partial x^i}{\partial t} + \frac{\partial x^i}{\partial x^\alpha}\dot{x}^\alpha \tag{3.2.5}$$

Since the two expressions, (3.2.4) and (3.2.5) are equivalent in Cartesian coordinates, it is unnecessary and inconvenient, as pointed out by Truesdell and Toupin (1960), to use both terms. Moreover, since in a stationary coordinate system Eq. (3.2.4) reduces anyway to (3.2.5), which is a much simpler term, we shall adopt this term when needed.

The superposed dot, indicating differentiation with respect to time, substitutes for d/dt which is known as the *material derivative*, and the partial differentiation with respect to time $\partial/\partial t$ is known as the *space derivative*. The material derivative is quite different from the space derivative, for, while the material derivative is the measure of the rate of change following the point, the space derivative is the measure of the rate of change as it appears to one located on the point, as we observed earlier.

If the coordinate system is a fixed frame of reference, where X is held constant during the motion, the equation of velocity (3.2.5) will be simplified considerably

$$v^i = \frac{dx^i}{dt} \equiv \frac{\partial x^i}{\partial t} \tag{3.2.6}$$

The magnitude of the velocity v^i is the *speed*

$$v = \dot{x} = \sqrt{(\dot{x}_\alpha \dot{x}_\alpha)} \tag{3.2.7}$$

The point where the speed vanishes, $v = 0$, is the *stagnation point*, and a motion where $v = \text{const}$ with time is a *steady* motion, the velocity becoming a function of space alone.

3.3 Acceleration

The rate of change of velocity, as it appears to the observer stationed on the moving point is the *acceleration*

$$a^i = \frac{dv^i}{dt} = \frac{d\dot{x}^i}{dt} = \ddot{x} = \frac{\partial \dot{x}^i}{\partial t} + x^i_{,\alpha}\, \dot{x}^\alpha \tag{3.3.1}$$

where $\partial \dot{x}^i/\partial t$ is the *local* acceleration, representing the change of the velocity field apparent to the observer stationed on the point, while the second term $x^i_{,\alpha}\, \dot{x}^\alpha$ is the *convective* acceleration, representing the change in the velocity as it appears to an observer situated outside the coordinate system.

In mechanics of the continuum, expressions are often given in terms of velocity or acceleration. Other implications of the velocity and acceleration will be discussed as they arise.

3.4 Material Derivatives of Displacement Gradients

Following Eq. (3.1.1) we may write (Truesdell and Toupin 1960)

$$\frac{d(dx^i)}{dt} = \dot{x}^i_{,\alpha}\, dx^\alpha = v^i_{,\alpha}\, dx^\alpha \tag{3.4.1}$$

By multiplying both sides of Eq. (3.4.1) by $1/\mathrm{d}X^K$ we obtain the material derivative of the displacement gradient $x^i_{,K}$

$$\frac{\mathrm{d}x^i_{,K}}{\mathrm{d}t} = \dot{x}^i_{,\alpha}\, x^\alpha_{,K} = v^i_{,\alpha}\, x^\alpha_{,K} \tag{3.4.2}$$

Also from Eq. (3.4.1) it follows that

$$\frac{\mathrm{d}(\mathrm{d}s^2)}{\mathrm{d}t} = \frac{\mathrm{d}(g_{\alpha\beta}\mathrm{d}x^\alpha \mathrm{d}x^\beta)}{\mathrm{d}t} = g_{\alpha\beta}\, \mathrm{d}x^\alpha \frac{\mathrm{d}}{\mathrm{d}t}(\mathrm{d}x^\beta) + g_{\alpha\beta}\mathrm{d}x^\alpha \frac{\mathrm{d}}{\mathrm{d}t}(\mathrm{d}x^\beta)$$

$$= g_{\alpha\beta}\, v_{\alpha,\gamma}\, \mathrm{d}x^\beta\, \mathrm{d}x^\gamma + g_{\alpha\beta} v_{\alpha,\gamma}\, \mathrm{d}x^\beta\, \mathrm{d}x^\gamma = 2f_{\alpha\beta}\, \mathrm{d}x^\alpha\, \mathrm{d}x^\beta \tag{3.4.3}$$

where f_{ij} is the *deformation rate tensor* or *flow tensor*, defined

$$f_{ij} = \tfrac{1}{2}(\dot{x}^i_{,j} + \dot{x}^j_{,i}) = \tfrac{1}{2}(v^i_{,j} + v^j_{,i}) \tag{3.4.4}$$

The flow tensor is thus a symmetric tensor. Replacing dx in the last part of Eq. (3.4.3) by dX we get

$$\frac{\mathrm{d}s^2}{\mathrm{d}t} = 2f_{\alpha\beta}\, \mathrm{d}x^\alpha\, \mathrm{d}x^\beta = 2f_{\alpha\beta} x^\alpha_{,\gamma} x^\beta_{,\rho}\, \mathrm{d}X^\gamma\, \mathrm{d}X^\rho = 2\dot{\overset{\mathrm{L}}{\varepsilon}}_{\gamma\rho}\, \mathrm{d}x^\gamma\, \mathrm{d}x^\rho \tag{3.4.5}$$

where $\dot{\overset{\mathrm{L}}{\varepsilon}}_{LM}$ is the strain rate of the Lagrangian strain tensor. If the strain rate vanishes, it is an indication of a rigid body motion, and, conversely, a necessary and sufficient condition for a motion to be a rigid body motion is $f_{ij} = 0$.

The first invariant of the flow tensor, a second-order symmetric tensor, when evaluated in view of Eq. (A.17.3) and applied to Eq. (3.4.4) results in

$$\mathrm{I}_f = \dot{x}_{\alpha,\alpha} = v_{\alpha,\alpha} = \dot{\overline{\log \mathrm{d}v}} \tag{3.4.6}$$

called the *expansion*. If the expansion vanishes, $\mathrm{I}_f = 0$, the motion is *isochoric*, namely, the motion proceeds without any change in volume.

From the material derivative of the displacement gradient the *spin tensor* w_{ij} may also be derived, as follows

$$w_{ij} = \tfrac{1}{2}(\dot{x}_{i,j} - \dot{x}_{j,i}) = \tfrac{1}{2}(v_{i,j} - v_{j,i}) \tag{3.4.7}$$

The spin tensor is skew-symmetric, thus $w_{ij} = -w_{ji}$. By adding the flow tensor to the spin tensor we obtain

$$f_{ij} + w_{ij} = \dot{x}_{i,j} = v_{i,j} \tag{3.4.8}$$

From a skew-symmetric tensor it is always possible to obtain an axial vector associated with the skew-symmetric tensor, see Eq. (A.17.13). Similarly, the spin tensor may be related to the *vorticity* vector w_k

$$w_k = e_{k\alpha\beta}\, w_{\alpha\beta} = \tfrac{1}{2}e_{k\alpha\beta}(\dot{x}_{\alpha,\beta} - \dot{x}_{\beta,\alpha}) \quad \text{or} \quad w_k \equiv \operatorname{curl}\mathbf{v} \tag{3.4.9}$$

acting in the direction normal to the plane in which the rotation occurs.

3.5 Strain Rates

The velocity and acceleration have been defined as the material derivatives of the deformation vector x^i and the displacement vector u^i. On the other hand, it has been demonstrated that the Cauchy deformation tensor, (2.1.14), is also a function of the same displacement vector. Before deriving the strain rates, the deformation tensor rates will be derived. For the Green deformation tensor C_{LM} the calculation is easy, since $\mathrm{d}X = 0$, and according to Eq. (2.1.13) we get

$$\dot{\overline{\mathrm{d}s^2}} = \dot{C}_{\alpha\beta}\,\mathrm{d}X^\alpha\,\mathrm{d}X^\beta \tag{3.5.1}$$

From (3.4.5) and (2.1.8) we have

$$\dot{\overline{\mathrm{d}s^2}} = 2f_{\alpha\beta}\,\mathrm{d}x^\alpha\,\mathrm{d}x^\beta = 2f_{\alpha\beta}\,x^\alpha_{,\gamma}\,x^\beta_{,\rho}\,\mathrm{d}X^\gamma\,\mathrm{d}X^\rho \tag{3.5.2}$$

By comparing Eqs. (3.5.1) and (3.5.2) we get

$$\dot{C}_{LM} = 2f_{\alpha\beta}\,x^\alpha_{,L}\,x^\beta_{,M} = 2\dot{\overset{E}{\varepsilon}}_{LM} \tag{3.5.3}$$

Similarly, differentiating Eq. (2.1.12) we get

$$0 = \dot{c}_{\alpha\beta}\,\mathrm{d}x^\alpha\,\mathrm{d}x^\beta + c_{\alpha\beta}\,(\dot{x}^\alpha_{,\gamma}\,\mathrm{d}x^\gamma\,\mathrm{d}x^\beta + \dot{x}^\beta_{,\gamma}\,\mathrm{d}x^\alpha\,\mathrm{d}x^\gamma) \tag{3.5.4}$$

and since $\mathrm{d}x^i$ is arbitrary, it follows that

$$\dot{c}_{ij} = -(c_{\alpha i}\,\dot{x}^\alpha_{,j} + c_{\alpha j}\,\dot{x}^\alpha_{,i}) \tag{3.5.5}$$

The rate of strain of the various strain measures can now be easily deduced from (3.5.3) and (3.5.5). The rates of strain of the Lagrangian and Eulerian strain tensors, for instance, can immediately be observed

$$\dot{\overset{L}{\varepsilon}}_{ij} = \tfrac{1}{2}\dot{C}_{ij} = f_{\alpha\beta}\,x^\alpha_{,i}\,x^\beta_{,j} \tag{3.5.6}$$

$$\dot{\overset{E}{\varepsilon}}_{ij} = f_{\alpha\beta} - \overset{E}{\varepsilon}_{\alpha i}\,\dot{x}^\alpha_{,j} + \overset{E}{\varepsilon}_{\alpha j}\,\dot{x}^\alpha_{,i} \tag{3.5.7}$$

However, not all rates of strain are of great significance in our study. Of particular interest for us is the rate of strain of the Hencky measure, $\overset{H}{\varepsilon}_{ij}$, and for the sake of completeness, also the measure of the infinitesimal strain, to which all strain measures reduce when the displacement vector u^i and its derivatives with respect to the coordinates are infinitesimal, as discussed in Sect. 2.5. The rate of strain of the Hencky measure is obtained from Eq. (2.7.7) and has been derived in Eq. (2.8.2)

$$\dot{\overset{\mathrm{H}}{\varepsilon}} = \overline{\ln \sqrt{\mathbf{C}(d)}}^{\,\bullet} = \tfrac{1}{2}\overline{\ln \mathbf{C}}^{\,\bullet} = \tfrac{1}{2}\frac{\dot{\mathbf{C}}}{\mathbf{C}} = \overline{\ln \lambda}^{\,\bullet} \tag{3.5.8}$$

The rate of strain of the infinitesimal measure is obtained from Eq. (2.5.5)

$$\dot{\overset{\mathrm{inf}}{\varepsilon}}_{ij} = \tfrac{1}{2}(\dot{u}^{i}_{,j} + \dot{u}^{j}_{,i}) = \tfrac{1}{2}(\dot{x}^{i}_{,j} + \dot{x}^{j}_{,i}) = f_{ij} \tag{3.5.9}$$

Finally we shall formulate also the rate of volumetric strain for both the finite Hencky measure and the infinitesimal measure. The rate of volumetric strain for the Hencky measure is derived from Eq. (2.7.7)

$$\dot{\overset{\mathrm{H}}{\varepsilon}}_{ij} = \tfrac{1}{2}\overline{(\ln \mathbf{C})_{ij}}^{\,\bullet} = \tfrac{1}{2}\frac{\partial(\ln \mathbf{C})_{ij}}{\partial C_{\alpha\beta}}\dot{C}_{\alpha\beta} \tag{3.5.10}$$

for which the rate of strain matrix is obtained considering Eqs. (2.7.7)–(2.7.9)

$$|\dot{\overset{\mathrm{H}}{\varepsilon}}_{ij}| = \begin{vmatrix} \dot{\lambda}_1/\lambda_1 & 0 & 0 \\ 0 & \dot{\lambda}_2/\lambda_2 & 0 \\ 0 & 0 & \dot{\lambda}_3/\lambda_3 \end{vmatrix} \tag{3.5.11}$$

Fitzgerald (1980) indicates that Eq. (3.5.11) has certain limitations and it does not always diagonalize.

The volumetric strain for both the Hencky and the infinitesimal strain measures may be derived from Eqs. (3.5.11) and (3.5.9), respectively

$$\dot{\overset{\mathrm{H}}{\varepsilon}}_{ij} = \mathrm{I}_{\mathbf{H}} = \overline{\ln \sqrt{\mathbf{C}}}^{\,\bullet} = \tfrac{1}{2}\frac{\dot{\mathrm{III}}_C}{\mathrm{III}_C} = \overline{\ln \lambda_{\alpha\alpha}}^{\,\bullet} \tag{3.5.12}$$

$$\dot{\overset{\mathrm{inf}}{\varepsilon}}_{\alpha\alpha} = \dot{\overset{\mathrm{inf}}{\mathrm{I}}}_{\varepsilon} = \dot{u}_{\alpha,\alpha} = \dot{x}_{\alpha,\alpha} = \mathrm{I}_f \tag{3.5.15}$$

which shows that the rate of volumetric strain of the Hencky measure is the first invariant of the strain rate, that is, the sum of the diagonals of matrix (3.5.11), while in the case of the infinitesimal strain the rate of volumetric strain is equal to the first invariant of the flow tensor.

3.6 The Fundamental Theorem of Deformations

In view of the many test modes employed in the study of soils, the following sections, which describe the different forms of deformation and deformation rates, would justify, perhaps, a separate chapter. On the other hand, they are a direct continuation of the present sections and the previous chapter – the reason they have been included here, as their extension.

The understanding of the basic types of deformation and motion is indispensable to the theoretical and experimental study of mechanics of materials, among them soils. The progress of geotechnical engineering study relies, however, not so much on theoretical developments, but mostly on laboratory and field tests of various kinds, which simulate field conditions of consistency, geometry and straining. This is due, on one hand, to the immense variation of soils from location to location, from depth to depth, and from one consistency to another, and on the other hand to the lack of well-developed constitutive equations and boundary conditions for soil problems corresponding to practical field conditions. It is, therefore, instructive to get acquainted with the specific theoretical background of deformations related to laboratory and field tests, and their implications. A selection of the fundamental and basic deformations and motions relevant to the study of the mechanics of soils is given in the forthcoming sections, and the conditions of their existence are outlined and discussed. A much broader selection and a more comprehensive study is found in the works of Truesdell and Toupin (1960) and Eringen (1962).

A theorem called the *fundamental theorem* (Truesdell and Toupin 1960) asserts that *every deformation, at any point, may be regarded as resulting from a translation, a rigid rotation of the principal axes of strains, and the stretches along these axes.* That is to say that every deformation that carries a neighborhood of $\mathbf{X}$ into a neighborhood of $\mathbf{x}$ is specified locally by the translation of $\mathbf{X}$ to $\mathbf{x}$, along the axis A_i, the angles of rotation ϑ_i of the respective axes from one set of axes into the other set, and the three principal stretches λ_i.

We shall decribe the various important deformations and motions, and we shall emphasize whether, and to what extent, these deformations cause a change of volume or a distortion.

3.7 Rigid Deformation and Motion

A deformation is considered *rigid* if the distance between two points remains unaltered throughout the motion. The necessary and sufficient condition for a deformation to be rigid is that the deformation tensor retain the value of unity, as we have already seen in Sect. 2.1.

$$X^K_{,lm} = U^K_{,lm} \equiv x^k_{,LM} = u^k_{,LM} \equiv C_{LM,K} = c_{lm,k} = 0 \tag{3.7.1}$$

Further equivalent conditions are

$$\lambda_i^2 = 1; \quad \mathrm{I}_C = \mathrm{II}_C = 3; \quad \mathrm{III}_C = 1 \tag{3.7.2}$$

Consequently, the conditions for a rigid motion may be derived from Eqs (3.7.1) and (3.7.2)

$$\dot{C}_{LM} = \dot{c}_{lm} \equiv \dot{\lambda}_i^2 \equiv \dot{\text{I}}_C = \dot{\text{II}}_C = \dot{\text{III}}_C = 0 \tag{3.7.3}$$

3.8 Homogeneous Strain

A deformation that carries a straight line into another straight line, an ellipse into an ellipse, an ellipsoid into an ellipsoid is a *homogeneous strain*. A homogeneous strain is possible if the displacement tensor maintains a constant value over the deformed body. A formal expression of this condition of invariance throughout the space is given by

$$X^K_{,lm} = U^K_{,lm} \equiv x^k_{,LM} = u^k_{,LM} \equiv C_{LM,K} = c_{lm,k} = 0 \tag{3.8.1}$$

A number of particular cases of homogeneous strains are forthcoming in the subsequent sections.

3.9 Pure Strain

A deformation that preserves the direction of the principal axes is a *pure strain*. The necessary and sufficient condition for pure strain is

$$\mathbf{R} = \mathbf{1} \tag{3.9.1}$$

where $\mathbf{R}$ is the *mean rotation tensor*.

It should be remarked that the fact that the principal axes do not rotate, does not prevent the rotation of any other line element between two points.

3.10 Isochoric Deformation and Motion

A deformation of a body is *isochoric* if its volume remains unaltered. The necessary and sufficient conditions for the deformation to be isochoric are derived from Eq. (2.4.10), by setting $\Lambda(X) = \Lambda(x) = 1$. Several of the many alternative forms are given here

$$|J| = 1; \quad \text{III}_C = 1; \quad \text{III}_c = 1; \quad \lambda_1^2\lambda_2^2\lambda_3^2 = 1 \tag{3.10.1}$$

Two cases are of particular interest. One relates to the infinitesimal strain and the second to the Hencky strain. In both these cases the first invariant vanishes identically for isochoric deformation

$$\overset{\text{inf}}{\mathrm{I}}_{\varepsilon} = U_{\alpha,\alpha} = u_{\alpha,\alpha} = 0 \tag{3.10.2}$$

$$\overset{\mathbf{H}}{\mathrm{I}}_{\varepsilon} = \ln \lambda_1 \lambda_2 \lambda_3 = 0 \tag{3.10.3}$$

A motion is isochoric if the volume that occupies any region remains unaltered with time, irrespective of the changes that the region undergoes. The condition for isochoric motion is

$$\dot{U}_{\alpha,\alpha} = \dot{X}_{\alpha,\alpha} \equiv \dot{u}_{\alpha,\alpha} = \dot{x}_{\alpha,\alpha} = f_{\alpha\alpha} = \mathrm{I}_f = 0 \tag{3.10.4}$$

from which it follows that the velocity field is *solenoidal*.

Another condition, rather important, is derived by differentiating Eq. (2.7.12) with respect to time, which yields

$$\overset{\dot{\mathbf{H}}}{\mathrm{I}}_{\varepsilon} = \frac{\dot{\lambda}_1}{\lambda_1} + \frac{\dot{\lambda}_2}{\lambda_2} + \frac{\dot{\lambda}_3}{\lambda_3} \tag{3.10.5}$$

The importance of this equation is in the fact that it shows that the rate of the first invariant of the Hencky strain is a natural indicator of isochoric motion.

3.11 Irrotational Motion

A motion in which the spin tensor w_{ij} vanishes is *irrotational*.

$$w_{ij} = \tfrac{1}{2}(\dot{x}_{i,j} - \dot{x}_{j,i}) = \tfrac{1}{2}(\dot{u}_{i,j} - \dot{u}_{j,i}) = 0 \tag{3.11.1}$$

A necessary and sufficient condition for a motion to be irrotational is that the velocity field be *lamellar*, that is, that there exists a scalar function $\Phi_{\mathfrak{v}}$ called the *velocity potential*, whose gradient is equal to the velocity field

$$\dot{x}_i = -\Phi_{\mathfrak{v},i} \tag{3.11.2}$$

An irrotational motion is also isochoric, if it satisfies Eq. (3.10.4)

$$\dot{x}_{\alpha,\alpha} = \Phi_{\mathfrak{v},\alpha\alpha} = 0 \tag{3.11.3}$$

Thus in an irrotational isochoric motion the Laplacian of the velocity potential vanishes identically.

3.12 Laminar Motion

If the *stream lines* of the velocity field $\dot{x}_i$ are orthogonal to the vorticity vector w_i, that is, according to Eq. (A.11.3), Appendix A

$$\dot{x}_\alpha w_\alpha = 0 \tag{3.12.1}$$

then the motion is a *laminar motion*.

Expressing Eq. (3.12.1) in terms of the spin tensor, the condition for laminar motion will be

$$e_{\alpha\beta\gamma} w_{\gamma\beta} \dot{x}_\alpha = 0 \tag{3.12.2}$$

The vorticity vector that appears in the expression of the laminar motion has many implications in the theories of flow. One of the well-known classical phenomena caused by the vorticity vector is what is called the *teapot effect* (Reiner 1960).

3.13 Spherical Deformation

A homogeneous deformation in which the principal stretches are all identical $\lambda_1 = \lambda_2 = \lambda_3 = \lambda$ is a *spherical deformation* or *isotropic deformation*. In mechanics literature the term *uniform dilatation* is often used. For such a deformation the stretch matrix may be written

$$|\lambda_{ij}| = \begin{vmatrix} \lambda & 0 & 0 \\ 0 & \lambda & 0 \\ 0 & 0 & \lambda \end{vmatrix} = \lambda|\delta_{ij}| \quad -\infty < \lambda < \infty \tag{3.13.1}$$

Since $C_i = 1/c_i = \lambda_i^2$, the invariants of the deformation tensor for spherical deformation are

$$\mathrm{I}_C = \bar{\mathrm{I}}_C = 3\lambda^2; \quad \mathrm{II}_C = \overline{\mathrm{II}}_C = 3\lambda^4; \quad \mathrm{III}_C = \overline{\mathrm{III}}_C = 3\lambda^6 \tag{3.13.2}$$

$$\Pi_C = 0 \tag{3.13.3}$$

and the invariants of the Hencky measure defined in Sect. 2.7 are

$$\overset{\mathrm{H}}{\mathrm{I}}_\varepsilon = \overset{\mathrm{H}}{\bar{\mathrm{I}}}_\varepsilon = 3\ln\lambda; \quad \overset{\mathrm{H}}{\mathrm{II}}_\varepsilon = \overset{\mathrm{H}}{\overline{\mathrm{II}}}_\varepsilon = 3(\ln\lambda)^2;$$

$$\overset{\mathrm{H}}{\mathrm{III}}_\varepsilon = (\ln\lambda)^3; \quad \overset{\mathrm{H}}{\overline{\mathrm{III}}}_\varepsilon = 3(\ln\lambda)^3 \tag{3.13.4}$$

$$\overset{\mathrm{H}}{\Pi}_{\varepsilon} = 0 \tag{3.13.5}$$

Eqs. (3.13.3) and (3.13.5), representing the octahedral invariant, have to be noticed in particular. For spherical deformation, where only volume changes occur, this invariant must always vanish. It is, therefore, an additional condition for spherical deformation.

Here again a positive stretch is a decrease in volume, a *compression*, whereas a negative stretch is an increase in volume, a *decompression* or *dilation*.

3.14 Simple Straining

A homogeneous deformation in which two principal stretches are equal is a *simple straining*. If the straining is an increase of stretch in the third direction it is a *simple extension*, while if it is a decrease of stretch in the third direction it is a *simple compression*. Simple straining is of great interest in soil engineering. In laboratory tests such as unconfined compression tests, triaxial tests, cell tests, etc. the soil samples, considered volume elements, undergo simple straining.

Consider rectangular coordinates with axes in the direction of the principal stretches, then the stretch matrix is

$$|\lambda| = \begin{vmatrix} B\lambda & 0 & 0 \\ 0 & B\lambda & 0 \\ 0 & 0 & B\lambda \end{vmatrix} \qquad \begin{matrix} -\infty < B < \infty \quad B \neq 1 \\ -\infty < \lambda < \infty \end{matrix} \tag{3.14.1}$$

where B is the *coefficient of transverse stretching*. The sign of λ controls whether the straining is an extension or a compression, while the sign of B controls whether the transverse stretching is in the same direction as λ or in the opposite direction. In materials, in general, B goes in the direction opposite to λ. If λ indicates an extension, the transverse stretching will be a contraction, and vice versa. There are, however, exceptions to that rule, where there might be either an extension or a contraction in all three directions. For $B = 1$ the simple straining becomes a spherical deformation.

By denoting with ν the ratio of the transverse dilatations to that of the main dilatation, we get

$$\nu = \frac{\delta_{11}}{\delta_{33}} = \frac{\delta_{22}}{\delta_{33}} = \frac{1 - B\lambda}{1 - \lambda} \tag{3.14.2}$$

$$B = \frac{(1 - \nu + \lambda\nu)}{\lambda} \tag{3.14.3}$$

In the theory of elasticity ν is recognized as *Poisson's ratio*.

The components of the deformation tensor for simple straining are

$$C_{11} = C_{22} = B^2\lambda^2; \quad C_{33} = \lambda^2; \quad C_{LM} = 0 \quad \text{for } L \neq M \tag{3.14.4}$$

and its invariants

$$\mathrm{I}_C = \lambda^2(1 + 2B^2); \quad \mathrm{II}_C = B^2\lambda^4(2 + B^2); \quad \mathrm{III}_C = B^2\lambda^6 \tag{3.14.5}$$

$$\bar{\mathrm{I}}_C = \lambda^2(1 + 2B^2); \quad \overline{\mathrm{II}}_C = \lambda^4(1 + 2B^4); \quad \overline{\mathrm{III}}_C = \lambda^6(1 + 2B^6) \tag{3.14.6}$$

$$\Pi_C = \tfrac{1}{2}\lambda^4(1 - B^2)^2 \tag{3.14.7}$$

The first invariant of the Hencky strain might also be of interest, as it indicates the volume change of the body

$$\overset{\mathrm{H}}{\mathrm{I}}_\varepsilon = \overset{\mathrm{H}}{\bar{\mathrm{I}}}_\varepsilon = 2\ln B + 3\ln\lambda \tag{3.14.8}$$

For $B = 1$ the simple straining reduces to a spherical deformation, the octahedral invariant of the deformation tensor, Eq. (3.14.5), vanishes, and Eqs. (3.14.3) and (3.14.8) reduce to the form of Eq. $(3.13.4)_1$.

3.15 Uniaxial Straining

A particular case of the simple straining, where $B = 1/\lambda$ is a *uniaxial straining* and the stretch tensor becomes

$$|\lambda_{ij}| = \begin{vmatrix} 1 & 0 & 0 \\ 0 & 1 & 0 \\ 0 & 0 & \lambda \end{vmatrix} \qquad -\infty < \lambda < \infty \tag{3.15.1}$$

The one-dimensional consolidation test is an example of uniaxial straining with λ positive. The deformation tensor matrix becomes then

$$|C_{LM}| = \left|\frac{1}{c_{lm}}\right| = \begin{vmatrix} 1 & 0 & 0 \\ 0 & 1 & 0 \\ 0 & 0 & \lambda^2 \end{vmatrix} \tag{3.15.2}$$

and its invariants are

$$\mathrm{I}_C = 2 + \lambda^2; \quad \mathrm{II}_C = 1 + 2\lambda^2; \quad \mathrm{III}_C = \lambda^2 \tag{3.15.3}$$

$$\bar{I}_C = 2 + \lambda^2; \quad \overline{II}_C = 2 + \lambda^4; \quad \overline{III}_C = 2 + \lambda^6 \tag{3.15.4}$$

$$\Pi_C = \tfrac{1}{2}(1 - 2\lambda^2 + \lambda^4) = \tfrac{1}{2}(1 - \lambda^2)(1 + \lambda^2) \tag{3.15.5}$$

3.16 Plane Strain

A particular case of simple straining, in which the deformation in the direction of one of the axes is zero, while the deformations in the direction of the two other axes are functions of the coordinates, is a *plane strain*. A plane strain is characterized by the following equations:

$$x^l = x(X^L, X^M)^l; \quad x^m = x(X^L, X^M)^m; \quad x^n = X^N \tag{3.16.1}$$

The deformation tensor will be represented by the matrix

$$|C_{LM}| = \begin{vmatrix} C_{LL} & C_{LM} & 0 \\ C_{ML} & C_{MM} & 0 \\ 0 & 0 & 1 \end{vmatrix} \tag{3.16.2}$$

The N axis is a principal axis, the L and M axes are not. The invariants of the deformation tensor $\mathbf{C}$ are

$$\mathrm{I}_C = C_{LL} + C_{MM} + 1; \quad \mathrm{II}_C = C_{LL}C_{MM} + C_{LL} + C_{MM} - C_{LM}^2$$

$$\mathrm{III}_C = C_{LL}C_{MM} - C_{LM}^2 \tag{3.16.3}$$

The roots of the characteristic equation, (A.15.5) Appendix A, are

$$C_1 = C_2 = \tfrac{1}{2}[\mathrm{I}_C - 1 \pm \sqrt{(\mathrm{I}_C^2 + 2\mathrm{I}_C - 4\mathrm{II}_C - 3)}]; \quad C_3 = 1 \tag{3.16.4}$$

C_1, C_2, and C_3 are components of the deformation tensor in the directions of the principal axes. The directions of the axes are determined by solving Eq. (A.15.10) for each C_n.

3.17 Simple Shear

A homogeneous strain defined by the equations

$$x = X; \quad y = Y + SZ; \quad z = Z \quad -\infty < S < \infty \tag{3.17.1}$$

and described by a deformation matrix in which the diagonal components are unity and one non-diagonal component is non-zero, is a *simple shear*, Fig. 3.17.1.

$$|x_{i,j}| = \begin{vmatrix} 1 & 0 & 0 \\ 0 & 1 & S \\ 0 & 0 & 1 \end{vmatrix} \tag{3.17.2}$$

Since it conforms to Eq. (3.16.1), it is a case of plane strain.

There is no relative displacement between points in the direction of X, and all planes normal to the direction Z maintain their shape, but slide in linear proportion to the coordinate Z and are called *shearing planes*. At the same time the planes $y = \text{const}$ rotate rigidly about their intersection with the plane $Z = 0$ through an angle called the *angle of distortion*, related to the coefficient of proportionality S by

$$\tan \alpha = S \tag{3.17.3}$$

that is, each shearing plane $Z = \text{const}$ slides in the direction Y by an amount proportional to its distance from the plane $Z = 0$. The stretch suffered by lines in the Z direction is $\sqrt{(1 + S^2)}$. So a point (0,0,1) is carried into a point (0,S,1) and a point (0,$-S/2$,1) into a point (0,$+S$,1). It follows that in simple shear there are two families of planes that are carried rigidly from their undeformed positions to their deformed positions. These are the planes normal to the shearing planes. The *angle of rotation* γ is equal to half the angle of distortion

$$\tan \gamma = S/2 \tag{3.17.4}$$

as evident from the displacement tensor, Eq. (3.17.2). From the displacement tensor the deformation tensor can be derived

$$|C_{LM}| = \begin{vmatrix} 1 & 0 & 0 \\ 0 & 1 + S^2 & 0 \\ 0 & 0 & 1 \end{vmatrix} \tag{3.17.5}$$

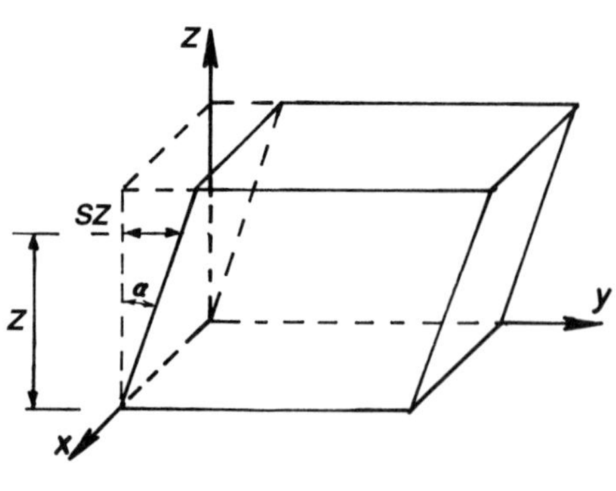

Fig. 3.17.1. Simple shear.

and its invariants are

$$\mathrm{I}_C = \bar{\mathrm{I}}_C = \mathrm{II}_C = \bar{\mathrm{II}}_C = 3 + S^2; \quad \mathrm{III}_C = \bar{\mathrm{III}}_C = 1 \tag{3.17.6}$$

$$\Pi_C = 0 \tag{3.17.7}$$

From Eqs. $(3.10.1)_2$, (3.17.5) and (3.17.6) it is evident that a simple shear is an isochoric deformation. Since the stretch tensor of the undeformed coordinates is equal to the square root of the deformation tensor, the principal stretches can be evaluated as the proper numbers of the stretch determinant. It follows then

$$\begin{aligned} \lambda_{11} &= 1 \\ \lambda_{22} &= \sqrt{\left[1 + \frac{S^2}{2} + S\sqrt{\left(1 + \frac{S^2}{4}\right)}\right]} \\ \lambda_{33} &= \sqrt{\left[1 + \frac{S^2}{2} + S\sqrt{\left(1 + \frac{S^2}{4}\right)}\right]} \end{aligned} \tag{3.17.8}$$

These results are consistent with previous findings in two respects: there is no displacement in the direction X, $\lambda_{XX} = \lambda_{11} = 1$; and the deformation is isochoric, hence $\lambda_{22} = 1/\lambda_{33}$.

The principal strain components in the Hencky measure are

$$\begin{aligned} &\overset{\mathrm{H}}{\varepsilon}_{11} = 0; \quad \overset{\mathrm{H}}{\varepsilon}_{22} = \tfrac{1}{2}\ln\left[1 + \frac{S^2}{2} + S\sqrt{\left(1 + \frac{S^2}{4}\right)}\right]; \\ &\overset{\mathrm{H}}{\varepsilon}_{33} = \tfrac{1}{2}\ln\left[1 + \frac{S^2}{2} - S\sqrt{\left(1 + \frac{S^2}{4}\right)}\right] \end{aligned} \tag{3.17.9}$$

Consequently, the first invariant of the Hencky strain tensor vanishes, another indication of isochoric deformation

$$\overset{\mathrm{H}}{\mathrm{I}} = 0 \tag{3.17.10}$$

If in a simple shear the shear displacement is a linear function of time, then the motion is defined *simple shearing*, Fig. 3.17.2, and we may write

$$\dot{x} = 0; \quad \dot{y} = SZ; \quad \dot{z} = 0 \tag{3.17.11}$$

and the flow tensor f_{LM} is then

$$|f_{LM}| = \tfrac{1}{2}\begin{vmatrix} 0 & 0 & 0 \\ 0 & 0 & S \\ 0 & S & 0 \end{vmatrix} \tag{3.17.12}$$

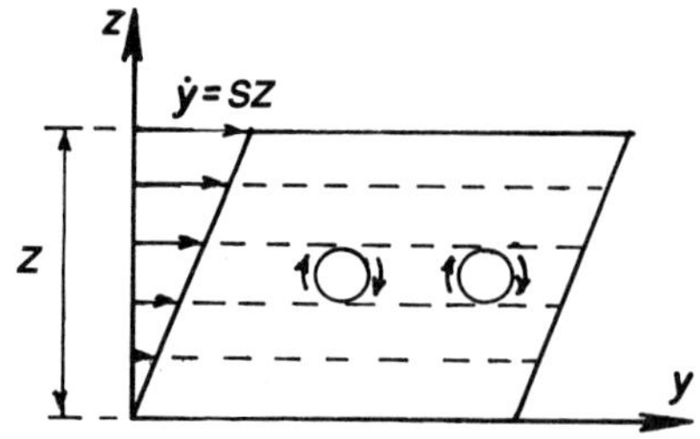

Fig. 3.17.2. Shearing in simple shear.

The simple shearing is an isochoric motion, since Eq. (3.17.11) corresponds to Eq. (3.10.4), and it is also a laminar motion since it corresponds to Eq. (3.12.2). Thus the motion is composed of a relative movement of laminae in which every particle travels along the stream lines in a straight line and with the same velocity, and a rotational motion within each lamina is expressed by the vorticity vector w_i or the spin tensor w_{ij}

$$w_i = w_i\,(-B, 0, 0); \quad |w_{ij}| = \frac{1}{2}\begin{vmatrix} 0 & 0 & 0 \\ 0 & 0 & S \\ 0 & -S & 0 \end{vmatrix} \tag{3.17.13}$$

The vorticity vector is in the direction normal to the shearing, and its magnitude $\surd(w_\alpha w_\alpha)$ represents the *amount of shearing*.

The conditions necessary and sufficient to satisfy a simple shearing are

$$\mathrm{I}_f = \mathrm{III}_f = 0 \tag{3.17.14}$$

w_i is in the direction of the principal axes with zero stretching

$$w^2 = w_\alpha w_\alpha = -4\mathrm{II}_f \tag{3.17.15}$$

3.18 Simple Torsion of a Circular Cylinder

A simple shear, in cylindrical orthogonal coordinates, expressed by

$$r = R; \quad \theta = \Theta + BZ; \quad z = Z \tag{3.18.1}$$

where R, Θ and Z are the cylindrical coordinates of a material point before deformation and r, θ and z are the cylindrical coordinates of the same point after deformation, is a *simple torsion*, and is produced by twisting a prism along its length, oriented in the direction $z = Z$, Fig. 3.18.1. The deformation and the deformation gradient, given in Eqs. (3.17.1) and (3.17.2), will remain the same, and B is then the *twist per unit length*. However, the components of the fundamental tensor will be $G_{11} = 1$, $G_{22} = R^2$, $G_{33} = 1$

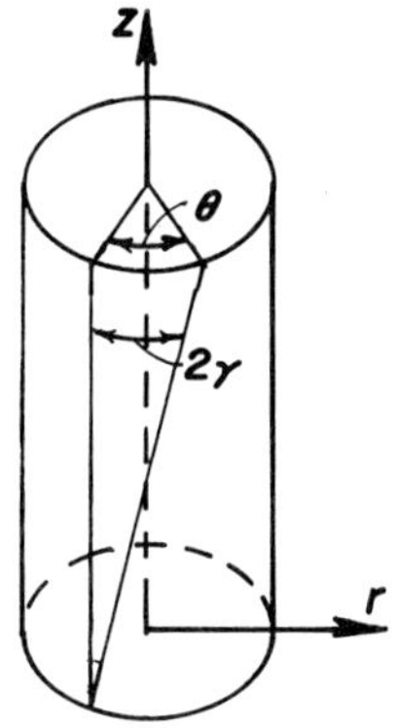

Fig. 3.18.1. Simple torsion.

while all the other components vanish, and the deformation tensor and its invariants are

$$|C_{LM}| = \begin{vmatrix} 1 & 0 & 0 \\ 0 & R^2 & BR^2 \\ 0 & BR^2 & 1 + B^2R^2 \end{vmatrix} \tag{3.18.2}$$

$$\mathrm{I}_C = \mathrm{II}_C = 3 + B^2R^2; \quad \mathrm{III}_C = 1 \tag{3.18.3}$$

$$\bar{\mathrm{I}}_C = 3 + B^2R^2; \quad \overline{\mathrm{II}}_C = 3 + 4B^2R^2 + B^4R^4;$$

$$\overline{\mathrm{III}}_C = 3 + (3 + B^2R^2)^2B^2R^2 \tag{3.18.4}$$

$$\Pi_C = B^2R^2(3 + B^2R^2) \tag{3.18.5}$$

from which it is evident that a simple torsion is isochoric when the planes $Z = \text{const}$ remain planar and are twisted relatively towards each other. The angle of rotation, analogously to (3.17.4) is

$$\tan\gamma = \tfrac{1}{2}BR^2 \tag{3.18.6}$$

and the principal stretches, analogously to Eq. (3.17.8) are

$$\lambda_{11} = 1; \quad \lambda_{22} = \sec\gamma + \tan\gamma; \quad \lambda_{33} = \sec\gamma - \tan\gamma \tag{3.18.7}$$

If the torsion proceeds proportional to time, the motion is a *simple torsional shearing*, satisfying Eqs. (3.17.14) and (3.17.15). The vorticity vectors are normal to the axes of the twist, Fig. 3.18.2. The motion is defined similarly to that of simple shear, Eq. (3.17.11)

$$r = R; \quad \theta = \Theta + AZt, \quad z = Z \tag{3.18.8}$$

$$\dot{r} = 0; \quad \dot{\theta} = AZ; \quad \dot{z} = 0 \tag{3.18.9}$$

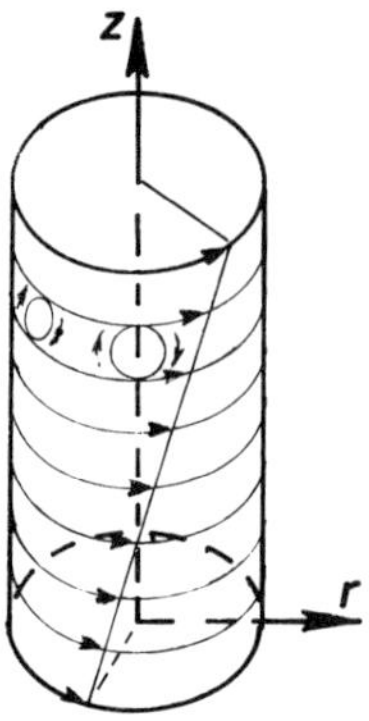

Fig. 3.18.2. Torsional shearing.

$$|f_{ij}| = \frac{1}{2}\begin{vmatrix} 0 & 0 & 0 \\ 0 & 0 & A \\ 0 & A & 0 \end{vmatrix} \tag{3.18.10}$$

Eqs. (3.18.9) and (3.18.10) indicate that the motion is isochoric, since it corresponds to Eq. (3.10.4), and laminar, since it corresponds to Eq. (3.12.2). The relative movement of the laminae normal to the direction z is a torsional motion, where the angular velocity $\dot{\theta}$ of a lamina is proportional to the distance from the plane $Z = 0$. The stream lines in this case are in the direction θ. A vorticity vector is also present, and its direction is R, the radial direction of the cylinder, normal to the direction of the stream lines and to the direction in which no stretching occurs. This vorticity vector and spin tensor are similar to the vorticity vector and spin tensor expressed in Eq. (3.17.13), with the single difference that they refer to cylindrical coordinates. They also satisfy the condition of simple shearing in Eq. (3.17.15), for which reason the motion is termed simple torsional shearing.

3.19 Telescoping Deformation

A deformation defined in cylindrical orthogonal coordinates by

$$R = R; \quad \theta = \Theta; \quad z = Z + K(R) \tag{3.19.1}$$

$$|x_{i,j}| = \begin{vmatrix} 1 & 0 & 0 \\ 0 & 1 & 0 \\ \dfrac{\partial K}{\partial R} & 0 & 1 \end{vmatrix} \tag{3.19.2}$$

is a *general telescopic deformation*, Fig. 3.19.1. Its application is important in the theories of flow in sealed conduits.

The deformation tensor and its invariants are

$$|C_{LM}| = \begin{vmatrix} 1 & 0 & \dfrac{\partial K}{\partial R} \\ 0 & 1 & 0 \\ \dfrac{\partial K}{\partial R} & 0 & 1 + \left(\dfrac{\partial K}{\partial R}\right)^2 \end{vmatrix} \tag{3.19.3}$$

$$\mathrm{I}_C = 3 + \left(\frac{\partial K}{\partial R}\right)^2; \quad \mathrm{II}_C = 3 + \left(\frac{\partial K}{\partial R}\right)^2; \quad \mathrm{III}_C = 1 \tag{3.19.4}$$

$$\bar{\mathrm{I}}_C = 3 + \left(\frac{\partial K}{\partial R}\right)^2; \quad \overline{\mathrm{II}}_C = 3 + 4\left(\frac{\partial K}{\partial R}\right)^2 + \left(\frac{\partial K}{\partial R}\right)^4;$$

$$\overline{\mathrm{III}}_C = 3 + 9\left(\frac{\partial K}{\partial R}\right)^2 + 6\left(\frac{\partial K}{\partial R}\right)^4 + \left(\frac{\partial K}{\partial R}\right)^6 \tag{3.19.5}$$

$$\Pi_C = \frac{1}{2}\left[3\left(\frac{\partial K}{\partial R}\right)^2 + \left(\frac{\partial K}{\partial R}\right)^4\right] \tag{3.19.6}$$

which indicates that the telescopic deformation is isochoric. All points lying on the cylinder with $R = \text{const}$ have equal displacement in the direction of the cylinder axis.

If the telescopic deformation is also a function of time $K = K(R, t)$, the motion is a *telescopic shearing*, defined by the equations

$$r = R; \quad \theta = \Theta; \quad z = Z + K(R, t) \tag{3.19.7}$$

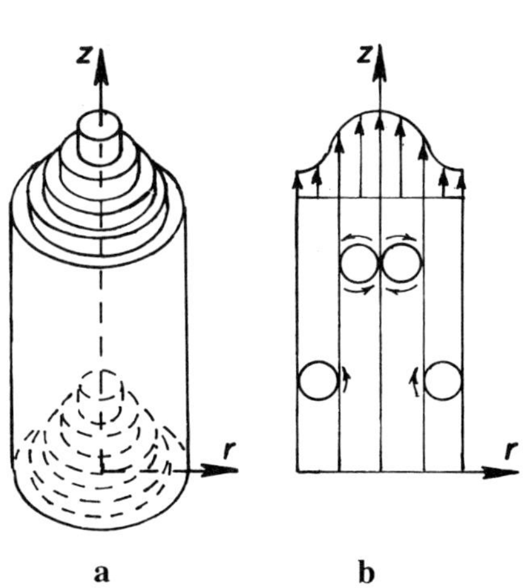

Fig. 3.19.1. Telescopic deformation and shearing. **a** Laminar. **b** Continuous laminar.

$$\dot{r} = 0; \quad \dot{\theta} = 0; \quad \dot{z} = \dot{K}(R, t) \tag{3.19.8}$$

$$\|f_{ij}\| = \frac{1}{2}\begin{vmatrix} 0 & 0 & \dfrac{\partial \dot{K}}{\partial R} \\ 0 & 0 & 0 \\ \dfrac{\partial \dot{K}}{\partial R} & 0 & 0 \end{vmatrix} \tag{3.19.9}$$

$$w_i = w\left(0, -\frac{\partial \dot{K}}{\partial R}, 0\right)_i; \quad \|w_{ij}\| = \begin{vmatrix} 0 & 0 & -\dfrac{\partial \dot{K}}{\partial R} \\ 0 & 0 & 0 \\ \dfrac{\partial \dot{K}}{\partial R} & 0 & 0 \end{vmatrix} \tag{3.19.10}$$

It may be seen that the telescopic shearing is an isochoric motion, whose stream lines are in the direction z and the vorticity vector is in the direction θ, normal to the stream lines and to the direction r of zero stretching.

We shall mention here only the particular case $K = At + BRt$, where A is the velocity of the motion at $R = 0$ and $B = A/R$. This motion is a simple telescoping shearing, Fig. 3.19.2, used by Hagen (1839) in an early study to explain the flow of water in narrow pipes, and has a conical distribution of velocity.

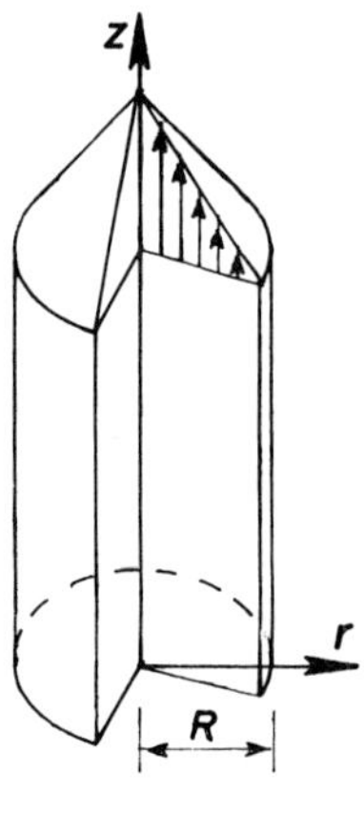

Fig. 3.19.2. Simple telescoping shearing.

3.20 Rotational Deformation

A motion described in cylindrical orthogonal coordinates by the deformation and deformation gradient

$$r = R; \quad \theta = \Theta + B(R); \quad z = Z \tag{3.20.1}$$

$$|x_{i,j}| = \begin{vmatrix} 1 & 0 & 0 \\ \dfrac{\partial B}{\partial R} & 1 & 0 \\ 0 & 0 & 1 \end{vmatrix} \tag{3.20.2}$$

is a *rotational deformation*. Fig. 3.20.1. Its deformation tensor and its invariants are

$$|C_{LM}| = \begin{vmatrix} 1 & \dfrac{\partial B}{\partial R}R^2 & 0 \\ \dfrac{\partial B}{\partial R} & 1 + \left(\dfrac{\partial B}{\partial R}\right)^2 R^2 & 0 \\ 0 & 0 & 0 \end{vmatrix} \tag{3.20.3}$$

$$\mathrm{I}_C = 3 + \left(\frac{\partial B}{\partial R}\right)^2 R^2; \quad \mathrm{II}_C = 3 + \left(\frac{\partial B}{\partial R}\right)^2 R^2; \quad \mathrm{III}_C = 1 \tag{3.20.4}$$

$$\bar{\mathrm{I}}_C = 3 + \left(\frac{\partial B}{\partial R}\right)^2 R^2; \quad \overline{\mathrm{II}}_C = 3 + 4\left(\frac{\partial B}{\partial R}\right)^2 R^2 + \left(\frac{\partial B}{\partial R}\right)^4 R^4$$

$$\overline{\mathrm{III}}_C = 3 + 9\left(\frac{\partial B}{\partial R}\right)^2 R^2 + 6\left(\frac{\partial B}{\partial R}\right)^4 R^4 + \left(\frac{\partial B}{\partial R}\right)^6 R^6 \tag{3.20.5}$$

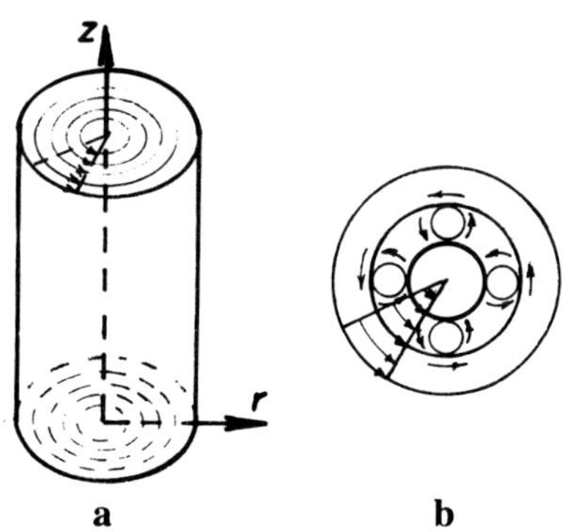

Fig. 3.20.1. Rotational deformation and shearing. **a** Isometric view. **b** Top view.

$$\Pi_C = \frac{1}{2}\left(\frac{\partial B}{\partial R}\right)^2 R^2\left[3 + \left(\frac{\partial B}{\partial R}\right)^2 R^2\right] \tag{3.20.6}$$

which is, again, an indicator that the deformation is isochoric. All points on the cylindrical surface with $R = \text{const}$ have equal rotational displacements. If $\partial B/\partial R = \text{const}$ the rotational motion also produces a simple shear, as seen when Eq. (3.20.3) is compared with Eq. (3.18.2).

If the rotation is a function of time, it is a *rotational shearing* or *vorticity*, defined as

$$r = R; \quad \theta = \Theta + B(R, t); \quad z = Z \tag{3.20.7}$$

$$\dot{r} = 0; \quad \dot{\theta} = \dot{B}(R, t); \quad \dot{z} = 0 \tag{3.20.8}$$

$$|f_{ij}| = \frac{1}{2}\begin{vmatrix} 0 & \dfrac{\partial \dot{B}}{\partial R} & 0 \\ \dfrac{\partial \dot{B}}{\partial R} & 0 & 0 \\ 0 & 0 & 0 \end{vmatrix} \tag{3.20.9}$$

$$w_i = w\left(0, 0, \frac{\partial \dot{B}}{\partial R}\right) \tag{3.20.10}$$

$$|w_{ij}| = \begin{vmatrix} 0 & -\dfrac{\partial \dot{B}}{\partial R} & 0 \\ \dfrac{\partial \dot{B}}{\partial R} & 0 & 0 \\ 0 & 0 & 0 \end{vmatrix} \tag{3.20.11}$$

The motion is laminar, since all particles belonging to the same radius describe circles with the same angular velocity. Thus the stream lines are in the direction θ, the direction of zero stretching is direction R and the vorticity vector is therefore in the direction Z, as seen from Eq. (3.20.10). Considering Sect. A.18 in Appendix A, the circulation of the circle with $R = \text{const}$ is given by

$$\int_0^{2\pi} R\,\dot{\theta}\,R\,\mathrm{d}\theta = \int_0^{2\pi} \overline{\dot{B}(R, t)}\,R^2\,\mathrm{d}\theta = 2\pi R^2\,\overline{\dot{B}(R, t)} \tag{3.20.12}$$

The angular velocity necessarily vanishes at $R = 0$. The particular case $B(R, t) = bRt$, where $b = \text{const}$, represents a motion with angular velocity increasing linearly with R, Fig. 3.20.1, and a steady vorticity exists in the cylindrical laminae or annuli.

Another particular case is obtained when the angular velocity is a function of time only, $\theta = B(t)$, in which case Eqs. (3.20.10) and (3.20.11) vanish. Although no vorticity or spin exist, the motion is still rotational, and it is a rigid rotation.

Couette viscometers, of rotating coaxial cylinders, produce rotational deformations and motions.

3.21 Steady Motion

A motion is defined *steady*, if the velocity field is not a function of time at any given point of the space

$$\dot{x}_i = \dot{x}(R)_i \tag{3.21.1}$$

4 Balance Equations for Homogeneous Media

4.1 Mass

In order to permit the study of the mechanical behavior of multi-phase mixtures, we shall first review the equations that govern the mechanical behavior of homogeneous single-phase media. For simplicity, but without impairing the generality, our exposure will be in the Cartesian coordinates unless otherwise stated. Hence, all tensorial indices will be subscripts.

The quantity affixed to the volume element, which expresses the amount of matter, is defined by the concept of *mass*. The volume element becomes then associated with the *mass particle*. This definition of the mass draws its essence from the following assertions.

1. The mass of a volume is the sum of the mass particles of all volume elements.

$$m = \int_V \mathrm{d}m = \int_V \rho \,\mathrm{d}V \tag{4.1.1}$$

where m is the mass and ρ is a scalar named the *density* of the material element. Thus the mass is a *measure* of the quantity of matter, in addition to the two other measures already encountered, the measure of length and the measure of time.

2. The mass is independent of the coordinate system used or of the motion of the particle. We say that the mass is invariant of both the space and time.

3. While the mass is related to the volume of the material, it is not determined by the size of it.

The mass invariance brings forth the important law of *conservation of mass*. The concept of the continuum makes it possible to define the *density* as a scalar function of space and time, thus $\rho = \rho(x^i, t)$. From Eq. (4.1.1) the following can be deduced

$$\rho \,\mathrm{d}V = \rho_0 \,\mathrm{d}V_0 \tag{4.1.2}$$

where ρ_0 and V_0 are *initial density* and *initial volume* elements, respectively, and Eq. (4.1.2) is the *material continuity equation*.

It follows from this equation that if the volume undergoes an *isochoric* motion, a motion without change in volume, then $\rho = \rho_0 = \text{const}$. Such a motion, in which the density remains unaltered in space and time is termed *homochoric*.

Now consider a volume V of a material element bounded by its surface S. The mass m contained in this volume is given by Eq. (4.1.1), where the density ρ is a function of x^i and t. Thus the rate of increase of mass is

$$\frac{\mathrm{d}m}{\mathrm{d}t} = \int_V \frac{\partial \rho}{\partial t} \,\mathrm{d}V$$

Let n_i denote the direction cosine of the normal to any surface element $\mathrm{d}S$, then the rate of mass flow outward across that surface element is $\rho n_\alpha v^\alpha \,\mathrm{d}S$. The rate of increase of mass is expressed by $-\int_S \rho n_\alpha v^\alpha \,\mathrm{d}S$. We have, therefore

$$\int_V \frac{\partial \rho}{\partial t} \,\mathrm{d}V + \int_S \rho n_\alpha v^\alpha \,\mathrm{d}S = 0 \tag{4.1.3}$$

which is another form of the equation of conservation of mass, implying an equilibrium between the mass flux across the bounding surface and the change of mass within the volume. Applying Green's transformation to the surface integral, Eq. (A.19.1), Appendix A, we obtain

$$\int_V (\rho v^\alpha)_{,\alpha} \,\mathrm{d}V + \int_V \frac{\partial \rho}{\partial t} \,\mathrm{d}V = 0 \tag{4.1.4}$$

By eliminating the integral over the volume $\mathrm{d}V$ and considering Eq. (3.2.2), we have

$$(\rho v^\alpha)_{,\alpha} + \frac{\partial \rho}{\partial t} = (\rho \dot{\overline{x}}^{\alpha})_{,\alpha} + \frac{\partial \rho}{\partial t} = \frac{\mathrm{d}\rho}{\mathrm{d}t} + \rho \dot{\overline{x}}^{\alpha}_{\ ,\alpha} = 0 \tag{4.1.5}$$

which is the *spatial continuity equation*. Another form given to Eq. (4.1.5) is

$$\frac{1}{\rho}\frac{\mathrm{d}\rho}{\mathrm{d}t} + \dot{\overline{x}}^{\alpha}_{,\alpha} = \frac{\dot{\rho}}{\rho} + \mathrm{I}_f = \dot{\overline{\ln \rho}} + \mathrm{I}_f = 0 \tag{4.1.6}$$

If the motion is a steady motion, it is also a motion with *steady density* so that $\partial\rho/\partial t = 0$. Therefore, from (4.1.5) we obtain

$$(\rho v^\alpha)_{,\alpha} = (\rho \dot{x}^\alpha)_{,\alpha} = 0 \tag{4.1.7}$$

In a homochoric motion, as we have seen earlier, $\rho = \text{const}$, therefore

$$v^{\alpha}_{,\alpha} = \dot{\overline{x^{\alpha}_{,\alpha}}} = 0 \tag{4.1.8}$$

Eqs. (4.1.3) and (4.1.8) are both valid for curvilinear coordinates as well, if vectors v^i and x^i are substituted by the contravariant vectors v_i and x_i, respectively.

Since Eq. (4.1.2) implies $(\mathrm{d}/\mathrm{d}t)(\rho\,\mathrm{d}V) = 0$, the conservation of mass, we may derive the following equation, frequently used

$$\frac{\mathrm{d}}{\mathrm{d}t}\int_V \rho\mathcal{F}\,\mathrm{d}V = \int_V \rho\dot{\mathcal{F}}\,\mathrm{d}V \equiv \frac{\mathrm{d}}{\mathrm{d}t}\int_M \rho\mathcal{F}\,\mathrm{d}m = \int_M \rho\dot{\mathcal{F}}\,\mathrm{d}m \tag{4.1.9}$$

where $\mathcal{F}$ represents a quantity related to a volume mass.

4.2 The General Balance Equation

When extensive quantities of a volume element, like mass, moment, energy, entropy, etc. do not change with time, the material is said to be in a *steady state of equilibrium*. Most often, however, these quantities do change with time, and the nature of these changes is sought. Any change of such quantities within a volume element is due either to transfer of quantities through the surface $\mathrm{d}S$ of the element in the form of *influx* or *efflux*, or to a *supply* or *loss* of quantities of the element through an internal source or sink, say, an internal reaction. Be Ξ an extensive quantity of the volume, defined by the quantity ξ per unit mass of the element

$$\Xi = \int_V \rho\xi\,\mathrm{d}V \tag{4.2.1}$$

then any change in Ξ is caused, as said, by an influx $\Phi(\xi)_i$ of that quantity and by a supply $\sigma(\xi)$ of it, and so

$$\frac{\mathrm{d}}{\mathrm{d}t}\Xi = \frac{\mathrm{d}}{\mathrm{d}t}\int_V \rho\xi\,\mathrm{d}V = -\int_V \Phi_\alpha n_\alpha\,\mathrm{d}S + \int_V \rho\sigma\,\mathrm{d}V$$

$$= -\int_V \Phi_{\alpha,\alpha}\,\mathrm{d}V + \int_V \rho\sigma\,\mathrm{d}V \tag{4.2.2}$$

Eq. (4.2.2) defines a relation between the quantities Φ_i and σ and is called the *general balance* or the *general conservation law* (Truesdell and Toupin 1960).

From Eq. (4.2.2) the *local* or *spatial balance* equation for volumes locally constant can be derived in the differential form

$$\frac{\partial(\rho\xi)}{\partial t} + \Phi_{\alpha,\alpha} - \rho\sigma = 0 \tag{4.2.3}$$

One can relate the balance equation to a velocity field v_i by which the volume elements and their surfaces move. The influx Φ_i, in Eqs. (4.2.2) and (4.2.3), is then naturally substituted by $\Phi_i - \rho\xi v_i$, formulating thus the *general material balance* equation

$$\frac{\mathrm{d}}{\mathrm{d}t}\Xi = \frac{\mathrm{d}}{\mathrm{d}t}\int_M \xi \mathrm{d}m = \int_M \frac{1}{\rho}(\Phi_\alpha - \rho v_\alpha)_{,\alpha}\mathrm{d}m + \int_M \sigma \mathrm{d}m \tag{4.2.4}$$

where $\rho\mathrm{d}V = \mathrm{d}m$, according to Eq. (4.1.1), and is considered constant with time. It should be remarked that the volume elements and their surfaces move along with the velocity field $v(X, t)_i$. Consequently, the differentiation affects only ξ on the left-hand side of Eq. (4.2.4), and therefore it can be written

$$\frac{\mathrm{d}}{\mathrm{d}t}\int_M \xi \mathrm{d}m = \int_M \frac{\mathrm{D}\xi}{\mathrm{D}t}\mathrm{d}m \tag{4.2.5}$$

The notation $\mathrm{D}/\mathrm{D}t$ refers to a material derivative, and Eq. (4.2.4), will therefore become

$$\rho\left(\frac{\mathrm{D}\xi}{\mathrm{D}t}\right) + (\Phi_\alpha - \rho\xi v_\alpha)_{,\alpha} - \rho\sigma = 0 \tag{4.2.6}$$

representing the differential form of the general material balance equation.

Eliminating σ from Eqs. (4.2.2) and (4.2.6) and considering Eq. $(4.1.5)_1$, the relation between the spatial and material balance equations is obtained

$$\frac{\mathrm{D}\xi}{\mathrm{D}t} = \frac{\partial\xi}{\partial t} + v_\alpha \xi_{,\alpha} \tag{4.2.7}$$

In a differential operator form, Eq. (4.2.7) reads

$$\frac{\mathrm{D}}{\mathrm{D}t} = \frac{\partial}{\partial t} + v_\alpha \frac{\partial}{\partial X_\alpha} \tag{4.2.8}$$

and may be applied to any tensorial quantity ξ, of whatever order.

Another form for the general material balance equation is obtained by substituting Eq. (4.2.3) into (4.2.8)

$$\frac{\mathrm{D}\rho\xi}{\mathrm{D}t} + \Phi_{\alpha,\alpha} - v_\alpha(\rho\xi)_{,\alpha} - \rho\sigma = 0 \tag{4.2.9}$$

By choosing properly the functions ξ, Φ_i and σ, the specific balance equations for the density, linear moment, moment of momentum, energy, etc. can be derived. This will be carried out in the forthcoming sections.

4.3 Density Balance

Considering $\xi = 1$, the simplest function representing the specific mass, Eqs. (4.2.3) and (4.2.9), yield

$$\frac{\partial \rho}{\partial t} + (\rho v_\alpha)_{,\alpha} - \rho\sigma = 0 \tag{4.3.1}$$

$$\frac{D\rho}{Dt} + \rho v_{\alpha,\alpha} - \rho\sigma = 0 \tag{4.3.2}$$

representing the *spatial* and *material continuity equations*, respectively, also known as the *spatial* and *material density balance*. Here we find that the density flux is $\Phi_i = v_i$.

In materials where no internal reactions go on, the more simple equations are known

$$\frac{\partial \rho}{\partial t} + (\rho v_\alpha)_{,\alpha} = 0 \tag{4.3.3}$$

$$\frac{D\rho}{Dt} + \rho v_{\alpha,\alpha} = 0 \tag{4.3.4}$$

If the motion is steady, it is also a motion with *steady density*, so that $\partial\rho/\partial t = 0$. Therefore, from (4.3.3) we have

$$(\rho v_\alpha)_{,\alpha} = (\rho x_\alpha)_{,\alpha} = 0 \tag{4.3.5}$$

In a homochoric motion $\rho = \text{const}$, therefore

$$v_{\alpha,\alpha} = \dot{u}_{\alpha,\alpha} = 0 \tag{4.3.6}$$

Eq. (4.3.6) is also valid for curvilinear coordinates.

4.4 Forces Acting on Deformable Bodies

If a material body or volume is undeformed, the particles and their interaction are said to be in a state of thermal equilibrium. In other words, the resultant force acting on any portion of the body is zero.

When a deformation occurs, the original state of equilibrium has been apparently disturbed and the order and interaction of particles changed. *Internal* forces arise to counteract the motion, and return the body to equilibrium. The deformation and, consequently, the internal equilibrium disturbance have been caused by *external* forces or constraints imposed on the material body. Gravity, inertia, magnetic forces, heat, etc., are, in this regard, external forces, for, despite the fact that they act on the individual mass points or volume elements within the body, their sources are the potential fields external to the body.

A translational motion of the body is caused when the mass center of the body lies in the line of action of the external forces or the resultant force acting on the body, and a rotation of the body will occur when the external force or the resultant force is not passing through the mass center of the body. The rotation depends on the *torque* acting on the body.

The force and the torque are vectorial quantities, defined

$$F_i = \int_V \mathrm{d}F_i \tag{4.4.1}$$

$$M_i = \int_V (e_{i\alpha\beta} x_\alpha \, \mathrm{d}F_\beta + \mathrm{d}M_i) \tag{4.4.2}$$

where F_i is the result of the forces acting on volume V and M_i is the torque with respect to the origin, known as the *moment*. It consists of two parts, $\int_V e_{i\alpha\beta} x_\alpha \, \mathrm{d}F_\beta$ – the moment of the forces with respect to the origin of the position vector x_i, and a *force couple* $\int_V \mathrm{d}M_i$.

It can also be shown that, in certain instances, the force is equal to the time derivative of the impulse, and the torque is equal to the time derivative of the angular momentum with respect to the origin

$$\begin{aligned} F_i &= \frac{\mathrm{d}}{\mathrm{d}t}\int_M \dot{x}_i \, \mathrm{d}m = \frac{\mathrm{d}}{\mathrm{d}t}\int_V \rho \dot{x}_i \, \mathrm{d}V \\ &= \frac{\mathrm{d}}{\mathrm{d}t}\int_V \rho v_i \, \mathrm{d}V = \frac{\partial}{\partial t}\int_V \rho v_i \, \mathrm{d}V + \int_S \rho v_i v_\alpha n_\alpha \, \mathrm{d}S \end{aligned} \tag{4.4.3}$$

$$\begin{aligned} M_i &= \frac{\mathrm{d}}{\mathrm{d}t}\int_M e_{i\alpha\beta} x_\alpha \dot{x}_\beta \, \mathrm{d}m = \frac{\mathrm{d}}{\mathrm{d}t}\int_V \rho e_{i\alpha\beta} x_\alpha v_\beta \, \mathrm{d}V \\ &= \frac{\partial}{\partial t}\int_V \rho e_{i\alpha\beta} x_\alpha v_\beta \, \mathrm{d}V + \int_S \rho e_{i\alpha\beta} x_\alpha v_\beta v_\gamma n_\gamma \, \mathrm{d}S \end{aligned} \tag{4.4.4}$$

Eqs. (4.4.3) and (4.4.4) represent the *balance of momentum* and the *balance of moment of momentum*, respectively.

4.5 Tractions and Body Forces

In order to specify the state of stress at a point in a body or a volume element, let us regard the body as continuous and given in a Cartesian coordinate system and in equilibrium under the action of several forces, Fig. 4.5.1. Let us bisect it by an imaginary closed diaphragm S. The orientation of such a section can be described by the unit vector n_i normal to the area S, regarded positive if directed from the section outward. An internal force F_{in} is acting on each section of the body,* such that the pair of forces F_{in} are equal and opposite in direction.

The existence of a vector field t_{in} is asserted by the following equations

$$F_i = \int_S t_{in}\,dS + \int_M \sigma_i\,dm = \int_S t_{in}\,dS + \int_V \rho\sigma_i\,dV = \frac{d}{dt}\int_M \dot{x}_i\,dm \qquad (4.5.1)$$

$$\begin{aligned} M_i &= \int_S e_{i\alpha\beta}(m_{\alpha\beta n} + x_\alpha t_{\beta n})\,dS + \int_M e_{i\alpha\beta}(l_{\alpha\beta} + x_\alpha\sigma_\beta)\,dm \\ &= \int_S e_{i\alpha\beta}(m_{\alpha\beta n} + x_\alpha t_{\beta n})\,dS + \int_V e_{i\alpha\beta}(\rho l_{\alpha\beta} + x_\alpha\rho\sigma_\beta)\,dV \\ &= \frac{d}{dt}\int_V e_{i\alpha\beta}x_\alpha\dot{x}_\beta\,dV \end{aligned} \qquad (4.5.2)$$

where t_{in} is the *traction* or *stress vector*, σ_i is the *body force* per unit volume acting on the body material, m_{ijn} are *surface couples* and l_{ij} is an internal *body moment* per unit mass.

The traction $\mathbf{t} \equiv t_{in}$ is a force that represents the intensity of the surface loading or the distribution of the total force F_{in} over the area S. Its magnitude and direction, and the magnitude and direction of F_{in} depend on the position of area S, determined by its normal unit vector n_i. The traction is a function of both the position vector X_i and the normal vector n_i.

The projection of $\mathbf{t} \equiv t_{in}$ on n_i is called the *normal traction* $\mathbf{t}_n \equiv t_{nn}$, and its projection on the surface normal to n_i is called the *shear traction* or *tangential traction* $\mathbf{t}_t \equiv t_{nt}$.

By rotating the plane normal to n_i so that n_i will, in turn, coincide with each of the directions $\imath$ of the coordinates, the tractions acting on the coordinate planes will be t_{ii}.

While t_{ji} are acting on the planes normal to the directions $\imath$, they are not necessarily normal to those planes.

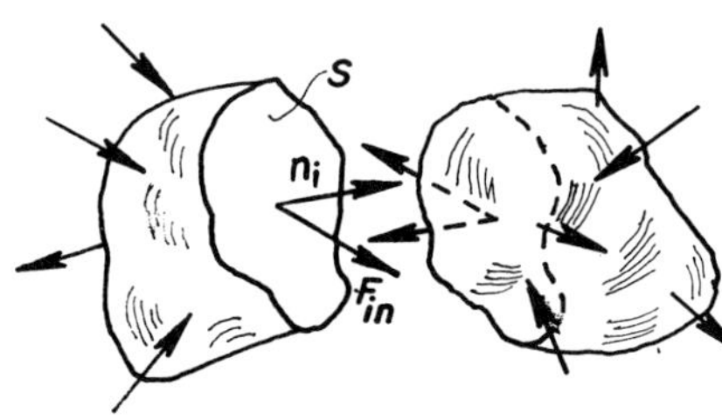

Fig. 4.5.1. Forces acting on a continuous body in equilibrium.

* The subscript n indicates that the vector acts on an area normal to the unit vector n_i.

Analogously to t_{in}, each of the tractions t_{ji} can be resolved into the three directions of the coordinates, one normal $t_{ii} \equiv t_{ii}$ and two shear components $t_{ji} \equiv t_{ji}$, a total of nine components t_{ij}. These nine components form a tensor of second order, the *stress tensor*, where t_{ij} represents the *stress components*, Fig. 4.5.2.

The stress tensor t_{ij}, being a second-order tensor, has two indices. The first index stands for the direction normal to the plane in which the component acts, and the second index indicates the direction in which the component acts. As such, the components with the same indices, of the form t_{ii}, are the *normal stress components* of the stress, while the components with two different indices, of the form t_{ij}, $i \neq j$, are the *shear stress components* of the stress.

It remains to show that the stress tensor t_{ij} serves to determine the state of stress at any point. For that purpose we shall consider three mutually orthogonal planar elements, parallel to the coordinate planes passing through a point O and forming a tetrahedron, with a fourth plane normal to a direction n_i and at a small distance h from O, Fig. 4.5.3. See also Sect. A.20, Appendix A. Let the area of the inclined plane be S, the area of the orthogonal planes will then be $-Sn_i$.

The forces acting on the tetrahedron are t_{in}^*, t_{ij}^* and σ_i^*, the stress vector, the stress tensor, and the body force, respectively. Thus, from the equilibrium of the forces acting on the tetrahedron we have

$$t_{in}^* S - t_{\alpha i}^* n_\alpha S + \tfrac{1}{3}(\rho \sigma_i^* h S) \tag{4.5.3}$$

Dividing (4.5.3) by S at the limit $h \to 0$ we get

$$t_{in} = t_{\alpha i} n_\alpha \tag{4.5.4}$$

which is the equation of equilibrium at the point O, since

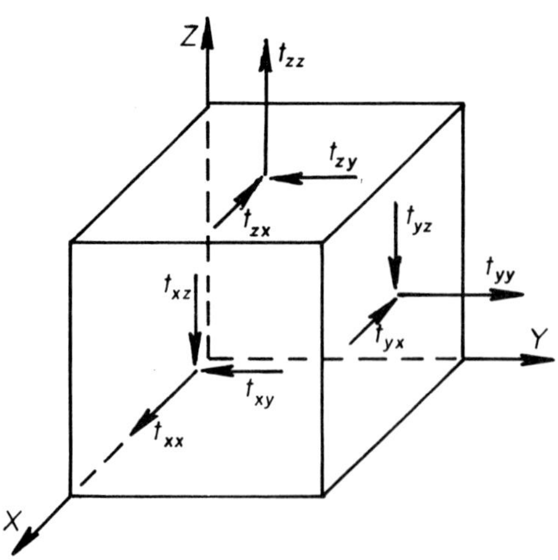

Fig. 4.5.2. Stress components.

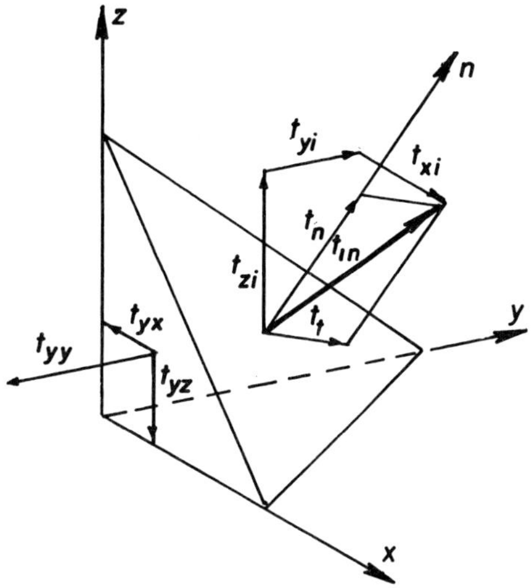

Fig. 4.5.3. Stress tetrahedron.

$$\lim_{h\to 0} t^*_{in} = t_{in}, \quad \lim_{h\to 0} t^*_{ij} = t_{ij} \text{ and } \lim_{h\to 0} \sigma^*_i = \sigma_i.$$

It should be noted that the direction outward of the surface is positive, which implies

$$t_{in} = -t_{i-n} \tag{4.5.5}$$

that is, the tractions acting on opposite sides of the same surface are equal in magnitude but opposite in sign. Consequently, the stress acting on the negative side of the plane normal to n_i and equivalent to the traction t_{in} is

$$t_{in} \equiv t_{\alpha i} n_\alpha \tag{4.5.6}$$

Parallel to that presentation we may assume the existence of *couple tractions* in the form of a skew-symmetric tensor m_{ijn} acting next to the traction t_{in} on the surface normal to n_i, and obtain equations analogous to (4.5.4) and (4.5.5).

$$m_{in} = m_{i\alpha} n_\alpha \tag{4.5.7}$$

$$m_{in} = -m_{i-n} \tag{4.5.8}$$

In terms of the equivalent skew-symmetric tensor, we have

$$m_{ijn} = m_{ij\alpha} n_\alpha \tag{4.5.9}$$

where m_{ijk} is the *couple stress*, acting on the volume element at point O. Being a skew-symmetric tensor, its indices do not commute directly, thus

$$m_{ijk} = m_{jki} = m_{kij} = -m_{ijk} = -m_{jki} = -m_{kij} \tag{4.5.10}$$

In general curvilinear coordinates, both the stress and the couple stress tensors are contravariant tensors of the second and third order, respectively, t^{ij} and m^{ijk}.

4.6 Balance of Linear Momentum

If in Eq. (4.5.1) the value of t_{in} from Eq. (4.5.4) is substituted, we obtain, according to Green's transformation, Eq. (A.18.1), Appendix A

$$F_{in} = \int_S t_{\alpha i} n_\alpha \, \mathrm{d}S + \int_M \sigma_i \, \mathrm{d}m = \int_V t_{\alpha i,\alpha} \, \mathrm{d}V + \int_V \rho\sigma_i \, \mathrm{d}V \tag{4.6.1}$$

Equating (4.6.1) with (4.4.3) the equation of *conservation of linear momentum* or the *balance of linear momentum* is obtained

$$\begin{aligned}\int_V t_{\alpha i,\alpha} \, \mathrm{d}V + \int_V \rho\sigma_i \, \mathrm{d}V &= \frac{\mathrm{d}}{\mathrm{d}t} \int_V \rho v_i \, \mathrm{d}V + \int_S \rho v_i v_\alpha n_\alpha \, \mathrm{d}S \\ &= \frac{\mathrm{d}}{\mathrm{d}t} \int_V \rho v_i \, \mathrm{d}V + \int_V \rho v_{i,\alpha} v_\alpha \, \mathrm{d}V \end{aligned} \tag{4.6.2}$$

By eliminating the integration over the volume $\mathrm{d}V$, Eq. (4.6.2) reduces to

$$t_{\alpha i,\alpha} + \rho\sigma_i = \frac{\mathrm{d}}{\mathrm{d}t}(\rho v_i) + \rho v_{i,\alpha} v_\alpha = \rho\dot{v}_i = \rho\ddot{x}_i \tag{4.6.3}$$

which is known as *Cauchy's first law of motion.*

If the body is considered in equilibrium, there is no motion and Eq. (4.6.3) yields

$$t_{\alpha i,\alpha} + \rho\sigma_i = 0 \tag{4.6.4}$$

The balance of linear momentum can be derived, without going through all that labor, from the general balance equations (4.2.3) and (4.2.6) by inserting in those equations the following values for ξ, Φ and σ: $\xi = v_i$, $\Phi = \rho v_i v_j - t_{ij}$ and $\sigma = \sigma_i$

$$\frac{\mathrm{d}}{\mathrm{d}t}(\rho v_i) - t_{\alpha i,\alpha} + (\rho v_i v_\alpha)_{,\alpha} - \rho\sigma_i = 0 \tag{4.6.5}$$

$$\rho\dot{v}_i - t_{i\alpha,\alpha} - \rho\sigma_i = 0 \tag{4.6.6}$$

representing the spatial and material balance equations of linear momentum, respectively, the latter being identical to Eq. (4.6.3).

4.7 Balance of Moment of Momentum

By substituting the values of t_{in} from Eq. (4.5.6) and the values of m_{ijn} from Eq. (4.5.9) into Eq. (4.4.2), we obtain, by applying again Green's transformation

$$M_i = \int_V e_{i\alpha\beta}(m_{\alpha\beta\gamma} + x_\alpha t_{\beta\gamma})_{,\alpha}\,\mathrm{d}V + \int_V e_{i\alpha\beta}(\rho l_{\alpha\beta} + x_\alpha \rho\sigma_\beta)\,\mathrm{d}V$$

$$= \frac{\mathrm{d}}{\mathrm{d}t}\int_V e_{i\alpha\beta}\rho x_\alpha \dot{x}_\beta\,\mathrm{d}V \tag{4.7.1}$$

By eliminating the integration over the volume $\mathrm{d}V$ and the permutation tensor e_{ijk}, Eq. (4.7.1) reduces to the differential equation

$$m_{ij\alpha,\alpha} + (x_i t_{j\alpha})_{,\alpha} + \rho l_{ij} + x_i \rho\sigma_j =$$

$$= m_{ij\alpha,\alpha} + \rho l_{ij} - \tfrac{1}{2}(t_{ij} - t_{ji}) + (x_i t_{j\alpha,\alpha} + x_i\rho\sigma_j) = \rho x_i \dot{x}_j \tag{4.7.2}$$

and in view of Eq. (4.6.3) the term $x_i t_{j\alpha,\alpha} + x_i\rho\sigma_j$ cancels out with the term $\rho x_i \dot{x}_j$, resulting in what is known as *Cauchy's second law of motion*

$$m_{ij\alpha,\alpha} + \rho l_{ij} - \tfrac{1}{2}(t_{ij} - t_{ji}) = 0 \tag{4.7.3}$$

Equation (4.7.3) in its present form serves us very little, since no couple stresses nor couple moments are considered in soils at the level of our study. If, however, the study of soils reduces to the level of the particles where electronic dipoles of the ion, permanent or induced, and magnetic moments of electrons are considered, Eq. (4.7.3) might come in handy. Therefore, by eliminating the couple stress and the couple moment we are left with

$$e_{i\alpha\beta} t_{\alpha\beta} = 0 \tag{4.7.4}$$

which can be expanded to yield three equations of the following form

$$e_{ijk} t_{jk} = e_{ikj} t_{kj} = 0 \tag{4.7.5}$$

Considering the permutation sequences as being i, j, k, $e_{ijk} = 1$ and $e_{ikj} = -1$, Eq. (4.7.5) reduces to

$$t_{ij} = t_{ji} \tag{4.7.6}$$

So far nothing has been said or assumed about the symmetry or asymmetry of tensor t_{ij}. It turns out, however, that the balance of moment of momentum dictates the symmetry of the stress tensor t_{ij} as a necessary and sufficient condition, whenever couple stresses or polar forces are ignored. Thus the stress tensor t_{ij} is a symmetric second-order tensor.

4.8 The Pressure Tensor

The directions of the stress tensor components acting on the volume element, Fig. 4.5.2, have been defined by *convention*. Accordingly, the directions outward of the surface of the element are the directions designated as positive, while the directions inward are considered negative. This convention is commonly accepted in mechanics. In several disciplines of mechanics and in particular in those related to fluid mechanics, such as hydrostatics, hydrodynamics, mechanics of soils, etc., a different convention is accepted. According to this convention, a stress vector is defined positive when it acts on a surface inward and negative when it acts on a surface outward. The origin of such a convention may rest in the fact that the hydrostatic pressure is regarded as a basis for directions and labeled positive. As stated before, we shall adhere to the convention accepted in soil mechanics and adopt it in our study and, accordingly, we shall define the *pressure tensor* p_{ij} as a tensor opposite in direction to the stress tensor t_{ij}, Fig. 4.8.1.

$$p_{ij} = -t_{ij} \tag{4.8.1}$$

The same can be said of the direction of the body force σ_i, which is replaced by $-B_i$ and is of an opposite direction

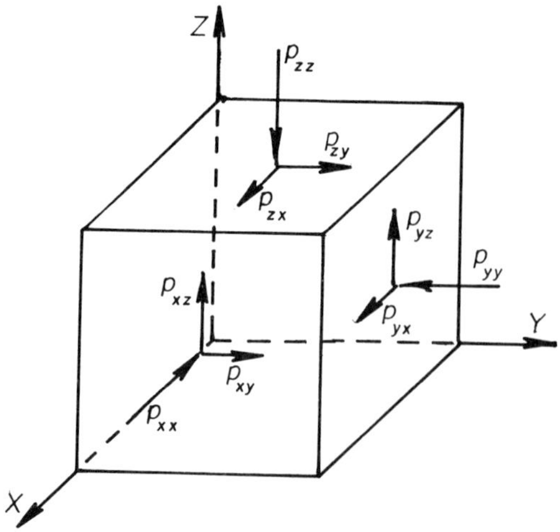

Fig. 4.8.1. Pressure components.

$$B_i = -\sigma_i \tag{4.8.2}$$

Still, in the following treatment of p_{ij} we shall use mainly the term "stress tensor", as is generally accepted.

4.9 The Stress Tensor

The stress tensor p_{ij} is a symmetric second-order tensor, and has, therefore, two indices. The first index stands for the direction normal to the plane on which the component acts and the second index indicates the direction in which the component acts. The components with the same indices, of the form p_{ii}, are the *normal stress components* of the stress tensor, while the components with the indices of the form p_{ij}, where $i \neq j$, are the six *shear stress components* of the stress tensor, which, in accord with Eq. (4.7.6) form three pairs of identical components, rendering the stress tensor symmetric.

As a second-order symmetric tensor, the stress tensor p_{ij} can be resolved into two tensors, as follows

$$p_{ij} = \tfrac{1}{3}p_{\alpha\alpha}\delta_{ij} + s_{ij} = p_{\mathrm{m}}\delta_{ij} + s_{ij} = \tfrac{1}{3}\mathrm{I}_p\delta_{ij} + s_{ij} \tag{4.9.1}$$

where $p_{\mathrm{m}}\delta_{ij}$ is the *spherical pressure tensor* and s_{ij} is the *deviatoric stress tensor*, while I_p is the first invariant of the stress tensor, as we shall see in the next section.

p_{m} is also known as the *equivalent hydrostatic pressure*, or *spherical pressure* since, as we shall see later, it has the effect of an all around pressure.

If all non-diagonal components of the stress tensor vanish, $p_{ij} = 0$ for $i \neq j$, the stress is said to be expressed by its *principal normal stress components*, denoted p_{11}, p_{22}, p_{33}. On the other hand, if the diagonal components vanish, $p_{ij} = 0$ for $i = j$, the stress is expressed by its *principal shear stress components*, denoted s_{11}, s_{22}, s_{33}.

The spherical stress is the mean pressure multiplied by the Kronecker unit tensor. The mean pressure is a scalar, obtained as the mean value of the normal stress components, and is equal to one third of the value of the first invariant of the stress tensor. It is also known as the *isotropic pressure*

$$p_{\alpha\alpha} = p_{ii} + p_{jj} + p_{kk} = p_{11} + p_{22} + p_{33} = \mathrm{I}_{\mathrm{p}} \tag{4.9.2}$$

If the stress tensor is expressed by the principal stress components which are all equal, $p_{11} = p_{22} = p_{33} = p$, the state of stresses is *hydrostatic* and p is called the *hydrostatic pressure*. Conversely, a hydrostatic stress is free of shear stresses, $s_{ij} = 0$

$$p_{ij} = p\delta_{ij}; \quad s_{ij} = 0 \tag{4.9.3}$$

The stress is a *pure shear* or *pure deviator*, if and only if the spherical tensor vanishes, $p = 0$, and then

$$p_{ij} = s_{ij}; \quad p_{\alpha\alpha} = \mathrm{I}_p = 0 \tag{4.9.4}$$

Thus Eq. (4.9.4) is the necessary and sufficient condition that the stress tensor be a pure shear tensor.

4.10 The Stress Invariants

The stress tensor p_{ij}, like the strain tensor, being a second-order symmetric tensor permits the definition of its invariant quantities.

The *principal stress invariants* are

$$\mathrm{I}_p = p_{\alpha\beta}\delta_{\alpha\beta} = p_{ii} + p_{jj} + p_{kk} = p_{11} + p_{22} + p_{33} \tag{4.10.1}$$

$$\begin{aligned}\mathrm{II}_p &= \frac{1}{2!}\delta_{\alpha\beta}\delta_{\gamma\rho}p_{\alpha\beta}p_{\gamma\rho} = p_{ii}p_{jj} + p_{jj}p_{kk} + p_{kk}p_{ii} - p_{ij}^2 - p_{jk}^2 - p_{ki}^2 \\ &= p_{11}p_{22} + p_{22}p_{33} + p_{33}p_{11}\end{aligned} \tag{4.10.2}$$

$$\begin{aligned}\mathrm{III}_p &= \frac{1}{3!}e_{\alpha\beta\gamma}e_{\rho\eta\mu}p_{\alpha\beta}p_{\gamma\rho}p_{\eta\mu} \equiv |p_{ij}| \\ &\equiv p_{ii}p_{jj}p_{kk} - p_{ii}p_{jk}^2 - p_{jj}p_{ik}^2 - p_{kk}p_{ij}^2 + 2p_{ij}p_{jk}p_{ki} \\ &= p_{11}p_{22}p_{33}\end{aligned} \tag{4.10.3}$$

where $\|p_{ij}\|$ is the determinant formed from the stress tensor matrix. Any function of the principal invariants is also an invariant. Such are the moment invariants and the octahedral invariant, see Sect. A.17, Appendix A. The *moment invariants of the stress* are

$$\bar{\mathrm{I}}_p = \mathrm{I}_p = p_{\alpha\alpha} = p_{ii} + p_{jj} + p_{kk} = p_{11} + p_{22} + p_{33} \tag{4.10.4}$$

$$\begin{aligned}\overline{\mathrm{II}}_p &= p_{\alpha\beta}p_{\alpha\beta} = p_{ii}^2 + p_{jj}^2 + p_{kk}^2 + 2p_{ij}^2 + 2p_{jk}^2 + 2p_{ki}^2 \\ &= p_{11}^2 + p_{22}^2 + p_{33}^2\end{aligned} \tag{4.10.5}$$

$$\begin{aligned}\overline{\mathrm{III}}_p &= p_{\alpha\beta}p_{\beta\gamma}p_{\gamma\rho} = p_{ii}^3 + p_{jj}^3 + p_{kk}^3 + 3p_{ij}^2(p_{ii} + p_{jj}) \\ &\quad + 3p_{jk}^2(p_{jj} + p_{kk}) + 3p_{ki}^2(p_{kk} + p_{ii}) + 6p_{ij}p_{jk}p_{ki} \\ &= p_{11}^3 + p_{22}^3 + p_{33}^3\end{aligned} \tag{4.10.6}$$

and the *octahedral stress invariant*, Π_p, introduced by Nadai (1933), which, along with the octahedral strain invariant is one of the important invariants used in the theories of plastic flow, is

$$\Pi_p = \tfrac{1}{2}(\overline{\mathrm{II}}_p - \mathrm{II}_p) = \tfrac{1}{2}(\mathrm{I}_p^2 - 3\mathrm{II}_p) = \tfrac{1}{4}(3\overline{\mathrm{II}}_p - \mathrm{I}_p^2)$$

$$= \tfrac{1}{4}[(p_{ii} - p_{jj})^2 + (p_{jj} - p_{kk})^2 + (p_{kk} - p_{ii})^2 + 6(p_{ij}^2 + p_{jk}^2 + p_{ki}^2)]$$

$$= \tfrac{1}{2}(p_{11}^2 + p_{22}^2 + p_{33}^2 - p_{11}p_{22} - p_{22}p_{33} - p_{33}p_{11}) \qquad (4.10.7)$$

For pure shear, $p_{\alpha\alpha} = 0$, the following relations between invariants are obtained from Eqs. (A.18.9) and (A.18.10), Appendix A

$$\bar{\mathrm{I}}_{\mathfrak{s}} = \mathrm{I}_{\mathfrak{s}} = 0 \qquad (4.10.8)$$

$$\overline{\mathrm{II}}_{\mathfrak{s}} = -2\mathrm{II}_{\mathfrak{s}} = s_{ii}^2 + s_{jj}^2 + s_{kk}^2 + 2s_{ij}^2 + 2s_{jk}^2 + 2s_{ik}^2$$

$$= (s_{11}^2 + s_{22}^2 + s_{33}^2) \qquad (4.10.9)$$

$$\overline{\mathrm{III}}_{\mathfrak{s}} = 3\mathrm{III}_{\mathfrak{s}} = s_{ii}^3 + s_{jj}^3 + s_{kk}^3 + 3s_{ij}^2(s_{ii} + s_{jj}) + 3s_{jk}^2(s_{jj} + s_{kk})$$

$$+ 3s_{ik}^2(s_{ii} + s_{kk}) + 6s_{ij}s_{jk}s_{ki} = s_{ii}^3 + s_{jj}^3 + s_{kk}^3 \qquad (4.10.10)$$

where Eqs. (4.10.9) and (4.10.10) serve as necessary and sufficient conditions for a stress tensor to be a pure shear. Of special importance is the octahedral invariant of the deviator

$$\Pi_{\mathfrak{s}} = -\tfrac{3}{2}\mathrm{II}_{\mathfrak{s}} = \tfrac{3}{4}\overline{\mathrm{II}}_{\mathfrak{s}} = \Pi_p = \tfrac{3}{4}(s_{ii}^2 + s_{jj}^2 + s_{kk}^2 + 2s_{ij}^2 + 2s_{jk}^2 + 2s_{ik}^2)$$

$$= \tfrac{3}{4}(s_{11}^2 + s_{22}^2 + s_{33}^2) \qquad (4.10.11)$$

Note that the octahedral invariant of the traceless stress tensor is equal to the octahedral invariant of the stress tensor. We shall see later that the octahedral invariant is responsible for the properties of the material at failure.

5 Energetics

5.1 Energy Considerations

This chapter is a continuation of the previous chapter on balance equations, since it discusses the balance of the internal energies. A separate chapter is devoted to it because of its extent and its great importance in general, and its importance for soils in particular.

In the previous chapters we have discussed *strains* – deformations and motions that are the *consequences* of the disturbance of equilibrium in matter, and *stresses* – the external forces and constraints that are the immediate *cause* of the strain.

Since one is the consequence of the other, there must be an established relationship expressing this causality. Energetics is the medium in which this relationship between stress and strain should be discussed, and its analysis will be the purpose of the following sections.

As we consider the mechanism of cause and consequence, it is clear that *power* is used, as the process begins in one state and ends in another within a finite time. This power is used while *energy* is invested during the time the process is going on.

There are many ways of investing energy in a material. The motion of a mass point along a path, caused by a force acting during a period of time, results in work done, and the energy invested equals the force multiplied by the path, two vectorial quantities. The strain of a volume element caused by stresses acting on it for a period of time is also work, but here the quantities involved are two second-order tensors. Still another way of expending energy is a deformation caused by a change in temperature where a first- or second-order tensor (deformation or strain, respectively) is related to a scalar (temperature). Many other ways of investing energy exist, but the outcome is the same, work is done as the energy invested proceeds for a length of time. We find the energy to be a scalar, a most reasonable and convenient tensorial order to account for quantities derived from tensorial terms of various orders. A bar subjected to a uniform temperature increment will undergo an extension; it may undergo the same extension if subjected to a tension. What is certain is that in both cases the same amount of internal energy has been invested in the process.

To illustrate the matter with a more appropriate example, let us take a clay sample and dry it slowly in an oven or in the air. The sample will gradually reduce in volume, it will shrink. The same shrinkage, or reduction in volume, can be attained by subjecting the sample to a spherical pressure in a triaxial cell, and we would call it then consolidation. In both instances an equivalent amount of energy must have caused the sample to change its volume. This is an internal energy, generated identically in both cases, in spite of the fact that one of the energies invested was thermal, the other mechanical. It is clear also that this is not all of the invested energy, since some of it was certainly lost by friction or by heat.

If this line of reasoning is further pursued by evaluating the changes that occur in the internal energy of the soil sample and inferring their magnitude, certain additional parameters have to be measured, such as the increase in temperature of the sample or its surroundings. In soils, the internal energy of the sample can be changed also through changes of energy in the diffuse double layer, as the clay particles are brought closer to one another. Thus we see that energy serves as a common denominator for investigation at different levels of aggregation, and so it is not only a binding factor of various thermodynamic processes, but also of the processes occurring in the various levels of aggregation.

The formulation of energetics presented here, known also as thermodynamics of matter, is based on developments in this field by Reik (1953), Meixner and Reik (1959) and Truesdell and Toupin (1960), considered the most lucid treatment that this subject has ever received.

The investigation of non-mechanical energy exchange concomitant to mechanical changes has always been difficult, mainly because of the necessity to measure small quantities of heat involved in the processes. Thermodynamics, a phenomenological study predominantly empirical, has advanced specifically in two areas: in chemistry, where chemical reactions are accompanied by mechanical and thermal energy transfer in rather ponderable quantities, and in the process of harnessing energy for steam engines, where the discharge of energy is concurrent with the phase changes of matter from liquid to vapor.

The mechanical deformation of a material is very seldom accompanied by one of these two forms of energy discharge, except for limited peripheral instances. It is no surprise, then, that engineers and scientists dealing with mechanical behavior of materials did not feel close to these processes, which were not directly related to their subjects of investigation, and that they did not, therefore, care to study them, in particular when their formulation lacked clarity and often showed discrepancies (Brønstedt 1955).

5.2 Kinetic Energy

The kinetic energy of a mass point dm having a velocity v_i is defined as $\frac{1}{2}v_\alpha v_\alpha\, dm$. When integrated over the volume V it reads

$$\mathfrak{K} = \frac{1}{2}\int_M v_\alpha v_\alpha \,\mathrm{d}m = \frac{1}{2}\int_M \mathsf{v}^2 \,\mathrm{d}m = \frac{1}{2}\int_M \dot{\mathsf{x}}\dot{\mathsf{x}}\,\mathrm{d}m \tag{5.2.1}$$

or in terms of density and volume

$$\mathfrak{K} = \frac{1}{2}\int_M v_\alpha v_\alpha \,\mathrm{d}m = \frac{1}{2}\int_V \rho v_\alpha v_\alpha \,\mathrm{d}V = \frac{1}{2}\int_V \rho\dot{\mathsf{x}}\dot{\mathsf{x}}\,\mathrm{d}V \tag{5.2.2}$$

Hence, $\mathfrak{K}$, the kinetic energy of a mass system is the sum of the kinetic energies of the mass points of the system.

In the particular case when the motion is homochoric, that is, there is no change in the density ρ, we obtain, from Eq. (5.2.2)

$$2\frac{\mathfrak{K}}{\rho} = \int_V v_\alpha v_\alpha \,\mathrm{d}V = \int_V \dot{\mathsf{x}}\dot{\mathsf{x}}\,\mathrm{d}V \tag{5.2.3}$$

Eliminating the integration over the volume V, we have, again from Eq. (5.2.2)

$$\epsilon_{\mathfrak{K}} = \tfrac{1}{2}\rho v_\alpha v_\alpha = \tfrac{1}{2}\rho\dot{\mathsf{x}}\dot{\mathsf{x}} \tag{5.2.4}$$

where $\epsilon_{\mathfrak{K}}$ is the *specific kinetic energy** or the kinetic energy per unit volume. From Eq. (5.2.4) we derive, considering Eqs. (4.1.5) and (4.1.2), the material derivative

$$\dot{\epsilon}_{\mathfrak{K}} = \frac{\mathrm{d}\epsilon_{\mathfrak{K}}}{\mathrm{d}t} = \frac{\mathrm{d}(\frac{1}{2}\rho v_\alpha v_\alpha)}{\mathrm{d}t} = \rho\dot{v}_\alpha v_\alpha + \rho v_{\alpha,\beta} v_\alpha v_\beta \tag{5.2.5}$$

where $\dot{\epsilon}_{\mathfrak{K}}$ is the *specific kinetic energy rate*.

From Eq. (4.6.3) we obtain

$$t_{\alpha i,\alpha} + \rho\sigma_i = \frac{\mathrm{d}(\rho v_i)}{\mathrm{d}t} + \rho v_{i,\alpha} v_\alpha = \rho\dot{v}_i \tag{5.2.6}$$

which, multiplied by v_i and contracted, yields

$$t_{\alpha\beta,\alpha} v_\beta + \rho\sigma_\alpha v_\alpha = \rho v_{\beta,\alpha} v_\beta v_\alpha + \rho\dot{v}_\alpha v_\alpha \tag{5.2.7}$$

In Eq. (5.2.5) we substitute $\rho\dot{v}_\alpha v_\alpha$ by its value from Eq. (5.2.7) and obtain the *kinetic energy rate* or *kinetic energy power*

$$\dot{\epsilon}_{\mathfrak{K}} = t_{\alpha\beta,\alpha} v_\beta + \rho\sigma_\alpha v_\alpha = \rho\sigma_\alpha v_\alpha + (t_{\alpha\beta} v_\beta)_{,\alpha} - t_{\alpha\beta} v_{\beta,\alpha} \tag{5.2.8}$$

* Neutral indices, as opposed to dummy indices, are the ones that do not originate from tensorial quantities, and their purpose is to award the term an additional distinction. All neutral indices are gothic indices.

Finally, substituting the pressure tensor $-p_{ij}$ for the stress tensor t_{ij} in Eq. (5.2.8) and the body force $-B_i$ for σ_i, we have

$$\dot{\epsilon}_{\mathfrak{K}} = -\rho B_\alpha v_\alpha - (p_{\beta\alpha} v_\alpha)_{,\beta} + p_{\beta\alpha} v_{\alpha,\beta} \tag{5.2.9}$$

Thus, the specific kinetic energy rate comprises three factors:

1. A decrease of the work done by the body force B_i acting on the volume element in motion.
2. A decrease of the work done by the pressure acting on the bounding area of the volume element.
3. An increase of the work done by the pressure acting on the volume element.

The term $p_{\beta\alpha} v_{\alpha,\beta}$ in Eq. (5.2.9) is also called the *specific stress work (pressure-work)* or *stress power* and it can be expressed in terms of the flow tensor f_{ij} and the spin tensor w_{ij}

$$\mathfrak{w}_{\mathfrak{s}} = p_{\alpha\beta} v_{\alpha,\beta} = p_{\alpha\beta} \dot{x}_{\alpha,\beta} = p_{\alpha\beta} f_{\alpha\beta} = p_{\alpha\beta}(\varepsilon_{\alpha\beta} + w_{\alpha\beta}) \tag{5.2.10}$$

If the term $p_{\alpha\beta} w_{\alpha\beta}$ in Eq. (5.2.10) vanishes, the stress-power is satisfied by the lamellar motion caused by the stress and is unaffected by the spin tensor.

By integrating Eq. (5.2.10) over the volume for a lamellar motion, the stress-power is obtained

$$\mathfrak{M}_{\mathfrak{s}} = \int_V \mathfrak{w}_{\mathfrak{s}}\, \mathrm{d}V = \int_V p_{\alpha\varphi} \dot{x}_{\alpha,\beta}\, \mathrm{d}V = \int_V p_{\alpha\beta} f_{\alpha\beta}\, \mathrm{d}V = -\int_V t_{\alpha\beta} f_{\alpha\beta}\, \mathrm{d}V \tag{5.2.11}$$

In motions with velocities of small order of magnitude, the entire kinetic energy becomes small and insignificant, for which reason it is then omitted.

5.3 Potential Energy

If the body force σ_i constitutes a steady lamellar field, then there exists a scalar Φ so that

$$\sigma_i = -B_i = \Phi_{,i}; \quad \Phi = \Phi(x, t) \tag{5.3.1}$$

where Φ, called the *force potential*, is a scalar function of space and time.

The *potential energy* is defined as the volume integral of the product of the force potential and the mass element

$$\mathfrak{P} = \int_M \Phi\, \mathrm{d}M = \int_V \rho\Phi\, \mathrm{d}V \tag{5.3.2}$$

The connected points $\Phi = \text{const}$ are called *equipotential lines*. From Eq. (5.3.1) it is evident that the body force B_i, that represents the intensity of the potential field, is inversely proportional to the distance between the equipotential lines and is opposite to the direction of the force potential. Crowded equipotentials mean great intensity of the force and sparce equipotentials mean slight intensity. Eq. (5.3.2) indicates that the potential energy increases with the increase of the position vector of the volume element.

The potential energy per unit volume will thus be obtained from Eq. (5.3.2)

$$\epsilon_p = \rho\Phi \tag{5.3.3}$$

where ϵ_p will be called the *specific potential energy*, from which we obtain the material derivative

$$\dot{\epsilon}_p = \frac{d\epsilon_p}{dt} = \rho\Phi_{,\alpha}v_\alpha \tag{5.3.4}$$

where $\dot{\epsilon}_p$ is the *specific potential energy rate*.

Eq. (5.3.4) can be written in terms of the body force by substituting into it Eq. (5.3.1)

$$\dot{\epsilon}_p = -\rho\sigma_\alpha v_\alpha = \rho B_\alpha v_\alpha \tag{5.3.5}$$

Thus, the specific potential energy rate is expressed by the decrease of work done by the body force B_i acting on the volume element in motion.

Substituting Eq. (5.3.5) into Eq. (5.2.8), it is evident that the sum of the kinetic and potential energy rates can be expressed entirely in terms of work done by the stress

$$\dot{\epsilon}_{\mathfrak{K}} + \dot{\epsilon}_p = -t_{\alpha\beta}f_{\alpha\beta} + (t_{\alpha\beta}v_\alpha)_{,\beta} = p_{\alpha\beta}f_{\alpha\beta} - (p_{\alpha\beta}v_\alpha)_{,\beta} \tag{5.3.6}$$

5.4 Internal Energy

The internal energy $\mathfrak{I}$ is the sum of the energies existing at any time in the mass. These energies include the supply and the influx or efflux of energy

$$\mathfrak{I} = \int_M \epsilon_i \, dm = \int_V \rho\epsilon_i \, dV \tag{5.4.1}$$

where ϵ_i is the *specific internal energy* or the internal energy per unit mass.

The material derivative of the internal energy from (5.4.1) yields the *internal energy rate*

$$\dot{\mathfrak{I}} = \int_V \rho\dot{\epsilon}_i \, dV \tag{5.4.2}$$

which, expressed in terms of its components, is

$$\dot{\mathfrak{I}} = \int_V \rho \dot{e}_i \, dV = -\mathfrak{M}_s = \dot{\mathfrak{E}} + \dot{\mathfrak{S}} \tag{5.4.3}$$

where $\mathfrak{M}_s$ is the *stress-power* of Eq. (5.2.11), $\dot{\mathfrak{E}}$ is the *energy efflux rate* through the surface which encloses the volume, and $\dot{\mathfrak{S}}$ is the *energy supply rate* to the volume, defined

$$\dot{\mathfrak{E}} = \int_S h_\alpha n_\alpha \, dS = \int_V h_{\alpha,\alpha} \, dV \tag{5.4.4}$$

$$\dot{\mathfrak{S}} = \int_M q \, dm = \int_V \rho q \, dV \tag{5.4.5}$$

where h_i is the *energy flux* and q is the *energy supply*.

The internal energy, as seen from Eq. (5.4.3), is opposite to the direction of the stress power.

The values of $\dot{\mathfrak{E}}$ and $\dot{\mathfrak{S}}$ from Eqs. (5.2.11), (5.4.4), and (5.4.5) are substituted into Eq. (5.4.3) omitting the volume integration, and the specific internal energy is so obtained

$$\rho \dot{e}_i = t_{\alpha\beta} f_{\alpha\beta} + h_{\alpha,\alpha} + \rho q \tag{5.4.6}$$

which is the differential equation of the *internal energy balance*.

5.5 The Total Energy Balance

Attempts to formulate an equation of energy balance which could be applicable in the study of energy exchange in matter has, so far, fallen behind expectations. This is particularly true for the study of energy exchange in chemically inert materials, whether they are homogeneous or heterogeneous.

Thermodynamics and energetics have been applied in the past to the study of materials in which the exchange of energy is manifested by chemical reactions, and changes in temperature, pressure, etc., have been measured in the course of laboratory experiments. This method is not readily applicable to materials in which chemical reactions are not commonly present. The existing laws of thermodynamics are inadequate also in the sense that they do not provide a criterion for distinction between conservative and non-conservative quantities.

In general, the existing laws of thermo-energetics are the product of empirical observations and conclusions, rather than the logical assessment of changes in matter. Quantities such as dissipative energy, mechanical energy and strain-power are presently so unrigorously defined, that we shall not attempt to include them here. Rather we shall discuss them later in our work, where terms such as stored energy and disbursed energy, not yet introduced, will also be discussed. These terms will be presented in connection with specific applications and defined as necessary, their universal merit not having been verified yet. Only the very general expressions of energy balance will be given here.

The *total energy* $\mathfrak{T}$ is defined as the sum of the internal energy $\mathfrak{I}$ and the kinetic energy $\mathfrak{K}$ on one hand, as discussed in Sects. 5.2–5.4, and the potential energy $\mathfrak{P}$ and the *dissipative energy* $\mathfrak{D}$, on the other hand

$$\mathfrak{T} = \mathfrak{I} + \mathfrak{K} = \mathfrak{P} + \mathfrak{D} \tag{5.5.1}$$

Its rate per unit volume, called the *specific total energy rate* $\dot{\epsilon}_{\mathrm{t}}$ is obtained from Eqs. (5.4.6) and (5.2.8)

$$\dot{\epsilon}_{\mathrm{t}} = \dot{\epsilon}_{\mathrm{i}} + \dot{\epsilon}_{\mathfrak{K}} = -\rho B_{\alpha} v_{\alpha} - (p_{\alpha\beta} v_{\alpha})_{,\beta} + h_{\alpha,\alpha} + \rho q \tag{5.5.2}$$

This is the rate of the specific total energy, which comprises: the energy decrease due to the work rate of the body force B_i, the decrease of the work done by the pressure p_{ij} over the boundary area, the efflux of the energies h_i, and the supply of any energies q to the mass.

Eq. (5.5.2) is considered the differential equation of the *total energy balance*.

If we agree that Eq. (5.5.2) accounts for all possible energy rates related to the state of the mass point, we will also be safe in saying that the total energy rate is equal to the *mechanical power* $\mathfrak{M}_{\mathrm{m}}$ plus the *non-mechanical power* $\mathfrak{M}_{\mathrm{n}}$ (thermal, chemical, etc.), and we can then write

$$\dot{\mathfrak{T}} = \dot{\mathfrak{I}} + \dot{\mathfrak{K}} = \dot{\mathfrak{P}} + \dot{\mathfrak{D}} = \mathfrak{M}_{\mathrm{m}} + \mathfrak{M}_{\mathrm{n}} \tag{5.5.3}$$

or, in specific energy rates per unit volume

$$\dot{\epsilon}_{\mathfrak{K}} = \dot{\epsilon}_{\mathrm{i}} + \dot{\epsilon}_{\mathrm{t}} = \dot{\epsilon}_{\mathfrak{p}} + \dot{\epsilon}_{\mathfrak{d}} = \mathfrak{w}_{\mathrm{m}} + \mathfrak{w}_{\mathrm{n}} \tag{5.5.4}$$

where $\mathfrak{w}_{\mathrm{m}}$ and $\mathfrak{w}_{\mathrm{n}}$ are the *specific mechanical* and *specific non-mechanical powers*, respectively.

The statement in Eq. (5.5.3), although correct, has only a scarce applicability in this form.

Again, the specific total energy balance can be derived from the general balance equations (4.2.3) and (4.2.6), along the lines discussed in Sect. 4.2, based on the treatment by Meixner and Reik (1959).

The *specific total energy per unit mass* ϵ_{t} is equal to the sum of the *specific kinetic energy* $\epsilon_{\mathfrak{K}} = \frac{1}{2} v_{\alpha} v_{\alpha}$ and the *specific internal energy* ϵ_{i}. To derive the balance of the specific total energy from Eqs. (4.2.3) and (4.2.6), it is assumed that the extensive quantity ξ is taken as the specific total energy $\xi = \epsilon_{\mathrm{t}} = \frac{1}{2} v_{\alpha} v_{\alpha} + \epsilon_{\mathrm{i}}$, Φ is taken as the sum of mechanical and non-mechanical flux $\Phi = \rho(\frac{1}{2} v_{\alpha} v_{\alpha} + \epsilon_{\mathrm{i}}) v_i + \ p_{i\beta} v_{\beta} - h_i$ and σ is taken as the exchange of mechanical and non-mechanical specific energy supply $\boldsymbol{\sigma} = \sigma_\alpha v_\alpha + q$. Thus, the spatial and material balance of the specific total energy rate is

$$\frac{\partial}{\partial t}\left(\tfrac{1}{2}\rho v_\alpha v_\alpha + \rho\epsilon_i\right) + \left(\tfrac{1}{2}\rho v_\alpha v_\alpha v_\beta + \rho\epsilon_i v_\beta - t_{\alpha\beta} v_\alpha + h_\beta\right)_{,\beta}$$

$$- \sigma_\alpha v_\alpha - \rho q = 0 \tag{5.5.5}$$

$$\rho v_\alpha \dot{v}_\alpha + \rho\epsilon_i - (t_{\alpha\beta} v_\alpha + h_\beta)_{,\beta} - \rho\sigma_\alpha v_\alpha - \rho q = 0 \tag{5.5.6}$$

Considering Eq. (4.6.3) and ignoring the effect of the kinetic energy, a much simpler form of the equations of balance of the specific total energy rate is obtained, in terms of the internal energy

$$\frac{\partial}{\partial t}\rho\epsilon_i + (\rho\epsilon_i v_\alpha)_{,\alpha} - t_{\alpha\beta} v_{\alpha,\beta} - h_{\alpha,\alpha} - \rho q = 0 \tag{5.5.7}$$

$$\rho\dot{\epsilon}_i - t_{\alpha\beta} v_{\alpha,\beta} - h_{\alpha,\alpha} - \rho q = 0 \tag{5.5.8}$$

Eq. (5.5.8) is exactly what we have already obtained in Eq. (5.4.6) for the specific internal energy rate.

5.6 Historical Notes on Irreversible Processes of the Continuum

The study of the frictional resistance by Newton (1687) is probably the earliest phenomenological description of irreversible processes, and perhaps it is not by accident that it followed, by less than a decade, the description of spring-like bodies by Hooke (1678). For nearly two centuries no major developments, except simple applications of the frictional resistance, were added to this subject. A prominent name in this period is that of Coulomb, who made contributions to the study of stability problems of frictional masses. It was not until the end of the nineteenth century that the classical thermodynamics began to take shape and was expressed in the treatise on thermodynamics by Bertrand (1887) and in the first clear formulation of the thermodynamics of fluids and substances and the equilibrium of heterogeneous substances by Gibbs (1873, 1875). A special mention should go to Caratheodory (1909) for his mathematical investigations of the basic axiomatics of thermodynamics, considered the theoretical foundation of thermodynamics.

The thermodynamical theory of irreversible processes was further developed in the works of Onsager (1931) on diffusion of substances and formulation of the reciprocal conditions through the calculus of variations, and a decade later Eckhart (1940), who gave a systematic modern treatment of irreversible thermodynamic processes.

Based on these two works, more light was shed on the phenomenological aspect of irreversible processes by Prigogine (1947), Meixner and Reik (1959), until it reached its present stage.

5.7 The Thermodynamic State

The thermodynamic state is the totality of the substates v_{m} which determine the specific internal energy ϵ_{i}. These substates are expressed in the mathematical formulation as the parameters v_{m} and their physical meaning is unspecified in the mathematical treatment.

The exposition of the subject in the next two sections follows very closely the studies by Prigogine, Reik and Meixner, mentioned before, all based on the elaborate work by Gibbs, differing slightly by the fact that the entropy is considered as one of the parameters v_{m} and is receiving the mathematical treatment as such. The definition and detailed discussion of entropy is given in Sects. 5.9–5.11.

The parameters v_{i} are tensor fields of any order, functions of the space x_i and the time t. No limit is imposed on the number of parameters, of different order or the same order but differing in their physical meaning, that can appear in the mathematical formulation.

The basic assumption of thermodynamics asserts the *sufficiency of the substates to determine the internal energy independently of time, place, motion and stress* (Truesdell and Toupin 1960).

There exists, therefore, a function such that

$$\epsilon_{\mathrm{i}} = \mathcal{F}(v_1, v_2, v_3, \ldots, v_{\mathfrak{M}}, x_i) = \epsilon_{\mathrm{i}}(v_{\mathrm{m}}, x_i) \tag{5.7.1}$$

where ϵ_{i} is the *specific internal energy per unit mass*, assumed to be one-to-one valued in the parameters. Eq. (5.7.1) is called the *caloric equation of state*.

Any given motion will, of course, be characterized by an equation of the form $\epsilon_{\mathrm{i}} = \mathcal{G}(x_i, t)$, where the functionality $\mathcal{G}$ will vary according to the particular motion. Consequently, the specific internal energy ϵ_{i} will not be dependent on the motion, but solely on the substate v_{m} and on the form of the function of Eq. (5.7.1), which is the *thermodynamic state*.

In a diagrammatical form this is seen in Fig. 5.7.1 in the two-dimensional Gibbs diagram, where each point corresponds to the thermodynamic state of the element designated by its position vector x_i, this state being determined at any time by $\epsilon_{\mathrm{i}} = \epsilon_{\mathrm{i}}(t)$, $v_{\mathrm{m}} = v_{\mathrm{m}}(t)$ and carried from a state ϵ_{i}, v_{m} at time t to a state ϵ_{i}^*, v_{m}^* at time t^*.

One state can be reached from another by many different paths, provided that at the different intervals of time the energy increments will be the same.

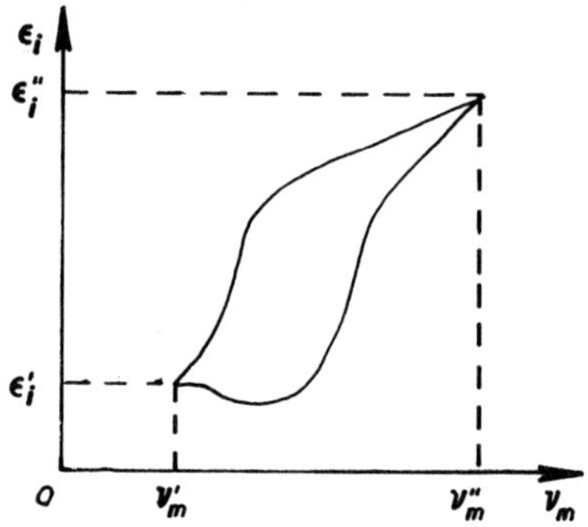

Fig. 5.7.1. Gibbs' diagram

Conversely, the thermodynamic state does not determine how any energy state has been reached.

The function $\mathcal{F}$, Eq. (5.7.1), is characteristic of the different *thermodynamic substates*. If the function $\mathcal{F}$ is independent of the position x_i of the element, the substate is *thermodynamically homogeneous*.

It is assumed that $\epsilon_{\mathfrak{i}}$ is a continuous and single-valued function of both the parameters $v_{\mathfrak{m}}$ and t. Consequently we may invert Eq. (5.7.1) and write

$$v_{\mathfrak{o}} = v_{\mathfrak{o}}(\epsilon_{\mathfrak{i}}, v_{\mathfrak{m}-\mathfrak{o}}, x_i) \tag{5.7.2}$$

where the convention on the notation $\mathfrak{m} - \mathfrak{o}$ indicates that all $\mathfrak{m}$ parameters, in our case $\mathfrak{m}$ substates, are considered, except for the $\mathfrak{o}$th parameter, in our case the $\mathfrak{o}$th substate.

5.8 Thermodynamic Tensions

Any change in the thermodynamic state is given by the total derivative of the internal energy

$$d\epsilon_{\mathfrak{i}} = \sum_{\mathfrak{m}=1}^{\mathfrak{M}} \tau_{\mathfrak{m}}\, dv_{\mathfrak{m}} \tag{5.8.1}$$

where $\tau_{\mathfrak{m}}$ are the *thermodynamic tensions*, defined

$$\tau_{\mathfrak{m}} \equiv \frac{\partial \epsilon_{\mathfrak{i}}}{\partial v_{\mathfrak{m}}} \tag{5.8.2}$$

and representing the slopes of the curve in Fig. 5.7.1.

From Eqs. (5.7.1) and (5.8.2) it is evident that the tensions are functions of the thermodynamic state.

$$\tau_{\mathfrak{m}} = \tau_{\mathfrak{m}}(v_{\mathfrak{m}}, x_i) \tag{5.8.3}$$

From Eq. (5.8.1) the time and space partial derivatives of the internal energy are obtained

$$\frac{\partial \epsilon_i}{\partial t} = \sum_{m=1}^{\mathfrak{M}} \tau_m \left(\frac{\partial v_m}{\partial t}\right) \tag{5.8.4}$$

$$\frac{\partial \epsilon_i}{\partial x_i} = \sum_{m=1}^{\mathfrak{M}} \tau_m \left(\frac{\partial v_m}{\partial x_i}\right) \tag{5.8.5}$$

The partial derivatives of ϵ_i from (5.7.1), with respect to any substate, say v_m, are

$$\tau_m = \left(\frac{\partial \epsilon_i}{\partial v_o}\right)_{v_{m-o}} = \left(\frac{\partial \mathcal{F}}{\partial v_o}\right)_{v_{m-o}} \tag{5.8.6}$$

Stipulating the conventions with respect to notation, the subscript outside the parentheses indicates the parameters held constant during differentiation.

The total derivative of any substate v_o can also be determined from Eq. (5.7.2)

$$\mathrm{d}v_o = \left(\frac{\partial v_o}{\partial \epsilon_i}\right)_{v_{m-o}} \mathrm{d}\epsilon_i + \sum_{m=1}^{\mathfrak{M}-o} \left(\frac{\partial v_o}{\partial v_{m-o}}\right)_{\epsilon_i} \mathrm{d}v_{m-o} \tag{5.8.7}$$

5.9 Entropy and Temperature

In Sects. 5.7 and 5.8 the thermodynamic state and its substates have been defined. In general, the substates are determined empirically, and for this reason they vary according to the particular material, the thermodynamic process and the phenomenon investigated. The specification of the substates is discussed in a later section.

There is, however, one substate present in all processes and phenomena and in any material, called the *entropy*. Clausius (1865) introduced the term entropy as the "measure of change that occurs in matter". Mathematically it is one of the parameters v_m representing a thermodynamic substate, denoted $v \equiv \eta$. Its corresponding tension $\tau \equiv T$ is the *temperature*, a new physical measure thus introduced. According to Eq. (5.8.2) we can write

$$T \equiv \frac{\partial \epsilon_i}{\partial \eta} \tag{5.9.1}$$

from which the *total entropy* H is defined as part of the additive set which represents the internal energy

$$H = \int_M \eta \,\mathrm{d}m = \int_V \eta \rho \,\mathrm{d}V \tag{5.9.2}$$

The specific internal energy per unit mass, Eq. (5.7.1), expressed in terms of the specific entropy η and the remaining substates ν_{m}, will be

$$\epsilon_{\mathrm{i}} = \mathcal{F}(\eta, \nu_1, \nu_2, \nu_3, \ldots, x_i) \tag{5.9.3}$$

where ϵ_{i} is a continuous and single-valued function of η as well, so that the following relations will also be valid

$$\eta = \eta(\epsilon_{\mathrm{i}}, \nu_{\mathrm{m}}, x_i) \tag{5.9.4}$$

$$\nu_{\mathrm{o}} = \nu_{\mathrm{o}}(\epsilon_{\mathrm{i}}, \eta, \nu_{\mathrm{m-o}}, x_i) \tag{5.9.5}$$

As we can see, the above two equations are of the form of Eq. (5.7.2).

The Gibbs diagram will become three-dimensional, the three coordinates being ϵ_{i}, η, ν_{m}, Fig. 5.9.1.

From Eqs. (5.9.3), (5.8.2) and (5.9.1) we obtain the following functions

$$T = T(\eta, \nu_{\mathrm{m}}) \tag{5.9.6}$$

$$\tau_{\mathrm{m}} = \tau_{\mathrm{m}}(\eta, \nu_{\mathrm{m}}) \tag{5.9.7}$$

The function (5.9.6), being single valued and continuous, can be inverted to obtain

$$\eta = \eta(T, \nu_{\mathrm{m}}) \tag{5.9.8}$$

From Eq. (5.8.1) we obtain the total derivative of the specific energy

$$d\epsilon_{\mathrm{i}} = T d\eta + \sum_{\mathrm{m}=1}^{\mathfrak{M}} \tau_{\mathrm{m}}\, d\nu_{\mathrm{m}} \tag{5.9.9}$$

The partial derivatives of time and space follow from Eqs. (5.8.4) and (5.8.5)

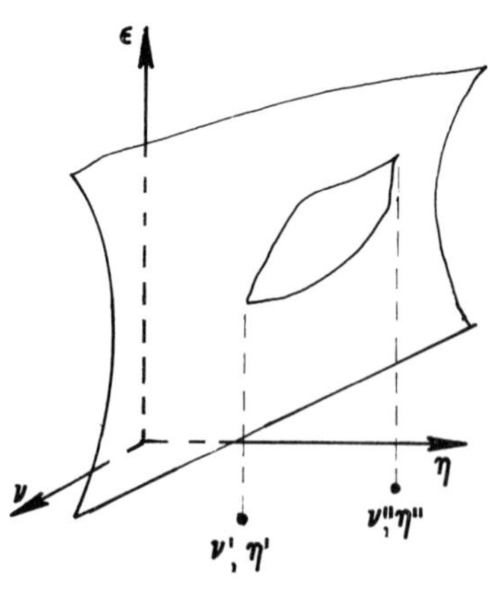

Fig. 5.9.1. Three-dimensional Gibbs' diagram

$$\frac{\partial \epsilon_{\mathrm{i}}}{\partial t} = \mathrm{T}\frac{\partial \eta}{\partial t} + \sum_{\mathrm{m}=1}^{\mathfrak{M}} \tau_{\mathrm{m}} \frac{\partial \nu_{\mathrm{m}}}{\partial t} \tag{5.9.10}$$

$$\frac{\partial \epsilon_{\mathrm{i}}}{\partial x_i} = T\eta_{,i} + \sum_{\mathrm{m}=1}^{\mathfrak{M}} \tau_{\mathrm{m}} \nu_{\mathrm{m},i} \tag{5.9.11}$$

Maintaining x_i constant, unless otherwise specified, we obtain from Eq. (5.9.1)

$$T \equiv \left(\frac{\partial \epsilon_{\mathrm{i}}}{\partial \eta}\right)_{\nu_{\mathrm{m}}} \tag{5.9.12}$$

From Eq. (5.9.4) we can obtain the total derivative of the entropy, and we substitute in it Eq. (5.9.9), so that

$$\mathrm{d}\eta = \left(\frac{\partial \eta}{\partial \epsilon_{\mathrm{i}}}\right)_{\nu_{\mathrm{m}}} \mathrm{d}\epsilon_{\mathrm{i}} + \sum_{\mathfrak{o}=1}^{\mathfrak{M}} \left(\frac{\partial \eta}{\partial \nu_{\mathfrak{o}}}\right)_{\epsilon_{\mathrm{i}},\nu_{\mathrm{m}-\mathfrak{o}}} \mathrm{d}\nu_{\mathrm{m}-\mathfrak{o}}$$

$$= \left(\frac{\partial \eta}{\partial \epsilon_{\mathrm{i}}}\right)_{\nu_{\mathrm{m}}} \mathrm{d}\epsilon_{\mathrm{i}} + \left(T\,\mathrm{d}\eta + \sum_{\mathrm{m}=1}^{\mathfrak{M}} \tau_{\mathrm{m}}\,\mathrm{d}\nu_{\mathrm{m}}\right) + \sum_{\mathfrak{o}=1}^{\mathfrak{M}} \left(\frac{\partial \eta}{\partial \nu_{\mathrm{m}}}\right)_{\epsilon_{\mathrm{i}},\nu_{\mathrm{m}-\mathfrak{o}}} \mathrm{d}\nu_{\mathfrak{o}} \tag{5.9.13}$$

By comparing the coefficients of the differentials, we obtain

$$\frac{1}{T} = \left(\frac{\partial \eta}{\partial \epsilon_{\mathrm{i}}}\right)_{\nu_{\mathrm{m}}} \tag{5.9.14}$$

$$\tau_{\mathfrak{o}} = -\left(\frac{\partial \eta}{\partial \nu_{\mathfrak{o}}}\right)_{\epsilon_{\mathrm{i}},\nu_{\mathrm{m}-\mathfrak{o}}} \tag{5.9.15}$$

Introducing Eq. (5.9.8) into Eq. (5.9.7), we obtain, owing to the single valued relation between the parameters

$$\tau_{\mathrm{m}} = \tau_{\mathrm{m}}(T, \nu_{\mathrm{m}}) \tag{5.9.16}$$

$$\nu_{\mathrm{m}} = \nu_{\mathrm{m}}(T, \tau_{\mathrm{m}}) \tag{5.9.17}$$

These are the *thermal equations of state*, the total derivatives of which are

$$\mathrm{d}\tau_{\mathrm{m}} = \left(\frac{\partial \tau_{\mathrm{m}}}{\partial T}\right)_{\nu_{\mathrm{m}}} \mathrm{d}T + \sum_{\mathfrak{o}=1}^{\mathfrak{M}} \left(\frac{\partial \tau_{\mathrm{m}}}{\partial \nu_{\mathrm{m}}}\right)_{T=\nu_{\mathrm{m}-\mathfrak{o}}} \mathrm{d}\nu_{\mathfrak{o}} \tag{5.9.18}$$

$$\mathrm{d}\nu_{\mathrm{m}} = \left(\frac{\partial \nu_{\mathrm{m}}}{\partial T}\right)_{\nu_{\mathrm{m}}} \mathrm{d}T + \sum_{\mathfrak{o}=1}^{\mathfrak{M}} \left(\frac{\partial \nu_{\mathrm{m}}}{\partial \tau}\right)_{T=\tau_{\mathrm{m}-\mathfrak{o}}} \mathrm{d}\tau_{\mathfrak{o}} \tag{5.9.19}$$

Eqs. (5.9.18) and (5.9.19) are more practical than (5.9.3) or (5.9.13), since their coefficients are measurable quantities during experiments, although not sufficient to determine the thermodynamic state.

5.10 The Thermodynamic Functions

We shall define here several other thermodynamic functions, developed by Gibbs (1875) and therefore known as the *Gibbs equations*, but more commonly known as *thermodynamic potentials*, which are quite widely used and are related to the specific internal energy per unit mass.

$$\psi = \epsilon_i - \eta T \tag{5.10.1}$$

$$\chi = \epsilon_i - \sum_{m=1}^{\mathfrak{M}} \tau_m \nu_m \tag{5.10.2}$$

$$\zeta = \chi - \eta T = \epsilon_i - \eta T - \sum_{m=1}^{\mathfrak{M}} \tau_m \nu_m \tag{5.10.3}$$

where ψ is the *free energy*, χ is the *enthalpy* and ζ is the *free enthalpy*, related by the equation

$$\epsilon_i - \psi + \chi - \zeta = 0 \tag{5.10.4}$$

From Eqs. (5.10.1)–(5.10.3) we obtain the total derivatives of the thermodynamic potentials

$$d\psi = -\eta\, dT + \sum_{m=1}^{\mathfrak{M}} \tau_m\, d\nu_m \tag{5.10.5}$$

$$d\chi = T\, d\eta - \sum_{m=1}^{\mathfrak{M}} \nu_m\, d\tau_m \tag{5.10.6}$$

$$d\zeta = -\eta\, dT - \sum_{m=1}^{\mathfrak{M}} \nu_m\, d\tau_m \tag{5.10.7}$$

Some simple relations are obtained from these three equations, in addition to the two relations already shown in Eqs. (5.8.6) and (5.9.12)

$$\eta = -\left(\frac{d\psi}{\partial T}\right)_{\nu_m} \tag{5.10.8}$$

$$\tau_m = \left(\frac{\partial \psi}{\partial \nu_o}\right)_{T, \nu_{m-o}} \tag{5.10.9}$$

$$T = -\left(\frac{\partial \chi}{\partial \eta}\right)_{\tau_m} \tag{5.10.10}$$

$$\nu_m = -\left(\frac{\partial \chi}{\partial \tau_o}\right)_{\eta, \tau_{m-o}} \tag{5.10.11}$$

$$\eta = -\left(\frac{\partial \zeta}{\partial T}\right)_{\tau_m} \tag{5.10.12}$$

$$\nu_o = -\left(\frac{\partial \zeta}{\partial \tau_o}\right)_{T,\nu_{m-o}} \tag{5.10.13}$$

which, when substituted into Eqs. (5.10.1)–(5.10.3), yield

$$\psi = \epsilon_i - T\left(\frac{\partial \psi}{\partial T}\right)_{\nu_m} \tag{5.10.14}$$

$$\chi = \epsilon_i - \sum_{m=1}^{\mathfrak{M}} \tau_o\left(\frac{\partial \chi}{\partial \tau_m}\right)_{\eta,\tau_{m-o}} \tag{5.10.15}$$

$$\zeta = \epsilon_i - T\left(\frac{\partial \zeta}{\partial \mathrm{T}}\right)_{\tau_m} + \sum_{m=1}^{\mathfrak{M}} \tau_o\left(\frac{\partial \zeta}{\partial \tau}\right)_{T,\tau_{m-o}} \tag{5.10.16}$$

It is evident from Eqs. (5.9.9) and (5.10.5)–(5.10.7) that the functions ϵ_i, ψ, χ and ζ from Eqs. (5.9.9), (5.10.14), (5.10.15) and (5.10.16) are functions of the following independent variables, respectively

$$\eta,\ \nu_m; \quad T,\ \nu_m; \quad \eta,\ \tau_m; \quad T,\ \tau_m \tag{5.10.17}$$

The thermodynamic potentials, expressed in terms of these sets, represent the fundamental equations of thermodynamics. These equations allow the evaluation of the variables ν_m, τ_m, η, T, ϵ_i called the *thermodynamic variables*. The fundamental equations are

$$\epsilon_i = \epsilon_i(\eta, \nu_m) \tag{5.10.18}$$

$$\psi = \psi(T, \nu_m) \tag{5.10.19}$$

$$\chi = \chi(\eta, \tau_m) \tag{5.10.20}$$

$$\zeta = \zeta(T, \tau_m) \tag{5.10.21}$$

5.11 The Production of Entropy

In Sects. 5.4 and 5.5 we discussed the internal and total energy balance, and in the subsequent sections we have discussed, independently, the thermodynamic state, thermodynamic tensions, entropy, etc. It is not by accident that all this material has been presented in the same chapter. As pointed out in Sect. 5.1, there is an interrelationship between the mechanical work that the material is subjected to and the thermal energy exchange.

In an attempt to investigate this relationship, let us first consider the equation of internal energy balance, Eq. (5.4.6), composed of the specific stress power $t_{\alpha\beta}f_{\alpha\beta}$ and the energy flux and supply $h_{\alpha,\alpha} + \rho q$. All three are imposed externally and affect the material, the specific stress power as a mechanical work and the energy flux and supply as a non-mechanical work.

On the other hand, let us consider the specific internal energy, Eq. (5.9.10), which, multiplied by the density ρ yields the equivalent of the internal energy of the thermodynamic state, or the internal thermodynamic exchange occurring in the element. This equation consists of a mechanical part, the thermodynamic tensions

$$\rho \sum_{m=1}^{\mathfrak{M}} \tau_m \dot{v}_m$$

and a non-mechanical part, the entropy $\rho T\dot{\eta}$.

By equating these two expressions, we assume that the externally imposed energies balance with the internal exchange of thermal energy

$$\rho\dot{\epsilon}_i = t_{\alpha\beta}f_{\alpha\beta} + h_{\alpha,\alpha} + \rho q = \rho T\dot{\eta} + \rho \sum_{m=1}^{\mathfrak{M}} \tau_m \dot{v}_m \tag{5.11.1}$$

or

$$\begin{aligned}\rho\dot{\eta} &= \frac{1}{T}\left(t_{\alpha\beta}f_{\alpha\beta} - \rho \sum_{m=1}^{\mathfrak{M}} \tau_m \dot{v}_m + h_{\alpha,\alpha} + \rho q\right) \\ &= \frac{1}{T}\left(t_{\alpha\beta}f_{\alpha\beta} - \rho \sum_{m=1}^{\mathfrak{M}} \tau_m \dot{v}_m + \rho q\right) + \left(\frac{h_\alpha}{T}\right)_{,\alpha} + \frac{h_\alpha T_{,\alpha}}{T^2} \\ &= \frac{1}{T}\left(t_{\alpha\beta}f_{\alpha\beta} - \rho \sum_{m=1}^{\mathfrak{M}} \tau_m \dot{v}_m + \rho q\right) + \frac{h_\alpha}{T}(\ln T)_{,\alpha} + \left(\frac{h_\alpha}{T}\right)_{,\alpha}\end{aligned} \tag{5.11.2}$$

which is the equation of the *production rate of specific entropy*.

It is evident from Eq. (5.11.2) that the entropy production rate per unit volume of matter is proportional to the energy rate imposed externally less the energy rate developed internally, and is inversely proportional to the temperature.

Substituting Eqs. (5.9.10) and (5.11.1) into (5.10.5)–(5.10.7), we obtain the specific rates of the free energy, enthalpy and free enthalpy, respectively

$$\rho\dot{\psi} = -\rho\dot{T}\eta + \rho \sum_{m=1}^{\mathfrak{M}} \tau_m \dot{v}_m = t_{\alpha\beta}f_{\alpha\beta} + h_{\alpha,\alpha} + \rho q - \rho T\dot{\eta} - \rho\dot{T}\eta \tag{5.11.3}$$

$$\rho\dot{\chi} = \rho T\dot{\eta} - \rho \sum_{m=1}^{\mathfrak{M}} v_m \dot{\tau}_m = t_{\alpha\beta}f_{\alpha\beta} - \rho \sum_{m=1}^{\mathfrak{M}} \tau_m \dot{v}_m + h_{\alpha,\alpha} + \rho q - \rho \sum_{m=1}^{\mathfrak{M}} v_m \dot{\tau}_m \tag{5.11.4}$$

$$\rho\dot{\zeta} = -\rho\eta\dot{T} - \rho \sum_{m=1}^{\mathfrak{M}} v_m \dot{\tau}_m$$

$$= t_{\alpha\beta} f_{\alpha\beta} - \rho \sum_{M=1}^{\mathfrak{M}} \tau_m \dot{v}_m - \rho \sum_{m=1}^{\mathfrak{M}} v_m \dot{\tau}_m + h_{\alpha,\alpha} + \rho q - \rho T\dot{\eta} - \rho\dot{T}\eta \tag{5.11.5}$$

We note from Eq. (5.11.3) that if the non-mechanical power vanishes, all the external power imposed as a stress power is directly converted into free energy

$$h_{\alpha,\alpha} + \rho q - \rho T\dot{\eta} - \rho\dot{T}\eta = 0 \tag{5.11.6}$$

Similarly, Eq. (5.11.4) shows that the vanishing of the mechanical power results in the conversion of all non-mechanical power into enthalpy

$$t_{\alpha\beta} f_{\alpha\beta} - \rho \sum_{m=1}^{\mathfrak{M}} \tau_m \dot{v}_m - \rho \sum_{m=1}^{\mathfrak{M}} v_m \dot{\tau}_m \tag{5.11.7}$$

$$\therefore \rho\chi = h_{\alpha,\alpha} + \rho q$$

Applying the time derivative to Eq. (5.9.2) and then substituting into it Eq. (5.11.2), we obtain the equation of the production rate of the total entropy

$$\dot{H} = \int_V \rho\dot{\eta}\, dV = \dot{H}' + \dot{H}'' = \int_V \rho\dot{\eta}'\, dV + \int_V \rho\dot{\eta}''\, dV$$

$$= \int_V \frac{1}{T}\left(t_{\alpha\beta} f_{\alpha\beta} - \rho \sum_{m=1}^{\mathfrak{M}} \tau_m \dot{v}_m\right) dV + \int_S \frac{h_\alpha}{T} n_\alpha\, dS$$

$$+ \int_V \frac{h_\alpha T_{,\alpha}}{T^2}\, dV + \int_V \frac{\rho q}{T}\, dV \tag{5.11.8}$$

where $\dot{H}'$, $\dot{H}''$, and $\dot{\eta}'$, $\dot{\eta}''$ are the two components of the production of the total and specific entropy rates, respectively, defined as follows

$$\dot{\eta}' = \frac{1}{\rho T} t_{\alpha\beta} f_{\alpha\beta} - \frac{1}{T} \sum_{m=1}^{\mathfrak{M}} \tau_m \dot{v}_m + \frac{h_\alpha T_{,\alpha}}{\rho T^2} \tag{5.11.9}$$

$$\dot{\eta}'' = \frac{1}{\rho}\left(\frac{h_\alpha}{T}\right)_{,\alpha} + \frac{q}{T} \tag{5.11.10}$$

It is clear, in particular from Eq. (5.11.9), that any entropy production is related to the temperature gradient.

Experience shows that:

1. If there are no sources of heat, $q = 0$, and the temperature is constant,

$T = \text{const}$, the mechanical power is not always recoverable, some of it being dissipative

$$t_{\alpha\beta} f_{\alpha\beta} - \rho \sum_{m=1}^{\mathfrak{M}} \tau_m \dot{v}_m \geq 0.$$

2. If no mechanical power is applied

$$t_{\alpha\beta} f_{\alpha\beta} - \rho \sum_{m=1}^{\mathfrak{M}} \tau_m \dot{v}_m = 0$$

and no heat is supplied, $q = 0$, the change of temperature is always from warm to colder and not vice versa $h_\alpha T_{,\alpha} \geq 0$.

These two observations, if applied to Eq. (5.11.9), yield the inequality

$$\rho T \dot{\eta}' = t_{\alpha\beta} f_{\alpha\beta} - \rho \sum_{m=1}^{\mathfrak{M}} \tau_m \dot{v}_m + \frac{h_\alpha T_{,\alpha}}{T} \geq 0. \tag{5.11.11}$$

a necessary condition called the *postulate of irreversibility* or the *Clausius–Duhem inequality*.

Assuming $T > 0$ we can apply the inequality (5.11.11) to Eq. (5.11.8) and we obtain the production of the total entropy rate as an inequality

$$\dot{H} \geq \oint_S \frac{h_\alpha n_\alpha}{T} \mathrm{d}S + \int_V \frac{\rho q}{T} \mathrm{d}V \tag{5.11.12}$$

which is another form of the postulate of irreversibility, expressing the production of the total entropy.

Eq. (5.11.12) together with Eq. (5.9.9) are also known as the *second law of thermodynamics*.

The assumption $T > 0$ is a valid assumption and is compatible with Eq. (5.9.1) where T was first introduced and defined, and with all that has been said about it in the following formulations. T is the *absolute temperature*, known also as the temperature of the *Kelvin scale*, and it can be demonstrated that the absolute temperature is bound by a lowest limit called the *absolute zero*.

5.12 Particular Cases of the Thermodynamic State

In Eq. (5.11.1) we have synthesized the external work and the internal thermodynamic exchange, by equating the internal energy obtained from the external work with its value obtained through considerations of the thermodynamic state. After rearranging this equation we have obtained the equation of the production rate of the specific entropy, Eq. (5.11.2), which is one of the basic equations controlling the behavior of matter as far as its thermodynamic state is concerned. Specifically, it governs the thermodynamics of irreversible processes. Most of the particular cases of the thermodynamic state can be deduced from this equation, under conditions pertaining to the thermodynamic variables.

Before discussing the various processes, we will define the forms of heat flux and heat supply, which constitute the non-mechanical external work. A process is *endothermic*, when heat flows into the element from the outside through its surface areas, or *exothermic*, when heat flows out of the element. A process is *thermogenic* when the supply of heat is generated within the element, or *thermo-dissipative* when it dissipates in the element.

1. *Isothermal process*. A thermodynamic process in which the temperature remains constant, $T = \text{const}$, is an *isothermal process*. The condition for such a process is one of the following:

$$T = \text{const}, \quad dT = 0 \tag{5.12.1}$$

Under these conditions the specific internal energy and energy rates, Eqs. (5.11.1) and (5.11.3)–(5.11.5), are

$$\rho \dot{\epsilon}_{\mathrm{i}} = \rho \sum_{\mathrm{m}=1}^{\mathfrak{M}} \tau_{\mathrm{m}} \dot{\nu}_{\mathrm{m}} + \rho T \dot{\eta} = t_{\alpha\beta} f_{\alpha\beta} + h_{\alpha,\alpha} + \rho q; \quad T = \text{const} \tag{5.12.2}$$

$$\rho \dot{\psi} = \rho \sum_{\mathrm{m}=1}^{\mathfrak{M}} \tau_{\mathrm{m}} \dot{\nu}_{\mathrm{m}} = t_{\alpha\beta} f_{\alpha\beta} + h_{\alpha,\alpha} + \rho q; \quad T = \text{const} \tag{5.12.3}$$

$$\rho \dot{\chi} = t_{\alpha\beta} f_{\alpha\beta} - \rho \sum_{\mathrm{m}=1}^{\mathfrak{M}} \tau_{\mathrm{m}} \dot{\nu}_{\mathrm{m}} + h_{\alpha,\alpha} + \rho q - \rho \sum_{\mathrm{m}=1}^{\mathfrak{M}} \dot{\tau}_{\mathrm{m}} \nu_{\mathrm{m}}; \quad T = \text{const} \tag{5.12.4}$$

$$\rho \dot{\zeta} = -\rho \sum_{\mathrm{m}=1}^{\mathfrak{M}} \dot{\tau}_{\mathrm{m}} \nu_{\mathrm{m}}; \quad T = \text{const} \tag{5.12.5}$$

It is worth noting that the specific free energy (5.12.3) and the free enthalpy rates (5.12.5) are functions of the rate of thermodynamic substates $\dot{\nu}_{\mathrm{m}}$ and the rate of the thermodynamic tensions $\dot{\tau}_{\mathrm{m}}$, respectively.

2. *Athermal process*. A process in which the non-mechanical external energy remains constant is an *athermal process*. In other words, the specific energy efflux and the specific energy supply balance one another and the entropy production will depend, according to Eq. (5.11.2), on the difference between the external and internal mechanical power.

$$h_{\alpha,\alpha} + \rho q = T \left(\frac{h_\alpha}{T} \right)_{,\alpha} + \frac{h_\alpha T_{,\alpha}}{T} + \rho q \tag{5.12.6}$$

$$\rho \dot{\eta} = \frac{1}{T} \left(t_{\alpha\beta} f_{\alpha\beta} - \rho \sum_{\mathrm{m}=1}^{\mathfrak{M}} \tau_{\mathrm{m}} \dot{\nu}_{\mathrm{m}} \right) \tag{5.12.7}$$

Eq. (5.12.6) indicates the condition of an athermal process. If the process is endothermic it must also be thermo-dissipative, and if the process is exothermic it must be thermogenic. The form of the energy flux h_i will uniquely determine the form of the supply function q, and vice versa.

The athermal process is always accompanied by a temperature gradient which satisfies this process, as seen from Eq. (5.12.6).

3. *Adiabatic process*. If the changes that occur in the matter do not involve any heat exchange and the individual components contributing to the non-mechanical external work vanish identically, $h_i = 0$, $q = 0$, the process is *adiabatic*. An adiabatic process is therefore neither endothermic nor exothermic, nor is it thermogenic or thermo-dissipative.

From Eq. (5.11.2) we can see that the adiabatic process does not prevent the production of entropy, even when it is isothermal, and the entropy production will depend on the difference between the external and internal mechanical power

$$\rho\dot{\eta} = \frac{1}{T}\left(t_{\alpha\beta}f_{\alpha\beta} - \rho\sum_{\mathrm{m}=1}^{\mathfrak{M}}\tau_{\mathrm{m}}\dot{v}_{\mathrm{m}}\right); \quad h_i = 0; \quad q = 0 \tag{5.12.8}$$

The adiabatic process can be considered a particular case of the athermal process, when it is specified that the flux h_i or the specific supply q vanish. This will also explain the identity of Eqs. (5.12.7) and (5.12.8).

4. *Isentropic process*. If the entropy remains constant during the thermodynamic process, $\eta = \text{const}$, the process is *isentropic*. In the isentropic process the mechanical power balances the non-mechanical power, and from Eq. (5.11.2) we obtain

$$-t_{\alpha\beta}f_{\alpha\beta} + \rho\sum_{\mathrm{m}=1}^{\mathfrak{M}}\tau_{\mathrm{m}}\dot{v}_{\mathrm{m}} = h_{\alpha,\alpha} + \rho q; \quad \eta = \text{const} \tag{5.12.9}$$

Thus, the rate of the specific enthalpy, Eq. (5.11.4), is dependent only on the power performed by the thermodynamic tensions

$$\dot{\chi} = \sum_{\mathrm{m}=1}^{\mathfrak{M}}\tau_{\mathrm{m}}\dot{v}_{\mathrm{m}}; \quad \eta = \text{const} \tag{5.12.10}$$

5. *Recoverable process*. Truesdell and Toupin (1960) have discussed under this name two particular cases, both characterized by the equivalence of the internal and external non-mechanical power

$$\rho T\dot{\eta} = h_{\alpha,\alpha} + \rho q \tag{5.12.11}$$

which is, at the same time, a necessary and sufficient condition for the equivalence of the external and internal mechanical power, given as

$$t_{\alpha\beta}f_{\alpha\beta} = \rho\sum_{\mathrm{m}=1}^{\mathfrak{M}}\tau_{\mathrm{m}}\dot{v}_{\mathrm{m}} \tag{5.12.12}$$

The last three processes identify and term individually three particular cases, by applying additional constraints to Eq. (5.12.11).

A process, in which the mechanical and the non-mechanical components of the power are balanced separately while the entropy remains constant $\eta = \text{const}$, is a *recoverable process*. Eq. (5.12.11) will vanish and the internal energy and energy potential rates, Eqs. (5.11.1) and (5.11.3)–(5.11.5), will change as follows

$$\rho\dot{\epsilon}_{\mathrm{i}} = t_{\alpha\beta} f_{\alpha\beta} - \rho \sum_{\mathrm{m}=1}^{\mathfrak{M}} \tau_{\mathrm{m}} \dot{v}_{\mathrm{m}}; \quad \eta = \text{const} \tag{5.12.13}$$

$$\rho\dot{\psi} = t_{\alpha\beta} f_{\alpha\beta} - \rho\dot{T}\eta; \quad \eta = \text{const} \tag{5.12.14}$$

$$\rho\dot{\chi} = -\rho \sum_{\mathrm{m}=1}^{\mathfrak{M}} \dot{\tau}_{\mathrm{m}} v_{\mathrm{m}}; \quad \eta = \text{const} \tag{5.12.15}$$

$$\rho\dot{\zeta} = -\rho \sum_{\mathrm{m}=1} \dot{\tau}_{\mathrm{m}} v_{\mathrm{m}} - \rho\dot{T}\eta; \quad \eta = \text{const} \tag{5.12.16}$$

We observe that if the process is recoverable, the internal energy is equal to the stress power, Eq. (5.12.13). The recoverable process has also been termed the *adiabatic elastic process* (Freudenthal and Geiringer 1958), although it would seem more correct to term it the *athermal elastic process*, since it is obtained as a particular case of the athermal process, by applying the same constraint, $\eta = \text{const}$ to Eq. (5.12.7).

6. *Reproducible process*. If the mechanical and non-mechanical components of the power are balanced separately under isothermal conditions, $T = \text{const}$, the process is a *reproducible process*. In this case we shall obtain, for internal energy and energy potential rates, Eqs. (5.11.1) and (5.11.3)–(5.11.5), the following results

$$\rho\dot{\epsilon}_{\mathrm{i}} = t_{\alpha\beta} f_{\alpha\beta} + \rho T\dot{\eta}; \quad T = \text{const} \tag{5.12.17}$$

$$\rho\dot{\psi} = t_{\alpha\beta} f_{\alpha\beta} = \rho \sum_{\mathrm{m}=1}^{\mathfrak{M}} \tau_{\mathrm{m}} \dot{v}_{\mathrm{m}}; \quad T = \text{const} \tag{5.12.18}$$

$$\rho\dot{\chi} = -\rho \sum_{\mathrm{m}=1}^{\mathfrak{M}} \dot{\tau}_{\mathrm{m}} v_{\mathrm{m}} + \rho T\dot{\eta}; \quad T = \text{const} \tag{5.12.19}$$

$$\rho\dot{\zeta} = -\rho \sum_{\mathrm{m}=1}^{\mathfrak{M}} \dot{\tau}_{\mathrm{m}} v_{\mathrm{m}}; \quad T = \text{const} \tag{5.12.20}$$

In the reproducible process the free energy is equal to the stress power. This process is also known as the *isothermal elastic process* and it can be obtained from Eqs. (5.12.2)–(5.12.5) by applying the additional condition given by Eq. (5.12.12).

7. *Reversible process*. The only process that may be properly called a

reversible process is not a real process, but a ficticious one. It would be a process in which all non-mechanical components vanish, while both entropy and temperature are constant.

The internal energy and energy potential rates, Eqs. (5.11.1) and (5.11.3)–(5.11.5), will be

$$\rho\dot{\epsilon}_{\rm i} = \rho\dot{\psi} = -p_{\alpha\beta}f_{\alpha\beta} = \rho\sum_{\rm m=1}^{\mathfrak{M}}\tau_{\rm m}\dot{\nu}_{\rm m};\quad \eta = 0 \tag{5.12.21}$$

$$\rho\dot{\chi} = \rho\dot{\xi} = \rho\sum_{\rm m=1}^{\mathfrak{M}}\dot{\tau}_{\rm m}\nu_{\rm m};\quad \eta = 0 \tag{5.12.22}$$

Thus, in a reversible process, which could also be called the *ideal elastic process*, the internal energy and free energy rates are identical and equal to the stress power, and the enthalpy and free enthalpy rates are also identical.

Another possible form for the fully reversible process can be obtained from Eqs. (5.12.21) and (5.12.22)

$$\sum_{\rm m=1}^{\mathfrak{M}}\tau_{\rm m}\nu_{\rm m} = \epsilon_{\rm i} + \chi = \psi + \zeta = \epsilon_{\rm i} + \zeta = \psi + \chi; \tag{5.12.23}$$

for $\eta =$ const; $T =$ const; $\eta' =$ const which is consistent with Eq. (5.10.4) and is a necessary and sufficient condition for ideal reversibility.

Many details of energetics formulation, which would belong in this chapter if the subject matter were theoretical mechanics, have been omitted here, in order that they be discussed in places where a direct relevancy and a possibility of application exists. In this chapter we have tried to concentrate the discussion around the fundamentals required to present the formulation and the method of reasoning; the details can always be deduced and studied.

6 Multi-phase Mixtures

6.1 Extensive and Intensive Variables

In a multi-phase mixture with the phases homogeneously distributed it is convenient to adopt the distinction between the thermodynamic variables introduced by Maxwell (1876):

Mass, volume, strain, energy, entropy, charges, thermodynamic substates, etc. are *extensive variables*. The magnitude of each variable in the mixture is addable, and is equal to the sum of its magnitudes in the parts of the mixture, or in the phases.

Variables such as density, stress, temperature, thermodynamic tensions, etc. are *intensive variables*. Their value is independent of the amount of matter in the phase or in any part of the material. Intensive variables are state functions indicating the intensity of the state shared by all the parts of the mixture.

6.2 Density, Volume, Mass and Weight of Constituents

The multi-phase nature of soils has been discussed to some extent in Sect. 1.8. It has been pointed out that, for practical purposes, the numerous components of soil can be grouped into three discernible classes, called phases, each of distinctly observable and measurable physical properties. Each phase is considered homogeneous or quasi-homogeneous, while the entire soil complex is regarded as multi-phase with the phases being homogeneously distributed throughout. This way of division and consideration of the material constitutes one of the fundamental assumptions of mechanics of multi-phase mixtures.

Before continuing the discussion of soils, however, the general theory of multi-phase mixtures will be expounded and it will be shown how it develops from the equations of balance.

The investigation of the properties of multi-phase mixtures goes back to the end of the nineteenth century, as investigators became interested in the problems of diffusion (Fick 1855; Stefan 1871) and motion of ground water (Slichter 1897). Later, the effects of chemical reactions and the dynamic theory of gases produced valuable studies relevant to the theory of multi-phase mixtures (Maxwell 1866, 1876; Helmholtz 1882; Jaumann 1911; Lohr 1917; Meissner 1938; Chapman and Cowling 1939; Eckhart 1940; Meixner 1941; Prigogine 1947; Hirschfelder et al. 1954).

The efforts of Meixner and Reik (1959), Truesdell and Toupin (1960), Green (1960) and others ended the sporadic study of multi-phase mixtures, and a systematic study continues to evolve in this subject. Valuable contributions have since been added and are still unfolding (Adkins 1963a,b; Green and Naghdi 1967, 1969; Bowen 1967, 1973; Muller 1968; Goodman and Cowin 1972; Passman 1974, 1977; Drew 1976; Kenyon 1976; Drumheller 1978; Bedford and Drumheller 1978, 1983; Nunziato and Walsh 1980; Murdoch 1985; Murdoch and Morro 1987), all based on the original notion that a mass-point can be occupied simultaneously by $\mathfrak{N}$ different constituents, as stated by Fick and Stefan (Fick 1855). This statement is known as the *equipresence of constituents* or *superposition of constituents*.

The application of the theory of multi-phase mixtures to soils offers great possibilities of advancement. The theory outlined further will be as general as possible. It will not include, for instance, the effect of interface interactions, although it will point out ways to consider such interactions.

If $\rho_\mathfrak{n}$ is the individual density of the $\mathfrak{n}$th constituent or phase, the statement of equipresence of constituents permits us to define the total density ρ of the mixture as the sum of the densities $\rho_\mathfrak{n}$ of the $\mathfrak{N}$ constituents.

$$\rho = \sum_{\mathfrak{n}=1}^{\mathfrak{N}} \rho_\mathfrak{n} \tag{6.2.1}$$

The *density concentration* $\alpha_\mathfrak{n}$ of the $\mathfrak{n}$th constituent is defined

$$\alpha_\mathfrak{n} = \frac{\rho_\mathfrak{n}}{\rho} \tag{6.2.2}$$

and satisfies Eq. (6.2.1) by

$$\sum_{\mathfrak{n}=1}^{\mathfrak{N}} \alpha_\mathfrak{n} = 1 \tag{6.2.3}$$

It is self-evident that $\rho_\mathfrak{n}$ and $\alpha_\mathfrak{n}$ can both be functions of space and time.

Since many changes are volume changes, volume relationships in a multi-phase mixture are further considered. The equation of conservation of volume in the mixture is

$$V = \sum_{\mathfrak{n}=1}^{\mathfrak{N}} V_\mathfrak{n} \tag{6.2.4}$$

where V and $V_{\mathfrak{n}}$ are the *total volume* and the *individual volume* of the $\mathfrak{n}$th constituent, respectively. Substituting Eqs. (6.2.1) and (6.2.4) into Eq. (4.1.1), the equation of conservation of mass for multi-phase mixtures follows

$$M = \int_V \rho \, \mathrm{d}V \equiv \sum_{\mathfrak{n}=1}^{\mathfrak{N}} \rho_{\mathfrak{n}} V = \rho \sum_{\mathfrak{n}=1}^{\mathfrak{N}} V_{\mathfrak{n}} = \sum_{\mathfrak{n}=1}^{\mathfrak{N}} m_{\mathfrak{n}} \tag{6.2.5}$$

where M and $m_{\mathfrak{n}}$ are the *total mass* and the *individual mass* of the $\mathfrak{n}$th constituent, respectively.

Multiplying both sides of Eq. (6.2.5) by the *gravitational acceleration* g, the analogous relationship between the *total weight* W and the *individual weights* $W_{\mathfrak{n}}$ of the constituents is obtained

$$W = Mg = g\sum_{\mathfrak{n}=1}^{\mathfrak{N}} m_{\mathfrak{n}} = gV\sum_{\mathfrak{n}=1}^{\mathfrak{N}} \rho_{\mathfrak{n}} = g\rho\sum_{\mathfrak{n}=1}^{\mathfrak{N}} V_{\mathfrak{n}} = \sum_{\mathfrak{n}=1}^{\mathfrak{N}} W_{\mathfrak{n}} \tag{6.2.6}$$

6.3 Diffusion Velocity and Barycentric Velocity

Since the constituents of a multi-phase mixture are in relative motion, it is required that the *individual velocities* $v_{\mathfrak{n}i}$ of the constituents be defined as functions of space and time. In this motional process, the individual velocities are not necessarily equal.

The *mean velocity* or *barycentric velocity* v_i of the mixture is defined as the sum of the individual mass transfers

$$\rho v_i = \sum_{\mathfrak{n}=1}^{\mathfrak{N}} \rho_{\mathfrak{n}} v_{\mathfrak{n}i} \tag{6.3.1}$$

and the *diffusion velocity* or *peculiar velocity* $\bar{v}_{\mathfrak{n}i}$ of the constituent is defined

$$\bar{v}_{\mathfrak{n}i} = v_{\mathfrak{n}i} - v_i \tag{6.3.2}$$

and is a velocity of the constituent *relative to the mean velocity*.

Introducing Eqs. (6.2.2) and (6.2.3) into Eq. (6.3.1) we obtain

$$v_i \sum_{\mathfrak{n}=1}^{\mathfrak{N}} \alpha_{\mathfrak{n}} = \sum_{\mathfrak{n}=1}^{\mathfrak{N}} v_{\mathfrak{n}i} \alpha_{\mathfrak{n}} \tag{6.3.3}$$

which is identical with the right-hand side of Eq. (6.3.2), when multiplied by $\alpha_{\mathfrak{n}}$ and summed over all $\mathfrak{N}$ contituents. Thus the left-hand side of Eq. (6.3.2) will yield

$$\sum_{\mathfrak{n}=1}^{\mathfrak{N}} \alpha_{\mathfrak{n}} v_{\mathfrak{n}i} - v_i \sum_{\mathfrak{n}=1}^{\mathfrak{N}} \alpha_{\mathfrak{n}} = \sum_{\mathfrak{n}=1}^{\mathfrak{N}} \alpha_{\mathfrak{n}} \bar{v}_{\mathfrak{n}i} = 0 \tag{6.3.4}$$

which indicates that the sum of the diffusive motion of the constituents vanishes.

In porous media like soils, where one is interested in the flow of the fluid or gas phase relative to the porous matrix, the relative velocity $v_{\mathrm{nm}i}$ is defined

$$\bar{v}_{\mathrm{nm}i} = v_{\mathrm{n}i} - v_{\mathrm{m}i} \tag{6.3.5}$$

where the index m represents the reference phase and the index n any one of the other phases.

If the velocities of all constituents increase or decrease by the same amount, say, $\eta_{\mathrm{n}i} = \eta_i$, and become $v'_{\mathrm{n}i} = v_{\mathrm{n}i} + \eta_{\mathrm{n}i}$, then the diffusion velocity remains unaltered, and thus, according to Eqs. (6.3.1) and (6.3.2), we obtain

$$\bar{v}_{\mathrm{n}i} = v'_{\mathrm{n}i} - v_i = v_{\mathrm{n}i} + \eta_i - \frac{1}{\rho}\sum_{\mathrm{n}=1}^{\mathfrak{N}} \rho_{\mathrm{n}}(v_{\mathrm{n}i} + \eta_i)$$

$$= v_{\mathrm{n}i} - \frac{1}{\rho}\sum_{\mathrm{n}=1}^{\mathfrak{N}} \rho_{\mathrm{n}} v_{\mathrm{n}i} + \eta_i - \frac{1}{\rho}\sum_{\mathrm{n}=1}^{\mathfrak{N}} \rho_{\mathrm{n}}\eta_i = v_{\mathrm{n}i} - \frac{1}{\rho}\sum_{\mathrm{n}=1}^{\mathfrak{N}} \rho_{\mathrm{n}} v_{\mathrm{n}i} = \bar{v}_{\mathrm{n}i} \tag{6.3.6}$$

since sometimes one is interested in the velocity of the constituents relative to one of the specific constituents, $v_{\mathrm{nm}i}$ as it has been discussed by De Groot and Mazur (1962).

6.4 The General Balance of Multi-phase Mixtures

In Sect. 4.2, the assumption was made that the general balance equation is valid for the mixture as a whole. When a multi-phase mixture is considered, the general balance equation must still hold for the mixture when it is summed up for the $\mathfrak{N}$ constituents; the single constituents, however, do not need to balance individually, since a transfer of the quantities of mass, momentum and energy from one constituent to another may occur. Thus, equating (4.2.3) and (4.2.9), the local and material balance, respectively, for the nth constituent, we obtain

$$\frac{\partial \rho_{\mathrm{n}}\xi_{\mathrm{n}}}{\partial t} + \phi_{\mathrm{n}\alpha,\alpha} - F_{\mathrm{n}} = 0 \tag{6.4.1}$$

$$\frac{\mathrm{D}\rho_{\mathrm{n}}\xi_{\mathrm{n}}}{\mathrm{D}t} - \frac{\partial \rho_{\mathrm{n}}\xi_{\mathrm{n}}}{\partial t} - (\rho_{\mathrm{n}}\xi_{\mathrm{n}})_{,\alpha} v_{\mathrm{o}\alpha}$$

$$= \frac{\mathrm{D}\rho_{\mathrm{n}}\xi_{\mathrm{n}}}{\mathrm{D}t} - \phi_{\mathrm{n}\alpha,\alpha} - (\rho_{\mathrm{n}}\xi_{\mathrm{n}})_{,\alpha} v_{\mathrm{o}\alpha} - F_{\mathrm{n}} = 0 \tag{6.4.2}$$

where ξ_{n}, $\phi_{\mathrm{n}i}$ and σ_{n}, are, respectively, the extensive quantity, the efflux and the specific supply rate of the $\mathfrak{n}$th constituent, and $F_\mathfrak{n} = \rho_\mathfrak{n}\sigma_\mathfrak{n}$ is the unbalanced supply of the quantity considered for the individual constituent, which may be transferred from one constituent to another, throughout the mixture.

The material derivative with respect to one of the constituents, say $\mathfrak{m}$, is, according to Truesdell and Toupin (1960)

$$\frac{\mathrm{D}\rho_\mathfrak{n}\xi_\mathfrak{n}}{\mathrm{D}_\mathfrak{m}t} - \frac{\partial\rho_\mathfrak{n}\xi_\mathfrak{n}}{\partial t} - (\rho_\mathfrak{n}\xi_\mathfrak{n})_{,\alpha}v_{\mathfrak{n}\alpha} = 0 \tag{6.4.3}$$

and therefore, when substituting Eq. (6.4.1) into (6.4.3) we obtain

$$\frac{\mathrm{D}\rho_\mathfrak{n}\xi_\mathfrak{n}}{\mathrm{D}_\mathfrak{m}t} + \phi_{\mathfrak{n}\alpha,\alpha} - (\rho_\mathfrak{n}\xi_\mathfrak{n})_{,\alpha}v_{\mathfrak{n}\alpha} - F_\mathfrak{n} = 0 \tag{6.4.4}$$

It should be noted that $v_{\mathfrak{n}i}$ is not necessarily the velocity of one of the constituents; it can be the velocity of any point related to a reference. As such, it could be the velocity of any one of the constituents, $v_{\mathfrak{n}i}$, with respect to which we consider the other constituents, or it could be the mean velocity, v_i, of the whole mixture. It will all depend on the nature of the problem we are facing.

6.5 Multi-phase Density Balance

Let $\xi_\mathfrak{n} = 1$ and the individual efflux and supply functions have the values $\phi_{\mathfrak{n}i} = \rho_\mathfrak{n}v_{\mathfrak{n}i}$, and $F_\mathfrak{n} = C_\mathfrak{n}(x, t)$, then, from Eq. (6.4.1) the local and material density balance for the $\mathfrak{n}$th constituent is obtained, respectively

$$\frac{\partial\rho_\mathfrak{n}}{\partial t} + (\rho_\mathfrak{n}v_{\mathfrak{n}\alpha})_{,\alpha} - C_\mathfrak{n} = \frac{\partial\rho_\mathfrak{n}}{\partial t} + (\rho_\mathfrak{n}\bar{v}_{\mathfrak{n}\alpha})_{,\alpha} + (\rho_\mathfrak{n}v_\alpha)_{,\alpha} - C_\mathfrak{n} = 0 \tag{6.5.1}$$

$$\frac{\mathrm{D}\rho_\mathfrak{n}}{\mathrm{D}t} - \frac{\partial\rho_\mathfrak{n}}{\partial t} - \rho_{\mathfrak{n},\alpha}v_{\mathfrak{n}\alpha} = \frac{\mathrm{D}\rho_\mathfrak{n}}{\mathrm{D}t} + \rho_\mathfrak{n}v_{\mathfrak{n}\alpha,\alpha} - C_\mathfrak{n} = 0 \tag{6.5.2}$$

Since only the density of the whole mixture need balance, and not the densities of the individual constituents $\mathfrak{n}$, $C_\mathfrak{n}$ represents the *supply of excess density rate of the individual constituent*, which, when summed up, should vanish

$$\sum_{\mathfrak{n}=1}^{\mathfrak{N}} C_\mathfrak{n} = 0 \tag{6.5.3}$$

Eq. (6.5.3) is a compatibility equation, necessary and sufficient for the balance of the whole mixture.

The material derivative with respect to the $\mathfrak{m}$th constituent is

$$\frac{\mathrm{D}\rho_{\mathfrak{n}}}{\mathrm{D}_{\mathfrak{m}}t} + (\rho_{\mathfrak{n}} v_{\mathfrak{n}\alpha})_{,\alpha} - \rho_{\mathfrak{n},\alpha} v_{\mathfrak{m}i} - C_{\mathfrak{n}}$$

$$= \frac{\mathrm{D}\rho_{\mathfrak{n}}}{\mathrm{D}_{\mathfrak{m}}t} + \rho_{\mathfrak{n}} v_{\mathfrak{n}\alpha,\alpha} + \rho_{\mathfrak{n},\alpha}(v_{\mathfrak{n}\alpha} - v_{\mathfrak{m}\alpha}) - C_{\mathfrak{n}} = 0 \qquad (6.5.4)$$

If the $\mathfrak{n}$th constituent becomes the $\mathfrak{m}$th constituent, the material derivative becomes

$$\frac{\mathrm{D}\rho_{\mathfrak{m}}}{\mathrm{D}_{\mathfrak{m}}t} + \rho_{\mathfrak{m}} v_{\mathfrak{m}\alpha,\alpha} - C_{\mathfrak{m}} = 0 \qquad (6.5.5)$$

which is equal to the second part of Eq. (6.5.2).

On the other hand, if the $\mathfrak{m}$th constituent is stationary, $v_{\mathfrak{m}i} = 0$, Eq. (6.5.4) becomes

$$\frac{\mathrm{D}\rho_{\mathfrak{n}}}{\mathrm{D}_{\mathfrak{m}}t} + \rho_{\mathfrak{n}} v_{\mathfrak{n}\alpha,\alpha} + \rho_{\mathfrak{n},\alpha} v_{\mathfrak{n}\alpha} - C_{\mathfrak{n}} = \frac{\mathrm{D}\rho_{\mathfrak{n}}}{\mathrm{D}_{\mathfrak{m}}t} + (\rho_{\mathfrak{n}} v_{\mathfrak{n}\alpha})_{,\alpha} - C_{\mathfrak{n}} = 0 \qquad (6.5.6)$$

which is different from Eq. (6.5.2).

Applying summation to both sides of Eqs. (6.5.1) and (6.5.2), and in view of Eqs. (6.3.1) and (6.5.3), we obtain

$$\sum_{\mathfrak{n}=1}^{\mathfrak{N}} \frac{\partial \rho_{\mathfrak{n}}}{\partial t} + \sum_{\mathfrak{n}=1}^{\mathfrak{N}} (\rho_{\mathfrak{n}} v_{\mathfrak{n}\alpha})_{,\alpha} - \sum_{\mathfrak{n}=1}^{\mathfrak{N}} C_{\mathfrak{n}} = \frac{\partial \rho}{\partial t} + (\rho v_{\alpha})_{,\alpha} = 0 \qquad (6.5.7)$$

$$\sum_{\mathfrak{n}=1}^{\mathfrak{N}} \frac{\mathrm{D}\rho_{\mathfrak{n}}}{\mathrm{D}t} - \sum_{\mathfrak{n}=1}^{\mathfrak{N}} \frac{\partial \rho_{\mathfrak{n}}}{\partial t} - \sum_{\mathfrak{n}=1}^{\mathfrak{N}} \rho_{\mathfrak{n},\alpha} v_{\mathfrak{n}\alpha} - \sum_{\mathfrak{n}=1}^{\mathfrak{N}} C_{\mathfrak{n}} = \frac{\mathrm{D}\rho}{\mathrm{D}t} + \rho v_{\alpha,\alpha} \qquad (6.5.8)$$

It is seen that what is obtained are Eqs. (4.3.3) and (4.3.4), provided that the sum of the density supplies vanishes, $C_{\mathfrak{n}} = 0$, which indicates that, although the individual supplies may not necessarily vanish, their sum should vanish identically. These equations represent the effect of chemical reactions and are, therefore, of marginal interest to mechanics of soils.

In a two-phase mixture, the assumption made by Fick (1855) considers the rate of diffusion of a density in one of the constituents proportional to the gradient of the density of that constituent, while it tends to restore uniformity

$$\rho_{\mathfrak{n}} v_{\mathfrak{n}} = -D\, \rho_{\mathfrak{n},i} \quad D > 0; \quad \mathfrak{n} = 1, 2, \qquad (6.5.9)$$

where the coefficient of proportionality D is called *diffusivity*.

Substituting Eq. (6.5.2) into Eq. (6.5.9) and assuming $\tau_{\mathrm{n},i} = 0$, the diffusion equation is obtained

$$\dot{\rho}_{\mathrm{n}} = -(\rho_{\mathrm{n}} v_{\mathrm{n}\alpha})_{,\alpha} + C_{\mathrm{n}} = -(\rho_{\mathrm{n}} v_{\mathrm{n}\alpha})_{,\alpha} - (\rho v_{\alpha})_{,\alpha} + C_{\mathrm{n}} = (D\rho_{\mathrm{n},\alpha})_{,\alpha} + C_{\mathrm{n}} \tag{6.5.10}$$

where the mean velocity v_i vanishes.

This equation is similar to Fourier's heat diffusion equation and we recognize in it the differential equation controlling the consolidation of a saturated soil. The more detailed discussion on this will be deferred to a later chapter.

Another frequent assumption found in the literature is $C_{\mathrm{n}} = 0$. Under this assumption Eq. (6.5.9) for a two-phase mixture, designated by n and m, becomes

$$-D\rho_{\mathrm{n},i} = \tfrac{1}{2}(\rho_{\mathrm{n}}\bar{v}_{\mathrm{n}i} - \rho_{\mathrm{m}}\bar{v}_{\mathrm{m}i}) = \frac{\rho_{\mathrm{n}}\rho_{\mathrm{m}}}{\rho}(v_{\mathrm{n}i} - v_{\mathrm{m}i}) = \frac{\rho_{\mathrm{n}}\rho_{\mathrm{m}}}{\rho}(\bar{v}_{\mathrm{n}i} - \bar{v}_{\mathrm{m}i}) \tag{6.5.11}$$

Thus the magnitude of the density gradient depends on the difference between the velocities of the constituents.

6.6 Multi-phase Balance of Linear Momentum

To develop an equation for the balance of linear momentum of a multi-phase mixture, Eqs. (4.6.5) and (4.6.6), in terms of the pressure tensor, should be written for the nth constituent

$$\frac{\partial \rho_{\mathrm{n}} v_{\mathrm{n}i}}{\partial t} - t_{\mathrm{n}i\alpha,\alpha} + (\rho_{\mathrm{n}} v_{\mathrm{n}i} v_{\mathrm{n}\alpha})_{,\alpha} - \rho_{\mathrm{n}}\sigma_{\mathrm{n}i} - F_{\mathrm{n}i}$$
$$= \frac{\partial \rho_{\mathrm{n}} v_{\mathrm{n}i}}{\partial t} - t_{\mathrm{n}i\alpha,\alpha} + (\rho_{\mathrm{n}} v_{\mathrm{n}i} v_{\mathrm{n}\alpha})_{,\alpha} - \rho_{\mathrm{n}}\sigma_{\mathrm{n}i} - \rho T_{\mathrm{n}i} = 0 \tag{6.6.1}$$

$$\rho_{\mathrm{n}}\frac{\mathrm{D}v_{\mathrm{n}i}}{\mathrm{D}t} - t_{\mathrm{n}i\alpha,\alpha} - \rho_{\mathrm{n}}\sigma_{\mathrm{n}i} - F_{\mathrm{n}i} = \rho_{\mathrm{n}}\frac{\mathrm{D}v_{\mathrm{n}i}}{\mathrm{D}t} - t_{\mathrm{n}i\alpha,\alpha} - \rho_{\mathrm{n}}\sigma_{\mathrm{n}i} - \rho T_{\mathrm{n}i} = 0 \tag{6.6.2}$$

where $F_{\mathrm{n}i}(x,t) = \rho T_{\mathrm{n}i}(x,t)$ represents here the *excess supply of linear momentum* of the constituent, a quantity with the properties of a body force within the individual constituent. Since momentum may be transferred from one constituent to another, the linear momentum in the constituent itself is not required to balance.

Applying summation to these equations, with the help of Eqs. (6.3.1) and (6.3.2), the following equations are obtained

$$\frac{\partial}{\partial t}\sum_{\mathrm{n}=1}^{\mathfrak{N}}\rho_{\mathrm{n}}v_{\mathrm{n}i}-\sum_{\mathrm{n}=1}^{\mathfrak{N}}(t_{\mathrm{n}i\alpha}-\rho_{\mathrm{n}}v_{\mathrm{n}i}v_{\mathrm{n}\alpha})_{,\alpha}-\sum_{\mathrm{n}=1}^{\mathfrak{N}}\rho_{\mathrm{n}}\sigma_{\mathrm{n}i}-\sum_{\mathrm{n}=1}^{\mathfrak{N}}F_{\mathrm{n}i}$$

$$=\frac{\partial}{\partial t}\sum_{\mathrm{n}=1}^{\mathfrak{N}}\rho v_i-\sum_{\mathrm{n}=1}^{\mathfrak{N}}[t_{\mathrm{n}i\alpha}-\rho_{\mathrm{n}}(v_{\mathrm{n}i}+\bar{v}_{\mathrm{n}i})(v_{\mathrm{n}\alpha}+\bar{v}_{\mathrm{n}\alpha})]_{,\alpha}$$

$$-\sum_{\mathrm{n}=1}^{\mathfrak{N}}\rho_{\mathrm{n}}\sigma_{\mathrm{n}i}-\rho\sum_{\mathrm{n}=1}^{\mathfrak{N}}T_{\mathrm{n}i}$$

$$=\frac{\partial}{\partial t}\sum_{\mathrm{n}=1}^{\mathfrak{N}}\rho v_i-\sum_{\mathrm{n}=1}^{\mathfrak{N}}(t_{\mathrm{n}\alpha}-\rho_{\mathrm{n}}\bar{v}_{\mathrm{n}i}\bar{v}_{\mathrm{n}\alpha})_{,\alpha}+(\rho v_i v_\alpha)_{,\alpha}$$

$$-\sum_{\mathrm{n}=1}^{\mathfrak{N}}\rho_{\mathrm{n}}\sigma_{\mathrm{n}i}-\rho\sum_{\mathrm{n}=1}^{\mathfrak{N}}T_{\mathrm{n}i}=0 \tag{6.6.3}$$

$$\sum_{\mathrm{n}=1}^{\mathfrak{N}}\rho_{\mathrm{n}}\frac{\mathrm{D}v_{\mathrm{n}i}}{\mathrm{D}t}-\sum_{\mathrm{n}=1}^{\mathfrak{N}}t_{\mathrm{n}i\alpha,\alpha}-\sum_{\mathrm{n}=1}^{\mathfrak{N}}\rho_{\mathrm{n}}\sigma_{\mathrm{n}i}-\sum_{\mathrm{n}=1}^{\mathfrak{N}}F_{\mathrm{n}i}$$

$$=\sum_{\mathrm{n}=1}^{\mathfrak{N}}\rho\frac{\mathrm{D}v_i}{\mathrm{D}t}-\sum_{\mathrm{n}=1}^{\mathfrak{N}}t_{\mathrm{n}i\alpha,\alpha}-\sum_{\mathrm{n}=1}^{\mathfrak{N}}\rho_{\mathrm{n}}\sigma_{\mathrm{n}i}-\rho\sum_{\mathrm{n}=1}^{\mathfrak{N}}T_{\mathrm{n}i}$$

$$=\sum_{\mathrm{n}=1}^{\mathfrak{N}}\rho\frac{\mathrm{D}v_i}{\mathrm{D}t}-\sum_{\mathrm{n}=1}(t_{\mathrm{n}i\alpha}-\rho_{\mathrm{n}}\bar{v}_{\mathrm{n}i}\bar{v}_{\mathrm{n}\alpha})_{,\alpha}-\sum_{\mathrm{n}=1}^{\mathfrak{N}}\rho_{\mathrm{n}}\sigma_{\mathrm{n}i}-\rho\sum_{\mathrm{n}=1}^{\mathfrak{N}}T_{\mathrm{n}i}=0 \tag{6.6.4}$$

which are identical to Eqs. (4.6.5) and (4.6.6), provided that

$$t_{ij}=\sum_{\mathrm{n}=1}^{\mathfrak{N}}(t_{\mathrm{n}ij}-\rho_{\mathrm{n}}\bar{v}_{\mathrm{n}i}\bar{v}_{\mathrm{n}j}) \tag{6.6.5}$$

$$\rho\sigma_i=\sum_{\mathrm{n}=1}^{\mathfrak{N}}\rho_{\mathrm{n}}\sigma_{\mathrm{n}i} \tag{6.6.6}$$

and

$$\sum_{\mathrm{n}=1}^{\mathfrak{N}}(T_{\mathrm{n}i}+C_{\mathrm{n}}\bar{v}_{\mathrm{o}i})=0 \tag{6.6.7}$$

where $t_{\mathrm{n}ij}$ is the *individual stress tensor* and $\alpha_{\mathrm{n}i}$ is the *individual body force* acting on the nth constituent.

This last equation, obtained after some mathematical manipulations, is another compatibility equation, necessary and sufficient for the balance of linear momentum in the mixture.

Note also that the term $\rho_{\mathrm{n}}\bar{v}_{\mathrm{n}i}\bar{v}_{\mathrm{n}j}$ in Eq. (6.6.5) is always symmetric, whereas $t_{\mathrm{n}ij}$ is not required to be symmetric. Thus, it is possible that t_{ij} be symmetric while $t_{\mathrm{n}ij}$ is not symmetric itself.

Also, it should be noted that the total stress t_{ij} is the sum of the stresses of the individual constituents, an important fact in soils as well as in other mixtures, and it depends on the diffusive velocities $v_{\mathfrak{n}i}$ of the constituents if dynamic excitations provide such velocities.

Finally, the following identity may be obtained from Eq. (6.6.4)

$$\rho \dot{v}_i - t_{i\alpha,\alpha} - \rho \sigma_i = \rho \sum_{\mathfrak{n}=1}^{\mathfrak{N}} (T_{\mathfrak{n}i} + C_{\mathfrak{n}} \bar{v}_{\mathfrak{n}i}) = 0 \tag{6.6.8}$$

which is identical to Eq. (4.6.6).

Eq. (6.6.7) does not vanish in all circumstances. In soils, where structural interactions are manifested in the form of interface stresses, discussed in Sects. 8.25 and 8.26, and electric potentials, discussed in Sects. 8.16–8.20, we have an alternative equation to (6.6.8)

$$\rho \dot{v}_i - t_{i\alpha,\alpha} - \rho \sigma_i - \rho \sum_{\mathfrak{n}=1}^{\mathfrak{N}} (T_{\mathfrak{n}i} - C_{\mathfrak{n}} \bar{v}_{\mathfrak{n}i}) = 0 \tag{6.6.9}$$

6.7 Multi-phase Balance of Moment of Momentum

The general balance of moment of momentum for multi-phase mixtures follows from Eq. (4.7.3). We have to assume that surface couples and body moments exist in each of the constituents. And so for the $\mathfrak{n}$th constituent we may write

$$m_{\mathfrak{n}ij\alpha,\alpha} + \rho_{\mathfrak{n}} l_{\mathfrak{n}ij} - \tfrac{1}{2}(t_{\mathfrak{n}ij} - t_{\mathfrak{n}ji}) = 0 \tag{6.7.1}$$

where $m_{\mathfrak{n}ijk}$ and $l_{\mathfrak{n}ij}$ are the *individual surface couple* and the *individual body moment* of the $\mathfrak{n}$th constituents, respectively. If Eq. (6.7.1) is summed over the $\mathfrak{n}$ constituents we obtain

$$\sum_{\mathfrak{n}=1}^{\mathfrak{N}} m_{\mathfrak{n}ij\alpha,\alpha} + \sum_{\mathfrak{n}=1}^{\mathfrak{N}} \rho_{\mathfrak{n}} l_{\mathfrak{n}ij} - \frac{1}{2} \sum_{\mathfrak{n}=1}^{\mathfrak{N}} (t_{\mathfrak{n}ij} - t_{\mathfrak{n}ji})$$
$$= m_{ij\alpha,\alpha} + \rho l_{ij} - \tfrac{1}{2}(t_{ij} - t_{ji}) = 0 \tag{6.7.2}$$

Since the substitution of Eq. (6.6.5) into the last part of Eq. (6.7.1) does not alter it, Eq. (6.7.2) is obtained, provided the following is satisfied

$$\sum_{\mathfrak{n}=1}^{\mathfrak{N}} m_{\mathfrak{n}ij\alpha,\alpha} + \sum_{\mathfrak{n}=1}^{\mathfrak{N}} \rho_{\mathfrak{n}} l_{\mathfrak{n}ij} - m_{\mathfrak{n}ij\alpha,\alpha} - \rho_{\mathfrak{n}} l_{\mathfrak{n}ij} = 0 \tag{6.7.3}$$

meaning that all individual surface couples and body moments have to balance out internally and vanish identically.

It should be remarked that Eq. (6.7.2) has not been solved for the case where the stresses $t_{\mathfrak{n}ij}$ of the constituents are non-symmetric but the overall stress t_{ij} of the mixture is symmetric. This case has relevance to soils, where charged clay particles would possess couple stresses and moments (see Chap. 8).

Surface couples and body moments will not be considered further in our study, unless specifically mentioned. Thus, if couple forces are ignored, then from Eq. (6.7.1) we have

$$t_{\mathrm{n}ij} = t_{\mathrm{n}ji} \tag{6.7.4}$$

meaning that the stress tensors of the individual constituents are all symmetric.

6.8 Multi-phase Balance of Internal Energy

When kinetic energy effects are ignored, the spatial and material balance of the specific internal energy for the nth constituents may be transcribed from Eqs. (5.5.7) and (5.5.8)

$$\frac{\partial}{\partial t}\rho_{\mathrm{n}}\epsilon_{\mathrm{n}} + (\rho_{\mathrm{n}}\epsilon_{\mathrm{n}}v_{\mathrm{n}\beta})_{,\beta} - t_{\mathrm{n}\alpha\beta}v_{\mathrm{n}\alpha,\beta} - h_{\mathrm{n}\alpha,\alpha} - \rho_{\mathrm{n}}q_{\mathrm{n}} - \rho E_{\mathrm{n}} = 0 \qquad ^{*}(6.8.1)$$

$$\rho_{\mathrm{n}}\dot{\epsilon}_{\mathrm{n}} - t_{\mathrm{n}\alpha\beta}v_{\mathrm{n}\alpha,\beta} - h_{\mathrm{n}\alpha,\alpha} - \rho_{\mathrm{n}}q_{\mathrm{n}} - \rho E_{\mathrm{n}} = 0 \tag{6.8.2}$$

where $E_{\mathrm{n}}(x, t)$ represents the *excess of internal energy supplies* of the individual constituents, not required to be balanced individually since transfer of internal energy from one constituent to another may occur. The balance of internal energy is required only for the whole mixture.

When summed over the $\mathfrak{N}$ constituents, Eqs. (6.8.1) and (6.8.2) yield, after some rearrangement

$$\begin{aligned}
&\sum_{\mathrm{n}=1}^{\mathfrak{N}} \frac{\partial}{\partial t}\rho_{\mathrm{n}}\epsilon_{\mathrm{n}} + \sum_{\mathrm{n}=1}^{\mathfrak{N}} (\rho_{\mathrm{n}}\epsilon_{\mathrm{n}}v_{\mathrm{n}\beta})_{,\beta} - \sum_{\mathrm{n}=1}^{\mathfrak{N}} t_{\mathrm{n}\alpha\beta}v_{\mathrm{n}\alpha,\beta} - \sum_{\mathrm{n}=1}^{\mathfrak{N}} h_{\mathrm{n}\alpha,\alpha} - \sum_{\mathrm{n}=1}^{\mathfrak{N}} \rho_{\mathrm{n}}q_{\mathrm{n}} - \rho\sum_{\mathrm{n}=1}^{\mathfrak{N}} E_{\mathrm{n}} \\
&\quad = \sum_{\mathrm{n}=1}^{\mathfrak{N}} \frac{\partial}{\partial t}\rho_{\mathrm{n}}\epsilon_{\mathrm{n}} + \sum_{\mathrm{n}=1}^{\mathfrak{N}} (\rho_{\mathrm{n}}\epsilon_{\mathrm{n}}(\bar{v}_{\mathrm{n}\beta} + v_{\beta}))_{,\beta} - \sum_{\mathrm{n}=1}^{\mathfrak{N}} t_{\mathrm{n}\alpha\beta}(\bar{v}_{\mathrm{n}\alpha} + v_{\beta})_{,\beta} \\
&\qquad - \sum_{\mathrm{n}=1}^{\mathfrak{N}} h_{\mathrm{n}\alpha,\alpha} - \sum_{\mathrm{n}=1}^{\mathfrak{N}} \rho_{\mathrm{n}}q_{\mathrm{n}} - \rho\sum_{\mathrm{n}=1}^{\mathfrak{N}} E_{\mathrm{n}} \\
&\quad = \frac{\partial}{\partial t}\rho\epsilon + (\rho\epsilon v_{\beta})_{,\beta} - t_{\alpha\beta}v_{\alpha,\beta} - h_{\alpha,\alpha} - \rho q \\
&\qquad - \rho\sum_{\mathrm{m}=1}^{\mathfrak{N}} [E_{\mathrm{n}} + T_{\mathrm{n}\alpha}\bar{v}_{\alpha} + C_{\mathrm{n}}(\dot{\epsilon}_{\mathrm{n}} + \tfrac{1}{2}\bar{v}_{\mathrm{n}\alpha}\bar{v}_{\mathrm{n}\alpha})] = 0
\end{aligned} \tag{6.8.3}$$

* In the following treatment the subscript i of the internal energy will be omitted, $\epsilon_{\mathrm{i}} \equiv \epsilon$.

$$\sum_{n=1}^{\mathfrak{N}} \rho_n \frac{D}{Dt} \epsilon_n - \sum_{n=1}^{\mathfrak{N}} t_{n\alpha\beta} v_{n\alpha,\beta} - \sum_{n=1}^{\mathfrak{N}} h_{n\alpha,\alpha} - \sum_{n=1}^{\mathfrak{N}} \rho_n q_n - \rho \sum_{n=1}^{\mathfrak{N}} E_n$$

$$= \sum_{n=1}^{\mathfrak{N}} \rho_n \frac{D}{Dt} (\epsilon_n + \tfrac{1}{2}(\bar{v}_{n\beta} + v_\beta)(\bar{v}_{n\alpha} + v_\alpha))_{,\beta}$$

$$- \sum_{n=1}^{\mathfrak{N}} t_{n\alpha\beta}(\bar{v}_{n\alpha} + v_\alpha)_{,\beta} - \sum_{n=1}^{\mathfrak{N}} h_{n\alpha,\alpha} - \sum_{n=1}^{\mathfrak{N}} \rho_n q_n - \rho \sum_{n=1}^{\mathfrak{N}} E_n$$

$$= \rho \frac{D}{Dt} \epsilon - t_{\alpha\beta} v_{\alpha,\beta} - h_{\alpha,\alpha} - \rho q$$

$$- \rho \sum_{n=1}^{\mathfrak{N}} [E_n + T_{n\alpha}\bar{v}_\alpha + C_n(\dot{\epsilon}_n + \tfrac{1}{2}\bar{v}_{n\alpha}\bar{v}_{n\alpha})] = 0 \tag{6.8.4}$$

provided the following equations are satisfied

$$\epsilon = \sum_{n=1}^{\mathfrak{N}} \alpha_n (\epsilon_n + \tfrac{1}{2}\bar{v}_{n\beta}\bar{v}_{n\beta}) \tag{6.8.5}$$

$$h_i = \sum_{n=1}^{\mathfrak{N}} [h_{ni} + t_{ni\beta}\bar{v}_{n\beta} - \rho_n(\epsilon_n + \tfrac{1}{2}\bar{v}_{n\beta}\bar{v}_{n\beta})\bar{v}_{ni}] \tag{6.8.6}$$

$$q = \sum_{n=1}^{\mathfrak{N}} \alpha_n(q_n - B_{n\alpha} v_{n\alpha}) \tag{6.8.7}$$

$$\rho \sum_{n=1}^{\mathfrak{N}} [E_n + T_{n\alpha}\bar{v}_\alpha + C_n(\dot{\epsilon}_n + \overline{\tfrac{1}{2}\bar{v}_{n\alpha}\bar{v}_{n\alpha}}^{\,\bullet})] = 0 \tag{6.8.8}$$

where E_n are the *excess interacting energy supplies*, T_{ni} the *excess linear momentum supplies* and C_n the *excess density rates*, as we have already seen. Eq. (6.8.8) is again a compatibility equation, obtained by mathematical manipulations and rearrangments, and it is necessary and sufficient to assert that the internal energy of the whole mixture is balanced, although the individual internal energies are not required to be balanced. Thus, Eq. (6.8.4) yields

$$\rho\dot{\epsilon} - t_{\alpha\beta} v_{\alpha,\beta} - h_{\alpha,\alpha} - \rho q = 0 \tag{6.8.9}$$

an equation identical to (5.5.8).

Here we see again that only under static or near static conditions will the total specific internal energy, the total energy efflux and the energy supply be equal to the sum of the specific internal energies, the energy effluxes, and the energy supplies of the individual constituents, respectively. Under dynamic conditions, the diffusion velocities v_{ni} must be considered.

The total internal energy ϵ is the sum of the *individual internal energies* of the constituents ϵ_{n} plus the kinetic energies of the diffusions $\frac{1}{2}(\rho_{\mathrm{n}}\bar{v}_{\mathrm{n}\alpha}v_{\mathrm{n}\alpha})$. The total energy flux h_i originates from three sources: the *individual fluxes* $h_{\mathrm{n}i}$, the *individual stress-work* against diffusion $t_{\mathrm{n}\alpha i}\bar{v}_{\mathrm{n}\alpha}$ and the *individual total energy fluxes* of diffusion $(\epsilon_{\mathrm{n}} + \frac{1}{2}\bar{v}_{\mathrm{n}\alpha}\bar{v}_{\mathrm{n}\alpha})\bar{v}_{\mathrm{n}i}$.

The total energy supply q appears as the sum of two components: the *individual energy supplies* q_{n} and the energy supply of the *individual body forces* working against diffusion $\alpha_{\mathrm{n}\alpha}\bar{v}_{\mathrm{n}\alpha}$.

While Eqs. (6.8.5) and (6.8.6) originate in the kinetic theory of gases by Maxwell (1866), the definitions differ among investigators. Here the treatment of Truesdell and Toupin (1960) is presented.

6.9 The Caloric Equations

The material balance of the specific entropy for a multi-phase mixture may be obtained when Eq. (5.9.10) is transcribed for the nth constituent

$$\rho_{\mathrm{n}} T_{\mathrm{n}} \dot{\eta}_{\mathrm{n}} - \rho_{\mathrm{n}} \dot{\epsilon}_{\mathrm{n}} + \rho_{\mathrm{n}} \sum_{\mathrm{n}=1}^{\mathfrak{N}} \tau_{\mathrm{mn}} \dot{v}_{\mathrm{mn}} = 0 \tag{6.9.1}$$

Substituting the specific internal energy rate, Eq. (6.8.4), one obtains the balance of entropy for the nth constituent.

$$\rho_{\mathrm{n}} T_{\mathrm{n}} \dot{\eta}_{\mathrm{n}} - t_{\mathrm{n}\alpha\beta} v_{\mathrm{n}\alpha,\beta} - h_{\mathrm{n}\alpha,\alpha} - \rho_{\mathrm{n}} q_{\mathrm{n}} + \rho_{\mathrm{n}} \sum_{\mathrm{m}=1}^{\mathfrak{M}} \tau_{\mathrm{mn}} \dot{v}_{\mathrm{mn}} - \rho H_{\mathrm{n}} = 0 \tag{6.9.2}$$

Summing over the $\mathfrak{N}$ constituents and considering Eq. (6.8.4) we obtain

$$\begin{aligned}
&\sum_{\mathrm{n}=1}^{\mathfrak{N}} \rho_{\mathrm{n}} T_{\mathrm{n}} \dot{\eta}_{\mathrm{n}} - \sum_{\mathrm{n}=1}^{\mathfrak{N}} t_{\mathrm{n}\alpha\beta} v_{\mathrm{n}\alpha,\beta} - \sum_{\mathrm{n}=1}^{\mathfrak{N}} h_{\mathrm{n}\alpha,\alpha} - \sum_{\mathrm{n}=1}^{\mathfrak{N}} \rho_{\mathrm{n}} q_{\mathrm{n}} + \sum_{\mathrm{n}=1}^{\mathfrak{N}} \rho_{\mathrm{n}} \sum_{\mathrm{m}=1}^{\mathfrak{M}} \tau_{\mathrm{mn}} \dot{v}_{\mathrm{mn}} - \rho_{\mathrm{n}} \sum_{\mathrm{n}=1}^{\mathfrak{N}} \eta_{\mathrm{n}} \\
&\quad = \rho_{\mathrm{n}} T_{\mathrm{n}} \dot{\eta}_{\mathrm{n}} - t_{\mathrm{n}\alpha\beta} v_{\mathrm{n}\alpha,\beta} - h_{\mathrm{n}\alpha,\alpha} - \rho_{\mathrm{n}} q_{\mathrm{n}} + \rho \sum_{\mathrm{n}=1}^{\mathfrak{N}} \tau_{\mathrm{mn}} \dot{v}_{\mathrm{mn}} - \rho \sum_{\mathrm{n}=1}^{\mathfrak{N}} \eta_{\mathrm{n}} = 0
\end{aligned} \tag{6.9.3}$$

where η_{n} are the *excess individual entropies*, T_{n} the *individual temperatures*, τ_{mn} the *individual thermodynamic tensions*, and v_{mn} the *individual thermodynamic substates* of the constituents, which balance outside the respective constituents but within the mixture. If the postulate is made that the temperatures of all constituents remain the same at any time and in any location

$$T_1 = T_2 = T_3 = \ldots = T_{\mathrm{n}} = \ldots = T_{\mathfrak{N}} = T \tag{6.9.4}$$

then the following equations are satisfied

$$\eta = \sum_{\mathrm{n}=1}^{\mathfrak{N}} \alpha_{\mathrm{n}} \eta_{\mathrm{n}} \tag{6.9.5}$$

$$\tau_{\mathrm{m}} = \sum_{\mathrm{n}=1}^{\mathfrak{N}} \alpha_{\mathrm{n}} \tau_{\mathrm{mn}} \tag{6.9.6}$$

$$\sum_{\mathrm{n}=1}^{\mathfrak{N}} \eta_{\mathrm{n}} = 0 \tag{6.9.7}$$

It can be seen from Eq. (6.9.3) that an analogy may be established between the work of the mth thermodynamic potential and the work of the nth constituent of each potential. In a sense they are interchangeable or at least they conform to the same rules.

Eq. (6.9.7) is a compatibility equation, necessary and sufficient to satisfy the balance of entropy production, if the use of the expression "balance" is at all proper for entropy production.

Truesdell and Toupin (1960) obtain a slightly different form for Eqs. (6.9.3) and (6.9.7), due to a different grouping of the expressions in those equations.

The rate of the thermodynamic functions or potentials, Eqs. (5.11.3)–(5.11.5), as defined in Sect. 5.10 by Eqs. (5.10.1)–(5.10.3), yields, for multi-phase mixtures, following Truesdell and Toupin (1960)

$$\begin{aligned} \rho\dot{\psi} &= \rho \sum_{\mathrm{n}=1}^{\mathfrak{N}} \alpha_{\mathrm{n}} \sum_{\mathrm{m}=1}^{\mathfrak{M}} \tau_{\mathrm{mn}} \dot{v}_{\mathrm{mn}} - \rho \sum_{\mathrm{n}=1}^{\mathfrak{N}} \alpha_{\mathrm{n}} \dot{T}_{\mathrm{n}} \eta_{\mathrm{n}} \\ &= \sum_{\mathrm{n}=1}^{\mathfrak{N}} t_{\mathrm{n}\alpha\beta} f_{\mathrm{n}\alpha\beta} + \sum_{\mathrm{n}=1}^{\mathfrak{N}} h_{\mathrm{n}\alpha,\alpha} + \sum_{\mathrm{n}=1}^{\mathfrak{N}} \rho_{\mathrm{n}} q_{\mathrm{n}} - \sum_{\mathrm{n}=1}^{\mathfrak{N}} \rho_{\mathrm{n}} T \dot{\eta}_{\mathrm{n}} - \sum_{\mathrm{n}=1}^{\mathfrak{N}} \rho_{\mathrm{n}} \dot{T} \eta_{\mathrm{n}} \end{aligned} \tag{6.9.8}$$

$$\begin{aligned} \rho\dot{\chi} &= -\rho \sum_{\mathrm{n}=1}^{\mathfrak{N}} \sum_{\mathrm{m}=1}^{\mathfrak{M}} \alpha_{\mathrm{n}} \dot{\tau}_{\mathrm{mn}} v_{\mathrm{mn}} - \rho \sum_{\mathrm{n}=1}^{\mathfrak{N}} \alpha_{\mathrm{n}} T_{\mathrm{m}} \dot{\eta}_{\mathrm{m}} \\ &= \sum_{\mathrm{n}=1}^{\mathfrak{N}} t_{\mathrm{n}\alpha\beta} f_{\mathrm{n}\alpha\beta} + \sum_{\mathrm{n}=1}^{\mathfrak{N}} h_{\mathrm{n}\alpha,\alpha} + \rho \sum_{\mathrm{n}=1}^{\mathfrak{N}} \alpha_{\mathrm{n}} q_{\mathrm{n}} - \rho \sum_{\mathrm{n}=1}^{\mathfrak{N}} \alpha_{\mathrm{n}} \sum_{\mathrm{m}=1}^{\mathfrak{M}} \tau_{\mathrm{mn}} \dot{v}_{\mathrm{mn}} \\ &\quad - \rho \sum_{\mathrm{n}=1}^{\mathfrak{N}} \alpha_{\mathrm{n}} \sum_{\mathrm{m}=1}^{\mathfrak{M}} \dot{\tau}_{\mathrm{mn}} v_{\mathrm{mn}} \end{aligned} \tag{6.9.9}$$

$$\begin{aligned} \rho\dot{\zeta} &= -\rho \sum_{\mathrm{n}=1}^{\mathfrak{N}} \alpha_{\mathrm{n}} \dot{T} \eta_{\mathrm{n}} - \rho \sum_{\mathrm{n}=1}^{\mathfrak{N}} \alpha_{\mathrm{n}} \sum_{\mathrm{m}=1}^{\mathfrak{M}} \dot{\tau}_{\mathrm{mn}} v_{\mathrm{mn}} \\ &= \sum_{\mathrm{n}=1}^{\mathfrak{N}} t_{\mathrm{n}\alpha\beta} f_{\mathrm{n}\alpha\beta} + \sum_{\mathrm{n}=1}^{\mathfrak{N}} h_{\mathrm{n}\alpha,\alpha} + \rho \sum_{\mathrm{n}=1}^{\mathfrak{N}} \alpha_{\mathrm{n}} q_{\mathrm{n}} - \rho \sum_{\mathrm{n}=1}^{\mathfrak{N}} \alpha_{\mathrm{n}} T \dot{\eta}_{\mathrm{m}} - \rho \sum_{\mathrm{n}=1}^{\mathfrak{N}} \alpha_{\mathrm{n}} \dot{T} \eta_{\mathrm{n}} \\ &\quad - \rho \sum_{\mathrm{n}=1}^{\mathfrak{N}} \alpha_{\mathrm{n}} \sum_{\mathrm{m}=1}^{\mathfrak{M}} \tau_{\mathrm{mn}} \dot{v}_{\mathrm{mn}} - \rho \sum_{\mathrm{n}=1}^{\mathfrak{N}} \alpha_{\mathrm{n}} \sum_{\mathrm{m}=1}^{\mathfrak{M}} \dot{\tau}_{\mathrm{mn}} v_{\mathrm{mn}} \end{aligned} \tag{6.9.10}$$

More useful expressions can be obtained from the caloric equation, discussed in Sects. 5.7–5.10. By using Eqs. (5.8.1) and (5.10.5)–(5.10.7), where practically no restrictions are imposed on the substates $\mathfrak{m}$ except that they be extensive variables, we obtain

$$d\epsilon = T\sum_{\mathfrak{n}=1}^{\mathfrak{N}} d\eta_{\mathfrak{n}} + \sum_{\mathfrak{m}=1}^{\mathfrak{M}}\sum_{\mathfrak{n}=1}^{\mathfrak{N}} \tau_{\mathfrak{mn}}\, dv_{\mathfrak{mn}} \tag{6.9.11}$$

$$d\psi = -\sum_{\mathfrak{n}=1}^{\mathfrak{N}} \eta_{\mathfrak{n}}\, dT + \sum_{\mathfrak{m}=1}^{\mathfrak{M}}\sum_{\mathfrak{n}=1}^{\mathfrak{N}} \tau_{\mathfrak{mn}} dv_{\mathfrak{mn}} \tag{6.9.12}$$

$$d\chi = T\sum_{\mathfrak{n}=1}^{\mathfrak{N}} d\eta_{\mathfrak{n}} - \sum_{\mathfrak{m}=1}^{\mathfrak{M}}\sum_{\mathfrak{n}=1}^{\mathfrak{N}} v_{\mathfrak{mn}} d\tau_{\mathfrak{mn}} \tag{6.9.13}$$

$$d\zeta = -\sum_{\mathfrak{n}=1}^{\mathfrak{N}} \eta_{\mathfrak{n}}\, dT - \sum_{\mathfrak{m}=1}^{\mathfrak{M}}\sum_{\mathfrak{n}=1}^{\mathfrak{N}} v_{\mathfrak{mn}} d\tau_{\mathfrak{mn}} \tag{6.9.14}$$

Of course all the restrictions and the particular cases discussed in Sect. 5.11 hold for multi-phase mixtures as well.

With respect to Eqs. (6.9.11)–(6.9.14) it is worth mentioning two points:

1. By the nature of the caloric equation (5.7.1), and since the only restrictions imposed on the thermodynamic substates are that they be continuous and many times differentiable extensive variables, there is no distinction in the second parts of the right-hand sides of Eqs. (6.9.11)–(6.9.14) between the indices $\mathfrak{m}$ and $\mathfrak{n}$, thus an interconvertibility exists between these indices. Consequently a single index, say $\mathfrak{n}$, and a single summation over that index is sufficient. The index could indicate either a process or a constituent.

2. Since changes in entropy are impossible to measure, Eqs. (6.9.12) and (6.9.14) are, from an experimental point of view, more useful.

6.10 The Production of Entropy

Many other forms of the production of entropy for multi-phase mixtures may be derived from Eqs. (5.11.8)–(5.11.10). The treatment adopted here follows closely the treatment of Truesdell (1957). Consequently, for the $\mathfrak{n}$th constituent we have

$$\dot{\eta}_{\mathfrak{n}} = \dot{\eta}'_{\mathfrak{n}} + \dot{\eta}''_{\mathfrak{n}} = \frac{t_{\mathfrak{n}\alpha\beta} f_{\mathfrak{n}\alpha\beta}}{\rho_{\mathfrak{n}} T} - \frac{1}{T}\sum_{\mathfrak{n}=1}^{\mathfrak{N}} \tau_{\mathfrak{nm}}\dot{v}_{\mathfrak{mn}} + \frac{h_{\mathfrak{n}\alpha} T_{,\alpha}}{\rho_{\mathfrak{n}} T^2} + \frac{1}{\rho_{\mathfrak{n}}}\left(\frac{h_{\mathfrak{n}\alpha}}{T}\right)_{,\alpha} + \frac{q_{\mathfrak{n}}}{T} \tag{6.10.1}$$

It is customary to require that the postulate of irreversibility, Eq. (5.11.12), in differential form, hold for each individual constituent

$$\dot{\eta}_{\rm n} \geq \frac{1}{\rho_{\rm n}} \left(\frac{h_{{\rm n}\alpha}}{T} \right)_{,\alpha} + \frac{q_{\rm n}}{T} \tag{6.10.2}$$

although no definite reason seems to support this requirement.

The practical implications of these equations for soils are presently very vague, due particularly to technical limitations in performing the measurements with respect to fluxes and exchanges of heat.

7 Constitutive Equations

7.1 Scope

In Chaps. 4 and 5 the eight balance or continuity equations derived for homogeneous media were presented. These equations, originally in integral form and reduced thereafter to differential form, constitute what is also known as the *field equations*. They are:

The density balance (one equation)

$$(4.3.1) \qquad \frac{\partial \rho}{\partial t} + (\rho v_\alpha)_{,\alpha} - \rho\sigma = 0$$

the balance of linear momentum (three equations)

$$(4.6.3) \qquad \rho\ddot{x} - \tfrac{1}{2}(t_{i\alpha,\alpha} + t_{\alpha i,\alpha}) - \rho\sigma_i = 0$$

the balance of moment of momentum (three equations)

$$(4.7.3) \qquad m_{\alpha ij,\alpha} + \rho l_{ij} - \tfrac{1}{2}(t_{ij} - t_{ji}) = 0$$

and the internal energy balance (one equation)

$$(5.4.6) \qquad \rho\dot{\varepsilon}_{\mathrm{i}} - (t_{\alpha\beta} - t_{\beta\alpha})f_{\alpha\beta} + m_{\alpha\beta\gamma}w_{\alpha\beta,\gamma} - h_{\alpha,\alpha} - \rho q = 0$$

Twenty-five variables are contained in these eight equations: ρ (one), v_i (three), t_{ij} (nine), m_{ijk} (nine) and h_i (three). f_{ij}, l_{ij}, q and $\dot{\varepsilon}_{\mathrm{i}}$ are supposed to be known variables, while w_{ij} is a function of v_i. Obviously, the eight equations are insufficient to uniquely determine the twenty-five variables. If the processes are not isothermal, the production of entropy, Eq. (5.11.1), should be added, with additional unknown variables: the entropy η and the thermodynamic substates v_{m}, one equation against $\mathfrak{M} + 1$ variables.

Matters may get further complicated if a heterogeneous medium is considered. In a material with $\mathfrak{N}$ distinguishable constituents we have $8\mathfrak{N}$ equations with $25\mathfrak{N}$ unknowns for the isothermal case and $9\mathfrak{N}$ equations with $(26 + \mathfrak{M})\mathfrak{N}$ unknowns for the non-isothermal case.

To compensate for the inadequate number of equations which are at our disposal, we use *constitutive equations* and *boundary conditions* to supplement the balance equations.

The balance equations express principles of mechanics and thermodynamics, general enough to refer to any continuous material irrespective of its nature. They are, in that sense, *universal*. Constitutive equations, however, are *specific* to the material or the constituents involved and represent the functional relations between the variables of the material as a whole or the variables of its constituents: they determine the differences between materials. Since material bodies of identically distributed mass and of the same geometry may respond differently to the same external excitations, it is fair to surmise that these responses reflect differences in the structural constitution of the materials. Constitutive equations define relations in terms of those variables which concern the intricate nature of the material, as closely as it can be approximated, indicating *ideal material* behavior. Fortunately many of these relations turn out to be very simple, such as the assumptions of rigid body motion, incompressibility of the material, isothermal process, etc. Others, as we shall see, are more complex. The need of many additional assumptions and equations for the definition of the relations between the variables makes this part of mechanics of materials sometimes weak and frequently unsolved. Mathematical speculations detached from experience, and experiments void of considerations of basic mechanics and physics hinder the progress of the study of constitutive equations.

Constitutive equations are *mechanical* if they define a relationship between the mechanical variables, *thermal* if they define a relationship of thermal variables, or *thermodynamical* if they define an interrelationship of mechanical and thermal variables.

Any assumption with respect to one of the variables is *a fortiori* a constitutive equation, for instance the assumption $\partial \rho / \partial t = 0$ limits the theory to incompressible materials.

Both the field and the constitutive equations must be satisfied at all points within the boundaries of the material and at all times of the process.

In general terms one defines *boundary conditions* as a known set of values which describe the state of the continuum along the boundaries enclosing it. It is assumed that once this state is defined at all points along the envelope, it is also defined at all points of the continuum within the envelope, but not at points outside it.

However, the values of the variables are not always given along an enclosed boundary. Any known value at a point on the boundary can serve as a condition in the evaluation of the variables. A boundary condition expressing velocities, heat-efflux, etc., can be referred to a specific time, as well as to a specific position in space, where the thermodynamic process (mechanical or thermal) is known.

7.2 Principles of Formulating Constitutive Equations

It has been pointed out that field equations enjoy universal acceptance, one of the reasons for this being that they hold when tested by the principles of physics.

In order to justify their validity, constitutive equations must satisfy the following mathematical and physical conditions or principles (Truesdell and Toupin 1960). In fact, not all constitutive equations in use are devised to conform to all of these principles.

1. *Consistency*. Constitutive equations must be compatible with the field equations and satisfy them.

2. *Coordinate invariance*. Constitutive equations have to be stated in such a way that they hold in any coordinate system. This can be achieved by expressing them through tensorial equations, which are independent of the coordinate systems.

3. *Dimensional invariance*. The coefficients of the constitutive equations must always have the same dimensions. This principle allows, and even supports, solutions through dimensional analysis, requiring dimensional homogeneity of the various terms in the equation.

4. *Uniqueness*. All variables of the field and constitutive equations are to be defined in such a way as to render single-valued solutions for each term, thus defining the phenomenon unequivocally. This principle seems to be disregarded in many studies, such as, for instance, theories of plasticity. Of course, it implies the introduction of hereditary properties, see Sect. 7.3.

5. *Material indifference*. The response of the material is independent of the observer. To overcome relative effects due to the observer's relative position, apparent forces and torques are introduced. D'Alambert and Coriolis forces are classical examples of preserving the principle of material indifference, also known as the principle of *objectivity*.

6. *Material identity*. The basic physical properties of the material are inherent to it, and once defined remain the same under any circumstances irrespective of coordinate changes, testing techniques or size of the material involved, as long as the material is homogeneous or quasi-homogeneous. If the properties of the material are defined by Ω material parameters, then any technique or circumstance will yield Ω and only Ω parameters. Moreover, if by one technique one obtains the Ω parameters $A_1, A_2, A_3, \ldots, A_\Omega$, and by another technique the parameters $B_1, B_2, B_3, \ldots, B_\Omega$ there must be a one to one correspondence between the two sets of parameters.

$$A_n = \mathcal{F}(B_1, B_2, \ldots, B_\Omega)$$

$$B_n = \mathcal{G}(A_1, A_2, \ldots, A_\Omega) \qquad (7.2.1)$$

This principle sustains the fact that the physical properties of a large sample are identical with those of a small sample, so long as the homogeneity is not violated.

7. *Isomorphism*. The properties of a material are divided into two groups: properties that do not exhibit preference with respect to direction and properties that exhibit preferences of direction. A material is *isotropic* if all its properties are of the kind that do not exhibit preference of direction. If a material contains one or more properties that are different in different directions, the material is *aeolotropic* or *anisotropic*. A material that is isotropic with respect to one property may be anisotropic with respect to another.

7.3 The Rheological Equation

Deformation, as defined in Chap. 2, is the result of the changes in the relative positions of the points in a material system. The relationship between the deformation and the mechanical variables upon which it depends, such as stress, rate of stress, rate of deformation, etc., describes the state of the continuum. At constant temperature this relationship is a mechanical constitutive equation, known as the *rheological equation* or the *stress–strain relationship*. These terms will be used interchangeably throughout this work.

Deformations by their nature contest the thermodynamical equilibrium of matter. The relative movement of particles produces frictional resistance, accompanied by a release of heat which increases with the increase of the movement, whether the movement is gradual or spontaneous. Hence the deformation is accompanied by an exchange of energy and eventually a heat flux within the material, inducing a change in the thermodynamic state of the continuum. If the thermal changes are assumed negligible, the deformation can be described by the rheological equation. A definition of rheology, according to Reiner (in a lecture series), is: *the branch of thermodynamics concerned with the changes in state of materials under isothermal conditions*.

For the sake of theoretical and experimental simplicity it is generally assumed that either the entropy η or the temperature T remain constant during the mechanical process, thus the process is either *isentropic* or *isothermal*. The most common and convenient assumption, without which the establishment of viscoelastic theories would not be possible (Freudenthal and Geiringer 1958), is $T = \text{const}$, which limits the study to rheology, a synonym to *viscoelasticity*.

In this context of rheology, a functional relationship is stipulated between the stress tensor p_{ij} and the deformation tensor $x_{i,j}$ and their time derivatives of any order. In a homogeneous medium this relationship retains the form

$$R'(p_{ij}, \dot{p}_{ij}, \ddot{p}_{ij}, \ldots, \varepsilon_{ij}, \dot{\varepsilon}_{ij}, \ddot{\varepsilon}_{ij}, \ldots, t) = 0 \tag{7.3.1}$$

where the strain tensor ε_{ij}, a function of the deformation tensor $x_{i,j}$ and defined according to any one of the known and admissible measures of strain, replaces the deformation tensor. Thus a stress–strain relationship is defined as including explicitly the time t, which indicates a *history effect*. No further restrictions have been imposed on the nature of the measure by which the strain is specified, or on the form of Eq. (7.3.1).

If history effects, also known as *hereditary properties*, are negligible, or if the phenomenon can be related to time $t = 0$ as a reference time, Eq. (7.3.1) is simplified and becomes

$$R(p_{ij}, \dot{p}_{ij}, \ddot{p}_{ij}, \ldots, \varepsilon_{ij}, \dot{\varepsilon}_{ij}, \dot{\varepsilon}_{ij}, \ldots) = 0 \tag{7.3.2}$$

This is the rheological equation, which serves as a basis for most phenomenological viscoelastic studies of materials.

The assumptions, not all explicitly stated, yet implied in the constitutive equation (7.3.2), will be summed up as follows:

1. Polar forces not present.
2. Isothermal processes always close to equilbrium.
3. Homogeneity or quasi-homogeneity of the material.
4. No previous history effects influence the phenomenon.

7.4 Linearity and Non-linearity of Constitutive Equations

One problem in the structure of constitutive equations is the fact that not always are the relations between the variables linear. Non-linear theories have so far been tackled only in simple problems and developed at a slow pace. On the other hand, the linear theories have been quite extensively developed and the results obtained are in general satisfactory. In many instances, however, some of the processes evolving can be proved to be non-linear, and even if those non-linearities are of a secondary nature they nevertheless play an important part in the understanding of many specific phenomena.

Being aware of the various sources of non-linearity in the constitutive equations, one can place the use of linear theories in its proper perspective, through the understanding of their limitations. Reiner (1958) discusses at length the three types of non-linearity that appear in constitutive equations.

1. *Physical non-linearity* is the most common and appears when the parameters representing the material properties, the moduli and the coefficients, assumed to be constant, do not satisfy the constitutive equation. Not only do they vary with temperature and electrical and electromagnetic field changes, but their numerical value changes even with the changes of the

rheological variables, such as density, velocity gradient, or internal heat dissipation.

It would seem that the existence of physical non-linearity or the fact that the parameters vary is inconsistent with regular concepts of study of the mechanical behavior of matter, according to which the behavior of the material is a direct consequence of its physical properties represented by constant parameters. The proper evaluation of this behavior may not be possible if the parameters are allowed to change during testing. The problem becomes more acute in soils, where changes in density occurring during tests continuously change their physical parameters. This difficulty is overcome by a step by step application of constraints and piecewise evaluation of parameters, so as to enable the changes in parameters to conform to the changes in conditions.

2. *Geometrical non-linearity* is the result of the products $\delta_{\alpha\beta}u_{\alpha i}u_{\beta j}$, and $\delta_{\alpha\beta}U_{\alpha \mathrm{L}}U_{\beta \mathrm{M}}$, which appear in the expression of the deformation tensor and the finite strain measures, Eqs. (2.1.14) and (2.1.15) respectively. These products cannot be neglected when the displacements increase, since they induce non-linearity in the strain measures.

It has been remarked in Sect. 2.8 that only when deformations are large, such as in highly compressible soil or in shear, will there be a discrepancy between the infinitesimal and the finite Hencky strain measures. And since we propose to use the Hencky measure as a strain measure in our treatment of soils, this type of non-linearity has to be coped with.

3. *Tensorial non-linearity* is a consequence of the postulate of isotropic relations among the stress and strain tensors, manifested by the appearance of the second powers $p_{\alpha i}p_{\alpha j}$ or $\varepsilon_{\alpha i}\varepsilon_{\alpha j}$ of the stress tensor p_{ij} or the strain tensor ε_{ij}, respectively, or their time derivatives, see Sects. 7.9, 7.10.

Occasionally some of the non-linearities cancel each other out, yet on other occasions they accumulate and the result is an increased non-linearity (Nadai 1937).

It is important to know which non-linearities could affect the study of phenomenological aspects of the mechanical behavior of soil. The physical non-linearity is perhaps the most critical.

It will be shown later that in soils both the geometrical non-linearity and the tensorial non-linearity may be overcome, to some extent, by focusing our attention upon one type of deformation at a time (Alfrey and Gurnee 1956; Klausner 1967, 1969a). The tensorial non-linearity, for instance, occurs in the shear stress–strain relationship only. The physical non-linearity, however, remains, and it can be proven that the volumetric stress–strain relationship is necessarily physically non-linear.

When the expressions of the stress and of the strain and their higher derivatives appearing in Eq. (7.3.2) are of the first order, the stress–strain relationship is linear, in other words Eq. (7.3.2) represents a *linear visco-elastic* constitutive equation. If, however, isotropy is requested from the continuum, it imposes certain constraints on the form of Eqs. (7.3.1) and (7.3.2). Reiner was the first to show (Reiner 1945) that when the stress tensor is expressed as a polynomial of the strain tensor and of its time derivatives, and if the Cayley–Hamilton relations are applied, then the stress–strain relationship, Eq. (7.3.2), contains terms of the stresses and of the strains up to the second order, and the equation becomes tensorially non-linear. In the following sections we shall present our argument first for the linear theory, then extend it to the non-linear theory.

7.5 The Dual Rheological Equation

In substituting Eqs. (2.3.12) and (4.9.1) of the strain and stress tensors, respectively, into Eq. (7.3.2), the latter may be resolved into a *dual stress–strain relationship*

$$\mathrm{R}_{\mathrm{i}}(p_{\alpha\alpha}, \dot{p}_{\alpha\alpha}, \ddot{p}_{\alpha\alpha}, \ldots, \varepsilon_{\alpha\alpha}, \dot{\varepsilon}_{\alpha\alpha}, \ddot{\varepsilon}_{\alpha\alpha}, \ldots) = 0 \tag{7.5.1}$$

$$\mathrm{R}_{\mathrm{b}}(s_{ij}, \dot{s}_{ij}, \ddot{s}_{ij}, \ldots, e_{ij}, \dot{e}_{ij}, \ddot{e}_{ij}, \ldots) = 0 \tag{7.5.2}$$

the *spherical* and *traceless* rheological equations, respectively.

Here the assumption was made that the *traceless stresses cannot cause spherical strains* and the *traceless strains cannot cause spherical stresses*, and that the functional relation between the traceless stresses and the traceless strains and their higher time derivatives is linear. Thus Eqs. (7.5.1) and (7.5.2) represent *linear stress–strain relations*.

The first, (7.5.1), is a scalar equation of the traceless stress and strain tensors and the second, (7.5.2), is a tensorial equation. $p_{\alpha\alpha}$ is the *spherical* or *isotropic stress*, $\varepsilon_{\alpha\alpha}$ is the *spherical* or *isotropic strain*, s_{ij} is the *traceless* or *deviatoric stress tensor*, and e_{ij} is the *traceless* or *deviatoric strain tensor*, defined as follows

$$p_{\alpha\alpha} = p_{ii} + p_{jj} + p_{kk} = p_{11} + p_{22} + p_{33} = \mathrm{I}_{\mathrm{p}} = 3p_{\mathrm{m}} \tag{7.5.3}$$

$$\varepsilon_{\alpha\alpha} = \varepsilon_{ii} + \varepsilon_{jj} + \varepsilon_{kk} = \varepsilon_{11} + \varepsilon_{22} + \varepsilon_{33} = \mathrm{I}_{\varepsilon} \tag{7.5.4}$$

$$s_{ij} = p_{ij} - p_{\mathrm{m}}\delta_{ij} \tag{7.5.5}$$

$$e_{ij} = \varepsilon_{ij} - \tfrac{1}{3}\varepsilon_{\alpha\alpha}\delta_{ij} \tag{7.5.6}$$

where I_{p} and I_{ε} are the first invariants of the stress and strain tensors, respectively, p_{ii}, p_{jj}, p_{kk} and ε_{ii}, ε_{jj}, ε_{kk} are the normal components of the stress and strain tensors, respectively, in any *ijk* coordinates, and p_{11}, p_{22}, p_{33} and ε_{11}, ε_{22}, ε_{33} are the principal components of the stress and strain tensors, respectively. p_{m} and s_{ij} have definite physical meaning, namely the *equivalent hydrostatic* or *isotropic pressure* and the *shear stress*, respectively, while $\varepsilon_{\alpha\alpha}$ and e_{ij} assume physical meaning, namely the *volumetric strain* and the *distortion*, respectively, in two cases:

1. When the strains are small and may be considered infinitesimal.
2. When the strains are defined by the Hencky strain measure.

In these cases we shall denote the spherical strain $\varepsilon_{\alpha\alpha}$ by a subscript, indicating that it represents a volumetric strain, ε_{m}.

Most important from the point of view of physics is the fact that one can relate the volumetric strain to the isotropic pressure, Eq. (7.5.1), and the distortions to the shear stresses, Eq. (7.5.2). Recently, a tensorial formulation for the Hencky measure for finite deformations was given (Fitzgerald 1980) and is presented in Sects. 2.7 and 2.8. This formulation permits and also facilitates the aforementioned physical interpretation of Eqs. (7.5.1) and (7.5.2).

Except for highly compressible materials, volumetric strains seldom exceed 15%–20% under reasonable isotropic pressures (up to 10 kg/cm^2). In structural engineering materials such as concrete, steel, bitumen and some soils, the infinitesimal strain measure would still be a fair approximation. Distortions, however, may exceed 50% and warrant the introduction of the Hencky measure in the shear stress–strain relationship, Eq. (7.5.2). This would be the case in many soils, certain plastics and plasticizers, as well as some rubberlike materials. Splitting Eq. (7.3.2) into two equations, (7.5.1) and (7.5.2) is not only mathematically and experimentally convenient, it also carries several physical justifications, presented here in the form of propositions.

Proposition 1. Spherical stresses and only spherical stresses will cause spherical strains.

This is evident from Eqs. (7.5.1) and (7.5.2), where the former contains only spherical components and the latter does not contain spherical components at all. Therefore the corollary of the above proposition exists as well.

Corollary 1. Spherical strains are caused by and only by spherical stresses.

We shall see later that neither proposition 1 nor its corollary will hold for non-linear stress–strain relationships, where isotropic relations between the stress and the strain are postulated.

Proposition 2. Traceless stresses and only traceless stresses may cause traceless strains.

The verification of this proposition and its corollary is similar to that of proposition 1 and evolves from Eqs (7.5.1) and (7.5.2).

Corollary 2. Traceless strains are caused by and only by traceless stresses.

The physical implications of these propositions and corollaries are apparent when the strain is defined as discussed, either through the infinitesimal or through the Hencky measure. The statements would then change to:

Proposition 1′. Isotropic pressures (tensions) and only isotropic pressures result in volumetric strains.

Corollary 1′. Volumetric strains result from and only from isotropic pressures.

Proposition 2′. Shear stresses and only shear stresses will cause distortions.

Corollary 2′. Distortions are caused by and only by shear stresses.

7.6 Dual Viscoelastic Models

Constitutive equations are idealizations of material behavior. Quite often constitutive equations of linear viscoelasticity are represented by what is known as viscoelastic models, which are analogous replicas of the mathematical equations they represent. The models may be mechanical if they are represented by springs and dashpots, or electrical if they are represented by electrical resistors, inductors and capacitors (Naslin 1965). And since there are two mathematical equations to be represented, any viscoelastic material has *two* viscoelastic models assigned.

Proposition 3. A necessary and sufficient condition in viscoelastic modeling is that it be represented by two models, one for the spherical and one for the traceless equation, or a volumetric and a shear model, respectively.

The two models are tangible representations of the dual equations (7.5.1) and (7.5.2) and are not required to be similar, as the functional relationships in these mathematical equations are not necessarily the same. While one model may be simple, the other can be quite complex.

Fallacy. A viscoelastic material is sufficiently described by a single viscoelastic model.

It is quite disturbing to find that some textbooks and publications relate the mechanical behavior of a material to one model only; others, when the complicated nature of the combined general equation becomes difficult to cope with, introduce different individual models for the three coordinate directions of the material element.

We shall present several examples of simple viscoelastic responses of material bodies, in order to make our point in proposition 3 more tangible.

1. *Elastically compressible elastic body*, defined by the two constitutive equations on which the entire theory of elasticity is based

$$p_{\mathrm{m}} = \varkappa \varepsilon_{\mathrm{m}}; \quad s_{ij} = 2Ge_{ij} \tag{7.6.1}$$

a volumetric and a shear stress–strain relationship, respectively, both with elastic response. $\varkappa$ is the *bulk modulus* or *compression modulus* and G is the *shear modulus* or *rigidity* (Stokes 1845) respectively, which we shall call the *physical* elastic coefficients. There are other known elastic coefficients, termed *technical* coefficients: the *Young modulus* E, *Poisson's ratio* ν and *Lamé's constant* λ. They evolve owing to the technical circumstances by which they are derived. The Young modulus and Poisson's ratio are obtained from the simple unidirectional compression test, while Lamé's constant appears in the equation of elastically compressible elastic materials corresponding to Eq. (7.3.1), which has the form

$$p_{ij} = \lambda \varepsilon_{\mathrm{m}} \delta_{ij} + 2G\varepsilon_{ij} \tag{7.6.2}$$

and is most effectively used in solutions where cylindrical coordinates are required. The technical coefficients do not come in addition to the physical coefficients; they can only replace them. In any case, irrespective of which coefficients are used in an elasticity problem, only two independent coefficients may appear. That is to say that any two of the above mentioned coefficients, or of any other coefficients that we may want to employ, are functions of two other coefficients. Table 7.6.1 illustrates this functional dependence (Gurtin 1972).

2. *Incompressible elastic body*, defined by the two constitutive equations

$$\varepsilon_{\mathrm{m}} = 0; \quad s_{ij} = 2Ge_{ij} \tag{7.6.3}$$

where the volumetric strain vanishes and the shear stress–strain relationship is elastic in its response. Such a material body is possible if the bulk modulus approaches infinity, $\varkappa = \infty$.

3. *Incompressible viscous body*, defined by equations

$$\varepsilon_{\mathrm{m}} = 0; \quad s_{ij} = 2\eta\dot{e}_{ij} \tag{7.6.4}$$

where the volumetric strain vanishes and the shear stress relates to the rate of strain through the *coefficient of viscosity* η. Here again the bulk modulus approaches infinity, $\varkappa = \infty$. The study of classical hydrodynamics is based on these two equations (7.6.4) with one coefficient.

Similarly to the Young modulus in elasticity, one can derive an analogous coefficient, the *viscous traction* λ_{T} (Trouton 1905, 1906), which relates the one-dimensional tension to the elongation $\lambda_{\mathrm{T}} = 3\eta$

4. *Elastically compressible viscous body*, defined by the equations

$$p = \varkappa\varepsilon_{\mathrm{m}}; \quad s_{ij} = 2\eta\dot{e}_{ij} \tag{7.6.5}$$

where the volumetric stress–strain relationship is elastic, while the shear stress–strain relationship is viscous. Most theoretical aerodynamics is based on such equations.

5. *General viscoelastic bodies*. We have seen some of the simple material bodies with one or two independent coefficients. Before we proceed we shall make two observations:

(i) The two equations, namely the volumetric stress–strain relationship and the shear stress–strain relationship do not have to be identical. The number of coefficients in one equation does not necessarily have to be equal to the number of coefficients in the other equation.

(ii) From experience we know that the shear models are usually less developed than the volumetric models, therefore the number of coefficients appearing in the volumetric equations is larger than in the shear equations. In most cases a two-coefficient equation satisfies the shear stress–strain relationship, while the volumetric stress–strain relationship may require as many as 4 to 6 coefficients.

A general viscoelastic material would, therefore, be represented by the following equations:

$$\varepsilon(t')_{\mathrm{m}} = \frac{p(t')_{\mathrm{m}}}{\varkappa_1} + \frac{1}{\mu_2}\int_0^{t'} p(t)_{\mathrm{m}}\,\mathrm{d}t + \sum_{\mathrm{i}=3}^{\mathfrak{Z}} \frac{1}{\mu_{\mathrm{i}}} \exp\frac{-t'}{T_{\mathrm{reti}}} \int_0^{t'} p(t)_{\mathrm{m}} \exp\frac{t}{T_{\mathrm{reti}}}\,\mathrm{d}t \tag{7.6.6}$$

$$e(t'')_{ij} = \frac{s(t'')_{ij}}{2G_1} + \frac{1}{2\eta_2}\int_0^{t''} s(t)_{ij}\,\mathrm{d}t + \sum_{\mathrm{i}=3}^{\mathfrak{Z}} \frac{1}{2\eta_{\mathrm{i}}} \exp\frac{-t''}{T'_{\mathrm{reti}}} \int_0^{t''} s(t)_{ij} \exp\frac{t}{T'_{\mathrm{reti}}}\,\mathrm{d}t \tag{7.6.7}$$

where $\varkappa_{\mathrm{i}}$ are the *volumetric elastic moduli*, μ_{i} the *volumetric viscous coefficients*, $T_{\mathrm{reti}} = \mu_{\mathrm{i}}/\varkappa_{\mathrm{i}}$ the *volumetric retardation times*, G_{i} the *shear moduli*, η_{i} the *shear viscous coefficients*, and $T'_{\mathrm{reti}} = \eta_{\mathrm{i}}/G_{\mathrm{i}}$ the *shear retardation times* of the ith order.

We shall now present a more general proposition concerning the coefficients involved in the viscoelastic equations.

Proposition 4. If a material is represented by a volumetric stress–strain relationship which contains $\mathfrak{n}$ *coefficients and a shear stress–strain relationship which contains* $\mathfrak{m}$ *coefficients, then there will always be a total of* $\mathfrak{n} + \mathfrak{m}$ *and only* $\mathfrak{n} + \mathfrak{m}$ *independent coefficients that fully represent this material.*

This proposition holds irrespective of any coordinate transformations or any changes in the equations involved, and satisfies the principle of material density.

From a practical point of view we believe that much simpler forms of Eqs. (7.6.6) and (7.6.7) sufficiently describe most familiar materials, including soils

$$\varepsilon(t')_{\mathrm{m}} = \frac{p(t')_{\mathrm{m}}}{\varkappa_1} + \frac{1}{\mu_2}\int_0^{t'} p(t)_{\mathrm{m}}\,\mathrm{d}t + \frac{1}{\mu_3} \exp\frac{-t'}{T_{\mathrm{ret3}}} \int_0^{t'} p(t)_{\mathrm{m}} \exp\frac{t}{T_{\mathrm{ret3}}}\,\mathrm{d}t \tag{7.6.8}$$

$$e(t'')_{ij} = \frac{s(t'')_{ij}}{2G_1} + \frac{1}{2\eta_2}\int_0^{t''} s(t)_{ij}\,\mathrm{d}t \tag{7.6.9}$$

Table 7.6.1. Interdependence of the elastic constants

	Lamé's constant λ	Shear modulus G	Young's modulus E	Poisson's ratio ν	Bulk modulus $\varkappa$
λ, G	λ	G	$\frac{(3\lambda + 2G)}{\lambda + G}$	$\frac{\lambda}{2(\lambda + G)}$	$\frac{3\lambda + 2G}{3}$
λ, E	λ	$\frac{(E - 3\lambda) + \surd((E - 3\lambda)^2 + 8\lambda}{4}$	E	$\frac{-(E + \lambda) + \surd((E + \lambda)^2 + 8\lambda^2}{4\lambda}$	$\frac{(3\lambda + E) + \surd((3\lambda + E)^2 - 4\lambda E)}{6}$
λ, ν	λ	$\frac{\lambda(1 - 2\nu)}{2\nu}$	$\frac{\lambda(1 + \nu)(1 - 2\nu)}{\nu}$	ν	$\frac{\lambda(1 + \nu)}{3\nu}$
$\lambda, \varkappa$	λ	$\frac{3(\varkappa - \lambda)}{2}$	$\frac{9\varkappa(\varkappa - \lambda)}{3\varkappa - \lambda}$	$\frac{\lambda}{3\varkappa - \lambda}$	$\varkappa$
G, E	$\frac{(2G - E)G}{E - 3G}$	G	E	$\frac{E - 2G}{2G}$	$\frac{GE}{3(3G - E)}$

Table 7.6.1. *Cont.*

	Lamé's constant λ	Shear modulus G	Young's modulus E	Poisson's ratio ν	Bulk modulus $\varkappa$
G, ν	$\frac{2G\nu}{(1-2\nu)}$	G	$2G(1+\nu)$	ν	$\frac{2G(1+\nu)}{3(1-2\nu)}$
$G, \varkappa$	$\frac{3\varkappa - 2G}{3}$	G	$\frac{9\varkappa G}{3\varkappa + G}$	$\frac{1}{2}\left[\frac{3\varkappa - 2G}{3\varkappa + G}\right]$	$\varkappa$
E, ν	$\frac{\nu E}{(1+\nu)(1-2\nu)}$	$\frac{E}{2(1+\nu)}$	E	ν	$\frac{E}{3(1-2\nu)}$
$E, \varkappa$	$\frac{3\varkappa(3\varkappa - E)}{9\varkappa - E}$	$\frac{3E\varkappa}{9\varkappa - E}$	E	$\frac{1}{2}\left[\frac{3\varkappa - E}{3\varkappa}\right]$	$\varkappa$
$\nu, \varkappa$	$\frac{3\varkappa\nu}{1+\nu}$	$\frac{3\varkappa(1-2\nu)}{2(1+\nu)}$	$3\varkappa(1-2\nu)$	ν	$\varkappa$

[After Gurtin 1972]

These stress–strain relationships, with a total of six coefficients, four for the volumetric behavior and two for the shear behavior, are believed to closely approximate the behavior of a wide range of materials. In materials such as plastics, rubberlike materials, soils, bitumen, etc. the volumetric model is more complex than the shear model.

We may evaluate the volumetric and shear *rates of strain*, respectively, as follows

$$\varepsilon(t')_{\mathrm{m}} = \frac{\dot{p}(t')_{\mathrm{m}}}{\varkappa_1} + \frac{p(t')_{\mathrm{m}}}{\mu_2} - \frac{1}{\mu_3 T_3} \exp \frac{-t'}{T_3} \int_0^{t'} p(t)_{\mathrm{m}} \exp \frac{t}{T_3} \, \mathrm{d}t + \frac{p(t')_{\mathrm{m}}}{\mu_3} \tag{7.6.10}$$

$$\dot{e}(t'')_{ij} = \frac{\dot{s}(t'')_{ij}}{2G_1} + \frac{s(t'')_{ij}}{2\eta_2} \tag{7.6.11}$$

Eqs. (7.6.8) and (7.6.9) and specifically (7.6.10) and (7.6.11) will be used in Chaps. 11 and 12.

7.7 Dual Volumetric Stress–Strain versus Shear Stress–Strain Relationship

While the resolution of Eq. (7.3.1) into the dual equations (7.5.1) and (7.5.2) has never before been strongly proposed, its necessity had already been felt by Stokes (1829, 1845), who developed the idea of the volumetric compression earlier observed (Poncelet 1839) and made the distinction between two kinds of elasticity, one resistant to compression and one to shear. Since then many investigators have noticed the effect of volume changes on the shearing resistance of materials and the effect of the isotropic pressure on their properties, and have reported experiments on a wide array of materials under isotropic pressures as high as 3×10^4 kg/cm^2 (Nadai 1950; Dow 1956).

The graphical presentation of the tri-dimensional stresses acting on materials is still somewhat deficient. First, there is the technical difficulty encountered in describing a tri-dimensional phenomenon by graphical means, second, the theoretical treatment is more involved than that of a two-dimensional representation. By reason of the difficulties involved in representing a tri-dimensional state, it is customary to resort to two-dimensional descriptions.

Let us now try to visualize a three-dimensional representation. We have to start from a coordinate system, and for the sake of simplicity we shall assume a Cartesian coordinate system, in which the coordinates are in the directions of the principal stresses. We have thus an isometric representation of the stresses, Fig. 7.7.1, and a hydrostatic pressure or tension would be marked along the *hydrostatic directrix*, that passes through the origin at equal

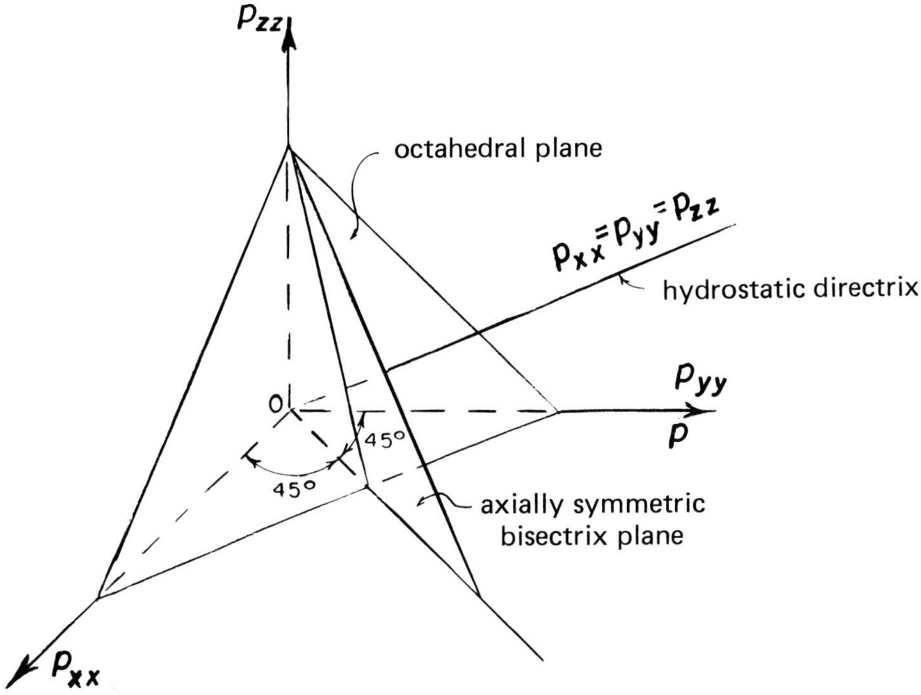

Fig. 7.7.1. Isometric representation of the stress coordinates.

distances from the coordinates and at angles with cosine $1/\sqrt{3} = 0.577$ (54°45′). Any deviation from the directrix indicates a shear stress, and any point of failure of the material lies on an envelope characteristic to the material, around and along the directrix, Fig. 7.7.2. A plane orthogonal to the directrix is the *octahedral plane*, which intersects the coordinates at equal distances and is the location of the shear stresses under equivalent hydrostatic pressure (tension), $p = \text{const}$. In failure theories, for instance, the failure envelopes are seen by looking at the octahedral plane in the direction of the

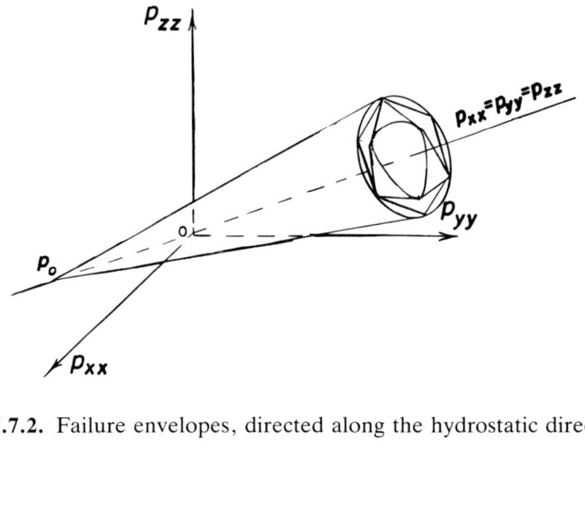

Fig. 7.7.2. Failure envelopes, directed along the hydrostatic directrix.

directrix, Fig. 7.7.3. When axial symmetric problems occur, the tri-dimensional state of stresses can be represented in two dimensions in the symmetric *bisectrix plane*, that is, the plane passing through the directrix and bisecting the $p_{11}-p_{22}$ plane, Fig. 7.7.4. The directrix is at an inclination of 35°15′ to the $p_{11}-p_{22}$ plane.

The graphical representations are very instructive, indicating that any point lying on the hydrostatic directrix is always removed from the failure envelope. Thus a failure, in the sense that mechanical failure is defined, can never occur by hydrostatic pressure (or tension).

The location of the stresses confined between the two extremes – the isotropic or zero shear state located on the directrix, and the failure state located on the failure envelope – may be considered a measure of the state of failure. One may say that between these two limits the material is directed towards failure. Since spherical stresses affect spherical strains and densities, which in turn affect the strength of materials, the interplay of shear and volumetric phenomena in materials can be crucial in many ways.

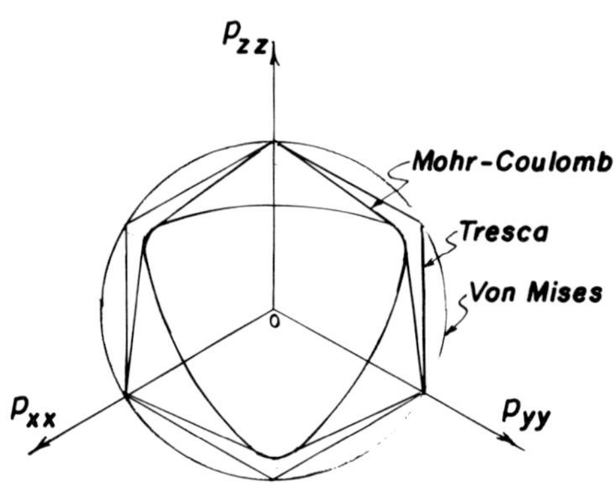

Fig. 7.7.3. Failure envelopes, looking on the octahedral plane.

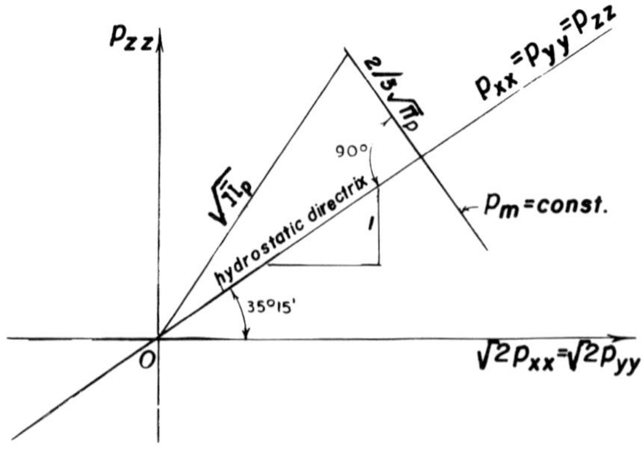

Fig. 7.7.4. The bisectrix for axially symmetric loading.

Proposition 5. Failure can never occur under hydrostatic pressures (tensions), only under shear stresses.

What may occur under hydrostatic pressure or tension is compression and contraction or decompression and expansion, but never failure of the material. Heat expansion is another volumetric phenomenon in materials that will not cause failure. Some very loose experimental evidence seems to support proposition 5 (Poulter 1932; Kármán 1913).

Except for marginal cases, the increase of spherical pressure densifies the material and increases its strength. A classical example is soils, which markedly increase their shearing resistance when consolidated. Consolidation is the densification of soil by squeezing out, under pressure, the water and air from the pores.

The increase in strength occurs, in fact, in any material. The strength of a material, which in general depends on its density and temperature, will depend under constant temperature on the density alone.

7.8 Energy Considerations in View of the Dual Equations

By applying hydrostatic pressure on the material element or by increasing its density, work is done. The energy invested is *stored* in the material and preserved there, enabling it to withstand higher shear stresses or strains and to defer the possibility of failure to higher shear energies. Thus a higher energy has to be *disbursed* in order to bring the material to failure. This process may have implications on the equation of the production rate of internal energy, which, for the non-polar case assumed to proceed close to equilibrium, is

$$\rho\dot{\epsilon} = -p_{\alpha\beta}f_{\alpha\beta} + h_{\alpha,\alpha} + \rho q = -p_{\alpha\beta}\dot{\varepsilon}_{\alpha\beta} + h_{\alpha,\alpha} + \rho q; \quad h_i \to 0; \quad q \to 0 \tag{7.8.1}$$

where ϵ is the specific internal energy, p_{ij} the stress tensor, ε_{ij} the strain tensor, f_{ij} the stretching tensor defined $f_{ij} = \frac{1}{2}(\dot{x}_{ij} + \dot{x}_{ji})$, h_i the energy flux and q the specific energy supply.

By resolving Eq. (7.8.1) into two equations, one originating in the spherical components and one in the shear components, we obtain

$$\rho\dot{\epsilon}_{\mathrm{i}} = -p_{\mathrm{m}}\dot{\varepsilon}_{\mathrm{m}} + h_{\alpha,\alpha} + \rho q \tag{7.8.2}$$

$$\rho\dot{\epsilon}_{\mathfrak{d}} = s_{\alpha\beta}\dot{\varepsilon}_{\alpha\beta} \tag{7.8.3}$$

an equation of the *stored internal energy* and an equation of the *disbursed internal energy*, respectively.

We maintain that any proper failure theory should consider the dependence of Eq. (7.8.3) on Eq. (7.8.2). In the case where the non-mechanical work vanishes, Eqs. (7.8.2) and (7.8.3) become simpler

$$\rho\dot{\epsilon}_{\mathrm{i}} = -p_{\mathrm{m}}\dot{\varepsilon}_{\mathrm{m}}; \quad h_{\alpha,\alpha} + \rho q = 0 \tag{7.8.4}$$

$$\rho\dot{\epsilon}_{\mathfrak{d}} = s_{\alpha\beta}\dot{\varepsilon}_{\alpha\beta} = s_{\alpha\beta}\dot{e}_{\alpha\beta} \tag{7.8.5}$$

Several illustrations of Eqs. (7.8.5) and (7.8.4) may be of interest and we shall mention two of them;

1. *Elastically compressible elastic case.* If $p = \varkappa\varepsilon_{\mathrm{m}}$, and $s_{ij} = 2Ge_{ij}$, then

$$\rho\dot{\epsilon}_{\mathrm{i}} = -p_{\mathrm{m}}\dot{\varepsilon}_{\mathrm{m}} = \varkappa\varepsilon_{\mathrm{m}}\dot{\varepsilon}_{\mathrm{m}} = \frac{p_{\mathrm{m}}\dot{p}_{\mathrm{m}}}{\varkappa} = \frac{\dot{\overline{p_{\mathrm{m}}^2}}}{2\varkappa} = \frac{\dot{\overline{\mathrm{I}_{\mathrm{p}}^2}}}{2\varkappa} \tag{7.8.6}$$

$$\rho\dot{\epsilon}_{\mathfrak{d}} = s_{\alpha\beta}\dot{e}_{\alpha\beta} = 2Ge_{\alpha\beta}\dot{e}_{\alpha\beta} = \frac{s_{\alpha\beta}\dot{s}_{\alpha\beta}}{2G} = \frac{\dot{\overline{s^2}}}{2G} = \frac{\dot{\mathrm{II}}_{\mathfrak{s}}}{4G} \tag{7.8.7}$$

where I_{p} and $\mathrm{II}_{\mathfrak{s}}$ are the first and second moment invariants of the spherical and shear stress-rates, respectively.

2. *Constant isotropic pressure*. The case of constant isotropic pressure $p = \text{const}$ is very common. Shear tests on soils are usually performed under constant isotropic pressure (tension), either as a torsion test under constant hydrostatic pressure, or a pure deviatoric test (Klausner 1964, 1967, 1970a–c) where $p_{11} = p_{22} = \frac{1}{2}p_{33}$. Then Eq. (7.8.4) becomes

$$p_{\mathrm{m}}\varepsilon_{\mathrm{m}} = \frac{\mathrm{I}_{\mathrm{p}}\dot{\mathrm{I}}_{\varepsilon}}{3} \tag{7.8.8}$$

Eq. (7.8.5) is the deviatoric part of the internal energy production. If we consider a compressible elastic material, where Eqs. (7.6.1) are valid, then, by inserting Eq. $(7.6.1)_1$ into Eq. (7.8.5) we obtain

$$\begin{aligned}\rho\dot{\epsilon}_{\mathfrak{d}} &= s_{\alpha\beta}\dot{e}_{\alpha\beta} = 2Ge_{\alpha\beta}\dot{e}_{\alpha\beta} = G\dot{\overline{e^2_{\alpha\beta}}} \\ &= \frac{\dot{\overline{s^2_{\alpha\beta}}}}{4G} = \frac{\dot{\overline{\mathrm{II}}}_{\mathfrak{s}}}{4G} = \frac{3\dot{\Pi}_{\mathrm{p}}}{G} = \frac{3(\dot{\mathrm{I}}_{\mathrm{p}} - 3\dot{\mathrm{II}}_{\mathrm{p}})}{2G}\end{aligned} \tag{7.8.9}$$

where $\overline{\mathrm{II}}_{\mathfrak{s}}$ is the *second moment invariant* of the deviatoric stress tensor p_{ij} and Π_{p} is the *octahedral invariant* of the stress tensor s_{ij}, defined $\Pi_{\mathrm{p}} = \Pi_{\mathfrak{s}} = \frac{1}{2}(\overline{\mathrm{II}}_{\mathrm{p}} - \mathrm{II}_{\mathrm{p}}) = \frac{1}{2}(\mathrm{I}_{\mathrm{p}}^2 - 3\mathrm{II}_{\mathrm{p}})$.

If, however, Eqs. (7.6.1) are substituted into Eq. (7.8.1) and the non-mechanical energy vanishes, or if Eqs. (7.8.6) and (7.8.8) are summed up, we obtain

$$\rho\dot{\epsilon}_{\mathrm{i}} = p_{\alpha\beta}\dot{\varepsilon}_{\alpha\beta} = 3p_{\mathrm{m}}\dot{\varepsilon}_{\mathrm{m}} + s_{\alpha\beta}\dot{e}_{\alpha\beta} = \frac{\dot{\overline{\mathrm{I}_{\mathrm{p}}^2}}}{2} + \frac{\dot{\mathrm{II}}_{\mathfrak{s}}}{4G} \tag{7.8.10}$$

It is interesting to note that several of the failure theories resemble either Eq. (7.8.9) or Eq. (7.8.10). The closest of all is the von Mises theory, which is expressed by the second moment invariant, that is, the first part of the right-hand side of Eq. (7.8.10). As it can be seen, the von Mises theory is not quite Eq. (7.8.10) nor Eq. (7.8.9), since it bears no inference whatsoever to the first invariant.

A proper theory for elastic materials should be expressed by the second octahedral invariant as in Eq. (7.8.9), and be related to the first invariant as in Eq. (7.8.6). A failure theory for other viscoelastic materials would be more complex, in any event it would have to relate the isotropic internal energy production, Eq. (7.8.3), to the deviatoric internal energy production, Eq. (7.8.2).

In soils, the octahedral invariant which represents the shear stress on the octahedral plane is indeed related to the first invariant. This relation is closest to the relation between Eqs. (7.8.3) and (7.8.2). It is always correct in the bisectrix plane; whether the relationship is maintained in any of the other planes is subject to further investigation.

The above can be summed up in the following proposition:

Proposition 6. Any failure theory properly defined has to relate the rate of internal energy production disbursed through the shearing process to the rate of internal energy production stored during the volumetric process.

7.9 The Isotropic Stress–Strain Relationship

A difficulty concerning the structure of constitutive equations is the fact that the relations between its variables are not always linear. Considering the various sources of non-linearity in the constitutive equations, Sect. 7.4, one can place the use of linear theories in its proper perspective by understanding their limitations.

In Sect. 7.4 we have discussed the possible non-linearities that may occur in constitutive equations. In this and the following sections we will be concerned mainly with the tensorial non-linearity arising from the postulated isotropic relations between the stresses and the strains.

When non-linear viscoelastic relationships due to tensorial non-linearity are considered, the resolution of Eq. (7.3.2) into the dual equations (7.5.1) and (7.5.2) results in further implications, in addition to those already discussed in the linear case. We aim to show that such a resolution can be maintained in equations with tensorial non-linearity as well.

We shall note that the strains admitted on account of their physical meaning, referred to in Eq. (7.3.2), are still either the infinitesimal strain or the finite strain of the Hencky measure.

The postulate of isotropy of the continuum imposes certain limitations on the form of Eq. (7.3.2). The tensorial formulation of the problem is due to

Rivlin and Ericksen (Rivlin, 1948; Rivlin and Ericksen 1955), after having been presented earlier by Reiner (1945) and observed experimentally (Poynting 1909, 1912; Weissenberg 1947, 1949). It has been shown that any matrix **P** which can be expressed as a matrix polynomial of any other matrix, say, **E** and of its higher order derivatives $\dot{\mathbf{E}}$, $\ddot{\mathbf{E}}$, . . ., can be further reduced to a polynomial function of the second order, if the *Cayley–Hamilton recursion equation* is applied (Rivlin and Ericksen 1955). The coefficients of the polynomial are functions of the invariants of the matrices **E**, $\dot{\mathbf{E}}$, ...

If the stress matrix **P** is defined as a polynomial function of the strain **E** and the strain rate $\dot{\mathbf{E}}$ matrices, then, since we have only six independent matrix components, only six linear homogeneous equations are possible (Rivlin and Ericksen 1955) and only six coefficients $\aleph_n$ can be obtained. Of the nine combinations six can be chosen at will, and the following form is one possible choice

$$\mathbf{P} = \aleph_0'\mathbf{1} + \aleph_1'\mathbf{E} + \aleph_2'\mathbf{E}^2 + \aleph_3'\dot{\mathbf{E}} + \aleph_5'(\mathbf{E}\dot{\mathbf{E}} + \dot{\mathbf{E}}\mathbf{E}) + \aleph_6'(\mathbf{E}^2\dot{\mathbf{E}} + \dot{\mathbf{E}}\mathbf{E}^2) \tag{7.9.1}$$

The only requirement regarding Eq. (7.9.1) is that the matrices **E**, $\mathbf{E}^2$, $\dot{\mathbf{E}}$, $(\mathbf{E}\dot{\mathbf{E}} + \dot{\mathbf{E}}\mathbf{E})$ and $(\mathbf{E}^2\dot{\mathbf{E}} + \dot{\mathbf{E}}\mathbf{E}^2)$ be linearly independent.

The strain matrix **E** can be expressed analogously as a function of the stress **P** and the stress-rate $\dot{\mathbf{P}}$ matrices

$$\mathbf{E} = \aleph_0\mathbf{1} + \aleph_1\mathbf{P} + \aleph_2\mathbf{P}^2 + \aleph_3\dot{\mathbf{P}} + \aleph_5(\mathbf{P}\dot{\mathbf{P}} + \dot{\mathbf{P}}\mathbf{P}) + \aleph_6(\mathbf{P}^2\dot{\mathbf{P}} + \dot{\mathbf{P}}\mathbf{P}^2) \tag{7.9.2}$$

Similarly, the strain rate matrix $\dot{\mathbf{E}}$ and the stress rate matrix $\dot{\mathbf{P}}$ may each be expressed as functions of the other matrices.

Eq. (7.9.1) as well as any other aforementioned equation similarly derived may be resolved into the dual equation, namely the isotropic and deviatoric equations, as follows

$$\pi = \beth_0 + \beth_1\varepsilon + \beth_2\varepsilon^2 + \beth_3\dot{\mathbf{E}} + \beth_5\varepsilon\dot{\varepsilon} + \beth_6\dot{\varepsilon}\varepsilon^2 + \daleth_2(\mathbf{e}^2)_{\mathrm{i}}$$

$$+ \daleth_5(\dot{\mathbf{e}}\mathbf{e} + \mathbf{e}\dot{\mathbf{e}})_{\mathrm{i}} + \daleth_6(\dot{\mathbf{e}}\mathbf{e}^2 + \mathbf{e}^2\dot{\mathbf{e}})_{\mathrm{i}} \tag{7.9.3}$$

$$\mathbf{s} = \text{ש}_1(\mathbf{e}) + \text{ש}_2(\mathbf{e}^2)_{\mathfrak{b}} + \text{ש}_3\dot{\mathbf{e}} + \text{ש}_5(\mathbf{e}\dot{\mathbf{e}} + \dot{\mathbf{e}}\mathbf{e})_{\mathfrak{b}} + \text{ש}_6(\mathbf{e}^2\dot{\mathbf{e}} + \dot{\mathbf{e}}\mathbf{e}^2)_{\mathfrak{b}} \tag{7.9.4}$$

where π and **s** are the trace and traceless matrix of the stress matrix **P**, respectively, defined $\mathbf{P} = \pi\mathbf{1} + \mathbf{s}$; ε and **e** are the trace and the traceless matrix of matrix **E**, respectively, defined $\mathbf{E} = \varepsilon\mathbf{1} + \mathbf{e}$; $\dot{\varepsilon}$ and $\dot{\mathbf{e}}$ are the trace and the traceless matrix of matrix $\dot{\mathbf{E}}$, respectively, defined $\dot{\mathbf{E}} = \dot{\varepsilon}\mathbf{1} = \dot{\mathbf{e}}$; $(\ )_{\mathrm{i}}$ are the isotropic components, and $(\ )_{\mathfrak{b}}$ are the deviatoric components, respectively, of the strain and strain rate matrices. It should be noticed that although **e** and $\dot{\mathbf{e}}$ are traceless matrices, their squared expressions yield matrices with non-vanishing diagonals. Consequently, these matrices containing squared expressions of **e** and $\dot{\mathbf{e}}$ can also be resolved into isotropic and deviatoric components. The expression in Eq. (7.9.3) with the coefficient $\daleth_2$ is the spherical component resulting from the squaring of the deviatoric stress **e**, whereas the expressions in Eq. (7.9.4) are all traceless deviatoric expressions.

With considerable labor it can be shown that the coefficients $ב_n$, $ד_n$, $ש_n$ appearing in Eqs. (7.9.3) and (7.9.4) are functions of the invariants of the stress and stress rate matrices, as it was shown with respect to the coefficients $\aleph_n$ in Eq. (7.9.1) (Rivlin and Ericksen 1955).

A case of particular interest to us is the case in which the strain matrix **E** is expressed in terms of the stress matrix **P** and the strain rate matrix $\dot{\mathbf{E}}$

$$\mathbf{E} = \aleph_0\mathbf{1} + \aleph_1\mathbf{P} + \aleph_2\mathbf{P}^2 + \aleph_3\dot{\mathbf{E}} + \aleph_5[\mathbf{P}\dot{\mathbf{E}} + \dot{\mathbf{E}}\mathbf{P}] + \aleph_6[\mathbf{P}^2\dot{\mathbf{E}} + \dot{\mathbf{E}}\mathbf{P}^2] \tag{7.9.5}$$

which can be resolved into isotropic and deviatoric equations

$$\begin{aligned}\varepsilon = {}& \aleph_0 + \aleph_1\pi + \aleph_2\pi^2 + \aleph_3\dot{\varepsilon} + 2\aleph_5\pi\dot{\varepsilon} + 2\aleph_6\pi^2\dot{\varepsilon} + [\aleph_2 + 2\aleph_6\dot{\varepsilon}][\mathbf{s}^2]_{\text{i}} \\ & + [\aleph_5 + 2\aleph_6\pi][\dot{\mathbf{e}}\mathbf{s} + \mathbf{s}\dot{\mathbf{e}}]_{\text{i}} + \aleph_6[\dot{\mathbf{e}}\mathbf{s}^2 + \mathbf{s}^2\dot{\mathbf{e}}]_{\text{i}}\end{aligned} \tag{7.9.6}$$

$$\begin{aligned}\mathbf{e} = {}& [\aleph_1 + 2\aleph_2\pi + 2\aleph_5\dot{\varepsilon} + 4\aleph_6\pi\dot{\varepsilon}][\mathbf{s}]_{\text{Ђ}} + [\aleph_2 + 2\aleph_6\dot{\varepsilon}][\mathbf{s}^2]_{\text{Ђ}} \\ & + [\aleph_3 + 2\aleph_5\pi + 2\aleph_6\pi^2][\dot{\mathbf{e}}]_{\text{Ђ}} + [\aleph_5 + 2\aleph_6\pi][\mathbf{s}\dot{\mathbf{e}} + \dot{\mathbf{e}}\mathbf{s}]_{\text{Ђ}} \\ & + \aleph_6[\mathbf{s}^2\dot{\mathbf{e}} + \dot{\mathbf{e}}\mathbf{s}^2]_{\text{Ђ}}\end{aligned} \tag{7.9.7}$$

The cumbersome notation and the elaborate form including multiplication of matrices render these equations quite difficult to handle, and in practice only short versions of Eq. (7.9.1) were used; today, however, computing facilities make this a simple task.

7.10 Dilatancy

The matrix equations resembling tensor equations indicate that the isotropic non-linear viscoelastic equations can also be resolved into dual equations. Moreover, by having the strains expressed either by the infinitesimal or the Hencky strain measures, the dual equations assume physical meaning in non-linear viscoelasticity as well as in linear viscoelasticity, inasmuch as they correlate isotropic pressure (tension) with volume changes and shear stresses with distortions. Almost all that has been said with respect to linear viscoelasticity, Sects. 7.3–7.8, is valid with respect to non-linear viscoelasticity as well, and therefore a short discussion of Eqs. (7.9.3) and (7.9.4) and the propositions put forward is in order.

Eq. (7.9.3) relates the volumetric stress p_{m} to the volumetric strain ε_{m} and to the volumetric strain rate $\dot{\varepsilon}_{\text{m}}$ and all their products. But it contains also volumetric strains resulting from the square power of the distortion tensor e_{ij}, the distortion rate tensor $\dot{e}_{ij}$ and their cross products. As was already mentioned, traceless tensors when squared result in products with non-vanishing traces. Therefore the product, although stemming from a traceless tensor, can be resolved into a volumetric strain and a distortion. The expressions in Eq. (7.9.3) preceded by the coefficients $\daleth_{\mathrm{n}}$ are all results of such squared cross products of the distortion or of the distortion rate tensors, and have been termed *dilatancy* by Reynolds (Reiner 1945). Thus, dilatancy is the volume change resulting from the squaring of the distortion or distortion rate tensors, whereas dilatation is the original volumetric strain.

This phenomenon of tensorial non-linearity is also called *cross viscoelasticity* or *second-order effect*, because of the cross products involved.

In the deviatoric equation (7.9.4) the traceless components of the strains, distortions and distortion rates are preceded by coefficients ש_{n}.

Eq. (7.9.1) expresses the stress tensor as a function of the strain and strain rate tensors. For a more comprehensive equation the stress rate may be added, as well as the strain acceleration or other higher-order time derivatives of the strain or stress tensor. However, this would prove too bulky to be handled.

In the same manner that the stress has been expressed in Eq. (7.9.1) as a function of the strain and strain rate, the strain can be expressed as a function of the stress and stress rate. The resolution of any of these equations into equations similar to Eqs. (7.9.3) and (7.9.4) is just as straightforward.

For viscoelastic materials with simple mechanical behavior and with postulated isotropic relations, Eqs. (7.9.3) and (7.9.4) simplify considerably. Two of these materials are discussed.

1. *Elastically compressible elastic.* For this material Eq. (7.9.1) will be

$$p_{ij} = \aleph_0\delta_{ij} + \aleph_1\varepsilon_{ij} + \aleph_2\varepsilon_{ij}^2 \tag{7.10.1}$$

which, when resolved into two equations, yields

$$p_{\mathrm{m}} = p_0 + \varkappa_1\varepsilon_{\mathrm{m}} + \varkappa_2\varepsilon'_{\mathrm{m}} = \beth_0 + \beth_1\varepsilon_{\mathrm{m}} + \beth_2\varepsilon_{\mathrm{m}}^2 + T_2(e_{ij}^2)_{\mathrm{i}} \tag{7.10.2}$$

$$s_{ij} = 2G_1e_{ij} + 2G_2e'_{ij} = \text{ש}_1e_{ij} + \text{ש}_2[2\varepsilon_{\mathrm{m}}e_{ij} + (e_{ij}^2)_{\mathfrak{d}}] \tag{7.10.3}$$

where ε_{ij}^2 is the tensor resulting from the squaring of the strain matrix, $(e_{ij}^2)_{\mathrm{i}}$ and $(e_{ij}^2)_{\mathfrak{d}}$ are the isotropic and deviatoric components, respectively, resulting from the squaring of the deviatoric strain matrix e_{ij}, and p_0 is the *cohesion*, an inherent spherical stress that holds together the material in its natural state and has to be overcome by a spherical tension in order to attain volumetric disintegration of the material. From Eqs. (7.10.2) and (7.10.3) the following identities may be derived: $\beth_0 = p_0$; $\beth_1 = \varkappa_1$; $\beth_2 = T_2 = \varkappa_2$; $\text{ש}_1 = 2G_1$; $\text{ש}_2 = 2G_2$ and $e'_{ij} = \varepsilon'_{\mathrm{m}} = \varepsilon_{\mathrm{m}}^2 + (e_{ij}^2)_{\mathrm{i}}$; $2\varepsilon_{\mathrm{m}}e_{ij} + (e_{ij}^2)_{\mathfrak{d}}$.

It should be noticed that while the volumetric strain ε_{m} can be either a volume increase or a volume decrease, the volumetric strain $\varepsilon'_{\mathrm{m}}$ resulting from the squaring of the deviatoric strain is *always* an increase in volume. In this context Reiner has drawn attention to the beach sand effect. When walking on a sandy beach close to the water, the water will seep away from under the feet due to shear stresses, resulting in a volume increase and a density decrease of the sand, and the closely surrounding area will appear dry.

The coupling of the deviatoric strain e_{ij} and the deviatoric component of its square under the one coefficient of the shear modulus is a necessary condition, since whenever a deviatoric strain is measured it is impossible to distinguish between its parts. In volumetric strains, however, which result on one hand from the isotropic stress and on the other hand from the squared deviatoric strain, the two parts can be distinguished and they warrant, therefore, separate volumetric coefficients.

In experiments directed towards the evaluation of the viscoelastic coefficients in soils and other materials the volumetric and shear properties should be treated separately if physical meaning is to be attributed. It means that a shear test – simple shear, torsion, triaxial shear, deviatoric shear – should be applied only after the sample has been exposed to a hydrostatic pressure.

2. *Elastically compressible viscous*. The transpose of Eq. (7.9.1) for such material would be

$$\rho_{ij} = \aleph_0 \delta_{ij} + \aleph_1 \varepsilon_{ij} + \aleph_2 \varepsilon_{ij}^2 + \aleph_3 \dot{\varepsilon}_{ij} \tag{7.10.4}$$

which decomposes into two equations

$$p_{\mathrm{m}} = p_0 + \varkappa_1 \varepsilon_{\mathrm{m}} + \varkappa_2 \varepsilon'_{\mathrm{m}} = \beth_0 + \beth_1 \varepsilon_{\mathrm{m}} + \beth_2 \varepsilon_{\mathrm{m}}^2 + T_2 (e_{ij}^2)_{\mathfrak{i}} + \beth_3 \dot{e}_{ij} \tag{7.10.5}$$

$$s_{ij} = 2G_1 e_{ij} + 2G_2 e'_{ij} + 2\eta \dot{e}_{ij} = \mathfrak{w}_1 e_{ij} + \mathfrak{w}_2 (e_{ij}^2)_{\mathfrak{d}} + \mathfrak{w}_3 \dot{e}_{ij} \tag{7.10.6}$$

from which the following identities are derived: $\beth_0 = p_0$; $\beth_1 = \varkappa_1$; $\beth_2 = T_2 = \varkappa_2$; $\beth_3 = 0$; $\mathfrak{w}_1 = 2G_1$; $\mathfrak{w}_2 = 2G_2$; $\mathfrak{w}_3 = 2\eta$; $\varepsilon'_{\mathrm{m}} = \varepsilon_{\mathrm{m}}^2$ and $e'_{ij} = (e_{ij}^2)_{\mathfrak{d}}$.

The remarks made with respect to the elastically compressible elastic materials are valid for the elastically compressible viscous materials as well.

7.11 Isotropic Non-linear versus Linear Viscoelasticity

It has been shown that the decomposition of Eq. (7.9.1), the general equation of the isotropic stress–strain relationship, into the dual equations (7.9.3) and (7.9.4) is possible, just as in linear viscoelasticity, where we obtain Eqs. (7.5.1) and (7.5.2).The physical meaning of these equations is also retained. All propositions and their corollaries, except for proposition 1, remain valid when isotropic relations are postulated and the stress–strain relationship becomes non-linear. Proposition 1, however, has to be modified to admit volume changes resulting from shear stresses, and, conversely, spherical pressures resulting from shear strains. We shall correct propositions 1 and 1′ as follows:

Proposition 1″. If isotropic relations are postulated in a viscoelastic material, then isotropic pressures, as well as deviatoric stresses, result in volumetric strains.

Isotropic pressures (tensions) result in first-order volumetric strains, contraction (dilatation), while deviatoric stresses result in second-order volumetric strains, dilatancy. Corollary 1 changes to 1″ as follows:

Corollary 1″. If isotropic relations are postulated in a viscoelastic material, then volumetric strains result either from isotropic pressures or from deviatoric stresses.

8 The Soil

8.1 Single-phase versus Multi-phase Considerations

The general single-phase and specific multi-phase theories of mechanics of the continuum have been derived in the previous chapters, without special and detailed reference to soils, except occasionally and in a general way. We shall now proceed to the bearing the outlined theories have on the mechanical behavior of soils.

There are two possible approaches to the investigation of soils, and they correspond to the theories of mechanics outlined above.

In one approach, the soil bulk is considered a single-phase entity, without dealing with the internal interactions of its constitutive phases. Most problems of failure and also some compression problems may be treated in such a way, which we shall call the *phenomenological approach* (see also Sect. 5). There are, however, problems where the application of the theories of multi-phase media is indispensable, specifically problems of flow, i.e. permeability of the soil matrix to water, to air, or to a mixture of the two. For the theory of consolidation, for example, it is essential to consider the interaction of the soil matrix with the water, thus implying, in fact, a multi-phase medium. Another example is the experimental assertion of the existence of pore pressure, which is a result of measurements performed on one of the phases, either the water or the air, in a multi-phase medium (soil). The proper way to deal with problems of this nature is by using the *structural approach*. Incidentally, the existence of pore pressure is not upheld by theoretical proof. In the following chapter such a theoretical proof is presented, based on the concept of multi-phase mixtures.

Experimentally the two approaches are manifested as follows: we may consider a soil sample, say a triaxial sample, subjected to certain confinements, and ask ourselves what are the overall parameters and coefficients responsible for the strains resulting from these confinements, as is done in some cases in soil mechanics and in the study of many other materials. Or, we may study the strains of the individual constituents of the sample and the interaction and relative motion of one constituent with respect to the other, by employing the theories of multi-phase media and applying boundary conditions as required; we will thus obtain not only better results, but also acquire more insight and understanding of the phenomena which support the motion-constraint relations of the sample.

We shall consider both approaches, and in order to do so we will have to divert our discussion, at some length, to the intricate nature of soils, through a third approach, the *molecular or submolecular* approach. Thus, before the mechanics of multi-phase media can be applied to soil, a detailed discussion is required concerning its constituents: soil particles, water and air, not necessarily in that order. The author would ask the reader to bear with him if he finds some of the details too elementary and superfluous.

8.2 Soil Constituents

The division of soils into three *phases*, solids, water and air, is only academic. The solid particles themselves are composed of several constituents, distinguished by their mineralogical nature, size and geometry. The water phase may come in a liquid or vapor state, with electrolyte concentration varying throughout the solution, and the gas phase may be composed of several gaseous constituents, be non-existent in soils known as *saturated*, or even be one of the phases in a two-phase system to which soils reduce if dry.

It is this division of soil into phases that permits its study as a multi-phase medium, once the limiting factors of the division are known and taken into consideration. The definition of soil as a three-phase system is, however, not a necessary condition. Cases where a four-phase system should be considered may well be envisaged, for instance when gravel has to be differentiated from the finer particles, just as a two-phase system is considered for a dry soil or a saturated soil. Except for the cases mentioned, in general a three-phase system seems a fair and reasonable assumption, inasmuch as it is congruent with the three major constituents of soils, solids, water and air. If it turns out that, for instance, clays have to be considered separately from the granular constituents, or a uniform sand has to be considered separately from the graded gravel constituting the solid phase, additional phases may be added. It should be kept in mind, however, that by adding phases the mathematics gets more and more intricate.

Classifications of soils, according to criteria satisfying specific engineering or agricultural needs, have been devised by scientists and engineers in order to reduce the large variety of soils to a reasonable number of groups of distinct properties. These classifications, however, relate mostly to the solid phase. They are either mineralogical, mechanical (grain size distribution), or technical (liquid and plastic limits, w_L and w_P respectively).

From a heuristic point of view the *solids phase* of soils is a particle system, at the extremes of which are the coarse grained granular particles and the fine grained clays. The concept of mechanics of materials, which assumes that the mechanical behavior of all materials is *qualitatively* similar, in spite of quantitative differences and in spite of the different mechanisms that trigger their behavior internally, can be applied to soils as well. In other words, all soils compress, some more some less, and perhaps at different rates; they all have a shearing resistance, again of various magnitudes; they are all, from the point of view of mechanics, cohesive; the cohesion of many granular soils is zero.

Between the two extremes, granular soils on one hand and fine clays on the other hand, there is a wide variety of soils with an array of quantitatively diverse properties, stemming from internal mechanisms. And although the internal mechanisms are different for the different soils, the overall end result, the deformation, is qualitatively the same.

When an external force is applied, the internal mechanism of compression in a granular material is a rearrangement of particles overcoming frictional resistance; in clays it is the equilibrium between charged particles in an ionic environment, as the oppositely charged ions are attracted and back-diffused at the same time.

The *water phase*, to the properties of which not much attention has been paid in soil mechanics, is known today to be responsible for many mechanical properties of clays. It is not an inert substance. The type and concentration of the electrolytes that the water phase accommodates affect all mechanical properties of fine grained soils, although they have no interactive effect in granular soils. Yet in a granular subsoil a subsidence will occur if the water table recedes due to excessive pumping, just as it does if, say, electrolytes of higher concentration or other chemicals seep into a clayey subsoil (Tschebotarioff 1948; Sampson 1950). Qualitatively the results are the same, although they are not necessarily due to the same cause. Obviously the water phase and its quality play very important roles in the mechanical behavior of all soils.

Finally, the *air phase* has a fairly well-known effect on all properties of soils: compression, shearing resistance and permeability. There is no evidence, however, as to the effect of the quality of air on these properties. The pore air pressure, whose effects are similar to those of the pore water pressure, has nevertheless different constitutive equations, derived in one of the following sections.

8.3 The Water

It is instructive to start with water since it is a rather simple structural material. Its molecule is composed of an oxygen atom (O^{2-}), with two free electrons that provide it with two valences of negative charge, and two hydrogen atoms (H^+), each single valent and positively charged. The opposite charges in the atoms, seemingly balanced, make possible the formation of the water molecule. The charges, however, are not concentrated in the center of the molecule, but revolve as electrons around the centers of the individual atoms, which are 0.957 Å apart. Moreover, the two hydrogen atoms are not symmetrically attached to the oxygen atom, Fig. 8.3.1, but are at an angle of about 105°. Thus the water molecule, although electostatically balanced, is not evenly charged at its envelope. There are spots around the hydrogen atoms, where the positive charges dominate, and spots around the oxygen atoms, where the negative charges dominate. These unevenly charged spots attract atoms of opposite charge from neighboring molecules, and they stick together to form the structure of water, Fig. 8.3.2. Water in its liquid form is a chain, or rather a cluster, of molecules that may extend from small drops to very large volumes such as oceans. Without the unevenly charged spots on the molecules, water would probably fall apart and occur in nature as individual molecules.

The wobbling and vibrating motion within the molecules, resulting from the revolving electrons, is what we call *heat*. As the temperature is increased the motion increases, and as temperature decreases the motion calms down. By further decreasing the temperature the motion not only slows steadily, but finally dies out while the various atoms recede to their determined locations.

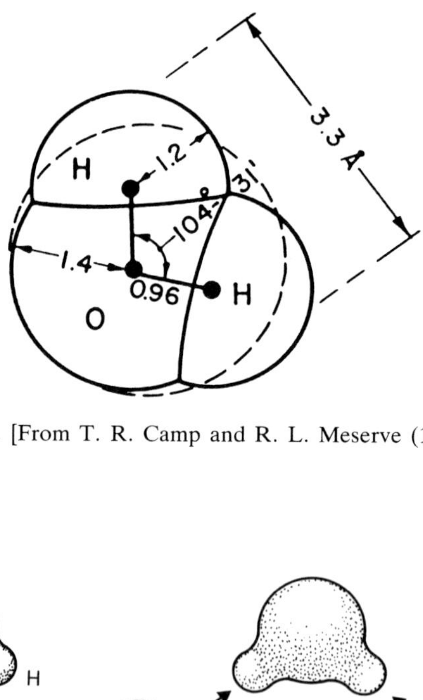

Fig. 8.3.1. Water molecule. [From T. R. Camp and R. L. Meserve (1974), courtesy of Dowden, Hutchison & Ross Inc.]

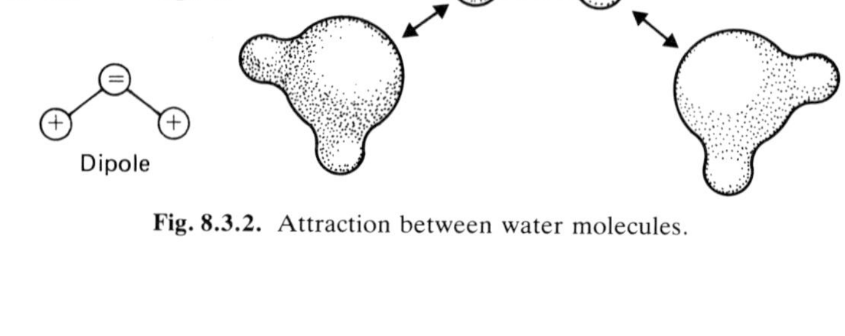

Fig. 8.3.2. Attraction between water molecules.

Fig. 8.3.3, an inaccurate two-dimensional picture of a three-dimensional phenomenon, shows what occurs at very low temperatures as the molecules lock into a new rigid pattern, the *ice*. The water becomes solid and a pattern of symmetry of hexagonal shape can be detected. This symmetry is observed

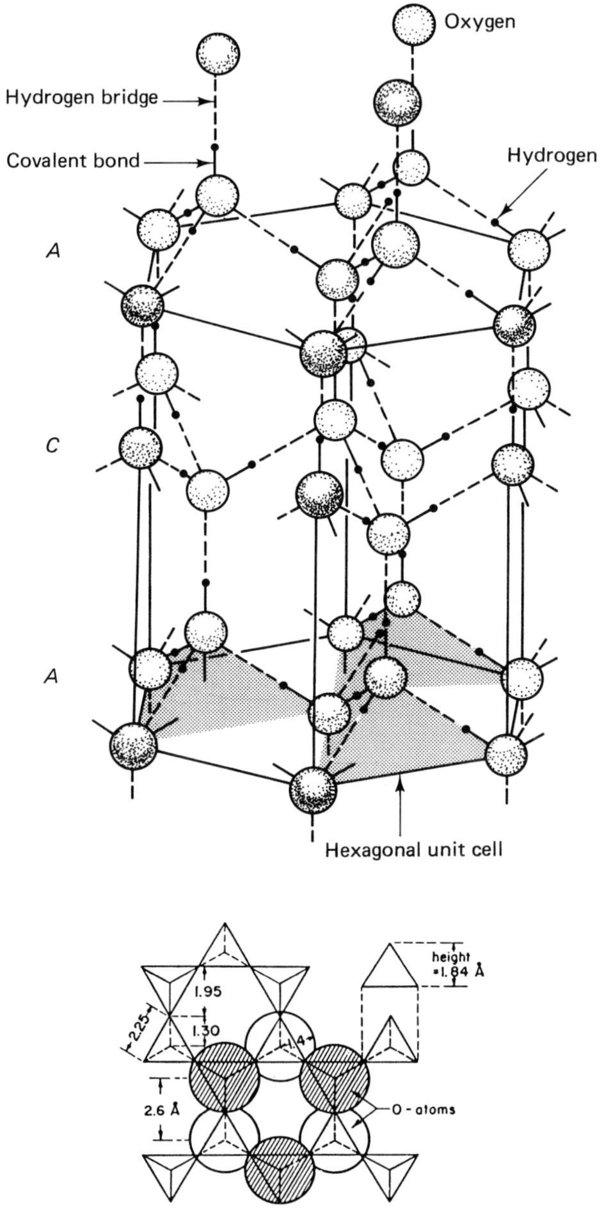

a

b

Fig. 8.3.3. Ice crystal. **a** Molecular structure. **b** Tetrahedral configuration of ice crystal. [From T. R. Camp and R. L. Meserve (1974), courtesy of Dowden, Hutchison & Ross Inc.]

in the numerous patterns of ice flakes, all hexagonal. The temperature at which water transforms from liquid to solid is the freezing temperature. By applying a force and pushing on one molecule, the whole solidified cluster moves as a unit, unlike water at room temperature, where by pushing on a molecule in one direction the other molecules may move in other directions and may even easily be disrupted into smaller volumes, behaving as a liquid. With further decrease of temperature a point is reached where the vibrational motion reduces to a minimum, and this is the *absolute zero* temperature. Between the absolute zero and the freezing point, ice has some heat and the wobbling motion in the atoms still goes on with limited amplitudes; as temperature increases, the amplitudes of the motion increase until the freezing point is reached again, and, as heat equal to the *heat of fusion* is provided, the molecules break loose and return to form again a liquid, the ice melts.

If the temperature of the water is further increased the wobbling motion increases and the volume of the water increases also, up to a point where the attraction between the molecules is not enough to hold them together and they fly apart. The molecules become separated forming steam or vapor, a gaseous substance in which the molecules are bouncing around and bumping against each other and against the walls of the vessel in which they are contained. The infinite impacts and jolts acting randomly in all directions enact an average pressure on the walls of the closed vessel, the *vapor pressure*. Thus, the vapor pressure is a function of temperature, Table 8.3.1. The vapor pressure increases slowly with the increase of the total pressure and it equals the *partial pressure* of water vapor in air saturated with moisture. The breaking of the bonds between the molecules requires additional energy called the *heat of vaporization*. As temperature increases without changing the volume of the vessel, the speed of the vibrations increases, the number of impacts increases and in turn the pressure increases. In reverse we can say that in order to confine a gas we must apply a pressure. The energy discharged by steam in a heated vessel, if tamed, produces work, this being the principle of the steam engine.

Table 8.3.1. Vapor pressure of pure water at atmospheric pressure

Temperature T (°C)	Vapor pressure p_v (atm)	Latent heat H (kcal/mole)	Free energy ψ (kcal/mole)
0	0.0060	10.73	2.770
10	0.0120	10.63	2.480
15	0.0168	10.58	2.340
20	0.0231	10.53	2.200
25	0.0313	10.48	2.054
30	0.0419	10.43	1.910
40	0.0728	10.33	1.630
50	0.1217	10.23	1.350
60	0.1965	10.13	1.080
80	0.4675	9.92	0.540
100	1.0000	9.70	0

The solid, liquid and gaseous states are *phases* or states of consistency of the water, but are characteristic of many other materials as well. Here the term "phases" relates to the various consistencies of a material, not to be confused with phases in a multi-phase material, where it relates to the different constituents within the material.

Now, let us turn and see what is happening on the free surface of the water contained in the vessel at ambient temperature. We will find there, of course, some water molecules in the form of vapor, that were kicked out from the liquid owing to the continuous vibrations within the molecules, particularly if the temperature of the air environment is lower than that of the water. Water vapors are always found on the surface of the water, owing to the equilibrium that exists between water and vapor. Other molecules are kicked out further into the air and fly away as gas. We say that the water *evaporates*. If we now cover the vessel, the evaporation continues and the water molecules reach the underside of the cover, get stuck together and form water drops that fall back into the vessel, the vapor *condenses*. The condensation is intensified if the cover becomes cooler. If evaporation continues at a higher rate so that the volume of air below the cover fills up rapidly with water vapor, an instance is reached when the air becomes *saturated* with vaporized water molecules, which have no other choice than to stick together and precipitate again as water drops. The instance of saturation depends on the temperature of the air. At higher temperatures more vapor can be contained in the air, since the vibrations in the vapor molecule is then more vigorous and molecules are less liable to join one another.

Besides water vapors on the free surface of water at ambient temperature, there are the molecules that constitute the air, such as oxygen molecules (formed by two oxygen atoms), nitrogen molecules, carbon dioxide, etc. Just as the water molecules are carried into the air, the oxygen and nitrogen molecules get into the water, and air is dissolved in water. We may observe this phenomenon by applying a vacuum on the top of the vessel and see bubbles of air rising from the water. Unless the water is treated by vacuum or heat, there is always dissolved air in the water, its amount being a function of the water temperature.

The compressibility of water under pressure is another property that we want to present here. Water is fairly compressible, as may be seen in Table 8.3.2, also presented graphically in Fig. 8.3.4, and this compressibility becomes important when investigating water layers in the proximity of clay particle surfaces.

8.4 Water Solutions

We have seen that water may contain air, but it can also contain solid solutes. Salts represent a class of solutes that dissolve in water and deserve our attention. Salts are crystallized solids where two *ions*, a metalic *cation* such as

Table 8.3.2. Compressibility of water under pressure

Temperature T [°C]	Pressure p_m [atm]	Compressibility $\varkappa_m$ [$(10^6$ atm$)^{-1}$]
0	1.0	50.4
10	1.0	47.8
20	1.0	45.8
20	12.8	49.6
20	197.4	43.6
20	394.8	41.5
20	493.5	39.5
25	1.0	45.7
25	1000.0	34.8
30	1.0	44.6
35	1.0	44.8
35	1000.0	34.2
40	1.0	44.1
40	493.5	38.5
40	987.0	33.4
40	11 884.0	9.1
45	1.0	44.1
45	1000.0	34.0
50	1.0	44.0
55	1.0	44.4
55	1000.0	34.0
60	1.0	44.3
65	1.0	44.8
65	1000.0	34.2
75	1.0	45.5
75	1000.0	34.7
85	1.0	46.5
85	1000.0	35.3
100	1.0	48.0

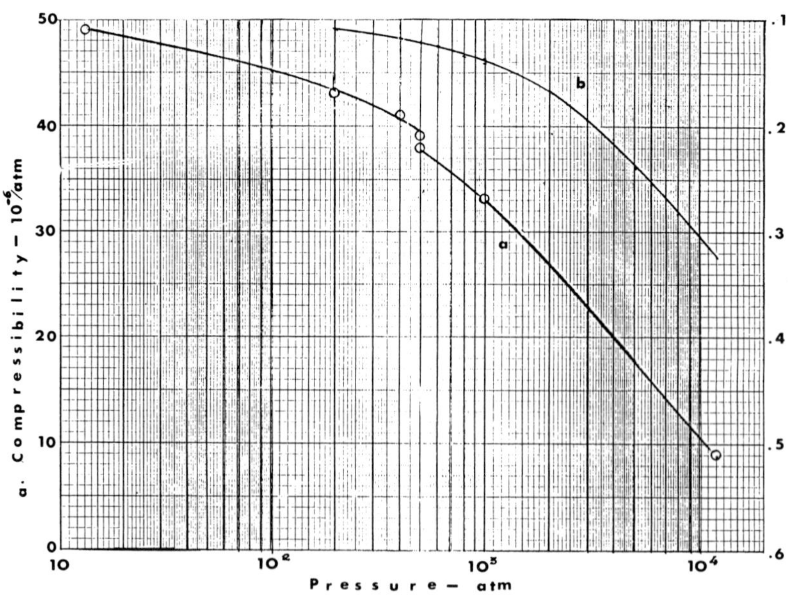

Fig. 8.3.4. Compressibility curve of water.

sodium, magnesium, potassium, calcium, etc., and an *anion* such as chlorine, sulfur, selenium, etc., join to form an organized array of ions, a *crystal*. An ion is an atom which has either a few extra electrons or a few missing ones. Those with the extra electrons like chlorine (anions) are negatively charged while those with missing electrons like sodium, calcium, magnesium (cations) are positively charged.

When placing salt crystals in water, the ions tend to break loose and to be attracted – the cations by the negatively charged oxygens and the anions by the positively charged hydrogens – until the whole crystal dissolves and the ions are evenly distributed in the water. The *ion concentration* n_∞ denotes the number of any particular type of ions contained in a cubic centimeter of a solution. The number of negatively charged ions is marked n_∞^- and the number of positively charged ions n_∞^+. Only in solutes of monovalent ions is the number of negatively charged ions equal to the number of positively charged ions, $n_\infty = n_\infty^+ = n_\infty^-$.

If several solutes, say i, are dissolved in water, the total *volume charge density* σ of the solution in e.s.u. per cm^3 is

$$\sigma = ez_\text{i}^+ n_\infty^+ + ez_\text{i}^- n_\infty^- \tag{8.4.1}$$

where e is the *electric charge* [4.8028×10^{10} e.s.u./electron] and z_i^+ and z_i^- are the valences of the positively and negatively charged ions, respectively.

Water may contain other substances and impurities as well, some even poisonous or radioactive. Most substances come in insignificant quantities, others in appreciable concentrations. Their sources are liquid waste of mines, industries, human sewage, and pesticides, both organic and inorganic. Air and other gases contained in water, in larger quantities, are of interest to us as they affect the compressibility and shear strength of soils. They are discussed in the next section.

8.5 Vapor Pressure

It was pointed out previously, specifically in Chap. 6, that soils will be considered a three-phase mixture. It was, however, emphasized that such division is arbitrary, and that additional phases present in the soil could be considered. We opted for three phases, the main constituents of the soil, in order not to complicate the already intricate formal presentation. We shall discuss the various manifestations of each phase, and in the present section the equilibrium between water and vapor is discussed.

Let us consider the classical example of a cylindrical vessel filled with water, with a snugly fitted piston, and a one-way valve which will release the air from the cylinder when the piston is lowered to come in contact with the water, and will be sealed off when it is pulled out of the cylinder. There is no air between the water and the piston and no air can enter between the piston and the wall of the cylinder.

If we pull the piston out of the cylinder under constant temperature, it becomes detached from the surface of the water and a *cavitation* occurs. We will notice boiling of the water, some of it evaporates and *steam* is built up in the space between the piston and the water surface. The quantity of steam that has been formed maintains a constant pressure, the *water vapor pressure*, independent of the volume of the water and a function of the temperature alone. By pulling the piston further out, the volume increases and the amount of steam increases proportionally. For any specific temperature there is, it appears, a constant amount of steam per volume that maintains that specific pressure. The volume containing the steam is said to be *saturated*. This is even more obvious when the motion of the piston is reversed. The volume of the steam decreases and, being compressed, the steam *precipitates*, i.e. transforms back to water, in a sufficient amount to maintain the volume containing the steam saturated at constant pressure. We say that there exists a dependence of the vapor pressure p_v on the temperature T

$$\mathcal{F}(p_v, T) = 0 \tag{8.5.1}$$

independent of the saturated volume. As the temperature of the liquid increases the mean kinetic energy of the molecules increases and a larger number of molecules get kicked out into the vapor phase.

In addition to the temperature there is another parameter on which the vapor pressure generally depends. It is the nature of the liquid involved. The general rule with respect to the various liquids is that in liquids in which the mutual attraction between the molecules is stronger their capacity to escape into the vapor phase is smaller. Since we are dealing only with water, this parameter is of lesser importance to us, nevertheless it is to be remembered that water solutes affect quantitatively the vapor pressure.

The values for the vapor pressure p_v as a function of temperature T, defined by Eq. (8.5.1), are given in Table 8.3.1 and shown in Fig. 8.5.1. The curve shown is often better presented by a semi-logarithmic representation, in which the logarithm of the pressure p_v is plotted against the reciprocal of the temperature T. Fig. 8.5.2 shows in a semi-logarithmic plot the same data shown in Fig. 8.5.1, the analytic expression of which is

$$\log p_v = \frac{\ln p_v}{2.303} = \frac{-\Delta h}{2.303\, RT} + C = \frac{-\Delta h}{4.58\, T} + C \tag{8.5.2}$$

where Δh is the heat required to transform one mole of water into vapor, $R = 1.987$ cal/(mole °K) is the gas constant and C is a constant depending on the units in which the vapor pressure is expressed.* An expression similar to Eq. (8.5.2) is obtained through thermodynamic considerations, and is known as the Clausius–Clapeyron equation.

* Eq. (8.5.2) is general for any liquid and its gassy phase, assuming it behaves as an ideal gas and Δh is independent of the temperature.

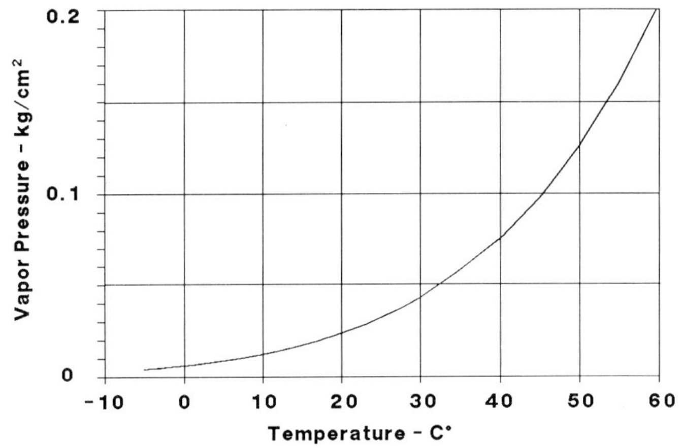

Fig. 8.5.1. Change of vapor pressure with temperature.

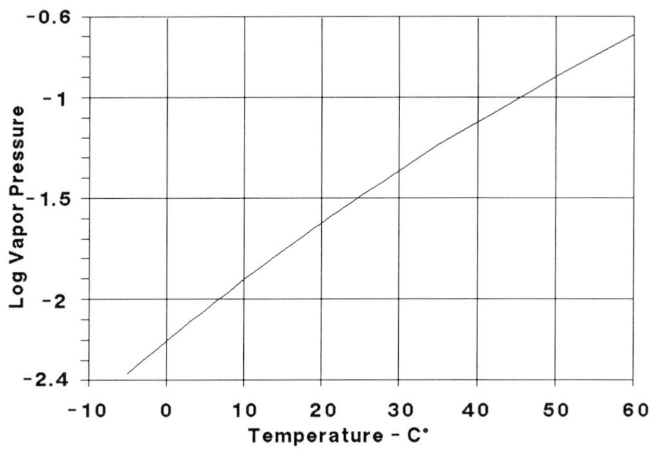

Fig. 8.5.2. Log vapor pressure versus reciprocal temperature.

The constant C is eliminated by evaluating Eq. (8.5.2) for two temperatures and subtracting one from the other

$$\log \frac{p'_v}{p''_v} = \log p'_v - \log p''_v = \frac{\Delta h}{4.58}\left(\frac{1}{T''} - \frac{1}{T'}\right) \tag{8.5.3}$$

For water at 100 °C ≡ 373 °K temperature, $\Delta h = 9700$ cal mole^{-1}, therefore

$$\log p_v = 5.605 - \frac{2090}{T} \quad \text{for } p_v \text{ in kg cm}^{-2} \text{ and } T \text{ in °K} \tag{8.5.4}$$

Eq. (8.5.4) gives the vapor pressure of water when saturated. Such a condition, however, very seldom occurs; in soil it occurs near the capillary zone. In general, the soil water is exposed to atmospheric conditions, particularly when the pores are continuous. In those pores, as in air in general, there is always moisture in the form of water vapor, in addition to the water in the form of liquid due to evaporation from water surfaces. These vapors act as any other gas and mix freely with other gases in the atmosphere, exchange energy by absorbing, retaining and releasing heat, and exert pressure. The moisture contained in the atmosphere can be conceived either in terms of *specific humidity*, which is the ratio of the weight of the water in grams, to the weight of the humid air containing the water in kilograms, or in terms of *relative humidity* R_H, which is the ratio of the partial pressure p exerted by the water vapor in the atmosphere to the vapor pressure of water p_v, for the same temperature. The relative humidity, expressed in percentage, is

$$R_H = \frac{p}{p_v} \times 100 \tag{8.5.5}$$

The relative humidity R_H approaches 100% as the partial vapor pressure p comes closer to the saturation vapor pressure p_v. Since the saturation vapor pressure p_v increases with temperature, the same amount of partial vapor pressure will result, in a higher temperature environment, in a lower degree of relative humidity. Thus, while at a given temperature the relative humidity may be less than 100%, if the temperature is lowered the relative humidity increases and can reach saturation, and the vapors condense and precipitate as liquid water.

8.6 The Air

The average composition of dry air is given in Table 8.6.1, but additional gases may be contained in water and consequently in soil. There are gases whose reaction with water is limited, like nitrogen (N_2), oxygen (O_2), hydrogen (H_2) and methane (CH_4). Others like carbon dioxide (CO_2), hydrogen sulfide (H_2S), sulfur dioxide (SO_2), and ammonia (NH_3) react with water and produce ions. The solubility of these and other gases in pure water at a pressure of 1 atm. is given in Table 8.6.2. The subscript T in Table 8.6.2

Table 8.6.1. Average composition of dry air

Gas	% by volume
Nitrogen (N_2)	78.08
Oxygen (O_2)	20.95
Carbon dioxide (CO_2)	0.03
Argon, etc.	0.94

[From T. R. Camp and R. L. Meserve (1974), courtesy of Dowden, Hutchison & Ross Inc.]

Table 8.6.2. Solubility of gases in pure water in contact with pure gas at a pressure of 1 atm

Temperature (°C)	Nitrogen 98.815% N_2 + 1.185% A		Oxygen O_2		Hydrogen H_2		Methane CH_4		Carbon dioxide CO_2		Hydrogen sulphide H_2S		Sulfur dioxide SO_2		Ammonia NH_3	
	$[N_2] \times 10^3$	mg/liter	$[O_2] \times 10^3$	mg/liter	$[H_2] \times 10^3$	mg/liter	$[CH_4] \times 10^3$	mg/liter	$[CO_2]_T \times 10^3$	mg/liter$_T$	$[H_2S]_T \times 10^3$	mg/liter$_T$	m_T	mg/liter$_T$	m_T	mg/liter$_T$
0	1.050	29.4	2.18	69.8	0.959	1.93	2.48	39.8	76.4	3360	208.3	7100	3.58	186 700	52.3	471 000
5	0.931	26.1	1.913	61.2	0.912	1.84	2.14	34.3	63.5	2790	177.4	6040	3.03	162 300	45.8	438 000
10	0.830	23.2	1.696	54.3	0.872	1.76	1.864	29.9	53.25	2345	151.6	5160	2.55	140 600	40.1	406 000
15	0.752	21.1	1.523	48.7	0.841	1.70	1.645	26.4	45.45	2000	131.4	4475	2.14	120 500	35.2	375 000
20	0.689	19.3	1.384	44.3	0.812	1.64	1.475	23.6	39.1	1720	115.2	3925	1.80	103 300	30.9	344 000
25	0.639	17.9	1.263	40.4	0.783	1.58	1.342	21.5	33.9	1495	101.8	3470	1.51	88 200	27.9	322 000
30	0.599	16.8	1.163	37.2	0.758	1.53	1.233	19.7	29.65	1305	90.9	3090	1.26	74 700	23.8	288 000
40	0.528	14.8	1.029	32.9	0.734	1.48	1.057	16.9	23.65	1040	74.1	2520	0.91	54 800	19.8	252 000
60	0.456	12.77	0.868	27.8	0.714	1.44	0.872	14.0	16.0	704	53.1	1810			14.0	192 000
80	0.427	11.96	0.786	25.1	0.714	1.44	0.79	12.7			40.9	1394			9.1	134 000
100	0.423	11.85	0.758	24.2	0.714	1.44	0.76	12.2			36.1	1230				

T signifies total solubility.
[From T. R. Camp and R. L. Meserve (1974), courtesy of Dowden, Hutchison & Ross Inc.]

indicates that this is the total solubility, including ionization products. At equilibrium with the atmosphere the solubility of the gases that do not react with pure water is proportional to the partial pressure of the gas according to *Henry's law*

$$\alpha'_g = p_g \mathcal{G}(T) \tag{8.6.1}$$

which states that at constant temperature *the quantity of solute gas absorbed by a liquid is proportional to the partial pressure of the free gas remaining above it*, where α'_g is the *concentration of the dissolved gas* and $\mathcal{G}(T)$ is the measure of solubility. The coefficient of proportionality is a measure of solubility of the gas, dependent only on the temperature (equal for the gas and the liquid), and is not affected by the presence of any other gases

$$\mathcal{G}(T) = \frac{\Delta h}{RT}; \quad \frac{\partial \mathcal{G}(T)}{\partial T} = -\frac{\Delta h}{RT^2} \tag{8.6.2}$$

where Δh is the heat absorbed during isothermal and isobaric evaporation of one gram molecule of the gas. For example, the solubility of O_2 in pure water at 20 °C exposed to dry air at barometric pressure of 1 atm. (Table 8.6.1) is $0.2095 \times 44.3 = 9.3$ mg/liter; and if the air is saturated with moisture (Table 8.3.1) the solubility is $(1 - 0.0231) \times 9.3 = 9.1$ mg/liter.

The principal source of dissolved nitrogen and oxygen in water is the air, but oxygen reaches the water also from plants through photosynthesis. The presence of hydrogen and methane in water is due primarily to decomposition of organic matter. Methane is present in swamps and coal mine drainage and persists in well waters in small quantities even after aeration.

8.7 Compressibility of Gases

The changes in the density, thus the compressibility of monatomic ideal gases has been the subject of scientific study since the days of Boyle and Gay-Lussac. The simple relation of the pressure p_g of a gas to its volume V_g and its temperature T is well established in the equation known as *Boyle's law*

$$p_g V_g = RT \tag{8.7.1.}$$

where $R = 8.314$ J mole^{-1} deg^{-1} is the *gas constant*.

As a result of theoretical and empirical developments, Eq. (8.7.1) has been further refined so as to take care of the singularities (Van der Waals 1873)

$$\left(p_g + \frac{a}{V_{gm}^2}\right)(V_{gm} - b) = RT \tag{8.7.2}$$

where $a = 27R^2T_c^2/(64p_{gc})$, $b = RT_c/(8p_{gc})$, V_{gm} is the *molal volume*, p_{gc} is the *critical pressure* and T_c is the *critical temperature* of the gas. The values of p_{gc}

and T_c are found in tables, for air $p_{gc} = 37.2$ atm. and $T_c = 132.5$ °K, while for water vapor $p_{wc} = 218.3$ atm. and $T_c = 647.4$ °K. The corresponding a and b values in Eq. (8.7.2) are $a_g = 0.5256\ \mathrm{N\,m^4\,mol^{-2}}$ $b_g = 3.6516 \times 10^{-5}\ \mathrm{m^3\,mol^{-1}}$ for air, and $a_w = 0.13583\ \mathrm{N\,m^4\,mol^{-2}}$ $b_w = 3.0417 \times 10^{-5}\ \mathrm{m^3\,mol^{-1}}$ for water vapor.

Like the Van der Waals equation, other equations have been proposed, partly theoretical partly empirical, some with six or eight parameters (Benedict et al. 1940; Redlich and Kwong 1949; Kennedy and Bhagia 1969), but none of them is applicable in the liquid region or in the liquid–gas boundary region.

An equation generally applicable to the gas region and more appropriate for engineering calculations at pressures above a few atmospheres and near the two-phase boundary is the *engineering gas law*, proposed by Standing and Katz (1942)

$$p_g V_{gm} = ZRT \tag{8.7.3}$$

where $Z = Z(p_{gr}, T_r)$ is the *compressibility factor*, a function of the *reduced pressure* p_{gr} and the *reduced temperature* T_r, defined $p_{gr} = p_g/p_{gc}$ and $T_r = T/T_c$, respectively. The review of the many attempts to correlate the compressibility factor, Eq. (8.7.3), with measured data (Leland and Chappelear 1968), reveals a similarity of behavior and of the ratio of critical pressures and temperatures for many fluid–gas phase changes. The compressibility factor chart by Viswanath and Su (1965), Fig. 8.7.1, is probably the best known chart used for simple gases. It predicts the compressibility of most common gases, with the exception of a very few like hydrogen, sulfur dioxide

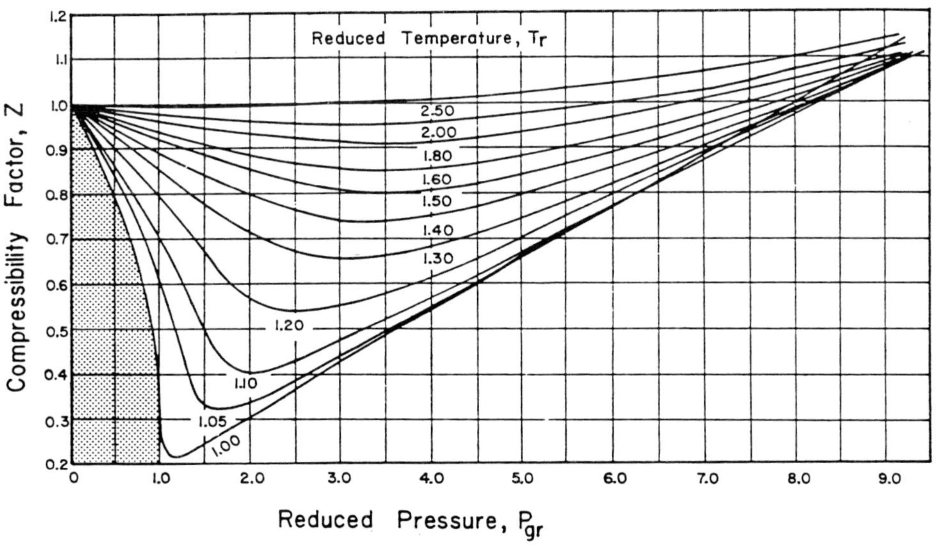

Fig. 8.7.1. Compressibility factor chart for pure gases. [From G. W. Govier and K. Aziz (1972), courtesy of Wadsworth Publishing Co.]

and hydrogen sulfite, within an average deviation of less than 2% and a maximum error of less than 10%.

8.8 Air-containing Pores

Soils containing air in their pores are *unsaturated*, that is, the pores are partly filled with air or gases produced by decomposition of organic matter. The literature developed since the late fifties treats unsaturated soils also under the name of *gassy* soils or the oxymoron *partly saturated* soils. Some really exciting investigations have been added in the past few years (Nageswaran 1983; Sills and Nageswaran 1984; Wheeler 1986; Thomas 1987) to previous studies (Aitchison 1956; Alpan 1961; Bishop and Blight 1963; Burland 1964; Barden 1965; Esrig and Kirby 1977; Fredlund 1975, 1985).

It has been mentioned in Sect. 8.3 that small amounts of air are usually dissolved in the water that fills the pores of the soil; it is, therefore, very hard to find completely saturated soils. Equally hard to find are dry fine-grained soils; as we shall see later, several molecular layers of water are always retained in most of the fine grained soils when dry. The energy required to extract this water in the course of drying the soil increases exponentially. In both instances the quantities are minute and in the range of 2%–3%. Between these two extremes of "completely" saturated soils and "completely" dry soils, that is, the pores of the soil are all filled up with water or with air, respectively, there are several possible configurations for the air to be accommodated along with the water within the pores.

1. The amount of air or gas in the gassy pockets is large enough not to be ignored, but not enough to form a continuous phase* throughout the soil, Fig. 8.8.1, and the bubbles formed are smaller than the voids, or pores, or

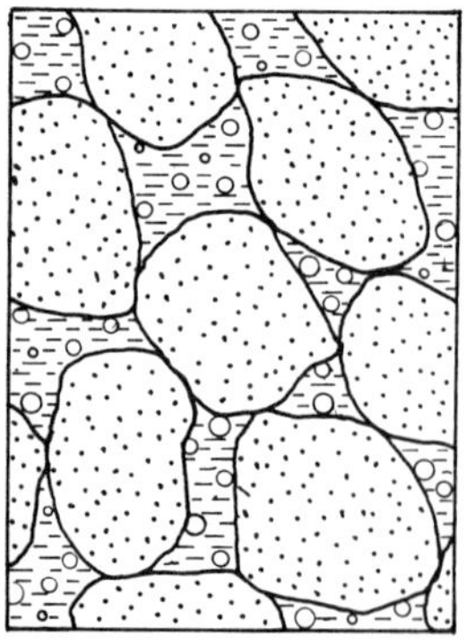

Fig. 8.8.1. Small gas bubbles.

* The fact that the gassy phase is not continuous does not mean that the soil–water–gas system cannot be considered a continuum; it means that a set of constitutive equations different from the one applied to a system with a continuous gassy phase will apply to it.

rather smaller than the constrictions of the pores. The air bubbles are spherical, and free to move with the water. The spherical shape of the bubbles allots the gas inclusions a minimal interface with the water for a given volume, and thus requires a minimal amount of energy in order to maintain that volume. If there is any movement of the water, however, these small bubbles collide with other small bubbles, because of the differences in their velocity due to their different sizes; this encounter of bubbles results sometimes in their fusion into larger size bubbles (see the next configuration). This phenomenon is found mostly in granular soils, where the considerable motion of water permits differences in the velocities of the bubbles.

2. The air inclusions are larger than the pore constrictions but still smaller than the distensions of the pores, so that the bubbles fit easily inside the pores and maintain still a spherical shape, but no longer pass the constrictions of the pores and thus cannot commute freely within the pores, Fig. 8.8.2. This type of air bubble configuration could be an original configuration of the soil, or a result of fusion of many smaller bubbles as they move in the pores. High pressures in the pore water may compress the bubbles to the size that will enable them to pass the pore constrictions, or, after a considerable time, they may be partly dissolved in the water. This latter possibility occurs usually in fine grained soils, where the time of dissolution of air in water relative to the time of water motion becomes larger.

3. The air bubbles are larger and the air contained in the voids is beyond 10%–15% of the porosity. They are not spherical in shape any more but follow, more or less, the contours of the voids, Fig. 8.8.3a. In granular soils the air inclusions may reach the surface of the grains and in fine grained soils they may be removed from the particle surfaces by many molecular layers of water, Fig. 8.8.3b. In spite of their larger size, the air inclusions do not yet form a continuous phase.

4. As soil becomes drier and its degree of saturation decreases, the air contained in the soil pores forms a continuous phase. This configuration is more likely to occur in granular soils and silts, at 10%–30% degree of saturation and slightly above the shrinkage limit. The water recedes to the points of contact of the particles, forming *menisci* at the air–water interface.

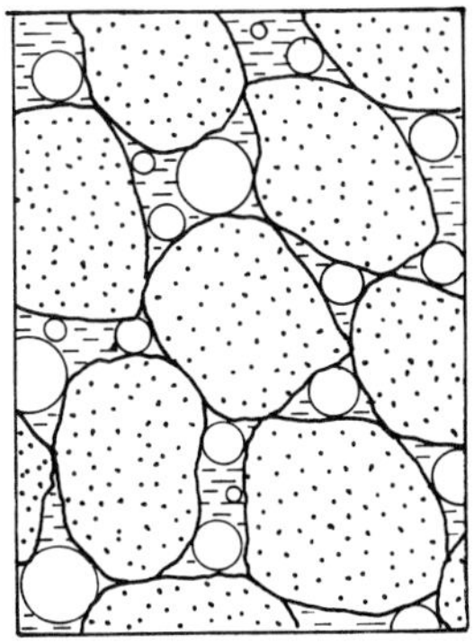

Fig. 8.8.2. Large gas bubbles.

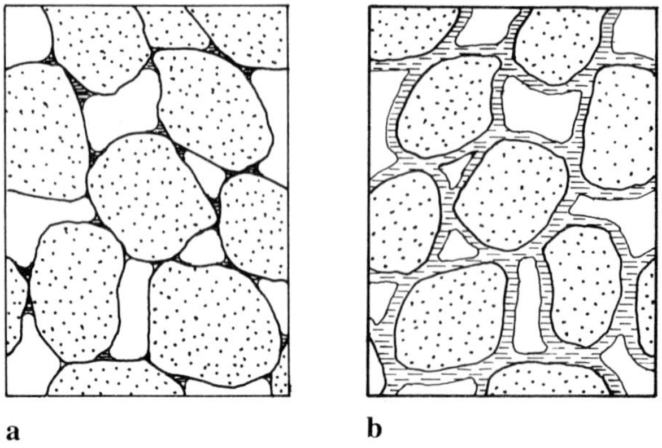

Fig. 8.8.3. Large gas inclusions; **a** Granular soils; **b** Fine grained cohesive soils.

In clayey soils such a configuration occurs at very low degrees of saturation, if at all. In some highly plastic clays, since the attractive forces between the clay particles and the water film attached to them is far larger than the tensional forces between the water molecules, a split occurs in the water long before it recedes to the narrow constrictions between the particles. This process tears the soil apart and results in a *crack*, which now forms a continuous air phase. Continuous air phases, including cracks, have the property that they allow free movement of air and water in the soil. Cracks, however, are different from the configuration of the continuous air phases in that they become larger with the decrease of the amount of water in the soil, and smaller with the absorption of water. In a field with a low water table, or no water table at all, cracks may go as deep as 6–8 meters below the soil surface, perhaps even more.

5. A seemingly unreasonable configuration but still a possible one, is that of large air inclusions surrounded by particles smaller than the air inclusion itself, Fig. 8.8.4. This configuration is encountered in cohesive soils, where the cohesive forces can prevent the particles from caving in. Compacted soils, or undisturbed soils where gas pressures maintain large inclusions, are examples of such a configuration.

In the above discussion on the different configurations, and particularly 1, 2 and 5, the impression may be that the air bubbles in any one configuration are all of the same size. In fact, within each configuration there is an array of sizes, statistically distributed throughout the voids. The determination of the distribution of the bubble sizes is not as simple as it would seem, nor is the implementation of the obtained values straightforward. The study of the microstructural behavior of air inclusions in unsaturated soils serves to understand the limitations of the phenomenological investigation, and to formulate as intelligently as possible the constitutive equations and boundary conditions of these soils.

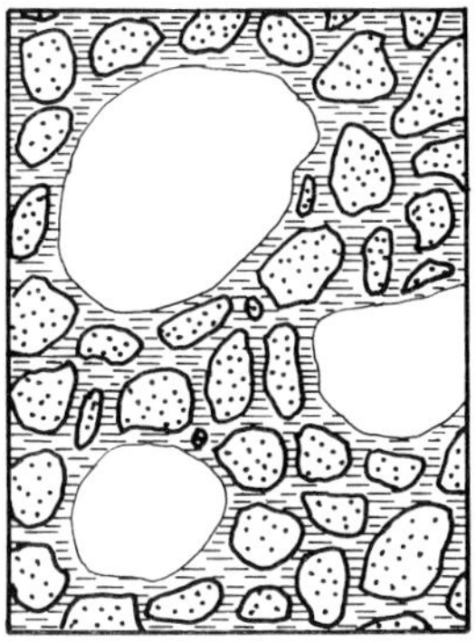

Fig. 8.8.4. Gas inclusions in compacted fine grained cohesive soils.

8.9 The Solid Particles

We shall not discuss here the origin and mineralogical composition of the various soils and their formation by mechanical or chemical weathering, their classification, or the tests used for classification. Those aspects of soils are well documented and we assume that the reader is adequately versed in the various publications.

We will discuss, however, one aspect of soil, namely its texture, and we will have to devote the necessary time to clay minerals, their physico-chemical interaction and their interaction with the aqueous surroundings, in order to understand the intrinsic forces that make soils so different from one another.

From the point of view of texture we distinguish two extremes, *granular* soils and *fines*. Between these two extremes there exists a whole array of soils of infinite variety, according to their origin, mineralogy, weathering, granular-clay content ratio and environment, forming a spectrum from gravels through sands of various coarseness, to silts and clays. It is customary to consider the division between the granular particles and the fine particles at the 74 μm size (no. 200 mesh), granular particles (gravels and sands) being larger than 74 μm and fine grained soils (silts and clays) below that size.

There are extreme differences between the mechanical behavior of granular soils and that of clays, and we maintain that all these differences are quantitative rather than qualitative. The permeability, for instance, may differ by up to ten orders of magnitude. Clays display explicitly non-linear effects, for instance a threshold gradient, however we may say that granular soils have a threshold gradient as well, except that it becomes reduced to zero. This assumption greatly simplifies the equations of granular soils, but that does not mean that these soils should be considered essentially different. The rate of consolidation is proportional to permeability; the bearing capacity of soils is a function of the rate of loading which again is a function of permeability. All these functional dependencies are expressible quantitatively, and in that respect all soils may be dealt with using the same physical equations.

There is, however, one particular property of soils that distinctly marks the difference between granular soils and clays, and it is related to the internal bonding between the particles of soils. Granular soils fail to exhibit bonding between their particles, and clay particles do interact by internal ties. We denote the internal forces that cause soil particles to be tied together as *cohesion*. Soils that lack these forces we call *non-cohesive* soils and soils that possess binding and adhesive forces we call *cohesive* soils.

We could surmise now that because of cohesion, the distinct property that differentiates between granular and clayey soils, the two types should be subjected to different mathematical treatments. We will show that the mathematical treatment is the same, and that for granular soils it becomes less complicated due to simpler assumptions. Thus, even of cohesion we can say that it is a property which exists in all soils, except that in granular soils its value diminishes and becomes zero.

In the few examples cited we see that the differences between non-cohesive (granular) and cohesive soils are quantitative. Cohesive soils such as clays are more complex in their mechanical behavior, and demand more complicated mathematical treatment than granular soils. Therefore, the behavior of the latter may always be derived from the behavior of clays, as a particular case with simplified conditions. Since this "complexity" is a result of cohesion, we will try in the following sections, after discussing granular soils, to reveal the origin of the internal forces responsible for the cohesion and the many other associated phenomena.

8.10 Specific Surface

All material particles, including soil particles, carry electric charges on their surfaces. The intensity of these charges is specific to the material and their *charge density* σ is measured in electrostatic units per area (e.s.u./m^2).* In general, these unbalanced charges are neutralized by ions from the surrounding environment. In the air, for instance, oppositely charged fine dust particles are attracted and attached to the material surfaces. In water, oppositely charged ions contained in the water solute are attracted to the surfaces.

Suppose we have a charged cubically shaped material with edges of one meter, and with a charge density σ e.s.u./m^2, then the total charge acting on the cube will be 6 σ e.s.u. If the cube is divided into 10 equal units in each direction, the total charge on all surfaces will increase tenfold, to 60 σ e.s.u., if the charge density remains the same. By dividing each cube again into ten equal sizes the total charge increases again by ten, to 600 σ e.s.u., and so on by each division into ten smaller units in each direction the total charge increases by ten. The total charge will thus reach 6×10^9 e.s.u. per cubic

* 1 e.s.u/m^2 = 0.333×10^{-9} coul/m^2.

meter of particles of the size of 10 Å and the same surface density. We see that as the particles become smaller the total surface retained in a certain volume of material increases, and the charge increases very fast. Thus, the total surface of particles contained in a unit weight or volume plays an important role in assessing the charge of the material. We define the *specific surface* as the total surface of all soil particles contained in one gram of soil. There are other definitions in the literature for specific surface; one of them, in which the total surface is related to a unit volume of soil particles (Scheidegger 1963), can be linked to our definition.

8.11 The Mineralogical Structure of Clays

Particles with diameters smaller than 1 μm (10^{-3} mm) but larger than molecular size are *colloidal*, and surface forces become dominant, as may be expected, and override gravitational forces. Clay particles fall within this range.

We mentioned that we shall not discuss the origin and the processes involved in the formation of different soils. All these topics are well covered by many textbooks and publications (Legget 1962; Gillott 1968; Lof 1983). Most of the material concerning the mineralogical aspects of clays, their interaction and the interaction with the surrounding water can also be found in the publications (Barshad 1964; Marshall 1964; Grim 1968; Jackson 1969). But in order not to impair the completeness of our study and since the further treatment of the topic relies on the basics of clay structure, the following sections will treat with some latitude the mineralogical aspect of clays and the forces acting in the clay–water system.

Evidence of the crystalline mineral structure of clays dates back to the early twenties. X-ray diffraction of clay particles was introduced in 1923, and clay mineralogy has since become a developing discipline, leading to systematic classification and description of the various clay minerals.

In general, the minerals contained in clay soils are alumino-silicates, while quartz (SiO_2), calcite ($CaCO_3$) and other minerals are found in granular soils. Two structural elements are fundamental in clay minerals: the silicon–oxygen (SiO_4) tetrahedron, Fig. 8.11.1a, and the aluminum–hydroxyl ($Al(OH)_6$) octahedron, Fig. 8.11.2a. The silicon has four positive valences and is bonded tetrahedrally to four oxygen atoms, each twice negatively valent, as shown in Fig. 8.11.1a. The single tetrahedrons, which are unbalanced elements, form a two-dimensional tetrahedral layer with a hexagonal array, Fig. 8.11.3, so that each oxygen at the base of the tetrahedron is shared by two silicons, and only the oxygens at the apex of the tetrahedrons, pointing upwards, remain unbalanced. The layer carries a net negative charge. Similarly, the charges of the single octahedrons, where the aluminum is tri-valent positive and the six hydroxyls are single valent negative, are unbalanced, and the octahedrons tend to form an octahedral layer, Fig. 8.11.3, so that each aluminum is shared

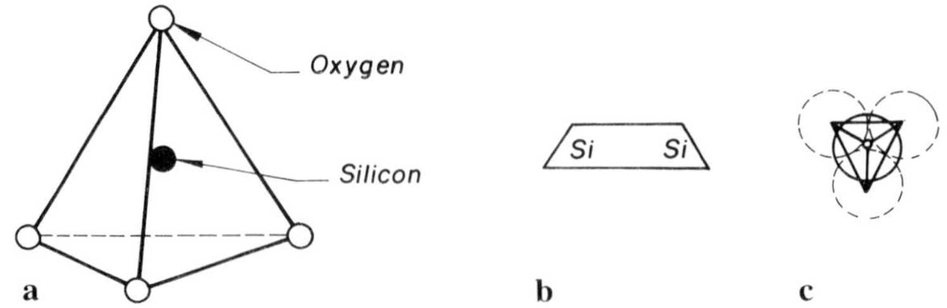

Fig. 8.11.1. Single silica tetrahedron. **a** Structure. **b** Schematic display. **c** Top view.

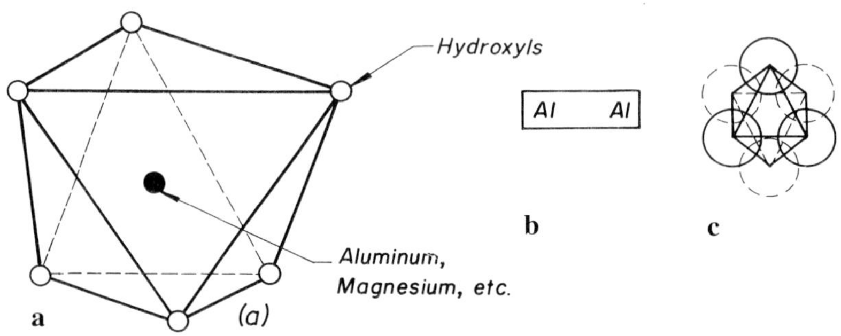

Fig. 8.11.2. Single aluminum and magnesium octahedron. **a** Structure. **b** Schematic display. **c** Top view.

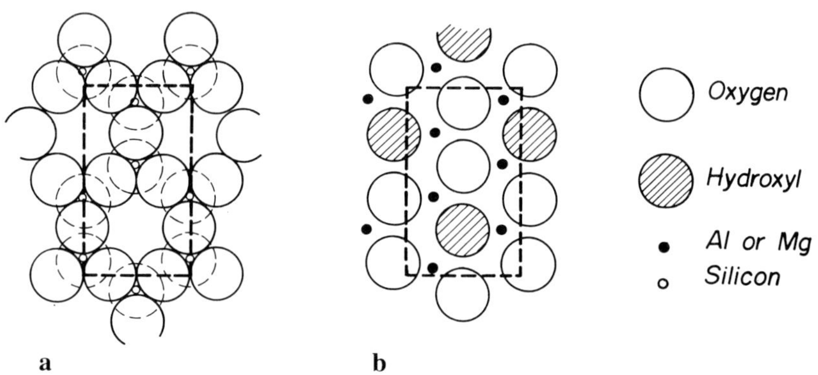

Fig. 8.11.3. Tetrahedral and octahedral layers. **a** Top view of the tetrahedral layer. **b** Top view of the octahedral layers.

by two hydroxyls. The octahedral layer is still unbalanced, having a net positive charge.

The two layers, the tetrahedral layer and the octahedral layer, form the building blocks or rather building layers of most common clays. The layers described here are *perfect* layers. The perfect tetrahedral layer is a *silica* crystal and the perfect octahedral layer is a *gibbsite* crystal, both *single-layer* crystals. Imperfections in the crystal lattice result from *isomorphous substitutions*, i.e. replacement of the quadri-valent silicon (Si^{4+}) in the tetrahedral layer by a tri-valent aluminum (Al^{3+}) and the tri-valent aluminum in the octahedral layer by iron (Fe^{3+}), magnesium (Mg^{2+}), chromium (Cr^{2+}), zinc

(Zn^{2+}), lithium (Li^{+}) or another ion. In a similar way, some of the hydroxyls in the octahedral layer may be substituted by oxygen atoms. Mostly an atom of lower positive valence replaces one of higher positive valence, resulting in a higher net negative charge of the layer. *Brucite* is another single-layer mineral where magnesium replaces all aluminums in the octahedral layer.

Isomorphous substitution not only changes the electric charges of the layers and their density, it also changes the affinity of the layers to other layers and ions, resulting in the combinations that form the different clays with different properties. Kaolinite, halloysite, serpentine are *two-layer* clays, a tetrahedral layer and an octahedral layer, whereas vermiculite, talc, illite, montmorillonite, etc. are *three-layer* clays with an octahedral layer sandwiched between two tetrahedral layers. In general, clay minerals are *platelets* of one, two, three or more layers. The forces that hold the layers together within the platelet are short-range forces of secondary bonds, also known as *Van der Waals forces*, that arise from internal dipoles of the molecules, and their range is from fractions of a kcal/mole up to about 10 kcal/mole for *hydrogen bonding*. Hydrogen bonding deserves special mention: unlike other atoms, the hydrogen proton is apparently not shielded enough by the single revolving electron, therefore it can be more strongly attracted to the electrons of other atoms.

The clay platelets have an approximate thickness of 3–14 Å, depending on the clay. Their length and width are 10–100 times their thickness (100–1000 Å).

There are clay minerals in which a number of platelets are piled up to form stacks of platelets, bound permanently together by secondary bonds due to low electric surface charges as in kaolinite, a two-layer mineral, and in talc, a three-layer mineral, or held together by interstitial ions that become part of the mineral. Muscovite and illite, for instance, have fixed potassium ions between their three-layer platelets and vermiculite has magnesium ions between its three-layer platelets.

For geotechnical engineering purposes clay minerals may roughly be classified into four groups:

1. *Kaolin* including kaolinite, halloysite, endellite, and some less frequent clays like allophane, anauxite, dickite, etc.
2. *Illite* including illite and muscovite clays.
3. *Montmorillonite* including montmorillonite, nontronite, talc, pyrophyllite, hectorite and some rarer minerals like beidellite, saponite, sauconite, etc.
4. *Miscellaneous minerals* including attapulgite, chlorite, vermiculite, sepiolite, sericite, glauconite, etc.

Kaolinite, illite, and montmorillonite are the three clay minerals most frequently encountered in soils in the field, appearing as mixtures in varying quantities. Other minerals occur only in traces except for specific locations where occasionally concentrated larger deposits are found and mined for commercial and engineering purposes. Table 8.11.1 shows a list of some of the clay minerals and their schematic structure.

Table 8.11.1. Data on clay minerals

Mineral	Isomorphous substitution (nature and amount)	Linkage between sheets (type and strength)	Specific surface (m^2/g)	$\frac{1}{\text{Charge density}}$ ($Å^2$/ion)	Potential exchange capacity (me/100 g)	Actual exchange capacity (me/100 g)	Particle shape	Particle size
Serpentine	None	H-bonding + secondary valence			1	1	Platy or fibrous	
Kaolinite	Al for Si, 1 in 400	H-bonding + secondary valence	10–20	83	3	3	Platy	$d = 0.3$ to $3\ \mu m$ thickness $= \frac{1}{3}$ to $\frac{1}{10}d$
Halloysite ($4H_20$)	Al for Si, 1 in 100	Secondary valence	40	55	12	12	Hollow rod	$OD = 0.07\ \mu m$ $ID = 0.04\ \mu m$ $L = 0.5\ \mu m$
Halloysite ($2H_2O$)	Al for Si, 1 in 100	Secondary valence	40	55	12	12	Hollow rod	$OD = 0.07\ \mu m$ $ID = 0.04\ \mu m$ $L = 0.5\ \mu m$
Talc	None	Secondary valence			1	1	Platy	
Pyrophyllite	None	Secondary valence			1	1	Platy	
Muscovite	Al for Si, 1 in 4	Secondary valence + K linkage			250	5–20	Platy	
Vermiculite	Al, Fe, for Mg; Al for Si	Secondary valence + Mg linkage	5–400	45	150	150	Platy	$t = \frac{1}{10}d$ to $\frac{1}{0.3}d$
Illite	Al for Si, 1 in 7; Mg, Fe for Al; Fe, Al for Mg	Secondary valence + K linkage	80–100	67	150	25	Platy	$d = 0.1$ to $2\ \mu m$ $t = \frac{1}{10}d$
Montmorillonite	Mg for Al, 1 in 6	Secondary valence + exchangeable ion linkage	800	133	100	100	Platy	$d = 0.1$ to $1\ \mu m$ $t = \frac{1}{100}d$

Table 8.11.1 *Cont.*

Mineral	Isomorphous substitution (nature and amount)	Linkage between sheets (type and strength)	Specific surface (m^2/g)	$\frac{1}{\text{Charge density}}$ ($Å^2$/ion)	Potential exchange capacity (me/100 g)	Actual exchange capacity (me/100 g)	Particle shape	Particle size
Nontronite	Al for Si, 1 in 6	Secondary valence + exchangeable ion linkage	800	133	100	100	Lath	$l = 0.4$ to $2\ \mu m$ $t = \frac{1}{100} l$
Chlorite	Al for Si, Fe; Al for Mg	Secondary valence + brucite linkage	5–50	700	20	20	Platy	

[After T. W. Lambe and R. V. Whitman (1969), courtesy of John Wiley & Sons, Inc.]

8.12 Electric Charges and Exchange Capacity

The mineral particles carry a net unbalanced electric charge arising from two sources:

1. Imperfections in the crystal lattice, due to isomorphous substitutions, i.e. ions of lower valency replacing ions of higher valency. Occasionally, in the tetrahedral layer tri-valent aluminum ions replace the quadri-valent silicon and in the octahedral layer magnesium, iron, as well as other metallic ions replace the aluminum, resulting in a negative charge uniformly distributed over the surface of the clay platelets.
2. The tetrahedral and the octahedral layers within the platelets, which are continuous in two directions, are disrupted at the edges of the platelets. Primary bonds being broken, the molecules near the edges are left unbalanced, resulting in edges that carry either positive or negative charges, depending on the pH of the water solution. As the pH of the water solution surrounding the platelets is high, the hydrogen ion concentration in the solution decreases and the unbalanced negative charges at the edges increase. By decreasing the pH of the water below a certain value the charges may even change to a positive value.

The unbalanced charges are compensated by the accumulation of an equivalent amount of ions of opposite sign from the solution surrounding the particles, keeping the whole assemblage *electroneutral*. These *counter-ions* can be replaced stoichiometrically by other ions of the same sign. The process of replacing the ions in the solution and those adsorbed on the particles is known as *ion exchange*, and the replaced ions as *exchangeable ions*.

Multiplying the charge density of the clay mineral by its specific surface per unit weight, the *exchange capacity* (EC) of the clay is obtained, and if the exchangeable ions are cations the *cation exchange capacity* (CEC) is obtained, that is, the amount of positively charged ions that a unit weight of the clay can attract from its aqueous surroundings, expressed in milli-equivalents per 100 grams (meq/100 g). The cation exchange capacity considers both charge deficiencies discussed here, the one caused by isomorphous substitution and the one caused by the disrupted edges. There is a fair correlation between the cation exchange capacity, the specific surface of the clay mineral and the particle size, Fig. 8.12.1.

Table 8.11.1 shows the cation exchange capacity of the clay minerals commonly encountered in soils. The cation exchange capacity of a soil containing several clay minerals in different quantities is the sum of the proportional amount of the individual exchange capacities of the constituent clay minerals.

Two clay minerals, kaolinite and montmorillonite, being located at the extremes of the CEC range, are the minerals most widely used by soil scientists in the basic research and investigation of clays.

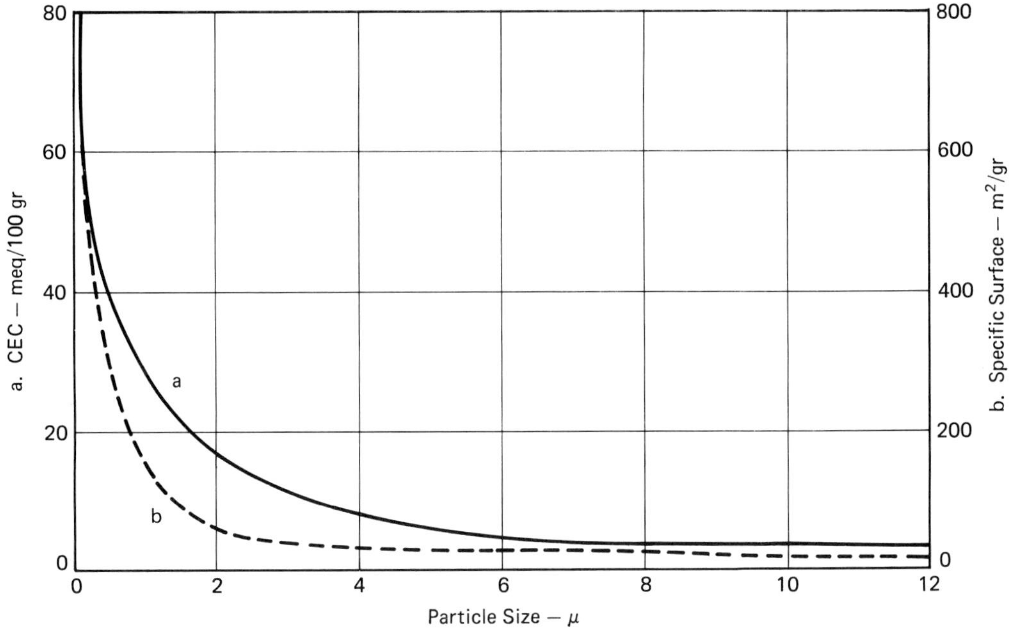

Fig. 8.12.1. Dependence of CEC and specific surface on the particle size.

8.13 The Diffuse Double Layer

Clay particles are always surrounded by a layer, however thin, of water, known as *adsorbed water*. The properties of the clay change greatly with the change in thickness of this layer. In the previous section it was pointed out that the clay particles have a net negative charge, counterbalanced by cations located on their surface. In the presence of water, these cations have a tendency to diffuse away from the surface, owing to differences in the concentration of the bulk solution. On the other hand, they are attracted electrostatically to the charged surfaces.

Thus, two competitive forces act on the cations in the water solution, in the neighborhood of the clay particles:

1. The electrostatical attraction between the negatively charged clay surfaces and the positively charged cations.
2. The tendency of the cations to back-diffuse from the clay particles into the solution, where the concentration of the solute is lower.

The result of this two-fold action is the distribution of cations in equilibrium in the form of a *diffuse electrical double layer*, their concentration decreasing away from the clay surface. Since this electric double layer is determined by the type and degree of the isomorphous substitution, there is a constant charge density on the mineral surface, independent of the presence of electrolytes in the solution.

Similarly, ions of negative charge are repelled by the clay surface and back-diffused from the bulk solution towards the clay surface, to a position where equilibrium is attained.

Gouy and Chapman (Gouy 1910, 1917; Chapman 1913) were the first to recognize this behavior of counter-ions and present a theory describing their distribution. This theory, known as the classical diffuse double-layer theory, yields a number of relationships useful in calculating the swelling pressure of a clay system, viscosity properties, movement of particles in dilute clay suspensions (*electrophoresis*) and that of a water solution in a concentrated clay (*electro-osmosis*). This theory was further developed by Langmuir (1938) and Verwey and Overbeek (1948), and refined by Stern (1924).

8.14 The Gouy–Chapman Double-layer Theory of Planar Surfaces

The Gouy theory, later extended by Chapman (Gouy 1910; Chapman 1913), considers the following factors which affect the two competitive forces acting on the counter-ions, the force of electrostatic attraction by the clay particles and the force of diffusion into the water solution:

1. The concentration of free electrolytes.
2. The valency of the ions.
3. The dielectric properties of the solvent (in the present case, of water).
4. The effect of the charge density per unit surface of the particles.

The theory was presented by Overbeek in great detail (Overbeek 1952) and only the essential concepts will be emphasized here, since this approach belongs to a different level of investigation of soils. A complete derivation of the theory is reproduced in several publications (Overbeek 1952; Van Olphen 1963; Yong and Warkentin 1966).

The equilibrium state, between the attractive and the repulsive forces acting on the counter-ions, is controlled by Poisson's equation

$$\frac{\mathrm{d}^2\Phi(x)_e}{\mathrm{d}x^2} = -\frac{4\pi\rho(x)}{D} \tag{8.14.1}$$

which relates the *charge density per unit volume of solution* $\rho(x)$, in e.s.u. per cm^3, defined

$$\rho(x) = z^+en(x)^+ - z^-en(x)^- \tag{8.14.2}$$

to the *electric potential* $\Phi(x)_e$ [e.s.u. of potentials], where x is the distance from the particle surface and D is the *dielectric constant* (dimensionless if charges are in e.s.u.).

The assumptions made in the theory are:

1. The charges on the particle surfaces are continuous and uniformly distributed.
2. The charged surface is of infinite extent.
3. The dielectric constant D is constant and uniform throughout the solution.
4. The distribution of the ions in the solution is according to *Boltzmann's distribution*

$$n(x)^{+} = n(\infty)^{+} \exp \frac{z^{+} e\Phi(x)_e}{kT} \tag{8.14.3}$$

$$n(x)^{-} = n(\infty)^{-} \exp \frac{z^{-} e\Phi(x)_e}{kT} \tag{8.14.4}$$

where $n(x)^+$ and $n(x)^-$ are the *concentrations* of the positive and negative ions, respectively, at a distance x from the particle surface (number of ions per cm^3), $n(\infty)^+$ and $n(\infty)^-$ are the concentrations farther out in the bulk solution (number of ions per cm^3), z is the *valency* of the ions, e is the *electronic charge* [4.8028×10^{-10} e.s.u. per electron], k is the *Boltzmann constant* [1.3805×10^{-16} erg per °K], and T is the *absolute temperature* [°K].

The validity of some of the above assumptions has been contested in the past (Bolt 1955; Low 1961; Babcock 1963; Shainberg and Kemper 1966a), nevertheless the theory has survived and has been extended and applied, and its ramifications yield fair correlation with experimental results.

In a homoionic and symmetrical electrolyte solution, $z^- = z^+ = z$, it is fair to assume that far out in the solution the potential vanishes, $\Phi(\infty)_e = 0$, and the concentration of the anions is equal to that of the cations, $n(\infty)^- = n(\infty)^+ = n(\infty)$.

Substituting Eqs. (8.14.2)–(8.14.4) into (8.14.1), the fundamental differential equation of the *Gouy–Chapman diffuse double-layer theory* is obtained

$$\frac{d^2\Phi(x)_e}{dx^2} = \frac{4\pi nze}{D}\left[\exp\frac{ze\Phi(x)_e}{kT} - \exp\frac{-ze\Phi(x)_e}{kT}\right]$$

$$= \frac{8\pi nze}{D}\sinh\frac{ze\Phi(x)_e}{kT} \tag{8.14.5}$$

Introducing the following dimensionless quantities

$$\xi = \sqrt{\left(\frac{8\pi e^2 z^2 n(\infty)}{DkT}\right)}\, x = \varkappa x \tag{8.14.6}$$

and

$$Y(\xi) = -\frac{ze\Phi(\xi)_e}{kT} \tag{8.14.7}$$

Eq. (8.14.5) can be rewritten in a more convenient dimensionless form

$$\frac{d^2 y}{d\xi^2} = \sinh y \tag{8.14.8}$$

$$\frac{1}{\varkappa} = \sqrt{\left(\frac{DkT}{8\pi e^2 z^2 n(\infty)}\right)}$$

may be looked upon as the *average thickness of the ionic double layer* measured in cm, a function of the nature of the electrolytes and their concentration alone.

To solve the differential equation (8.14.8), the following boundary conditions are known:

1. The electric potential at a distance far enough from the particle, say, at infinity vanishes; for $\xi = \infty$, $Y(\xi) = Y(\infty) = 0$.
2. The gradient of the electric potential at infinity vanishes as well; for $\xi = \infty$, $dy(\xi)/d\xi = dy(\infty)/d\xi = 0$.
3. At the particle surface the electric potential is at its highest value and depends on the clay and its surface charge density and on the properties of the electrolyte solution; for $\xi = 0$, $Y(\xi) = Y(0) = ze\Phi(0)_e/(kT) = Z$.

If the above boundary conditions are applied to Eq. (8.14.8), we obtain

$$\exp \tfrac{1}{2} Y = \frac{\exp \tfrac{1}{2} Z - 1 + (\exp \tfrac{1}{2} Z - 1) \exp - \xi}{\exp \tfrac{1}{2} Z + 1 + (\exp \tfrac{1}{2} Z - 1) \exp - \xi} \tag{8.14.9}$$

describing the decay of the electrical potential $Y(\xi)$ in the solution as a function of the distance ξ, provided the potential Z on the surface is known, Fig. 8.14.1.

By rearranging Eq. (8.14.9) and back transforming Eq. (8.14.7), after several simple algebraic manipulations Eq. (8.14.8) becomes

$$\frac{d^2\Phi_e}{dx^2} = \varkappa^2 \Phi_e \tag{8.14.10}$$

and the electric potential is obtained

$$\Phi_e(x) = \Phi_e(0) \exp(-\varkappa x) \tag{8.14.11}$$

the force field $F(x)$ per unit charge being

$$F(x) = \frac{d\Phi_e}{dx} = -\frac{2kT\varkappa}{ze} \sinh \frac{zex\Phi_e}{2kT} \tag{8.14.12}$$

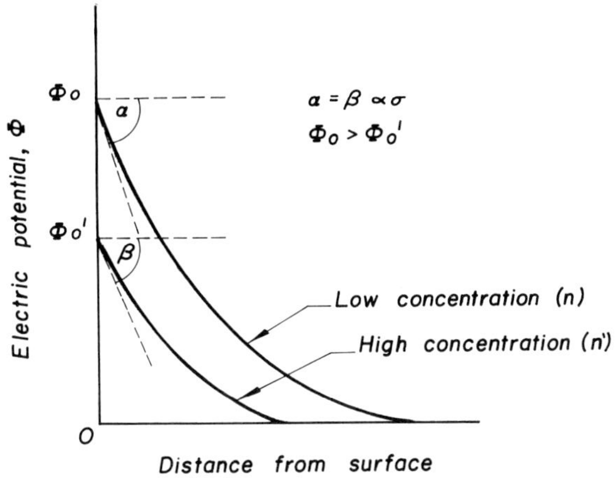

Fig. 8.14.1. Electric potential distribution in diffuse double layer with constant surface charge.

Also σ, the *total electric charge* per unit length of particle surface, is obtained from the *electric charge balance* equation

$$\sigma = -\int_0^\infty \rho(x)\,\mathrm{d}x = \frac{D}{4\pi}\int_0^\infty \frac{\mathrm{d}^2\Phi(x)_\mathrm{e}}{\mathrm{d}x^2}\,\mathrm{d}x = \frac{D}{4\pi}\left(\frac{\mathrm{d}\Phi(x)_\mathrm{e}}{\mathrm{d}x}\right)_{x=0}$$

$$= \frac{2n(\infty)DkT}{\pi}\sinh\tfrac{1}{2}Z = \frac{2n(\infty)DkT}{\pi}\sinh\frac{ze\Phi(x)_\mathrm{e}}{2kT} \qquad (8.14.13)$$

For small electric potentials Eq. (8.14.13) is approximated

$$\sigma \cong \left(\frac{n(\infty)D}{2\pi kT}\right)^2 ze\Phi(0)_\mathrm{e} = \frac{D\varkappa}{4\pi}\,\Phi(0)_\mathrm{e} \qquad (8.14.14)$$

Several conclusions can be drawn from Eqs. (8.14.13) and (8.14.14):

1. The electric potential at the particle surface increases approximately proportional to the surface charge density

$$\frac{\Phi(0)'_\mathrm{e}}{\Phi(0)''_\mathrm{e}} \cong \frac{\sigma'}{\sigma''} \qquad (8.14.15)$$

2. The electric potential at the particle surface increases inversely proportional to the square root of the concentration

$$\frac{\Phi(0)'_\mathrm{e}}{\Phi(0)''_\mathrm{e}} = \frac{\sqrt{n''}}{\sqrt{n'}} \qquad (8.14.16)$$

3. The electric potential at the particle surface increases approximately inversely to the valency of the counter ions

$$\frac{\Phi(0)'_e}{\Phi(0)''_e} \approx \frac{z''}{z'} \tag{8.14.17}$$

4. The slope of the electric potential at the surface is constant and depends on the surface charge density and the dielectric constant alone. From Eqs. (8.14.12) and (8.14.14) we have

$$\left(\frac{d\Phi_e}{dx}\right)_{X=0} = -\frac{2kT\varkappa}{ze}\sinh\frac{ze\Phi(0)_e}{2kT} = -\frac{8\pi\sigma}{D} \tag{8.14.18}$$

8.15 Limitations of the Gouy–Chapman Theory

It has been remarked in the previous section that the validity of the Gouy–Chapman theory has been challenged, since there are several circumstantial conditions that have not been taken into consideration. The limitations of the theory, as well as corrections intended to diminish or even to cancel them, are reviewed here.

1. The sizes of the atoms involved in the isomorphic substitutions introduce distortions that result in warping the mineral surface, which the theory does not account for.

2. The size of the ions in the solution has not been considered. The size poses a problem in the proximity of the particle surface, particularly in high electrolyte concentrations where the ions have to be accommodated. It should be added that the diffuse layer close to the particle surfaces is compressed due to the strong attractive forces and the ions tend to deform. It has been pointed out (Stern 1924) that the size of the ion determines the closest distance it can come to the clay particle surface, and a correction was suggested for the electric potential to take into account that distance. The cations likely to serve as counter ions were classified (Shainberg and Kemper 1966a) according to whether their energy of hydration is sufficient to sustain a molecule of water between the clay surface and the first layer of adsorbed cations, and a correction of the electric potential was proposed accordingly.

3. Bolt came to the conclusion (Bolt 1955) that "in the case of cations as counter ions, dehydration (of the ions next to the surfaces) does not seem probable, unless the dielectric constant of the colloid exceeds that of water." In other words cations in water are always hydrated, and therefore the clay particle surface is always hydrated. Consequently he proposed a correction to include a hydrated layer next to the clay surface.

4. The assumption made by the theory that the mineral surface and the charges on it are continuous and of infinite extent is unrealistic. The tetrahedral and octahedral layers are of finite length. At their edges they are disrupted and, therefore, positively charged. The edges attract either anions

or hydroxyls from the water molecules, or result in edge-to-face contacts between particles. All this is not taken care of by the theory.

5. Since the particle surfaces are negatively charged and compensated by an excess of positively charged cations, the negatively charged ions (the anions) are repelled by the surfaces and excluded from the solution next to the particle surface. The extent of this phenomenon, known as *anion exclusion*, has been thoroughly investigated (Schofield 1947; Bolt and Warkentin 1958; De Haan and Bolt 1963; Edwards *et al.* 1965), and fair agreement with the diffuse double layer theory for purified clays was found, except for low concentrations of electrolytes.

6. It has also been found that the distribution of exchangeable ions in the diffuse layer takes place only on the external surfaces of the mineral particles, where the ions are free to diffuse. So, for instance, the surfaces in kaolinite, where several platelets are joined together to form the mineral, or in illite, where the potassium ions are fixed between the three-layer platelets, are internal surfaces where no diffuse double layer can develop. Also in vermiculite, where the distance between the platelets is of two molecular layers of water, no diffuse layer is expected. In that respect the diffuse double-layer theory is ideal for montmorillonites, particularly in a mono-valent solution.

7. It has been noticed experimentally (Norrish and Quirk 1954; Warkentin *et al.* 1957; Blackmore and Miller 1961; Shainberg and Otoh 1968) that divalent ions behave differently from monovalent ions. It was found that Ca-montmorillonite forms *packets* or *tactoids*, a stack of several (4–6) platelets with a water film of approximately two molecular layers, 4.5 Å, in the internal surfaces between the platelets. The Ca ions adsorbed on the internal surfaces do not form a diffuse double layer, which is formed in a Ca-montmorillonite only on the external surfaces (Van Olphen 1959; Blackmore and Miller 1961; Shainberg 1968). Moreover, it was found (Shainberg and Otoh 1968) that in a solution with different ions, say Ca and Na, there are preferred locations for the Ca and for the Na. Na cations prefer the external surfaces whereas Ca cations prefer the internal surfaces of the tactoids. Only at approximately 20% Na do the tactoids begin to break up, and by 40% Na the tactoids have already broken up and a complete *demixing* takes place.

8. The negative charges, although evenly distributed on the clay mineral surfaces, are located at intervals of approximately 9 Å. At the spots with negative charges the electric potentials are maximal, and halfway between the spots the electric potentials are minimal. Cations adsorbed on the mineral surface have to overcome an energy barrier when moving along the surface from spots of higher electric potential to spots of lower electric potential.

9. The theory describes the interaction between the clay mineral and its environment of electrolyte solution. The assumption of the theory is that either the solution contains a single clay platelet, or the concentration of clay platelets in the solution is so low that the clay particles are in suspension and the diffuse layers of any two platelets do not interact. Of course, the theory presents in a very tangible way the importance and the effects of all factors

involved in the clay–electrolyte interaction: the charge density of the clay mineral surface, the concentration of the solution, the nature of ions in the solution, their valency, the temperature and the dielectric constant of the system. It cannot, however, describe correctly the clay–electrolyte interaction of an array of interacting clay particles. In the next section the main features of the extension of the theory to such an array are presented.

In spite of the reservations enumerated, the diffuse double-layer theory has been applied with considerable success to montmorillonite clays adsorbed with alkali ions most common in natural soils, correlating it with experimental results of swelling, consolidation, rheological properties, etc. (Shainberg 1968; Klausner 1970b). In surveying the assumptions and the shortcomings of the theory, Shainberg (1970) concludes that two of the points raised above, points 2 and 8, need to be reconsidered and further studied: the fact that the theory does not consider the size of the ions, while the affinity of the clay for cations of larger sizes has been demonstrated (Gast 1969; Gast *et al.* 1969), and the fact that the charges on the mineral surface are located discretely and not continuously as assumed, about 50% of the adsorbed ions being located within 3 Å of the surface.

8.16 Two Interacting Surfaces in Electrolyte Solution

The theory of the diffuse double layer derived in the previous section presents the interaction of a single clay particle and an electrolyte solution environment. It was noted above in point 9 that in order to make the theory of diffuse double layer useful in solving practical problems, it has to be extended to include the effect of an array of particles rather than that of a single particle. The extension of the theory was advanced by Verwey and Overbeek (1948) and reproduced in several publications (Overbeek 1952; Van Olphen 1963; Yong and Warkentin 1966; Klausner 1970a–c). The applications are far-reaching and concern problems of flow, consolidation, and strength.

Consider two parallel platelets of the same charge density, brought by an external constraint to such proximity that their potential fields overlap, Fig. 8.16.1, the distance between the two platelets being denoted $2d$, equal to the extent of the double layer $2/\varkappa$. The external constraint may be generated by a mechanical, thermal, chemical, electro-chemical, etc., source. In this case the two double layers begin to interfere, a superposition of the two takes place and neither double layer can develop completely. We are looking for the electric potential Φ_e between the two interacting surfaces.

Since the surface charges remain unchanged, the potentials Φ_e of the surfaces retain the same value $\Phi(0)_e$ as for free surfaces. Because of the symmetry and the dissipative nature of the potentials away from the particle surfaces, the superimposed potentials will result in a minimum value halfway between the surfaces, at $x = d$.

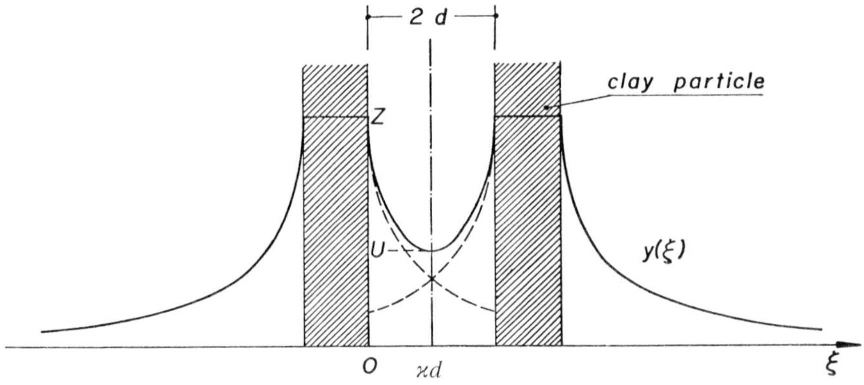

Fig. 8.16.1. Diffuse double layer between parallel platelets of sodium montmorillonite $N = 1$.

Starting with Poisson and Bolzmann equations, (8.14.1), (8.14.3) and (8.14.4.) and new boundary conditions

for $\xi = 0$ $\qquad Y(\xi) = Z = Y(0) = \dfrac{ze\Phi(0)_e}{kT}$

for $\xi = \varkappa d$ $\qquad Y(\xi) = U = Y(\varkappa d) = \dfrac{ze\Phi(d)_e}{kT}$

for $\xi = \varkappa d$ $\qquad \dfrac{dY}{d\xi} = 0$

where d is the *average half distance* between particles and U is the dimensionless electrical potential at $\xi = \varkappa d$ defined by

$$U = \frac{ze\Phi(d)_e}{kT} \tag{8.16.1}$$

one obtains, by integrating Eq. (8.14.5) and using the second and third boundary condition

$$\frac{dY(\xi)}{d\xi} = -\sqrt{(2\cosh Y - 2\cosh U)} \tag{8.16.2}$$

Separating the variables and integrating Eq. (8.16.2) between $0 < \xi < \varkappa d$, and then applying the first boundary condition, we obtain, after certain algebraic manipulations, the following relation (Overbeek 1952)

$$\begin{aligned} -\varkappa d &= -\int_0^d d\xi = \int_Z^U \sqrt{(2\cosh Y - 2\cosh U)}\, dY \\ &= 2\exp - \tfrac{1}{2}U\{F[\exp - U, \arcsin\exp\tfrac{1}{2}(U - Z)] \\ &\quad - F[\exp - U, \tfrac{1}{2}\pi]\} \end{aligned} \tag{8.16.3}$$

where

$$F(K, \varphi) = \int_0^{\psi} \frac{d\alpha}{\surd(1 - K^2 \sin \alpha^2)}$$

are elliptic integrals of the first kind, extensively tabulated. In Table 8.16.1 the values $\varkappa d$ are presented as functions of the midway potentials U, and the surface potentials Z. The computation of the distances, $\varkappa d$, for constant surface charges, is somewhat more elaborate.

The total surface charge, Eq. (8.14.13), is calculated from Eqs. (8.14.1), (8.16.1) and (8.16.2)

$$\sigma = -\int_0^d \rho(x)\, dx = -\frac{D}{4\pi}\left(\frac{d\Phi_e}{dx}\right)$$

$$= \sqrt{\left(\frac{n(\infty)DkT}{2\pi}\right)} \surd(2\cosh Z - 2\cosh U) \tag{8.16.4}$$

Since the surface charge in clays may fairly be assumed constant and since $DkT/2\pi$ is itself a constant, Eq. (8.16.4) yields

$$\surd n(\infty)\, \surd(2\cosh Z - 2\cosh U) = \text{const} \tag{8.16.5}$$

Eq. (8.16.3) has been tabulated by Verwey and Overbeek (1948) and reproduced (Overbeek 1952) for the midway potentials U and surface potentials Z, given surface charges σ, various electrolyte concentrations n, half distance between parallel platelets d, temperature T and dielectric constant D, see Table 8.16.2.

It was pointed out in Sect. 8.15 that calcium saturated montmorillonite consists of tactoids, a packet of 4–6 platelets with a distance of approximately 9 Å between them filled with water, rather than single platelets, Fig. 8.16.2. The tactoid formation reduces the amount of active interacting surfaces to the

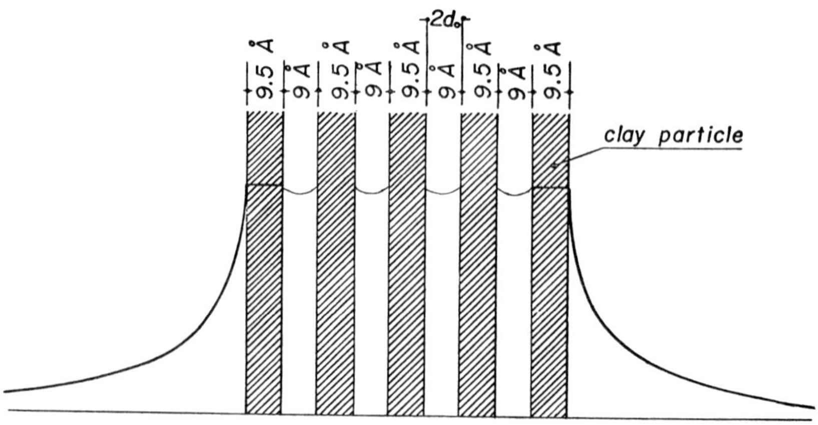

Fig. 8.16.2. Diffuse double layer between tactoids of calcium montmorillonite $N = 5$.

Table 8.16.1. $\varkappa d$ as a function of Z and U

$$Z = \frac{ze\psi(0)_e}{kT} \qquad U = \frac{ze\psi(d)_e}{kT}$$

	$Z - U = 0.1$	$Z - U = 0.3$	$Z - U = 0.6$	$U = 9$	$U = 8$	7	6	5	4	3	2	1	0.5	0.25	0.1
$Z = 10$	0.00434	0.00836	0.01337	0.02042	0.04374	0.08128	0.1429	0.2444	0.4117	0.6879	1.148	1.962	2.721	3.440	4.366
9	0.007324	0.01379	0.02208		0.03367	0.07211	0.1340	0.2356	0.4030	0.6792	1.139	1.953	2.712	3.431	4.357
8	0.01208	0.02273	0.03642			0.05551	0.1189	0.2210	0.3885	0.6647	1.124	1.939	2.698	3.417	4.343
7	0.0199	0.03748	0.06005				0.09154	0.1961	0.3644	0.6407	1.101	1.915	2.674	3.393	4.318
6	0.03275	0.06179	0.09900					0.1509	0.3232	0.6010	1.061	1.876	2.635	3.354	4.280
5	0.05410	0.10185	0.1632						0.2488	0.5333	0.9955	1.811	2.570	3.290	4.215
4	0.08915	0.1680	0.2692							0.4105	0.8837	1.702	2.462	3.181	4.107
3	0.1471	0.2774	0.4455								0.6813	1.518	2.280	2.998	3.924
2	0.2435	0.4643	0.7513									1.178	1.958	2.680	3.608
1	0.4353	0.8551	1.532										1.279	2.035	2.971
0.5	0.8706	1.558												1.309	2.241

From J. Th. G. Overbeek (1952), courtesy of Elsevier Science Publishers.

Table 8.16.2. Pairs of values of surface potential $-Z$ and midway potential $-U$ for three different values of the surface charge σ at various electrolyte concentrations $\varkappa$ and plate distances d.[a]

$\varkappa d$	$(dY/d\xi)_0 = 840$ $-U$	$-Z$
0.2555	5.0	13.47
0.1985	5.5	13.47
0.1540	6.0	13.47
0.1194	6.5	13.47
0.09249	7.0	13.47
0.07150	7.5	13.47
0.05516	8.0	13.47
0.04244	8.5	13.47
0.03253	9.0	13.48
0.02481	9.5	13.49
0.01881	10.0	13.50
0.01414	10.5	13.52
0.01052	11.0	13.55
0.00772	11.5	13.60
0.00557	12.0	13.67

$\varkappa d$	$(dY/d\xi)_0 = 560$ $-U$	$-Z$
0.2543	5.0	12.66
0.1973	5.5	12.66
0.1528	6.0	12.66
0.1182	6.5	12.66
0.09130	7.0	12.66
0.07032	7.5	12.66
0.05400	8.0	12.67
0.04126	8.5	12.67
0.03136	9.0	12.68
0.02366	9.5	12.70
0.01767	10.0	12.72
0.01304	10.5	12.77
0.00947	11.0	12.83
0.00674	11.5	12.93
0.00469	12.0	13.07

$\varkappa d$	$(dY/d\xi)_0 = 280$ $-U$	$-Z$
0.6943	3.0	11.27
0.5389	3.5	11.27
0.4181	4.0	11.27
0.3240	4.5	11.27
0.2507	5.0	11.27
0.1937	5.5	11.27
0.1493	6.0	11.27
0.1147	6.5	11.28
0.08776	7.0	11.28
0.06679	7.5	11.29
0.05049	8.0	11.31
0.03781	8.5	11.33
0.02799	9.0	11.37
0.02041	9.5	11.43
0.01460	10.0	11.52
0.01021	10.5	11.65
0.00697	11.0	11.84
0.00463	11.5	12.08
0.00301	12.0	12.39
0.00191	12.5	12.76
0.00120	13.0	13.16

$\varkappa d$	$(dY/d\xi)_0 = 187$ $-U$	$-Z$
0.8909	2.5	10.46
0.6907	3.0	10.46
0.5354	3.5	10.46
0.4145	4.0	10.46
0.3204	4.5	10.46
0.2472	5.0	10.47
0.1902	5.5	10.47
0.1458	6.0	10.47
0.1112	6.5	10.48
0.08428	7.0	10.49
0.06337	7.5	10.51
0.04713	8.0	10.54
0.03458	8.5	10.59
0.02493	9.0	10.67
0.01760	9.5	10.79
0.01213	10.0	10.95
0.00814	10.5	11.17

$\varkappa d$	$(dY/d\xi)_0 = 94$ $-U$	$-Z$
0.8803	2.5	9.09
0.6802	3.0	9.09
0.5248	3.5	9.09
0.4040	4.0	9.09
0.3099	4.5	9.10
0.2367	5.0	9.10
0.1798	5.5	9.11
0.1354	6.0	9.13
0.1010	6.5	9.16
0.07441	7.0	9.20
0.05390	7.5	9.27
0.03826	8.0	9.38
0.02653	8.5	9.53
0.01793	9.0	9.74

$\varkappa d$	$(dY/d\xi)_0 = 84$ $-U$	$-Z$
2.7112	0.5	8.86
1.9516	1.0	8.86
1.4792	1.5	8.86
1.1373	2.0	8.86
0.8778	2.5	8.86
0.6776	3.0	8.86
0.5223	3.5	8.87
0.4015	4.0	8.87
0.3074	4.5	8.87
0.2342	5.0	8.88
0.1773	5.5	8.90
0.1330	6.0	8.92
0.09871	6.5	8.95
0.07219	7.0	9.01
0.05184	7.5	9.09
0.03643	8.0	9.21
0.02497	8.5	9.39
0.01668	9.0	9.63
0.01087	9.5	9.92
0.06941	10.0	10.28
0.04357	10.5	10.68
0.02703	11.0	11.11

[a]The data are computed as follows:

(B.16.2) $dY/d\xi = -(2\cosh Y - 2\cosh U)^{1/2}$

For $Y = Z$ we obtain

$$\left(\frac{dY}{d\xi}\right)_0 = \frac{4\pi c v \sigma}{\epsilon k T \varkappa} = 1872\,\frac{\sigma}{\varkappa}$$

	$(dY/d\xi)_0 = 56$	
κd	$-U$	$-Z$
2.6993	0.5	8.05
1.9397	1.0	8.05
1.4673	1.5	8.05
1.1254	2.0	8.05
0.8659	2.5	8.05
0.6658	3.0	8.06
0.5105	3.5	8.06
0.3897	4.0	8.07
0.2958	4.5	8.08
0.2227	5.0	8.10
0.1660	5.5	8.13
0.1221	6.0	8.17
0.08834	6.5	8.24
0.06262	7.0	8.35
0.04334	7.5	8.51
0.02923	8.0	8.72
0.01923	8.5	8.99
0.01236	9.0	9.33

	$(dY/d\xi)_0 = 28$	
κd	$-U$	$-Z$
8.9158	0.001	6.67
6.6132	0.01	6.67
5.0031	0.05	6.67
3.6091	0.2	6.67
3.0366	0.35	6.67
2.6636	0.50	6.67
2.3826	0.65	6.67
1.9040	1.0	6.67
1.4317	1.5	6.67
1.0899	2.0	6.67
0.8305	2.5	6.68
0.6306	3.0	6.69
0.4756	3.5	6.71
0.3554	4.0	6.73
0.2623	4.5	6.77
0.1905	5.0	6.84
0.1357	5.5	6.94
0.09445	6.0	7.08
0.06409	6.5	7.28
0.04329	7.0	7.54
0.02739	7.5	7.86
0.01736	8.0	8.23
0.01084	8.5	8.65
0.00670	9.0	9.09
0.002513	10.0	10.03

	$(dY/d\xi)_0 = 18.7$	
κd	$-U$	$-Z$
8.8804	0.001	5.86
6.5778	0.01	5.86
4.2728	0.1	5.86
3.0013	0.35	5.86
2.4337	0.6	5.86
2.0521	0.85	5.86
1.8687	1.0	5.87
1.3965	1.5	5.87
1.0549	2.0	5.87
0.7959	2.5	5.98
0.5965	3.0	5.91
0.4423	3.5	5.95
0.3234	4.0	6.00
0.2322	4.5	6.09
0.1631	5.0	6.21

	$(dY/d\xi)_0 = 9.4$	
κd	$-U$	$-Z$
8.7760	0.001	4.50
6.4734	0.01	4.50
3.2424	0.25	4.51
2.5240	0.5	4.51
2.0868	0.75	4.51
1.7650	1.0	4.52
1.5080	1.25	4.52
1.2939	1.5	4.53
0.9540	2.0	4.56
0.6979	2.5	4.61
0.5029	3.0	4.69
0.3551	3.5	4.80
0.2449	4.0	4.96
0.1646	4.5	5.18
0.1079	5.0	5.47

	$(dY/d\xi)_0 = 8.4$	
κd	$-U$	$-Z$
15.659	10^{-6}	4.28
13.356	10^{-5}	4.28
11.054	10^{-4}	4.28
8.7513	10^{-3}	4.28
6.4487	0.01	4.28
4.1436	0.1	4.28
2.4993	0.5	4.29
1.7406	1.0	4.30
1.2700	1.5	4.32
0.9309	2.0	4.36
0.6760	2.5	4.41
0.4827	3.0	4.51

	$(dY/d\xi)_0 = 5.6$	
κd	$-U$	$-Z$
15.545	10^{-6}	3.51
13.242	10^{-5}	3.51
10.940	10^{-4}	3.51
8.637	10^{-3}	3.51
6.335	0.01	3.51
4.030	0.1	3.51
2.386	0.5	3.51
1.6292	1.0	3.54
1.1621	1.5	3.58
0.8289	2.0	3.66
0.5824	2.5	3.78
0.4003	3.0	3.94
0.2684	3.5	4.17
0.1756	4.0	4.45
0.1124	4.5	4.80

Furthermore,

$$\int_Z^U (2\cosh Y - 2\cosh U)^{-1/2}\, dY = -\kappa d \tag{B.16.3}$$

	$(dY/d\xi)_0 = 2.8$			$(dY/d\xi)_0 = 1.87$			$(dY/d\xi)_0 = 0.94$	
κd	$-U$	$-Z$	κd	$-U$	$-Z$	κd	$-U$	$-Z$
15.741	10^{-7}	2.28	15.476	10^{-7}	1.67	14.906	10^{-7}	0.91
15.231	10^{-6}	2.28	14.965	10^{-6}	1.67	14.396	10^{-6}	0.91
12.928	10^{-5}	2.28	12.663	10^{-5}	1.67	12.093	10^{-5}	0.91
10.625	10^{-4}	2.28	10.360	10^{-4}	1.67	9.790	10^{-4}	0.91
8.323	10^{-3}	2.28	8.058	10^{-3}	1.67	7.488	10^{-3}	0.91
6.020	10^{-2}	2.28	6.448	0.005	1.67	5.185	10^{-2}	0.91
4.410	0.05	2.28	5.755	0.01	1.67	3.576	0.05	0.91
3.715	0.1	2.28	4.145	0.05	1.67	2.883	0.1	0.91
2.077	0.5	2.30	3.450	0.1	1.67	1.303	0.5	1.02
1.3351	1.0	2.38	1.8213	0.5	1.72			
0.8936	1.5	2.52	1.1075	1.0	1.86			
0.5948	2.0	2.73						

external surfaces alone. The above theory of two interacting surfaces was extended and investigated experimentally for interacting tactoids (Klausner and Shainberg 1967, 1971).

8.17 The Work of Interacting Surfaces

Let $\Phi(x)_e$ be the electric potential of the diffuse double layer of two interacting surfaces at any point x, and $\rho(x)$ be the charge per unit volume, defined, according to Eqs. (8.14.2)–(8.14.4)

$$\rho(x) = 2zen(\infty)\sinh\frac{ze\Phi_e}{kT} \qquad (8.17.1)$$

where a symmetric electrolyte solution has been assumed with $z^+ = z^- = z$ and $n(\infty)^+ = n(\infty)^- = n(\infty)$. The work $d\mathfrak{w}$ per unit volume against the potential field along the path dx will be

$$d\mathfrak{w} = \rho(x)\, d\Phi(x)_e \qquad (8.17.2)$$

But the work per unit volume is also equal to an equivalent spherical

Pairs of values of U and Z are evaluated for the following conditions:

Normality of 1 – 1 electrolyte	0.964×10^{-5}	0.867×10^{-4}	0.964×10^{-3}	0.867×10^{-2}	0.964×10^{-1}	0.867
$\varkappa$ (cm^{-1})	10^5	3×10^5	10^6	3×10^6	10^7	3×10^7
$(dY/d\xi)_0$ for						
$\sigma = 1.5 \times 10^4$	280	94	28	9.4	2.8	0.94
$\sigma = 3.0 \times 10^4$	560	187	56	18.7	5.6	1.87
$\sigma = 4.5 \times 10^4$	840	280	84	28	8.4	2.8
e.s.u./cm^2						

for a series of values of d between 6 and 240 Å.

From Y. Van Olphen (1957), courtesy of John Wiley & Sons Inc.

pressure p existing at x and acting in all directions. In order to determine the function $\mathfrak{w}$, we shall resort to the transformations (8.14.6), (8.14.7) and (8.16.1) already used, as well as to Eq. (8.16.2). Consequently, Eq. (8.17.2) results in

$$\mathfrak{w} = -p(x) = -\int \rho(x)\, \mathrm{d}\Phi(x)_e = 2zen(\infty) \int \sinh(e/kT)\, \mathrm{d}\Phi(x)_e$$

$$= 2n(\infty)kT \int \sinh Y\, \mathrm{d}Y = 2n(\infty)kT \cosh Y + C$$

$$= 2n(\infty)kT \cosh(ze\Phi(x)_e/kT) + C \tag{8.17.3}$$

Another form of this equation is obtained by the substitution of Poisson's equation (8.14.1) in Eq. (8.14.12)

$$\mathfrak{w} = -p(x) = -\int \frac{D}{4\pi} \frac{\mathrm{d}^2\Phi(x)_e}{\mathrm{d}x^2} \mathrm{d}\Phi(x)_e = -\frac{D}{8\pi}\left(\frac{\mathrm{d}\Phi(x)_e}{\mathrm{d}x}\right)^2 + C \tag{8.17.4}$$

In both equations, (8.17.3) and (8.17.4), the difference between the pressure p which is a hydrostatic pressure and the stress generated by the electrostatic potential is a constant. This constant, according to Eq. (8.17.4), is equal to the pressure $p(d)_r$ midway between the two surfaces, where the potential gradient must be zero. This is a repulsive pressure, known as the *osmotic pressure* p_o, denoted in soil science literature by π. The osmotic pressure is equal to the increase of the repulsive pressure between two platelets, when the distance between them decreases from infinity to $2d$.

The repulsive pressure p_r at midplane between the two surfaces is obtained from Eq. (8.17.3), at the distance of $2d$ between the surfaces

$$-p(d)_r = 2n(\pi)kT \cosh \frac{ze\Phi(d)_e}{kT} + C = 2n(\infty)kT \cosh U + C \tag{8.17.5}$$

as well as for two surfaces infinitely apart $\Phi(\infty)_e = 0$

$$-p(\infty)_r = 2n(\infty)kT + C \tag{8.17.6}$$

The osmotic pressure p_o becomes, therefore

$$-p_o - [p(d)_r - p(\infty)_r] = 2n(\infty)kT[\cosh U - 1] \tag{8.17.7}$$

which represents also the work per unit volume required in order to bring two particles from infinity to a proximity of $2d$.

The work per unit surface of the particle is equal to the work of the excess repulsive pressure of two surfaces along the thickness of the diffuse double layer

$$\mathfrak{W}_r = \frac{p_o}{\varkappa} = \frac{2n(\infty)kT}{\varkappa}(\cosh U - 1) \tag{8.17.8}$$

The force per unit volume $f(x)$, or the pressure gradient in the diffuse double layer is

$$f(x) = \frac{\mathrm{d}\mathfrak{w}}{\mathrm{d}x} = \frac{\mathrm{d}Y}{\mathrm{d}\xi} = -2n(\infty)kT\varkappa \sinh Y(\xi) \times \sqrt{(2\cosh Y(\xi) - 2\cosh U)}$$

$$= -2n(\infty)kT\varkappa \sinh \frac{ze\Phi(0)_e}{kT} \times \sqrt{\left(2\cosh \frac{ze\Phi(0)_e}{kT} - 2\cosh \frac{ze\Phi(0)_e}{kT}\right)} \tag{8.17.9}$$

This gradient next to the particle surface follows from Eq. (8.17.8) and can be expressed also in terms of the surface charge σ, as given by Eq. (8.16.4)

$$f(0) = -2n(\infty)kT\varkappa \sinh \frac{ze\Phi(0)_e}{kT} \sqrt{\left(2\cosh \frac{ze\Phi(0)_e}{kT} - 2\cosh \frac{ze\Phi(0)_e}{kT}\right)}$$

$$= -2n(\infty)kT\varkappa \sinh Z \sqrt{(2\cosh Z - 2\cosh U)}$$

$$= -\frac{8\pi n(\infty)ez}{kT}\sigma \sinh Z \tag{8.17.10}$$

Eqs. (8.17.9) and (8.17.10) indicate change of pressure gradient across the diffuse double layer in the direction x, which may be responsible for many phenomena in the mechanical behavior of the clayey soil, and in the first place for the mobility of the water layers, containing adsorbed ions, in the soil pores.

8.18 Osmotic Pressure and Consolidation

Sects. 8.11–8.17 deal with soils in the molecular and submolecular level. The detailed study of this level is beyond the scope of the present work; it seemed nevertheless instructive to examine the forces and the resulting phenomena that emanate from this level, as an introduction to the study of the structural and phenomenological levels, which is our main purpose. In the following sections the relevant conclusions reached at the molecular and sub-molecular levels will be discussed.

The osmotic pressure p_o, Eq. (8.17.7), is a net repulsive pressure acting between two clay particles, a result of an electrostatic attraction and a ionic diffusion. Two particles at infinite distance between them have a zero osmotic pressure and zero repulsive pressure; if the distance between the particles is brought from infinity to $2d$, an osmotic repulsive pressure p_o will develop, corresponding to the distance $2d$. In order for that to happen some work has to be done on the particles.

When the distance between the particles decreases, water must escape from between the particles and the volume of the soil–water system decreases. By exerting an external constraint, which may be mechanical, thermal, chemical, electro-chemical, etc., the particles in a soil–water system are brought from one inter-particle distance $2d'$ to another inter-particle distance $2d''$, where $d' > d''$. We have already established that the process of getting from a position $2d'$ to a position $2d''$ involves a volume change due to the exit of water, squeezed out or extracted from the spaces between the particles, the pores. The result is a volume decrease of the clay–water system, a *compression*. This process is time consuming and by the time the particles advance from $2d'$ to $2d''$ their osmotic repulsive pressure gradually increases, until it reaches equilibrium at $2d''$ for the particular constraint.

Let p_{m} be a confining pressure applied at $2d' \equiv \infty$ at time $t = 0$, then the osmotic pressure p_{o} increases with the time $t \geq 0$ from $p_{\mathrm{o}} = 0$ when the particles are at infinite distance to $p_{\mathrm{o}} = p_{\mathrm{m}}$ when they are at the distance $2d''$, and attains equilibrium. The diffuse double-layer theory is an equilibrium theory, and as such it describes the soil only at equilibrium states and not between them. Introducing soil mechanics terminology, p_{m} is defined as the *consolidation pressure*, while $p_{\mathrm{o}} = p_{\mathrm{m}}$ is known as the *effective pressure* when the soil comes to equilibrium. If an additional constraint is applied, a new equilibrium position, perhaps at $2d''$, is attained, again a time-dependent process. The increase in the osmotic pressure as the particles move from $2d'$ to $2d''$, is derived from Eq. (8.17.7)

$$-\Delta p_{\mathrm{m}} = -[p(d'')_{\mathrm{m}} - p(d')_{\mathrm{m}}] = 2n(\infty)kT(\cosh U'' - \cosh U') \qquad (8.18.1)$$

The meaning of this equation is that, owing to a certain consolidation pressure increment Δp_{m}, two particles are drawn from a distance d' to a distance d'' and an osmotic-repulsive pressure increment develops, increasing the effective pressure by an effective pressure increment Δp_{o}.

In order to avoid a very common misunderstanding, the term "intergranular pressure" should not be taken literally. It serves to bring under a common denominator volumetric phenomena that occur both in granular soils and in fine grained clays. As is evident from the double-layer theory, clay particles can subsist under pressure without coming in contact with one another, since they are surrounded by several layers of water molecules, unlike in granular soils where the contact between particles is a real physical contact. Thus the osmotic repulsive pressure, which permits particles to reach proximity, acts as an "intergranular pressure", which we are so familiar with in granular soils, and has also the same affect.

The consolidation pressure p_{m}, and the effective pressure p_{o} are not always equal. The theory of diffuse double layer deals only with states of equilibrium, as explained before, and not with the processes that the system undergoes when passing from one state of equilibrium to another. In spite of the fact that the diffuse double-layer theory defines only equilibrium states, we may want to know the value of the osmotic pressure existing between the equilibrium states. Once an excitation is applied to the system, e.g. when a consolidation pressure p_{m} or, rather, a consolidation pressure increment Δp_{m}

is applied, the particles are drawn together. But this is a kinematic process, which is time-dependent, the distance $2d$ between particles decreasing monotonically. Only at equilibrium does the effective pressure or effective pressure increment become equal to the applied consolidation pressure or pressure increment, respectively. Because of the monotonically decreasing nature of the distance $2d$, such an equilibrium is reached theoretically only at a time that tends to infinity.

Experimentally, equilibrium is reached much faster and in practice it is determined, by convention, at a stage when the effective pressure increment rate is considered negligible. Fig. 8.18.1 presents a typical consolidation test showing curves of volumetric strain versus time of consolidation for a series of incremental loadings, in regular scale in Fig. 8.18.1a and in semi-logarithmic scale in Fig. 8.18.1b.

8.19 Pore Water Pressure

The consolidation process is an act of drawing together two particles, made possible by an outflow of water solution from the space between the clay platelets and accompanied by a volume decrease. Such an outflow occurs if a pressure gradient is established in the water solution, or can be postulated to exist in the water solution.

The difference between the consolidating pressure $p_{\mathfrak{m}}$ and the effective pressure $p_{\mathfrak{e}}$ during the consolidation process can be taken as the source of such a pressure gradient. Consequently, from Eq. (8.17.7) we obtain

$$p(d)_{\mathfrak{w}} = p_{\mathfrak{m}} - p(d)_{\mathfrak{e}} = p_{\mathfrak{m}} - 2n(\infty)kT(\cosh U - 1) \quad (d'' < d < d') \tag{8.19.1}$$

where $p(d)_{\mathfrak{w}}$ is the *pore water pressure* between the clay platelets.

Assuming the consolidating pressure $p_{\mathfrak{m}}$ constant, the effective pressure $p_{\mathfrak{e}}(d, t)$ is time dependent, going from zero to $p_{\mathfrak{m}}$ as consolidation proceeds from one equilibrium state to another. Consequently the pore pressure $p_{\mathfrak{w}}(d, t)$ is also time dependent, decreasing from the value $p_{\mathfrak{w}}(d', t) = p_{\mathfrak{m}}$ to $p_{\mathfrak{w}}(d'', t) = 0$, as consolidation progresses.

The pore water pressure is the actual driving pressure behind the process of consolidation. In the case of parallel interacting platelets, the double-layer theory does not consider states other than states of equilibrium. It is, therefore, a one-dimensional theory and it is also time independent. In order to enable the consolidation between two states of equilibrium to proceed, a pore pressure gradient must be postulated in the direction of the flow. The pore pressure gradient, being a vectorial quantity, can have one or two non-vanishing components in the direction parallel to the platelets, while the component normal to the platelets is zero. The pressure gradient has the form

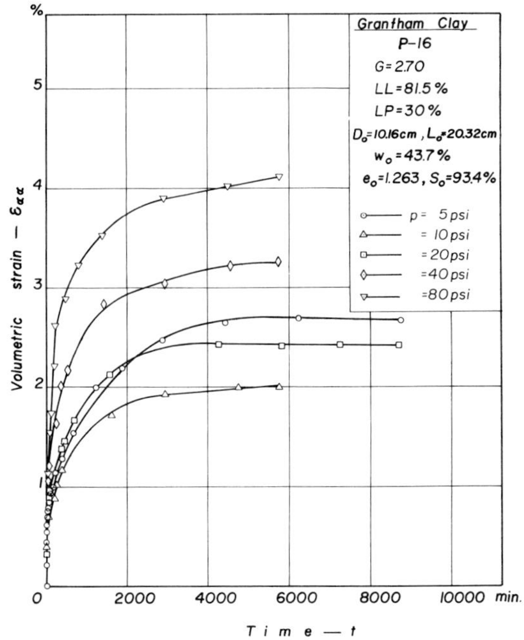

a

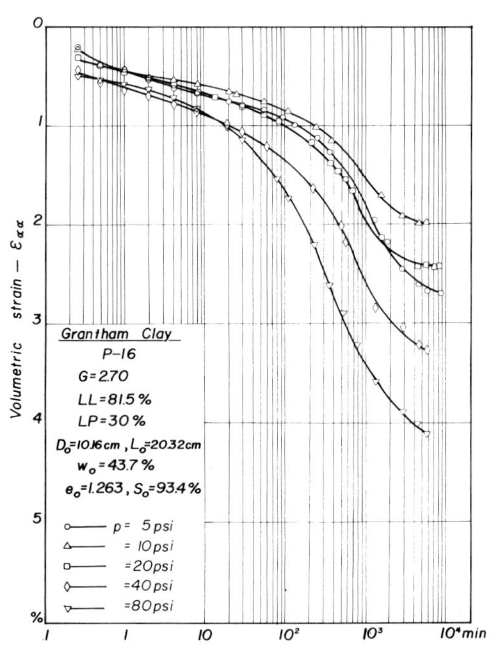

b

Fig. 8.18.1. Triaxial consolidation. **a** Volumetric strain versus time. **b** Volumetric strain versus log time.

$$p_{w,i} = \frac{dp_w}{dx_i} = 0 \quad \text{for } x_i = z \quad \text{and } p_{w,i} = 0 \quad \text{for } d'' < d < d' \qquad (8.19.2)$$

This extension of the theory has not been further developed as yet.

The diffuse double-layer theory is a two-phase theory concerning the interaction of clay particles and water solution. So far, no double-layer theory concerning a three-phase system and including the air phase has been proposed. Only in the structural level, discussed in the next chapter, can the air pore pressure be included.

Although the double-layer theory, as derived above, is a one-dimensional theory, i.e. a plane stress theory, the osmotic pressure, Eqs. (8.17.3)–(8.17.7), the effective pressure, Eq. (8.19.1), derived from it, and the pore pressure derived in turn from the effective pressure, are all scalar quantities: they are all spherical pressures. Since we have already pointed out that the end result of consolidation is a change in volume of the clay–water system, we may conclude by stating that *consolidation is a volumetric phenomenon*. A more rigorous proof is provided later in the study at the phenomenological level.

8.20 Swelling Pressure

If the outside constraint on the clay–water system or on a part of it is removed, the osmotic repulsive pressure, previously in equilibrium with that constraint, tends to repulse the clay platelets, increasing the distance $2d$ between them. If water solution is supplied from outside the system, it is sucked up to fill the growing interparticle space. The system increases its volume, it swells. Therefore, the osmotic pressure p_o is sometimes called the *swelling pressure*.

The double-layer theory does not distinguish between consolidation and swelling except for their opposite sign. From a thermodynamic point of view the process described by the theory is a reversible process, see Sect. 5.12. In reality, consolidation of clays is an irreversible process and part of the energy invested transforms into heat. Fig. 8.20.1 shows volumetric strain versus hydrostatic pressure in the consolidation and swelling processes, for pressures up to 80 p.s.i. See also Fig. 8.18.1 which shows volumetric strain versus time of consolidation for the same sample.

Several causes can be responsible for this loss of energy:

1. The resilience of the adsorbed water layer next to the clay surfaces, subjected to pressures up to a thousand atmospheres and pressure gradients up to 10^8 atm. per cm (Zaslavsky *et al.* 1967). Evidence points to the fact that the water next to the particle surfaces is of a semi-crystalline nature and exhibits non-Newtonian behavior (Low 1961).

2. Variations in the electric potentials along the clay surfaces, as pointed

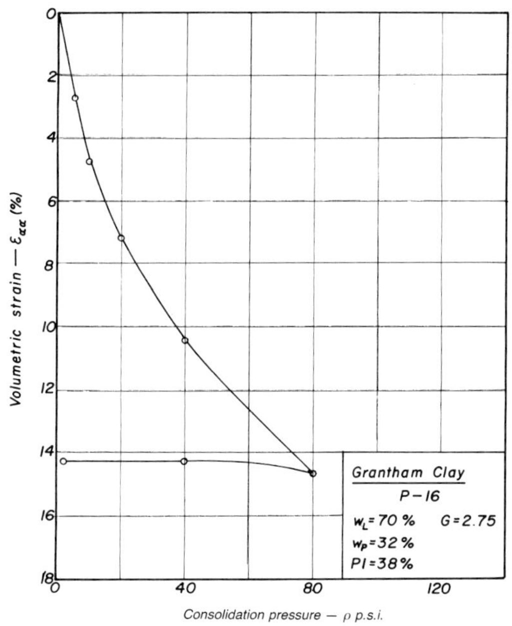

Fig. 8.20.1. Irreversibility of triaxial consolidation.

out in Sect. 8.15 point 8, impeding the normal flow of water solution along the surfaces.

3. Variations in the flatness of the clay surfaces due to isomorphic substitutions, as larger molecules usually replace smaller ones and warp the clay mineral surface, hampering the flow.

4. Lack of parallelness of the platelets, as assumed in the theory, and edge-to-face contacts between particles as explained in Sect. 8.15 point 4, causing *tortuosity* and obstructing the even flow of water in the pores.

5. Rotation of particles not in the direction of the flow, thus blocking its path.

6. Thixotropic gain, the strengthening of the clay with time.

Practically, the consolidation-swelling cycle is dissipative, where mechanical, chemical, thermal, etc., energies dissipate during the process by the production of heat. Usually the *compression*, the volume decrease by consolidation, is larger than the *expansion*, the volume increase by swelling. The difference between the amount of compression and the amount of swelling due to loss of energy for the same pressure increment is known as *hysteresis*. Fig. 8.20.2 shows volumetric strain versus time, in a consolidation test followed by a swelling test, for a specific pressure increment/decrement of 40 p.s.i. Notice that the volumetric strain computed from measured drained water is lagging behind the total volumetric strain by approximately 1.2%.

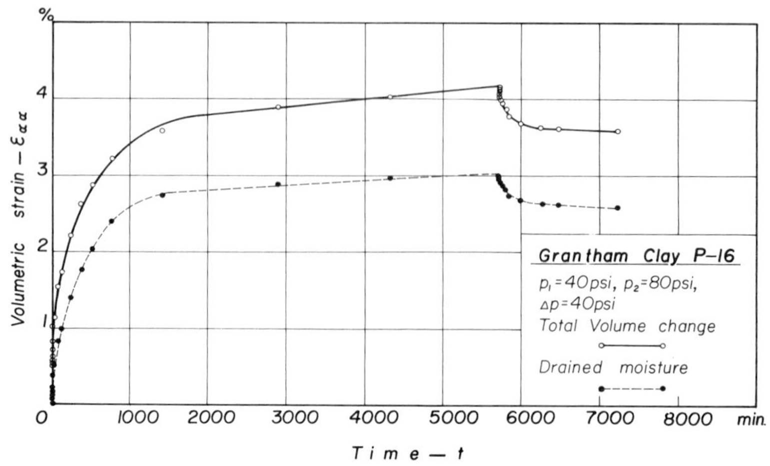

Fig. 8.20.2. Rebound cycle of triaxial consolidation.

8.21 Factors Affecting the Behavior of Clays in Consolidation

The consolidation of clays, as derived from the double-layer theory, depends on several factors, such as the surface charge density of the clay, the concentration of the bulk water solution, the temperature T, the dielectric constant D, the valence of the ions in suspension z, and an additional factor, not obvious from the theory: the nature of the exchangeable ions and their preferred location. These factors will be reviewed briefly, in the same order.

1. *Surface charge density*. If all factors are kept constant, a particle system with a lower surface charge density will compress to a higher density, i.e. a smaller d, for the same consolidation pressure p_{m}. This may be seen from Eqs. (8.16.3), (8.16.4) and (8.17.7). From Eq. (8.17.7) it is evident that for the same consolidation pressure the same *midplane potential* U is expected in clay, if the temperature and concentration are kept constant

$$\cosh U = 1 + \frac{p_{\mathrm{m}}}{2n(\infty)kT} \tag{8.21.1}$$

Consequently, from Eq. (8.16.4) two clays with surface charge densities σ' and σ'' will relate

$$\frac{\sigma'}{\sigma''} = \sqrt{\left(\frac{\cosh Z' - \cosh U}{\cosh Z'' - \cosh U}\right)} = \sqrt{\left(\frac{2n(\infty)kT(\cosh Z' - 1) - p_{\mathrm{m}}}{2n(\infty)kT(\cosh Z'' - 1) - p_{\mathrm{m}}}\right)} \tag{8.21.2}$$

The corresponding average distances d' and d'' can be found from Eq. (8.16.3)

$$\frac{d'}{d''} = \frac{\int_{Z'}^{U} \sqrt{(2\cosh Y - 2\cosh U \, \mathrm{d}Y)}}{\int_{Z''}^{U} \sqrt{(2\cosh Y - 2\cosh U \, \mathrm{d}Y)}} \tag{8.21.3}$$

which yields elliptic integrals, easily found in tables.

2. *Concentration of water solution*. From Eqs. (8.14.4) and (8.16.4) we can determine the distribution of the ionic concentration across the diffuse double layer between two parallel particles for two different ionic concentrations, and the distribution of their potential, Fig. 8.21.1.

It can be seen from Eqs. (8.14.2) and (8.14.13) that for identical ionic solutions, where only the concentrations are different, the following equation is obtained

$$\int_0^d (n'^+ + n'^-) \, \mathrm{d}x = \int_0^d (n^+ - n^-) \, \mathrm{d}x \tag{8.21.4}$$

Thus, the areas ADC and $A'D'C'$ in Fig. 8.21.1 are equal, and depend on the surface charge density of the particles, irrespective of the distance d between the particles. If the concentration of the ionic solution is lower, the diffuse double layer is necessarily more developed, and swelling–shrinkage phenomena are more evident.

From Eq. (8.17.7) it is seen that for the same osmotic pressure two ionic concentrations in the bulk solution relate inversely to the magnitude of their potential in the midplane

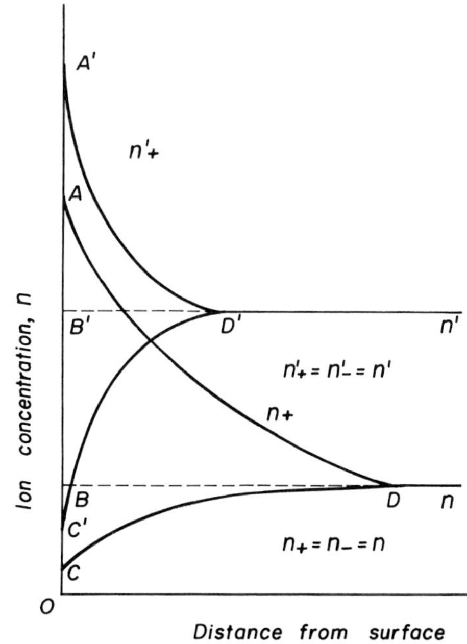

Fig. 8.21.1. Charge distribution in the diffuse double layer of two electrolyte concentrations with constant surface charge.

$$\frac{n'}{n''} = \frac{\cosh U'' - 1}{\cosh U' - 1} \tag{8.21.5}$$

The distances d' and d'' relate according to Eq. (8.16.3) and considering (8.21.5)

$$\frac{d'}{d''} = \sqrt{\frac{n''}{n'}} = \frac{\int_{Z'}^{U} \sqrt{(2\cosh Y - 2\cosh U dY)}}{\int_{Z''}^{U} \sqrt{(2\cosh Y - 2\cosh U dY)}} \tag{8.21.6}$$

The effect of Na^+ montmorillonite in NaCl solutions on the ionic concentration has been investigated by Warkentin and Schofield (1962) and extensive investigations with Ca^{2+} montmorillonite and mixtures of Na^+ and Ca^{2+} were conducted (Shainberg and Kemper 1966b; Shainberg and Otoh 1968; Bar-On et al. 1970; Klausner and Shainberg 1971; Shainberg et al. 1971).

3. *Temperature*. Since the midplane potential $U = \Phi(d)_e$ is a function of temperature, two clays, differing in temperature and under the same pressure, will relate

$$\frac{T'}{T''} = \frac{\cosh U'' - 1}{\cosh U' - 1} \tag{8.21.7}$$

The inter-particle distances d' and d'' depend on the temperature, because the electronic potentials Z and U and the dielectric constant D are functions of temperature

$$D - 1 \cong \frac{1}{T} \tag{8.21.8}$$

and therefore $\varkappa \cong \sqrt{[1/(T + A)]}$, where A is a function dependent on the dipole of the water. Consequently

$$\frac{d'}{d''} = \sqrt{\left(\frac{T' + A}{T'' + A}\right)} = \frac{\int_{Z'}^{U} \sqrt{(2\cosh Y - 2\cosh U dY)}}{\int_{Z''}^{U} \sqrt{(2\cosh Y - 2\cosh U dY)}} \tag{8.21.9}$$

Yong *et al.* (1963) investigated the effect of temperature on Na^+ montmorillonite clays and found it insignificant and hardly detectable experimentally.

4. *Dielectric constant*. The dielectric constant D does not affect the pressure directly. Therefore, for any pressure, the interparticle distances of two systems, identical in all details except the dielectric constant of the solution, relate as the square roots of their dielectric constants.

$$\frac{d'}{d''} = \sqrt{\frac{D'}{D''}} \tag{8.21.10}$$

When water is replaced with alcohol, which has a lower dielectric constant, the extent of the diffuse layer effect is reduced, without detrimental effects to the soil structure itself. This technique is occasionally employed in the investigation of loess soil, saturating it with alcohol without collapse of its structure, which would be expected if it were saturated with water.

5. *Valency*. When considered according to the diffuse double-layer theory, the valency has an effect similar to that of the ionic concentration. When the valency increases the extent of the diffuse layer is diminished and the swelling–shrinkage properties of the soil change. For the same osmotic pressure, according to Eq. (8.17.7) we have

$$\cosh U' = \cosh U''; \quad \Phi(d')_e = \Phi(d'')_e \tag{8.21.11}$$

and the interparticle distances will relate, according to Eq. (8.16.3), inversely proportional to the valencies

$$\frac{d'}{d''} = \frac{z''}{z'} = \frac{\int_{Z'}^{U} \sqrt{(2\cosh Y - 2\cosh U \mathrm{d}Y)}}{\int_{Z''}^{U} \sqrt{(2\cosh Y - 2\cosh U \mathrm{d}Y)}} \tag{8.21.12}$$

This result is correct only in general terms, and a further correction of the theory concerning the nature of ions, discussed in the next point, may cancel the deviations from the theory.

6. *Nature of ions*. It has been noticed that when clays, particularly montmorillonite clays, are adsorbed with polyvalent ions, the assumption of platelet distribution at random with average interparticle spacing of $2d$ is not observed in the system. While montmorillonite clays, saturated with monovalent ions such as Na^+ and Li^+, conform to the theory and are dispersed as single platelets, polyvalent ion-saturated montmorillonites consist of tactoids, packets of platelets, rather than of single platelets, Fig. 8.21.2.

The tactoid formation in polyvalent ion-saturated clays reduces the size of active sufaces that interact according to the osmotic model, thereby further reducing the shrinkage–swelling phenomena for all osmotic pressures. The studies of van Olphen and Blackmore and Miller (1961) indicated that the number of single clay platelets $\mathfrak{N}$ per tactoid for calcium montmorillonite varies between 4 and 6 and up to 9 in air-dried clay, that the internal surfaces of tactoids were more than three times that of the external surfaces for an electrolyte concentration of 10^{-3}M, and that the calculated affinity will

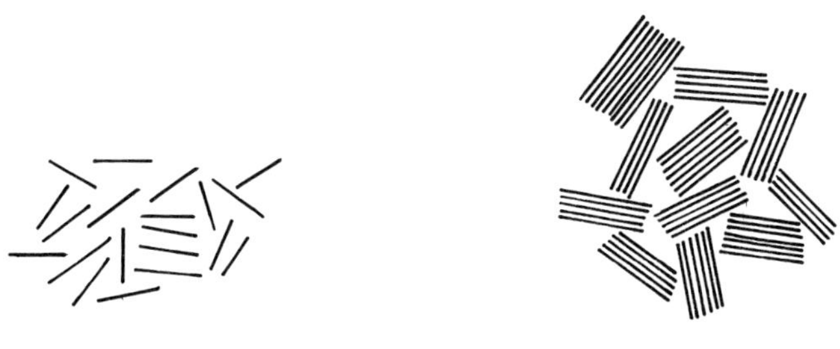

a b

Fig. 8.21.2. Array of single platelets **a** and of tactoids **b**.

increase if the electrolyte concentration is decreasing. The average half space of the platelets in the internal surfaces, as determined by the X-ray measurements, is $d_0 = 4.5$ Å. Experimental evidence seems to indicate that monovalent and bivalent ions in a tactoid of montmorillonite clay prefer certain locations, for instance Ca ions prefer the internal surfaces, while Na ions prefer external surfaces (Mering and Glaeser 1961; Shainberg and Otoh 1968; Klausner and Shainberg 1971).

8.22 Properties of Clays, as Predicted by the Diffuse Double Layer

Macrophenomena, the effects of physical and mechanical properties, evolve from microphenomena, that is, the interaction between the clay particles, the electrolyte solution and the air.

Among the numerous properties of clays some are of particular interest.

1. *Viscosity*. Perhaps the most important physical property, which depends on the diffuse layer of a clay–water solution system is viscosity. Viscosity is affected by all the factors enumerated in the previous section: electrolyte concentrations, surface charge, temperature, valency, dielectric constant and the nature of the ions, and in turn has a strong influence on the flow properties of the electrolyte solution, whether the solution flows through the matrix of the clay solids, or forms a dilute clay suspension.

Eq. (8.17.9) indicates that a pressure gradient exists in the diffuse double layer close to the clay surfaces, dissipating towards the bulk suspension. While in the bulk suspension this pressure may be of the order of 1 atm., it can reach, close to the surfaces, values of as high as 10–20 atm. (R. H. Kemper 1969, personal communication) and the electric fields developed between the electrons involved may result in internal pressures as high as 1000 atm., and pressure gradients of 10^8 atm./cm (Zaslavsky et al. 1967). Such extremes would certainly affect viscosity or apparent viscosity of the solution in the layers close to the clay surfaces, where the properties of the water are believed to be different from those of the water far away from the surfaces (Low 1961). Shainberg and Otoh (1968) have shown the effect of the nature of the ions on the viscous properties of a montmorillonite clay suspension.

2. *Permeability*. The viscosity of the electric solution flowing through a clay matrix is related to permeability, an important technical property of soils in general, which controls many mechanical phenomena in clays. The properties of the water solution in the first few layers next to the clay surfaces may differ significantly from those of the bulk water further away from the surfaces, as has already been noted (Low 1961; Zaslavsky et al. 1967). This would cause permeability of clays to deviate from the classical Darcian flow and assume non-linear characteristics. It has been shown theoretically

(Klausner and Kraft 1965, 1966) that if a pressure gradient $f(x)$ such as given in Eq. (8.17.9) exists in a porous medium or in clay, threshold gradients may develop and non-linear flow is possible. Non-Darcian flow has been detected experimentally (Lutz and Kemper 1959; Hansbo 1960; Miller and Low 1962), but the results were severely contested on the grounds of inaccuracy of measurements. A review of this topic has been presented by Kutilek (1969).

3. *Atterberg limits*. The liquid limit w_L and the plastic limit w_P, so extensively used in geotechnical engineering, are moisture contents characteristic of two specific stresses in the soil, a shear stress and a volumetric stress, respectively. The Atterberg limits represent a certain state of equilibrium and are related to electric potentials, related, in turn, to volume changes. Thus there is a possibility of tracing the Atterberg limits to shear and volumetric behavior via the diffuse double layer. Experiments of a limited nature on the effect of Na and Ca ions and their concentration have been carried out by White (1949), Warkentin (1961) and others (Yong and Warkentin 1966). A comprehensive study of the effect of electrolyte solution on the Atterberg limits is in progress by the author and his colleagues.

4. *Compressibility*. The theory of the diffuse double layer, specifically its extension to two interacting surfaces, which predicts the average half-space d between particles for an osmotic pressure p_0, as shown in Sect. 8.17, is in fact a compressibility theory. It has been experimentally shown (Bolt 1956; Warkentin et al. 1957; Yong and Warkentin 1960; Klausner and Shainberg 1967b; Klausner and Shainberg 1971; Shainberg et al. 1971) that the theory can be applied to consolidation, and can successfully predict the compressibility, e-log p curve, of Na and Ca montmorillonite clays. It is thus a three-dimensional consolidation theory, which has to be verified for clays other than montmorillonite, that is, clays with lower cation exchange capacity.

5. *Shear strength*. So far, no shear strength theory has been derived from the theory of the diffuse double layer. It is the author's belief that such a theory can be inferred from the compressibility theory, by investigating the sources of shear strength and its relation either to the fixed ionic array in the electrolyte solution and its mobility, or to the orientation of particles.

8.23 The Structure of Clays

In Sect. 8.13 most of the limitations of the double-layer theory have been exposed, some of them touching upon the most debated question of particle arrangement in a clay–water solution system, i.e. their orientation and distribution, called also the "fabric" or "architecture" of the clay suspension. Terms like "card house structure", "flocculated structure", "dispersed structure" have become common expressions in soil mechanics since the late forties and have been copied periodically from one textbook to another, illustrations included, and the topic is still intensely debated (Shainberg et al. 1987).

Let us assume that we are dealing with a clay suspension in water and that the concentration of the clay platelets is sufficiently low to allow their free movement. Even in high concentrations the clay particles never really come in contact with one another, however close they might come, since there are always water molecules and solute ions that separate the particles (Bolt 1955).

According to the double-layer theory, the platelets are parallel, extend infinitely in two directions, the charge density and the external constraints are evenly distributed over the surfaces, and what is known as *face to face* interaction between particles occurs. The question arises, is there a preferred direction in which the platelets orient themselves? Is there any reason for the platelets to prefer a particular direction? It was also pointed out that in clays not all of the assumptions discussed in Sect. 8.14 are satisfied. Owing to isomorphic substitution, different ions (Al^{3+}, Fe^{3+}, Mg^{2+}, Cr^{2+}, etc.), are located along the platelets, introducing both deformation and warping of the platelets as well as uneven charge densities along the surfaces. Since the platelets have finite length and are disrupted at the edges, some *edge to face* attraction occurs. At the edges, where the tetrahedral silica and the octahedral alumina or magnesia are exposed and excess positive charges dominate, the charges are balanced by hydroxyl $(OH)^-$ ions, resulting in *edge to edge* interaction of particles, to which a theory similar to the diffuse double-layer theory is applied. On account of all this the platelets are far from parallel, even in a state of equilibrium; moreover, in montmorillonites, where the double-layer theory works best, the thickness of the platelets is $\frac{1}{100}$ to $\frac{1}{1000}$ of their length or width, and they are flexible enough to be easily bent.

Clays with low cation exchange capacity (CEC) in general have a thickness to length ratio of less than $\frac{1}{10}$. The edge to face attraction between particles in clays with lower CEC is more developed and dominant. The double-layer theory for clays with lower CEC is of a more limited applicability, but it is by the same token also of lesser importance, since at that stage the soil ceases to behave like a clay.

Considering all circumstances, and the fact that there is no reason for the particles to prefer any particular orientation, the answer must inevitably be that in a low concentration of particle suspension at equilibrium the orientation is, at best, a *preferred random* orientation, determined by the electric potential of the diffuse layer and other forces acting at equilibrium. Is this preferred random orientation the equivalent of a "dispersed" structure since the platelets are randomly dispersed and oriented? Or is it a "flocculated" structure since it is, nevertheless, determined by the electric potential which holds the particles tightly together?

The "flocculation value", an arbitrarily defined technique, (Arora and Coleman 1979) is not a scientific term and, as has been pointed out, it "may not be suitable for flocculation determination under certain conditions" (Shainberg et al. 1987).

If at equilibrium the orientation of the platelets is preferred random, $2d$ refers to an average distance between two platelets, at one end of the platelet the distance being perhaps less than $2d$ and at the other end more than $2d$.

The hydrostatic repulsive pressure (the effective pressure) $p_r(x, t) \equiv$

$p(d, t)_c$ and the pore water pressure $p(d, t)_w$ are derived from Eqs. (8.17.3) and (8.17.4) which express the work per unit volume, and are average values over the whole surface of the platelet, being thus scalar quantities without specific direction. The pore pressure gradient (8.19.2) is, however, a vector, in the direction of the flow. The existence of a pressure gradient directed along the platelets results in a deflection of the platelets, the distance $2d$ between the platelets changing as consolidation proceeds.

There are misconceptions in the literature concerning the direction of the platelets during the consolidation process.

Fallacy. The clay particles orient themselves perpendicular to the pressure applied.

This statement is illustrated by a technical description and certainly not a scientific term, the "card house structure", which, when subjected to a force, collapses and becomes flat, with all cards parallel and perpendicular to the force. Let us check this concept.

Suppose we subject a remolded cylindrical clay sample to consolidation under a hydrostatic pressure in a triaxial cell, where it drains freely at the base. At the outset of the test the particles are randomly oriented due to remolding, for instance, by compaction. The random orientation due to remolding must not be confused with the preferred random orientation due to internal forces in the clay, which brings the particles to a state of equilibrium. Since the pressure is equal in all directions, would it make any sense for the particles to become oriented in one particular direction? Indeed it would. But this direction could not possibly be perpendicular to the applied pressure, since the pressure was assumed equal in all directions. We maintain that the particles would tend to direct themselves in the direction of the flow path of the draining water, so as to provide the least resistance to this flow. This proposition was never proved; its verification would have to be made by inference and by confirming its applicability to other circumstances. In Sect. 8.19 it was shown that the pressure gradient is everywhere in the direction of the flow. Since the pore water pressure which drives the flow is perpendicular to the pressure gradient and also acts perpendicular to the particle surfaces, we may infer that the average direction of the particles at any specific location is parallel to the flow in that location. We also see that particles are oriented perpendicular to the pore water pressures acting in the sample, and not to the pressures applied externally.

Proposition. If flow occurs in a clay–water system, particles will tend, mobility conditions permitting, to orient themselves parallel to the flow path of the water. Otherwise they will tend to persist in their preferred random orientation.

Once the equilibrium is upset, flow commences and the particles, as stated in the proposition, tend to rotate in the direction of the flow. This rotation is resisted by the forces that participated in the earlier equilibrium (ionic adsorbtion and back diffusion, edge to face attraction, etc.), and by friction (of the particles, the ions, and the water molecules near the particle surfaces, owing to their rearrangement and to the densified environment). At the end of the process a new equilibrium is attained, the driving pore water pressure diminishes, and the particles gradually return to their preferred random orientation, as far as frictional resistance permits. The frictional resistance acts in both directions, and in the process it tranforms into heat resulting in hysteresis, discussed in Sect. 8.20.

The gradual return of the particles to their preferred random orientation after the flow ceases is a very slow, monotonically decreasing and time-dependent process, subject to the possible mobility of the particles, water molecules and ions. The preferred random orientation is partly responsible for the *thixotropic regain* of sensitive clays.

When random orientation is introduced in a clay by remolding (kneading or compaction), the particles tend, in time, to reorient and reach the preferred random orientation. We observe a change in density, strength, etc. of the sample, beginning right after remolding. This change can be in both directions: increase or decrease of the density or strength, depending on whether the remolding energy was originally lower or higher than the preferred internal energy of the interacting particles in the clay–electrolyte system.

A quantitative approach to the thixotropic regain was given by Skempton and Bishop (1954) and was linked to the *sensitivity* of the clay, which is defined as the loss in strength when the soil passes from the original undisturbed state to its remolded state. The strength is measured by the unconfined compression test, Sect. 12.7

$$S_t = \frac{p_u}{p_{ur}} \tag{8.23.1}$$

where S_t is the sensitivity of the clay as defined by Terzaghi (1944), p_u is the *unconfined compression strength of the undisturbed* clay sample and p_{ur} is the *unconfined compression strength of the remolded* clay sample.

Since the final strength of the undisturbed and the remolded samples is not reached at the same strains, Fig. 8.23.1, Tschebotarioff (1952) proposes an alternative equation, in which the strength ratio of the unconfined compression test is taken at the same strain for both the undisturbed and the remolded samples

$$S'_t = \frac{p_u}{p'_{ur}} \tag{8.23.2}$$

where p'_{ur} is the unconfined compression stress of the remolded clay, at the strain at which the final strength of undisturbed unconfined compression is reached.

According to Skempton, for most clays the regain curves, Fig. 8.23.2, indicate "that thixotropic hardening does not increase the strength by more than about 200%". In other words, the sensitivity is not more than 3. The question still remains how to explain high sensitivities of up to 150.

A clay in its preferred random orientation is necessarily isotropic, whereas clays in which the particles become oriented due to flow are anisotropic and

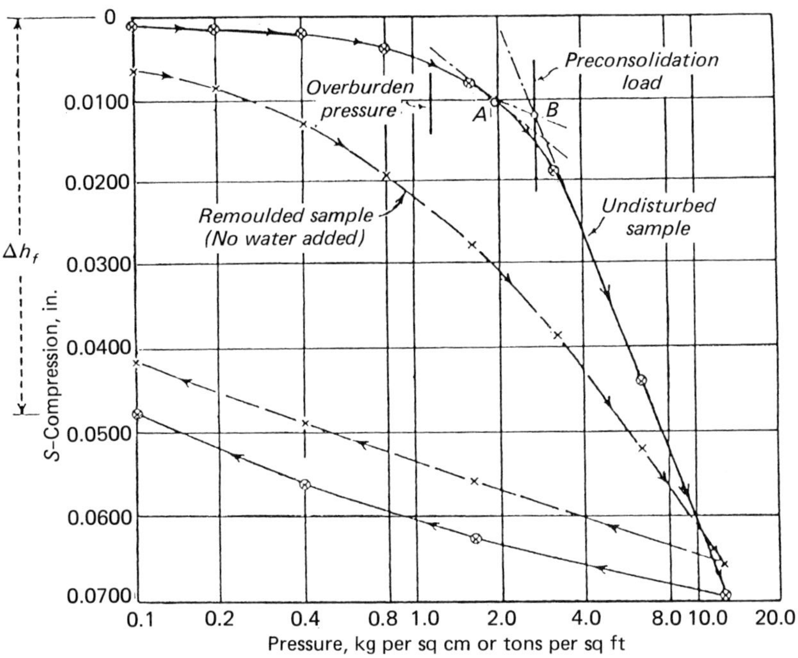

Fig. 8.23.1. Comparison between undisturbed and remolded stress–strain curve of unconfined compression test. [From G. P. Tschebotarioff (1952), courtesy of McGraw-Hill Book Co. Inc.]

acquire different strength properties in the different directions. The shear strength of clays in the direction parallel to the flow has been found to be lower than that in the direction normal to the flow (Ho 1962). This is plausible, considering the fact that in the first case the failure plane passes mostly through the water phase between the particles, while in the latter it cuts through the particles themselves. It was also found that the shear strength of a compacted clay with random orientation is higher than that of a clay with orientation parallel to the direction of the flow and lower than that of a clay with orientation normal to it.

There are further implications to the matter of orientation not discussed here. It should be noted that the water in clays is in continuous dynamic change owing to dessication, precipitation, consolidation and underground or other flow, which makes the clays likely to have an oriented structure and be anisotropic. The orientation of natural deposits is less likely to be deterred from its preferred direction and equilibrium.

8.24 Interfacial Forces

So far it has been implied that in a multi-phase mixture the completely homogeneous phases are bounded by sharply defined geometrical surfaces.

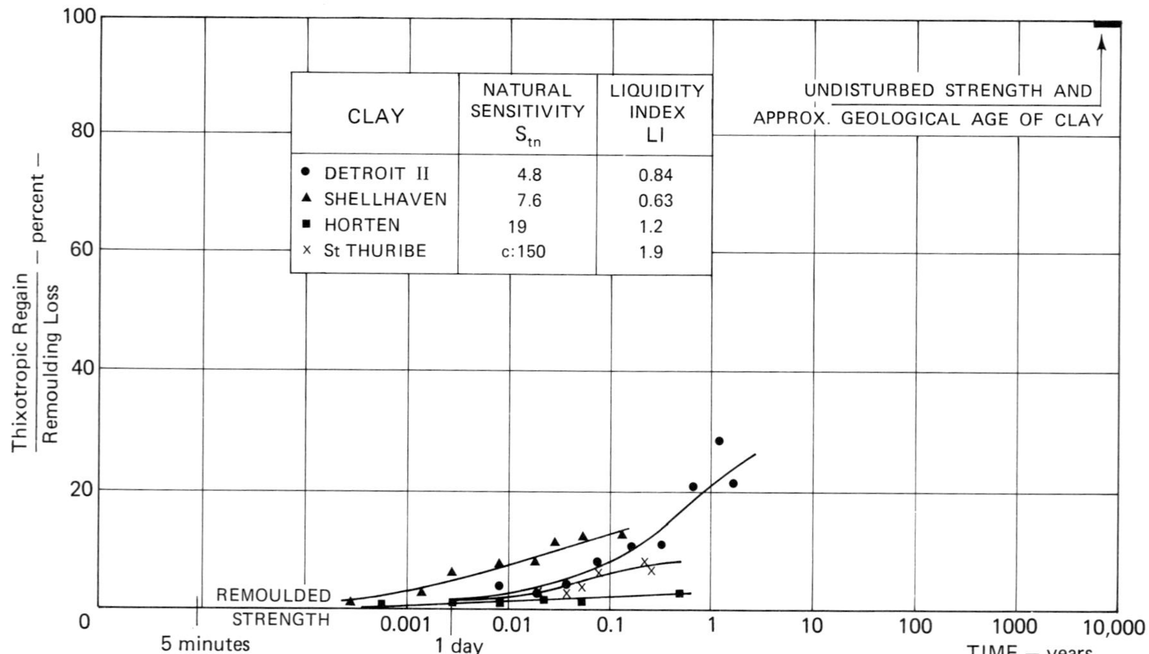

Fig. 8.23.2. Thixotropic regain in sensitive clays. [From A.W. Skempton and A. W. Bishop (1954), courtesy of North Holland Publ. Co.]

This, however, is an oversimplification, since the interface between two phases is rather a thin layer, across which the physical properties vary, changing gradually from the properties of one phase to those of the other phase. In the case of two interacting phases with different densities, surface charges and perhaps other properties (dielectric, electro-magnetic, etc.), a local discontinuity of the properties is never possible, and an *interface* is always present. In fact, the diffuse double layer, already discussed, is such an interface.

The surface energy of the separate phases can be defined, generally speaking, as an additional potential energy per unit surface, when the particles of the surface layer lack neighbors on their external side. The surface molecules attract other matter whose molecules or ions have greater freedom of movement, such as liquids or gases. Thus, an important characteristic feature of the interface is *adsorbtion*.

Since ionic or molecular mobility of solids is very small the adhesive attraction between solids is barred, while the mobility in liquids and gases allows a gradual change of ion concentration at the interface. Gases, being free to move and mix, average out their mixture and thus do not form interfaces among themselves; liquids form interfaces with solids and gases, and with other liquids if they are *immiscible*; gases form interfaces with solids.

The interface, however thin, has the properties of a catenary membrane whose tensional strength depends on the materials involved. This membrane behaves like a three-dimensional catenary, and deforms according to the force polygon acting in the direction normal to it. In cases where liquids and gases are involved the forces are spherical (hydrostatic). The solid–gas interface differs from the solid–liquid interface in that the gas, by its nature, cannot transfer tensional forces through its molecules, only pressure forces.

In soils we are dealing basically with three constituents: the solids, which can transfer normal stresses, pressures and tensions, as well as shear stresses; water, which transfers only normal stresses, both pressures and tensions, and air which transfers only pressures.

Considering interfaces in soils, we can distinguish two basic states, while further subdivisions of these states are always possible:

1. The soil is *saturated*, and there is only a *solids–water* interface, provided that the small amounts of gas or water vapor, always present, are insignificant and therefore ignored.

In a saturated soil the solids–water interface is the diffuse double layer of fine grained soils, discussed in Sects. 8.10–8.15. In granular soils the solids–water interface draws from the same source of surface charges as fine grained soils, but is inconsequential since the specific surface of granular soils is several orders of magnitude smaller than that of fine grained soils, and therefore insignificant. The tensional strength of the solids–water interface, both in granular and fine grained soils, is weak compared to the strength of the grains which it borders.

2. The soil is *unsaturated* and in addition to solids and water variable

amounts of gases are contained in it. The various degrees of unsaturation were discussed in Sect. 8.8. In the unsaturated state there is an *air–water* interface and an *air–solids* interface, in addition to the solids–water interface. The air–solids interface is insignificant and irrelevant to our study. The air–water interface is discussed in Sect. 8.25.

There is, however, a third, most important state. It is the boundary state at the fringes of the saturated soil, between its saturated part and its unsaturated part, the topic of Sect. 8.26.

8.25 Air–Water Interface

The simplest way to introduce this subject is to consider a spherical air bubble with a radius R, small enough to be contained in the pore water when the system is in equilibrium and the influence of gravitation is neglected. Let the pressure in the water be p_w and in the air bubble p_g, and let us assume that we cut the globule with an imaginary plane at a circular angle α, Fig. 8.25.1. The forces acting on each part of the sections are now the pressure p_w of the water acting on the outer curvature of the sections, the pressure p_g of the air acting on the inner curvature of the sections, and a tensional force of the air–water interface T acting along the circumference of the circular sections and tangential to the spherical envelope. The components of these forces in the direction normal to the partitional plane are in equilibrium, therefore we may write

$$(p_w - p_g)2\pi R^2 \sin\alpha + T2\pi R \sin\alpha = 0 \qquad (8.25.1)$$

from which the *Kelvin relation* (Thomson 1858) is obtained

$$p_i = p_w - p_g = -\frac{2T}{R} \qquad (8.25.2)$$

where p_i is the *air–water interfacial stress*, directly proportional to the *tensional force* of the water T measured per unit length $[KM^{-1}]$, and inversely proportional to the radius of the spherical bubble. It should be remarked that T is in fact the *ultimate surface tension* of the water. Unlike a rubber membrane, it is assumed that the interface membrane does not by itself possess elasticity of form. Its resistance against deformation is furnished solely by the tensional force around the circumference of the bubble, tangential to it. In general, other assumptions can be made with respect to the deformation characteristics of the interface. T is temperature dependent, and its value at 20 °C at atmospheric pressure is 72.7 dynes per cm. A wider range of temperatures is given in Table 8.25.1.

The radius R of bubbles for given interfacial pressures p_i may be computed from Eq. (8.25.2), $R = -2T/p_i$, and is given in Table 8.25.2.

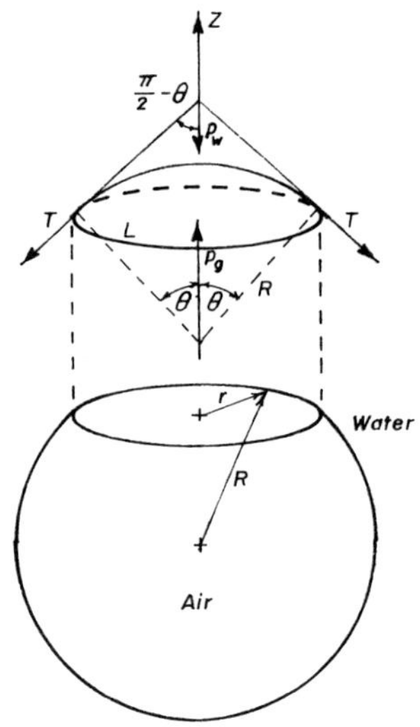

Fig. 8.25.1. Equilibrium of forces at the interface between immiscible fluids. [After H. J. Morel-Seytoux (1961) A. W. Skempton and A. W. Bishop (1954), courtesy of North Holland Publ. Co.]

Table 8.25.1. Surface tension T of air–water interface at different temperatures T

Temperature T (°C)	Surface tension T (dynes/cm)
−5	76.4
0	75.6
5	74.9
10	74.22
15	73.49
20	72.75
25	71.97
30	71.18
40	69.56
50	67.91
60	66.18
70	64.4
80	62.6
90	60.8
100	58.9

A more refined approach (Kondo 1956) seeks a particular fictitious dividing spherical surface with a radius R^* defined by

$$\left[\frac{\partial T}{\partial R}\right]_{R=R^*} = 0 \tag{8.25.3}$$

which yields

$$p_i = p_w - p_g = -\frac{2T^*}{R^*} \tag{8.25.4}$$

Table 8.25.2. Capillary head h_c and capillary tension p_c functions of the air pore radii R

Pore radius R [μm]	Capillary head h_c [cm H_2O]	p_i Pressure head atm.	Capillary tension p_c ($10^{-3}\times$ bar)	($dynes/cm^2$)	Suction potential pF [log h_c]
148.320	10	0.010	9.81	9.81×10^3	1
24.720	60	0.058	58.9	5.89×10^4	1.78
14.400	103	0.101	101.0	1.01×10^5	2.01
4.310	344	0.333	337.5	3.38×10^5	2.54
2.875	516	0.5	506	5.06×10^5	2.71
1.435	1033	1	1013	1.01×10^6	3.01
0.720	2066	2	2027	2.03×10^6	3.32
0.480	3099	3	3040	3.04×10^6	3.49
0.275	5165	5	5067	5.07×10^6	3.71
0.145	10330	10	10134	1.10×10^7	4.01
0.095	15495	15	15201	1.52×10^7	4.19

[After D. Hillel (1980), courtesy of Academic Press.]

where

$$T = \frac{R^{*2}T^*}{3R^2} + \frac{2T^*R}{3R^*} \tag{8.25.5}$$

Since T must always be positive, it has a minimum value in Eq. (8.25.5) at $R = R^*$ as $T = T^*$, meaning that the surface tension is independent of the location of the interface as long as it is within its thickness.

If an external energy, such as drying or suction, is applied to the soil in order to extract some water from its pores, we will say that external "stresses" have been applied to it. Extraction of the water from the soil will result in an increase of the air phase in the pores at the expense of the water phase. The bubbles will grow, "touching" the pore walls, and their shape will cease to be spherical, gradually assuming the shape of the pores. The water recedes to the interparticle constrictions and the radius of the air–water interface decreases correspondingly.

We intend to investigate the air–water interface of the interstitial water of arbitrarily shaped air voids in the interparticle constrictions. It is convenient to follow Gauss' approach (Gauss 1830), who proceeds from the minimum principle of the surface energy. Accordingly, our starting point is the existence of an interfacial pressure $p_i = p_w - p_g$ indicating a pressure difference between the water and the air across the interface, similar to that assumed earlier for the spherical bubble. Our intention is to show that the interfacial pressure may be derived from this minimum principle. We also stipulate, as before, that in the interface the water has a surface tension $T = T^*$, equal everywhere and tangential to the interface surface.

We restrict ourselves to a simple case, assuming a cylindrical air–water interface or a cylindrical membrane. The tension equilibrium of such a membrane is similar to that of a vibrating string, whose shape coincides, at any time, with a cylindrical profile. Let u be the displacement of the string between two neighboring points and Δs their distance, a cylindrical segment of the radius R, with the tensions $\mathbf{T}$ and $\mathbf{T}'$ acting at its ends in the tangential directions of the string, Fig. 8.25.2.** $\mathbf{T}$ and $\mathbf{T}'$ are equal in magnitude but not parallel to each other, and an angular difference of $\Delta\theta$ exists between them. The resulting force $\Delta\mathbf{T} = \mathbf{T}\Delta\theta$ in the force triangle is perpendicular to the arc element Δs and it counteracts the reinstating force $\mathbf{p}_i$ acting on the arc element Δs of the string, and so we have

$$\mathbf{p}_i\Delta s = -\Delta\mathbf{T} = -\mathbf{T}\Delta\theta \tag{8.25.6}$$

On the other hand, the curvature $1/R$ is given by

$$\pm\frac{1}{R} = \lim_{\Delta s\to 0}\frac{\Delta\theta}{\Delta s} = \frac{\partial^2 u/\partial x^2}{[1 + (\partial u/\partial x)^2]^{3/2}} \tag{8.25.7}$$

where u is the deflection of the string at x, the abscissa along the string. In the denominator of the last member, the positive value of the square root is taken, if s increases with x, as assumed.

Eqs. (8.25.6) and (8.25.7) determine the magnitude of $\mathbf{p}_i$

$$\mathbf{p}_i = -\frac{\mathbf{T}}{R} \tag{8.25.8}$$

with $\mathbf{p}_i$ pointing always to the concave side of the string, or, for sufficiently small deflection

$$\mathbf{p}_i = \frac{\partial^2 u}{\partial x^2}\mathbf{T} \tag{8.25.9}$$

If we now pass from the vibrating string to the cylindrical membrane by transcribing the variables $\mathbf{p}_i$ and $\mathbf{T}$ of the string from Eq. (8.25.8) to the

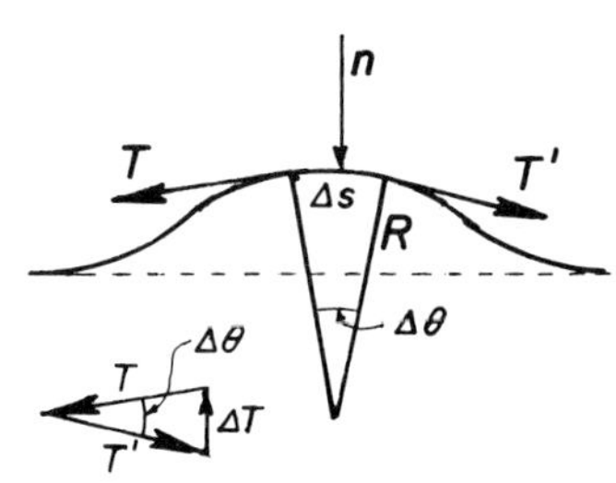

Fig. 8.25.2. Forces acting in a vibrating string and in an air–water interface.

** Note that the dimension of $\mathbf{T}$ is that of a force rather than a force per length, in view of the fact we are considering a string.

$\mathbf{p}$, defined later, has the dimension of a force per length rather than that of a stress.

corresponding variables p_i and T of the interface membrane, respectively, we obtain

$$p_i = -\frac{T}{R} \tag{8.25.10}$$

an equation similar to Eq. (8.25.2) of the spherical bubbles, indicating that the air–water pressure interface is dependent on the radius of the meniscus but not on the shape of the air voids.

When the air–water interface pressure, Eq. (8.25.10), acts on a membrane with a double curvature and two principal radii R_1 and R_2 at the surface point of consideration, Figure 8.25.3, we may write

$$p_i = p_w - p_g = -\left(\frac{1}{R_1} + \frac{1}{R_2}\right)T \tag{8.25.11}$$

Other known forms given to this equation are

$$p_i = p_w - p_g = -2CT = -\frac{2T}{R^*} = -K(R_1 + R_2)T \tag{8.25.12}$$

where $C = \frac{1}{2}(1/R_1 + 1/R_2) = 1/R^*$ is the *mean curvature* of the surface as defined by Minkowski in his carefully written article on capillarity (Minkowski 1898–1935), while the measure of curvature $K = 1/R_1R_2$ is *Gauss's total curvature* (Gauss 1830).

While in general the normals of two infinitesimally distant surface points do not necessarily intersect on the same line, here the normals do intersect and

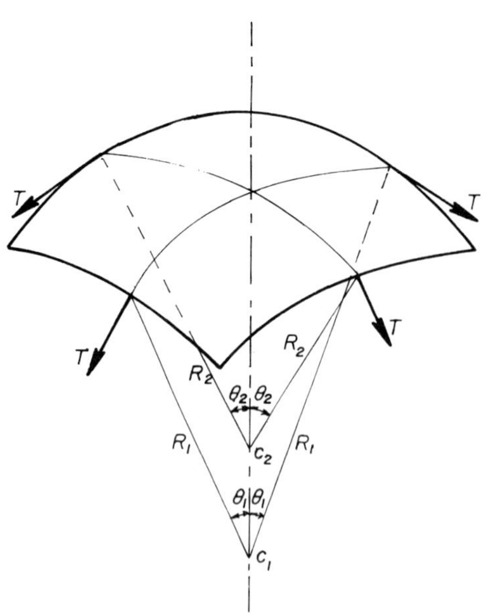

Fig. 8.25.3. Capillary interface with double curvature on the same side.

define the principal directions. The curvatures C_1 and C_2 are the reciprocals of the radii R_1 and R_2, respectively, and denote the centers of curvature in Fig. 8.25.3. We see that the points C_1 and C_2 fall on the same line normal to the surface. If R_1 and R_2 are the same, $R_1 = R_2$, Eq. (8.25.8) becomes the same as Eq.(8.25.2), of the spherical bubbles.

R_1 and R_2 can be of different signs, and then the centers of their curvature, C_1 and C_2 respectively, are located on the normal line in opposite directions, Fig. 8.25.4, and Eq. (8.25.8) becomes

$$p_i = p_w - p_g = -\left(\frac{1}{R_1} - \frac{1}{R_2}\right)T \tag{8.25.13}$$

The menisci of unsaturated soils can have centers of curvature on opposite directions on the line normal to the interface.

If the expression in the parentheses containing the curvatures is positive, p_i is a tension, which is the general case in natural soils. If the expression is negative, as it may be in compacted or remolded soils with air inclusions, p_i is a compression, and will remain so until an equilibrium of the soil--water–air mixture is reached.

When the air–water interface is transformed into a spherical stress we obtain

$$p_i\delta_{ij} = (p_w - p_g)\delta_{ij} = -\left(\frac{1}{R_1} - \frac{1}{R_2}\right)T\delta_{ij} \tag{8.25.14}$$

indicating that the air–water interface acts in all directions like a hydrostatic pressure and is therefore spherical.

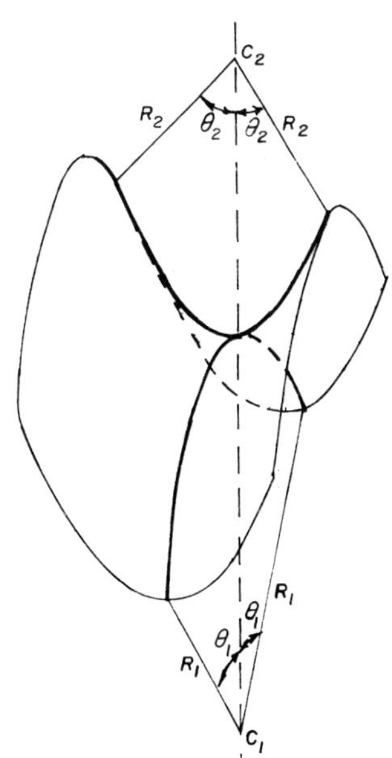

Fig. 8.25.4. Capillary interface with double curvature on opposite sides.

8.26 Capillarity

The soil–water potentials, such as the suction potential, the osmotic potential, the gravitational potential, etc. usually expressed in terms of pressure (tension) $[p] \equiv [GL^{-2}]$, may be interpreted on one hand as an energy per unit volume, $[GL]/[L^3] = [GL^{-2}]$, and on the other hand it may be expressed in terms of a pressure head when divided by the unit weight of water at 4 °C, $\gamma_{w0} = 1$, $[L] = [GL^{-2}]/[GL^{-3}]$. The suction p_i expressed as a pressure head is

$$h_i = \frac{p_i}{\gamma_{w0}} = -\frac{2T}{\gamma_{w0}R^*} \tag{8.26.1}$$

where h_i is the *head* of the corresponding tension p_i.

From Eq. (8.26.1) and considering also Eq. (8.25.12) the equivalent mean radius of curvature of the water meniscus in the soil pore R^* can be evaluated from the head h_i.

At an elevation in the soil where the tension vanishes, $h_i \equiv p_i = 0$, there is no air–water interface, the soil is necessarily saturated and the pore water pressure is zero. In the field it indicates the elevation where the *water table* exists and the hydrostatic pressure of the soil water is zero. At any depth d below the water table the hydrostatic pressure is $p_\delta = \gamma_{w0}d$, a positive pressure increasing linearly with the depth, Fig. 8.26.1.

Above the water table there still exists a saturated zone up to an elevation $h = h_c$. h_c not only defines a suction and an air–water interface, but also a *capillary tension* in the soil water, and is called the *capillary head*

$$h_c = \frac{p_c}{\gamma_{w0}} = \frac{-2T}{\gamma_{w0}R^*} \tag{8.26.2}$$

At any elevation h which is less than the capillary head, $0 \le h \le h_c$, when the soil is saturated the hydrostatic pressure is negative, namely, a suction, linearly increasing up to $h = h_c$.

This is an in-between zone, the *capillary zone*, between the saturated zone and the unsaturated zone, called by Childs (1969) the *capillary fringe*. The

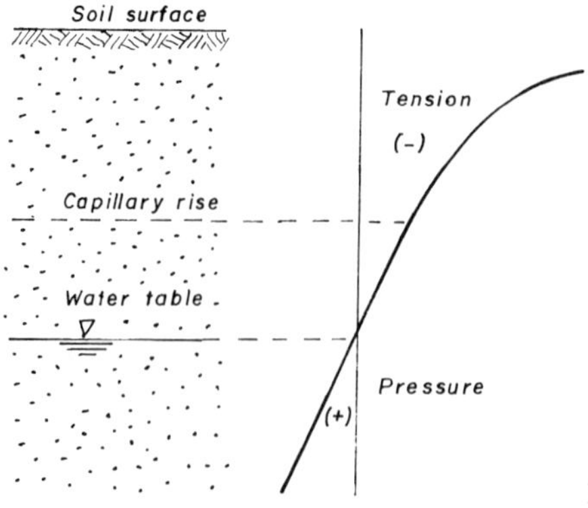

Fig. 8.26.1. Suction profile.

water head in the capillary zone is similar to the water rising in a capillary tube and it seems to be suspended from a sort of membrane funicular, the meniscus. *Pendular water*, another name for the water in the capillary zone, is thus appropriate. Table 8.25.2 relates the radius of curvature of the menisci to the capillary head and capillary suction.

There are two ways to distinguish between the zone of the soil below the water table and the capillary zone, which are both saturated:

1. By installing tensiometers, thermistors or any other suction measuring devices at various depths and monitoring the pressures in the soil water.

2. By drilling a hole in the ground, going down below the water table. Well conditions will develop in the surroundings of the hole, and after some time, depending on the permeability of the soil, an equilibrium state will be reached when the elevation of the water in the borehole equals the elevation of the water table.

Right above the elevation of the capillary head, at $h \geq h_c$, the suction in the water exceeds the tensional strength of the water, and *cavitation bubbles* develop in the soil. Since most of the pores are filled with water and the bubbles are remote from the atmospheric environment, air from the atmosphere cannot gain access to the bubbles, thus vacuum conditions and vapor pressure develop. The vapor pressure itself, as discussed in Sect. 8.5, is mainly a function of temperature, $p_v(T)$, and is given in Table 8.3.1. In the bubbles an interface is generated between the water and the vapors and it determines the suction in the water, which is greatly intensified and is no more a linear function $\mathcal{F}$ of the elevation

$$p_b = p_w - p_v = \mathcal{F}(h, T) \tag{8.26.3}$$

As the elevation h increases beyond the capillary head, $h > h_c$, and comes closer to the ground surface, the suction increases and the soil moisture and the degree of saturation of the soil decrease, until the elevation reaches a point h_a where continuous air voids evolve and atmospheric environment prevails in them.

Fig. 8.26.1 is a suction profile, representing a homogeneous ground profile and is typical of the ground suction conditions of the soil in many parts of the world. It shows the various zones discussed above, as well as the ground profile close to the surface, subject to seasonal variations.

8.27 Suction and Shrinkage

In an unsaturated natural soil, the tensional forces T of the menisci in the constrictions between the particles exert a tension stress p_c, which acts in the water and is carried over to the soil particles, Fig. 8.27.1. The net effect of

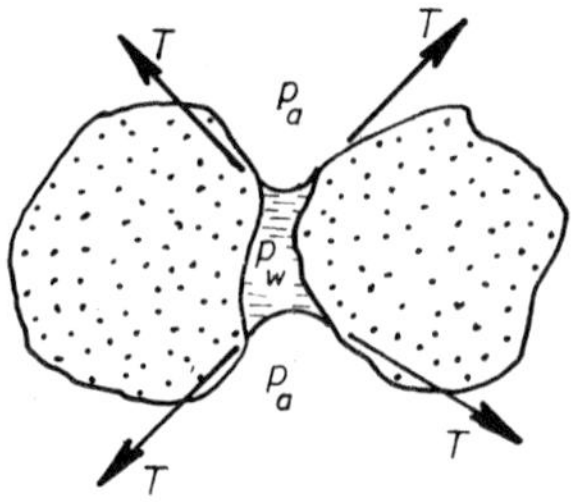

Fig. 8.27.1. Schematic display of capillary forces acting on soil particles.

these stresses acting on all sides of the particles is a resultant attraction force between the particles, which is a function of the stress p_c acting inwardly and pulling them closer together.

If the interstitial water layers and the packing of the particles allow the distances between the particles to decrease, the result is a volume decrease of the soil bulk. As the soil becomes drier, that is, it undergoes a *desorption*, the water recedes further between the interstitial voids, the radii of the menisci decrease, and the attraction forces between the particles increase further. Thus, with the drying of the soil higher interparticle stresses develop, the result being an additional volume decrease, a *shrinkage* of the soil.

In fine grained soils these forces are more distinct and the interstices permit menisci of smaller radii. This causes further increase in the stresses and more volume changes, to a point where the packing of the particles cannot tolerate any further movements and rearrangements, and additional volume changes are obstructed. The volume changes gradually cease, despite the continual decrease in moisture and water content. The soil has attained its minimum volume and maximum density, for this particular state and particular initial condition. Fig. 8.27.2 shows a typical *shrinkage curve* for a soil, and the water content at which the volume change of the soil ceases is defined as the *shrinkage limit* w_S.

When energy is invested in the soil, in this case by drying, the concomitant changes are on one hand changes in the water content w (when related to weights) or changes in the moisture content θ (when related to volumes), and on the other hand volume changes. This phenomenon is characteristic for each particular soil, and it represented by two characteristic curves, one known as the *suction curve* and the other as the shrinkage curve. Fig. 8.27.3 shows both curves for an unidentified heavy clay, presented by Croney and Coleman (1958, 1961) and redrawn so as to correspond to our presentation. The significance and important implications of these curves are discussed in more detail in Chaps. 11 and 13. Here only a brief description of the curves and the technique by which they are obtained will follow.

The shrinkage curve, in its narrow sense, is the basis for the shrinkage limit test, used widely in soil mechanics as one of the consistency tests that characterize each specific soil. By definition, the shrinkage limit w_S of a soil is the moisture content beyond which no more changes in volume occur when the moisture content decreases, in a saturated soil cake of a standardized size. A typical shrinkage curve is shown in Fig. 8.27.2. The curve is in general

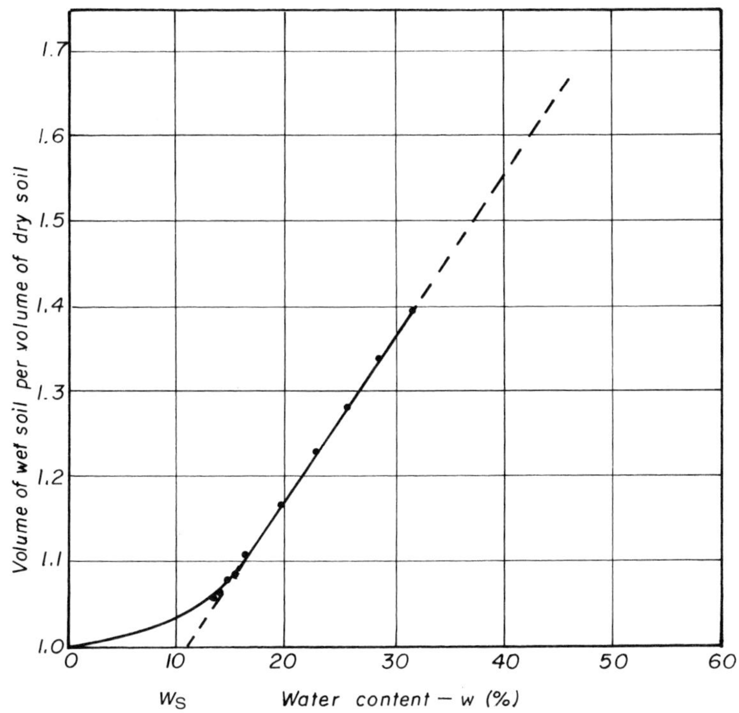

Fig. 8.27.2. Typical shrinkage curve. [After D. Croney and J. D. Coleman (1961).]

presented as $\Delta V/V_\delta$ versus w, where V_δ is the volume of the dry sample. It can be seen that in the range where volume changes occur, a linear relationship exists between the volume changes and the corresponding moisture content.

An early study (Haefeli and Amberg 1948) pointed out the important features of the shrinkage curve and the shrinkage limit, which has since been accepted as a test of decisive importance in other disciplines and industries where control of volume changes is imperative.

The suction curve, less known and applied in soil mechanics, has been the subject of intense studies and investigations in soil science serving agriculture. Early studies by Gardner (1920a), Gardner (1920b), Lebedev (1927) and Richards (1928) are still amended today by valuable bits of information on suction curves for various soils, techniques to obtain better curves, and conditions that affect the suction.

The suction curve represents a functional relationship between the capillary potential, which is the energy necessary to remove water from the soil, and the moisture content. At a known applied suction, moisture equilibrium is established in the water. The energy may be expressed in different ways: energy per unit weight, energy per unit volume, equivalent pressure, equivalent pressure head, etc. Table 8.27.1 shows practical equivalent energy levels of soil water. Since the range of energies for certain soils may span as much

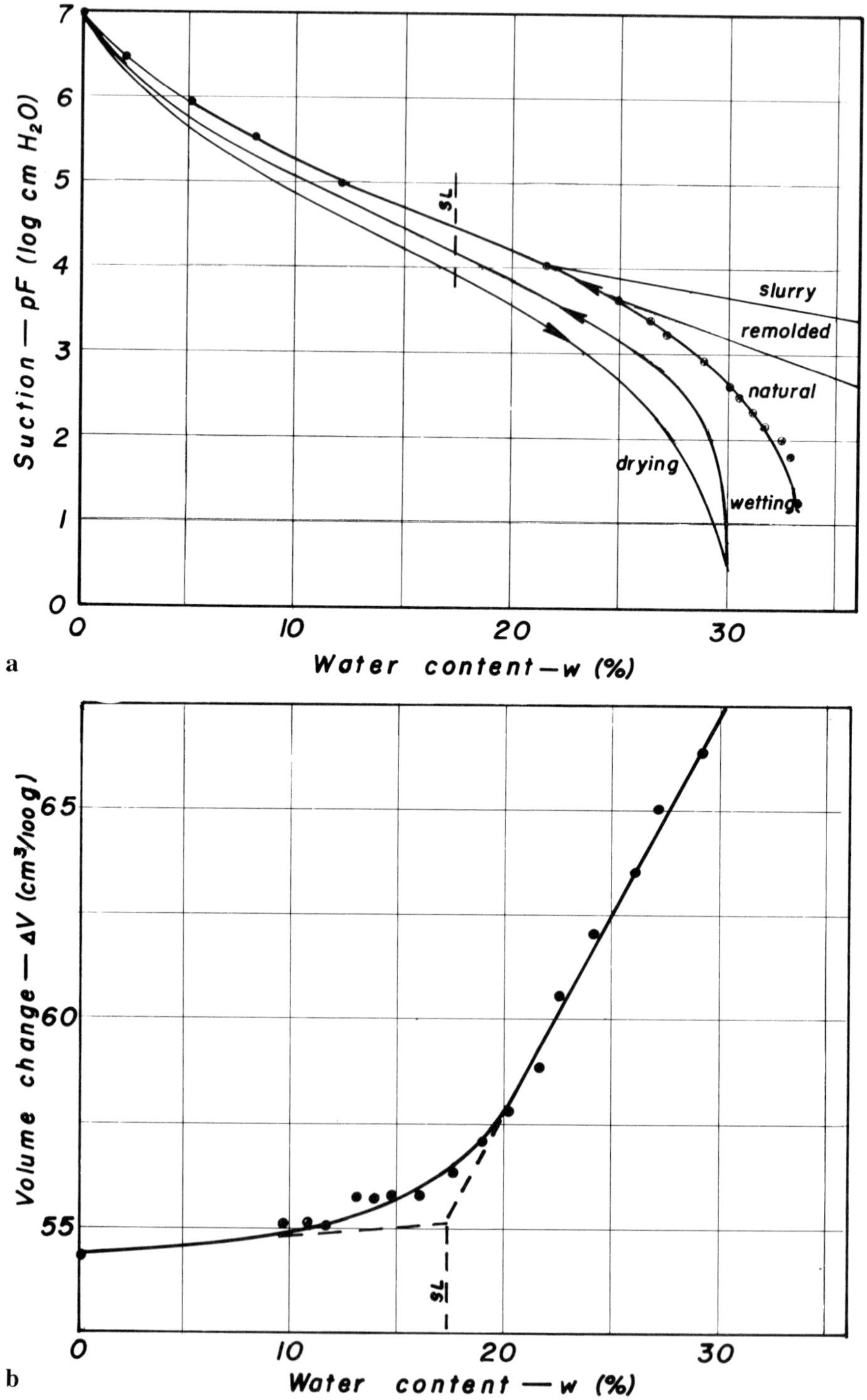

Fig. 8.27.3. **a** Suction and **b** shrinkage curves of a London clay. [After D. Croney and J. D. Coleman (1961).]

Table 8.27.1. Equivalent energy levels of soil water

Soil–water potential				Soil–water suction[a]		Vapor pressure (torr) 20 °C (%)	Relative humidity[b] at 20 °C (%)
Per unit mass		Per unit volume		Pressure (bar)	Head (cm H_2O)		
(erg/gm)	(joule/kg)	(bar)	(cm H_2O)				
0	0	0	0	0	0	17.5350	100.00
-1×10^4	−1	−0.01	−10.2	0.01	10.2	17.5349	100.00
-5×10^4	−5	−0.05	−51.0	0.05	51.0	17.5344	99.997
-1×10^5	−10	−0.1	−102.0	0.1	102.0	17.5337	99.993
-2×10^5	−20	−0.2	−204.0	0.2	204.0	17.5324	99.985
-3×10^5	−30	−0.3	−306.0	0.3	306.0	17.5312	99.978
-4×10^5	−40	−0.4	−408.0	0.4	408.0	17.5299	99.971
-5×10^5	−50	−0.5	−510.0	0.5	510.0	17.5286	99.964
-6×10^5	−60	−0.6	−612.0	0.6	612.0	17.5273	99.956
-7×10^5	−70	−0.7	−714.0	0.7	714.0	17.5260	99.949
-8×10^5	−80	−0.8	−816.0	0.8	816.0	17.5247	99.941
-9×10^5	−90	−0.9	−918.0	0.9	918.0	17.5234	99.934
-1×10^6	−100	−1.0	−1020	1.0	1020	17.5222	99.927
-2×10^6	−200	−2	−1040	2	1040	17.5089	99.851
-3×10^6	−300	−3	−3060	3	3060	17.4961	99.778
-4×10^6	−400	−4	−4080	4	4080	17.4833	99.705
-5×10^6	−500	−5	−5100	5	5100	17.4704	99.637
-6×10^6	−600	−6	−6120	6	6120	17.4572	99.556

[From D. Hillel (1980), courtesy of Academic Press.]
[a] In the absence of osmotic effects (soluble salts), soil–water suction equals matric suction; otherwise, it is the sum of matric and osmotic suctions.
[b] Relative humidity of air in equilibrium with the soil at different suction values.

as 7–8 orders of magnitude, as seen in Fig. 8.27.4, Schofield (1935) suggested the use of the term *pF of soil water* to express the energy by which water is held by the soil. *pF* is the logarithm of the tension expressed in centimeters of water. A *pF* of 1 represents a tension head of 10 cm of water, thus a *pF* of 3 is a tension head of 1000 cm or close to 1 atm.

The capillary potential may be determined by a series of different techniques, such as calculations from freezing point depression, dilatometer and vapor-pressure measurements, tensiometers, centrifuge techniques, sorbtion

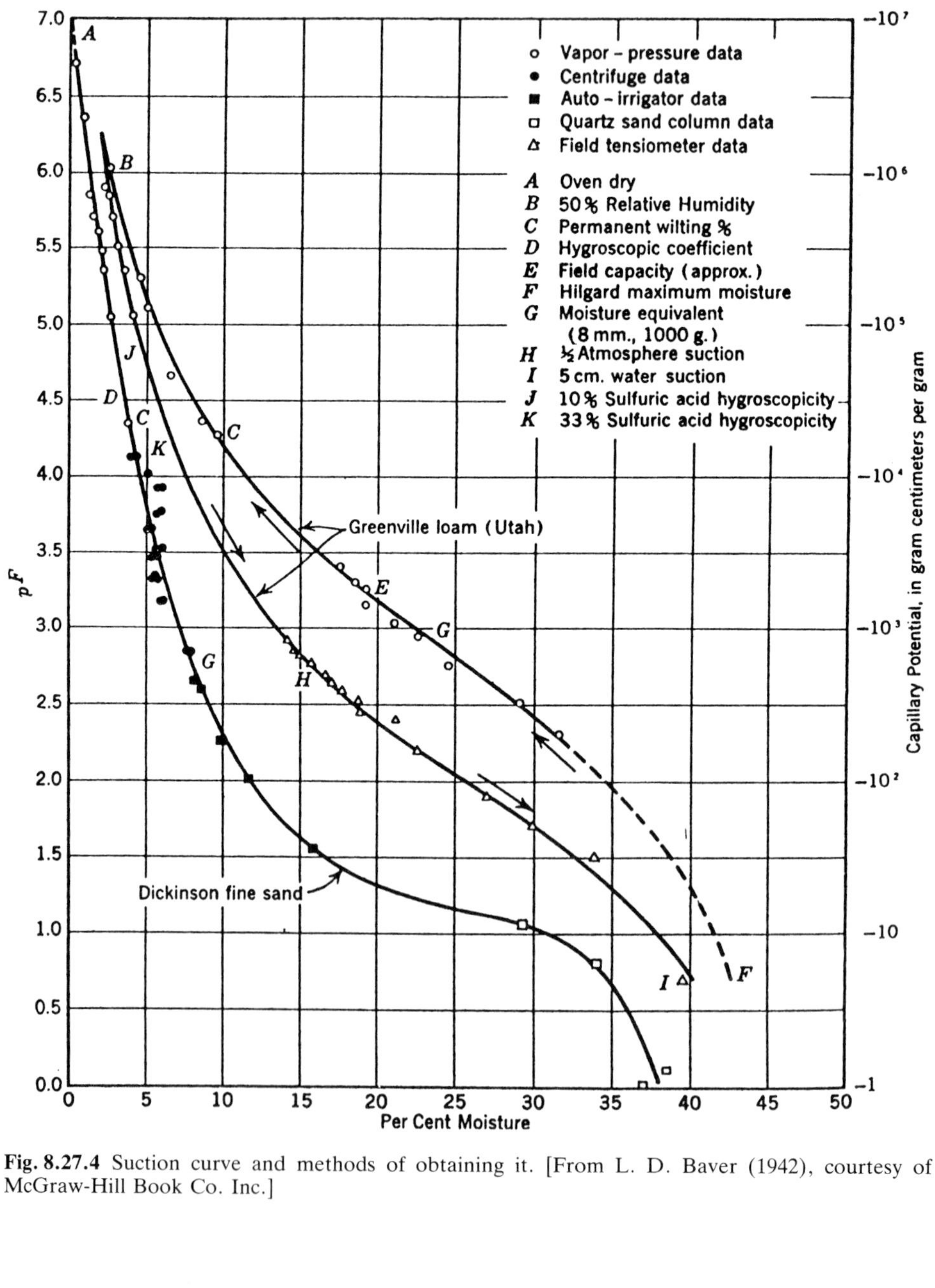

Fig. 8.27.4 Suction curve and methods of obtaining it. [From L. D. Baver (1942), courtesy of McGraw-Hill Book Co. Inc.]

balance, pressure membranes and others. A short review of the various techniques is presented by Croney and Coleman (1958, 1961). Not all methods will provide the full range of tensions; some may work better in a certain range, others in another range, but all ranges can be covered, while some methods even overlap. Fig. 8.27.4 indicates the various techniques by which the suction curves were obtained.

In spite of the fact that the suction curve is obtainable and reproducible to a high degree of accuracy and there is a considerable consistency between the various techniques, the suction is sensitive to the initial conditions in the soil, such as remolding, initial moisture, etc., particularly in the low-suction range.

A capillary potential curve, similar to the one obtained by drying, *desorption*, can be obtained by gradual *wetting* of the soil, *sorption*, proceeding from its initial dry state and successively correlating between the suction potential and its corresponding equilibrated moisture content. The sorption curve is, however, different from the desorption curve, Fig. 8.27.4. An interesting feature is involved here. The suction potential p_c, is not a single-valued function of the moisture content w. In mathematical terms, the integral $\oint w \, dp_c$ around one complete cycle is *not* equal to zero. The loop generated by the cycle is a *hysteresis loop* while the area enclosed in the loop is the *hysteresis loss*, an energy loss attributed to the inability of the water to move freely in the pores. Some of the obstructions to the free flow are the following:

1. The irregularities in the cross-sections of the void passages, known as the "ink-bottle" effect.

2. The effect of the contact angle, which is greater in an advancing meniscus than in a receding one.

3. Entrapped air or a vacuum condition, with advancing or receding meniscus, respectively, and corresponding pressure or suction, respectively.

4. Thixotropic regain or aging due to the wetting and drying history of the soil.

9 Soil as a Multi-phase Mixture

9.1 Introduction

The previous chapters introduce the basic concepts of mechanics, as they are related to soils. Chaps. 2–7 outline the fundamentals of physics and continuum mechanics, including the basics of multiphase media and stressing their relevance to our main topic, and Chap. 8 presents a fairly detailed discourse on soils, unfolding its molecular and structural system. The former introduce us to the kinematics, dynamics and thermodynamics of matter in general, presenting the balance equations and outlining the directives for the formulation of constitutive equations, while the latter is concerned specifically with the soil and its constituents, describing as closely as possible most aspects of its behavior.

Considering the principles and propositions discussed in Chap. 7, our objective now is to formulate the two constitutive equations for soils, a spherical and a deviatoric constitutive equation, and to determine the interrelation of the two.

The present chapter is a transition, introducing the terminology of soil mechanics so that concepts of physics, continuum mechanics and the theory of mixtures can be readily applied to soils.

9.2 Volume and Weight Relations in Soils

From Eqs. (6.2.1)–(6.2.6) we derive the corrcsponding equation for soils, considering it a three-phase mixture

$$\rho = \rho_{\mathfrak{s}} + \rho_{\mathfrak{w}} + \rho_{\mathfrak{g}} = \rho(\alpha_{\mathfrak{s}} + \alpha_{\mathfrak{w}} + \alpha_{\mathfrak{g}}) \quad \rho_{\mathfrak{g}} \equiv \alpha_{\mathfrak{g}} \cong 0 \tag{9.2.1}$$

where $\rho_{\mathfrak{s}}$, $\rho_{\mathfrak{w}}$ and $\rho_{\mathfrak{g}}$ indicate the solids, water and gaseous constituents, respectively, and relate the individual densities to the total density.

The density $\rho_{\mathfrak{g}}$ of the air is in general very small when compared to that of the water $\rho_{\mathfrak{w}}$ or the solids $\rho_{\mathfrak{s}}$. Their approximate magnitude, given in [FLT] units, is:

$$\rho_{\mathfrak{s}} \cong 20\text{–}300 \text{ kg s}^2/\text{m}^4 \text{ for various soils}$$

$$\rho_{\mathfrak{w}} \cong 100 \text{ kg s}^2/\text{m}^4 \text{ for de-aired water}$$

$$\rho_{\mathfrak{g}} \cong 0.1293 \text{ kg s}^2/\text{m}^4 \text{ for air at } 0\,°\text{C and 1 atm.}$$

which shows that the density of the air is of the order of magnitude of one promil or less with respect to the other constituents. Unless velocities or impulses related to the air phase are high, $\rho_{\mathfrak{g}}$ can be neglected, a common practice in soil mechanics. Consequently, the following equations are used

$$V = V_{\mathfrak{s}} + V_{\mathfrak{w}} + V_{\mathfrak{g}} = V_{\mathfrak{s}} + V_{\mathfrak{v}} \tag{9.2.2}$$

$$m = m_{\mathfrak{s}} + m_{\mathfrak{w}} + m_{\mathfrak{g}}; \quad m_{\mathfrak{g}} \cong 0 \tag{9.2.3}$$

$$W = W_{\mathfrak{s}} + W_{\mathfrak{w}} + W_{\mathfrak{g}}; \quad W_{\mathfrak{g}} \cong 0 \tag{9.2.4}$$

We recognize the sum of the volume of the water $V_{\mathfrak{w}}$ and of the air $V_{\mathfrak{g}}$ as the *volume of the voids* $V_{\mathfrak{v}}$. The familiar relations in soil mechanics between volumes and weights, Fig. 9.2.1, can now be defined.

Porosity n is the relation of the volume of the voids $V_{\mathfrak{v}}$ to the total volume V of the soil mixture

$$n = \frac{V_{\mathfrak{v}}}{V} \tag{9.2.5}$$

Void ratio e is the ratio of the volume of the voids $V_{\mathfrak{v}}$ to the volume of the solids $V_{\mathfrak{s}}$

$$e = \frac{V_{\mathfrak{v}}}{V_{\mathfrak{s}}} \tag{9.2.6}$$

The compressibility of the soil, under the assumption that the solid particles

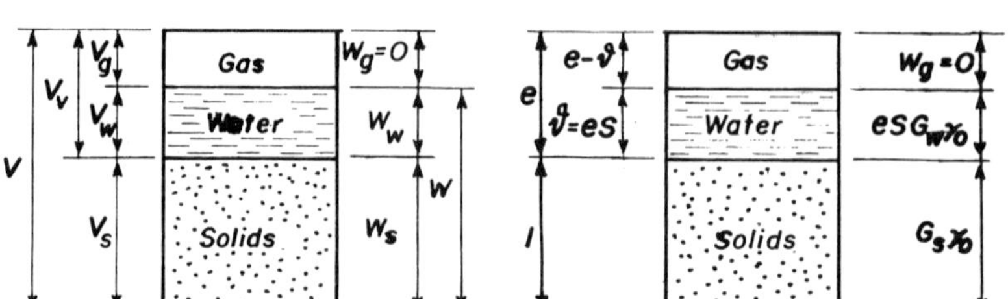

Fig. 9.2.1. Elementary volume of soil, showing weight and volume components.

are incompressible, is conveniently described by the change of the void ratio. Since all changes in volume occur in that case in the volume of the voids, the denominator of Eq. (9.2.6) remains constant and any change in density is reflected by the changes of the numerator. The assumption of incompressibility of solids is not necessarily always correct.

Degree of saturation S is the ratio of the volume of the voids occupied by water V_w to the total volume of voids V_v

$$S = \frac{V_w}{V_v} \tag{9.2.7}$$

The degree of saturation indicates how much of the voids is filled with water, and it varies from zero for dry soil to 1.0 for saturated soil.

Water content w is the ratio of the weight of the free water contained in the voids W_w to the weight of the solids W_s

$$w = \frac{W_w}{W_s} \tag{9.2.8}$$

The water content is conventionally expressed as a percentage and is extensively used in soil mechanics.

Volumetric solids fraction or *solids content* θ_s is the ratio of the volume of the solids V_s to the total volume V

$$\theta_s = \frac{V_s}{V} = 1 - n = \frac{1}{1 + e} \tag{9.2.9}$$

Volumetric moisture fraction or *moisture content* θ_w is the ratio of the volume of the water contained in the voids V_w to the total volume V

$$\theta_w = \frac{V_w}{V} = Sn = \frac{Se}{1 + e} \tag{9.2.10}$$

The moisture content concept is widely used by soil physicists.

Volumetric air fraction or *air content* θ_g, a term similar to that used by Philip (1969a), is the ratio of the volume of the air V_g to the total volume V

$$\theta_g = \frac{V_g}{V} = (1 - S)n = \frac{(1 - S)e}{1 + e} \tag{9.2.11}$$

It follows from Eqs. (9.2.9)–(9.2.11) that

$$\sum_{n=1}^{3} \theta_n = \theta_s + \theta_w + \theta_g = 1 \tag{9.2.12}$$

Moisture ratio ϑ is the ratio of the volume of the water V_w to the volume of the solids V_s

$$\vartheta = \frac{V_w}{V_s} = \frac{\theta_w}{\theta_s} = \theta_w (1 + e) = eS = \frac{nS}{1 - n} = \frac{\gamma_s}{\gamma_w} w \tag{9.2.13}$$

introduced (Philip and Smiles 1969) to represent the concentration of the water in the voids.

The ratio of weight to volume is defined as *unit weight*, and the unit weights of the non-vanishing constituents of the soil can be defined. The unit weight establishes a relationship between the geometrical quantity, volume, and the mechanical quantity, weight, of the element.

Total unit weight is the ratio of the total weight W to the total volume V

$$\gamma = \frac{W}{V} = \frac{gm}{V} = g\rho \tag{9.2.14}$$

Unit weight of solids γ_s is the ratio of the weight of the solids W_s to their volume V_s

$$\gamma_s = \frac{W_s}{V_s} = \frac{g\rho_s}{\theta_s} = \frac{g\rho_s}{1 - \mathrm{n}} = g\rho_s (1 + e) \tag{9.2.15}$$

Unit weight of water γ_w is the ratio of the weight of the water W_w contained in the voids to its volume V_w

$$\gamma_w = \frac{W_w}{V_w} = \frac{g\rho_w}{\theta_w} = \frac{g\rho_w}{Sn} = \frac{g\rho_w (1 + e)}{Se} \tag{9.2.16}$$

The unit weight of water is greatly affected by changes in temperature and varies accordingly. The unit weight of distilled deaired water at 4 °C has special technical significance: it serves to establish the weight measure of the c.g.s. system, its value being exactly 1 gram per cubic centimeter. In the foot-pound system its value is 62.4 pounds per cubic foot. It has also a special notation γ_{w0}.

Unit weight of air γ_g is the ratio of the weight of the air W_g contained in the voids not filled with water to its volume V_g

$$\gamma_g = \frac{W_g}{V_g} = \frac{g\,\rho_g}{\theta_g} = \frac{g\,\rho_g}{(1 - S)\,n} = \frac{g\,\rho_g(1 + e)}{(1 - S)e} \tag{9.2.17}$$

Dividing the numerator as well as the denominator by V_s and substituting Eqs. (9.2.15), (9.2.16) and (9.2.17) into Eq. (9.2.14), γ is obtained in terms of γ_s, γ_w and γ_g

$$\gamma = \frac{W}{V} = \frac{\dfrac{W_s}{V_s} + \dfrac{W_w V_w V_v}{V_s V_w V_v} + \dfrac{W_g}{V_s}\dfrac{(V_v - V_w)V_v}{V_g\, V_v}}{\dfrac{V_s}{V_s} + \dfrac{V_v}{V_s}} =$$

$$= \frac{\gamma_s + eS\gamma_w + (1 - S)e\gamma_g}{1 + e} = g\rho \tag{9.2.18}$$

The dry unit weight $\gamma_{\mathfrak{d}}$ is the ratio of the weight of the solids $W_{\mathfrak{s}}$ to the total volume V

$$\gamma_{\mathfrak{d}} = \frac{W_{\mathfrak{s}}}{V} = \frac{\gamma}{1 + \omega} = g\rho_{\mathfrak{s}} \tag{9.2.19}$$

The magnitude of the unit weight of a substance is not the same in various measuring systems. It will have a different value when expressed in the c.g.s. system, for example, than when expressed in the foot-pound system. If the unit weight of water at 4 °C is used as a reference term to relate the unit weights of other substances, their *specific gravity* is obtained. So we define the *specific gravity of solids* $G_{\mathfrak{s}}$

$$G_{\mathfrak{s}} = \frac{\gamma_{\mathfrak{s}}}{\gamma_{\mathfrak{w}0}} \tag{9.2.20}$$

and the *specific gravity of water* $G_{\mathfrak{w}}$ which is temperature dependent

$$G_{\mathfrak{w}} = \frac{\gamma_{\mathfrak{w}}}{\gamma_{\mathfrak{w}0}} \tag{9.2.21}$$

Finally we should be interested in the interdependence between the water content w, defined as a relation of weights, and the moisture content $\theta_{\mathfrak{w}}$, defined as a relation of volumes

$$w = \frac{W_{\mathfrak{w}}}{W_{\mathfrak{s}}} = \frac{W_{\mathfrak{w}}}{W_{\mathfrak{s}}}\frac{V_{\mathfrak{w}}}{V_{\mathfrak{w}}}\frac{V_{\mathfrak{s}}}{V_{\mathfrak{s}}}\frac{V}{V}\frac{\gamma_{\mathfrak{w}0}}{\gamma_{\mathfrak{w}0}} = \theta_{\mathfrak{w}}\frac{(1 + e)G_{\mathfrak{w}}}{G_{\mathfrak{s}}} = \frac{eSG_{\mathfrak{w}}}{G_{\mathfrak{s}}} \tag{9.2.22}$$

If the specific gravity of the water is considered unity, $G_{\mathfrak{w}} \cong 1$, $G_{\mathfrak{w}}$ is omitted from the equation.

9.3 Density Balance

In the following sections the balance equations of mixtures are applied to soils. By introducing a number of assumptions, great simplifications are achieved in the balance equations, this being done either because the equations are not applicable to soils, or in order not to complicate the study at the outset. It will be assumed that there are no transfers of supply of density, momentum, or energy from one constituent of the soil to another, since chemical reactions are seldom responsible for the changes in the mechanical behavior of soils, and whenever such changes occur, the general balance equations of mixtures can be used. The effect of interfacial forces has not been included in the balance equations, but such forces can be considered, when required, through the equations of state. In most instances the processes occurring in soils will be assumed mechanical, being thus either isothermal or isentropic, or both.

Considering soil as a three-phase mixture and substituting $\mathfrak{s}$, $\mathfrak{w}$ and $\mathfrak{g}$, the solids, the water and the air phases, respectively, for $\mathfrak{n}$ in Eq. (6.5.2), the density balance equations of the constituents of the soil are obtained

$$\frac{D\rho_{\mathfrak{s}}}{Dt} + \rho_{\mathfrak{s}} v_{\mathfrak{s}\alpha,\alpha} - C_{\mathfrak{s}} = 0 \quad C_{\mathfrak{s}} \cong 0 \tag{9.3.1}$$

$$\frac{D\rho_{\mathfrak{w}}}{Dt} + \rho_{\mathfrak{w}} v_{\mathfrak{w}\alpha,\alpha} - C_{\mathfrak{w}} = 0 \quad C_{\mathfrak{w}} \cong 0 \tag{9.3.2}$$

$$\frac{D\rho_{\mathfrak{g}}}{Dt} + \rho_{\mathfrak{g}} v_{\mathfrak{g}\alpha,\alpha} - C_{\mathfrak{g}} = 0 \quad C_{\mathfrak{g}} \cong 0 \tag{9.3.3}$$

where $\rho_{\mathfrak{s}}$, $\rho_{\mathfrak{w}}$ and $\rho_{\mathfrak{g}}$ are the densities of the solids, the water and the air, respectively, $v_{\mathfrak{s}i}$, $v_{\mathfrak{w}i}$ and $v_{\mathfrak{g}i}$ are the individual velocities and $C_{\mathfrak{s}}$, $C_{\mathfrak{w}}$ and $C_{\mathfrak{g}}$ are the *individual density supplies* of the solids, the water and the air, respectively. In view of Eqs. (6.3.1), (6.5.2), (6.5.7) and (9.2.1) we obtain

$$\frac{D\rho}{Dt} = -(\rho_{\mathfrak{s}} v_{\mathfrak{s}\alpha,\alpha} + \rho_{\mathfrak{w}} v_{\mathfrak{w}\alpha,\alpha} + \rho_{\mathfrak{g}} v_{\mathfrak{g}\alpha,\alpha})$$

$$= \frac{\partial\rho}{\partial t} + \rho_{\mathfrak{s},\alpha} v_{\mathfrak{s}\alpha} + \rho_{\mathfrak{w},\alpha} v_{\mathfrak{w}\alpha} + \rho_{\mathfrak{g},\alpha} v_{\mathfrak{g}\alpha} \tag{9.3.4}$$

The motion is always expressed with respect to a point of reference. In a multiphase mixture, where each of the constituents is in motion relative to the other constituents, the density balance can be expressed in more than one manner. One expression considers the *mean velocity of the mixture* $\bar{v}_i$ as the reference velocity of the constituents (Truesdell and Toupin 1960; Bowen 1973), another expression considers the motion of the constituents $v_{\mathfrak{n}i}$ with respect to one particular constituent considered fixed. Still another reference could be the main velocity v_i. We will see later that many of the known laws used in soil mechanics can be deduced from the balance equations, in particular the density balance, by using the solids phase as a frame of reference. In problems of permeability and consolidation the solid phase constitutes a steady matrix, with respect to which the other constituents, the water and the air phases, move. In other problems, such as those with electrophoretic phenomena, the water phase is considered as the steady phase. In consolidation problems the bottom of the consolidating layer is occasionally used as a reference.

In general, in a soil mixture with $\mathfrak{N}$ constituents, if the solids are considered as the reference constituent, the following two equations are obtained from Eq. (6.5.4)

$$\frac{D\rho_{\mathfrak{s}}}{D_{\mathfrak{s}}t} + \rho_{\mathfrak{s}} v_{\mathfrak{s}\alpha,\alpha} = 0 \tag{9.3.5}$$

$$\frac{D\rho_{\mathfrak{n}}}{D_{\mathfrak{s}}t} + [\rho_{\mathfrak{n}}(v_{\mathfrak{n}\alpha} - v_{\mathfrak{s}\alpha})]_{,\alpha} + \rho_{\mathfrak{n}} v_{\mathfrak{s}\alpha,\alpha} = 0 \tag{9.3.6}$$

where Eq. (9.3.5) is the density balance equation of the solids and Eq. (9.3.6) is the density balance equation of the other $\mathfrak{N} - 1$ constituents of the soil mixture, both equations with coordinates attached to the solids as reference and $\mathfrak{s}$ substituted for $\mathfrak{m}$.

If the value of $v_{\mathfrak{s}\alpha,\alpha} = -(1/\rho_{\mathfrak{s}})(D\rho_{\mathfrak{s}}/D_{\mathfrak{s}}t)$ from Eq. (9.3.5) is substituted in (9.3.6), the *general flow equations* for the $\mathfrak{N}$ constituents of soils are obtained

$$\frac{D\rho_{\mathfrak{n}}}{D_{\mathfrak{s}}t} - \frac{\rho_{\mathfrak{n}}}{\rho_{\mathfrak{s}}}\frac{D\rho_{\mathfrak{s}}}{D_{\mathfrak{s}}t} = \rho_{\mathfrak{s}}\frac{D\dfrac{\rho_{\mathfrak{n}}}{\rho_{\mathfrak{s}}}}{D_{\mathfrak{s}}t} = -[\rho_{\mathfrak{n}}(v_{\mathfrak{n}\alpha} - v_{\mathfrak{s}\alpha})]_{,\alpha} \tag{9.3.7}$$

Actually, Eq. (9.3.7) is the flow equation of the $\mathfrak{n}$th constituent for a soil of $\mathfrak{n}$ constituents, derived from the density balance equation. In a $\mathfrak{N}$ constituent soil, $\mathfrak{N} - 1$ constituents flow relative to the one remaining constituent. For practical reasons soils are envisaged as three-phase mixtures, $\mathfrak{N} = 3$ (solids, water and air), though more refined approaches with additional constituents are possible.

Eq. (9.3.7) considers the flow of the $\mathfrak{N} - 1$ constituents relative to the solids, and here, as well as in the following sections, this flow is our main concern. Specifically, for $\mathfrak{N} = 3$ we obtain two equations: when $\mathfrak{m}$ in Eq. (6.5.4) is replaced by $\mathfrak{s}$, and $\mathfrak{n}$ is replaced in turn by $\mathfrak{w}$ or $\mathfrak{g}$, the flow equations of water or air, respectively, are obtained with respect to the solids

$$\frac{D\rho_{\mathfrak{w}}}{D_{\mathfrak{s}}t} - \frac{\rho_{\mathfrak{w}}}{\rho_{\mathfrak{s}}}\frac{D\rho_{\mathfrak{s}}}{D_{\mathfrak{s}}t} = \rho_{\mathfrak{s}}\frac{D\dfrac{\rho_{\mathfrak{w}}}{\rho_{\mathfrak{s}}}}{D_{\mathfrak{s}}t} = -[\rho_{\mathfrak{w}}(v_{\mathfrak{w}\alpha} - v_{\mathfrak{s}\alpha})]_{,\alpha} \tag{9.3.8}$$

$$\frac{D\rho_{\mathfrak{g}}}{D_{\mathfrak{s}}t} - \frac{\rho_{\mathfrak{g}}}{\rho_{\mathfrak{s}}}\frac{D\rho_{\mathfrak{s}}}{D_{\mathfrak{s}}t} = \rho_{\mathfrak{s}}\frac{D\dfrac{\rho_{\mathfrak{g}}}{\rho_{\mathfrak{s}}}}{D_{\mathfrak{s}}t} = -[\rho_{\mathfrak{g}}(v_{\mathfrak{g}\alpha} - v_{\mathfrak{s}\alpha})]_{,\alpha} \tag{9.3.9}$$

Rather than expressing the flow equations (9.3.8) and (9.3.9) only in terms of the densities $\rho_{\mathfrak{n}}$, there are three additional options to substitute for the densities: the void ratio e, the porosity n and the volumetric fractions $\theta_{\mathfrak{n}}$, defined in Sect. 9.2. The degree of saturation S or the moisture ratio ϑ in the flow equations, which will be discussed in Chap. 10, are additional parameters characterizing the flow in unsaturated soil. The void ratio e is the more acknowledged parameter in soil mechanics; in soil physics, however, the volumetric fractions $\theta_{\mathfrak{n}}$ have greater appeal. On other occasions the porosity, where the volumes are related to the total volume, appears to be valuable. With the help of Eqs. (9.2.15)–(9.2.17) almost all derivations are convertible from any one to any other parameter. Consequently Eqs. (9.3.8) and (9.3.9) may be expressed

$$\frac{\gamma_{\mathfrak{s}}}{1+e}\,\frac{\mathrm{D}\dfrac{\gamma_{\mathfrak{w}} eS}{\gamma_{\mathfrak{s}}}}{\mathrm{D}_{\mathfrak{s}}t} = \gamma_{\mathfrak{s}}(1-n)\,\frac{\mathrm{D}\dfrac{\gamma_{\mathfrak{w}} nS}{\gamma_{\mathfrak{s}}(1-n)}}{\mathrm{D}_{\mathfrak{s}}t} = \gamma_{\mathfrak{s}}\theta_{\mathfrak{s}}\,\frac{\mathrm{D}\dfrac{\gamma_{\mathfrak{w}}\theta_{\mathfrak{w}}}{\gamma_{\mathfrak{s}}\theta_{\mathfrak{s}}}}{\mathrm{D}_{\mathfrak{s}}t}$$
$$= -[\gamma_{\mathfrak{w}}(v_{\mathfrak{w}\alpha} - v_{\mathfrak{s}\alpha})]_{,\alpha} \tag{9.3.10}$$

$$\frac{\gamma_{\mathfrak{s}}}{1+e}\,\frac{\mathrm{D}\dfrac{\gamma_{\mathfrak{g}} e(1-S)}{\gamma_{\mathfrak{s}}}}{\mathrm{D}_{\mathfrak{s}}t} = \gamma_{\mathfrak{s}}(1-n)\,\frac{\mathrm{D}\dfrac{\gamma_{\mathfrak{g}} n(1-S)}{\gamma_{\mathfrak{s}}(1-n)}}{\mathrm{D}_{\mathfrak{s}}t} = \gamma_{\mathfrak{s}}\theta_{\mathfrak{s}}\,\frac{\mathrm{D}\dfrac{\gamma_{\mathfrak{g}}\theta_{\mathfrak{g}}}{\gamma_{\mathfrak{s}}\theta_{\mathfrak{s}}}}{\mathrm{D}_{\mathfrak{s}}t}$$
$$= -\,[\gamma_{\mathfrak{g}}(v_{\mathfrak{g}\alpha} - v_{\mathfrak{s}\alpha})]_{,\alpha} \tag{9.3.11}$$

The material balance equation for linear momentum (6.6.4) and Eqs. (6.6.5) and (6.6.6) that evolve from it are of particular interest in the study of soils. We shall substitute the pressure tensor p_{ij} for the stress tensor t_{ij}, and the gravity force B_i and an acceleration α_i for the body force σ_i, according to Eqs. (4.8.1) and (4.8.2). If the diffusion velocities of the constituents are small enough, so that the dynamic components $\rho_{\mathfrak{n}}\bar{v}_{\mathfrak{n}i}\bar{v}_{\mathfrak{n}j}$ may be ignored, Eq. (6.6.5) becomes

9.4 Balance of Linear Momentum

The material balance equation for linear momentum (6.6.4) and Eqs. (6.6.5) and (6.6.6) that evolve from it are of particular interest in the study of soils. We shall substitute the pressure tensor p_{ij} for the stress tensor t_{ij}, and the gravity force B_i and an acceleration α_i for the body force σ_i, according to Eqs. (4.8.1) and (4.8.2). If the diffusion velocities of the constituents are small enough, so that the dynamic components $\rho_{\mathfrak{n}}\bar{v}_{\mathfrak{n}i}\bar{v}_{\mathfrak{n}j}$ may be ignored, Eq. (6.6.5) becomes

$$p_{ij} = \sum_{\mathfrak{n}=1}^{\mathfrak{N}} p_{\mathfrak{n}ij}; \quad \bar{v}_{\mathfrak{s}i} = \bar{v}_{\mathfrak{w}i} = \bar{v}_{\mathfrak{g}i} \cong 0 \tag{9.4.1}$$

where $p_{\mathfrak{n}ij}$ represents the static stress tensor acting on each individual constituent.

However, the dynamic component cannot always be ignored. Unless special care is taken to apply the load slowly or in small increments, loadings could be accompanied by a dynamic component, due to the velocity or acceleration at which the applied load excites the mass of the soil constituents, solids, water and air alike.

Since the water and air phases do not transfer shear stresses, the following quantities may be substituted for the constituents of soil: $p_{\mathfrak{s}ij} = p_{\mathfrak{s}ij}$, $p_{\mathfrak{w}ij} = p_{\mathfrak{w}}\delta_{ij}$, $p_{\mathfrak{g}ij} = p_{\mathfrak{g}}\delta_{ij}$ and equation (6.6.5) will become

$$p_{ij} = p_{\mathfrak{s}ij} + p_{\mathfrak{w}}\delta_{ij} + p_{\mathfrak{g}}\delta_{ij} + \rho_{\mathfrak{s}}\bar{v}_{\mathfrak{s}i}\bar{v}_{\mathfrak{s}j} + \rho_{\mathfrak{w}}\bar{v}_{\mathfrak{w}i}\bar{v}_{\mathfrak{w}j} + \rho_{\mathfrak{g}}\bar{v}_{\mathfrak{g}i}\bar{v}_{\mathfrak{g}j} \tag{9.4.2}$$

where $p_{\mathfrak{s}ij}$, $p_{\mathfrak{w}}\delta_{ij}$ and $p_{\mathfrak{g}}\delta_{ij}$ are the *individual static stress tensors* of the three constituents of soils, and $\rho_{\mathfrak{s}}\bar{v}_{\mathfrak{s}i}\bar{v}_{\mathfrak{s}j}$, $\rho_{\mathfrak{w}}\bar{v}_{\mathfrak{w}i}\bar{v}_{\mathfrak{w}j}$ and $\rho_{\mathfrak{g}}\bar{v}_{\mathfrak{g}i}\bar{v}_{\mathfrak{g}j}$ are the *individual dynamic stresses* generated by the diffusional motion.

The body forces expressed in Eq. (6.6.6) are gravity forces represented by the gravity acceleration g_i, or centrifugal and other forces represented by any other acceleration, say α_i

$$\begin{aligned}
\sum_{\mathfrak{n}=1}^{\mathfrak{N}} \rho_{\mathfrak{n}}\sigma_{\mathfrak{n}} &= -\sum_{\mathfrak{n}=1}^{\mathfrak{N}} \rho_{\mathfrak{n}} B_{\mathfrak{n}} = -\rho_{\mathfrak{s}} B_{\mathfrak{s}i} - \rho_{\mathfrak{w}} B_{\mathfrak{w}i} - \rho_{\mathfrak{g}} B_{\mathfrak{g}i} \\
&= -\rho_{\mathfrak{s}}(g_i + a_{\mathfrak{s}i}) - \rho_{\mathfrak{w}}(g_i + a_{\mathfrak{w}i}) - \rho_{\mathfrak{g}}(g_i + a_{\mathfrak{g}i}) \\
&= -\rho_{\mathfrak{s}}(g_i + a_{\mathfrak{s}i}) - \rho_{\mathfrak{w}}(g_i + a_{\mathfrak{w}i}) \qquad (9.4.3) \\
&= -\rho_{\mathfrak{g}}(g_i + a_{\mathfrak{g}i}) \cong 0
\end{aligned}$$

where $B_{\mathfrak{s}i}$, $B_{\mathfrak{w}i}$, $B_{\mathfrak{g}i}$ are the *individual body forces per unit mass*, acting on the constituents of the soil.

Eq. (6.6.2), when applied to the three phases of soils, reads as following

$$\rho_{\mathfrak{s}}\dot{v}_{\mathfrak{s}i} + p_{\mathfrak{s}i\alpha,\alpha} + \rho_{\mathfrak{s}} B_{\mathfrak{s}i} = 0 \tag{9.4.4}$$

$$\rho_{\mathfrak{w}}\dot{v}_{\mathfrak{w}i} + p_{\mathfrak{w}i\alpha,\alpha} + \rho_{\mathfrak{w}} B_{\mathfrak{w}i} = \rho_{\mathfrak{w}}\dot{v}_{\mathfrak{w}i} + p_{\mathfrak{w},\alpha}\delta_{i\alpha} + \rho_{\mathfrak{w}} B_{\mathfrak{w}i} = 0 \tag{9.4.5}$$

$$\rho_{\mathfrak{g}}\dot{v}_{\mathfrak{g}i} + p_{\mathfrak{g}i\alpha,\alpha} + \rho_{\mathfrak{g}} B_{\mathfrak{g}i} = \rho_{\mathfrak{g}}\dot{v}_{\mathfrak{g}i} + p_{\mathfrak{g},\alpha}\delta_{i\alpha} + \rho_{\mathfrak{g}} B_{\mathfrak{g}i} = 0 \tag{9.4.6}$$

and represents the nine balance equations of linear momentum for soils.

Eq. (9.4.2) indicates that in the absence of dynamic stresses, *the sum of the internal stresses in the constituting phases is equal to the macroscopic stress acting on the soil mixture.* The corollary also exists: *The total stress applied over the soil mixture is carried by and distributed among its constituents*

$$p_{ij} = p_{\mathfrak{s}ij} + p_{\mathfrak{w}}\delta_{ij} + p_{\mathfrak{g}}\delta_{ij} \quad \text{for } \bar{v}_{\mathfrak{n}i} \cong 0 \tag{9.4.7}$$

In Eq. (9.4.7) we recognize the *total pressure* p_{ij}, which is the sum of the *effective pressure tensor* $p_{\mathfrak{s}ij}$ and the *pore water* and *pore air pressure tensors* $p_{\mathfrak{w}}\delta_{ij}$ and $p_{\mathfrak{g}}\delta_{ij}$, respectively. This equation will be discussed further in Sect. 12.2.

The stress tensor of Eq. (9.4.7) may be decomposed, according to Eq. (4.9.1), into two parts: a spherical stress $\frac{1}{3}p_{\alpha\alpha}$ and a deviator stress s_{ij}, as follows

$$\tfrac{1}{3}p_{\alpha\alpha} = \tfrac{1}{3}p_{\mathfrak{s}\alpha\alpha} + p_{\mathfrak{w}} + p_{\mathfrak{g}} = p_{\mathfrak{m}} + p_{\mathfrak{p}} \tag{9.4.8}$$

$$
\begin{aligned}
s_{ij} &= p_{ij} - \tfrac{1}{3} p_{\alpha\alpha} \delta_{ij} = p_{ij} - \tfrac{1}{3} p_{\mathfrak{s}\alpha\alpha} \delta_{ij} - p_{\mathfrak{w}} \delta_{ij} - p_{\mathfrak{g}} \delta_{ij} \\
&= p_{\mathfrak{s}ij} + p_{\mathfrak{w}} \delta_{ij} + p_{\mathfrak{g}} \delta_{ij} - \tfrac{1}{3} p_{\mathfrak{s}\alpha\alpha} \delta_{ij} - p_{\mathfrak{w}} \delta_{ij} - p_{\mathfrak{g}} \delta_{ij} \\
&= p_{\mathfrak{s}ij} - \tfrac{1}{3} p_{\mathfrak{s}\alpha\alpha} \delta_{ij} = s_{\mathfrak{s}ij}
\end{aligned} \tag{9.4.9}
$$

where $s_{ij} = s_{\mathfrak{s}ij}$ is the *shear stress* of the soil equal to the shear stress of the solids, $p_{\mathfrak{s}ij} = \frac{1}{3} p_{\mathfrak{s}\alpha\alpha} \delta_{ij} + s_{\mathfrak{s}ij}$ is the *effective pressure*, and the shear stresses of the water and the air vanish, $s_{\mathfrak{w}ij} = s_{\mathfrak{g}ij} = 0$.

A close look at Eq. (9.4.8) reveals that we have here the much sought-after formal proof of the existence of pore pressures, an essential concept widely used in soil engineering. We recognize here several other soil mechanics concepts: p_{ij} the *total stress*, $p_{\mathfrak{s}ij}$ the *effective pressure*, $p_{\mathfrak{w}}$ the *pore–water pressure* and $p_{\mathfrak{g}}$ the *pore–air pressure*. It should be observed that the spherical stress $\frac{1}{3} p_{\alpha\alpha}$ introduced in Sect. 4.9 is carried by the equivalent effective stress $\frac{1}{3} p_{\mathfrak{s}\alpha\alpha}$, the pore–water pressure $p_{\mathfrak{w}}$ and the pore–air pressure $p_{\mathfrak{g}}$ in an unsaturated soil considered as a three-phase mixture. For saturated soils the last term on the right-hand side of Eq. (9.4.7) vanishes, and then the total stress is carried by the equivalent effective stress and the pore–water pressure alone.

Eq. (9.4.9) confirms our original assumption that shear stresses are carried only by the solid phase. Eqs. (9.4.7) and (9.4.8) are discussed at length in Chaps. 11 and 12.

If the dynamic phenomena cannot be ignored, the stress tensor of Eq. (9.4.2) can be similarly decomposed, according to Eq. (4.9.1), into two parts: a spherical stress $\frac{1}{3} p_{\alpha\alpha}$ and a deviator stress s_{ij}, as follows

$$
\tfrac{1}{3} p_{\alpha\alpha} = \tfrac{1}{3} p_{\mathfrak{s}\alpha\alpha} + p_{\mathfrak{w}} + p_{\mathfrak{g}} + \tfrac{1}{3} (\rho_{\mathfrak{s}}\, \bar{v}_{\mathfrak{s}\alpha} \bar{v}_{\mathfrak{s}\alpha} + \rho_{\mathfrak{s}}\, \bar{v}_{\mathfrak{s}\alpha} \bar{v}_{\mathfrak{s}\alpha} + \rho_{\mathfrak{s}}\, \bar{v}_{\mathfrak{s}\alpha} \bar{v}_{\mathfrak{s}\alpha}) \tag{9.4.10}
$$

$$
\begin{aligned}
s_{ij} &= p_{ij} - \tfrac{1}{3} p_{\alpha\alpha}\, \delta_{ij} = p_{\mathfrak{s}ij} - \tfrac{1}{3} p_{\mathfrak{s}\alpha\alpha} \delta_{ij} + \rho_{\mathfrak{s}} (\bar{v}_{\mathfrak{s}i} \bar{v}_{\mathfrak{s}j} - \tfrac{1}{3} \bar{v}_{\mathfrak{s}\alpha}\, \bar{v}_{\mathfrak{s}\alpha} \delta_{ij}) \\
&\quad + \rho_{\mathfrak{w}}\, (\bar{v}_{\mathfrak{w}i} \bar{v}_{\mathfrak{w}j} - \tfrac{1}{3} \bar{v}_{\mathfrak{w}\alpha} \bar{v}_{\mathfrak{w}\alpha} \delta_{ij}) + \rho_{\mathfrak{g}} (\bar{v}_{\mathfrak{g}i} \bar{v}_{\mathfrak{g}j} - \tfrac{1}{3} \bar{v}_{\mathfrak{g}\alpha} \bar{v}_{\mathfrak{g}\alpha} \delta_{ij})
\end{aligned} \tag{9.4.11}
$$

where $s_{\mathfrak{s}ij}$ may be substituted for $p_{\mathfrak{s}ij} - \frac{1}{3} p_{\mathfrak{s}\alpha\alpha} \delta_{ij}$ in Eq. (9.4.11).

Dynamic forces such as oscillatory loadings or transient loadings are represented in Eq. (9.4.10) by the squared values of the diffusive velocities $\bar{v}_{\mathfrak{s}i}$, $\bar{v}_{\mathfrak{w}i}$ and $\bar{v}_{\mathfrak{g}i}$. Since the mobility of the solids is limited in many instances, $\bar{v}_{\mathfrak{s}i}$ is small and the effect of $\rho_{\mathfrak{s}} \bar{v}_{\mathfrak{s}i} \bar{v}_{\mathfrak{s}j}$ is negligible. The velocities $\bar{v}_{\mathfrak{w}i}$ and $\bar{v}_{\mathfrak{g}i}$ of the water and the air, however, are more mobile and susceptible to dynamic excitations. Being squared in Eqs. (9.4.10) and (9.4.11) they have always positive values, irrespective of their direction, and consequently they contribute to an increase of the pore pressures and to a reduction of the equivalent effective pressure in the soil. Fig. 9.4.1 shows peak pore-water pressures in a sample close to saturation at the instant of applying the constant pressure increment, as recorded at several points at various distances from the drainage boundaries (Schmid 1962). The curves show the dynamic

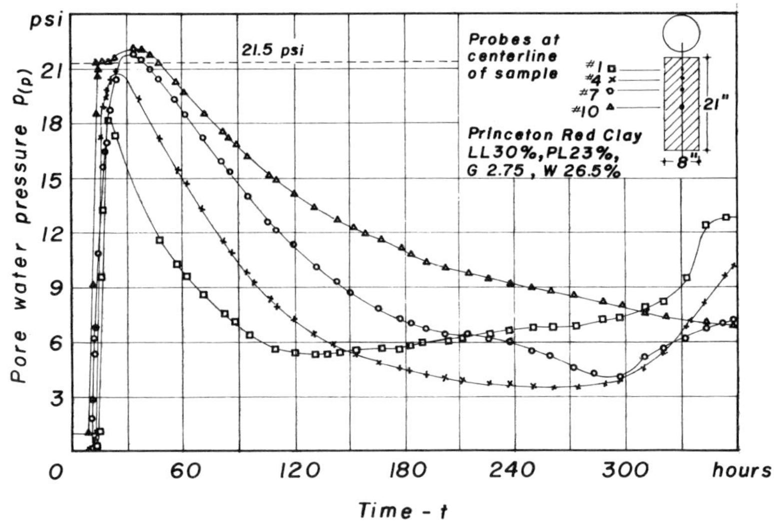

Fig. 9.4.1. Pore pressure measurement in large cell test.

effect of the transient loading on the pore pressures, compared to the magnitude of the applied pressure.

This increase of the pore-water and possibly the pore-air pressure, if carried beyond a certain limit, may reduce drastically the equivalent effective pressure, the main cause for the phenomenon known as *liquefaction*. Many studies, mostly experimental, have been presented along the years on the phenomenon of liquefaction, and criteria have been suggested for its occurrence (Arthur et al. 1979; Seed et al. 1981; Selig and Chang 1981). In most of these studies, however, the results cannot be evaluated on the basis of Eqs. (9.4.10) and (9.4.11), since many relevant data, such as information about the velocities or accelerations of the individual constituents are not disclosed. Moreover, the volumetric and shear phenomena dealt with are superimposed, and this prevents their discriminate consideration. A re-evaluation of the results could most efficiently be done when the necessary data are provided. It is instructive to quote a pertinent sentence from a discussion (Pyke 1979) on one of these publications on liquefaction: "It is more important to understand the process involved than it is to come up with a dictionary-type definition."

The dynamic component of the spherical stress in Eq. (9.4.10) has the effect of compressing the material, the latter resists the compression and a wave analogous to a *sound wave*, also called *normal wave* or *longitudinal wave*, can be generated. If such a wave is generated, the dynamic component in Eq. (9.4.10) will change slightly, displaying a cyclic expression.

Eq. (9.4.11) reveals two effects of the dynamic component contained in the shear stresses: a decrease of the shear stress due to residuals of the spherical component $\rho_{\mathrm{n}}\bar{v}_{\mathrm{n}\alpha}\bar{v}_{\mathrm{n}\alpha}\delta_{ij}$ which is always positive, and a "decrease" of the shear stress related to the diffusive velocities $\bar{v}_{\mathrm{n}i}$ of the components $\rho_{\mathrm{n}}\bar{v}_{\mathrm{n}i}\bar{v}_{\mathrm{n}j}$. The sign of this second "decrease" is conditional and depends on the signs of the two velocities $\bar{v}_{\mathfrak{n}i}$ and $\bar{v}_{\mathfrak{n}j}$ which appear in the equation. If $\bar{v}_{\mathfrak{n}i}$ and $\bar{v}_{\mathfrak{n}j}$ are opposite in sign, the effect is a decrease, and $\rho_{\mathfrak{n}}\bar{v}_{\mathfrak{n}\alpha}\bar{v}_{\mathfrak{n}\alpha}\delta_{ij}$ adds up with $\rho_{\mathfrak{n}}\bar{v}_{\mathfrak{n}i}\bar{v}_{\mathfrak{n}j}$. But if $\bar{v}_{\mathfrak{n}i}$ and $\bar{v}_{\mathfrak{n}j}$ have the same sign, the effect is an increase in the shear stress.

In the shear stresses in soils dynamic effects in the form of waves could be present as well. A wave known as *shear wave* or *transversal wave* can develop and bounce back and forth affecting the shear stresses. As we have just seen, the dynamic components may increase or decrease the shear stresses, depending on which of the components is predominant. In general, the term depending on the individual velocities and residuals of the pore pressures dominates, and the result is a decrease in the shearing resistance of the soil.

The differences between Eqs. (9.4.10) and (9.4.11) should be noted: Eq. (9.4.10) is a scalar equation. If the dynamic component results in a compressive wave, such a wave would act with the same intensity in all directions, and, as we have already remarked, would be in this respect similar to a sound wave (sound has no polarization, it is just a pressure wave). Eq. (9.4.11), however, is a tensorial equation, with components different in the different directions. Thus, if shear waves were to be generated, they would be somewhat more analogous, so far as their polarization is concerned, to light waves.

Waves also travel at a certain speed: longitudinal waves travel at the same speed in all directions, while shear waves propagate at a characteristic speed in each direction, their speed is always less than the speed of longitudinal waves, and they dissipate faster.

All these concepts are most important for problems of soil dynamics, but will not be explored much further in the present work.

From Eq. (6.6.4) we obtain the balance equation for linear momentum applied to soils

$$\rho_{\mathfrak{s}}\frac{\mathrm{D}v_{\mathfrak{s}i}}{\mathrm{D}t} + \rho_{\mathfrak{w}}\frac{\mathrm{D}v_{\mathfrak{w}i}}{\mathrm{D}t} + \rho_{\mathfrak{g}}\frac{\mathrm{D}v_{\mathfrak{g}i}}{\mathrm{D}t} + (p_{\mathfrak{s}i\alpha} + \rho_{\mathfrak{s}}v_{\mathfrak{s}i}v_{\mathfrak{s}\alpha})_{,\alpha}$$

$$+ (p_{\mathfrak{w}i\alpha} + \rho_{\mathfrak{w}}v_{\mathfrak{w}i}v_{\mathfrak{w}\alpha})_{,\alpha} + (p_{\mathfrak{w}i\alpha} + \rho_{\mathfrak{w}}v_{\mathfrak{w}i}v_{\mathfrak{w}\alpha})_{,\alpha}$$

$$+ \rho_{\mathfrak{s}}(g_i + a_{\mathfrak{s}i}) + \rho_{\mathfrak{w}}(g_i + a_{\mathfrak{w}i}) + \rho_{\mathfrak{w}}a_{\mathfrak{w}i} = 0; \quad \rho_{\mathfrak{w}}g_i \cong 0 \qquad (9.4.12)$$

For the non-polar case, Eq. (6.7.4) yields

$$s_{ij} = s_{ji} \qquad (9.4.13)$$

which results in

$$p_{\mathfrak{s}ij} = p_{\mathfrak{s}ji}; \quad s_{\mathfrak{s}ij} = s_{\mathfrak{s}ji} \qquad (9.4.14)$$

since $\rho_{\mathfrak{n}}v_{\mathfrak{n}i}v_{\mathfrak{n}j}$ are always symmetric.

9.5 Balance of the Internal Energy

The balance of internal energy in soils, Eq. (6.8.2), is obtained when the proper substitutions are made for the indices designating the constituents, assuming that energy is not transferred from one constituent to another

$$\rho_{\mathfrak{s}}\dot{\epsilon}_{\mathrm{i}\mathfrak{s}} + p_{\mathfrak{s}\alpha\beta}f_{\mathfrak{s}\alpha\beta} - h_{\mathfrak{s}\alpha,\alpha} - \rho_{\mathfrak{s}}q_{\mathfrak{s}} = 0 \tag{9.5.1}$$

$$\rho_{\mathfrak{w}}\dot{\epsilon}_{\mathrm{i}\mathfrak{w}} + p_{\mathfrak{w}\alpha\beta}f_{\mathfrak{w}\alpha\beta} - h_{\mathfrak{w}\alpha,\alpha} - \rho_{\mathfrak{w}}q_{\mathfrak{w}} = 0 \tag{9.5.2}$$

$$\rho_{\mathfrak{g}}\dot{\epsilon}_{\mathrm{i}\mathfrak{g}} + p_{\mathfrak{g}\alpha\beta}f_{\mathfrak{g}\alpha\beta} - h_{\mathfrak{g}\alpha,\alpha} - \rho_{\mathfrak{g}}q_{\mathfrak{g}} = 0 \tag{9.5.3}$$

where, according to Eqs. (6.8.5)–(6.8.7), the following identities are valid

$$\epsilon_{\mathrm{i}} = \alpha_{\mathfrak{s}}(\epsilon_{\mathrm{i}\mathfrak{s}} + \tfrac{1}{2}\bar{v}_{\mathfrak{s}\alpha}\bar{v}_{\mathfrak{s}\alpha}) + \alpha_{\mathfrak{w}}(\epsilon_{\mathrm{i}\mathfrak{w}} + \tfrac{1}{2}\bar{v}_{\mathfrak{w}\alpha}\bar{v}_{\mathfrak{w}\alpha}) + \alpha_{\mathfrak{g}}(\epsilon_{\mathrm{i}\mathfrak{g}} + \tfrac{1}{2}\bar{v}_{\mathfrak{g}\alpha}\bar{v}_{\mathfrak{g}\alpha}) \tag{9.5.4}$$

$$\begin{aligned} h_i = {} & h_{\mathfrak{s}i} + h_{\mathfrak{w}i} + h_{\mathfrak{g}i} - p_{\mathfrak{s}\beta i}\bar{v}_{\mathfrak{s}\beta} - p_{\mathfrak{w}\beta i}\bar{v}_{\mathfrak{w}\beta} - p_{\mathfrak{g}\beta i}\bar{v}_{\mathfrak{g}\beta} \\ & - \rho_{\mathfrak{s}}(\epsilon_{\mathrm{i}\mathfrak{s}} + \tfrac{1}{2}\bar{v}_{\mathfrak{s}\alpha}\bar{v}_{\mathfrak{s}\alpha})\bar{v}_{\mathfrak{s}i} - \rho_{\mathfrak{w}}(\epsilon_{\mathrm{i}\mathfrak{w}} + \tfrac{1}{2}\bar{v}_{\mathfrak{w}\alpha}\bar{v}_{\mathfrak{w}\alpha})\bar{v}_{\mathfrak{w}i} - \rho_{\mathfrak{g}}(\epsilon_{\mathrm{i}\mathfrak{g}} + \tfrac{1}{2}\bar{v}_{\mathfrak{g}\alpha}\bar{v}_{\mathfrak{g}\alpha})\bar{v}_{\mathfrak{g}i} \end{aligned} \tag{9.5.5}$$

$$\begin{aligned} q = {} & \alpha_{\mathfrak{s}}q_{\mathfrak{s}} + \alpha_{\mathfrak{w}}q_{\mathfrak{w}} + \alpha_{\mathfrak{g}}q_{\mathfrak{g}} \\ & - \alpha_{\mathfrak{s}}(g_\alpha + a_{\mathfrak{s}\alpha})\bar{v}_{\mathfrak{s}\alpha} - \alpha_{\mathfrak{w}}(g_\alpha + a_{\mathfrak{w}\alpha})\bar{v}_{\mathfrak{w}\alpha} - \alpha_{\mathfrak{g}}(g_\alpha + a_{\mathfrak{g}\alpha})\bar{v}_{\mathfrak{g}\alpha} \end{aligned} \tag{9.5.6}$$

It was pointed out in Sect. 6.8 that under static or near static conditions the total specific internal energy, the total efflux and the total energy supply will be equal to the sum of the specific internal energies, the energy effluxes and the energy supplies of the constituents, respectively. Under dynamic conditions, however, the diffusion velocities have to be considered.

When Eqs. (9.5.1)–(9.5.3) are summed up, we obtain the internal energy rate from Eq. (6.8.4)

$$\begin{aligned} \rho\dot{\epsilon}_{\mathrm{i}} = {} & - p_{\mathfrak{s}\alpha\beta}f_{\mathfrak{s}\alpha\beta} - p_{\mathfrak{w}}\delta_{\alpha\beta}f_{\mathfrak{w}\alpha\beta} - p_{\mathfrak{g}}\delta_{\alpha\beta}f_{\mathfrak{g}\alpha\beta} + (h_{\mathfrak{s}\alpha} + h_{\mathfrak{w}\alpha} + h_{\mathfrak{g}\alpha})_{,\alpha} \\ & + \alpha_{\mathfrak{s}}q_{\mathfrak{s}} + \alpha_{\mathfrak{w}}q_{\mathfrak{w}} + \alpha_{\mathfrak{g}}q_{\mathfrak{g}} = 0 \end{aligned} \tag{9.5.7}$$

provided the equivalent of equation (6.8.8) is satisfied, that is

$$\begin{aligned} & \rho_{\mathfrak{s}}E_{\mathfrak{s}} + \rho_{\mathfrak{w}}E_{\mathfrak{w}} + \rho_{\mathfrak{g}}E_{\mathfrak{g}} + T_{\mathfrak{s}\alpha}\bar{v}_{\mathfrak{s}\alpha} + T_{\mathfrak{w}\alpha}\bar{v}_{\mathfrak{w}\alpha} + T_{\mathfrak{g}\alpha}\bar{v}_{\mathfrak{g}\alpha} + C_{\mathfrak{s}}(\dot{\epsilon} + \overline{\tfrac{1}{2}\bar{v}_{\mathfrak{s}\alpha}\bar{v}_{\mathfrak{s}\alpha}}^{\,\cdot}) \\ & \quad + C_{\mathfrak{w}}(\dot{\epsilon}_{\mathfrak{w}} + \tfrac{1}{2}\,\overline{\bar{v}_{\mathfrak{w}\alpha}\bar{v}_{\mathfrak{w}\alpha}}^{\,\cdot}) + C_{\mathfrak{g}}(\dot{\epsilon}_{\mathfrak{g}} + \tfrac{1}{2}\,\overline{\bar{v}_{\mathfrak{g}\alpha}\bar{v}_{\mathfrak{g}\alpha}}^{\,\cdot}) = 0 \end{aligned} \tag{9.5.7'}$$

The thermodynamic potentials, or the Gibbs functions, as they are known, are of interest with respect to soils, particularly the free energy and the enthalpy. From Eqs. (6.9.8)–(6.9.10) the corresponding equations for soils are obtained

$$\begin{aligned}
\rho\dot{\psi} = &- p_{\mathfrak{s}\alpha\beta} f_{\mathfrak{s}\alpha\beta} - p_{\mathfrak{w}} \delta_{\alpha\beta} f_{\mathfrak{w}\alpha\beta} - p_{\mathfrak{g}} \delta_{\alpha\beta} f_{\mathfrak{g}\alpha\beta} + (h_{\mathfrak{s}\alpha} + h_{\mathfrak{w}\alpha} + h_{\mathfrak{g}\alpha})_{,\alpha} \\
&+ \rho_{\mathfrak{s}} q_{\mathfrak{s}} + \rho_{\mathfrak{w}} q_{\mathfrak{w}} + \rho_{\mathfrak{g}} q_{\mathfrak{g}} - \rho_{\mathfrak{s}} T\dot{\eta}_{\mathfrak{s}} - \rho_{\mathfrak{w}} T\dot{\eta}_{\mathfrak{w}} - \rho_{\mathfrak{g}} T\dot{\eta}_{\mathfrak{g}} \\
&- \rho_{\mathfrak{s}} \dot{T}\eta_{\mathfrak{s}} - \rho_{\mathfrak{w}} \dot{T}\eta_{\mathfrak{w}} - \rho_{\mathfrak{g}} \dot{T}\eta_{\mathfrak{g}} \\
&= \rho_{\mathfrak{s}} \sum_{\mathrm{m}=1}^{\mathfrak{M}} \tau_{\mathrm{m}\mathfrak{s}} \dot{v}_{\mathrm{m}\mathfrak{s}} + \rho_{\mathfrak{w}} \sum_{\mathrm{m}=1}^{\mathfrak{M}} \tau_{\mathrm{m}\mathfrak{w}} \dot{v}_{\mathrm{m}\mathfrak{w}} + \rho_{\mathfrak{g}} \sum_{\mathrm{m}=1}^{\mathfrak{M}} \tau_{\mathrm{m}\mathfrak{g}} \dot{v}_{\mathrm{m}\mathfrak{g}} \\
&- \rho_{\mathfrak{s}} \eta_{\mathfrak{s}} \dot{T} - \rho_{\mathfrak{w}} \eta_{\mathfrak{w}} \dot{T} - \rho_{\mathfrak{g}} \eta_{\mathfrak{g}} \dot{T}
\end{aligned} \tag{9.5.8}$$

$$\begin{aligned}
\rho\chi = &- p_{\mathfrak{s}\alpha\beta} f_{\mathfrak{s}\alpha\beta} - p_{\mathfrak{w}} \delta_{\alpha\beta} f_{\mathfrak{w}\alpha\beta} - p_{\mathfrak{g}} \delta_{\alpha\beta} f_{\mathfrak{g}\alpha\beta} + (h_{\mathfrak{s}\alpha} + h_{\mathfrak{w}\alpha} + h_{\mathfrak{g}\alpha})_{,\alpha} \\
&+ \rho_{\mathfrak{s}} q_{\mathfrak{s}} + \rho_{\mathfrak{w}} q_{\mathfrak{w}} + \rho_{\mathfrak{g}} q_{\mathfrak{g}} - \rho_{\mathfrak{s}} \sum_{\mathrm{m}}^{\mathfrak{M}} \dot{\tau}_{\mathrm{m}\mathfrak{s}} v_{\mathrm{m}\mathfrak{s}} - \rho_{\mathfrak{w}} \sum_{\mathrm{m}=1}^{\mathfrak{M}} \dot{\tau}_{\mathrm{m}\mathfrak{w}} v_{\mathrm{m}\mathfrak{w}} - \rho_{\mathfrak{g}} \sum_{\mathrm{m}=1}^{\mathfrak{M}} \dot{\tau}_{\mathrm{m}\mathfrak{g}} v_{\mathrm{m}\mathfrak{g}} - \\
&- \rho_{\mathfrak{s}} \sum_{\mathrm{m}=1}^{\mathfrak{M}} \tau_{\mathrm{m}\mathfrak{s}} \dot{v}_{\mathrm{m}\mathfrak{s}} - \rho_{\mathfrak{w}} \sum_{\mathrm{m}=1}^{\mathfrak{M}} \tau_{\mathrm{m}\mathfrak{w}} \dot{v}_{\mathrm{m}\mathfrak{w}} - \rho_{\mathfrak{g}} \sum_{\mathrm{m}=1}^{\mathfrak{M}} \tau_{\mathrm{m}\mathfrak{g}} \dot{v}_{\mathrm{m}\mathfrak{g}} \\
&= \rho_{\mathfrak{s}} \dot{\eta}_{\mathfrak{s}} T + \rho_{\mathfrak{w}} \dot{\eta}_{\mathfrak{w}} T + \rho_{\mathfrak{g}} \dot{\eta}_{\mathfrak{g}} T - \rho_{\mathfrak{s}} \sum_{\mathrm{m}=1}^{\mathfrak{M}} \dot{\tau}_{\mathrm{m}\mathfrak{s}} v_{\mathrm{m}\mathfrak{s}} - \rho_{\mathfrak{w}} \sum_{\mathrm{m}=1}^{\mathfrak{M}} \dot{\tau}_{\mathrm{m}\mathfrak{w}} v_{\mathrm{m}\mathfrak{w}} - \rho_{\mathfrak{g}} \sum_{\mathrm{m}=1}^{\mathfrak{M}} \dot{\tau}_{\mathrm{m}\mathfrak{g}} v_{\mathrm{m}\mathfrak{g}}
\end{aligned} \tag{9.5.9}$$

$$\begin{aligned}
\rho\dot{\zeta} = &-p_{\mathfrak{s}\alpha\beta} f_{\mathfrak{s}\alpha\beta} - p_{\mathfrak{w}} \delta_{\alpha\beta} f_{\mathfrak{w}\alpha\beta} - p_{\mathfrak{g}} \delta_{\alpha\beta} f_{\mathfrak{g}\alpha\beta} + (h_{\mathfrak{s}\alpha} + h_{\mathfrak{w}\alpha} + h_{\mathfrak{g}\alpha})_{,\alpha} \\
&- \rho_{\mathfrak{s}} \sum_{\mathrm{m}=1}^{\mathfrak{M}} \dot{\tau}_{\mathrm{m}\mathfrak{s}} v_{\mathrm{m}\mathfrak{s}} - \rho_{\mathfrak{w}} \sum_{\mathrm{m}=1}^{\mathfrak{M}} \dot{\tau}_{\mathrm{m}\mathfrak{w}} v_{\mathrm{m}\mathfrak{w}} - \rho_{\mathfrak{g}} \sum_{\mathrm{m}=1}^{\mathfrak{M}} \dot{\tau}_{\mathrm{m}\mathfrak{g}} v_{\mathrm{m}\mathfrak{g}} - \rho_{\mathfrak{s}} \sum_{\mathrm{m}=1}^{\mathfrak{M}} \tau_{\mathfrak{s}} \dot{v}_{\mathfrak{s}} \\
&- \rho_{\mathfrak{w}} \sum_{\mathrm{m}=1}^{\mathfrak{M}} \tau_{\mathrm{m}\mathfrak{w}} \dot{v}_{\mathrm{m}\mathfrak{w}} - \rho_{\mathfrak{g}} \sum_{\mathrm{m}=1}^{\mathfrak{M}} \tau_{\mathrm{m}\mathfrak{g}} \dot{v}_{\mathrm{m}\mathfrak{g}} + \rho_{\mathfrak{s}} q_{\mathfrak{s}} + \rho_{\mathfrak{w}} q_{\mathfrak{w}} + \rho_{\mathfrak{g}} q_{\mathfrak{g}} \\
&- \rho_{\mathfrak{s}} \eta_{\mathfrak{s}} \dot{T} - \rho_{\mathfrak{w}} \eta_{\mathfrak{w}} \dot{T} - \rho_{\mathfrak{g}} \eta_{\mathfrak{g}} \dot{T} - \rho_{\mathfrak{s}} \dot{\eta}_{\mathfrak{s}} T - \rho_{\mathfrak{w}} \dot{\eta}_{\mathfrak{w}} T - \rho_{\mathfrak{g}} \dot{\eta}_{\mathfrak{g}} T \\
&= - \rho_{\mathfrak{s}} \sum_{\mathrm{m}=1}^{\mathfrak{M}} \dot{\tau}_{\mathrm{m}\mathfrak{s}} v_{\mathrm{m}\mathfrak{s}} - \rho_{\mathfrak{w}} \sum_{\mathrm{m}=1}^{\mathfrak{M}} \dot{\tau}_{\mathrm{m}\mathfrak{w}} v_{\mathrm{m}\mathfrak{w}} - \rho_{\mathfrak{g}} \sum_{\mathrm{m}=1}^{\mathfrak{M}} \dot{\tau}_{\mathrm{m}\mathfrak{g}} v_{\mathrm{m}\mathfrak{g}} \\
&- \rho_{\mathfrak{s}} \eta_{\mathfrak{s}} \dot{T}_{\mathfrak{s}} - \rho_{\mathfrak{w}} \eta_{\mathfrak{w}} \dot{T}_{\mathfrak{w}} - \rho_{\mathfrak{g}} \eta_{\mathfrak{g}} \dot{T}_{\mathfrak{g}}
\end{aligned} \tag{9.5.10}$$

It may be seen that if the non-mechanical energy rates in Eq. (9.5.8) vanish, then all the external work, namely the first three terms referring to the work of the effective pressure, the pore water pressure and the pore air pressure, is converted into free energy.

Concluding the energy balance equations for soils, the three inequality equations of the entropy production are obtained from Eq. (6.10.2)

$$\left.\begin{aligned}
\eta_{\mathfrak{s}} &\geq \frac{1}{\rho_{\mathfrak{s}}}\left[\frac{h_{\mathfrak{s}\alpha}}{T_{\mathfrak{s}}}\right]_{,\alpha} + \frac{q_{\mathfrak{s}}}{T_{\mathfrak{s}}} \\
\eta_{\mathfrak{w}} &\geq \frac{1}{\rho_{\mathfrak{w}}}\left[\frac{h_{\mathfrak{w}\alpha}}{T_{\mathfrak{w}}}\right]_{,\alpha} + \frac{q_{\mathfrak{w}}}{T_{\mathfrak{w}}} \\
\eta_{\mathfrak{g}} &\geq \frac{1}{\rho_{\mathfrak{g}}}\left[\frac{h_{\mathfrak{g}\alpha}}{T_{\mathfrak{g}}}\right]_{,\alpha} + \frac{q_{\mathfrak{g}}}{T_{\mathfrak{g}}}
\end{aligned}\right\} \qquad (9.5.11)$$

10 Flow in Soils

10.1 Introduction

From this chapter on we shall deal with the application of the fundamental concepts of physics and continuum mechanics to soils, based on the notions expressed by Eqs. (7.5.1), (7.5.2) and (7.9.3), (7.9.4), which show that the stress–strain relationship can be resolved into a spherical stress–strain relationship and a shear stress–strain relationship. Eqs. (7.5.3)–(7.5.6) indicate further how the stress and strain tensors can each be split into a spherical and shear tensor.

When the concept of stress was extended to mixtures in general and it was shown that the macroscopic stress is equal to the sum of the stresses of the constituents, Eq. (6.6.5), it was also possible to show that in soil, assumed to be a three-phase mixture and provided that no dynamic effects are present, the total macroscopic stress p_{ij} is equal to the sum of the stresses of its constituents: the stress in the solids $p_{\mathfrak{s}ij}$, the stress in the water $p_{\mathfrak{w}}\delta_{ij}$ and the stress in the air $p_{\mathfrak{g}}\delta_{ij}$, Eq. (9.4.7). Two constitutive equations were implemented, and they show that the water and the air phases do not carry shear stresses, since the shear moduli of the water and the air both vanish, $G_{\mathfrak{w}} = 0$ and $G_{\mathfrak{g}} = 0$, respectively. As a result, only spherical stresses exist in the water and the air phases, while in the solids the stresses decompose into spherical and shear stresses and Eqs. (9.4.8) and (9.4.9) are obtained.

Analogously, the strain tensor can be resolved into volumetric strain and distortion, according to Eqs. (7.5.4) and (7.5.6), respectively. The total volumetric strain is therefore a result of the volumetric strains of the three constituents of the soil: the volumetric strain of the solids $\varepsilon_{\mathfrak{s}\alpha\alpha}$, the volumetric strain of the water $\varepsilon_{\mathfrak{w}\alpha\alpha}$ and the volumetric strain of the air $\varepsilon_{\mathfrak{g}\alpha\alpha}$, resulting from the compressibility of the respective phases, while the overall distortion of the soil is produced by the distortion of the solids alone. It should be remarked here that although there are no distortions in the water phase, distortion rates of the first order exist in the water and distortion rates of higher order might also exist.

Spherical stresses in the water exert a decompressive inflating pressure on the solids and result in a decrease in the effective stresses, an overall increase in the volume of the soil if the environment permits it, and a compressive stress on the air inclusions resulting in a decrease of their volume. If the spherical stresses in the water are concurrent with spherical stress gradients, flow of water will occur in the soil. The present chapter deals with this topic, namely flow of water in soils.

Spherical stresses and stress gradients in the water and in the air are generated in two ways: either by applying the stress directly, as just explained, or by applying a spherical stress or stress gradient on the solids, which, when compressed, induce spherical stresses and stress gradients in the water and air.

Like the pore water pressure, the pore air pressure decreases the effective stress of the solids as well, and adds to the overall volume increase of the soil. It will, also, either increase or decrease the flow of both water and air and increase the condensation and solubility of the air. Unfortunately, in spite of its importance in the stability and settlement problems, particularly of offshore deposits (Benyaako et al. 1974; Coleman and Prior 1978; Whelan and Lester 1980) and in the mechanics of unsaturated soils in general, the flow of air in soil has not commanded the appropriate attention. The reason is perhaps the foreseen experimental impediments, but most probably the fact that, unlike in the flow of water, the common pressure potentials we are familiar with such as gravity, gravitation, velocity, osmotic, etc. have minimal or no effect on the air, and that air bubbles and air pressures are sometimes generated by chemical reactions, biogenical decomposition or air locking. The flow of air in soil remains an open field for investigation and research.

The following chapters will consider the spherical stresses and the spherical stress–strain relationship (consolidation), the shear stresses and the shear stress–strain relationship (shear), and the mutual effects of the two (failure).

10.2 Force Fields

Besides the true complexity of the topics of flow and diffusion in porous media, there is an added handicap in the fact that the topic is being treated by such diverse disciplines as agriculture, civil engineering, geophysics, refractory, metallurgy, chemical engineering, rheology, biodynamics, mechanics, etc., each in its own way and from different standpoints. Terminology, notation and clarity are the immediate casualties. Moreover, there is very little common language throughout the disciplines, and the literature is too extensive to be mastered. Thermodynamics, the unifying discipline which should provide the necessary answers, is still involved in trying to understand, clarify, and formulate its own basic concepts; grappling with the feedback from the various disciplines, thermodynamics directs the investigation of the specific problems back to the disciplines themselves, which are, apparently, more adept to deal with them. Indeed, studies on flow of water and air in soils, started at the beginning of the century (Buckingham 1907; Gardner 1920a; Edlefsen and Anderson 1943) and continuously advanced and promoted, rest on thermodynamics concepts. Still, there is much more to accomplish, to mention just two problems yet unsolved: the problem of unsaturated soils and the problem of hysteresis.

Before proceeding with the problems of flow, it is proper that we should gain some insight into the nature of Newtonian attraction forces, acting at all points of a space rather than at fixed and isolated ones.

Let **f** be a *force field*, a tensor of rank one defined at all points of the space. We are familiar with many force fields, such as that of gravitation, velocity of moving matter, heat flow, electric currents in conductors, magnetic fields, etc. A force field is an assemblage of points in space on which individual forces are acting, and it can be imagined as needles placed in various points of the space, each pointing in the direction of the force, and having a length proportional to the magnitude of the force. Their direction changes gradually across the space, and the impression of *force lines* is perceived. Now let us take any small closed contour, Fig. 10.2.1, and draw the force lines that pass through every point within the contour. These lines form thin tubes, the *force tubes*, with varying cross-sections, their generators at any point effectively parallel. The *strength* of the tube, called the *flux*, is defined as the integral of the product of the force field **f** and the sectional area $\mathrm{d}\mathsf{S}$ through the force field

$$\int_{\mathsf{S}} \mathbf{f}\,\mathrm{d}\mathsf{S} \equiv \int_{\mathsf{S}} f_\alpha\,\mathrm{d}S_\alpha = \int_{\mathsf{S}} f_\alpha n_\alpha\,\mathrm{d}\mathsf{S} \tag{10.2.1}$$

where $\mathrm{d}S_i$ is the vector representing the projection of the area normal to the force field and its magnitude is equal to that of the cross-section, and n_i is a unit vector in the direction of the force field, normal to the cross-section $\mathrm{d}S_i$.

Force tubes, like force lines, are everywhere parallel to the direction of the flux **f**. Therefore two force lines can never meet, for that would mean that there are two distinct directions of **f** at a common point. Exceptions are points where $\mathbf{f} = 0$, which are singular, meaning that at such points the resultant forces of the field are in *equilibrium*.

From Eq. (10.2.1) it is evident that the formula containing $\mathbf{f}\,\mathrm{d}\mathsf{S}$ equally holds, even when $\mathrm{d}S_i$ is not normal to the field. Since $\mathrm{d}S_i$ may vary along the

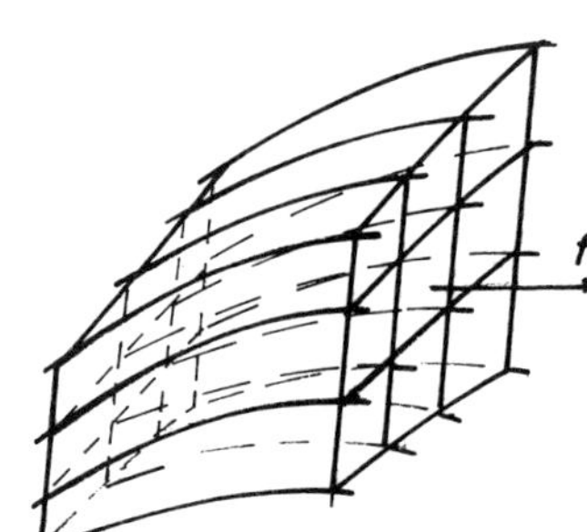

Fig. 10.2.1. Force field.

force tube while the flux is the same, we may write

$$\int_S f_\alpha n_\alpha \, dS = \int_V f_{\alpha,\alpha} \, dV \tag{10.2.2}$$

where the flux was observed at two cross-sections, dS_α^a and dS_α^b. From Eq. (10.2.2) it is seen that the intensity of the force field at any of the cross-sections varies inversely to the area of the cross-section

$$\frac{f_i^a}{f_i^b} = \frac{dS_i^b}{dS_i^a} \tag{10.2.3}$$

According to Gauss' theorem, Appendix A.19, we have

$$\int_S f_\alpha n_\alpha \, dS = \int_V f_{\alpha,\alpha} \, dV \equiv \int_V \operatorname{div} \mathbf{f} \, dV \tag{10.2.4}$$

which indicates that the flux of a force field, Eq. (10.2.1), is equal to the divergence of the force field **f**. All properties of force fields relate to the divergence, namely, *the integral of the divergence of a force field (vector field) over a region of space is equal to the integral over the surface of that region of the component of the field in the direction of the outward directed normal to the surface*. The left-hand side of Eq. (10.2.4) is the *flux* of the force across the surface, while the terms on the right-hand side of the equation are the divergence. A divergence indicates the existence of either a source or a sink, an excess of the outward over the inward flow which diverges, or an excess of the inward over the outward flow which converges, respectively. The sign of the divergence is, accordingly, positive (+) for source, negative (−) for sink.

If the flux of the force field across every closed surface vanishes, the field is solenoidal. A necessary and sufficient condition for the field to be solenoidal is that the divergence vanish everywhere, provided the surface is continuous and everywhere differentiable. Eq. (10.2.4) becomes then

$$\operatorname{div} \mathbf{f} \equiv f_{\alpha,\alpha} = 0 \tag{10.2.5}$$

for, if the divergence vanishes everywhere, the flux of the force across the surface vanishes, and vice versa. If such a vanishing divergence of a force field exists, it means that there are no sources or sinks at any part of the surface. The absence of divergence in a velocity field is a necessary and sufficient condition that the matter flowing be incompressible.

A force field **f** is *lamellar* or *conservative* if the line integral about every closed path in space vanishes

$$\oint_C \mathbf{f} \, d\mathbf{x} \equiv \oint_C f_\alpha \, dx_\alpha = 0 \tag{10.2.6}$$

It follows that in order to satisfy Eq. (10.2.6) the force field **f** is continuous and differentiable throughout the space and there exists a scalar Φ, called the *potential* of **f**, such that

$$f_i = -\Phi_{,i} = -\text{grad}\,\Phi \tag{10.2.7}$$

Another aspect of the lamellar field is highlighted, if Eq. (10.2.6) is written

$$\oint_C \mathbf{f}\,\mathrm{d}x = \int_{\mathfrak{A}}^{\mathfrak{B}} \mathbf{f}\,\mathrm{d}x_{\mathfrak{a}} + \int_{\mathfrak{B}}^{\mathfrak{A}} \mathbf{f}\,\mathrm{d}x_{\mathfrak{b}} = 0 \tag{10.2.8}$$

where the path followed from $\mathfrak{A}$ to $\mathfrak{B}$, marked $\mathfrak{a}$, is different from the one followed from $\mathfrak{B}$ to $\mathfrak{A}$, marked $\mathfrak{b}$, and in fact both are unspecified and arbitrary. The second part of Eq. (10.2.8) may be written

$$\int_{\mathfrak{A}}^{\mathfrak{B}} \mathbf{f}\,\mathrm{d}x_{\mathfrak{a}} = -\int_{\mathfrak{B}}^{\mathfrak{A}} \mathbf{f}\,\mathrm{d}x_{\mathfrak{b}} = \int_{\mathfrak{A}}^{\mathfrak{B}} \mathbf{f}\,\mathrm{d}x_{\mathfrak{b}} = \int_{\mathfrak{A}}^{\mathfrak{B}} \mathbf{f}\,\mathrm{d}x \tag{10.2.9}$$

This shows that *in a lamellar field the line integral of* **f** *from* $\mathfrak{A}$ *to* $\mathfrak{B}$ *is independent of the path followed*, subject to the requirements mentioned earlier, field continuity and differentiability. The right-hand side of Eq. (10.2.9) depends only upon the initial point $\mathfrak{A}$ and the final point $\mathfrak{B}$ of the path, that is

$$\int_{\mathfrak{A}}^{\mathfrak{B}} \mathbf{f}\,\mathrm{d}x = \Phi_{\mathfrak{B}} - \Phi_{\mathfrak{A}} \tag{10.2.10}$$

If the two points $\mathfrak{A}$ and $\mathfrak{B}$ are taken very close together, we have

$$\mathbf{f} = -\frac{\mathrm{d}\Phi}{\mathrm{d}x_i} = -\,\text{grad}\,\Phi = f_i = -\frac{\mathrm{d}\Phi}{\mathrm{d}x} = -\Phi_{,i} \tag{10.2.11}$$

that is, **f** must be a gradient of some scalar function Φ, as has been suggested in Eq. (10.2.7).

According to Stokes' theorem, Eq. (A.19.5), Eq. (10.2.6) may be written

$$\oint_C \mathbf{f}\,\mathrm{d}x \equiv \oint_C f_\alpha\,\mathrm{d}x_\alpha = e_{\alpha\beta\gamma}\int_S f_{\alpha,\beta}n_\gamma\,\mathrm{d}S \equiv \int_S \text{curl}\,\mathbf{f}\,\mathbf{n}\,\mathrm{d}S = 0 \tag{10.2.12}$$

Since the right-hand side of Eq. (10.2.12) vanishes for any surface S and any normal direction **n**, it implies, quite generally, that

$$\text{curl}\,\mathbf{f} \equiv f_{i,j} - f_{j,i} \equiv \text{curl}\,\text{grad}\,\Phi = 0 \tag{10.2.13}$$

The curl indicates in general a rotatory element in the field or a character of vortex motion. In a force field, the left-hand side integral in Eq. (10.2.12) is called the *circulation* along the closed loop curve. If this integral vanishes, as in our case, the motion is said to be *irrotational*. Thus a field is irrotational if and only if it is *lamellar*, that is, if it has a potential through which it is defined. The necessary and sufficient condition for an irrotational motion is Eq. (10.2.13). If the field is a velocity field, this condition reduces to Eq. (3.11.1) discussed in Sect. 3.11.

If Eq. (10.2.11) is substituted into Eq. (10.2.4)

$$\int_V f_{\alpha,\alpha}\, \mathrm{d}V = \int_V \Phi_{,\alpha\alpha}\, \mathrm{d}V = \int_S f_\alpha n_\alpha\, \mathrm{d}S \tag{10.2.14}$$

the general differential form of the divergence equation for a lamellar non-solenoidal force field, known as *Poisson's equation*, is obtained

$$f_{\alpha,\alpha} = \Phi_{\alpha,\alpha} \equiv \operatorname{div}\mathbf{f} = \operatorname{div}\operatorname{grad}\Phi = \nabla^2\Phi = 0 = \mathcal{F} \tag{10.2.15}$$

where ∇^2 is the *Laplacian differential operator* (see Appendix A, Sect. A.16), and $\mathcal{F}$ is a function depending on the nature of the force field. In an electromagnetic force field Φ is the electrostatic potential and $\mathcal{F} = -4\pi\sigma/\varepsilon$; in a gravitational field Φ is the gravitational potential and $\mathcal{F} = -4\pi$; while in a velocity field Φ is the flow potential and $\mathcal{F}$ is a function related to the compressibility of the field.

If any of those force fields is solenoidal, Eq. (10.2.15) reduces to the Laplace differential equation presented in Sect. 3.11

$$f_{\alpha,\alpha} = \Phi_{,\alpha\alpha} \equiv \operatorname{div}\mathbf{f} = \operatorname{div}\operatorname{grad}\Phi \equiv \nabla^2\Phi = 0 \tag{10.2.16}$$

10.3 Flow Potentials

In Sect. 5.5 we have seen that the rate of change of the total energy, that is, the sum of the internal and kinetic energy rates, is fully determined in terms of the work done by the stresses and strains involved, Eq. (5.4.6). This total energy, and specifically the internal energy, is the energy contained in the material, and it can take the form of potential energy and be released in the form of work. When work, mechanical or non-mechanical, is done on the material, the internal energy increases, and together with the kinetic energy invested we have the total energy of the material. This total energy generates potential energy, which is able to do work, again mechanical or non-mechanical, on the surroundings, and when the energy contained in the material is partly or fully released the total energy decreases and the internal and kinetic energies decrease correspondingly. In a non-conservative, or irreversible, system there is some loss of energy which turns into heat, either when it works on the material or on the surroundings.

The kinetic energy of the material may be computed at any instance from known velocities. To evaluate the internal energy, however, is harder. One has to know the amount of work invested at any time in the increase of the internal energy, an almost impossible task. There is a way to circumvent it, however, by means of evaluating the potential energy.

The method is the *spring balance method*. One may not know the weight hanging on a spring, but knowing the spring's coefficient of elasticity by calibration and measuring its extension, one should be in a position to evaluate the weight at any instance. Some energy may be lost through heat, but since our aim is to know the weight, the calibration of the spring provides an equitable answer. In soils we have a similar situation. The internal energy of the soil may not be known but it is a function of several potentials which, if known, may reveal it. The pressure potential $\Phi_{\mathfrak{p}}$, the gravitational potential $\Phi_{\mathfrak{g}}$, the electro-chemical potential $\Phi_{\mathfrak{e}}$, the capillary potential $\Phi_{\mathfrak{c}}$, the overburden potential Φ_{γ}, the heat potential $\Phi_{\mathfrak{h}}$ are some of the known potentials active in soils. Other potentials, less explored for their effect in soils, but significant in materials other than soils, are the chemical potential, the vaporization potential, the thermo-electric potential (Thomson effect), the electro-magnetic potential, etc. Some of these have already been mentioned in the previous chapters, others will be discussed in the appropriate sections.

It is not so easy to discriminate between the various potentials. Some of them are superimposed and when the suction potential, discussed in Sect. 8.27, is measured, the measurement includes most potentials mentioned above. In different circumstances one or the other potential prevails. So, for instance, the capillary potential is dominant in unsaturated, particularly non-cohesive, soils and the electro-chemical and osmotic potentials are dominant in cohesive saturated soils.

Following Philip, (1969a) the *total potential* $\Phi_{\mathfrak{t}} \equiv \mathbf{\Phi}$, or the *flow potential* which drives the water from a location of higher potential to a location of lower potential can be expressed as the sum of three distinct potentials, using Philip's notation:

1. The *moisture potential* $\Phi_{\mathfrak{w}}$ is the sum of most of the potentials mentioned earlier, such as the osmotic potential, the capillary potential, the electro-chemical potential, etc., which arise from the internal interaction of the soil–water system and are lumped together in the measurements obtained by well-established techniques. The moisture potential is in fact the suction potential discussed in Sect. 8.27, and is often called so in the literature. Other names for the same are the *matric potential, swelling potential*, etc. In our further discussion the moisture potential will be denoted simply by $\mathbf{\Psi} \equiv \Phi_{\mathfrak{w}}$.

2. The *gravitational potential* $\Phi_{\mathfrak{g}}$ is the pressure head above a specified datum elevation. It appears therefore only where differences in the elevations referred to the datum elevation contribute to the pressure head such as in vertical columns. It is convenient to consider the datum elevation at the water table level measuring the positive direction downwards, so that capillary rises above the water table are negative. The gravitational potential, denoted $\Phi_{\mathfrak{g}}$, is defined

$$\Phi_{\mathfrak{g}} = -z; \quad \Phi_{\mathfrak{g},i} = -\frac{\partial z}{\partial Z} \tag{10.3.1}$$

and is obtained by simple calculations once the elevations are known.

A more general way to formulate the *gravity potential* denoted by ζ, which includes the gravitational potential $\Phi_g = \zeta$ as well, is

$$\zeta = \frac{1}{g}\int_0^X (g_i + a_i)\,\mathrm{d}X = \Phi_g + \frac{a}{g}\int_0^{X^i} \mathrm{d}X^i = -z + \frac{a}{g}X^i \tag{10.3.2}$$

where $-z$ is the gravitational potential from Eq. (10.3.1) and $(a/g)\,X^i$ is a gravity potential dependent on a body force such as a centrifugal force.

3. The *overburden potential* Φ_γ is the excess pressure head expected in the field. It is a consequence of the overburden pressure in the soil column which decreases the void ratio of the soil, causing a decrease of the moisture content with depth in a saturated soil, and an increase of the degree of saturation with depth in an unsaturated soil. The overburden pressure is a result of the weight of the soil column and any additional load acting on the surface of the soil. Attention to that phenomenon was drawn by Croney and Coleman (1958), who presented a theoretical equilibrium curve, Fig. 10.3.1, showing the moisture content distribution with depth, calculated from the suction curve similar to the one in Fig. 8.27.3a. Since then others have observed it. The distribution of the overburden pressure with depth, if $z_0 = 0$ represents the surface of the ground, is

$$p(z) = p(z_0) + \int_{z_0}^{z} \gamma\,\mathrm{d}z \tag{10.3.3}$$

where $p(z)$ is the overburden pressure at depth z, $p(z_0)$ is a distributed load over the ground surface and γ is the total unit weight of the soil as defined by Eq. (9.2.14). The last part of Eq. (10.3.3) presents the commonly adopted technique of replacing the load distributed over the ground surface by an equivalent weight of the soil layer.

The overburden potential, denoted $\mathbf{\Omega}$, is defined as the change in the volume of the voids, resulting from the change in the volume of the water under the overburden pressure

$$\Phi_\gamma = \mathbf{\Omega} = \frac{\mathrm{d}V_e}{\mathrm{d}V_w}\,p(z) = \frac{\mathrm{d}e}{\mathrm{d}\vartheta}\,p(z) = \frac{\mathrm{d}e}{\mathrm{d}\vartheta}\left[\frac{p(z_0)}{\gamma_w} + \int_{z_0}^{z}\mathrm{d}z\right] \tag{10.3.4}$$

where e is the void ratio and ϑ is the moisture ratio defined in Eqs. (9.2.6) and (9.2.13).

The total potential at any depth Φ, therefore, is the sum of the three potentials, the *matric potential* $\mathbf{\Psi}$, the *gravity potential* ζ and the *overburden potential* $\mathbf{\Omega}$

$$\mathbf{\Phi} = \mathbf{\Psi} + \zeta + \mathbf{\Omega} \tag{10.3.5}$$

Since the total potential is the pressure acting in the water, it is equivalent

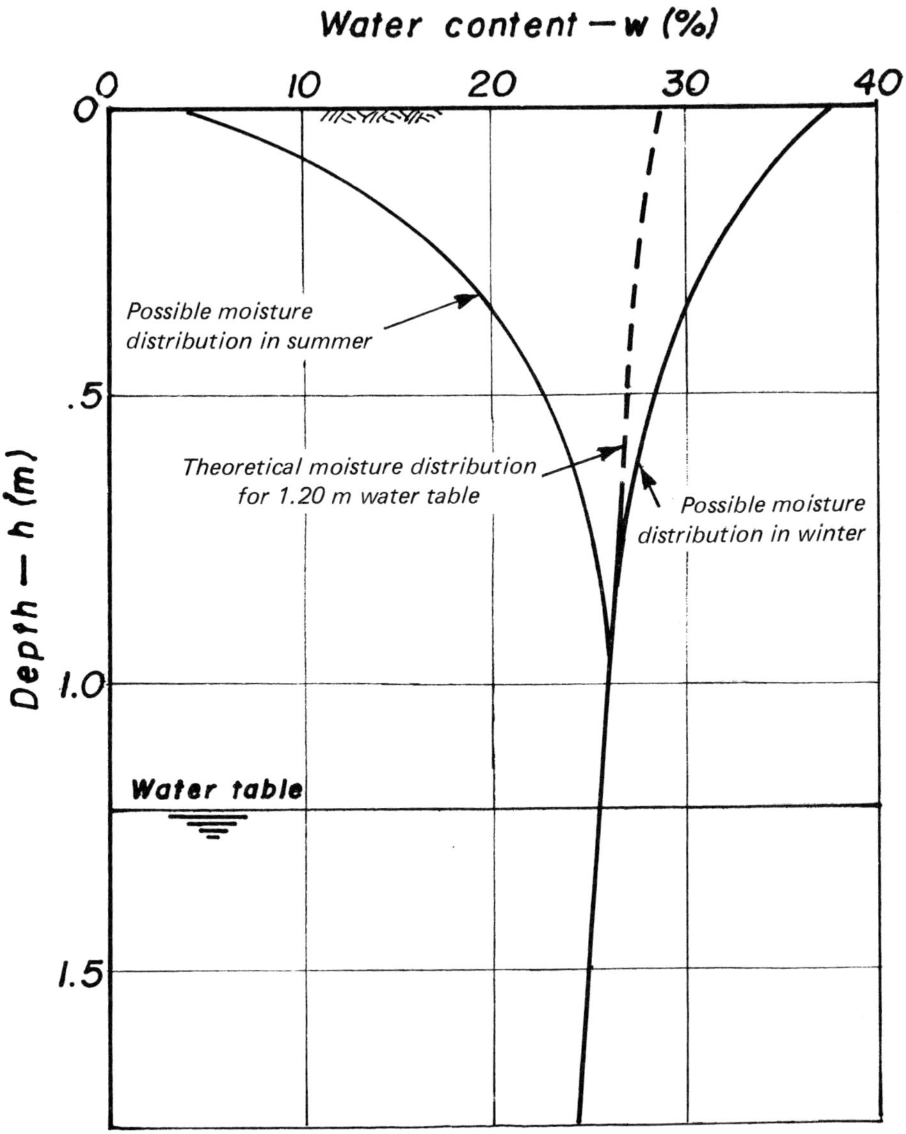

Fig. 10.3.1. Water content distribution in the field for saturated compressible soil.

to the pore water pressure p_w

$$\Phi \equiv p_w \tag{10.3.6}$$

Considering only the gravitational part of the gravity potential $-z$, we obtain the total potential used by Philip and most other investigators

$$\Phi = \Psi + \zeta + \Omega = \Psi - z + \frac{de}{d\vartheta}\left[\frac{p(z_0)}{\gamma_w} + \int_{z_0}^{z} dz\right] \tag{10.3.7}$$

If the soil is saturated, $de/d\vartheta = 1$, Eq. (10.3.7) becomes

$$\mathbf{\Phi} = \mathbf{\Psi} + \zeta + \mathbf{\Omega} = \mathbf{\Psi} - z + \left[\frac{p(z_0)}{\gamma_w} + \int_{z_0}^{z} dz\right] \tag{10.3.8}$$

If the soil is a rigid non-swelling soil, the change of the void ratio in the last term of Eq. (10.3.7) is zero, $de = 0$, the term containing de vanishes and the equation simplifies further

$$\mathbf{\Phi} = \mathbf{\Psi} + \zeta = \mathbf{\Psi} - z \quad \text{for } de = 0 \tag{10.3.9}$$

For horizontal columns $\zeta = 0$, and Eq. (10.3.7) becomes even simpler

$$\mathbf{\Phi} = \mathbf{\Psi} \quad \text{for } \zeta = 0 \tag{10.3.10}$$

10.4 Review of Linear and Non-linear, Saturated and Unsaturated Flow

The most outstanding manifestation of the theory of mixtures, when it is applied to soils, is the phenomenon of flow and diffusion. We have noted in Chap. 8 the complexity of the constituents of the soil, each in its own category, solids as a structured lacy matrix, water as a liquid and air as a multi-component gas. When combined, the complexity of each is multiplied by that of the others. The movement of water in the solid matrix, conditioned by restrictions imposed by the gas phase within that matrix, as well as the movement of water in many other porous systems, was termed "the most consequential phenomenon with respect to the life and activities of the human race" (Philip 1958a–c). In geotechnical engineering, flow and diffusion of water in soils are the most crucial phenomena. Except for the fact that they relate directly to problems of seepage and water transfer in the soil, as in the case of wells, dams, wetting and drying of the subgrades of pavement structures and others, they control the phenomena of consolidation, shrinkage and swelling, affecting, in turn, shearing resistance and stability.

Most efforts in soil mechanics have been devoted to the study of saturated soils, based on Darcy's law of linear relationship between the flow discharge and the pressure gradient. The non-linear relationships, between the flow discharge and the void ratio (porosity) (Howe and Hudson 1927; Lambe 1954; Michaels and Lin 1954; Schmid 1957), or the flow discharge and the pressure gradient (Fatt and Davis 1952; Derjagin and Krylov 1944; Li 1960) are documented in experimental studies, but no suitable application has been worked out, although several analytical speculations have been advanced.

The study of flow in unsaturated soils has been promoted since the nineteen fifties by soil scientists (Wooltorton 1950; Philip 1955, 1968a, b,

1969a, b; Smiles and Rosenthal 1968; Childs 1972; Klute 1972) and applied mainly in agriculture. Some of these studies penetrated into soil mechanics (Aitchison 1956, 1961a; Jennings 1960; Jennings and Burland 1962; Alpan 1961; Donald 1963a, b; Barden 1965), but the interest was localized and limited to several small groups of researchers, without much impact on the discipline as a whole. The topic recently surfaced, and, although limited again to localized groups (Fredlund 1975, 1979, 1985; Esrig and Kirby 1977; Wood 1979; Lloret and Alonso 1980, 1985; Nageswaran 1983; Sills and Nageswaran 1984; Wheeler 1986; Thomas 1987), it produced most valuable experimental work, evaluated, however, through the concepts of classical soil mechanics. Of particular interest is the work of Helm (1987), who introduces concepts of continuum mechanics of mixtures into the geotechnical literature, even though continuum mechanics is not specifically mentioned.

Problems of flow and diffusion, where motions relative to a reference are observed and where transfer of matter occurs within any arbitrary volume element, are necessarily controlled by the density balance equation. This equation accounts for the volume of matter leaving the volume element in order to make space for another volume of matter to enter the volume element. Moreover, it accounts for the change in density in the volume element, if the volume leaving the element is not equal to the volume entering it. The volume element may be that of a single-phase material or of a multi-phase mixture, and it can be either locked in a stationary position, or on any moving constituent of the multi-phase mixture.

Through the volume element there is a *discharge* Q_i of the constituents

$$Q_i = Q_{\mathfrak{s}i} + Q_{\mathfrak{w}i} + Q_{\mathfrak{g}i} \tag{10.4.1}$$

where $Q_{\mathfrak{s}i}$, $Q_{\mathfrak{w}i}$ and $Q_{\mathfrak{g}i}$ are the *individual discharges* of the solids, the water and the air constituents, respectively.

The *flux* or *specific discharge* q_i is defined as the discharge Q_i per cross-section S

$$\begin{aligned} q_i &= \frac{Q_i}{S} = \frac{Q_{\mathfrak{s}i}}{S} + \frac{Q_{\mathfrak{w}i}}{S} + \frac{Q_{\mathfrak{g}i}}{S} = q_{\mathfrak{s}i} + q_{\mathfrak{w}i} + q_{\mathfrak{g}i} \\ &= \frac{1}{(1+e)}\left(v_{\mathfrak{s}i} + Sev_{\mathfrak{w}i} + (1-S)ev_{\mathfrak{g}i}\right) \end{aligned} \tag{10.4.2}$$

where $q_{\mathfrak{s}i}$, $q_{\mathfrak{w}i}$ and $q_{\mathfrak{g}i}$ are the *individual fluxes* of the solid, the water and the air constituents, respectively, within a streaming volume, and $v_{\mathfrak{s}i}$, $v_{\mathfrak{w}i}$ and $v_{\mathfrak{g}i}$ are the individual velocities of the constituents of the soil, as defined in Sect. 6.3. From Eq. (10.4.2) we have the relations between the individual velocities and the specific discharges of the constituents

$$v_{\mathfrak{s}i} = (1+e)q_{\mathfrak{s}i}; \quad v_{\mathfrak{w}i} = \frac{(1+e)}{Se}q_{\mathfrak{w}i}; \quad v_{\mathfrak{g}i} = \frac{(1+e)}{(1-S)e}q_{\mathfrak{g}i} \tag{10.4.3}$$

In the following sections the density balance equations will be explored for the saturated and unsaturated, uniformly and non-uniformly distributed, steady and unsteady, compressible and incompressible constituents. For the latter it should be added that in geotechnical engineering incompressibility of the water and certainly of the solids is generally assumed for reasons of simplicity. In fact, all constituents are compressible to some degree: the compressibility of the solids is very difficult to assess; the compressibility of the water, which is by at least two orders of magnitude higher than that of the solids, can be characterized by its bulk modulus, which, for instance, at atmospheric pressure and 20 °C is $\varkappa = 2.2 \times 10^4$ kg/cm^2; the compressibility of the air is a function of temperature.

The balance equations (9.3.8)–(9.3.11) and (9.2.17) in terms of the void ratio e are commonly used in geotechnical engineering but are not necessarily the simplest.

The balance equations ensure that the built-in conditions in which the flow occurs are properly specified. The compressibility of the constituents was discussed in Chap. 8. If the densities of the constituents, which depend on their compressibility, remain unaltered through the flow process, or are considered steady, then the terms containing the time derivatives of the corresponding unit weights, $\dot{\gamma}_{\mathfrak{s}}$, $\dot{\gamma}_{\mathfrak{w}}$ and $\dot{\gamma}_{\mathfrak{g}}$ will vanish. If the constituents are homogeneous and the unit weights do not vary from place to place, then their respective space derivatives will vanish, $\gamma_{\mathfrak{s},i} = \gamma_{\mathfrak{w},i} = \gamma_{\mathfrak{g},i} = 0$. If the void ratio e or the porosity n are uniformly distributed throughout the porous medium, the spatial derivatives $e_{,i}$ and $n_{,i}$, respectively, vanish, and if the void ratio e or the porosity n remain unaltered through the flow process, $\dot{e}$ and $\dot{n}$, respectively, vanish. Analogously to $e_{,i}$ and $\dot{e}$ will $\theta_{,i}$ and $\dot{\theta}$ vanish in Eqs. (9.3.10) and (9.3.11).

10.5 Darcy's Law

Let us consider a water saturated soil, $S = 1$, which will then have only two constituents: the solids and the water.

The flux q_i for saturated soils is obtained from Eq. (10.4.2) by setting $S = 1$

$$q_i = q_{\mathfrak{s}i} + q_{\mathfrak{w}i} = \frac{1}{(1 + e)} (v_{\mathfrak{s}i} + e v_{\mathfrak{w}i}); \quad S = 1 \tag{10.5.1}$$

The total flux q_i is composed, as observed from Eq. (10.5.1), from the flux of the solids $q_{\mathfrak{s}i}$ and the flux of the water $q_{\mathfrak{w}i}$.

The flow through a saturated porous medium like that of a saturated soil with a steady solid matrix, is assumed to be controlled by *Darcy's Law*, which maintains that the specific water discharge through a porous medium is linearly proportional to the gradient driving the flow

$$\frac{Q_{wi}}{S} = q_{wi} = \frac{e}{(1+e)}\, v_{wi} = \theta_w v_{wi} = k_{i\alpha}\frac{\partial \mathbf{\Phi}}{\partial X} = k_{i\alpha}\mathbf{\Phi}(X)_{,\alpha} \qquad (10.5.2)$$

where q_{wi} is the specific water discharge defined in Eq. (10.4.2), $\mathbf{\Phi}(X)$ is a potential function of X_i taken as the appropriate form of one of Eqs. (10.3.4)–(10.3.8), and k_{ij} is the *coefficient of permeability* or *hydraulic conductivity*.

The expression $\mathbf{\Phi}(X)_{,i}$ is the *hydraulic gradient*, defined

$$\mathbf{\Phi}(X)_{,i} = \mathbf{\Psi}_{,i} + \zeta_{,i} + \mathbf{\Omega}_{,i} = \left[\frac{p_w}{\gamma_w}\right]_{,i} + \frac{B_i}{g} + \frac{\mathrm{d}e}{\mathrm{d}\vartheta}\left(\frac{p(z_0)}{\gamma_w} + \int_{z_0}^{z}\mathrm{d}z\right)$$

$$= \left[\frac{p_w}{\gamma_w}\right]_{,i} + \frac{g_i + a_i}{g} + \frac{\mathrm{d}e}{\mathrm{d}\vartheta}\left[\frac{p(z_0)}{\gamma_w} + \int_{z_0}^{z}\mathrm{d}z\right] \qquad (10.5.3)$$

where $\mathbf{\Phi}(X) = p_w/\gamma_w$ is the *pressure potential* or *pressure head*, $\mathbf{\Phi}(X)_{,i} = [p_w/\gamma_w]_{,i}$ is recognized as the *hydrostatic pressure head gradient* related to a pressure p_w within the water, $\Phi_g = -zg(Z)/g$ is the *gravitational potential* which, when acting in the direction Z is equal to the elevation $-z$ above the reference point, and a_i/g is a *body force potential* that may be acting in the water in any direction.

The coefficient of permeability k_{ij} is a second-order symmetric tensor, $k_{ij} = k_{ji}$. When rotated in its principal directions, the coefficients k_{11}, k_{22}, k_{33} are not necessarily equal, therefore the permeability coefficient k_{ij} is not necessarily isotropic. Only in the particular case when $k_{11} = k_{22} = k_{33}$ is the flow considered isotropic and Darcy's equation becomes

$$\frac{Q_{wi}}{S} = q_{wi} = \frac{e}{(1+e)}\, v_{wi} = \theta_w v_{wi} = k\frac{\partial \Phi}{\partial X_i} = k\Phi(X)_{,i} \qquad (10.5.4)$$

Eq. (10.5.1) is of course correct, provided the solids phase is steady, $v_{\mathfrak{s}i} = 0$.

Gersevanov (1934) and subsequently Biot (1956) have remarked that in a two-phase porous medium, where one of the constituents is a solid phase and the other a fluid phase like water, and where both constituents are in motion, $v_{\mathfrak{s}i} \neq 0$, the flow of water has to be considered relative with respect to the solids matrix, $v_{wi} - v_{\mathfrak{s}i}$. This relative motion of the water is called the *seepage velocity*. Applying it to Darcy's law, Eq. (10.5.2), we have

$$\overline{q_{wi}} = \frac{e}{1+e}\,[v_{wi} - v_{\mathfrak{s}i}] = \theta_w[v_{wi} - v_{\mathfrak{s}i}] = q_i - v_{\mathfrak{s}i} = k_{i\alpha}\Phi(X)_{,\alpha} \qquad (10.5.5)$$

Eq. (10.5.5) makes use of Eq. (6.3.5), which defines the *diffusion velocity* as a relative velocity with respect to a reference in a saturated soil. In this case $\overline{q}_{wi}$ is defined with respect to the solids.

A more general expression, including unsaturated soils as well, would be

$$\overline{q}_{\mathfrak{w}i} = \frac{eS}{1+e}(v_{\mathfrak{w}i} - v_{\mathfrak{s}i}) = \theta_{\mathfrak{w}}(v_{\mathfrak{w}i} - v_{\mathfrak{s}i}) = k(\theta_{\mathfrak{w}})_{i\alpha}\Phi(\theta_{\mathfrak{w}})_{,\alpha} \tag{10.5.6}$$

Here, however, the coefficient of permeability $k(\theta_{\mathfrak{w}})$ as well as the potential $\Phi(\theta_{\mathfrak{w}})$ are functions of the moisture content $\theta_{\mathfrak{w}}$.

10.6 Flow in Saturated Soils

In Sect. 9.3 the general flow equation for soils of $\mathfrak{N}$ constituents, (9.3.7), was derived from the density balance equation, and explicitly implemented for $\mathfrak{N} = 3$ constituents, Eqs. (9.3.8) and (9.3.9).

It was pointed out that Eqs. (9.3.8) and (9.3.9), which are stated in terms of the densities $\rho_{\mathfrak{n}}$, may also be stated in terms of e, n or $\theta_{\mathfrak{n}}$, Eqs. (9.3.10) and (9.3.11).

In a saturated soil, where only two constituents are present, $\mathfrak{N} = 2$, the density balance of air, Eq. (9.3.11), vanishes, while the density balance of water, Eq. (9.3.10), except for the substitution of $S = 1$, remains unaltered

$$\rho_{\mathfrak{s}} \frac{\mathrm{D}\dfrac{\rho_{\mathfrak{w}}}{\rho_{\mathfrak{s}}}}{\mathrm{D}_{\mathfrak{s}}t} = \frac{\gamma_{\mathfrak{s}}\theta_{\mathfrak{s}}}{g} \frac{\mathrm{D}\dfrac{\gamma_{\mathfrak{w}}\theta_{\mathfrak{w}}}{\gamma_{\mathfrak{s}}\theta_{\mathfrak{s}}}}{\mathrm{D}_{\mathfrak{s}}t} = -[\rho_{\mathfrak{w}}(v_{\mathfrak{w}\alpha} - v_{\mathfrak{s}\alpha})]_{,\alpha}$$

$$= -\left[\frac{\gamma_{\mathfrak{w}}\theta_{\mathfrak{w}}}{g}(v_{\mathfrak{w}\alpha} - v_{\mathfrak{s}\alpha})\right]_{,\alpha}$$

$$= \left(\frac{\gamma_{\mathfrak{w}}}{g}[k_{\alpha\beta}\phi_{,\alpha}]\right)_{,\beta} \tag{10.6.1}$$

From Eq. (9.2.12) we obtain

$$\theta_{\mathfrak{s}} + \theta_{\mathfrak{w}} = 1 \tag{10.6.2}$$

and hence

$$\frac{\mathrm{D}\theta_{\mathfrak{s}}}{\mathrm{D}_{\mathfrak{s}}t} = -\frac{\mathrm{D}\theta_{\mathfrak{w}}}{\mathrm{D}_{\mathfrak{s}}t}; \quad \frac{\partial\theta_{\mathfrak{s}}}{\partial t} = -\frac{\partial\theta_{\mathfrak{w}}}{\partial t}; \quad \theta_{\mathfrak{s},i} = -\theta_{\mathfrak{w},i}; \ S = 1 \tag{10.6.3}$$

The left-hand side of Eq. (10.6.1) is the measure of the relative change with time of the water with respect to the solids, and accounts for the changes in the densities of the solids and the water. These changes depend on the compressibility of the materials involved which affects their unit weight, and on the packing of the materials, expressed by the void ratio, the porosity, or the volumetric fractions.

On the right-hand side we have the mass flux of the water relative to the solids $\gamma_{\mathfrak{w}}\theta_{\mathfrak{w}}(v_{\mathfrak{w}i} - v_{\mathfrak{s}i})$ and the changes that occur in that flux throughout the space, expressed by the space derivative of the flux. The former, consistent with Eq. (10.5.5), indicates that if the flow of the water is evaluated with respect to the solids when the solids are in motion, this motion should obviously be considered. The latter accounts for possible unhomogeneity of the saturated soil.

Several assumptions can be made with respect to Eq. (10.6.1), and accordingly a variety of flow modes are possible in a saturated soil. The assumptions are:

1. *Incompressibility of the constituents*. If incompressibility of the solids is assumed, $\gamma_{\mathfrak{s}}$ is independent of time and the left-hand side of Eq. (10.6.1) becomes

$$\theta_{\mathfrak{s}} \frac{\mathrm{D}\dfrac{\gamma_{\mathfrak{w}}\theta_{\mathfrak{w}}}{\theta_{\mathfrak{s}}}}{\mathrm{D}_{\mathfrak{s}}t} = \theta_{\mathfrak{w}} \frac{\mathrm{D}\gamma_{\mathfrak{w}}}{\mathrm{D}_{\mathfrak{s}}t} + \frac{\gamma_{\mathfrak{w}}}{\theta_{\mathfrak{s}}} \frac{\mathrm{D}\theta_{\mathfrak{w}}}{\mathrm{D}_{\mathfrak{s}}t} = \theta_{\mathfrak{w}} \frac{\mathrm{D}\gamma_{\mathfrak{w}}}{\mathrm{D}_{\mathfrak{s}}t} - \frac{\gamma_{\mathfrak{w}}}{\theta_{\mathfrak{s}}} \frac{\mathrm{D}\theta_{\mathfrak{s}}}{\mathrm{D}_{\mathfrak{s}}t}$$

$$= \frac{1}{1+e} \frac{\mathrm{D}\gamma_{\mathfrak{w}}e}{\mathrm{D}_{\mathfrak{s}}t} = n \frac{\mathrm{D}\gamma_{\mathfrak{w}}}{\mathrm{D}_{\mathfrak{s}}t} + \frac{\gamma_{\mathfrak{w}}}{1-n} \frac{\mathrm{D}n}{\mathrm{D}_{\mathfrak{s}}t} \tag{10.6.4}$$

Here Eqs. (10.6.2) and $(10.6.3)_1$ have been used, and we see that the flow equation for incompressible solids in a saturated soil can be expressed both in terms of $\theta_{\mathfrak{s}}$ and $\theta_{\mathfrak{w}}$.

If the solids and the water are postulated incompressible, the unit weights $\gamma_{\mathfrak{s}}$ and $\gamma_{\mathfrak{w}}$ are independent of the time *t*. $\gamma_{\mathfrak{w}}$ is also independent of its location in space and that affects the right-hand side of Eq. (10.6.1), which ultimately becomes

$$-\frac{\gamma_{\mathfrak{w}}}{\theta_{\mathfrak{s}}} \frac{\mathrm{D}\theta_{\mathfrak{s}}}{\mathrm{D}_{\mathfrak{s}}\mathrm{t}} = \frac{\gamma_{\mathfrak{w}}}{1+e} \frac{\mathrm{D}e}{\mathrm{D}_{\mathfrak{s}}\mathrm{t}} = \frac{\gamma_{\mathfrak{w}}}{1-n} \frac{\mathrm{D}n}{\mathrm{D}_{\mathfrak{s}}\mathrm{t}}$$

$$= -\gamma_{\mathfrak{w}}[\theta_{\mathfrak{w}}(v_{\mathfrak{w}\alpha} - v_{\mathfrak{s}\alpha})]_{,\alpha} \tag{10.6.5}$$

It should be noted that when Eqs. (10.6.4) and (10.6.5) are expressed in terms of the densities $\rho_{\mathfrak{s}}$ and $\rho_{\mathfrak{w}}$ the assumption of incompressibility is not distinguishable, since the densities are phenomenological parameters and reflect the phenomenological process of flow. When they are expressed in terms of the void ratio *e*, the porosity *n* or the volumetric fractions $\theta_{\mathfrak{n}}$, which reflect the structural processes and thus reveal the structural interactions, the assumption is apparent in the equations.

2. *Rigid motion of constituents*. If a rigid motion of the solids is postulated, it means that the void ratio *e*, the porosity *n* or the volumetric fraction $\theta_{\mathfrak{s}}$ remain constant during the flow. From equation (10.6.2) it follows that if $\theta_{\mathfrak{s}}$ is constant $\theta_{\mathfrak{w}}$ is also constant. Eq. (10.6.1) for rigid solids motion becomes

$$\gamma_{\mathfrak{s}}\theta_{\mathfrak{w}} \frac{\mathrm{D}(\gamma_{\mathfrak{w}}/\gamma_{\mathfrak{s}})}{\mathrm{D}_{\mathfrak{s}}t} = \gamma_{\mathfrak{s}}n \frac{\mathrm{D}(\gamma_{\mathfrak{w}}/\gamma_{\mathfrak{s}})}{\mathrm{D}_{\mathfrak{s}}t} = \gamma_{\mathfrak{s}} \frac{e}{1+e} \frac{\mathrm{D}(\gamma_{\mathfrak{w}}/\gamma_{\mathfrak{s}})}{\mathrm{D}_{\mathfrak{s}}t}$$

$$= -\,[\gamma_{\mathfrak{w}}\theta_{\mathfrak{w}}(v_{\mathfrak{w}\alpha} - v_{\mathfrak{s}\alpha})]_{,\alpha} \tag{10.6.6}$$

If the motion in a saturated soil is rigid, no volume changes are involved in the process and any changes in influx and efflux are revealed by changes of the densities $\gamma_{\mathfrak{s}}$ and $\gamma_{\mathfrak{w}}$.

3. *The solids are fixed to the coordinate system*. If the solids are attached to the coordinate system, as is convenient in many instances, there will be no distinction between material and local time derivatives, and the local time derivative can replace the material time derivative, $D/Dt = \partial/\partial t$. If the motion is also a rigid motion the deformation gradient becomes $\partial x_i/\partial X_j = 1$.

On the right-hand side of Eq. (10.6.1) the velocity of the solids $v_{\mathfrak{s}i}$ does not necessarily vanish, except at the points defined by boundary conditions, or when the flow proceeds as a rigid motion. The assumption of fixed coordinates affects the right-hand side of Eq. (10.6.1) when $v_{\mathfrak{s}i}$ is omitted

$$[\gamma_{\mathfrak{w}}\theta_{\mathfrak{w}}v_{\mathfrak{w}\alpha}]_{,\alpha} = \left[\frac{\gamma_{\mathfrak{w}}e}{1+e}v_{\mathfrak{w}\alpha}\right]_{,\alpha} = [\gamma_{\mathfrak{w}}nv_{\mathfrak{w}\alpha}]_{,\alpha} \tag{10.6.7}$$

4. *Homogeneity of the soil*. The assumption of homogeneity in a saturated soil affects the right-hand side of Eq. (10.6.1). If homogeneity is assumed, $\theta_{\mathfrak{w}}$ is no more a function of the space, yet the density of the water is likely to change from one place of the space to another and Eq. (10.6.1) becomes

$$\gamma_{\mathfrak{s}}\theta_{\mathfrak{s}}\frac{D\dfrac{\gamma_{\mathfrak{w}}\theta_{\mathfrak{w}}}{\gamma_{\mathfrak{s}}\theta_{\mathfrak{s}}}}{Dt} = -\rho_{\mathfrak{w}}(v_{\mathfrak{w}\alpha} - v_{\mathfrak{s}\alpha})_{,\alpha} = -\theta_{\mathfrak{w}}[\gamma_{\mathfrak{w}}(v_{\mathfrak{w}\alpha} - v_{\mathfrak{s}\alpha})]_{,\alpha} \tag{10.6.8}$$

10.7 Modes of Saturated Flow

The assumptions listed in the previous section are frequently linked, and appear so in many theoretical studies. The following examples illustrate this point.

1. *Rigid solids in a homogeneous soil*. The assumption of rigid solids in saturated soils inevitably leads, according to Eq. (10.6.2), to rigid motion of the water as well. If we follow the solids with the coordinate system, as it is convenient, the velocity $v_{\mathfrak{s}i}$ vanishes. This is achieved by replacing the material derivative by the local derivative and we obtain

$$\frac{\partial\gamma_{\mathfrak{w}}}{\partial t} - \gamma_{\mathfrak{w}}\log\gamma_{\mathfrak{s}} = -[\gamma_{\mathfrak{w}}v_{\mathfrak{w}\alpha}]_{,\alpha} \tag{10.7.1}$$

Note that the moisture content $\theta_{\mathfrak{w}}$ disappeared from Eq. (10.7.1), due to the assumption of homogeneity.

2. *Rigid and incompressible solids*. If, in addition to being in a rigid array,

the solids are also incompressible, the second term on the left-hand side of Eq. (10.7.1) vanishes and we obtain

$$\frac{\partial \gamma_{\mathfrak{w}}}{\partial t} = -[\gamma_{\mathfrak{w}} v_{\mathfrak{w}\alpha}]_{,\alpha} \tag{10.7.2}$$

3. *Incompressible solids and water with rigid solids*. When incompressible solids, incompressible water, and a rigid array of solids are assumed, then all three terms on the left-hand side of Eq. (10.6.6) vanish, and the right-hand side, according to Eq. (10.5.5), results in the equation of *steady state* flow

$$[\theta_{\mathfrak{w}}(v_{\mathfrak{w}\alpha} - v_{\mathfrak{s}\alpha})]_{,\alpha} = \left[\frac{e}{1+e}(v_{\mathfrak{w}\alpha} - v_{\mathfrak{s}\alpha})\right]_{,\alpha} = k_{\alpha\beta}\Phi_{,\alpha\beta} = 0 \tag{10.7.3}$$

also known as the general *Laplace equation*, with the constant coefficient of permeability k_{ij}, to be discussed in Sect. 10.8.

As noticed earlier, in a rigid motion the array of solids may move rigidly with a velocity $v_{\mathfrak{s}i}$, as seen in Eqs. (10.7.1)–(10.7.3). If, however, the array is fixed in space, the velocity $v_{\mathfrak{s}i}$ vanishes.

4. *Incompressibility of solids*. If incompressibility of solids is postulated, $\gamma_{\mathfrak{s}}$ is independent of the time t, $\dot{\gamma}_{\mathfrak{s}} = 0$, and Eq. (10.6.1) yields

$$\theta_{\mathfrak{s}} \frac{\mathrm{D}\dfrac{\gamma_{\mathfrak{w}}\theta_{\mathfrak{w}}}{\theta_{\mathfrak{s}}}}{\mathrm{D}_{\mathfrak{s}}t} = \theta_{\mathfrak{w}} \frac{\mathrm{D}\gamma_{\mathfrak{w}}}{\mathrm{D}_{\mathfrak{s}}t} - \gamma_{\mathfrak{w}} \log \theta_{\mathfrak{s}} = \frac{e}{1+e}\frac{\mathrm{D}\gamma_{\mathfrak{w}}}{\mathrm{D}_{\mathfrak{s}}t} - \gamma_{\mathfrak{w}} \log \frac{1}{1+e}$$

$$= n\frac{\mathrm{D}\gamma_{\mathfrak{w}}}{\mathrm{D}_{\mathfrak{s}}t} - \gamma_{\mathfrak{w}} \log(1-n) = -[\gamma_{\mathfrak{w}}\theta_{\mathfrak{w}}(v_{\mathfrak{w}\alpha} - v_{\mathfrak{s}\alpha})]_{,\alpha} \tag{10.7.4}$$

If in Eq. (10.7.4) rigidity of solids is assumed as well, the term containing the logarithm vanishes. If, in addition, the coordinate system moves along with the solids, we may replace the material derivative with the local derivative and Eq. (10.7.2) is obtained again.

5. *Incompressibility of solids and water*. When the assumption of incompressibility of solids and water is considered, Eq. (10.6.6) reduces further, with the help of Eqs. (10.6.2) and (10.6.3)

$$\gamma_{\mathfrak{w}}\theta_{\mathfrak{s}} \frac{\mathrm{D}(\theta_{\mathfrak{w}}/\theta_{\mathfrak{s}})}{\mathrm{D}_{\mathfrak{s}}t} = \frac{\gamma_{\mathfrak{w}}}{\theta_{\mathfrak{s}}}\left[\theta_{\mathfrak{s}}\frac{\mathrm{D}\theta_{\mathfrak{w}}}{\mathrm{D}_{\mathfrak{s}}t} - \theta_{\mathfrak{w}}\frac{\mathrm{D}\theta_{\mathfrak{s}}}{\mathrm{D}_{\mathfrak{s}}t}\right]$$

$$= \frac{\gamma_{\mathfrak{w}}}{\theta_{\mathfrak{s}}}\left[-\theta_{\mathfrak{s}}\frac{\mathrm{D}\theta_{\mathfrak{s}}}{\mathrm{D}_{\mathfrak{s}}t} - (1-\theta_{\mathfrak{s}})\frac{\mathrm{D}\theta_{\mathfrak{s}}}{\mathrm{D}_{\mathfrak{s}}t}\right] = -\gamma_{\mathfrak{w}}\log\theta_{\mathfrak{s}}$$

$$= \frac{\gamma_{\mathfrak{w}}}{1-\theta_{\mathfrak{w}}}\frac{\mathrm{D}\theta_{\mathfrak{w}}}{\mathrm{D}_{\mathfrak{s}}t} = \frac{\gamma_{\mathfrak{w}}}{1+e}\frac{\mathrm{D}e}{\mathrm{D}_{\mathfrak{s}}t} = \gamma_{\mathfrak{w}} n \frac{\mathrm{D}n}{\mathrm{D}_{\mathfrak{s}}t}$$

$$= -\gamma_{\mathfrak{w}}[\theta_{\mathfrak{w}}(v_{\mathfrak{w}\alpha} - v_{\mathfrak{s}\alpha})]_{,\alpha} \tag{10.7.5}$$

Following the motion with the coordinates, the material derivatives are replaced by the local derivatives, and keeping the velocity of the solids fixed, $v_{\mathfrak{s}i} = 0$, we obtain

$$\frac{1}{1-\theta_{\mathfrak{w}}}\frac{\partial \theta_{\mathfrak{w}}}{\partial t} = \frac{1}{1+e}\frac{\partial e}{\partial t} = n\frac{\partial n}{\partial t} = -[\theta_{\mathfrak{w}}(v_{\mathfrak{w}\alpha} - v_{\mathfrak{s}\alpha})]_{,\alpha} \tag{10.7.6}$$

Eq. (10.7.6) will be further discussed in Chap. 11, as we identify in it Terzaghi's consolidation theory.

10.8 The Coefficient of Permeability

Darcy's law, Eq. (10.5.2), is universally accepted as a valid law controlling the flow in a porous medium. It owes its popularity to its simplicity, to its being linear, and to the fact that it lends itself easily to additional improvements and can be extended to cases beyond its original limits. Nevertheless, its limitations are known, and several comments are here in order.

The law, developed empirically in the middle of the nineteenth century by Darcy (1856), a city engineer in Dijon, while experimenting on the flow of water through sand columns, was formulated in very simple terms. Darcy found that the discharge Q through the column is directly proportional to the cross-section S of the column and to the piezometric head H, and is inversely proportional to the length of the column L

$$Q = -kS\frac{H}{L} \tag{10.8.1}$$

where k is the coefficient of proportionality. When the data obtained was plotted on a Q/S versus H/L diagram, Fig. 10.8.1, it indicated a linear relationship between the specific discharge $q = Q/S$ and the pressure gradient H/L. The slope β of the line going through the origin was

$$k = \tan\beta = -\frac{QL}{SH} \tag{10.8.2}$$

It may be seen that Eq. (10.8.1) is a particular case of Eqs. (10.5.4) and (10.5.5) in that the flow is one-directional, the porous sand matrix is stationary and the discharge measured contains only water. The fact that the flow is one-directional is reflected in the permeability coefficient k as a constant scalar value. In its general form k_{ij} is a symmetric second-order tensor and it indicates anisotropy in the flow properties of the soil, occurring in sedimentary deposits and varved clays, in compacted fills and in structurally oriented clays. Actually, if in addition to the one-directional permeability the other directions had been checked, different values would have been

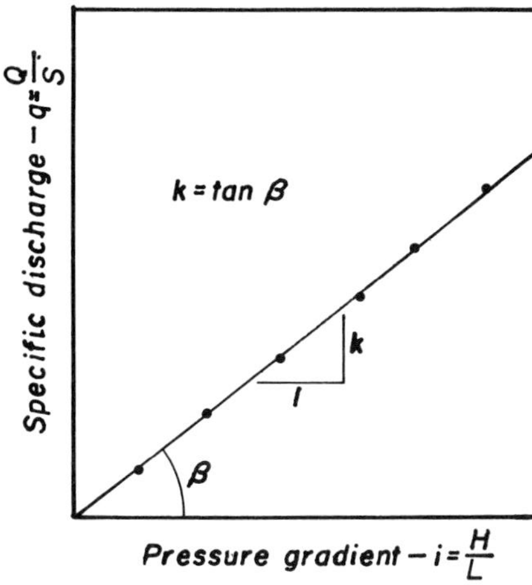

Fig. 10.8.1. Permeability results.

obtained. If the soil exhibits identical permeabilities in its principal directions, $k_{11} = k_{22} = k_{33} = k$, the coefficient reduces to a scalar.

k_{ij} has the dimensions of velocity, $[LT^{-1}]$; it is, however, a technical quantity and, considering Eq. (10.8.1), it seems to be dependent only on the geometry of the solids matrix. Still, there are other factors that affect the permeability coefficient:

1. *Homogeneity*. The assumption of homogeneity, both of the solids matrix and the water, is inherent in Darcy's law. However, like in any physical process, quasi-homogeneity is acceptable.

2. *Grain size distribution*. The grain size distribution is strongly related to the size and distribution of the pores (Morel-Seytoux 1969), through which the flow occurs. Therefore, permeability is ordinarily considered to be related to the effective diameter D_e of the grains. The permeability decreases with the decrease of the effective diameter.

3. *Specific surfaces*. The flow is generally related to the frictional resistance generated and expressed by the hydraulic radius, $R_{\mathfrak{H}}$, which is the ratio of the volume of the voids to the grain envelope. The permeability decreases with the decrease of the hydraulic radius.

4. *Shape of grains*. It is obvious that the different shapes of the grains affect the permeability differently. In a soil with sphere-like grains the hydraulic radius is minimal, as compared with that of a soil with plate-like grains. Since flow and permeability decrease with the decrease of the square of the hydraulic radius, it seems consistent that in a sphere-like grain soil, where the contours of the pores are sharper and constrictions and enlargements follow one another more closely, resistance to flow is higher.

5. *Texture of grains*. Grains of angular and irregular surface have greater frictional resistance and greater resistance to flow than grains of rounded and smooth texture.

6. *Packing of grains*. Soils, particularly granular soils, may have the same effective granular diameter and the same granular shape and texture, but be

packed differently (Dallavalle 1943; Farouki 1963), and thereby result in different permeability coefficients. In fine grained soils the packing effect is manifested in the orientation of the particles, as explained in Sect. 8.23.

The effects of shape, texture, and packing, when lumped together, are what is known as the *tortuosity* of the pores.

7. *Charge density.* In Sect. 8.10 we have seen that the surface charges divided by the specific surface define the charge density. In the sections following (specifically 8.12, 8.21 and 8.22) the effects of the surface charges, particularly in fine grained soils, have been discussed and stressed in relation with the mobility of water next to the particle surfaces. An increase of the charge density reduces the permeability of fine grained soils.

8. *Viscosity of the permeating fluid.* It is known that the coefficient of permeability, as formulated by Darcy's law, seems to depend on the properties of the solids matrix only. A closer scrutiny indicates that it depends also on the concealed properties of the permeating fluid (Nutting 1930; Wykoff et al. 1933; Sullivan and Hertel 1942). The relevant fluid properties are the fluid density ρ_w and the *viscosity* η

$$K_{ij} = \frac{k_{ij}\gamma_w}{\eta} = \frac{k_{ij}g\rho_w}{\eta} = \frac{k_{ij}g}{\nu} \tag{10.8.3}$$

where K_{ij} is the *intrinsic permeability* or *conductivity*, with the dimensions $[L^2]$ and dependent on the porous matrix, and $\nu = \eta/\rho_w$ is the *kinematic viscosity*. The dimension of ν is $[L^2T^{-1}]$ while that of η is $[FL^{-1}T^{-1}]$.

It is evident from Eq. (10.8.3) that the permeability is dependent on the viscosity η or the kinematic viscosity ν of the permeating fluid. But if the permeating fluid is water, its viscosity will depend on the concentration of the electrolytes and their nature. As the number and size of the ions increase, they interfere with the flow and decrease the permeability.

9. *Density of the permeating fluid.* The various concentrations of electrolytes in the water, as well as the nature and size of ions alter the density of the water. Eq. (10.8.3) indicates that the permeability is inversely proportional to the density. There is evidence that in fine grained soils with a developed diffuse double layer the density of the water next to the particle surface is considerably higher than the density in the bulk water (Low 1961, 1968). The density is also affected by the temperature.

10. *Temperature.* The temperature of the fluid affects greatly its viscosity: as the temperature of the water decreases the viscosity increases, and so does the permeability. We should keep in mind that the temperature also affects the density of the permeating water and the diffuse double layer, thus influencing again the permeability.

11. *Ionic concentration of the permeating water.* In fine grained soils the ionic concentration affects the permeability in another way as well. A decrease of the ionic concentration in the water solution results in a more developed diffuse double layer and in turn in a reduction of the mobility of the water, or a decrease of its permeability.

12. *Pressure gradient.* It has been found that for any particular soil there is an upper limit and a lower limit, between which the permeability is maintained according to Darcy's law. For granular soils there may not be a lower limit or *threshold gradient* and the Q/S versus H/L line would go through the origin. There could, however, be an upper limit, the *critical gradient* (Terzaghi 1925, 1943), beyond which the stability of a granular soil is jeopardized. The stability could also be threatened under high discharge flows within the soil, when the critical *Reynolds number* becomes large $R_e = q_{wi} D_e \gamma_w / g\nu < 1$ (Harr 1962), where D_e is the *average effective diameter* of the soil particles.

In fine grained soils, on the other hand, the upper limit would require a high pressure gradient in order to reach either a critical gradient or the critical Reynolds number. Because of both the cohesive interaction that holds together the particles, and the relatively narrow inter-particle pores that impede high velocities of solutes, such an upper limit is reached only under high pressure gradients. Threshold gradients have been anticipated in soils (Low 1960; Hansbo 1960; Rosenqvist 1961; Blackmore 1969), observed and recorded (Howe and Hudson 1927; Franzini 1949; Lambe 1954; Michaels and Lin 1954; Schmid 1957; Lutz and Kemper 1959; Li 1960; Miller and Low 1962; Swartzendruber 1962; Kutilek 1969), studied and explained (Schmid 1957; Klausner and Kraft 1965, 1966; Basak and Madhav 1976). It has been found that as pressure decreases in a fine grained soil, a threshold gradient is reached gradually, below which no flow can occur, Fig. 10.8.2. Such a phenomenon would indicate a gradually decreasing coefficient of permeability $k(\Phi)_{ij}$, which is a function of the gradient $\Phi_{,i}$ and attains zero for Φ_0.

13. *Entrapped gas.* Saturated soils, even undisturbed samples, are likely to contain small amounts of entrapped air or other gases, or the water may contain dissolved air, so that practically it is quite impossible to obtain

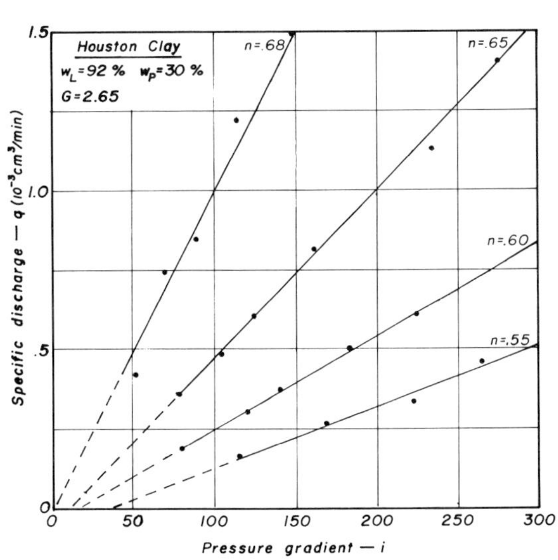

Fig. 10.8.2. Permeability threshold gradient. [After Li Ping Seung (1960).]

completely saturated soils. In embankments and fills, the entrapped air may reach over 10% of the volume, even when compacted on the wet side of the optimum moisture content. The permeability would be greatly affected by the entrapped air and change continuously, either increasing as the air is dissolved in the water and drained, or decreasing when air accumulates in the voids, forming bubbles larger than the constrictions of the pores and blocking the flow.

14. *Other Factors*. Unsteady flow due to either a transient potential or a dynamic potential, be it periodic or aperiodic, will induce corresponding changes in the density of the water and the solid matrix, respective to their compressibility. The effect of density changes will be considered later.

Time has certainly an effect on permeability. Remolded and rolled fills of fine grained soils, in particular, will exhibit an added time effect, due to particle rearrangement and thixotropic phenomena. Particle rearrangement may occur in undisturbed soils as well, since the particles tend to orient themselves in the direction of the flow; yet, when the mobility of the particles is limited, their orientation may not exactly coincide with the direction of the flow generated. Reversal or change of the flow direction may result in a time lag due to reorientation of particles, even in homogeneous soils. Such a time lag could be of considerable length if the pressure gradients are small.

It has been pointed out earlier that the coefficient of permeability is a technical *calculated* quantity, obtained from other *measured* quantities: the flow discharge Q, the hydraulic head H and the geometry of the solid matrix – its cross-section S and length L. It should also be kept in mind that while the dimensions of the specific water discharge $q_{\mathfrak{w}i}$ are the same as those of the velocity $v_{\mathfrak{w}i}$, $[LT^{-1}]$, $v_{\mathfrak{w}i}$ is a measurable quantity but $q_{\mathfrak{w}i}$ is calculated.

10.9 Seepage in Saturated Soils

It has been shown that the general equation of flow of a saturated soil, Eq. (10.6.1), reduces to Eq. (10.6.5) if the solids and the water are considered incompressible. If, in addition, we assume that the solids are steady in their position and do not move, $v_{\mathfrak{s}i}$ in Eq. (10.6.5) vanishes and the equation reduces further

$$k_{\alpha\beta}\Phi_{,\alpha\beta} = 0 \qquad (10.9.1)$$

to the *Laplace differential equation of seepage*, that controls the flow in a saturated homogeneous soil with anisotropic permeability. In a soil with isotropic permeability, $k_{11} = k_{22} = k_{33} = k$, we have the even simpler and more familiar Laplace differential equation

$$\Phi_{,\alpha\alpha} \equiv \operatorname{div}\operatorname{grad}\Phi = \nabla^2\Phi = 0 \qquad (10.9.2)$$

which occurs in various branches of physics, where Φ represents: the gravitational potential in a medium where no other potentials or attractive forces are present, the electrostatic potential in a uniform dielectric in electrostatics (diffuse double layer), the magnetic potential in free space, the electric potential in the theory of steady flow of electric currents in solid conductors, the temperature in the theory of thermal equilibrium. In spite of the disciplinal diversity of the subjects just mentioned, the mathematical treatment is much the same for all of them. Some of the theories concerning the potentials mentioned above serve also as analogous methods in the empirical solution of Eq. (10.9.2) for problems of seepage with known boundary conditions.

Eq. (10.9.2) is a linear, homogeneous, partial differential equation of second order, where $\Phi(X, Y, Z)$ is a function of the coordinates X, Y, Z. Specific problems, which involve the solution of differential field equations, are subject to specified conditions. Usually these conditions are at the geometrical boundaries of the specific problem and are known, therefore, as *boundary conditions*, and such problems as *boundary value* problems. Eq. (10.9.2), being a second-order differential equation and Φ being a function of three variables, there are six boundary conditions.

A powerful tool, but not the only one, to solve many important linear, homogeneous, differential equations is the method of separation of variables. Frequently the boundary is of a simple geometrical form, e.g. a rectangle or a circle, we may thus find appropriate the use of orthogonal or curvilinear coordinates. The Laplace equation, (10.9.2), after transformation to the new coordinates, say x, y, z, usually yields the solution

$$\Phi(x, y, z) = \Phi(x)_1\Phi(y)_2\Phi(z)_3 \tag{10.9.3}$$

The problem now is to obtain a system of ordinary differential equations for the functions $\Phi(x)_1$, $\Phi(y)_2$ and $\Phi(z)_3$, and an appropriate linear combination for the solution of Eq. (10.9.3), that will also satisfy the boundary conditions.

There are few problems for which Eq. (10.9.3) can be solved directly, and even for those problems the solutions are mostly numerical. In engineering, however, most practical problems requiring the solution of Eq. (10.9.3) can be formulated as problems of two variables, where Φ can be considered as being independent of one of its variables, say z. Owing either to symmetry or to the fact that the variation of the physical field in one direction is zero or can be neglected in comparison with the variations in the other directions, the dependence of Φ on that direction is cancelled

$$\Phi(X, Y)_{,\alpha\alpha} \equiv \nabla^2\Phi(X, Y) = \frac{\partial^2\Phi}{\partial Z^2} = 0 \tag{10.9.4}$$

For example, many problems can be formulated as planar two-dimensional problems, with the flow being the same in parallel planes, $\partial\Phi/\partial Z = 0$, although the similarity is limited to a finite length in the third direction.

Other problems, formulated in cylindrical or spherical coordinates, display axial symmetry $\partial\Phi/\partial\theta = 0$. In such cases, where the problem becomes two-dimensional, the required boundary conditions reduce to four. The two-dimensional Laplace equation has an important property, that any analytic function, like the *complex potential*

$$F(\mathscr{Z}) = \phi(X, Y) + i\psi(X, Y) \tag{10.9.5}$$

is a solution of it. According to Eq. (10.2.13), the *Cauchy–Riemann equations* follow

$$\phi_{,i} = \psi_{,j} \quad \text{and} \quad \phi_{,j} = -\psi_{,i} \tag{10.9.6}$$

indicating that we have two sets of functions ϕ and ψ orthogonal one to the other, Fig. 10.9.1. One of them is the potential lines, representing points of equal potentials, the other the force lines that were mentioned in Sect. 10.2, representing here flow lines. A final solution depends on the specific problem with a given geometry and on four boundary conditions, two with respect to known potential characteristics and two with respect to known flow line characteristics, Fig. 10.9.1. Solutions for a variety of boundary-value problems, as well as for layered soils and soils with anisotropic permeability may be found in the numerous textbooks.

10.10 Unsaturated Flow of Multi-phase Fluids

In Sects. 10.4–10.9 the flow in saturated soils was discussed. However, most frequently encountered in practice is the flow in unsaturated soils, which is much more complex and ramified, yet its study is less developed and less conclusive. Infiltration from sources of various geometrical characteristics (surface, line, point), water table draw-down, both in non-swelling and swelling soils, are some of the typical problems posed by this type of flow.

Part of the pores in a soil may be filled with liquid, while the rest of the pores are filled with a gas or a gaseous mixture, say air. We have discussed in Sect. 8.1 the existence of water vapor in the pores, and in Sects. 8.6–8.8 the

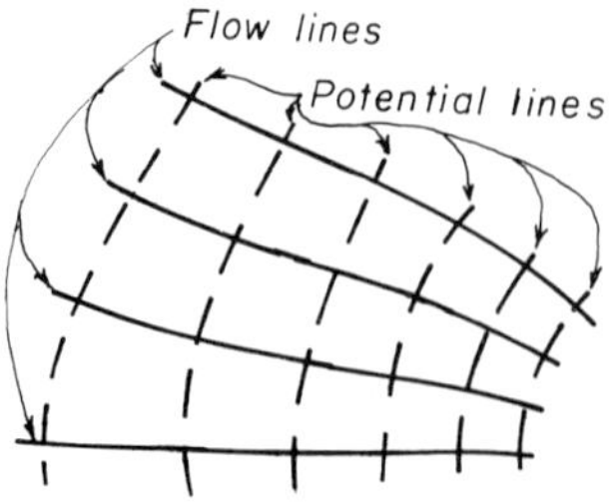

Fig. 10.9.1. Orthogonal flow lines and potential lines.

various forms, shapes and sizes of the air inclusions in soils. It was shown that air could be in the form of small or large bubbles or inclusions that fill out the pores, except for the regions of the constrictions where the liquid bound by the menisci prevails.

The pores may be filled with several *immiscible* liquids, or several immiscible liquids and air. If the pores contain, say, $\mathfrak{F}$ immiscible liquids, then the degree of saturation $S_{\mathfrak{f}}$ of the $\mathfrak{f}$th immiscible liquid is defined

$$S_{\mathfrak{f}} = \frac{V_{\mathfrak{f}}}{V_{\mathfrak{p}}} \tag{10.10.1}$$

and we may write

$$\sum_{\mathfrak{f}=1}^{\mathfrak{F}} S_{\mathfrak{f}} = 1; \quad V_{\mathfrak{g}} = 0 \tag{10.10.2}$$

for the case where no gaseous phase is contained in the pores.

According to Eq. (10.10.2) a porous medium with several immiscible liquids in its pores may have all its pores filled and still be unsaturated, since by definition the saturation is considered with respect to each of the liquids individually. If there are air inclusions in the pores in addition to the $\mathfrak{F}$ liquids, Eq. (10.10.2) is replaced by

$$\sum_{\mathfrak{f}=1}^{\mathfrak{F}} S_{\mathfrak{f}} = 1 - \frac{V_{\mathfrak{g}}}{V_{\mathfrak{p}}}; \quad V_{\mathfrak{g}} \neq 0 \tag{10.10.3}$$

The flow of several immiscible liquids in a porous medium is another aspect of unsaturated flow and forms a disciplinary subdivision called *hydrodynamic dispersion*. It is covered by extensive literature (Saffman 1959; Scheidegger 1961; Dagan 1964; Morel-Seytoux 1969; Baer 1972) and will not be pursued further in this work, except for certain aspects considered basic to the understanding of the mechanical behavior of soils.

10.11 Flow in Unsaturated Soils

In a very general way, the flow in unsaturated soils can be sorted into two sets (Philip 1970): flow in non-swelling soils and flow in swelling soils.

Flow and diffusion, the transport phenomena in an unsaturated soil conceived as a three-phase mixture, are controlled by two equations, (9.3.10) and (9.3.11), derived from the density balance equations which govern the flow of water and air, respectively. With respect to a three-phase soil it is important to consider the pertinent equation (9.2.12) as well, from which we may write

$$\frac{D\theta_{s}}{Dt} + \frac{D\theta_{w}}{Dt} + \frac{D\theta_{g}}{Dt} = 0; \quad \frac{\partial\theta_{s}}{\partial t} + \frac{\partial\theta_{w}}{\partial t} + \frac{\partial\theta_{g}}{\partial t} = 0;$$

$$\theta_{s,i} + \theta_{w,i} + \theta_{g,i} = 1 \tag{10.11.1}$$

The derivation of the proper equations for the many modes of flow is subject to the assumptions made with respect to the properties of the constituents and the properties of the motion, and to a treatment similar to that of saturated soils, discussed in Sects. 10.6 and 10.7.

10.12 Flow in Unsaturated Non-swelling Soils

The assumption of rigidity of solids infers the absence of soil swelling. If the solids are packed in a rigid array and in addition incompressibility of solids is postulated, then θ_{s}, γ_{s}, e and n in Eqs. (9.3.8)–(9.3.11) become independent of time and we obtain

$$\frac{D\rho_{w}}{D_{s}t} = \frac{D\gamma_{w}\theta_{w}}{D_{s}t} = \frac{e}{1+e}\frac{D\gamma_{w}S}{D_{s}t} = n\frac{D\gamma_{w}S}{D_{s}t} = -[\gamma_{w}\theta_{w}(v_{w\alpha} - v_{s\alpha})]_{,\alpha} \tag{10.12.1}$$

$$\frac{D\rho_{g}}{D_{s}t} = \frac{D\gamma_{g}\theta_{g}}{D_{s}t} = \frac{e}{1+e}\frac{D\gamma_{g}(1-S)}{D_{s}t}$$

$$= n\frac{D\gamma_{g}(1-S)}{D_{s}t} = -[\gamma_{g}\theta_{g}(v_{g\alpha} - v_{s\alpha})]_{,\alpha} \tag{10.12.2}$$

If incompressibility of the water is also required and Eq. (10.5.5) is introduced, Eq. (10.12.1) simplifies to

$$\frac{D\rho_{w}}{D_{s}t} = \gamma_{w}\frac{D\theta_{w}}{D_{s}t} = \frac{\gamma_{w}e}{1+e}\frac{DS}{D_{s}t} = \gamma_{w}n\frac{DS}{D_{s}t} = -[\gamma_{w}\theta_{w}(v_{w\alpha} - v_{s\alpha})]_{,\alpha}$$

$$= -[\gamma_{w}\bar{q}_{w\alpha}]_{,\alpha} = [\gamma_{w}(k_{\alpha\beta}\Phi_{,\alpha})]_{,\beta} \tag{10.12.3}$$

If, in addition, homogeneity of the soil is postulated, γ_{w} on the right-hand side becomes independent of its location in space. Since $v_{si} = 0$ and also $v_{s\alpha,\alpha} = 0$, it is convenient to transfer from material coordinates to local coordinates, and Eq. (10.12.3) reduces further to

$$\frac{1}{\gamma_{w}}\frac{\partial\rho_{w}}{\partial t} = \frac{\partial\theta_{w}}{\partial t} = \frac{e}{1+e}\frac{\partial S}{\partial t} = n\frac{\partial S}{\partial t} = -\bar{q}_{w\alpha,\alpha} = [k_{\alpha\beta}\Phi_{,\alpha}]_{,\beta} \tag{10.12.4}$$

For unsaturated soils, the coefficient of permeability can no longer be considered constant, since it is a function of the moisture content $k(\theta_{w})_{ij}$ and evidently varies as the moisture content changes. From Sects. 8.27 and 10.3 we may conclude that the moisture potential Φ_w is also a function of the moisture content θ_w. Consequently Eq. (10.12.4) may be written

$$\frac{\partial \theta_w}{\partial t} = \frac{e}{1+e}\frac{\partial S}{\partial t} = n\frac{\partial S}{\partial t} = -\bar{q}_{w\alpha,\alpha} = [k(\theta_w)_{\alpha\beta}\Phi_{,\alpha}]_{,\beta}$$

$$= [k(\theta_w)_{\alpha\beta}\frac{\partial \Phi(\theta_w)}{\partial \theta_w}\theta_{w,\alpha}]_{,\beta} \tag{10.12.5}$$

If Eq. (10.3.7) is substituted for $\Phi(\theta)$ in Eq. (10.12.5) we obtain

$$\frac{\partial \theta_w}{\partial t} = [k(\theta_w)_{\alpha\beta}\frac{\partial \Phi(\theta_w)}{\partial \theta_w}\theta_{w,\alpha}]_{,\beta} - \frac{\partial k(\theta_w)_{\alpha\beta}}{\partial \theta_w}\theta_{w,\alpha}\left(1 - \left[\frac{p(z_0)}{\gamma_w z} + 1\right]\right)n_\beta$$

$$= [D(\theta_w)_{\alpha\beta}\theta_{w,\alpha}]_{,\beta} - \frac{\partial k(\theta_w)_{\alpha\beta}}{\partial \theta_w}\theta_{w,\alpha}\left(1 - \left[\frac{p(z_0)}{\gamma_w z} + 1\right]\right)n_\beta \tag{10.12.6}$$

where

$$D(\theta_w)_{ij} = k(\theta_w)_{ij}\frac{\partial \Phi(\theta_w)}{\partial \theta_w} \tag{10.12.7}$$

is the *moisture diffusivity*, a function of the coefficient of permeability $k(\theta_w)_{ij}$ and of the derivative of the moisture potential $\partial\Phi(\theta_w)/\partial t$, both functions of the moisture content. Its dimensions are $[D(\theta_w)_{ij}] \equiv [L^2T^{-1}]$.

Since in soil engineering the water content w is a more familiar term than the moisture content θ_w, Eqs. (10.12.5) and (10.12.7) may be expressed in terms of w. By substituting the value of θ_w from Eq. (9.2.22) with its value in terms of w, we obtain

$$\frac{\partial w}{\partial t} = \frac{eG_w}{G_s}\frac{\partial S}{\partial t} = \frac{nG_w}{(1-n)G_s}\frac{\partial S}{\partial t} = \frac{(1+e)G_w}{G_s}\left[k(w)_{\alpha\beta}\frac{\partial \Phi(w)}{\partial w}w_{,\alpha}\right]_{,\beta}$$

$$= \frac{(1+e)G_w}{G_s}\left[(D(w)_{\alpha\beta}w_{,\alpha})_{,\beta} - \frac{\partial k(w)_{\alpha\beta}}{\partial w}w_{,\alpha}\left(1 - \left[\frac{p(z_0)}{\gamma_w z} + 1n_\beta\right)\right]\right] \tag{10.12.8}$$

$$D(w)_{ij} = k_{ij}(w)\frac{\partial \Phi(w)}{\partial w} = \frac{G_s}{(1+e)G_w}D(\theta_w)_{ij} \tag{10.12.9}$$

where $D(w)_{ij}$ is the *water diffusivity*, a function of the coefficient of permeability $k(w)_{ij}$ and of the moisture potential derivative $\partial\Phi(w)/\partial w$, both in terms of the water content w. Figs. 10.12.1–10.12.4 show the coefficients of permeability $k(w)$, the moisture potentials $\Phi(w) = \Phi_w$, the moisture potential derivatives $\partial\Phi(w)/\partial w$ and the diffusivities $D(w)_{ij}$, respectively, reproduced from data presented by Gardner (1958) for three representative soils: Chino clay, Indio silt and Pachappa sandy silt.

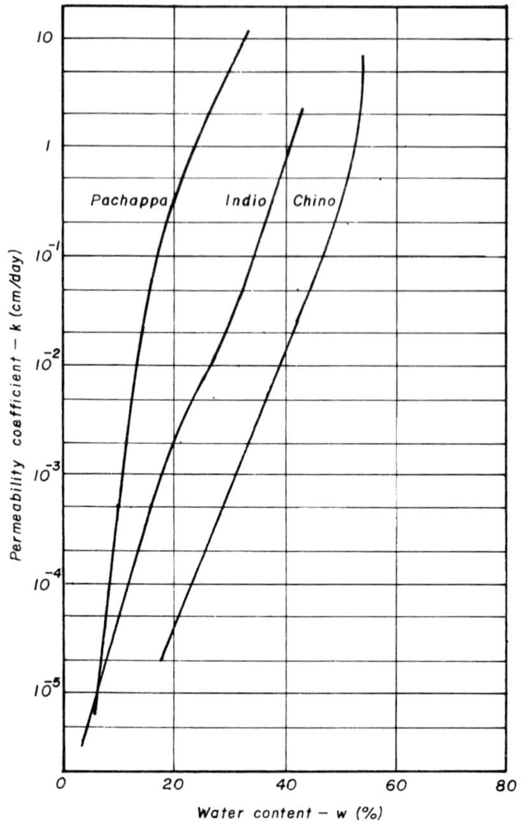

Fig. 10.12.1. Coefficients of permeability versus water content. [After W. R. Gardner (1958).]

Eqs. (10.12.6) and (10.12.7) serve as basic equations for the solution of boundary-value problems such as infiltration and sorption (Bruch and Zyvoloski 1974; Knight and Philip 1974; Nielsen et al. 1972; Kunze and Nielsen 1982; Parlange 1971a–c; Philip 1957a–d, 1966a, b, 1967, 1968a, 1973), in general for the isotropic soil where $k_{ij} = k\delta_{ij}$.

The flow of water in unsaturated soils seems to be independent of the flow of air that occurs in the soil simultaneously and next to it. At first glance, Eqs. (10.12.1) and (10.12.2) are apparently independent of one another. On a closer look, however, θ_w and θ_g, and v_{wi} and v_{gi} are interdependent, according to Eqs. (9.2.12) and (6.3.1), respectively. Moreover, the moisture potential Φ, which is the pressure head of the water, is definitely dependent on the pressure of the air p_g. The solution of any of Eqs. (10.12.1), (10.12.6) and (10.12.8) is nevertheless possible, independent of what goes on in the air phase, if the moisture potential curve is experimentally available.

Eq. (10.12.2) has hardly been explored in the study of flow of air in soils, and there is much to be done in that respect. It could shed light on phenomena related to the many facets of air in soils, and its further study could enhance, perhaps, analytical solutions of the flow of water in unsaturated soils. The flow of air through porous media, including soil, has been of

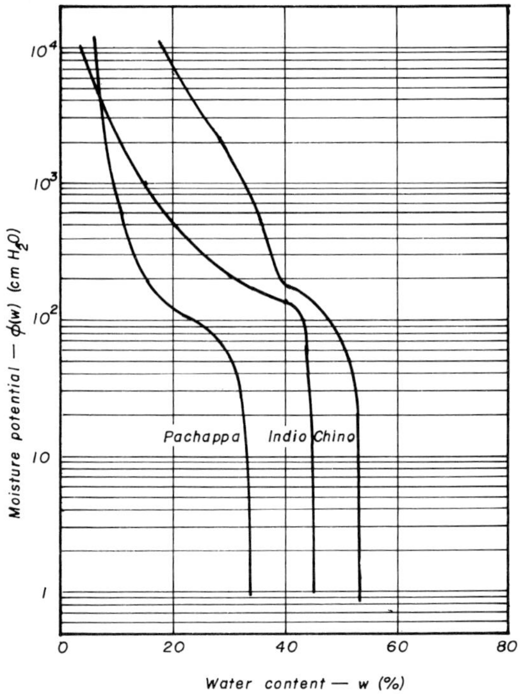

Fig. 10.12.2. Moisture potentials versus water content. [After W. R. Gardner (1958).]

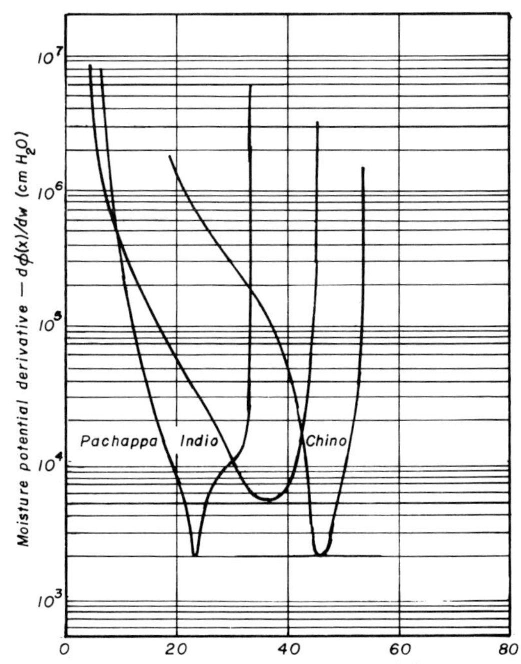

Fig. 10.12.3. Moisture potential derivatives versus water content. [After W. R. Gardner (1958).]

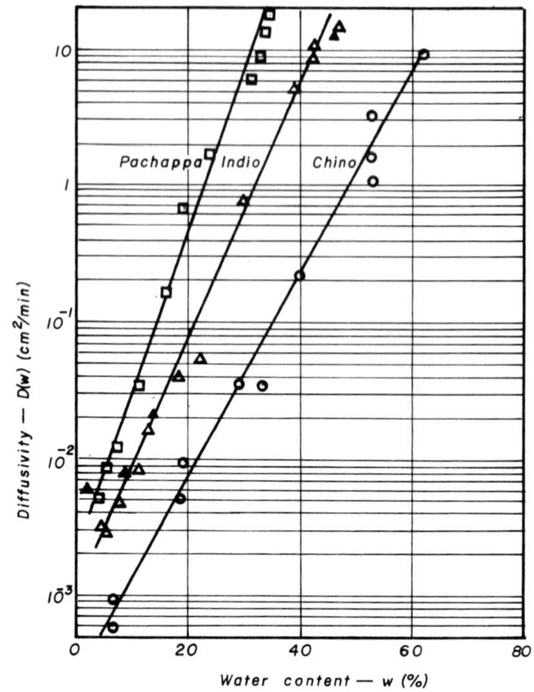

Fig. 10.12.4. Diffusivities versus water content. [After W. R. Gardner (1958).]

continuous interest for many investigators (Wykoff and Botset 1936; Carman 1956; Alpan 1962; Le Van Phuc and Morel-Seytoux 1972; Fredlund 1979; Parlange et al. 1982; De Jong et al. 1983), the study being, however, mostly empirical or semi-empirical.

To further the exploration of Eq. (10.12.2), let us consider the same assumptions made with respect to the flow of water, namely, incompressibility of both solids and water and rigidity of the solids, and we obtain

$$-\frac{D\rho_g}{D_s t} = -\frac{D\gamma_g\theta_g}{D_s t} = -\frac{e}{1+e}\frac{D\gamma_g(1-S)}{D_s t} = -n\frac{D\gamma_g(1-S)}{D_s t}$$

$$= [\gamma_g\theta_g(v_{g\alpha} - v_{s\alpha})]_{,\alpha} \qquad (10.12.10)$$

We introduce the *bubbling velocity* $\bar{q}_{gi}$, a relative motion of the air and an expression analogous to the seepage velocity in Eq. (10.5.6), defined

$$\bar{q}_{gi} = \frac{(1-S)e}{1+e}[v_{gi} - v_{si}] = \theta_g[v_{gi} - v_{si}] = q_i - v_{si} \qquad (10.12.11)$$

Since the discharge gradient of the air, the right-hand side of Eq. (10.12.10), relates to the solids as well, it is also convenient to consider local derivatives and we obtain

$$-\frac{\partial \rho_g}{\partial t} = -\frac{\partial \gamma_g \theta_g}{\partial t} = -\frac{e}{1+e}\frac{\partial \gamma_g (1-S)}{\partial t} = -n\frac{\partial \gamma_g (1-S)}{\partial t} = [\gamma_g \bar{q}_{g\alpha}]_{,\alpha} \tag{10.12.12}$$

Considering the change in time and space of the volume of the air V_g, according to Boyle's law at constant temperature, we obtain from Eq. (8.7.1)

$$V_g = \frac{RT}{p_g} \tag{10.12.13}$$

$$\frac{\partial V_g}{\partial t} = RT\frac{\partial \frac{1}{p_g}}{\partial t} = \frac{-RT}{p_g^2}\frac{\partial p_g}{\partial t}; \quad V_{g,i} = RT\left(\frac{1}{p_g}\right)_{,i} = \frac{-RT}{p_g^2}p_{g,i} \tag{10.12.14}$$

Since the volume V_g is proportional to the volumetric air fraction θ_g, the proportionality equation for rigid solids may be written

$$V_g \approx \theta_g \approx S_g \tag{10.12.15}$$

If Eqs. (10.12.13) and (10.12.14) are substituted into Eq. (10.12.10), the flow equation of air is obtained

$$\frac{\partial \gamma_g}{\partial t} - \frac{\gamma_g}{p_g}\frac{\partial p_g}{\partial t} = \frac{\gamma_g v_{g\alpha}}{p_g}p_{g,\alpha} - (\gamma_g v_{g\alpha})_{,\alpha} \tag{10.12.16}$$

Eq. (10.12.16) will be useful in the following chapter.

Here we have explored the flow of water and air in unsaturated non-swelling soils under the assumption of incompressibility of solids and water. A more restrictive assumption considering the compressibility of water is also possible, however it is more laborious and will not be pursued here.

10.13 The Boltzmann Transformation Solution

Since Gardner (1956) first suggested the method of calculating D and k from experimental data, several approaches to the solution of the equation of flow in unsaturated soils have been presented. Bruce and Klute (1956) used the method of separation of variables known as the *Boltzmann transformation*, suggested sixty years earlier (Boltzmann 1894). This solution is outlined here because of its simplicity. In order to facilitate the solution of the differential equation, it is assumed that the flow is one-directional and the moisture potential is a single-valued monotonic function of the moisture content θ_w, which is a single-valued function of χ

$$\theta_w = \mathcal{F}(\chi) \tag{10.13.1}$$

where χ is the Boltzmann transformation function, defined

$$\chi = xt^{-1/2} \tag{10.13.2}$$

the space and time derivatives of which are

$$\frac{\partial \chi}{\partial x} = t^{-1/2} \tag{10.13.3}$$

$$\frac{\partial \chi}{\partial t} = -\frac{1}{2} xt^{-3/2} \tag{10.13.4}$$

If Eqs. (10.13.2)–(10.13.4) are substituted into Eq. (10.12.5) we obtain

$$-\chi \frac{d\theta_w}{d\chi} = 2 \frac{d}{d\chi}\left[D(\theta_w) \frac{d\theta_w}{d\chi}\right] \tag{10.13.5}$$

assuming that $D(\theta_w)$ remains a function of θ_w, as already defined in Eq. (10.12.7).

The boundary conditions with respect to θ_w can now be introduced, to solve a number of problems of unsaturated flow with the help of the Boltzmann transformation.

θ_{wi} and θ_{ws} being the initial moisture content and the moisture content at saturation, respectively, the boundary conditions for a horizontal flow are:

$$\begin{aligned}
\theta(x, t)_w &= \theta_{wi} \quad \text{for } t = 0, \quad x > 0, \quad \chi = \infty \\
\theta(x, t)_w &= \theta_{wi} \quad \text{for } t > 0, \quad x = 0, \quad \chi = \infty \\
\theta(x, t)_w &= \theta_{ws} \quad \text{for } t > 0, \quad x = 0, \quad \chi = 0 \\
d\theta_w/d\chi &= 0 \quad \text{for } \theta_w = \theta_{wi}
\end{aligned} \tag{10.13.6}$$

By multiplying both sides of Eq. (10.13.5) by $d\chi$ and integrating from the initial moisture content θ_{wi} to the moisture content θ_{wr} at a distance x we obtain

$$-\int_{\theta_{wi}}^{\theta_{wr}} \chi d\theta_w = 2\int_{\theta_{wi}}^{\theta_{wr}} d\left(D(\theta_w)\frac{d\theta_w}{d\chi}\right) = 2\left[D(\theta_{wr})\frac{d\theta_w}{d\chi}\bigg|_{\theta_{wr}} - D(\theta_{wi})\frac{d\theta_w}{d\chi}\bigg|_{\theta_{wi}}\right] \tag{10.13.7}$$

where the last term is zero on account of the boundary condition $(10.13.6)_4$, and therefore the following equation is obtained from (10.13.7)

$$D(\theta_{wr}) = -\frac{1}{2}\frac{d\chi}{d\theta_w}\bigg|_{\theta_{wr}} \int_{\theta_{wi}}^{\theta_{wr}} \chi d\theta_w \tag{10.13.8}$$

This equation can be evaluated, by semi-analytical methods, from the θ_{w} versus χ curve obtained by experiment.

The Boltzmann transformation method serves as a starting point for some of the other methods, since it allows the solution of a large number of problems such as: infiltration, sorption, desorption, drying, vertical flow, horizontal flow, etc.

10.14 Flow in Unsaturated Swelling Soils

If we deal with a swelling soil, rigidity of the solids array is not required, and the assumption of incompressibility of the solids affects its unit weight alone, for which we require $\dot{\theta}_{s}$. Accordingly, Eqs. (9.3.8)–(9.3.11) become

$$\frac{D\rho_{w}}{D_{s}t} = \frac{D\gamma_{w}\theta_{w}}{D_{s}t} = \frac{1}{1+e}\frac{D\gamma_{w}eS}{D_{s}t} = \frac{D\gamma_{w}nS}{D_{s}t} = -[\gamma_{w}\theta_{w}(v_{w\alpha} - v_{s\alpha})]_{,\alpha} \tag{10.14.1}$$

$$\frac{D\rho_{g}}{D_{s}t} = \frac{D\gamma_{g}\theta_{g}}{D_{s}t} = \frac{1}{1+e}\frac{D\gamma_{g}e(1-S)}{D_{s}t} = \frac{D\gamma_{g}n(1-S)}{D_{s}t}$$

$$= -[\gamma_{g}\theta_{g}(v_{g\alpha} - v_{s\alpha})]_{,\alpha} \tag{10.14.2}$$

Here we see that the flow of water or of air in unsaturated swelling soils are functions not only of the degree of saturation S as in non-swelling soils, but also of the void ratio e, or the porosity n. Following Philip (1969a), we shall introduce the moisture ratio ϑ defined by Eq. (9.2.13) into Eq. (10.14.1). We will come to appreciate the advantage of having the flow expressed by the moisture ratio, which is the void ratio multiplied by the degree of saturation, $\vartheta = eS$. The void ratio and the degree of saturation may increase or decrease, and we shall thus have nine combinations of changes of eS. The same may be achieved by three possible changes obtained from the moisture ratio: $\partial\vartheta > 0$, $\partial\vartheta = 0$, $\partial\vartheta < 0$. At this stage we also stipulate incompressibility of the water and follow the solids by moving to the local coordinate system, and so we obtain

$$\frac{1}{\gamma_{w}}\frac{\partial\rho_{w}}{\partial t} = \frac{\partial\theta_{w}}{\partial t} = \frac{1}{1+e}\frac{\partial eS}{\partial t} = \frac{1}{1+e}\frac{\partial\vartheta}{\partial t}$$

$$= \frac{\partial nS}{\partial t} = -\bar{q}_{w\alpha,\alpha} = [k(\theta_{w})_{\alpha\beta}\Phi(\theta_{w})_{,\alpha}]_{,\beta}$$

$$= \left[k(\theta_{w})_{\alpha\beta}\frac{\partial\Phi(\theta_{w})}{\partial\theta}\theta_{w,\alpha}\right]_{,\beta} = [D(\theta_{w})_{\alpha\beta}\theta_{w,\alpha}]_{,\beta} \tag{10.14.3}$$

which is the flow equation of unsaturated swelling soils, in terms of the moisture content θ_w.

If we substitute the value of $\Phi(\theta_w)$ from Eq. (10.3.7) into Eq. (10.14.3), we obtain

$$\frac{\partial \theta_w}{\partial t} = \frac{1}{1+e}\frac{\partial \vartheta}{\partial t} = \left[k(\theta_w)_{\alpha\beta}\frac{\partial \Phi(\theta_w)}{\partial \theta_w}\theta_{w,\alpha}\right]_{,\beta} - \frac{\partial k(\theta_w)_{\alpha\beta}}{\partial \theta_w}\theta_{w,\alpha}\left(1-\left[\frac{p(z_0)}{\gamma_w z}+1\right]\right)n_\beta$$

$$= [D(\theta_w)_{\alpha\beta}\theta_{w,\alpha}]_{,\beta} - \frac{\partial k(\theta_w)_{\alpha\beta}}{\partial \theta_w}\theta_{w,\alpha}\left(1-\frac{de}{d\vartheta}\left[\frac{p(z_0)}{\gamma_w z}+1\right]\right)n_\beta \quad (10.14.4)$$

where the value of $D(\theta_w)_{ij}$ is given in Eq. (10.12.7).

The solution of the flow equation, (10.14.3), is not so straightforward. Its non-linear nature, the dependence of the diffusivity and of the potential on the moisture content, and the hysteresis involved in the drying–wetting cycle require the use of additional assumptions.

By substituting ϑ for θ_w in Eq. (10.14.3), as follows from Eq. (9.2.13)

$$\theta_w = \frac{1}{1+e}\vartheta \quad (10.14.5)$$

we obtain the diffusion equation of water in terms of the moisture ratio

$$\frac{\partial \vartheta}{\partial t} = (1+e)\left[k(\vartheta)_{\alpha\beta}\frac{\partial \Phi(\vartheta)}{\partial \vartheta}\vartheta_{,\alpha}\right]_{,\beta} = (1+e)[D(\vartheta)_{\alpha\beta}\vartheta_{,\alpha}]_{,\beta} \quad (10.14.6)$$

where $D(\vartheta)_{ij}$, the diffusivity in terms of the moisture ratio, relates to the diffusivity $D(\theta_w)_{ij}$ in terms of the moisture content

$$D(\vartheta)_{ij} = \frac{1}{1+e}D(\theta_w) \quad (10.14.7)$$

Application of these equations may be found in the literature mentioned above and many other works, and concern some very practical problems such as dynamics of absorption, infiltration from various sources (point, line, surface section or infinite surface), capillary rise, various stages of drying, etc. The accuracy of the application depends on the original assumptions and on the further assumption to be made with respect to the boundary conditions.

The flow in unsaturated soils was discussed only briefly in order to present a complete picture of the present state of art. The study outlined rests on the assumption that the flow of air in soil is similar to the flow of water and it can be characterized by a permeability function like that of the water, although it was not specifically stated, and that the flow of the air or the water is not upset by any interaction or interdependence of the two. Unfortunately the problem of flow of air–water mixtures in micropores is more complicated and is still under investigation. Even for the flow of air–water mixtures in pipes and channels, which can be more readily investigated, only empirical results are available.

11 Volumetric Stress–Strain Phenomena

11.1 The Volumetric Stress–Strain Relationship

In Sect. 7.5 it was shown how the general rheological equation, (7.3.2), which is the general constitutive equation of materials, can be resolved into two equations, (7.5.1) and (7.5.2), to relate the spherical component $p_{\alpha\alpha}$ of the stress tensor p_{ij} to the spherical component $\varepsilon_{\alpha\alpha}$ of the strain tensor ε_{ij}, and the traceless component s_{ij} of the stress tensor p_{ij} to the traceless component e_{ij} of the strain tensor ε_{ij}. The first of these two equations is the spherical stress–strain relationship, and the second is the traceless stress–strain relationship. It was also noted that beyond the mathematical possibility of resolving Eq. (7.3.2) into Eqs. (7.5.1) and (7.5.2), the latter two equations assume physical meaning under either of the two following conditions:

1. The strain ε_{ij} is an infinitesimal strain, $\overset{\text{inf}}{\varepsilon}_{ij}$.
2. The strain ε_{ij} is a finite strain and is defined by the Hencky strain measure, $\overset{\mathbf{H}}{\varepsilon}_{ij}$.

If such a resolution is performed, the spherical and traceless stress–strain relationships become volumetric and shear stress–strain relationships, respectively.

In our further discussion the simple notation ε_{ij} will be used, with the understanding that only the permissible strains are considered; when the Hencky strain measure is specifically referred to, the full notation $\overset{\mathbf{H}}{\varepsilon}_{ij}$ will be used.

It was also pointed out that the resolution of Eq. (7.3.2) into Eqs. (7.5.1) and (7.5.2) is straightforward, if Eq. (7.3.2) represents a linear relationship between the stress p_{ij} and the strain ε_{ij}. If isotropic relations are postulated then the relationship between the stress tensor p_{ij} and the strain tensor ε_{ij} is not linear, and the general stress–strain relationship can be resolved by the superposition of the traceless stress–strain relationship over the spherical stress–strain relationship. The physical meaning is still preserved, albeit with the added provision of coaxiality of the principal directions of the stress and strain tensors and their time derivatives.

This chapter will explore the phenomena related to the volumetric stress–strain relationship of soils, both in the structural level and the phenomenological level.

11.2 Volume Changes in Soils

Volume changes in soils are induced by stresses from a variety of sources: mechanical loading by application or removal of surcharges from the surface or from any depth in the soil, thermal stresses such as desiccation, hydrological constraints such as water supply to the soil by rain or by lowering or raising the water table, domestic environs of garden waterings or damaged sewer systems percolating water into the ground, electrochemical changes due to the electrolytes in the industrial refuse percolating into the soil, ecological influence of vegetation and forestry continuously draining the water from the ground, and more. All of the above are physical circumstances. Some of the changes are decreases, others are increases in volume, and in terms of deformations they result in the subsidence and settlement of the soil or in its heaving, respectively. The subsidence of soils resulting from a volume decrease is known in geotechnical engineering as *consolidation*, while the heaving of soils, a volume increase, is known as *swelling*.

As was stated in Chap. 7, if linear viscoelasticity is assumed *volume decreases and increases are results of spherical stresses and only of spherical stresses*, i.e. *compression or decompression*. When soils are stressed, they undergo volume changes of two kinds:

1. *Compression-dilation* of the constituents, the solids, the water and the air, discussed in detail in Sects. 8.4–8.7. This compression-dilation, as in any material, depends on the bulk modulus $\varkappa_\mathfrak{n}$ of the constituent affecting its unit weight $\gamma_\mathfrak{n}$ and its specific gravity $G_\mathfrak{n}$.

2. *Dissipation-sorption* of water and air from or into the pores and simultaneous rearrangement, i.e. densification or expansion, of the solids matrix. The dissipation of water and air from the pores and the sorption of water and inflow of air into the pores is a result of the changes in the existing pore-water pressure and pore-air pressure. The effect is a decrease or increase, respectively, in the moisture content $\theta_\mathfrak{w}$ and air content $\theta_\mathfrak{g}$, thus a net volume decrease or increase. Irrespective of the total stresses applied to the soil, and since we have already seen that water and air are not receptive of shear stresses but of spherical stresses only, the driving dissipative forces must be, in our present case, the pore-water pressure or pore-air pressure.

A fair assumption with respect to the compressibility of the solids and the water is that the solids and the water are elastic and thus compressible. When the dissipative-sorptive volume changes of the entire soil system are large relative to the compression-dilation volume changes, then the solids, and occasionally the water, may be assumed incompressible. The dissipation of water and air through the pores, however, is more complex and unfolds as a time-dependent phenomenon.

In granular soils the effect of the compressibility of the constituents is, in most cases, more dominant than the dissipative effect. In fine grained clayey or silty soils the effect of the dissipative phenomenon overrides that of compressibility of the constituents, which is sometimes negligible compared to the volume changes due to the dissipation of water and air from the pores. On the other hand, in some loosely packed granular soils considerable time-dependent volume changes are measured, mostly the effect of dissipation. Soils in which the dissipation is greater than the compressibility command added attention because of the extent of the volume changes and the time element involved.

In the following sections of this chapter a discussion of the phenomena of consolidation and swelling of saturated and unsaturated soils will be presented, in both the phenomenological and the structural levels.

The equations controlling consolidation and swelling are the same density balance equations that control flow in soils. In saturated soils the equations of interest are Eqs. (10.6.1), (10.6.2) and (9.4.2), while in unsaturated soils Eqs. (10.12.1), (10.12.2), (9.2.12), (10.11.1) and (9.4.2).

11.3 Consolidation of Saturated Soils

According to our definition *consolidation* is *the decrease in volume that occurs in the soil, of whatever cause.*

The general material equation that controls the consolidation of saturated soils is identical to the density balance equations of a mixture of two constituents, Eqs. (10.6.1) and (10.6.2), which, as we have seen, control the flow of water in saturated soils and are reproduced again here

$$(10.6.1) \qquad \rho_{\mathfrak{s}} \frac{D(\rho_{\mathfrak{w}}/\rho_{\mathfrak{s}})}{D_{\mathfrak{s}}t} = \gamma_{\mathfrak{s}}\theta_{\mathfrak{s}} \frac{D[\gamma_{\mathfrak{w}}\theta_{\mathfrak{w}}/(\gamma_{\mathfrak{s}}\theta_{\mathfrak{s}})]}{D_{\mathfrak{s}}t}$$

$$= \gamma_{\mathfrak{s}}(1-n)\frac{D\{\gamma_{\mathfrak{w}}n/[\gamma_{\mathfrak{s}}(1-n)]\}}{D_{\mathfrak{s}}t}$$

$$= \frac{\gamma_{\mathfrak{s}}}{1+e}\frac{D(\gamma_{\mathfrak{w}}e/\gamma_{\mathfrak{s}})}{D_{\mathfrak{s}}t} = -[\gamma_{\mathfrak{w}}\theta_{\mathfrak{w}}(v_{\mathfrak{w}\alpha} - v_{\mathfrak{s}\alpha})]_{,\alpha}$$

$$= [\gamma_{\mathfrak{w}}(k_{\alpha\beta}\phi_{,\alpha})]_{,\beta}$$

$$(10.6.2) \qquad \theta_{\mathfrak{s}} + \theta_{\mathfrak{w}} = 1$$

Eq. (10.6.1) expresses the fact that the density flux of the water through a material point represented by an infinitesimal volume is equal to the changes in time of the amounts of solids and water. Like in the flow of water in a saturated soil, the driving force of these changes is the hydraulic gradient $\Phi(X)_{,i}$, defined in Eq. (10.5.3).

We see that the general equation of consolidation, given here in material coordinates moving along with the solids phase, can be expressed in terms of any one of the four parameters: the densities $\rho_{\mathfrak{w}}$ and $\rho_{\mathfrak{s}}$, the volume fractions $\theta_{\mathfrak{w}}$ and $\theta_{\mathfrak{s}}$, the porosity n and the void ratio e. Note that if the consolidation equation is expressed in terms of the volume fractions one can distinguish and separate between the part of the volume change that is due to the compressibility of the constituents involved as $\gamma_{\mathfrak{w}}$ and $\gamma_{\mathfrak{s}}$ vary, and the part of the volume change that is due to the dissipation of the water from the pores as $\theta_{\mathfrak{w}}$ varies. In saturated soils $\theta_{\mathfrak{s}}$ is dependent on $\theta_{\mathfrak{w}}$ through Eq. (10.6.2).

When we apply Eq. (4.3.2) to the material derivatives of Eq. (10.6.1), keeping in mind that (10.6.1) refers to the solids and considering Eq. (9.3.5), we obtain the *general spatial differential equation controlling the consolidation of saturated soils*

$$\frac{\partial \rho_{\mathfrak{w}}}{\partial t} = \frac{\partial \gamma_{\mathfrak{w}} \theta_{\mathfrak{w}}}{\partial t} = \frac{\partial [\gamma_{\mathfrak{w}} e/(1+e)]}{\partial t} = \frac{\partial \gamma_{\mathfrak{w}} n}{\partial t}$$

$$= -(\gamma_{\mathfrak{w}} \theta_{\mathfrak{w}} v_{\mathfrak{s}\alpha})_{,\alpha} - [\gamma_{\mathfrak{w}} \theta_{\mathfrak{w}} (v_{\mathfrak{w}\alpha} - v_{\mathfrak{s}\alpha})]_{,\alpha}$$

$$= -\left(\gamma_{\mathfrak{w}} \frac{e}{1+e} v_{\mathfrak{s}\alpha}\right)_{,\alpha} + [\gamma_{\mathfrak{w}} (k_{\alpha\beta} \Phi_{,\alpha})]_{,\beta} \tag{11.3.1}$$

Eq. (10.6.1) of the material density balance and Eq. (11.3.1) of the spatial density balance serve as the basic equations for the theory of consolidation.

The simplest boundary condition considers a soil layer of $2H$ thickness and infinite extent in the X and Y directions (in cylindrical coordinates in the R direction), intercalated between two draining layers. To Eqs. (10.6.1) or (11.3.1) we add assumptions in the form of constitutive equations and boundary conditions that correspond to field and laboratory tests. In some studies the assumptions made were simplistic, or the equations used were different from Eqs. (10.6.1) and (11.3.1).

Special interest will be directed towards the solution of the consolidation problem in the consolidometer test and in the triaxial cell test.

11.4 Terzaghi's Theory of Consolidation

The theory of consolidation by Terzaghi relates to the third part and the last part of Eq. (11.3.1), namely

$$\frac{\partial \gamma_{w} e/(1+e)}{\partial t} = -\left(\gamma_{w} \frac{e}{1+e} v_{\bar{s}\alpha}\right)_{,\alpha} + [\gamma_{w}(k_{\alpha\beta}\Phi_{,\alpha})]_{,\beta} \tag{11.4.1}$$

It is instructive to see how Terzaghi's theory of consolidation (Terzaghi 1925), which opened the era of classical soil mechanics, evolves from Eq. (11.3.1). The underlying assumptions of the theory with respect to the constitutive equations and the boundary conditions, are:

1. Homogeneity or quasi-homogeneity of the soil.
2. Complete saturation of the soil.
3. Incompressibility of the solids. This corresponds to a constitutive equation whereby the bulk modulus of the solids is equal to infinity, $\varkappa_{\bar{s}} = \infty$, implying $dV_{\bar{s}} = 0$. The term $1/(1+e)\,dV$ in Eq. (11.4.1) represents the volume of the solids in the volume element dV. If the solids are incompressible, the terms $1/(1+e)$ and $\gamma_{\bar{s}}$ are independent of time and space and can be removed from the differentiation on both sides of Eqs. (10.6.1) and (11.4.1).
4. Incompressibility of the water. The assumption of incompressibility of the water corresponds to a constitutive equation where the bulk modulus of the water is equal to infinity, $\varkappa_{w} = \infty$. Therefore no changes in the density of the water are possible, thus γ_{w} is independent of time and space and can also be removed from the differentiation on both sides of Eqs. (10.6.1) and (11.4.1).
5. The processes occurring during consolidation are slow enough so that the velocity and the velocity gradient of the solids are negligible, $v_{\bar{s}i} \cong 0$ and $v_{\bar{s}\alpha,\alpha} \cong 0$. This assumption represents a constitutive equation and permits the removal of $v_{\bar{s}i}$ and $v_{\bar{s}\alpha,\alpha}$ from Eqs. (10.6.1) and (11.4.1).
6. The force driving the consolidation is the moisture potential Φ_{w}, by means of its gradient $\Phi_{w,i}$. We have already seen that the moisture potential, in the absence of the gravitational and the overburden potentials, is proportional to the pore-water pressure p_{w}, and this indicates that the driving force in consolidation is the pore-water pressure. Although the theory of Terzaghi originally considered the pore-water pressure as the direct cause of consolidation and the moisture potential was not named, it was soon realized that the theory can be extended to the total potential, including the overburden and gravitational potentials.
7. A $2H$ thick soil layer extending to infinity in the X and Y directions (R in cylindrical coordinates) and intercalated between two permeable incompressible layers, Fig. 11.4.1, consolidates due to a pore pressure gradient $p_{w,i}/\gamma_{w} = \Phi_{,i}$ defined in the previous assumption. Since the consolidating layer can drain only towards the two permeable layers the drainage path of the system and the potential gradient are in the Z direction, $\Phi_{,k} = p_{w,k}/\gamma_{w}$, and thus all processes in the soil occur in the Z direction. At the top and bottom of the consolidating layer there is complete drainage, and in this respect this assumption is a boundary condition. One-dimensional flow is not an assumption as such; it is, however, inherent in the geometry of the specific theory, thus $\Phi_{,z} \neq 0$ and $\Phi_{,x} = \phi_{,y} = 0$.

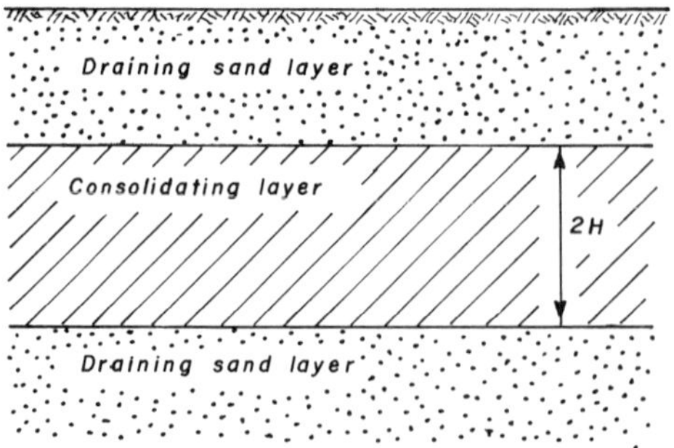

Fig 11.4.1. Consolidating layer.

8. The validity of Darcy's law. The law assumes constant value for the coefficient of permeability, $k_{ij} = \text{const.}$ or $k_{i\alpha,\alpha} = 0$, thus k_{ij} can be removed from the brackets.*

9. A linear relationship between the change in the void ratio and the change in the pore-water pressure,** presenting another constitutive equation

$$de = a_v \, dp_w \tag{11.4.2}$$

where a_v is the *coefficient of compressibility*† assumed to be constant. Based on Eq. (11.4.2) the following relation may be written

$$U_z = \frac{\Delta V}{\Delta V_{t=\infty}} = \frac{e_0 - e}{e_0 - e_{t=\infty}} = \frac{p_{w0} - p_w}{p_{w0}} = 1 - \frac{p_w}{p_{w0}} \tag{11.4.3}$$

where U_z is the *degree of consolidation*, indicating the amount of consolidation that has already been accomplished up to the time t out of the total consolidation that occurs at the time $t = \infty$ which is the end of consolidation, e_0 is the initial void ratio corresponding to the initial pore-water pressure p_{w0}, and $e_{t=\infty}$ is the void ratio at the end of the consolidation at $t = \infty$ when the pore-water pressure has dissipated to zero, $p_{wt=\infty} = 0$.

When the above assumptions are applied to Eq. (11.4.1), the following simple form is obtained

$$\frac{\partial p_w}{\partial t} = \left(\frac{k(1 + e)}{a_v \gamma_w} \right) p_{w,\alpha\alpha} = c_v p_{w,\alpha\alpha} = c_v \frac{\partial^2 p_w}{\partial Z^2} \tag{11.4.4}$$

* Darcy's law does not exclude anisotropy of permeability. Darcy performed his experiments on one-dimensional columns and formulated his theory accordingly. Therefore it cannot reflect multi-directional flow.

** In early textbooks it was customary to formulate this assumption (Taylor 1948; Terzaghi and Peck 1948; Tschebotarioff 1952) by stating that a linear relationship exists between the change in the void ratio e and the change in the applied pressure p, which in turn is equal to the change in the pore-water pressure p_w, $de = -a'_v \, dp_g = a'_v \, dp_w$. In others (Terzaghi 1943) and later textbooks (Scott 1965) the pressure p is replaced by the effective pressure p_g, $de = -a''_v \, dp_g = a''_v \, dp_w$. Notice that $a'_v = 3a_v/(1 + 2K_0)$ while $a''_v = a_v$.

†'Coefficient of compressibility' is unfortunate terminology, since it seems to refer to the compressibility, whereas a_v relates to the dissipative part of the consolidation.

known in general as the *diffusion equation** governing the dissipation of the pore-water pressure in soils, where c_v is the *coefficient of consolidation*, a constant value defined

$$c_v = \frac{k(1 + e)}{a_v \gamma_w} \tag{11.4.5}$$

Several boundary conditions are implied by some of the listed assumptions. Assumption 7, for instance, implies that at any time $0 < t \leq \infty$ there is complete drainage at $Z = 0$ and $Z = 2H$ and therefore the pore-water pressure at these boundaries vanishes, $p_w = 0$.

The boundary condition specifying the magnitude and form of the moisture potential, i.e. pore-water pressure, which drives the consolidation process, is not included in the above assumptions since it depends on the specific boundary value problem.

The Terzaghi consolidation theory assumes a transient loading condition at $t = 0$ with an initial value of the pore-water pressure, $p_w = p_{w0}$, which may vary throughout the thickness of the layer. However, a transient loading is never a natural condition, and in engineering practice an "instantaneous" stress load is applied to the soil over a finite time period. Other loading conditions are possible.

Another boundary condition results from the symmetry with respect to the plane $Z = H$, where the pore-pressure gradient necessarily vanishes, $\partial p_w / \partial Z = 0$.

The solution of Eq. (11.4.4) for the above boundary conditions is straightforward and is achieved by separation of variables and by means of Fourier's series,** and may be found in many textbooks. Only the final solution which gives the pore-water pressure p_w as a function of the elevation Z in the consolidating layer at any time $0 < t \leq \infty$ is reproduced here

$$p_w = \sum_{n=1}^{n=\infty} \frac{2p_{w0}}{n\pi} (1 - \cos n\pi) \left(\sin \frac{n\pi Z}{2H}\right) \exp\left(\frac{-n^2\pi^2 c_v t}{4H^2}\right) \tag{11.4.6}$$

where n is any integer. When n is even, $1 - \cos n\pi$ vanishes; when n is odd, it becomes 2.

Eq. (11.4.6) can be further simplified. If the initial pore-water pressure is constant throughout the layer thickness, namely it is not a function of Z, then $p_{wi} = p_{w0}$. It is also convenient to make the following substitutions

* Eq. (11.4.4) is perhaps the partial differential equation next in importance to Laplace's equation in applied mathematics. It governs the heat flow and the distribution of temperature in solids, and the current-density flow for high electric conductivities when Maxwell's electromagnetic equation reduces to the skin effect equation. It is the diffusion equation of the current-density, and the propagation equation of electric potential and current along electrical cables. It is, of course, also the diffusion equation which controls the transport of material by diffusive motion along surfaces and grain boundaries and through the volume of solids, the latter being based, the same as our Chap. 6, on Fick's law.

** Another way of solving Eq. (11.4.4) is by the Laplace transform, a more general method for the solution of time-dependent equations.

$$M = \tfrac{1}{2}\pi\,(2m + 1) \qquad (11.4.7)$$

$$T_v = \frac{c_v t}{H^2} \qquad (11.4.8)$$

in which m in an integer and T_v is a dimensionless time element called the *time factor*. Eq. (11.4.6) becomes

$$p_w = \sum_{m=1}^{m=\infty} \left(\frac{2p_{w0}}{M} \sin \frac{MZ}{H}\right) \exp\left(-M^2 T_v\right) \qquad (11.4.9)$$

If Eq. (11.4.9) is inserted into Eq. (11.4.4) we obtain

$$U_z = 1 - \sum_{m=1}^{m=\infty} \frac{2}{M}\left(\sin \frac{MZ}{H}\right) \exp\left(-M^2 T_v\right) \qquad (11.4.10)$$

The progress of consolidation is illustrated in the diagram, Fig. 11.4.2, in which the degree of consolidation U_z versus Z is presented by curves, *isochrones*, for various time intervals T_v. We note that the pore-water pressure distribution is symmetrical about the center of the consolidating layer. If the soil layer is only half as thick and drains on one side only, as in the case when the layer rests on an impervious base, then the consolidation progresses as shown in the upper half of the diagram in Fig. 11.4.2.

Frequently it becomes necessary to find the average degree of consolidation U for the layer of soil. This is obtained when the average value of the pore-water pressure,

$$\overline{p_w} = \frac{1}{2}\int_0^{2H} p_w \, dZ$$

is substituted into Eq. (11.4.3)

$$U = 1 - \frac{\int_0^{2H} p_w \, dZ}{2H\, p_{w0}} = 1 - \sum_{m=1}^{m=\infty} \frac{2}{M^2} \exp\left(-M^2 T_v\right) \qquad (11.4.11)$$

Fig. 11.4.3 is the graphical representation of U versus T, a curve of practical importance, whereas the curves in Fig. 11.4.2 are mainly illustrative and explanatory of the consolidation process.

11.5 Discussion of Terzaghi's Theory of Consolidation

Applying a set of simplified assumptions, Terzaghi's theory of consolidation provides a solution for the simplest geometry and boundary conditions corresponding to a particular field condition. It was shown in the previous section how this theory, considered at the structural level, evolves from the

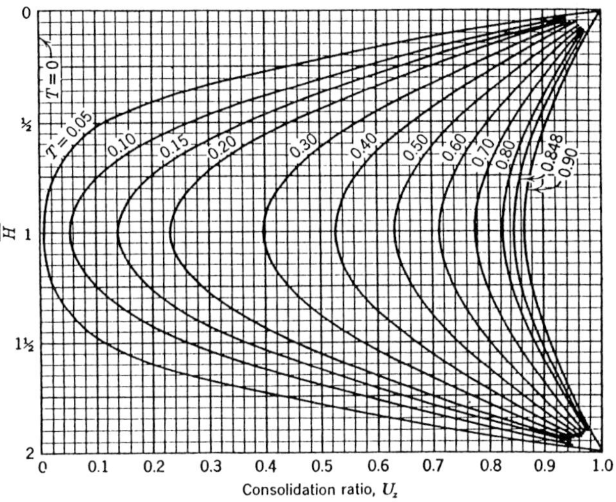

Fig 11.4.2. Degree of consolidation U_z versus depth Z. [From D. W. Taylor (1948) courtesy of John Wiley and Sons Inc.]

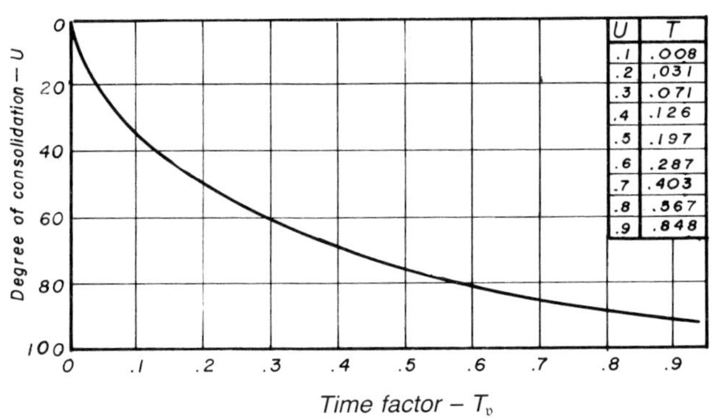

U	T
.1	.008
.2	.031
.3	.071
.4	.126
.5	.197
.6	.287
.7	.403
.8	.567
.9	.848

Fig 11.4.3. Average consolidation U versus time factor T_v. [From D. W. Taylor (1948) courtesy of John Wiley and Sons Inc.]

theory of mixtures based on continuum mechanics and applied to soils. It is indeed at the structural level that Terzaghi's theory was formulated; however, rather than deducing it from the general theory, Terzaghi considers the interaction between the constituents of the soil, solids and water, applying to it a balance equation.

It has been amply demonstrated that the correlation between Terzaghi's theory and experimental data, either from the laboratory or the field, is only approximate. Discrepancies between measured settlements and those predicted by the theory indicate that the theory provides only a good first approximation of the time curve, and the agreement between theory and measurements is valid only for part of that curve. Beyond this part the theoretical curve flattens out, becoming asymptotic to a horizontal line, while the measurements continue for a long period of time even under a very small excess of pore-water pressure. The part of the volume changes accounted for by the theory is referred to as the *primary consolidation*, while the second part, not explained by it, is called *secondary consolidation*.

The attempts to improve on Terzaghi's theory of consolidation or present an alternate theory go back almost to the time of its disclosure (Buisman 1936; Taylor and Merchant 1940; Biot 1956, 1963; Tan 1957b; Lo 1959; J. B. Hansen 1961; Gibson and Lo 1961; Schiffman et al. 1964; Christie 1965; Gibson et al. 1967; Barden 1968; Lee and Sills 1979; Schiffman 1980), and the continuous efforts to explain its inconsistencies and the discrepancies between theory and observations concern almost all of the assumptions listed in the previous section.

Let us review critically those assumptions:

1. It must be realized that the assumption of homogeneity, discussed in Sect. 1.7, is inherent to almost all theoretical studies, while in practice most materials are inhomogeneous to a great extent. This is particularly important when laboratory tests are performed or measuring instrumentation is installed in the field. In order that the variations in the soil remain within a constant statistical average, care must be taken that the test samples, with respect to their number and size, be representative of the field, and that the field instrumentation be compatible with those variations.

2. It is well known that significant amounts of entrapped air may be found even in saturated soils. Smith and Browning (1942) found the average air entrapment in 200 "saturated" samples to be 9% of the soil volume, with a maximum of 22% at complete saturation. These data seem exaggerated, since even for remolded and compacted soils higher degrees of saturation have been reached (Casagrande and Hirschfeld 1960; Schmertmann and Osterberg 1960). In compacted "saturated" clays a degree of saturation above 94% was found (Klausner 1987a–d) and in a series of 16 compacted samples of montmorillonite the degree of saturation was found to be above 97%.

3. The assumption of incompressibility of solids does not introduce any big errors; for pressures used in soil mechanics the compressibility of solids can indeed be considered negligible. Volume changes of solids reported (Gibbs et al. 1960) for soils with $e = 1$ and for pressures of $7\ \mathrm{kg/cm^2}$ (100 p.s.i.) were approximately 0.002%, which amounts to a bulk modulus of $\varkappa_{\mathfrak{s}} = 3.5 \times 10^5\ \mathrm{kg/cm^2}$.

4. The compressibility of the water, however, is somewhat higher than that of the solids and in marginal cases it may have to be considered. In a soil similar to that considered above, with a void ratio of $e = 1$, pressures of $7\ \mathrm{kg/cm^2}$ and a bulk modulus of $\varkappa_{\mathfrak{w}} = 2.1 \times 10^4\ \mathrm{kg/cm^2}$, a volume change of 0.033% is obtained, about 17 times the volume change of the solids, see Table 8.3.2. In clay pastes such as Na-montmorillonites, where a void ratio as high as 25–30 can easily be obtained, volume changes become considerable, amounting approximately to 4%. Notice that whatever consolidation pressure is applied, as the consolidation proceeds the volume change of the water

diminishes and pore-water pressure dissipates, and therefore when computing the volume changes at the end of the process, the compressibility of the water may be ignored.

5. The velocity and velocity gradient of the solids, $v_{\mathfrak{s}i}$ and $v_{\mathfrak{s}\alpha,\alpha}$, respectively, are not negligible, particularly at the beginning of the consolidation process and near the application of the transient load, when the pore-water pressure is high, the effective pressure low and the mobility of the solid particles increased by the velocity gradient of the water.

There are also indications, not yet fully substantiated, that at least part of the secondary consolidation and the volume increase occasionally observed at the end of the consolidation process with the termination of the velocity gradient of the water is related to the reorientation of the particles to a random array.

6. Terzaghi's theory of consolidation considers the moisture potential alone. However, it is no problem to incorporate in it the gravitational and the overburden potentials as well. Taylor (1948) already treated this subject considering the gravitational and overburden potentials and presented the corresponding average time curve $T_{\mathfrak{o}}$. Gibson et al. (1967) also included gravitational potential in their study.

7. This assumption is actually a boundary condition, and as such no improvement can be added. With respect to the thickness $2H$ of the consolidating layer, however, it should be remarked that if the layer is thin, the effect of the overburden potential and even that of the gravitational potential in the field is negligible compared with the effect of the applied pressures. In the laboratory the influence of the friction between the sample and the consolidometer ring is minimal.

8. The consolidation theory, which assumes Darcy's linear law, can be improved greatly be assuming a non-linear relationship between the coefficient of permeability k_{ij} and the void ratio e, that is, dependence of the permeability on the void ratio, $k(e)_{ij}$ (Schmid 1957; 1958) or existence of interactive forces between solids and water controlling the flow (Klausner and Kraft 1966), discussed in Sects. 8.21 and 8.22. Gibson et al. (1967), improved the theory of consolidation by including in it the dependence of the permeability on the void ratio.

The assumption of functional dependence of the permeability on the void ratio is equivalent to functional dependence of the permeability on the moisture content $\theta_{\mathfrak{w}}$, $k(e)_{ij} \equiv k(\theta_{\mathfrak{w}})_{ij}$, see Eqs. (10.5.6) and (10.12.5).

9. The relationship between the change in void ratio and the change in pore-water pressure is by no means linear as shown in Eq. (11.4.2), which is responsible for the discrepancy between the theory and experience manifested by what is known as secondary consolidation; it is possible that an assumption of a non-linear relationship will greatly improve the agreement between theory and experience in consolidation.

A description of the laboratory consolidation test and equipment and a discussion of the test results, specifically the relationship between the void ratio and pressure, is appropriate here.

11.6 The Consolidometer

The laboratory determination of the compressibility of soils and the rate at which compression occurs is performed in a device illustrated in Fig. 11.6.1, known as a *consolidometer*, or by its earlier name, an *oedometer*.

A circular undisturbed or compacted soil sample (2), depending on the objectives of the test, is carefully and tightly fitted into a solid ring (1) by using special trimming tools. The ring with the soil sample is then fastened to a solid base plate (11) provided with a porous filter plate (3), by means of a cover ring (9), a sealing gasket (10) and screws. Channel cavities in the base plate under the porous plate lead the water which drains from the sample to a standpipe (8). The cover ring is provided with a collar capable of holding the water covering the sample. A second porous filter plate (3) is then placed on top of the sample and on it rests a circular load-distributing plate (6).

The axial vertical load is applied externally to the sample through a loading yoke (4) and transmitted centrally from the loading yoke through a steel ball (5) located in a groove specially made in the distributing plate. The loads are applied as dead weights or by pneumatic or hydraulic loading jacks, controlled through proving rings, pressure gages or electronic load cells.

The vertical deformations of the soil sample are measured by a dial gage (7) with the traveling pin resting on the yoke. The dial gage, clamped rigidly to the base plate, is considered an independent reference point outside the

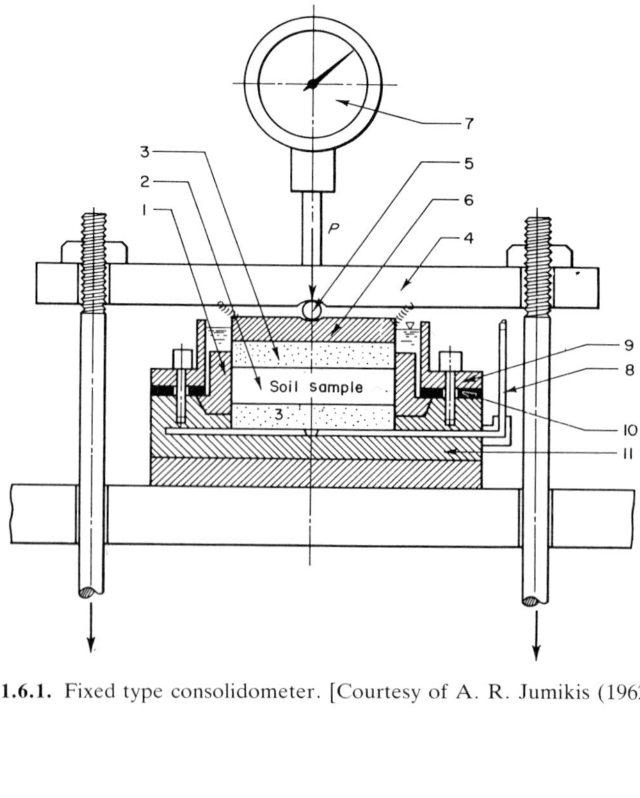

Fig 11.6.1. Fixed type consolidometer. [Courtesy of A. R. Jumikis (1962).]

soil sample. Dial gages used in standard consolidometers read to an accuracy of 10^{-4} in. or 10^{-3} mm.

Standard sample sizes are 2.5 in., 3 in. and 4 in. in diameter or 30 cm^2 and 100 cm^2 area, and sample thickness varies between $\frac{1}{2}$ in. and 1 in. or 1 cm and 2.5 cm. Other sizes are also in use.

The consolidometer described above is known as the *fixed type* consolidometer, while another variation of the consolidometer is the *floating type*, Fig. 11.6.2, where the sample ring is a thin ring not fastened to the base plate and the soil sample can compress at both sides. Although this type of device has the advantage of low cost, it cannot be used for the consolidation of soft clays or loose sandy soils. It is maintained that frictional resistance in the floating type consolidometer is lower than in the fixed type. This, however, has not been proved.

Unlike the floating type, the fixed type consolidometer is a fairly versatile piece of equipment, adaptable to other tests as well such as:

1. *Swelling test*, in which the amount of swelling of a soil sample is measured through the change of its height under any specific fixed vertical load.

2. *Swelling pressure test*, in which the swelling pressure is measured while the height of the soil sample is kept unchanged.

3. *Permeability test*, in which the standpipe is connected to a falling-head permeameter graduated tube.

4. *Lateral earth pressure coefficient test*, by which the lateral earth pressure coefficient K can be evaluated if the ring containing the sample is fitted with a load cell and if the friction between the sample and the ring is ignored.

5. *Pore pressure measurements* can be made at the bottom porous plate by connecting the standpipe to a pore-pressure measuring device. In this case the sample drains through the top porous plate, and its height is considered a half layer, H.

11.7 The Consolidation Test

After the soil sample is placed in the consolidometer and the device is assembled the loads are applied successively so that each increment is equal

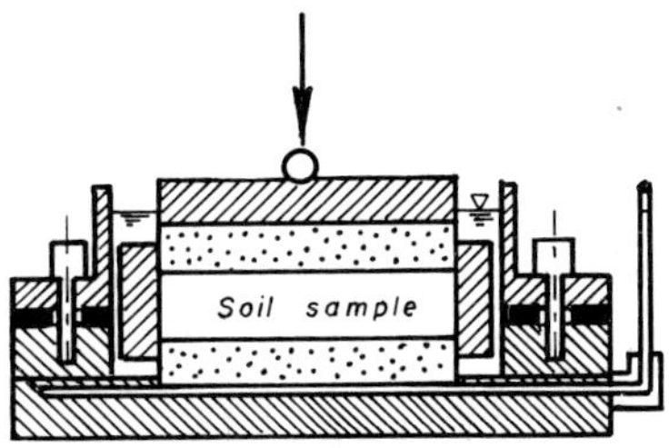

Fig 11.6.2. Floating type consolidometer.

to the previous total load. So, for instance, if the first initial load produced a vertical pressure p_{zz} of 0.01 kg/cm^2, the successive loads should raise the pressure to 0.02, 0.04, 0.08. 0.16, 0.32, 0.64, 1.28, 2.56, 5.12 kg/cm^2 and so on. Such incremental loading is convenient, since the results are related to the pressure p_{zz} and the data will appear at equal distances if presented in a $\log p_{zz}$ plot. But, of course, any other loadings are acceptable.

For each load increment the soil is allowed to consolidate. A time of 24 hours is, in general, sufficient for the consolidation of most of the common soils tested at the standard thicknesses mentioned in the previous section. However, thicker samples and certain soils require a consolidation time of more than 24 hours. Careful continuous dial readings are recorded at 0.25, 0.5, 1, 2, 4, 8, 15, 30, 60, 120, 240, 480 and 1440 minutes, representing the times elapsed from the beginning of the consolidation. Here again the reason is convenience, as the dial readings d are plotted against the logarithm of the time. The curves d–$\log t$ are called *time curves*. Fig. 11.7.1 shows typical time curves of a consolidation test.

After the largest load required for the test has been applied the load is removed in decrements, to provide data for the *expansion* curve of the soil.

The vertical deformation Δ of the soil sample at any time can be computed from the dial reading d at that time and the known initial dial reading d_0, $\Delta = d - d_0$. If we know the total weight of the sample W, the initial water content w_0, the specific gravity of the soil G_s and the cross-section area of the soil sample S, then the void ratio e corresponding to the dial reading d can be computed

$$e = \frac{SG_s(1 + w_0)(2H - \Delta)}{W} - 1 \tag{11.7.1}$$

Also the vertical stretch λ_z can be found

$$\lambda_z = \frac{\Delta}{2H} \tag{11.7.2}$$

The void ratio e can thus be evaluated at every stage of the consolidation process. If evaluated at the end of the consolidation of each load increment or decrement and plotted against the total pressure applied, the void ratio e versus the vertical pressure p_{zz} curve is obtained. Fig. 11.7.2a shows such a curve for an undisturbed clay sample of high plasticity (CH). The same data is presented in a semi-logarithmic pressure scale e–$\log p_{zz}$. This presentation, Fig. 11.7.2b, has the advantage of a more evenly distributed pressure, which may range over 3–5 orders of magnitude, one reason for its wide acceptance and use in soil mechanics. Another reason is the fact that a portion of the e–$\log p_{zz}$ curve is believed to be linear and thus is described by an equation of a straight line

$$e = e_c - C_c \log \frac{p_{zz}}{p_{czz}} \tag{11.7.3}$$

where e_c is the void ratio at an arbitrary pressure p_{czz} and C_c is the

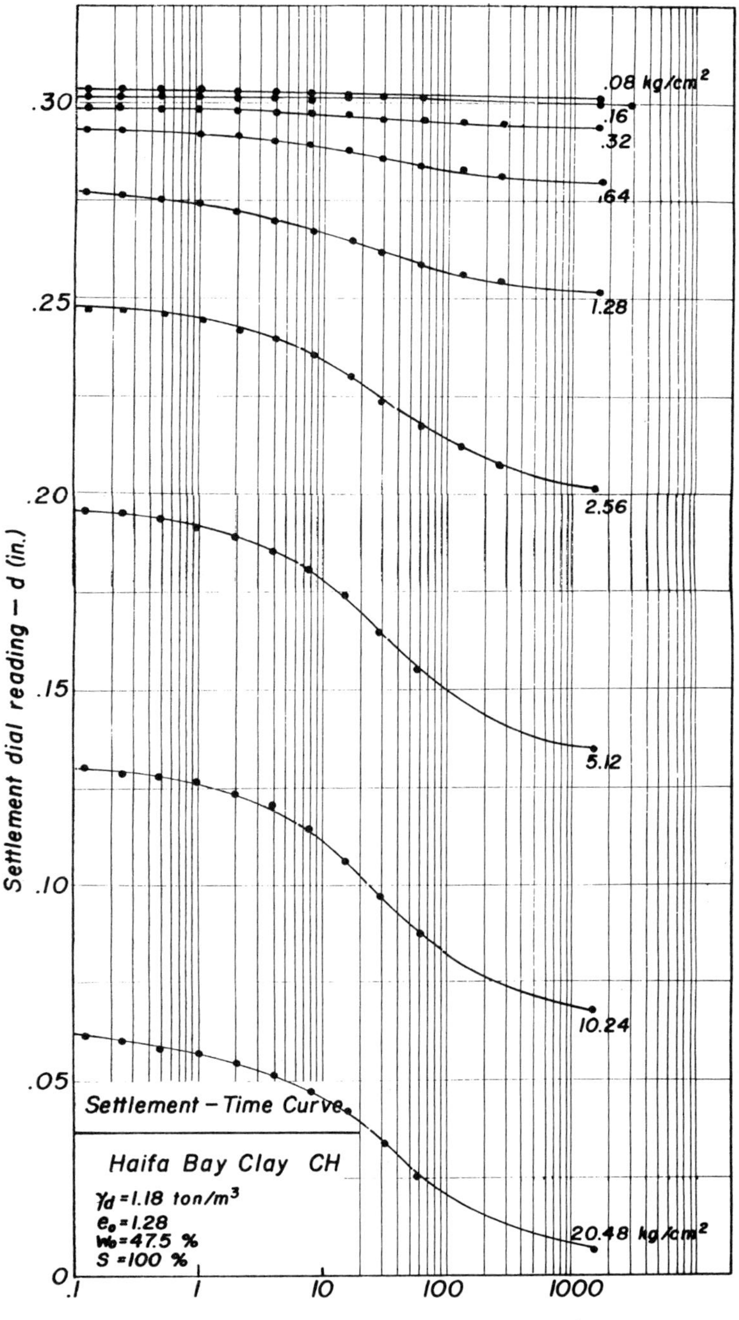

Fig 11.7.1. Time curves of consolidation test.

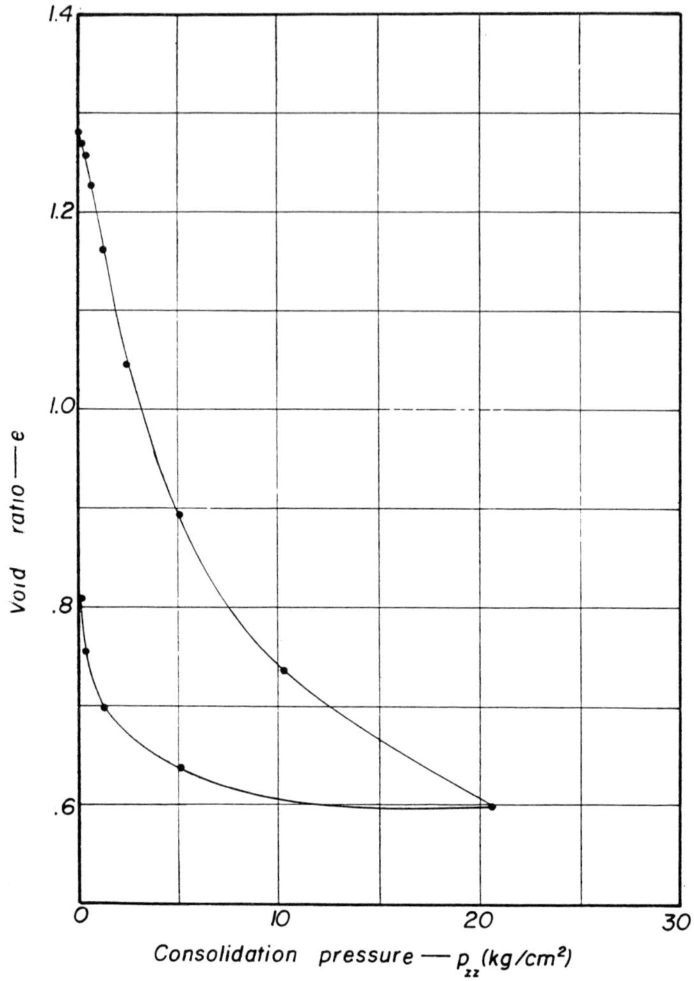

a

compression index. We will see in the next section that the linearity of e–$\log p_{zz}$ is only circumstantial and not necessarily an inherent property of the soil. The e–p_{zz} curve is not linear in any circumstance.

Further detailed descriptions of consolidation tests of soils are found in ASTM, *Procedures of Testing Soils*, April, 1958.

11.8 The Void Ratio–Pressure Dependence

To assist the discussion of the dependence of the void ratio on the pressure, we shall present several specific e–p_{zz} and e–$\log p_{zz}$ curves from the literature.

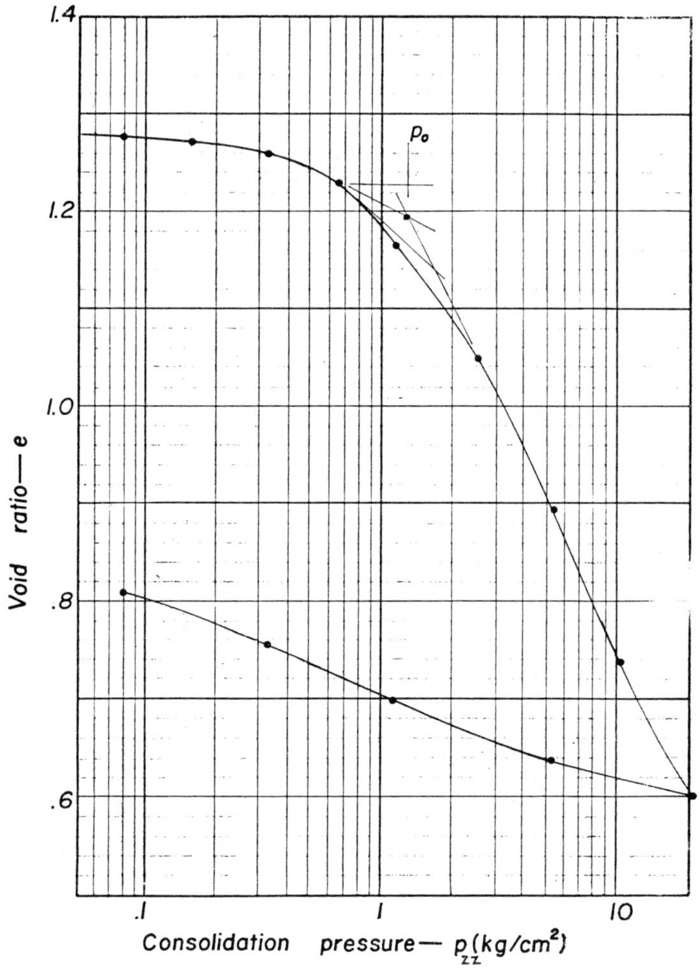

b

Fig 11.7.2. Void ratio e versus vertical stress p_{zz}. **a** e versus p_{zz}. **b** e versus $\log p_{zz}$.

Taylor (1948) presented consolidation test results of an undisturbed Boston blue clay, Fig. 11.8.1, displayed in the e versus p_z curve (a) and e versus $\log p_z$ curve (b), similar to the test results presented in Fig. 11.7.2. Taylor's curves have an additional loading sequence creating a loading loop.

If the soil sample is reloaded, the void ratio will again decrease and the loading part of the curve will follow closely the unloading part but will usually lie above it, as seen in Fig. 11.8.1. As the reloading curve approaches the maximum value applied earlier it reverses its curvature in the $e-p_{zz}$ diagram and intensifies it by decreasing its radius of curvature in the $e-\log p_{zz}$ diagram. Thus it blends into the initial curve, the *virgin curve*, and continues

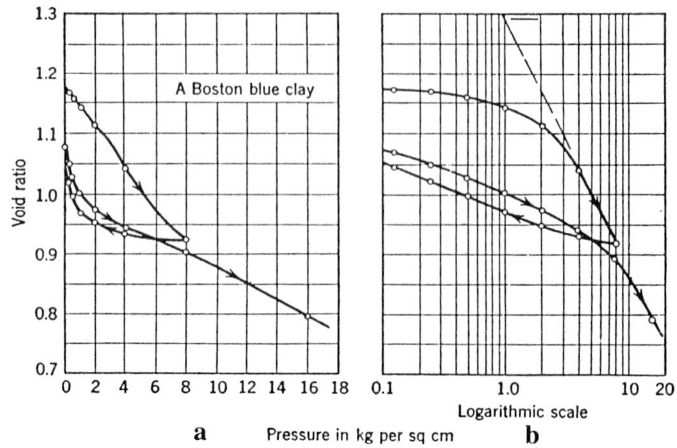

Fig 11.8.1. Void ratio e versus vertical stress p_{zz} of a Boston clay. [From D. W. Taylor (1948) courtesy of John Wiley and Sons Inc.] **a** e versus p_{zz}. **b** e versus $\log p_{zz}$.

as its extension. We observe that the extreme change in the curvature of the reloading sequence is an indication of an earlier loading that had affected the sample, and that the pressure applied in the past on the sample is somewhat beyond the pressure where the intense change in th curvature occurred. Referring to the reloading curves in Fig. 11.8.1, we could say that the initial maximum pressure applied to the sample was between 7 and 10 kg/cm^2. Analogously, by looking at the initial loading curve we can say that before the consolidation test the soil might have been already loaded and consolidated by a pressure of about 2–4 kg/cm^2. We say that the soil sample was *preloaded* or *preconsolidated*, and the pressure p_p corresponding to that preloading is called the *preconsolidation pressure* or *precompression pressure*. The preconsolidation pressure of the consolidation test data shown in Fig. 11.7.2 can be estimated at around 1.3 kg/cm^2.

For engineering purposes, a technical method of approximating the value of the preconsolidation pressure from the e–$\log p_z$ curve was developed and presented by Casagrande (1936). An undisturbed soil sample displaying a preconsolidation pressure may indicate that the soil layer in the field from where the sample originates has been consolidated in the past under that pressure. The pressure exerted on the soil layer from the existing overburden soil or fill can be computed and compared with the preconsolidation pressure found in the consolidation test of the sample taken from that layer. If the pressures match, the soil layer is said to be *normally consolidated*.

Soils may have been preconsolidated during the geologic past by the weight of ice or glaciers, by past overburden layers that subsequently eroded or were excavated, or by structures that were eventually torn down. The preconsolidation pressure found in the consolidation test may then indicate a higher preconsolidation pressure than expected from the existing overburden load. The soil layer is said to be *overconsolidated*. A saturated soil sample taken from the capillary zone or from a layer where a recent rise in the water table has taken place and the swelling of the layer is still in progress would also indicate an overconsolidated condition.

Occasionally the preconsolidation pressure is found to be lower than anticipated from the weight of the existing overburden load, in which case the layer is considered *underconsolidated*. Such a case may occur in a soil layer where the water table has receded recently but has not yet reached an equilibrium and the consolidation is still in progress.

Assessing the preconsolidation state of a soil layer is important in the evaluation of its load carrying capacity, in order to determine the amount of settlement that would result from the anticipated load. Nevertheless, because of the occasional low quality of undisturbed samples or variations in the soil structure the accuracy of determining the preconsolidation pressure could be impaired. Therefore, care must be exercised when preconsolidation pressures are applied, in spite of their practical value.

11.9 The Pressure and Strain Tensors in Uniaxial Consolidation

The vertically applied pressure p_{zz} on an element in the soil layer in the field, or on the laboratory sample, produces a total pressure p_{ij} which can be decomposed, according to Eq. (4.9.1), into a spherical pressure $\frac{1}{3}p_{\alpha\alpha}$ and a deviator stress s_{ij}

$$p_{ij} \equiv \begin{vmatrix} Kp_{zz} & 0 & 0 \\ 0 & Kp_{zz} & 0 \\ 0 & 0 & p_{zz} \end{vmatrix} \equiv \frac{1+2K}{3} p_{zz}\delta_{ij} + \frac{1-K}{3} p_{zz}\Xi_{ij} \tag{11.9.1}$$

where the spherical pressure is equal to

$$\tfrac{1}{3}p_{\alpha\alpha} = \frac{1+2K}{3} p_{zz} \tag{11.9.2}$$

and the deviatoric stress is

$$s_{ij} \equiv \frac{1-K}{3} p_{zz} \begin{vmatrix} -1 & 0 & 0 \\ 0 & -1 & 0 \\ 0 & 0 & 2 \end{vmatrix} \equiv \frac{1-K}{3} p_{zz}\Xi_{ij} \tag{11.9.3}$$

and where Ξ_{ij} is the *dimensionless deviatoric tensor*, discussed in Sect. A.21 of Appendix A, and K is the *coefficient of lateral earth pressure*, a dimensionless number defined as the ratio of the lateral horizontal pressure p_{xx} or p_{yy} to the vertical pressure p_{zz}

$$K = \frac{p_{xx}}{p_{zz}} = \frac{p_{yy}}{p_{zz}} \tag{11.9.4}$$

K varies during the consolidation process. For pressures higher than the preconsolidation pressure it may assume values between $K \leq 1$ at time $t = 0$, the beginning of the consolidation, and $K = K_0$ at time $t = \infty$, the end of consolidation. K_0 is the *coefficient of lateral earth pressure at rest*, that is, at equilibrium. K is an inherent property of the soil and depends on its consistency, density and moisture content. At pressures below the preconsolidation pressure the coefficient of lateral earth pressure at $t = 0$ will be much below unity and in most cases close to the coefficient of lateral earth pressure at rest, $1 \gg K \geq K_0$.

Since even "saturated" soils are seldom completely saturated, the coefficient of lateral earth pressure may start somewhat below unity, $1 > K$, and for pressures below the preconsolidation pressure it could be far below unity, $1 \gg K$.

While p_{zz} is a constant pressure, p_{xx}, p_{yy}, $p_{\alpha\alpha}$ are pressures varying with time.

In the laboratory test, unlike in the field, additional frictional forces act between the soil sample and the ring, which induce further deviatoric stresses and reduce the deformation. The effect of the friction weakens with the increase of the ratio of the diameter to the thickness of the sample $D/2H$.

In a normally consolidated saturated soil the moisture potential at the beginning of consolidation is equal to the spherical pressure. At the end of consolidation the moisture potential vanishes and the spherical pressure attains a finite value

$$\Phi(0)_w = p(0)_w = \tfrac{1}{3}p(0)_{\alpha\alpha} = \frac{1 + 2K(0)}{3} p_{zz} = p_{zz}; \quad \text{for } K \cong 1 \text{ at } t = 0$$

$$\Phi(\infty)_w = p(\infty)_w = 0; \quad \tfrac{1}{3}p(\infty)_{\alpha\alpha} = \frac{1 + 2K_0}{3} p_{zz}; \text{ for } K(\infty) = K_0 \text{ at } t = \infty \tag{11.9.5}$$

To satisfy the requirement imposed on the moisture potential we can write

$$\begin{aligned} \Phi(t)_w = p(t)_w &= p_{zz} - [p_{zz} - \tfrac{1}{3}p(t)_{\alpha\alpha}]\left(\frac{p_{zz}}{p_{zz} - \tfrac{1}{3}p(\infty)_{\alpha\alpha}}\right) \\ &= p_{zz}\left(1 - \frac{3p_{zz} - p(t)_{\alpha\alpha}}{3p_{zz} - p(\infty)_{\alpha\alpha}}\right) \\ &= p_{zz}\left(1 - \frac{1 - K(t)}{1 - K_0}\right) = p_{zz}\frac{K - K_0}{1 - K_0} \end{aligned} \tag{11.9.6}$$

where we see that $\Phi_w = p_w = p_{zz}$ for $t = 0$ and $\Phi_w = p_w = 0$ for $t = \infty$.* We also notice that the moisture potential is a function of the coefficient of lateral earth pressure K or its value at rest, K_0.

The shear stress s_{ij} does not exist at the beginning of the consolidation as $K \cong 1$, but appears, according to Eq. (11.9.3), right after the beginning of consolidation, increasing with time as K decreases. Towards the end of consolidation, while the spherical pressure vanishes and K approaches K_0, the shear stress attains a finite value

$$s(0)_{ij} \geq 0 \quad \text{for } t = 0; \quad s(\infty)_{ij} = \frac{1 - K_0}{3} p_{zz} \Xi_{ij} \quad \text{for } t = \infty \tag{11.9.7}$$

The infinitesimal strain tensor $\overset{\text{inf}}{\varepsilon}_{ij}$ has only one component, $\overset{\text{inf}}{\varepsilon}_{zz}$. All other components are zero, thus the strain tensor becomes, accordingly

$$\overset{\text{inf}}{\varepsilon}_{ij} \equiv \begin{vmatrix} 0 & 0 & 0 \\ 0 & 0 & 0 \\ 0 & 0 & \overset{\text{inf}}{\varepsilon}_{zz} \end{vmatrix} = \frac{\overset{\text{inf}}{\varepsilon}_{zz}}{3} \delta_{ij} + \frac{\overset{\text{inf}}{\varepsilon}_{zz}}{3} \Xi_{ij} \tag{11.9.8}$$

Similarly the Hencky strain measure can be derived from the stretch

$$\lambda_{zz} = (1 + \overset{\text{inf}}{\varepsilon}_{zz})$$

if it is substituted into $\ln \lambda_{zz}$

$$\overset{\mathbf{H}}{\varepsilon}_{ij} = \frac{\ln \lambda_{zz}}{3} \delta_{ij} + \frac{\ln \lambda_{zz}}{3} \Xi_{ij} \tag{11.9.9}$$

where the volumetric and shear strains in the infinitesimal and Hencky strain measures are, respectively

$$\overset{\text{inf}}{\varepsilon}_{zz} = \overset{\text{inf}}{\varepsilon}_{\alpha\alpha}; \quad \overset{\mathbf{H}}{\varepsilon}_{zz} = \overset{\mathbf{H}}{\varepsilon}_{\alpha\alpha} = \ln \lambda_{zz} = \ln (1 + \overset{\text{inf}}{\varepsilon}_{zz});$$

$$\overset{\text{inf}}{\varepsilon}_{xx} = \overset{\text{inf}}{\varepsilon}_{yy} = \overset{\mathbf{H}}{\varepsilon}_{xx} = \overset{\mathbf{H}}{\varepsilon}_{yy} = 0 \tag{11.9.10}$$

$$\overset{\text{inf}}{e}_{ij} = \frac{\overset{\text{inf}}{\varepsilon}_{zz}}{3} \Xi_{ij}; \quad \overset{\mathbf{H}}{e}_{ij} = \frac{\ln \lambda_{zz}}{3} \Xi_{ij} = \frac{\ln (1 + \overset{\text{inf}}{\varepsilon}_{zz})}{3} \Xi_{ij} \tag{11.9.11}$$

It can be seen that along the process of consolidation the strain has two components, which indicate that the uniaxial consolidation is not a pure consolidation but contains shear stresses as well and consequently also shear strains. It is different in the triaxial consolidation, as we will see in

* By now the reader is probably aware that the moisture potential Φ_w, Sect. 10.3, and the pore-water pressure p_w are equal, and therefore the terms, as well as the notations, are occasionally interchanged.

Sect. 11.10. Thus in the uniaxial consolidation there is a superposition of a spherical and a shear effect. No wonder then that the time curves indicate a "secondary consolidation", which is partly a shear strain following a fairly straight line and is finally balked by the frictional resistance induced by the consolidometer ring.

On the phenomenological level the following stress–strain relationships may be written from Eqs. (11.9.2), (11.9.3), (11.9.10) and (11.9.11)

$$\overset{\text{inf}}{\varepsilon}(t)_{\alpha\alpha} = \overset{\text{inf}}{\varepsilon}_{zz} = \mathcal{F}\left[\frac{1 + 2K(t)}{3}\right] p_{zz};$$

$$\overset{\mathbf{H}}{\varepsilon}(t)_{\alpha\alpha} = \ln \lambda_{zz} = \mathcal{F}'\left[\frac{1 + 2K(t)}{3}\right] p_{zz} \qquad (11.9.12)$$

$$\overset{\text{inf}}{\varepsilon}_{zz} = \mathcal{G}[1 - K(t)]\, p_{zz}; \quad \ln \lambda_{zz} = \mathcal{G}'[1 - K(t)] p_{zz} \qquad (11.9.13)$$

where $\mathcal{F}$, $\mathcal{F}'$, $\mathcal{G}$, $\mathcal{G}'$ are linear viscoelastic differential operators.

The *effective pressure tensor* $p_{\mathfrak{s}ij}$, acting during the consolidation of a saturated soil, can also be calculated from Eq. (9.4.7) where $p_{\mathfrak{g}} = 0$, and from Eq. (11.9.1)

$$p_{\mathfrak{s}ij} = p_{ij} - p_{\mathfrak{w}}\delta_{ij} = \left(\frac{1 + 2K}{3} - \frac{K - K_0}{1 - K_0}\right) p_{zz}\delta_{ij} + \frac{1 - K}{3} p_{zz}\Xi_{ij}$$

$$= \frac{1 - K + 2K_0 - 2KK_0}{3(1 - K_0)} p_{zz}\delta_{ij} + \frac{1 - K}{3} p_{zz}\Xi_{ij} \qquad (11.9.14)$$

From Eq. (11.9.14) it is seen that for normally consolidated soil, at the beginning of consolidation where $K = 1$ the effective pressure vanishes, $p_{\mathfrak{s}ij} = 0$, while at the end of consolidation, $t \to \infty$ and $K = K_0$, the effective pressure acts as a shear stress equal to Eq. (11.9.7) and equal also to the value of the total pressure

$$0 \le p(t)_{\mathfrak{s}ij} = \frac{1 - K(t) + 2K_0 - 2K_0K(t)}{3(1 - K_0)} p_{zz} \le \frac{1 + 2K_0}{3} p_{zz} = p(t)_{ij}$$

(11.9.15)

Terzaghi's consolidation theory of an infinite layer, in the field or in the laboratory sample, does not refer to a pure consolidation (as defined in Sect. 11.3), since the pressure acting on any soil element consists not only of a spherical pressure but includes a deviatoric component as well, meaning that shear stresses act in the soil. In Sect. 11.11 we will see that the triaxial consolidation is the closest to a spherical pure consolidation test. The uniaxial consolidation test,* however, is one of the many tests that simulates a field condition and provides technical parameters.

* The term uniaxial consolidation is used here for the one dimensionally loaded soil and the term triaxial consolidation for the three dimensionally loaded soil. Both terminologies are correct.

The internal driving force of consolidation in Terzaghi's theory, as derived in Sect. 11.4 from the density balance equation (10.6.1) and the assumption of saturation, Eq. (10.6.2), is the moisture potential gradient $\Phi_{w,i}$, proportional to the pore-water pressure gradient $p_{w,i}$. The external load applied is p_{zz} and the simplest way to present the results of a consolidation test is by the e versus p_{zz} curve or the e versus $\log p_{zz}$ curve. The presentation of the results in terms of effective pressures, e versus $p_{\mathfrak{s}zz}$ or e versus $\log\ p_{\mathfrak{s}zz}$, is unnecessary.

11.10 The Triaxial Testing Device

According to Eq. (11.9.2) the soil element in the uniaxial consolidation test, Fig. 11.10.1, is subjected to a spherical stress equal to $\frac{1}{3}p_{\alpha\alpha} = \frac{1}{3}[1 + 2K(t)]p_{zz}$ which varies with time as $K(t)$ varies, in spite of the fact that p_{zz} is constant. Actually the horizontal pressures depend on the coefficient of lateral earth pressure, according to Eq. (11.9.4), and are thus functions of time, $p_{xx} = p_{yy} = K(t)p_{zz}$. To obtain a state of stress in which the normal stresses acting on the soil element are equal in all directions is possible only by a *spherical consolidation test*, performed in the *triaxial testing device*, shown in Fig. 11.10.2 and in the schematic drawing in Fig. 11.10.3. The triaxial testing device was apparently first designed by the Prussian Waterways Experiment Station (Seifert 1933) for the purpose of studying the spherical consolidation of clays.

The triaxial testing device is one of the most perfect, sophisticated and versatile testing tools in mechanics, especially for testing compressible materials, and is also acknowledged as such in soil mechanics and engineering. It permits the application of both spherical and deviatoric stresses on a soil sample of cylindrical shape and the manipulation of these stresses to obtain a great variety of stress combinations.

The triaxial testing device has three major components:

1. The *triaxial cell* which contains the soil sample.

2. The auxiliary *measuring and control unit*, usually a panel accommodating valves, pressure gages, measuring burettes and dials.

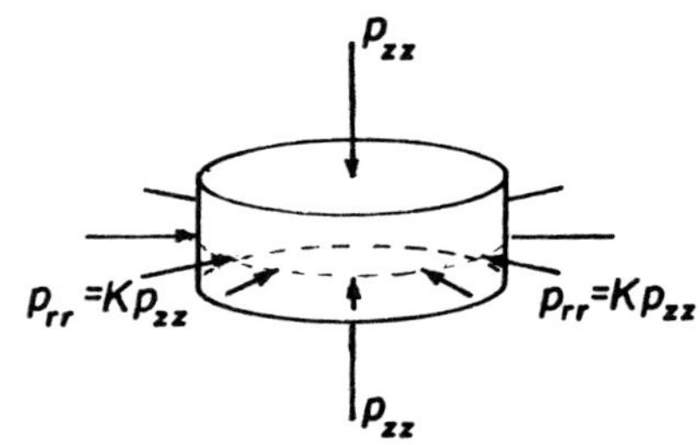

Fig 11.10.1. Stresses acting on a consolidating sample.

Fig 11.10.2. The triaxial cell and panel.

3. The *loading device* which applies the deviatoric stresses to the sample by a dead weight loading system, hydraulic or pneumatic ram.

Most up-to-date triaxial test devices have a connected computer facility, monitoring the stresses and measurements, evaluating the recorded data and providing readily available results.

At present we are interested in the first two components.

The main features of the triaxial cell, Fig. 11.10.3 are:

1. A *base plate*, made of solid metal, with an attached *pedestal*. The pedestal is provided with drainage grooves and a cavity leading the draining water to an outside connection. The base plate has a connection to the cell

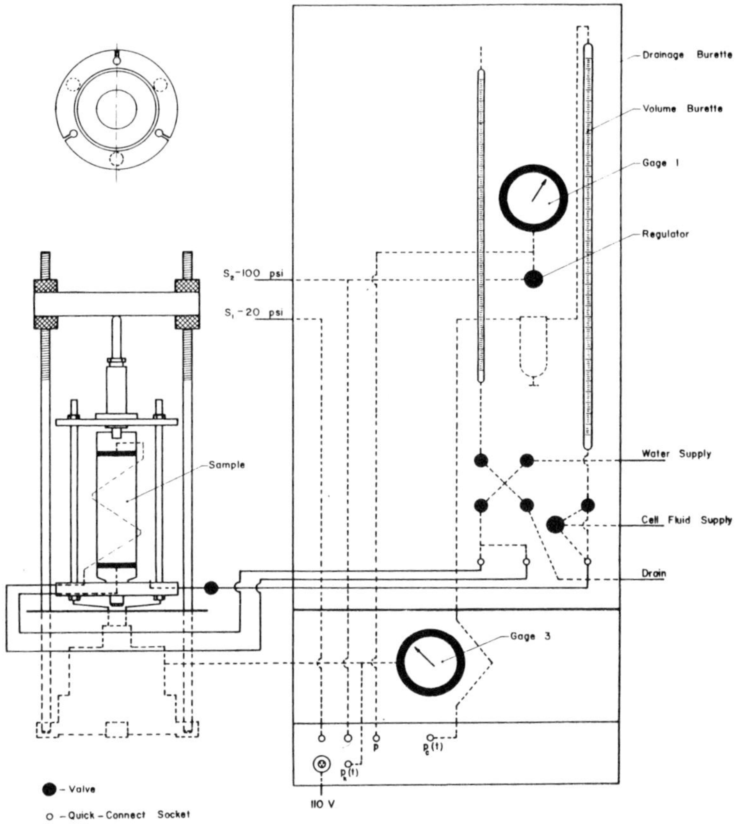

Fig 11.10.3. Schematic drawing of the triaxial testing device.

fluid and pressure supply and can have another connection to the eventual hookup to measuring facilities, to measure quantities such as pore pressures, strains, etc.

2. A *loading cap*, usually of Perspex, placed on top of the sample and provided with drainage grooves leading to a tube connector. A flexible thin tube leads from the top cap to an outside connection in the base plate.

3. *Porous plates*, placed between the soil sample and the pedestal and between the sample and the top cap.

4. A *top plate*, made of solid metal, which forms, with a *cylindrical tube* and the base plate, a closed airtight cell. The plates are tied together with three *tie bars*, and rubber gaskets at the end of the tube secure a leakproof cell. The top plate is fitted with a penetrating piston which rests during the test on the loading cap. It is required that the piston move with minimum frictional resistance and still prevent leaks along the piston. The cell tube for standard and small size cells is usually a Perspex tube that permits the continuous visual contact with the sample and the triaxial process. Large size cells have metal tube walls fitted with viewing portholes.

The control panel accommodates in general:

5. A *pressure gage*, that monitors the pressure in the triaxial cell, a *pressure regulator* that regulates it and usually a *toggle valve* that permits the adjusted pressure to be transferred instantaneously to the triaxial cell through an air–fluid interface.

6. A *volume change burette* between the pressure regulator and the triaxial cell, measuring the volume changes in the triaxial cell through the air–fluid interface level, moving up and down in the graduated burette. The confining pressure applied to the triaxial cell passes through the burette which is connected directly to the triaxial cell.

7. A *drainage burette*, a graduated burette connected to the drainage connectors of the pedestal and of the loading cap, measures the water drained from the sample or absorbed by the sample. Occasionally the loading cap and pedestal can be connected to separate burettes, so that permeability tests can be performed under pressure also. Samples, particularly of granular soils, can be saturated in the triaxial cell and the saturation monitored through these burettes.

8. A *vertical strain dial*, has a 5 cm travel with divisions to 10^{-2} mm or 10^{-4} in. and indicates the vertical deformations of the soil sample. The dial is fixed to an arm rigidly attached to the top plate, while its traveling pin touches the top of the loading bar that rests on the piston. It measures the penetration of the piston relative to the top plate to which it is attached. To obtain reliable results the piston must always be in touch with the loading cap.

9. A *pore-pressure measuring device* (not shown in the figures), may be of a variety of types: hydraulic, pneumatic, electronic, etc. The pore-pressure measurements are made either at the bottom of the sample through the pedestal drainage connector, which in this case is disconnected as a drainage source, or by probes inserted in the sample at chosen elevations which lead to the extra connectors provided in the base plate.

The set-up calls for an air pressure source, a water supply source (usually a distilled and de-aired water tank) and a confining fluid source. Water is commonly used as a confining fluid; however, in order to reduce leakage of the fluid from the triaxial cell, more viscous fluids such as oil, glycerine, Ucon lubricant, etc. are used. Air is also commonly used as a confining fluid, but then volume changes cannot be measured. A vacuum source is useful for removing the confining fluid from the triaxial cell, for saturating granular soil samples and other tasks.

A series of valves, pressure regulators, connectors, etc. facilitate the manipulation of the whole complex.

Further technical and conceptual details about the triaxial testing device, its calibration and testing techniques may be found in the literature (Marsal and Resines 1960; Warlam 1960; Bishop and Henkel 1962; Gibbs and Coffey 1969; ASTM D-2850 1970; B.S. 1377).

The testing starts with the preparation of the soil sample, an undisturbed or remolded cylindrical specimen enclosed in a thin latex membrane, which is placed on the porous plate of the base pedestal and is topped by a second porous plate and the loading cap. The membrane is fastened to the pedestal and to the loading cap and sealed by rubber bands or O rings.

There is no standard size of samples but some sizes are more popular than others. It is considered optimal if the height of the sample is twice the size of its diameter, $2H/D \cong 2$. Sample heights range from 10 cm to 1 m and diameters from 2.5 cm to 60 cm. The larger size samples are either for coarser granular material or for tests where pore pressure probes or other measuring installations are to be embedded.

When the soil sample is in place and all tubes connected, the triaxial cell is sealed and filled with the confining fluid, the measuring instrumentation is installed and connected, the drainage burette is de-aired, filled with water and adjusted to the required initial level, and the volume burette is adjusted to the required initial level. The piston is brought in contact with the loading cap and the vertical strain dial is set in place and adjusted to an initial reading. Now the triaxial testing device is ready for the spherical consolidation test.

11.11 The Spherical Consolidation

When a constant pressure p_m is applied to the confining fluid through the volume burette, say, by opening the toggle valve, the pressure is instantaneously transmitted to the soil sample as an all-round pressure and the process of consolidation commences.

There are three modes of consolidating a cylindrical soil sample in a triaxial cell, depending on the provided drainage path for the water.

1. Draining the sample at its ends is the most common, where the drainage path is in the vertical direction, allowing the flow of water toward the ends of the sample.
2. Radial drainage of the sample in the direction of its radius R from the center outward, where the vertical flow of water is directed toward drainage strips on the envelope of the sample, connected to the drainage outlet.
3. Radial drainage of the sample in the direction of its radius R towards the center, where a porous drainage element is provided.

In many instances vertical drainage is combined with outward radial drainage in order to accelerate the consolidation of the sample. Such a combination cannot provide reliable information on the time required for consolidation, but on the amount of consolidation only.

The pressure tensor p_{ij} of the confining pressure is

$$p_{ij} = p_m \equiv \begin{vmatrix} p_m & 0 & 0 \\ 0 & p_m & 0 \\ 0 & 0 & p_m \end{vmatrix} = p_m \begin{vmatrix} 1 & 0 & 0 \\ 0 & 1 & 0 \\ 0 & 0 & 1 \end{vmatrix} \qquad (11.11.1)$$

where $p_{\mathrm{m}} = p_{xx} = p_{yy} = p_{zz} = p_{11} = p_{22} = p_{33} = \frac{1}{3}p_{\alpha\alpha}$ and the assumption is that any volume element within the sample is identical to the sample as a whole and behaves like it. According to St Venant's principle, this assumption is fairly accurate as a first approximation, except in a small region close to the ends of the sample, since

$$p_{\mathrm{m}}S = \int_S p_{zz}\,\mathrm{d}S \tag{11.11.2}$$

and $p_{zz} = p_{\mathrm{m}}$.

Since the sample is saturated, the water which is squeezed out of the soil drains gradually in the vertical direction through the top and bottom porous plates. The expelled water rises in the drainage burette, and when recorded at proper time intervals the decrease in the volume of the voids of the soil is accounted for.

With the decrease in the volume of the voids the soil sample compresses and its total volume decreases. Fluid from the volume burette moves to replace the lost volume in the triaxial cell and descends accordingly. The changes in the volume burette represent the total changes that occur in the volume of the soil sample. The initial reading on the volume burette, however, has to be corrected for the volume changes due to the initial expansion of the triaxial cell and tubing and for the initial compression of the confining fluid under pressure. This correction is found by performing earlier calibration of the system for various pressures.

With the decrease in volume of the soil sample its height decreases correspondingly. If care has been taken to counterbalance the uplift pressure on the loading piston due to the pressure acting in the triaxial cell, then the piston will maintain contact and follow the loading cap and the strain dial will record the vertical deformations of the soil sample.

As the consolidation progresses, readings are taken on the volume burette, the drainage burette and the strain dial at predetermined time intervals, indicating the total volume changes of the soil sample, the changes of the volume of the voids and the vertical deformations of the soil sample, respectively (after the necessary calibration corrections). As in the uniaxial consolidation, the dial readings are carefully recorded after elapsed times of 0.25, 0.5, 1, 2, 4, 8, 15, 30, 60, 120, 240, 480, 1440 minutes and daily thereafter, as long as the consolidation proceeds. The time intervals are the times elapsed from the beginning of the consolidation. The succession of the readings are conveniently spaced at logarithmic intervals since the dial readings d are plotted against the logarithm of the time. Since a 24 hour period has been found sufficient for the consolidation of an 25 mm high soil sample in a uniaxial consolidation, and since, according to Eq. (11.4.8), the time of consolidation of a sample is directly proportional to the square of its height, it can be surmised that samples of the same height as the one used in triaxial consolidation would require several days to consolidate.

As in uniaxial consolidation, the pressures are successively applied so that each increment is equal to the previous pressure. So, for instance, if the first initial pressure p_m was 0.125 kg/cm^2, the successive pressures would be 0.25, 0.5, 1.0, 2.0, 4.0, 8.0, 16.0 kg/cm^2 and so on, up to the required pressure or the capacity of the triaxial cell. Such incremental loading is convenient since the data, as related to the pressure p_m, will appear at equal distances if presented in a $\log p_m$ plot. But, of course, any other pressures or initial pressures are acceptable as well.

For each pressure increment the soil is allowed to consolidate for any time required by the nature of the soil. Long term consolidation tests, however, may turn out to be damaging to the test itself if no precautions are taken and leaks along the piston, from the triaxial cell, from any of the connections or through the latex membrane occur. Also a "freeze" of the piston or the dial gage may be detected.

Consolidation time curves for the kaolinite clay, representing the volumetric strain versus time are given in Fig. 11.11.1a and Fig. 11.11.1b, in regular time scale and in log time scale, respectively. They show the changes with time t of the total volumetric strain $3\varepsilon_m = \varepsilon_{\alpha\alpha}$ and the volumetric strain due to the water expelled $\varepsilon_{w\alpha\alpha}$. The void ratio versus the spherical pressure is shown in the e versus p_m curve, Fig. 11.11.2a, and e versus $\log p_m$ curve Fig. 11.11.2b.

The overall macroscopic pressures acting on the sample in a triaxial consolidation test are assumed to be all equal, according to Eq. (11.11.1), and also equal to the spherical pressure, $p_{xx} = p_{yy} = p_{zz} = p_m = \frac{1}{3}p_{\alpha\alpha}$, while the pressures in a soil element within the sample are only nearly equal among themselves and to the spherical pressure. The exact evaluation of the pressures acting on a small element is possible through Eq. (6.6.7), however the theory is not presented in this work. The differences of stresses between the small elements in the sample are a result of additional internal frictional forces and affect the *inner part* of the stresses (Bowen 1967).

While the sample consolidates the volume changes and vertical deformations are recorded and the strains in all directions can be evaluated. Frictional restraints exist at the two ends of the sample, between the soil and the porous plates, and the horizontal strains are obstructed (Shockley and Ahlvin 1960; Kirkpatrick and Belshaw 1968; Perloff and Pombo 1969). The frictional restraint induces radial shear stresses at the ends of the sample, where the soil is consolidated right from the beginning of the consolidation process and its shearing resistance is highest. The shear stresses increase from the center of the sample outward and so does the shearing resistance. This effect of end restraint and shear stresses dies away quite fast and at a short distance from the ends it is not felt any more. It is important, therefore, to keep the height of the soil sample more than twice its diameter (Bishop and Green 1965). In Fig. 11.11.3 the "hourglass" contours of soil samples after triaxial consolidation tests are seen.

The horizontal deformations do not proceed at the same rate at different elevations: near the ends of the sample, where the consolidation has progressed more than at its mid-height, the horizontal deformations are more advanced. As a result, the resistance to shear stresses decreases towards the mid-height of the sample and increases as the consolidation proceeds. The soil is very sensitive in the central region, and if the vertical pressure

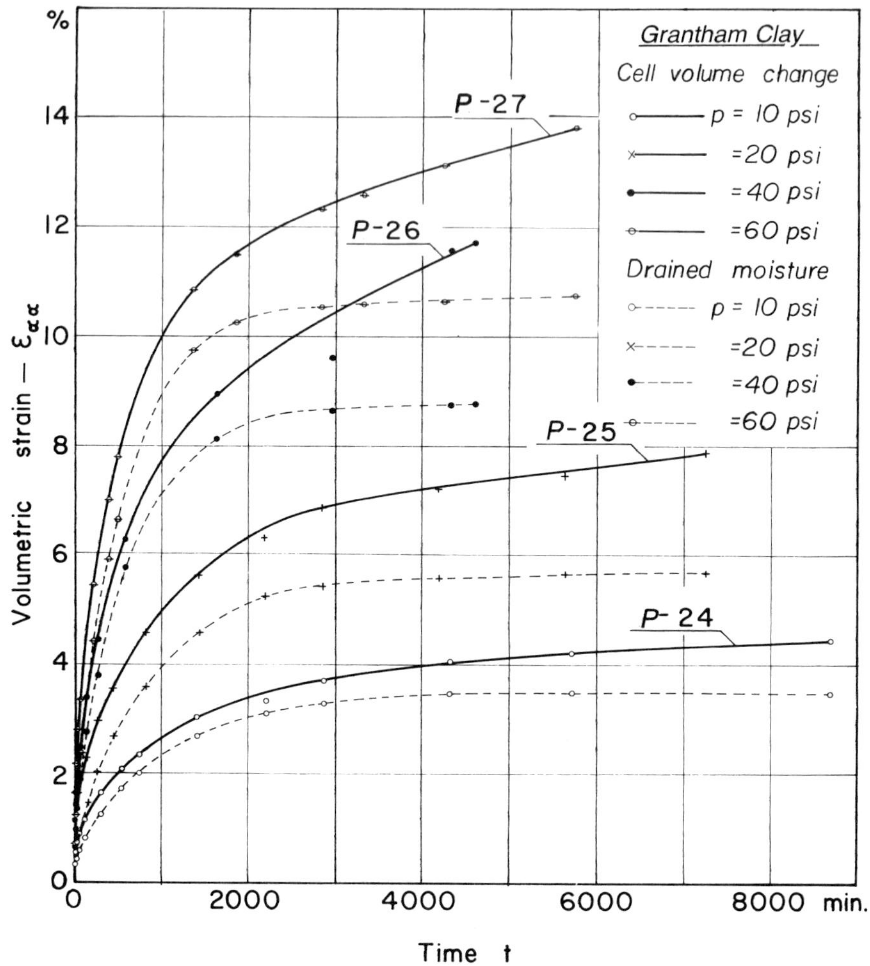

a

counterbalancing the piston uplift is not carefully administered and an excess vertical pressure applied, a bulging of the sample at mid-height is observed, a result of the presence of the shear stresses just where the pore-water pressures are highest.

11.12 The Pressure and Strain Tensors in the Triaxial Test

Given the volumetric stretch $\Lambda = V_1/V_0$ and the vertical stretch $\lambda_{zz} = 2H_1/2H$, and assuming a simple compression for the horizontal stretches $\lambda_{xx} =$

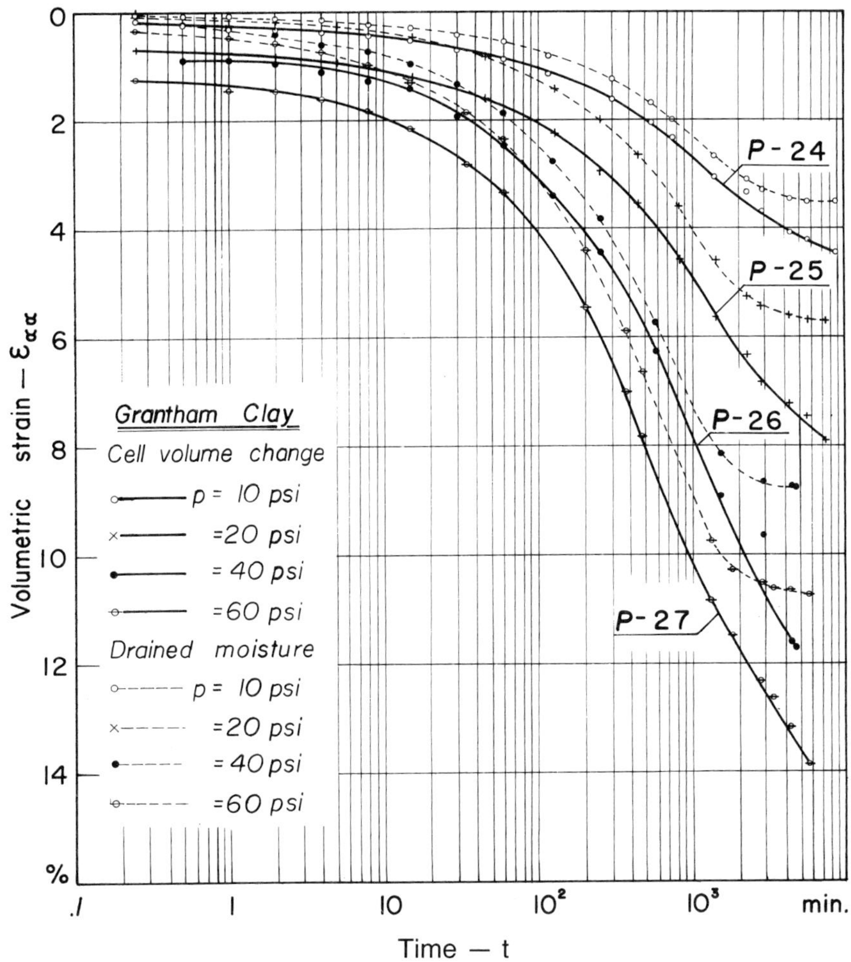

b

Fig 11.11.1. Spherical consolidation time curves. **a** Volumetric strain $\varepsilon_{\alpha\alpha}$ versus time t. **b** Volumetric strain $\varepsilon_{\alpha\alpha}$ versus log t.

$\lambda_{yy} = B\lambda_{zz}$, see Sect. 3.14, the infinitesimal strain tensor ε_{ij} and the Hencky strain tensor $\overset{\mathbf{H}}{\varepsilon}_{ij}$ are

$$\varepsilon_{ij} = \begin{vmatrix} \varepsilon_{xx} & 0 & 0 \\ 0 & \varepsilon_{yy} & 0 \\ 0 & 0 & \varepsilon_{zz} \end{vmatrix} = \begin{vmatrix} \nu\varepsilon_{zz} & 0 & 0 \\ 0 & \nu\varepsilon_{zz} & 0 \\ 0 & 0 & \varepsilon_{zz} \end{vmatrix}$$

$$= \begin{vmatrix} 1-\lambda_{xx} & 0 & 0 \\ 0 & 1-\lambda_{yy} & 0 \\ 0 & 0 & 1-\lambda_{zz} \end{vmatrix}$$

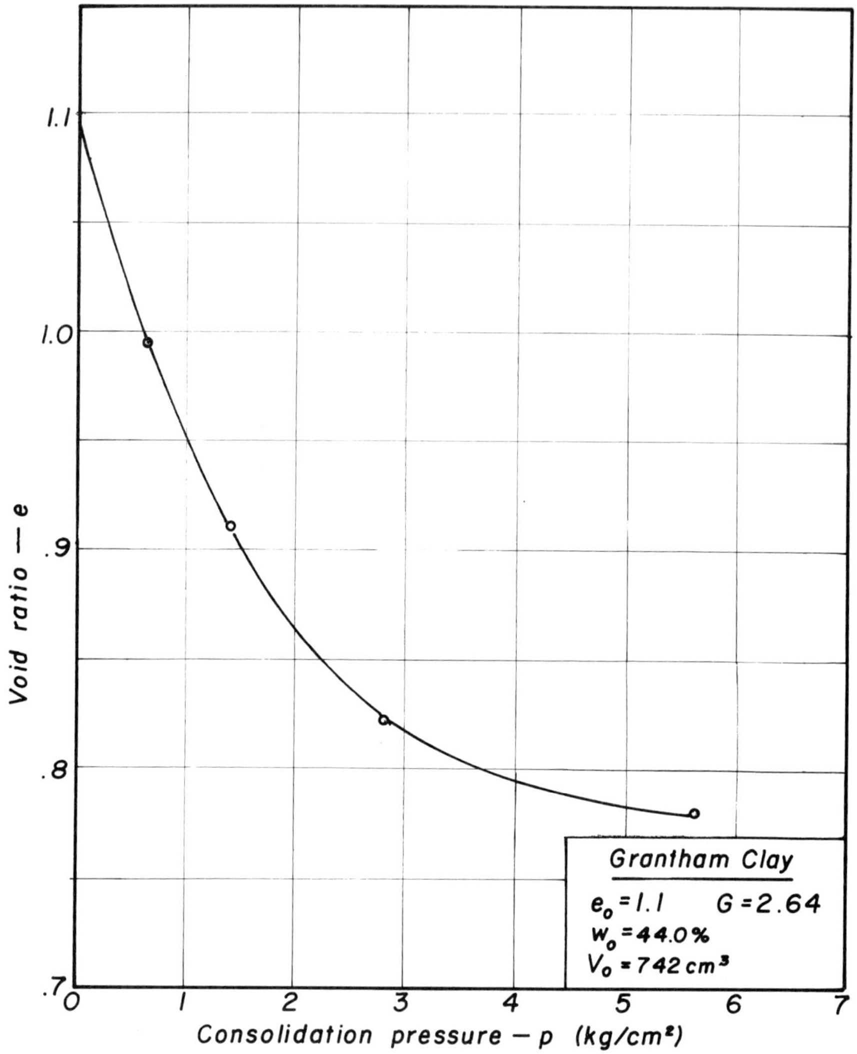

a

$$= \begin{vmatrix} 1 - B\lambda_{zz} & 0 & 0 \\ 0 & 1 - B\lambda_{zz} & 0 \\ 0 & 0 & 1 - \lambda_{zz} \end{vmatrix}$$

$$\equiv \left(1 - \frac{(1 + 2B)\lambda_{zz}}{3}\right)\delta_{ij} + \frac{(1 - B)\lambda_{zz}}{3}\,\Xi_{ij}$$

$$= \frac{\varepsilon_{\alpha\alpha}}{3}\,\delta_{ij} + e_{ij} \tag{11.12.1}$$

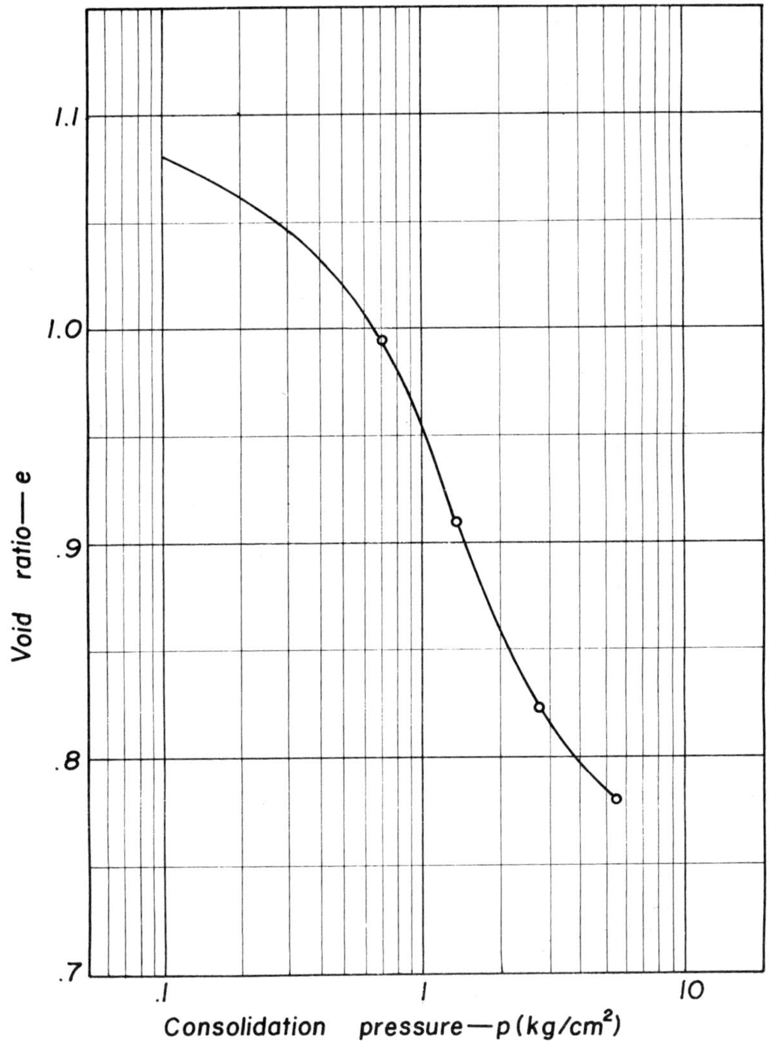

b

Fig 11.11.2. Void ratio e versus spherical pressure p_{m}. **a** e versus p_{m}. **b** e versus $\log p_{\mathrm{m}}$.

$$\overset{\mathrm{H}}{\varepsilon}_{ij} = \left|\begin{matrix} \overset{\mathrm{H}}{\varepsilon}_{xx} & 0 & 0 \\ 0 & \overset{\mathrm{H}}{\varepsilon}_{yy} & 0 \\ 0 & 0 & \overset{\mathrm{H}}{\varepsilon}_{zz} \end{matrix}\right| = \left|\begin{matrix} \ln \lambda_{xx} & 0 & 0 \\ 0 & \ln \lambda_{yy} & 0 \\ 0 & 0 & \ln \lambda_{zz} \end{matrix}\right|$$

$$= \left|\begin{matrix} \ln B\lambda_{zz} & 0 & 0 \\ 0 & \ln B\lambda_{zz} & 0 \\ 0 & 0 & \ln \lambda_{zz} \end{matrix}\right| \equiv \frac{\ln^2 B\lambda_{zz}^3}{3}\,\delta_{ij} + \frac{1}{\ln B^3}\,\Xi_{ij}$$

$$= \frac{\overset{\mathrm{H}}{\varepsilon}_{\alpha\alpha}}{3}\,\delta_{ij} + \overset{\mathrm{H}}{e}_{ij} \tag{11.12.2}$$

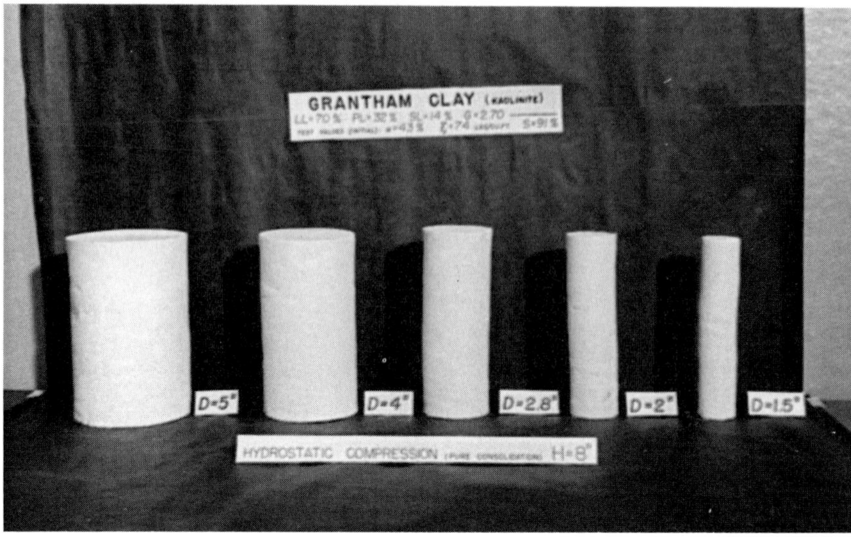

Fig 11.11.3. Hourglass shaped sample after spherical triaxial test.

where $B \leq 1$ is the *coefficient of transverse stretching* and $\nu \leq 1$ is the *ratio of transverse dilatation*.

The respective spherical strains are

$$\varepsilon_{\alpha\alpha} = \varepsilon_{xx} + \varepsilon_{yy} + \varepsilon_{zz} = 3 - \lambda_{xx} - \lambda_{yy} - \lambda_{zz} = 3 - (1 + 2B)\lambda_{zz} \tag{11.12.3}$$

$$\overset{\mathrm{H}}{\varepsilon}_{\alpha\alpha} = \overset{\mathrm{H}}{\varepsilon}_{xx} + \overset{\mathrm{H}}{\varepsilon}_{yy} + \overset{\mathrm{H}}{\varepsilon}_{zz} = \ln \Lambda = 2 \ln B\lambda_{zz} + \ln \lambda_{zz} = \ln B^2 \lambda_{zz}^3 \tag{11.12.4}$$

From Eqs. (11.12.1) and (11.12.2) it is seen that if the spherical compression is uniform, $B = 1$, then both the infinitesimal strain and the Hencky strain are spherical and do not have distortional components. If, however, the compression is not uniform, then the strain is *anisotropic* and internal distortions appear in the soil sample. The amount of distortion depends on the magnitude of B. In any event, the spherical part of Eqs. (11.12.1) and (11.12.2) is one third of the volumetric strains in both measures, as evident from Eqs. (11.12.3) and (11.12.4).

Another interesting feature of the Hencky strain in triaxial consolidation is that even if there are distortions, they do not depend on the strain of the sample but on the amount of anisotropy, namely on B, meaning that the anisotropy does not introduce rotation in the principal stresses acting on the soil sample.

Fig. 11.12.1 shows the various strains of the same soil sample, both in the infinitesimal and the Hencky measures.

Fig. 11.12.1. Comparison of the infinitesimal and the Hencky measures of strain. **a** Volumetric strain $\varepsilon_{\alpha\alpha}$ versus consolidation pressure p_{m}. **b** Vertical Strain ε_{zz} versus consolidation pressure p_{m}. **c** Transverse stretching B versus consolidation pressure p_{m}.

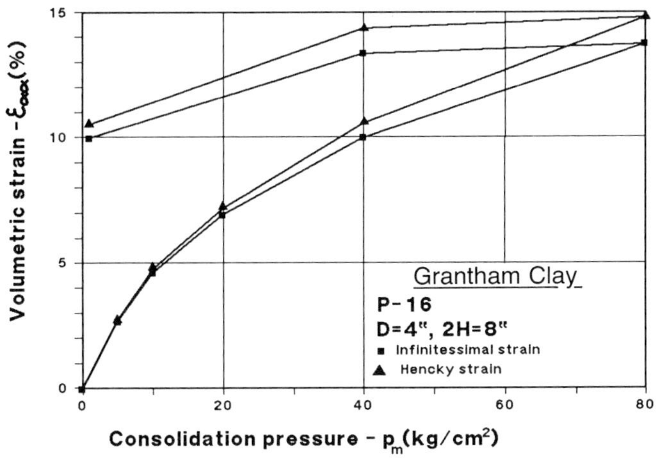
Volumetric strain
0
5
10
15
0
20
40
60
80
Consolidation pressure - p_m(kg/cm^2)
Grantham Clay
P-16
D=4", 2H=8"
■ Infinitessimal strain
▲ Hencky strain

a

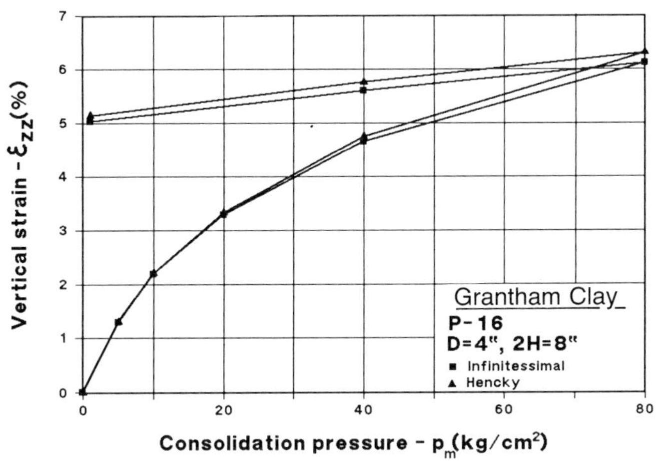
Vertical strain - ε_{ZZ}(%)
0
1
2
3
4
5
6
7
0
20
40
60
80
Consolidation pressure - p_m(kg/cm^2)
Grantham Clay
P-16
D=4", 2H=8"
■ Infinitessimal
▲ Hencky

b

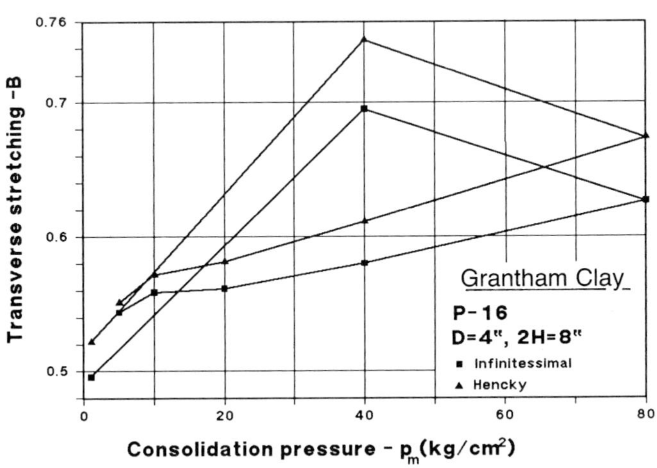
Transverse stretching -B
0.5
0.6
0.7
0.76
0
20
40
60
80
Consolidation pressure - p_m(kg/cm^2)
Grantham Clay
P-16
D=4", 2H=8"
■ Infinitessimal
▲ Hencky

c

While the stress tensor acting on the soil sample in a triaxial consolidation is, at least macroscopically, spherical, $p_{ij} = p_{\mathfrak{m}}\delta_{ij}$, the strain tensor hints at the internal microstructural shear distortions. In Fig. 11.12.1 we can detect an anisotropy which is emphasized in Fig. 11.12.1c, where the change in the coefficient of the transverse stretch B is presented.

Since the consolidation is driven by the water potential $\Phi_{\mathfrak{w}}$ or the equivalent pore-water pressure $p_{\mathfrak{w}}$, and since it is assumed that the all around pressure $p_{\mathfrak{m}}$ acting on the sample is equal to the total pressure, the effective pressure $p_{\mathfrak{m}} = \frac{1}{3}p_{\mathfrak{s}\alpha\alpha}$ acting at any time in a saturated sample can be found

$$p(t)_{\mathfrak{s}} = p_{\mathfrak{m}} - p(t)_{\mathfrak{w}} = \tfrac{1}{3}p_{\alpha\alpha} - p(t)_{\mathfrak{w}} \tag{11.12.5}$$

Since $p_{\mathfrak{m}}$ and $\frac{1}{3}p_{\alpha\alpha}$ are here assumed constant and the pore-water pressure $p(t)_{\mathfrak{w}}$ dissipates with time from an initial value $p(0)_{\mathfrak{w}} = p_{\mathfrak{m}}$ to a final value $p(\infty)_{\mathfrak{w}} = 0$, the effective pressure $p(t)_{\mathfrak{s}}$ is necessarily time dependent. If, however, the spherical pressure is time dependent as well, $p(t)_{\mathfrak{m}}$, then two possibilities are evident:

1. The applied spherical pressure $p(t)_{\mathfrak{m}}$ increases slower than the pore-water pressure $p(t)_{\infty}$ which has been building up can dissipate, $|\Delta p(t)_{\mathfrak{m}}| < |\Delta p(t)_{\mathfrak{w}}|$, in which case the effective pressure increases gradually

$$p(t)_{\mathfrak{s}} = p(t)_{\mathfrak{m}} - p(t)_{\mathfrak{w}} > 0 \tag{11.12.6}$$

2. The applied spherical pressure $p(t)_{\mathfrak{m}}$ increases faster than the pore-water pressure $p(t)_{\mathfrak{w}}$ dissipates, $|\Delta p(t)_{\mathfrak{m}}| > |\Delta p(t)_{\mathfrak{w}}|$, in which case there is a continuous net increase in the pore-water pressure

$$|p(t)_{\mathfrak{w}} - p(t)_{\mathfrak{m}}| > 0 \tag{11.12.7}$$

It is also possible to perform a triaxial consolidation on a cylindrical soil sample which drains radially. Here there are, in fact, two possibilities, both problematic.

1. A cylindrical soil sample drains inward, towards a cylindrical central draining core of a radius R_i, Fig. 11.12.2a. The problem that such a test poses is to find a suitable porous material stiff enough not to compress radially and compressible enough vertically so as not to prevent the compression of the soil sample, nor cause frictional resistance along its height.

2. A cylindrical soil sample drains outward, Fig. 11.12.2b. The problem of such a test is that the outside cylindrical envelope of the sample, which consolidates faster than its inside core by being closer to the drainage medium, forms stiff cylindrical tubes which repress radial deformation by arching, the result being an impediment to the consolidation and a reduced volumetric and radial strain. A combination of such a radial drainage with vertical drainage would not only reduce consolidation, but result in higher shear strains in the soil sample.

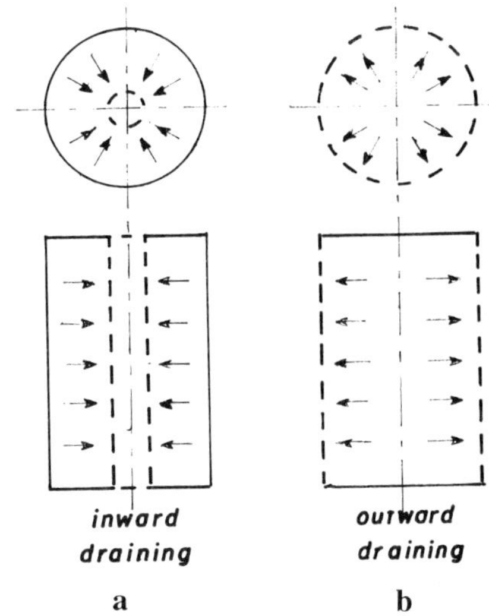

Fig 11.12.2. Radially draining cylindrical triaxial sample. a Inward draining. b Outward draining.

An interesting experiment of triaxial consolidation was made by consolidating a spherical soil sample with a central rigid spherical draining core (de Josselin de Jong and Verruijt 1965). The work, however, did not produce conclusive results and seems to have been discontinued.

Unfortunately, time curves, void ratio versus pressure curves and other relevant data of triaxial consolidation, which usually precede consolidated shear tests, are very seldom presented along with the shear data. It is, nevertheless, apparent from what has been published and from personal experience that triaxial consolidation time curves display less secondary consolidation than uniaxial consolidation time curves. This still remains to be substantiated. The fact that in triaxial consolidation the shear stresses and shear strains, as far as they exist, are not phenomenological but structural while in the uniaxial consolidation they are phenomenological as well, indicates that at least part of the secondary consolidation in uniaxial consolidation is made up of shear strains, a result of the shear stresses present.

The e versus $\log p_m$ curve of triaxial consolidation, as it appears in Fig. 11.11.2, as well as other similar curves, are not straight lines either. A more elaborate discussion of the e versus p_m curve in uniaxial and triaxial consolidation is the topic of the following section.

11.13 The Void Ratio–Pressure Curve

Fig. 11.13.1 is a selection of e–$\log p_{zz}$ curves of consolidation tests from a variety of soil types: sands of different densities and mineralogical compositions, silts and clays from different parts of the world, undisturbed and

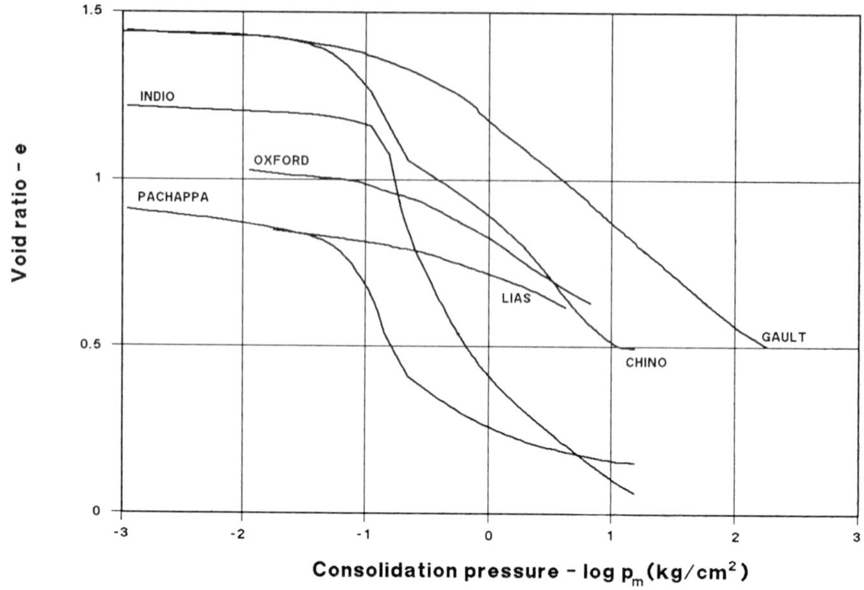

Fig 11.13.1. Suction curves converted to e versus $\log p_m$ curves.

remolded soils, ordinary and extrasensitive clays, slurries and slickenside clays, chemically or electrically treated soils, etc., found in the literature. The variety of shapes and ranges of void ratio indicates that the consolidation is an expression of diversified soil properties.

Several immediate observations should be made:

1. Even a casual look at the display of curves shows that neither the e versus $\log p_{zz}$ curve of the uniaxial consolidation nor the e versus $\log p_m$ curve of the triaxial consolidation are straight lines, as they are sometimes preferred to be thought. The linearity of the e versus p_{zz} curve, as assumed by Terzaghi's theory of consolidation (assumption 9 in Sect. 11.4) and expressed by Eq. (11.4.2), is even less obvious. It is true that some of the curves do show linear sections, occasionally of considerable extent, and sometimes the whole range of consolidation beyond the preconsolidation load appears linear. But this is due to the fact that the range of the consolidation test is only 10 kg/cm^2 or 20 kg/cm^2. Had the consolidation tests, uniaxial or triaxial, been extended to higher pressures, a curvature of the lines would have been evident.

2. In the suction curve a remolded soil reveals a slightly curved line, the *virgin curve*, which is close to a straight line and indicates no defined preconsolidation pressure, while undisturbed soils show some preconsolidation due to previous overburden pressure, Fig. 11.13.2.

3. For approximately the same pressure range, the range of void ratio in granular soils depends on the density of the soil: the denser the soil, the

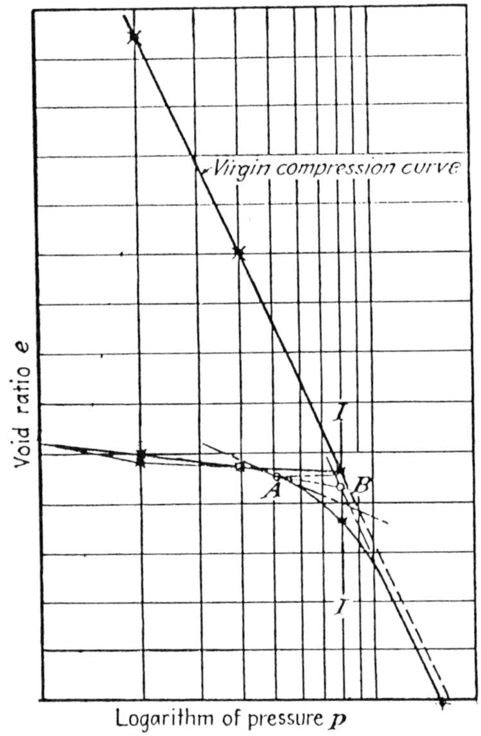

Fig 11.13.2. Consolidation curve with virgin line. [From G. P. Tschebotarioff (1952), courtesy of McGraw-Hill Book Co. Inc.]

smaller the range of void ratio. In fine grained soils it will depend on a number of parameters. We will say that the void ratio range increases mainly with the increase of the cation exchange capacity of the soils, the decrease of ionic concentration in the water solute and the nature of the ions, in the following order: Ca^{2+}, Mg^{2+}, K^+, Na^+, Li^+.

4. Let us have here a second look at the suction curves discussed in Sect. 8.27, for instance Fig. 8.27.4. The suction curve is represented by the logarithm of the suction pressure versus the moisture content $\theta_{\mathfrak{w}}$. Occasionally, instead of the moisture content $\theta_{\mathfrak{w}}$ the curve is presented in terms of the water content w, which requires a simple transformation

$$\theta_{\mathfrak{w}} = \frac{G_g w}{1 - G_g w}; \quad w = \frac{\theta_{\mathfrak{w}}}{G_g(1 + \theta_{\mathfrak{w}})} \tag{11.13.1}$$

By inverting the suction curve, that is, interchanging the ordinate with the abscissa and performing a simple transformation of the moisture content $\theta_{\mathfrak{w}}$ or the water content w into the void ratio e, according to Eqs. (9.2.10) and (9.2.22) for $S = 1$, respectively,

$$e = \frac{\theta_w}{1 - \theta_w}; \quad e = \frac{G_g}{G_w} w \tag{11.13.2}$$

we obtain a complete e versus $\log p_m$ curve, comprising sometimes a pressure range of seven orders of magnitude, Fig. 11.13.1. Likewise we can take the consolidation curves in Fig. 11.13.1 and show them in their original form as suction curves, Fig. 11.13.3.

The range of suction curves may extend to 10^4 kg/cm^2, some 500–1000 times the range of consolidation pressures, 10–20 kg/cm^2, at which saturated soils are usually consolidated.

5. The data for the different sections of the suction curve are obtained by different overlapping techniques, the common denominator of which is that a certain amount of energy must be invested in order to extract a certain amount of water from the soil; thus, in the consolidation process we invest an amount of energy by applying pressure, and squeeze out an amount of water from the soil. By applying a higher pressure an additional amount of water is squeezed out, and so on. We see that the e versus $\log p_m$ curve is equivalent to the suction curve and the two curves are directly interconvertible.

The nature of the energy directed to remove water from the soil is immaterial, and it can be of any kind: mechanical or hydraulic (pressure), vacuum (suction), thermal (drying), centrifugal, osmotic, electrical (electrophoresis), electrochemical, or any other energy form. Incidentally, one of the techniques applied to obtain a part of the suction curve, in the range of up to 10 kg/cm^2, is the pressure plate method, in which the water is subjected to a condition similar to that in consolidation.

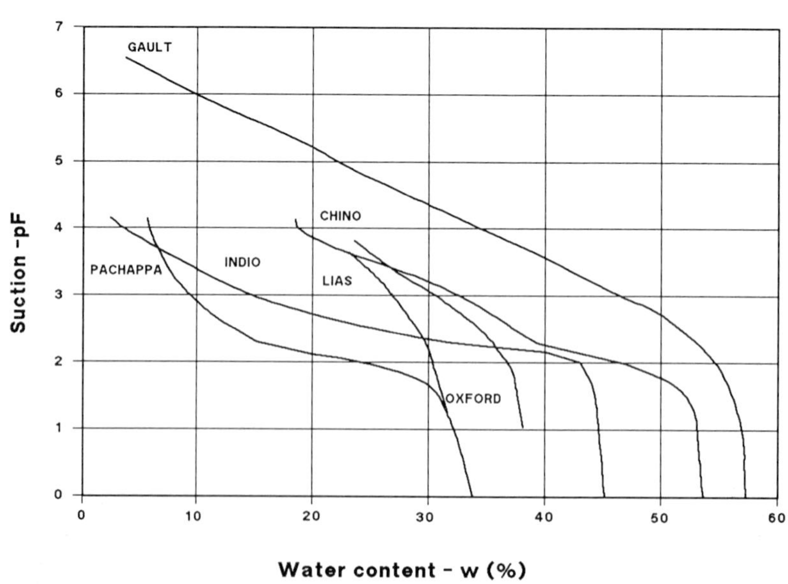

Fig 11.13.3. e versus $\log p_m$ curves from Fig. 11.11.1 converted to suction curves.

6. There is an equivalence between the suction acting in the pore-water and sustaining a moisture content and the spherical consolidation pressure resulting in the same moisture content. In other words, when a soil sample is consolidated by a spherical pressure, say p_m, then, with the removal of the pressure a suction head of p_m/γ_m will take control over the pore-water. If the soil sample is subjected to a series of incremental pressures so that an e versus $\log p_m$ curve can be drawn, Fig. 11.13.4, then an inverted suction versus water content curve can also be presented, Fig. 11.13.5. If the corresponding points are now projected on the shrinkage curve of the soil, Fig. 11.13.6, then the points corresponding to the consolidation pressures will all lie on the straight line of the shrinkage curve.

To illustrate further the above point, the relation of the shrinkage curve to the suction curve, discussed in Sect. 8.27, should be mentioned. The shrinkage is in linear relationship with the water content: as the water content in a soil sample decreases the volume of the sample decreases, that is, the sample shrinks, the decrease in volume of the soil sample being equal to the volume of the water lost. The linear relation between the shrinkage of the sample and the water content continues until the shrinkage limit is reached, and from there on the volume change stops, in spite of the continual reduction of the

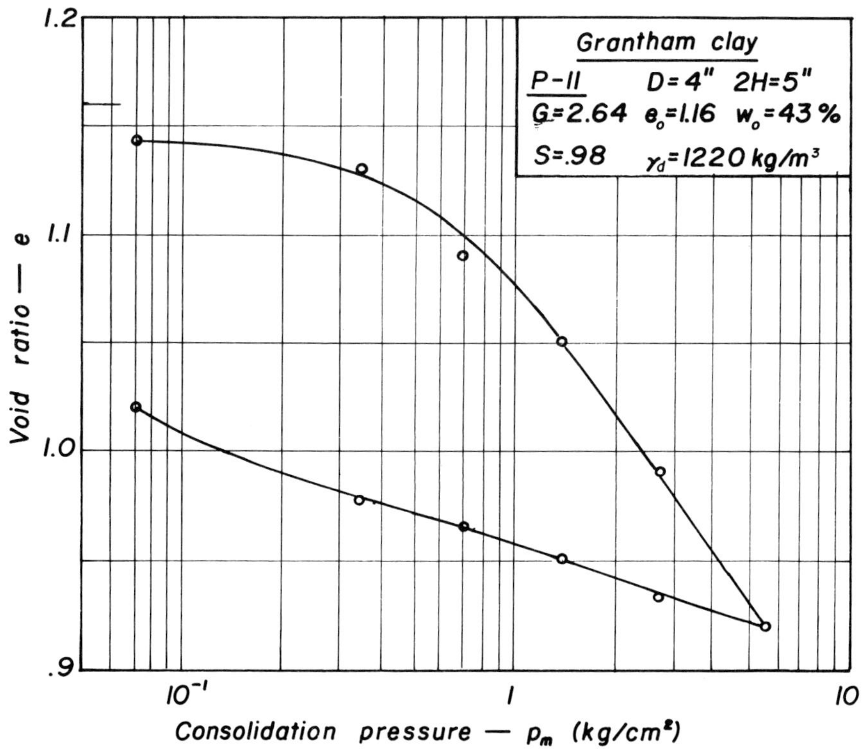

Fig 11.13.4. e versus $\log p_m$ curve of a spherical consolidation test with hysteresis.

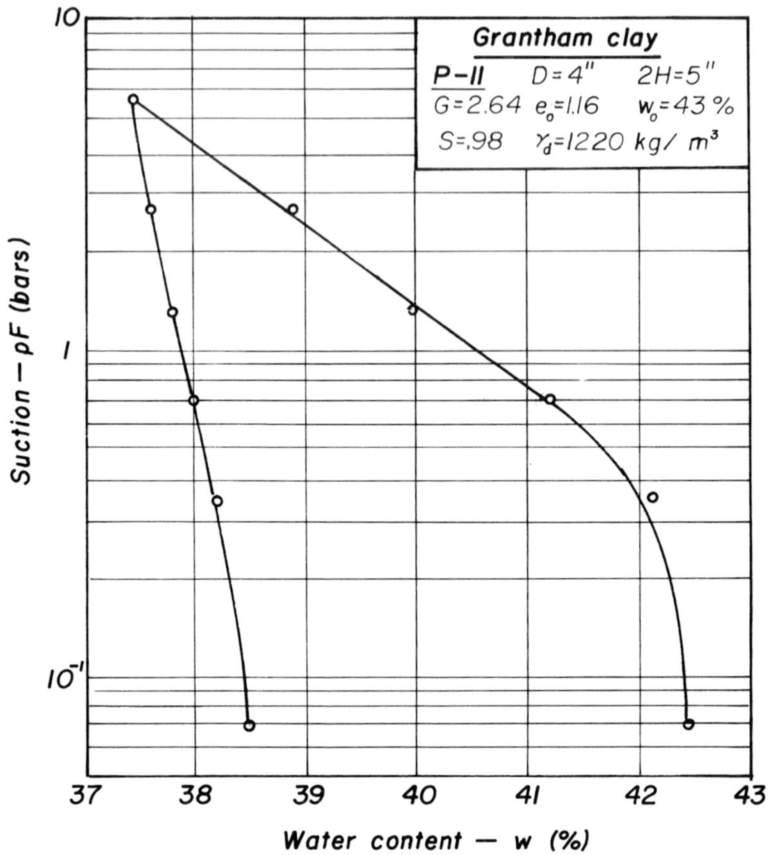

Fig 11.13.5. Suction curve inverted from e versus $\log p_m$ consolidation curve shown in Fig. 11.13.4.

water content.* This linear relation exists as long as the soil sample is saturated, and below the shrinkage limit the soil sample becomes unsaturated.

7. The parallel changes that occur in the suction forces of the soil sample are seen in the adjacent suction curve, Fig. 11.13.5. The suction potential starts in the range of capillary tension, at zero suction, increasing continuously as the water content decreases and the suction curve passes through the point representing the shrinkage limit. While the shrinkage curve shows a dramatic change at the shrinkage limit when the soil sample becomes unsaturated, in the suction curve the change from the saturated to the unsaturated state of the soil is not directly distinguishable and the curve rises continuously.

* The formal definition of the shrinkage limit w_S is the water content at which the linear relation between the volume change ΔV and the water content w of a soil sample ends. Actually the linear relation ends somewhat above the shrinkage limit, from which point on there is a non-linear gradual decrease of the volume change with the change in the water content.

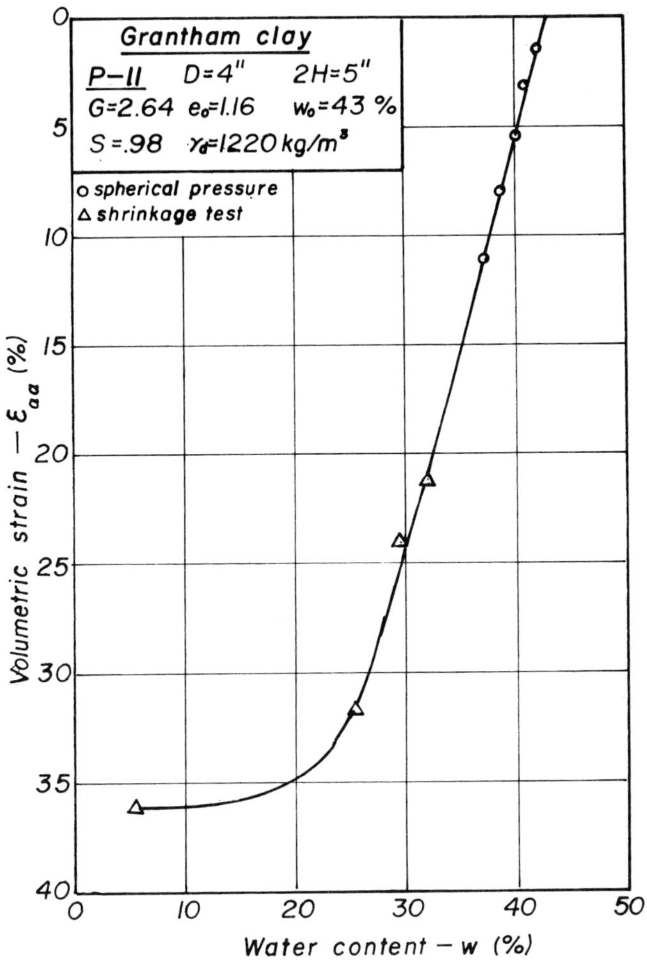

Fig 11.13.6. Shrinkage curve from Figs. 11.13.4 and 11.13.5.

Summarizing these observations it should be clear that the e versus $\log p_{zz}$ curve and the suction curve are two different presentations of the same results of the volumetric behavior of a soil.

11.14 Normally Consolidated and Overconsolidated Soils

In Sect. 11.8 the difference between a remolded soil and an overconsolidated soil was noted and in Fig. 8.27.3 several suction curves of a London clay were shown.

These curves show a natural undisturbed clay sample in the drying process, from an initial natural water content $w_0 = 33.2\%$ to zero moisture content, the suction potential increasing from practically zero suction to a suction of 10^{-7} cm of water. Then the sample is wetted and dried again and a wetting and drying loop is obtained: with the wetting, the suction potential decreases and the water content increases, but remains far below the water content of the initial suction curve; with the repeated drying of the sample, a new suction curve is obtained, again with lower water content than that of the initial suction curve, but higher than that of the wetting curve. Thus the suction is not a unique function of the moisture content, but indicates a hysteretic phenomenon.

The suction curves of two remolded samples are also shown: one at an initial water content of 90%, on which the results of the plastic limit test and the values of the liquid limit test for different numbers of blows are marked (in the original publication), the other in a slurried state with an initial water content of 106.7%. Both curves are linear down to a water content of approximately 55%. The difference between the two curves is probably in the amount of air entrapped in the samples, in the slurried sample the amount is closer to zero and the curve can be considered as the *virgin curve*.

In the slurried sample two wetting and drying cycles were performed at 57.6% and 47.7% water contents, represented by two loops attached to the suction curve (Croney et al. 1958). When wetted, the sample reaches water contents of 66.4% and 60.0%, respectively, far below the initial 106.7%. It is also apparent that the rewetting sections of the loops join the virgin curve at the points of the previously applied suctions, and the increase in the curvature is an indication of preconsolidation pressures. By checking the loops of the e–$\log p_m$ curve of this sample, preconsolidation pressures can be observed at about 0.4 and 1.0 kg/cm^2.

After having introduced the identity between the suction curve and the triaxial consolidation through comparison of their properties, two reservations are in order:

1. A slight discrepancy in the correspondence between the suction curve and the consolidation curve may exist in the lower water content ranges, due to the form of drainage, which at this stage is multidirectional and not unidirectional as in the triaxial test. This has not yet been experimentally verified, and it would require fairly high consolidation pressures and appropriate equipment to do it.

2. The suction curve and the consolidation curve are not uniquely determined, but depend on the initial moisture of the soil sample, its state of disturbance, its thixotropic regain and, most important, on the amount of entrapped air present in the soil and the size distribution of the bubbles. The latter has a strong effect on the shape of the consolidation curves. In the wetting and drying cycles these curves are subject to hysteretic behaviour, which, so far, has not been given a simple and satisfactory analytic treatment. Nevertheless, in practical applications the hysteretic effect is avoided, by referring to a specific curve or a specific cycle.

11.15 Consolidation of Unsaturated Soils

While the theory of consolidation was introduced in the nineteen twenties, and although most problems encountered in soil engineering concern unsaturated soils, the study of unsaturated soils by way of extending the theory of consolidation did not capture the interest of the soil mechanics community until the late fifties (Aitchison 1956, 1961a; Alpan 1959, 1961, 1963; Barden 1965; Bishop 1959; Bishop et al. 1960; Bishop and Blight 1963; Bishop and Donald 1961; Burland 1964; Fredlund 1975, 1985; Fredlund and Hasan 1979; Fredlund and Morgenstern 1976, 1977; Jennings 1960; Lloret and Alonso 1980; Matyas and Radhakrishna 1968; Philip and Smiles 1969; Sills 1987; Sills and Austin 1982; Sills and Nageswaran 1984; Sparks 1967; Thomas 1987; Wheeler 1986; Wood 1979; Yoshimi and Osterberg 1963). This impressive partial list of publications indicates a growing interest in the subject; still, many of the works relate to the shear strength of unsaturated soils, rather than to their volumetric properties. In soil physics the interest in unsaturated soils began somewhat earier (Childs and Collis-George 1950; Klute 1952; Philip 1954, 1957a–e, 1958a, b, 1968b, 1969b; Philip and Smiles 1969; Richards 1950; Smiles and Rosenthal 1968).

In fact unsaturated soil is the general case and saturation $S = 1$ is a particular case, in which the three equations (9.3.8), (9.3.9) and (9.2.12) are reduced to two equations (10.6.1) and (10.6.2).

As in the case of saturated soils, the consolidation of unsaturated soils is also controlled by the general equations of material density balance for a three-constituent mixture, Eqs. (9.3.8), (9.3.9) and (9.2.12), reproduced here

$$(9.3.8) \qquad \rho_{\mathfrak{s}} \frac{\mathrm{D}\dfrac{\rho_{\mathfrak{w}}}{\rho_{\mathfrak{s}}}}{\mathrm{D}_{\mathfrak{s}} t} = \frac{\mathrm{D}\rho_{\mathfrak{w}}}{\mathrm{D}_{\mathfrak{s}} t} = \frac{\mathrm{D}\gamma_{\mathfrak{w}}\theta_{\mathfrak{w}}}{\mathrm{D}_{\mathfrak{s}} t} = \frac{1}{1+e}\,\frac{\mathrm{D}\gamma_{\mathfrak{w}} eS}{\mathrm{D}_{y} t}$$

$$= \frac{\mathrm{D}\gamma_{\mathfrak{w}} nS}{\mathrm{D}_{\mathfrak{s}} t} = -[\gamma_{\mathfrak{w}}\theta_{\mathfrak{w}}(v_{\mathfrak{w}\alpha} - v_{\mathfrak{s}\alpha})]_{,\alpha}$$

$$(9.3.9) \qquad \rho_{\mathfrak{s}} \frac{\mathrm{D}\dfrac{\rho_{\mathfrak{g}}}{\rho_{\mathfrak{s}}}}{\mathrm{D}_{\mathfrak{s}} t} = \frac{\mathrm{D}\rho_{\mathfrak{g}}}{\mathrm{D}_{\mathfrak{s}} t} = \frac{\mathrm{D}\gamma_{\mathfrak{g}}\theta_{\mathfrak{g}}}{\mathrm{D}_{\mathfrak{s}} t} = \frac{1}{1+e}\,\frac{\mathrm{D}\gamma_{\mathfrak{g}} e(1-S)}{\mathrm{D}_{\mathfrak{s}} t} = \frac{\mathrm{D}\gamma_{\mathfrak{g}} nS}{\mathrm{D}_{\mathfrak{s}} t}$$

$$= -[\gamma_{\mathfrak{g}}\theta_{\mathfrak{g}}(v_{\mathfrak{g}\alpha} - v_{\mathfrak{s}\alpha})]_{,\alpha}$$

$$(9.2.12) \qquad \theta_{\mathfrak{s}} + \theta_{\mathfrak{w}} + \theta_{\mathfrak{g}} = 1$$

Eqs. (9.3.8) and (9.3.9) are accounts of the balance of flow of water and air, respectively, moving along with a material element, while Eq. (9.2.12) is the equation of equipresence of volumes in a three-constituent mixture. In the study of consolidation of unsaturated soils Barden (1965) used the same equation in terms of the porosity n. However, all terms are equivalent and we will use here the most customary term, the void ratio e, and follow the work of Philip and Smiles (1969), with minor changes.

The material derivatives on the left-hand side of Eqs. (9.3.8) and (9.3.9) may be transformed into spatial derivatives

$$\rho_s \frac{D\dfrac{\rho_w}{\rho_s}}{D_s t} = \rho_s \frac{\partial\dfrac{\rho_w}{\rho_s}}{\partial t} + \rho_w v_{s\alpha,\alpha} = \frac{\partial \rho_w}{\partial t} - \frac{\rho_w}{\rho_s}\frac{\partial \rho_s}{\partial t} + \rho_w v_{s\alpha,\alpha} = \frac{\partial \rho_w}{\partial t}$$

$$= \frac{1}{g}\frac{\partial\dfrac{\gamma_w Se}{1+e}}{\partial t} = -[\rho_w(v_{w\alpha} - v_{s\alpha})]_{,\alpha} \tag{11.15.1}$$

$$\rho_s \frac{D\dfrac{\rho_g}{\rho_s}}{D_s t} = \rho_s \frac{\partial\dfrac{\rho_g}{\rho_s}}{\partial t} + \rho_g v_{s\alpha,\alpha} = \frac{\partial \rho_g}{\partial t} - \frac{\rho_g}{\rho_s}\frac{\partial \rho_s}{\partial t} + \rho_g v_{s\alpha,\alpha} = \frac{\partial \rho_g}{\partial t}$$

$$= \frac{1}{g}\frac{\partial\dfrac{\gamma_w(1-S)e}{1+e}}{\partial t} = -[\rho_g(v_{g\alpha} - v_{s\alpha})]_{,\alpha} \tag{11.15.2}$$

In Eqs. (11.15.1) and (11.15.2) we considered the fact that for incompressible solids, $\partial\gamma_s/\partial t = 0$, the ratio of the rate of change of volume of the solids fraction to the volume of the solids fraction is equal to the spatial change of the velocity of the solids,

$$\frac{1}{\theta_s}\frac{\partial \theta_s}{\partial t} = \frac{1}{\rho_s}\frac{\partial \rho_s}{\partial t} = v_{s\alpha,\alpha}$$

We substitute the value of ϑ from Eq. (9.2.13) into Eq. (11.15.1) and obtain

$$\frac{\partial\dfrac{\gamma_w Se}{1+e}}{\partial t} = \frac{\partial\dfrac{\gamma_w \vartheta}{1+e}}{\gamma t} = -\left[\gamma_w \frac{\vartheta}{1+e}(v_{w\alpha} - v_{s\alpha})\right]_{,\alpha} \tag{11.15.3}$$

In the process of losing water, three ranges* of the degree of saturation S of soils are distinguishable:

1. The *range of saturation*, where the soil is saturated $S = 1$ and the saturation is maintained in spite of the change in the moisture content θ_w until the moisture content reaches the equivalent value of the water content w, slightly above the shrinkage limit. As already pointed out, in the range of

* The concept of ranges and the term *range* was introduced by Philip and Smiles (1969). The range of saturation, however, was named *normal range*.

saturation a linear relation exists between the water content w and the void ratio e, and the dependence of the void ratio e on the moisture content θ_w is obtained from Eq. (9.2.10)

$$e = \vartheta_{S=1} = \frac{\theta_w}{1 - \theta_w} \quad \text{for } S = 1 \tag{11.15.4}$$

for which we also have

$$\frac{de}{d\theta_w} = (1 - \theta_w)^{-2} \tag{11.15.5}$$

2. The *residual range* is the range where the soil is below saturation, $S < 1$, and it contains the shrinkage limit. A non-linear relation exists between the void ratio e and the water content w, and the void ratio and the moisture content θ_w

$$e = \frac{\theta_w}{S - \theta_w} > \frac{\theta_w}{1 - \theta_w} \quad \text{for } S < 1 \tag{11.15.6}$$

and

$$0 < \frac{de}{d\theta_w} = S(S - \theta_w)^{-2} = \frac{(1 + e)^2}{S} \tag{11.15.7*}$$

$$0 < \frac{d[(1 + e)\,\theta_w]}{d(1 + e)} = S \leq 1 \tag{11.15.8}$$

Eq. (11.15.8) states that any small change in the bulk volume of the soil, $(1 + e)$ per unit volume of solids, produces a change in the volume of the water, $(1 + e)\theta_w$ per unit volume of solids, equal to the degree of saturation itself.

3. The *zero range* where the soil does not change its volume with the change of the moisture content

$$e = \text{const} = e_{\min} = \left(\frac{\theta_w}{S - \theta_w}\right)_{\min} = \frac{G_g w}{G_w S} \tag{11.15.9}$$

from which it follows

$$\frac{de}{d\theta_w} = \frac{S}{(S - \theta_w)^2} = 0 \quad \therefore S = 0,\ \theta > 0 \text{ and } e > 0 \tag{11.15.10}$$

since the void ratio e is always positive.

Most points discussed in Sects. 11.4 and 11.5 with respect to saturated soils are relevant to unsaturated soils as well. Assuming that the soils and water are incompressible so that γ_g and γ_w are independent of the time t or the

* Philip and Smiles add also the constraint $de/d\theta_w < (1 + e)/(1 - \theta_w)$, which is unacceptable since that would lead to $S > 1$, while necessarily $S \leq 1$.

space x, and that the permeability $k(\theta_w)_{ij}$ is a function of the moisture content θ_w, we obtain

$$\frac{\partial \theta_w}{\partial t} = \frac{\partial \dfrac{Se}{1+e}}{\partial t} = -[\theta_w(v_{w\alpha} - v_{s\alpha})]_{,\alpha}$$

$$= -\left[\frac{\vartheta}{1+e}(v_{w\alpha} - v_{s\alpha})\right]_{,\alpha}$$

$$= [k(\theta_w)_{\beta\alpha}\,\Phi(\theta_w)_{,\alpha}]_{,\alpha} \tag{11.15.11}$$

Eq. (11.15.11) contains two parameters, the moisture content θ_w and the moisture potential $\Phi(\theta_w)$, and it is assumed that the moisture potential is a function of the moisture content and vice versa. Assuming that $\Phi(\theta_w)$ is a unique function of θ_w (no hysteresis), the transformation

$$\frac{\partial \theta_w}{\partial t} = \frac{\partial \Phi(\theta_w)_w}{\partial t}\,\frac{\partial \theta_w}{\partial \Phi(\theta_w)_w} \tag{11.15.12}$$

can be used in Eq. (11.15.11), and we obtain

$$\frac{\partial \Phi(\theta_w)_w}{\partial t} = \frac{\partial \Phi(\theta_w)_w}{\partial \theta_w}\,[k(\theta_w)_{w\beta\alpha}\,\Phi(\theta_w)_{w,\alpha}]_{,\alpha} \tag{11.15.13}$$

Opening the brackets and abandoning the functionality notation of the suction potential a simpler form of Eq. (11.15.13) can be obtained

$$\frac{\partial \Phi_w}{\partial t} = \frac{k_{\beta\alpha,\beta}}{\gamma_w}\,\Phi_{w,\alpha} + \frac{k_{\beta\alpha}}{\gamma_w}\,\Phi_{w,\alpha\beta} = D(\theta_w)'_\alpha\Phi_{w,\alpha} + D(\theta_w)''_{\alpha\beta}\Phi_{w,\alpha\beta} \tag{11.15.14}$$

$$\Phi_w = \Phi(\theta_w)_w; \quad k_{ij} = k(\theta_w)_{ij}$$

where $D(\theta_w)'_i$ and $D(\theta_w)''_{ij}$ are the diffusion coefficients, functions of the moisture content θ_w, defined

$$D(\theta)_w)'_i = \frac{k_{\beta i,\beta}}{\gamma_w}; \quad D(\theta_w)''_{ij} = \frac{k_{ij}}{\gamma_w} \tag{11.15.15}$$

Eq. (11.15.13) may be expressed in terms of the void ratio e by simple transformation, given in Eq. $(11.13.2)_1$

$$\frac{\partial \Phi(e)_w}{\partial t} = \frac{k_{\beta\alpha,\beta}}{\gamma_w}\,\frac{1}{\partial\theta_w/\partial e}\,\Phi(e)_{w,\alpha} + \frac{\kappa_{\beta\alpha}}{\gamma_w}\,\frac{1}{\partial\theta_w/\partial e}\,\Phi(e)_{w,\alpha\beta}$$

$$= D'_{e\alpha}\,\Phi_{w,\alpha} + D''_{e\alpha\beta}\Phi_{w,\alpha\beta} \tag{11.15.16}$$

$$\Phi_w = \Phi(e)_w; \quad k_{ij} = k(e)_{ij}$$

where D'_{ei} and D''_{eij} are the diffusion coefficients, functions of the void ratio e

$$D'_{ei} = \frac{k_{\beta i,\beta}}{\gamma_w} \frac{1}{\partial\theta_w/\partial e}; \quad D''_{eij} = \frac{k_{ij}}{\gamma_w} \frac{1}{\partial\theta_w/\partial e} \tag{11.15.17}$$

Eqs. (11.15.14) and (11.15.16) are the *general equations of consolidation* of soils in terms of the moisture content θ_w and the void ratio e, respectively, valid for saturated and unsaturated soils. They have the form of wave equations and their solution is by numerical methods; a solution in closed form is at present not in sight.

The solution of Eq. (11.15.14) is more straightforward than that of Eq. (11.15.16), when the suction curve $\Phi(\theta_w)$ and the coefficient of permeability curve $k(\theta_w)$ are available. Eq. (11.15.16), in terms of the void ratio e which is more familiar in soil mechanics, can also be solved, since $\Phi(e)$ and $k(e)$ are easily obtainable.

These equations encompass the whole range of degrees of saturation, as noted before.

The Saturation Range

In saturated soils the pore-water pressure is proportional to the moisture potential, $\rho(\theta_w)_w = \gamma_w\Phi(\theta_w)$, and by substituting the pore-water pressure in Eq. (11.15.14) we obtain, in terms of the moisture content θ_w

$$\frac{\partial \rho_w}{\partial t} = \frac{k_{\beta\alpha,\beta}}{\gamma_w} p_{w,\alpha} + \frac{k_{\beta\alpha}}{\gamma_w} p_{w,\alpha\beta} = D(\theta_w)'_\alpha p_{w,\alpha} + D(\theta_w)''_{\alpha\beta} p_{w,\alpha\beta} \tag{11.15.18}$$

$$p_w = p(\theta_w)_w; \quad k_{ij} = k(\theta_w)_{ij}$$

In terms of the void ratio e, Eq. (11.15.18) becomes

$$\frac{\partial p_w}{\partial t} = \frac{k_{\beta\alpha,\beta}}{\gamma_w} \frac{1}{\partial\theta_w/\partial e} p_{w,\alpha} + \frac{k_{\beta\alpha}}{\gamma_w} \frac{1}{\partial\theta_w/\partial e} p_{w,\alpha\beta}$$

$$= D(e)'_\alpha p_{w,\alpha} + D(e)''_{\alpha\beta} p_{w,\alpha\beta} \tag{11.15.19}$$

$$p_w = p(e)_w; \quad k_{ij} = k(e)_{ij}$$

where, applying Eq. (11.15.5), the diffusion coefficients will be

$$D(e)'_j = \frac{(1+e)^2}{\gamma_w} k_{\beta j,\beta}; \quad D(e)''_{ij} = \frac{(1+e)^2}{\gamma_w} k_{ij}; \quad k_{ij} = k(e)_{ij} \tag{11.15.20}$$

It should be noted that although the soil is saturated the coefficient of permeability $k(e)_{ij}$ can vary, since the soil can be in any of the states of the capillary zone.

If Eqs. (11.15.16) or (11.15.19) are applied to either uniaxial consolidation or triaxial consolidation where the drainage path is unidirectional in the direction Z, the coefficient of permeability becomes $k(\mathrm{e})_{ij} = k(\mathrm{e})_{zz} = k(\mathrm{e})$.

The Residual Range

In the residual range, when the soil is unsaturated $0 < S < 1$, Eq. (11.15.16) holds, and if Eq. (11.15.7) is substituted into Eqs. (11.15.17) the diffusion coefficients become

$$D'_{\mathfrak{e}j} = \frac{(1+e)^2}{S\gamma_{\mathfrak{w}}} k_{\beta j,\beta}; \quad D''_{\mathfrak{e}ij} = \frac{(1+e)^2}{S\gamma_{\mathfrak{w}}} k_{ij}; \quad k_{ij} = k(e)_{ij} \tag{11.15.21}$$

The Zero Range

In the zero range where $\partial e/\partial \theta_{\mathfrak{w}} = 0$, it is evident from Eq. (11.15.16) that

$$\frac{\partial \Phi_{\mathfrak{w}}}{\partial t} = 0 \tag{11.15.22}$$

meaning that the change of moisture potential is zero.

The fact that the equations of consolidation of unsaturated soil seem to be independent of the air content of the soil is interesting. While they do not include explicitly the parameters of the air and their effect on consolidation, in the suction potentials $\Phi(\theta_{\mathfrak{w}})$ or $\Phi(e)$ and in the permeability functions $k(\theta_{\mathfrak{w}})$ or $k(e)$ the effects of air inclusions in the soil are lumped together.

The equations presented in this section can help solve a series of problems of swelling and consolidation when proper boundary conditions are applied (Philip 1968a,b, 1969a,b; Smiles and Rosenthal 1968; Philip and Smiles 1969; Smiles and Poulos 1969). It must be realized, however, that while the soil passes through the different ranges of moisture content the boundary conditions vary as well, and in most cases only numerical solutions can surmount the computational dificulties.

11.16 Hysteresis

In Sect. 11.13 and earlier in Sect. 8.27 we introduced the concept of *hysteresis*, defining the relation between suction and moisture content in soil, or, respectively, between the spherical pressure and the volume strain. Repeated stressing and relaxing cycles result in *hysteretic loops*. Hysteretic loops are always connected with energy dissipation, which cannot be uniquely described by one to one valued functions. Similar phenomena occur in ferroelectricity and ferromagnetism, but the most familiar is the hysteresis observed when a material is subject to repeated loadings in a series of loading–unloading cycles. In terms of a stress–strain diagram, loops formed by consecutive segments of the diagram enclose areas equal to the energy consumed during the respective cycles. Hysteresis always implies irreversibility in the sequence of states, which is in general traced to the molecular structure of the material.

Fallacy. Hysteresis occurs when shear stresses are applied.

Shear stresses and strains are by nature completely dissipative, therefore they cannot sustain hysteretic loops. On the other hand, all hysteretic phenomena that we have discussed are related to spherical phenomena, encountered in soils and materials in general. The confusion in many textbooks is a result of inappropriate testing methods, such as unidirectional or tridirectional compression or tension tests that are in all cases unseparable superpositions of spherical and deviatoric phenomena. This may be the main reason for the lack of breakthrough in solving the problem of hysteresis in a closed mathematical form. The existing theories and solutions based on the assumption of a *domain* model (Enderby 1955; Poulovassilis 1962, 1970; Topp 1969, 1971; Talsma 1970; Mualem 1974, 1984a; Mualem and Dagan 1975) have rarely been considered, due mainly to the following three reasons (Mualem 1984b): the complexity of the hysteresis theories, the large amount of data needed to calibrate the models and the complexity of the numerical solutions of unsaturated flow problems, greatly increased by the changing boundary conditions along the ranges of moisture changes, see Sect. 11.15.

The study of hysteresis requires perhaps a more original insight, directed to thermodynamic concepts, the only approach that can provide a working basis. The predicament which we find ourselves confronted with when handling hysteresis is the fact that the functional relationship between two parameters is not a single-valued relationship. For instance, in the volumetric strain ε_{m} versus spherical pressure p_{m} curves shown in Fig. 11.16.1, for every value of p_{m} we have several values of ε_{m} and vice versa. Such a situation not only puts us in an awkward and exigent position, but presents us also with a quandary as to the deterministic character of nature, which is contradictory to all our scientific endeavor. It seems that the answer lies in the problem itself. What is actually the problem? A compressive spherical pressure p_{m} is applied on a volume element of a material, resulting in a decrease in size and an increase in negative volumetric strain. By reducing the pressure, the volume increases and the strain decreases, and we have thus a *recovery* of the volume together with the decrease in strain. This strain decrease is not commensurate, however, with the strain increase during the compression. The deficiency in the recovery is attributed to energy dissipation due to frictional resistance during the compression and the decompression stages, and because of this energy dissipation there is no justification to expect an equivalence in strains for comparable stresses. If the energy dissipation is considered, what appears to be a multi-valued relation between the stresses and the strains in Fig. 11.16.1 is in reality a single-valued relationship. The functional relationship exists between three parameters – stress, strain and dissipative energy,

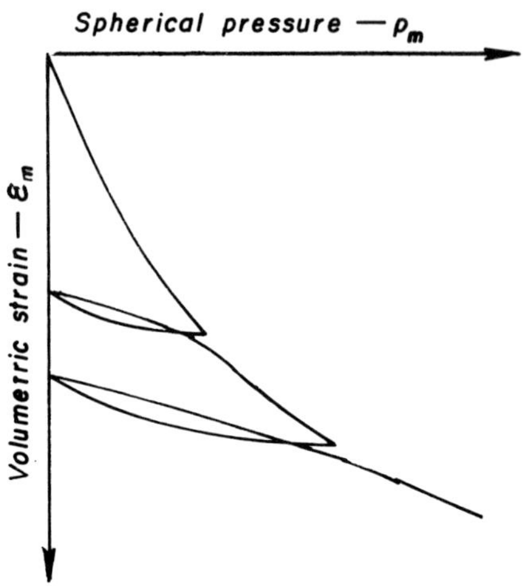

Fig 11.16.1. Hysteresis curve in two dimensions, of volumetric strain ε_m versus spherical pressure p_m.

rather than between the two parameters, stress and strain alone. Fig. 11.16.2 shows the same volumetric strain ε_m versus the spherical pressure p_m in a three-dimensional display, where the third dimension is the *entropy production* ηT. It is evident in Fig. 11.16.2 that the ε_m versus p_m curve in three dimensions is a single-valued spiraling curve, while Fig. 11.16.1 is its projection on the ε_m–p_m plane. Further investigation of the hysteresis should concentrate on the relationship between the three parameters, and the means of determining numerically the energy dissipation.

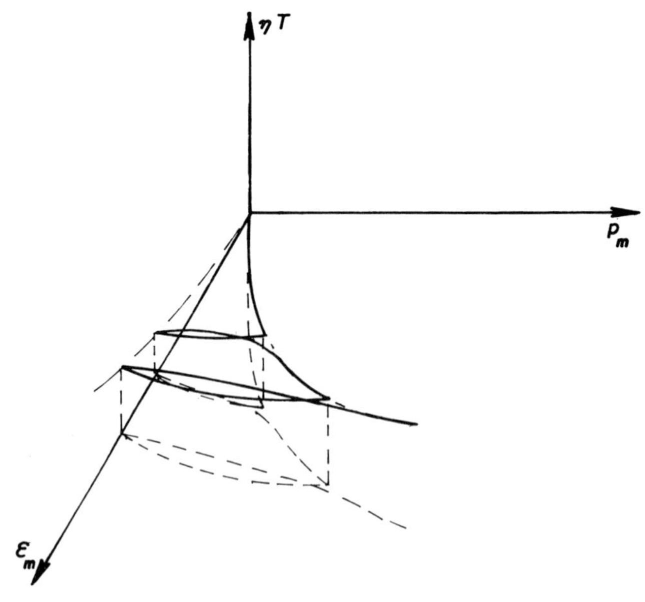

Fig 11.16.2. Hysteresis curve in three dimensions, of volumetric strain ε_m versus spherical pressure p_m shown in Fig. 11.16.1.

Meanwhile it is preferable to assume no hysteresis, as was done in a previous section, or to apply the assumption of hysteresis piecewise to the segments of the hysteresis curve.

11.17 Phenomenological Linear Volumetric Stress–Strain Relationship

It has already been shown that the general *phenomenological* or *macroscopic* stress–strain relationship, as expressed in the rheological equation, (7.3.2), can be studied by assuming either a linear or a non-linear relationship between the stresses and strains and their higher time derivatives. It was also shown that if we assume a linear stress–strain relationship, Eq. (7.3.2) can be resolved into two equations, (7.5.1) and (7.5.2), which is straightforward provided that the requirements with respect to the strain tensor, discussed in Sect. 11.1, are satisfied. Thus a general volumetric stress–strain relationship in the form of Eq. (7.6.6) represents a volumetric viscoelastic behavior.

Eq. (7.6.6) can be portrayed by a *model* of $\mathfrak{I}$ volumetric *Kelvin bodies* coupled in series, Fig. 11.17.1. A short review of rheological modeling can be found in Appendix C. Actually Eq. (7.6.6) was derived so as to correspond to the model in Fig. 11.17.1, which is a replica of the strains occurring in an ideal viscoelastic material. The model contains two degenerated Kelvin bodies, one resulting in a Hookean spring and the other in a Newtonian dashpot.

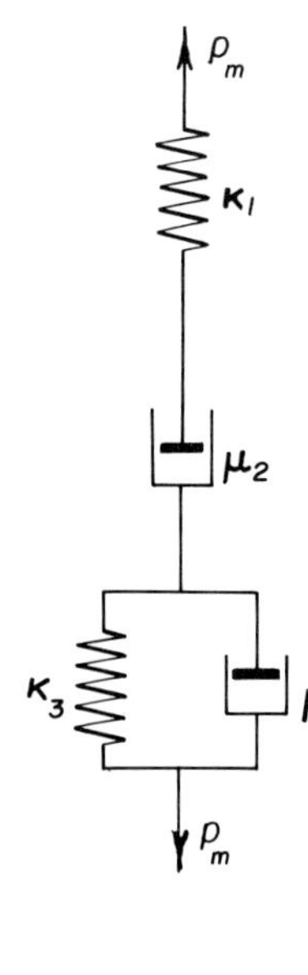

Fig 11.17.1. A three-element degenerated Kelvin series.

11.18 Modeling the Linear Volumetric Stress–Strain Relationship

Observing the volumetric strains of soils, we can say that if a constant spherical pressure $p_m = \frac{1}{3} p_{\alpha\alpha} = \text{const}$ is applied to a soil volume the volume will undergo a strain, made up of an *instantaneous elastic* strain, a *time-dependent unrecoverable* strain and a *time-dependent recoverable* strain, all of volumetric nature. It should be clear that the volumetric stress–strain relationship affects the density of the soil: when a spherical pressure is applied, the volume of the soil decreases while the density increases, and when the pressure is removed the volume rebounds and increases while the density decreases. In other words, the loading–unloading cycle results in an increase in the density of the soil, which is, however, only partly recoverable. We assert also that for a finite pressure such as p_m a finite volumetric strain $\varepsilon_{\alpha\alpha}$ is bound to exist. Phenomenologically, there are no shear stresses involved and consequently there are no distortions; the sample simply compresses or distends, in thermodynamic terms its internal energy increases or decreases, respectively.

When applying a rheological model to describe the volumetric strain of soils there are five requirements:

1. The total strain must contain an instantaneous recoverable strain.
2. The total strain must contain a time-dependent recoverable strain.
3. The total strain must contain a time-dependent unrecoverable strain.
4. The total strain must be bounded by an upper strain limit, for an applied constant stress.
5. In a loading–unloading stress cycle the strain manifests a hysteretic behavior. (This can also be deduced from point 3).

In rheological modeling we distinguish between the actual behavior of the soil and that of the model, which is an idealization of the actual behavior. We understand that the model which conforms with the five requirements outlined above is a *visco-elasto-plastic* model, Fig. 11.18.1, composed of a Burgers body of four coefficients, coupled in parallel with a St Venant element, see Appendices C.5 and C.9. The four coefficient Burgers body coupled with the St Venant element is the absolute minimal model necessary to represent volumetric stress–strain relations in soils, as well as in any other material.

The visco-elasto-plastic model includes a Hookean element with instantaneous strain and a viscoelastic Kelvin body which provides a time-dependent strain, and is the basic model of consolidation. These two elements are fully recoverable, which compels us to include the Newtonian viscous element. This element, however, provides a continuous strain increment not bounded by an upper limit, as required by point 4 above, even when a constant spherical stress is applied and such a limit is anticipated. To remedy this we include the St Venant element, which dissipates part of the energy of the

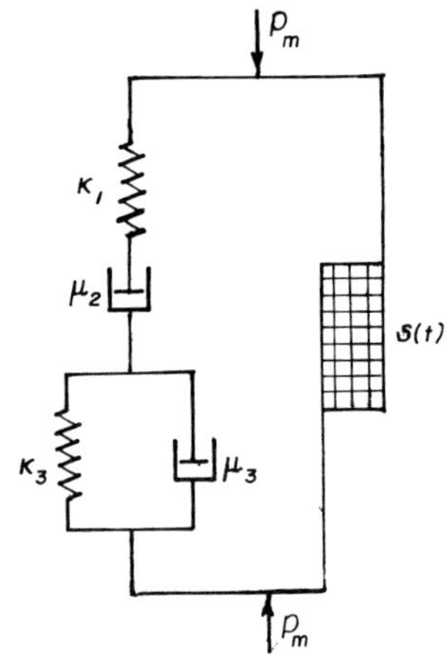

Fig 11.18.1. A three-element degenerated Burgers model coupled with a St Venant element.

applied stress, reducing it to an *activating stress*, and keeps the strain bounded. It also reduces the instantaneous and viscoelastic strains, so that the recovered strains are smaller than the initial strain, thus inducing *hysteresis*, as actually observed. If properly chosen, the St Venant element provides for the plastic restraint and also for the thixotropic regain with time.

If we substitute $[p(t)_{\mathrm{m}} - p_0\vartheta(t)]$ for the pressure $p(t)_{\mathrm{m}}$, Eqs. (7.6.8) and (7.6.10) will look as follows

$$\begin{aligned}\varepsilon(t)_{\alpha\alpha} &= \frac{[p(t)_{\mathrm{m}} - p_0\vartheta(t)]}{\varkappa_1} + \frac{1}{\mu_2}\int_0^t [p(\tau)_{\mathrm{m}} - p_0\vartheta(\tau)]\,\mathrm{d}\tau \\ &\quad + \frac{1}{\mu_3}\exp\frac{-t}{T_3}\int_0^t [p(\tau)_{\mathrm{m}} - p_0\vartheta(\tau)]\exp\frac{\tau}{T_3}\,\mathrm{d}\tau \\ &= \frac{p(t)_{\mathrm{m}}}{\varkappa_1}\left(1 - \frac{p_0}{p(t)_{\mathrm{m}}}\,\vartheta(t)\right) + \frac{1}{\mu_2}\int_0^t p(\tau)_{\mathrm{m}}\left(1 - \frac{p_0}{p(\tau)_{\mathrm{m}}}\,\vartheta(\tau)\right)\mathrm{d}\tau \\ &\quad + \frac{1}{\mu_3}\exp\frac{-t}{T_3}\int_0^t p(\tau)_{\mathrm{m}}\left(1 - \frac{p_0}{p(\tau)_{\mathrm{m}}}\,\vartheta(\tau)\right)\exp\frac{\tau}{T_3}\,\mathrm{d}\tau \end{aligned} \tag{11.18.1}$$

$$\begin{aligned}\dot{\varepsilon}(t)_{\alpha\alpha} &= \frac{1}{\varkappa_1}[\dot{p}(\mathrm{t})_{\mathrm{m}} - p_0\dot{\vartheta}(t)] + \frac{p(t)_{\mathrm{m}}}{\mu_2}\left(1 - \frac{p_0}{p(t)}\,\vartheta(t)\right) \\ &\quad + \frac{p(t)_{\mathrm{m}}}{\mu_3}\left(1 - \frac{p_0}{p(t)_{\mathrm{m}}}\,\vartheta(t)\right) \\ &\quad - \frac{1}{\mu_3 T_3}\exp\frac{-t}{T_3}\int_0^t p(\tau)_{\mathrm{m}}\left(1 - \frac{\dot{p}_0}{p(\tau)_{\mathrm{m}}}\,\vartheta(\tau)\right)\exp\frac{\tau}{T_3}\,\mathrm{d}\tau \end{aligned} \tag{11.18.2}$$

where $T_3 = \mu_3/\varkappa_3$ is the volumetric retardation time of the time-dependent recoverable strain rate.

The time-dependent recoverable part of the strain rate is

$$\dot{\varepsilon}(t)_{\mathrm{p}} = \frac{p(t)_{\mathrm{m}}}{\mu_3}\left(1 - \frac{p_0}{p(t)_{\mathrm{m}}}\,\vartheta(t)\right)$$

$$- \frac{1}{\mu_3 T_3}\exp\frac{-t}{T_3}\int_0^t p(\tau)_{\mathrm{m}}\left(1 - \frac{p_0}{p(\tau)_{\mathrm{m}}}\,\vartheta(\tau)\right)\exp\frac{\tau}{T_3}\,\mathrm{d}\tau \tag{11.18.3}$$

from which the corresponding pore pressures p_{p} can be computed

$$p_{\mathrm{p}} = \mu_3\dot{\varepsilon}(t)_{\mathrm{p}} = -\frac{1}{T_3}\exp\frac{-t}{T_3}\int_0^t p(\tau)_{\mathrm{m}}\left(1 - \frac{p_0}{p(\tau)_{\mathrm{m}}}\,\vartheta(\tau)\right)\exp\frac{\tau}{T_3}\,\mathrm{d}\tau \tag{11.18.4}$$

It is evident that only the Kelvin element contributes to the pore pressure.

11.19 Constant Spherical Pressure

The total volumetric strain, which results from a spherical pressure $p_{\mathrm{m}} = \mathcal{P} = \mathrm{const}$ and corresponds to the above model, is obtained from Eqs. (7.6.8) and (11.18.1) and will be at any time t

$$3\varepsilon(t)_{\mathrm{m}} = \varepsilon(t)_{\alpha\alpha} = \varepsilon(t)_{\alpha\alpha 1} + \varepsilon(t)_{\alpha\alpha 2} + \varepsilon(t)_{\alpha\alpha 3}$$

$$= \frac{\mathcal{P} - p_0\vartheta(t)}{\varkappa_1} + \frac{\mathcal{P}}{\mu_2}\int_0^t\left(1 - \frac{p_0}{\mathcal{P}}\,\vartheta(\tau)\right)\mathrm{d}\tau$$

$$- \frac{\mathcal{P}}{\mu_3}\exp\frac{-t}{T_3}\int_0^t\left(1 - \frac{p_0}{\mathcal{P}}\,\vartheta(\tau)\right)\exp\frac{\tau}{T_3}\,\mathrm{d}\tau$$

$$p_{\mathrm{m}} = \mathcal{P} = \mathrm{const} \tag{11.19.1}$$

where $\varepsilon_{\alpha\alpha 1}$, $\varepsilon_{\alpha\alpha 2}$ and $\varepsilon_{\alpha\alpha 3}$ are the *elastic*, *viscous* and *viscoelastic* parts of the total strain, respectively, $\varkappa_1$ and $\varkappa_3$ are *volumetric bulk moduli*, μ_2 and μ_3 are the *volumetric viscous coefficients* or the *bulk coefficients of viscosity*, $T_3 = \mu_3/\varkappa_3$ is the *volumetric retardation time* or the *bulk retardation time*, $\vartheta(t)$ is a time-dependent *plastic restraint* and p_0 is the *coefficient of plastic restraint*.

Fig. 11.19.1 shows the individual contributions of the elastic, viscous and viscoelastic strains to the total strain, which is due to the particular constant spherical pressure of a Burgers model, that is, the visco-elasto-plastic model without the plastic restraint. It is seen that the contribution of the Hookean elastic element is a strain which remains constant for the entire duration of the pressure, of the Newtonian viscous element a strain increasing linearly

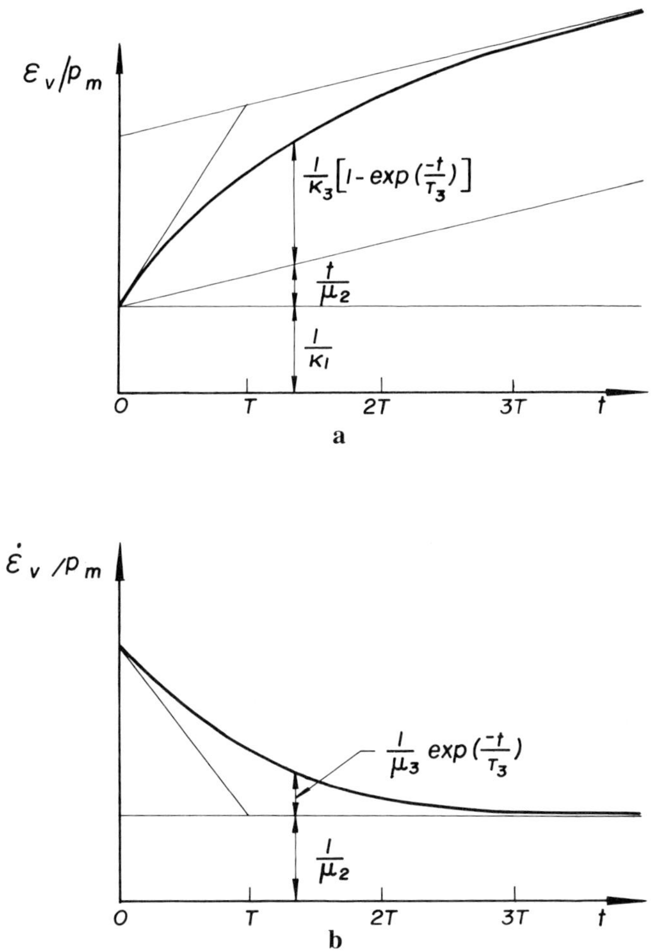

Fig 11.19.1. Response of a Burgers model. **a** Volumetric strain versus time. **b** Volumetric strain rate versus time.

with time, and of the Kelvin viscoelastic body a strain increasing exponentially with time. If the spherical pressure is not constant but a function of time t, the volumetric strain is obtained through Eq. (7.6.8).

With the removal of the pressure p_m in the Burgers model a partial recovery is taking place. The elastic strain is recovered instantaneously and the viscoelastic strain is also fully recovered, asymptotically with time at $t = \infty$; the viscous strain, however, is retained and remains permanent, having reached its final value just before the removal of the pressure.

We believe, with certain reservations to be discussed later, that at least a four coefficient model is necessary to describe properly the volumetric stress–strain relationship of soils. An additional Kelvin body with two additional coefficients will, perhaps, give better results and add another recoverable strain,

$$\varepsilon_{\alpha\alpha 4} = \frac{1}{\mu_4} \exp \frac{-t}{T_4} \int_0^t p(\tau)_{\mathrm{m}} \left[1 - \frac{p_0}{p(\tau)_{\mathrm{m}}} \vartheta(\tau)\right] \exp \frac{\tau}{T_4} \, \mathrm{d}\tau$$

It is worth noting that while the elastic strain $\varepsilon_{\alpha\alpha 1} = p_{\mathrm{m}}/\varkappa_1$ is instantaneous and remains unchanged, and the viscoelastic strain is bounded and approaches asymptotically the value $\varepsilon_{\alpha\alpha 3} = p_{\mathrm{m}}/\varkappa_3$ at time $t = \infty$, the viscous strain continues to increase indefinitely, proportional with time. A model of only an elastic element and a viscoelastic Kelvin body coupled in series will always recover its strain after the removal of the load. This is the reason why a viscous element is also needed in the model. However, when we include the viscous element we see that the strain continues indefinitely, and this is in discord with the fourth requirement, of a bounded strain for a finite pressure.

A closer observation of experimental results reveals further facts of the volumetric stress–strain relationship. Actually, neither the instantaneous elastic strain nor the viscoelastic strain is fully recoverable, and therefore it can be surmised that in addition to the energy dissipated due to viscosity, part of the energy produced by the pressure p_{m} dissipates as well. Since p_{m} is a "pressure" of dissipative nature, more like a plastic resistance, it has been termed *volumetric plastic restraint* and denoted $p_{\mathrm{m}}\vartheta(t)$ (Klausner 1961), where $\vartheta(t)$ is the coefficient of plastic restraint, the exact form of which is yet to be found.

The volumetric strain rate is derived from Eq. (11.19.1)

$$3\dot{\varepsilon}(t)_{\mathrm{m}} = -\frac{p_0}{\varkappa_1}\dot{\vartheta}(t) - \left(\frac{\mathcal{P}}{\mu_2} + \frac{\mathcal{P}}{\mu_3}\right)\left(1 - \frac{p_0}{\mathcal{P}}\vartheta(t)\right) + \frac{\mathcal{P}}{\mu_3 T_3}\exp\frac{-t}{T_3}\int_0^t \left(1 - \frac{p_0}{\mathcal{P}}\vartheta(\tau)\right)\exp\frac{\tau}{T_3}\,\mathrm{d}\tau \qquad (11.19.2)$$

The plastic restraint is a function of time, and therefore it induces non-steady elements in the equations. Thus a creep experiment with a constant pressure $\mathcal{P}$, which seems to be a steady-state problem, turns out to be a non-steady problem varying with time, $\mathcal{P} - p_0\vartheta(t)$.

Since in triaxial consolidation the sample is under constant hydrostatic pressure, Eqs. (11.19.1) and (11.19.2) which consider such a pressure are the practical equations for the strain and strain rate in the phenomenological approach.

11.20 General Spherical Pressure

For the sake of completeness, the general case of a four parameter volumetric stress–strain relation, where the spherical stress applied is a function of time, $p(t)_{\mathrm{m}}$, is also presented.

Including the coefficient of plastic restraint in Eq. (7.6.6), we obtain

$$3\varepsilon(t)_{\mathrm{m}} = \varepsilon(t)_{\alpha\alpha} = \varepsilon(t)_{\alpha\alpha 1} + \varepsilon(t)_{\alpha\alpha 2} + \varepsilon(t)_{\alpha\alpha 3} = \frac{p(t)_{\mathrm{m}} - p_0\vartheta(t)}{\varkappa_1} +$$

$$+ \frac{1}{\mu_2}\int_0^t (p(\tau)_{\mathrm{m}} - p_0\vartheta(\tau))\,\mathrm{d}\tau$$

$$+ \sum_{i=3}^{\mathfrak{I}} \frac{1}{\mu_i}\exp\frac{-t}{T_i}\int_0^t (p(\tau)_{\mathrm{m}} - p_0\vartheta(\tau))\exp\frac{\tau}{T_i}\,\mathrm{d}\tau$$

$$= \frac{p(t)_{\mathrm{m}}}{\varkappa_1}\left(1 - \frac{p_0}{p(t)_{\mathrm{m}}}\vartheta(t)\right) + \frac{1}{\mu_2}\int_0^t p(\tau)_{\mathrm{m}}\left(1 - \frac{p_0}{p(\tau)_{\mathrm{m}}}\vartheta(\tau)\right)\mathrm{d}\tau$$

$$+ \sum_{i=3}^{\mathfrak{I}} \frac{1}{\mu_i}\exp\frac{-t}{T_i}\int_0^t p(\tau)_{\mathrm{m}}\left(1 - \frac{p_0}{p(\tau)_{\mathrm{m}}}\vartheta(\tau)\right)\exp\frac{\tau}{T_i}\,\mathrm{d}\tau \quad (11.20.1)$$

and the strain rate becomes

$$3\dot{\varepsilon}(t)_{\mathrm{m}} = \frac{1}{\varkappa_1} + \frac{p(t)_{\mathrm{m}}}{\mu_2}\left(1 - \frac{p_0}{p(t)}\vartheta(t)\right)$$

$$+ \sum_{i=3}^{\mathfrak{I}} \frac{p(t)_{\mathrm{m}}}{\mu_i}\left(1 - \frac{p_0}{p(t)_{\mathrm{m}}}\vartheta(t)\right)$$

$$- \sum_{i=3}^{\mathfrak{I}} \frac{1}{\mu_i T_i}\exp\frac{-t}{T_i}\int_0^t p(\tau)_{\mathrm{m}}\left(1 - \frac{\dot{p}_0}{p(\tau)_{\mathrm{m}}}\vartheta(\tau)\right)\exp\frac{\tau}{T_i}\,\mathrm{d}\tau \quad (11.20.2)$$

The medium thus defined is the *visco-elasto-plastic continuum*, a linear medium, since it was derived from the linear visco-elastic continuum.

11.21 The Volumetric Plastic Restraint

The introduction of the plastic restraint provides an expression for an additional dissipative energy in the process of consolidation. This should enable us to render the volumetric strain as a bounded function, possessing a non-recoverable part.

In order to support the introduction of the plastic restraint $p_0\vartheta(t)$, two explanations are given:

1. The work of compression does not proceed in a steady environment, say, of constant internal energy. As the material is being compressed (consolidated) its properties continuously change. The internal energy increases and the relative contribution and effect of the pressure $p(t)_{\mathrm{m}}$ is reduced accordingly. Therefore, $p_0\vartheta(t)$ is the part of the pressure by which $p(t)_{\mathrm{m}}$ is reduced, and $p(t)_{\mathrm{m}} - p_0\vartheta(t)$ becomes the *activating pressure* that actually strains the material. This activating pressure decreases with time and creates a non-steady-state condition.

2. The plastic restraint $p_0\vartheta(t)$, acting as a *fudge* function, reduces the effect of the pressure $p(t)_m$ applied to the element as the time of consolidation proceed, and compels the volume strain to satisfy the requirements of irreversibility and of confined limits.

It is necessary to remark here that although $p_0\vartheta(t)$ has the dimensions of pressure, it is not a pressure which can do work. It is aroused only upon the application of the pressure $p(t)_m$ and is merely a resistance to it.

The mathematical form of the plastic restraint is still open to speculation. It should be derived and deduced from experimental evidence, and correspond also to the physical concepts, which indicated, as we have seen, that the plastic restraint controls the mathematical formulation of consolidation. Thus it is clear that there exists a linkage between the degree of consolidation and the plastic restraint, the latter increasing as the consolidation and the effective pressures increase.

The minimal requirements of the functional form of the plastic restraint can be derived from the above considerations. For normally consolidated soils, the plastic restraint should start from zero at the outset of the stress application and increase monotonically and asymptotically to 1 at the end of the consolidation process

$$\vartheta(0) = 0; \quad \vartheta(\infty) = 1; \quad \frac{d\vartheta(\infty)}{dt} = 0 \tag{11.21.1}$$

When the load is removed at time $t = t_1$, the activating pressure left to produce the recovery is $p(t_1)_m - p_0\vartheta(t_1)$ and the recovery will proceed according to this pressure.

Several functional forms satisfying the above requirements are suggested, see Appendix C.14. It would be desirable if $\vartheta(t)$, in addition to satisfying the mathematical requirements and providing the physical interpretation through heuristic considerations at the phenomenological level, could also be derived from thermodynamic considerations of the intricate interactions in the soil system, at the structural or microscopic level.

11.22 Effective Pressure

From the total pressure, Eq. (9.4.8), derived from the linear momentum balance equation of the soil–water–air system, we obtain, assuming the absence of kinetic effects

$$\tfrac{1}{3}p_{\mathfrak{s}\alpha\alpha} = \tfrac{1}{3}p_{\alpha\alpha} - p_{\mathfrak{w}} - p_{\mathfrak{g}} = p_m - p_{\mathfrak{w}} - p_{\mathfrak{g}} \tag{11.22.1}$$

where $p_m = \frac{1}{3}p_{\alpha\alpha}$ is the *total spherical pressure*, $\frac{1}{3}p_{\mathfrak{s}\alpha\alpha}$ is the *quasi-effective pressure*, carried by the solids, $p_{\mathfrak{w}}$ is the *quasi-pore-water pressure*, carried by the water and $p_{\mathfrak{g}}$ is the *quasi-pore-air pressure*, carried by the air. This is a simple equation of the *quasi-effective pressure*, derived from the theory of continuum mechanics of multi-phase media applied to soils, not including inertial effects, interfacial effects, vapor pressures and non-mechanical stresses, and subject to several constraints. We will see later how Eq. (11.22.1) can be extended to obtain the effective pressure, and generalized to include electrochemical effects and suction stresses.

For saturated soils the pore-air pressure p_g vanishes and Eq. (11.22.1) simplifies to

$$\tfrac{1}{3}p_{\mathfrak{s}\alpha\alpha} = \tfrac{1}{3}p_{\alpha\alpha} - p_w \tag{11.22.2}$$

where $\tfrac{1}{3}p_{\mathfrak{s}\alpha\alpha}$ is the actual *effective pressure* (Terzaghi 1923) which depends only on the pore-water pressure. The only existing interfacial stresses between the solids and the water are negligible, and non-mechanical stresses are ignored.

Terzaghi introduced the concept of effective pressure for saturated soils as a logical consequence of his theory of consolidation for saturated soils, confirmed experimentally years later (Rendulic 1937; Taylor 1944; Bishop and Eldin 1950). His formulation, which corresponds to Eq. (11.22.2) in the conventional notation, was

$$\sigma' = \sigma - u_w \tag{11.22.3}$$

where σ is the *total pressure* applied to the soil, σ' is the *effective pressure* and u_w is the *pore water pressure*. The difference between Eqs. (11.22.2) and (11.22.3) is that Eq. (11.22.3) is given in terms of the applied pressure, while Eq. (11.22.2) is given in terms of its spherical component. The concept of effective pressure was soon extended to unsaturated soils as well (Hilf 1956; Bishop 1959; Alpan 1959, 1963; Bishop et al. 1960; Lambe 1960b; Aitchison 1961a, b; Jennings 1960, 1961; Bishop and Donald 1961; Blight 1967, 1983; Fredlund and Morgenstern 1977; Kassif and Ben-Shalom 1971), and the studies resulted in diversified opinions as to the form of its expression and the factors that affect it. Schuurman (1966) presented a most penetrating study on the relations between the air and water pressures in soils with degrees of saturation of 85% and above. A review of the studies revolving around the effective pressure in unsaturated soils was presented by Blight (1983). It can be shown that most of them are special cases of Bishop's equation, formulated and given in the conventional notation

$$\sigma' = (\sigma - u_a) + \chi(u_a - u_w) \tag{11.22.4}$$

where u_a and u_w are the pore-air pressure and the pore-water pressure, respectively and χ is a *pore pressure parameter* dependent on the degree of saturation.

The discussion of the effective pressure is introduced at the end of this chapter because, as we intend to show, it is a volumetric concept, although its implications are apparent only when shear stresses are applied. As such, it is a proper introduction to the next chapter, which will deal with shear stress–strain relations.

While the pore-water pressure u_w in Eqs. (11.22.3) and (11.22.4) is a hydrostatic pressure which is unquestionably a spherical pressure and thus a scalar, the total pressure σ and the effective pressure σ' could be tensorial pressure entities, which can always be decomposed into spherical and shear components. The pore-water pressure u_w affects only the spherical component of the total pressure σ, leaving its shear component intact; thus the effective pressure is different from the total pressure only in its spherical component and not in its shear component.

If we maintain that the effective pressure, rather than the total pressure, controls the shear properties of the soil, it could not possibly be on account of the shear component, which has not been affected, but of the changes that occur in the spherical component. Moreover, if the shear component of the total pressure did not exist, that is, it would be equal to zero, there would still be an effective pressure dependent on the spherical pressure, while if the spherical component of the total pressure vanished there would not be any effective pressure to consider. Consequently, one must realize that the spherical pressure components are the ones that matter, and if only the spherical components of Eq. (11.22.3) are considered it leads necessarily to Eq. (11.22.2).

In Sect. 8.25 the air–water interfacial pressure p_i existing in the solids matrix of the soils was studied. We could consider the air–water interface as a fourth constituent of the soil, and even the water vapor in the air phase and the air bubbles dissolved in the water could be considered as separate constituents and the pressures existing in each of them can then be considered individually. If they are not considered as separate constituents, they can be included in any other constituent and their pressures will be subtracted from the total pressure in Eq. (11.22.1). In any case they affect the effective pressure in the same manner.

In order not to complicate the study too much, we opted for a three-constituent soil and considered the interfacial pressure according to Eq. (6.6.4), as an excess linear momentum supply which emerges as an interconstituent momentum transfer. This was not considered in Eq. (9.4.3). The vapor and air bubbles affect mainly the volume strains of the soil and their effect on the pressures is very limited. Similarly to the interfacial stresses the changes in the osmotic pressure $p_\mathfrak{b}$, discussed in Sects. 8.16–8.18 and 8.21, also affect the total pressure and contribute to the effective pressure as an interconstituent momentum transfer.

The general equation which replaces Eq. (11.22.1) and corresponds more accurately to Eq. (6.6.4) is the *general effective pressure* equation, written

$$
\begin{aligned}
\tfrac{1}{3}p_{\mathfrak{s}\alpha\alpha} &= \tfrac{1}{3}p_{\alpha\alpha} + \rho_\mathfrak{s}\bar{v}_{\mathfrak{s}\alpha}\bar{v}_{\mathfrak{s}\alpha} + \rho_\mathfrak{w}\bar{v}_{\mathfrak{w}\alpha}\bar{v}_{\mathfrak{w}\alpha} + \rho_\mathfrak{g}\bar{v}_{\mathfrak{g}\alpha}\bar{v}_{\mathfrak{g}\alpha} - p_\mathfrak{w} - p_\mathfrak{g} + p_\mathrm{i} - p_\mathfrak{b} \\
&= \tfrac{1}{3}p_{\alpha\alpha} + \rho_\mathfrak{s}\bar{v}_{\mathfrak{s}\alpha}\bar{v}_{\mathfrak{s}\alpha} + \rho_\mathfrak{w}\bar{v}_{\mathfrak{w}\alpha}\bar{v}_{\mathfrak{w}\alpha} + \rho_\mathrm{m}\bar{v}_{\mathfrak{g}\alpha}\bar{v}_{\mathfrak{g}\alpha} \\
&\quad - p_\mathfrak{w} - p_\mathfrak{g} - \left(\frac{1}{R_1} + \frac{1}{R_2}\right)T - p_\mathfrak{b} \qquad (11.22.5)
\end{aligned}
$$

where $\frac{1}{3}p_{\alpha\alpha}$ is a mean spherical pressure of the externally applied total pressure tensor p_{ij}, $\frac{1}{3}p_{\mathfrak{s}\alpha\alpha}$ is the mean effective pressure, $p_{\mathfrak{w}}$ is the pore-water pressure, $p_{\mathfrak{g}}$ is the pore-air pressure, $\rho_{\mathfrak{s}}\bar{v}_{\mathfrak{s}\alpha}\bar{v}_{\mathfrak{s}\alpha}$, $\rho_{\mathfrak{w}}\bar{v}_{\mathfrak{w}\alpha}\bar{v}_{\mathfrak{w}\alpha}$ and $\rho_{\mathfrak{g}}\bar{v}_{\mathfrak{g}\alpha}\bar{v}_{\mathfrak{g}\alpha}$ are the dynamic pressures of the solids, pore-water and pore-air, respectively, $p_{\mathfrak{i}} = [(1/R_1) + (1/R_2)]T$ is the air–water interfacial pressure and $p_{\mathfrak{d}}$ is the osmotic pressure. All these components are spherical pressures and thus Eq. (11.22.5) defines the effective pressure as a spherical pressure.

Dynamic Pressures

The dynamic pressure components have diverse effects. The dynamic solids pressure $\rho_{\mathfrak{s}}\bar{v}_{\mathfrak{s}\alpha}\bar{v}_{\mathfrak{s}\alpha}$ increases the effective pressure as it adds to the pressure of the solids, while the dynamic pore-water and pore-air pressures $\rho_{\mathfrak{w}}\bar{v}_{\mathfrak{w}\alpha}\bar{v}_{\mathfrak{w}\alpha}$ and $\rho_{\mathfrak{g}}\bar{v}_{\mathfrak{g}\alpha}\bar{v}_{\mathfrak{g}\alpha}$ decrease the effective pressure by adding to the pore pressures. It has already been remarked that the directions of the velocities $\bar{v}_{\mathfrak{s}i}$, $\bar{v}_{\mathfrak{w}i}$ and $\bar{v}_{\mathfrak{g}i}$ are of no consequence, since the velocities are squared and therefore the dynamic pressures are always positive. Dynamic spherical pressures occur when dynamic oscillatory loadings or transient loadings are applied to the soil. Earthquakes are excellent examples of the effects of the dynamic spherical pressures. Granular soils, in general more stable in static loadings than highly wet fine grained soils since they have higher shear resistance, prove to be more vulnerable to earthquakes than fine grained soil. The pore-water and pore-air, attaining high accelerations and high mobility due to low permeability coefficients, can cause high momentary increases of the pore pressures. In spite of the small density of the air the dynamic pore-air pressure cannot be ignored, for the air may attain high velocity and its squaring may produce commensurable dynamic pore-air pressures.

Pore-water Pressure

The pore-water pressure $p_{\mathfrak{w}}$ and the other components of the effective pressure act concomitantly on the same surface area, or rather on the same volume element, according to the Fick–Stefan principle of equipresence of constituents discussed in Sect. 6.2. The idea implied by this principle, that the different pressures act on the same volume simultaneously, seemed so inconceivable that some investigators tried to link the *intergranular pressure* and *active pore pressures*, that act each on their own active area of contact, to the effective pressure. However, the pore pressures measured in soils are the ones defined by the Fick–Stefan principle.

Pore-air Pressure

Air dissolved in water comes in very small bubbles, $R \rightarrow \infty$, and from Eq. (8.25.2) we can write

$$p_i = p_w - p_g = -\left.\frac{2T}{R}\right|_{R\to\infty} \to 0 \quad \therefore \; p_g \cong p_w \tag{11.22.6}$$

namely, the pressure of the air dissolved in the water in the form of small bubbles can be considered equal to the pressure of the water.

In addition to small bubbles of air dissolved in the water air fills the voids to various degrees of saturation and at different pressures; so, for instance, air could be locked in a compacted soil at pressures higher than atmospheric pressure. When the air constituent forms a continuous phase throughout the soil, the pore-air pressure is necessarily atmospheric.

Interfacial Pressure

The interfacial pressure, or tension, varies in a wide range, from approximately 0.7 kg/cm^2 at atmospheric pressure to very high pressures. We have seen in Eq. (11.22.6) that if air bubbles are very small the interfacial pressure almost vanishes. As the bubbles increase their radius increases, and while the bubbles are still spherical the difference $p_w - p_g$ in Eq. (8.25.2) increases concurrently with the *air–water interfacial pressure*.

We have also seen that as the degree of saturation decreases and the air in the voids increases, the shape of the bubbles is not spherical any more but assumes the shape of the interparticle voids, and instead of having one radius like in a spherical bubble, it may develop two perpendicular radii. The air–water interfacial pressure p_i is evaluated according to Eq. (8.25.11)

(8.25.11) $$p_i = p_w - p_g = -\left(\frac{1}{R_1} + \frac{1}{R_2}\right)T$$

With further decrease of the degree of saturation and increase of the air inclusions, one of the radii, say R_2, increases to infinity, $R_2 \to \infty$, then inverts its sign to $R_2 \to -\infty$ and gradually decreases. The air–water interfacial pressure will be evaluated according to Eq. (8.25.13)

(8.25.13) $$p_i = p_w - p_g = -\left(\frac{1}{R_1} - \frac{1}{R_2}\right)T$$

It is worth noticing that at the same time R_1 also decreases gradually. Thus, as the degree of saturation changes the interfacial pressure undergoes changes as well. There is no knowledge, so far, as to exactly how the interfacial pressure varies as R_1 and R_2 vary independently.

Osmotic Pressure

In Sects. 8.18 and 8.19 it was shown that the osmotic pressure p_o, a result of the equilibrium between the attractive forces and the repulsive back-diffusion of solute ions and soil particles, regulates the particle interaction in fine grained soils in the molecular and submolecular levels. Since the osmotic pressure is part of the pore-water pressure, the factors that influence it discussed in Sect. 8.21 affect the pore-water pressure as well, thus enhancing or reducing the effective pressures. Flooding highly saline soils with water of low ionic concentration and leaching the salts to a greater depth is a common practice in arid regions to improve the agricultural quality of the soil. Such action induces an osmotic pressure equal to the pore-water pressure. Lambe (1960b) included the osmotic pressure in the formulation of the effective pressure, in the form of its two components, the "electrical repulsion between particles" less the "electrical attraction between particles".

Suction

At a low degree of saturation, the effect of pore-air pressure is negligible compared to the high negative tension of the pore-water pressure and the interfacial pressure, and if no dynamic and osmotic pressures are present Eq. (11.22.5) reduces to

$$\tfrac{1}{3}p_{\mathfrak{s}\alpha\alpha} = \tfrac{1}{3}p_{\alpha\alpha} + p_{\mathfrak{w}}; \qquad \rho_{\mathfrak{n}}\bar{v}_{\mathfrak{n}\alpha}\bar{v}_{\mathfrak{n}\alpha} = 0; \qquad p_{\mathfrak{o}} = 0; \quad p_{\mathfrak{g}} \ll p_{\mathfrak{w}} \qquad (11.22.7)$$

where $p_{\mathfrak{w}}$ is the suction pressure, a tension that can reach high values. As a tension, it becomes positive when added to the total pressure and increases the effective pressure. Dry soil samples, i.e. samples at low degrees of saturation indeed have higher strength than the same samples when wet, at higher degrees of saturation. Thus Eq. (11.22.5) considers effective pressures of the soil at any degree of saturation.

The velocities $\bar{v}_{\mathfrak{n}i}$ of the dynamic pressure should be available by computation from Eqs. (6.3.4) and (9.3.7)–(9.3.9). The pore-water pressure is measured by conventional techniques (Bishop and Henkel 1962), the pore-air pressure can be measured in limited form (Bishop 1960; Bishop et al. 1960; Fredlund and Morgenstern 1973), the interfacial pressure can be partly assessed from geometrical considerations (Schuurman 1966), the osmotic pressure should be available on the basis of tests and the physico-chemical considerations discussed in Chap. 8, and the suction pressure is found from tests (Croney and Coleman 1961). However, further extensive basic research, theoretical and experimental, is needed in all these topics in order to fully comprehend the intricacies of the phenomena in unsaturated soils.

11.23 Total Pressure

The total pressure has many facets and is not always exactly defined. The simplest definition is *pressure applied externally*. This includes forces that impart a pressure on the soil or soil sample, and also vacuum which is a negative spherical pressure of up to 1 atm. (approximately 1 kg/cm^2). A full vacuum (-760 mm mercury), for instance, applied to a soil sample is the equivalent of -1.0332 kg/cm^2 and has the same effect as an external confining pressure of 1.0332 kg/cm^2 as far as compression, consolidation and volume changes are concerned.

The pore-water pressure $p_{\mathfrak{w}}$ is either positive when it acts in a wet soil, or negative (a suction) when it acts in a dry soil. The effective pressure $p_{\mathfrak{e}} = \frac{1}{3}p_{\mathfrak{s}\alpha\alpha} - p_{\mathrm{m}}$ is defined in Eq. (9.4.8) as the mean total pressure $p_{\mathrm{m}} = \frac{1}{3}p_{\alpha\alpha}$ less the pore pressure $p_{\mathfrak{p}}$.

In a wet soil close to saturation, if the pore-air pressure can be assumed negligible $p_{\mathfrak{g}} \cong 0$, we have

$$p(t)_{\mathfrak{e}} = \tfrac{1}{3}p_{\mathfrak{s}\alpha\alpha} = \tfrac{1}{3}p_{\alpha\alpha} - p_{\mathfrak{w}} \tag{11.23.1}$$

In the presentation of shear stresses versus normal stresses of a soil element where pore-water pressure $p_{\mathfrak{w}}$ is acting in the water, Fig. 11.23.1, the effective pressure $p_{\mathfrak{e}}$ is marked by a point to the left of the total pressure p_{m}, the distance between them being equal to $p_{\mathfrak{w}}$. A Mohr circle is shifted to the left by the same distance $p_{\mathfrak{w}}$, since according to Eq. (9.4.7) the normal pressures p_{ij} where $i = j$ are similarly affected by $p_{\mathfrak{w}}$. In the octahedral presentation in Fig. 11.23.2 the mean total pressure p_{m} is represented by a point on the hydrostatic directrix line and the effective pressure $p_{\mathfrak{e}}$ by a point on the same

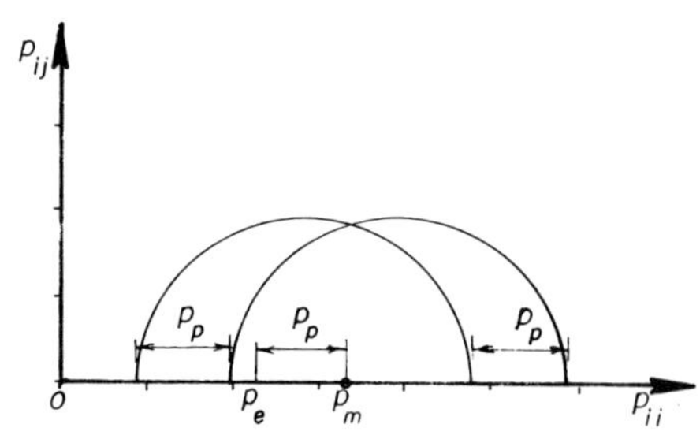

Fig 11.23.1. Shear stresses versus normal stresses in total and effective pressures representation.

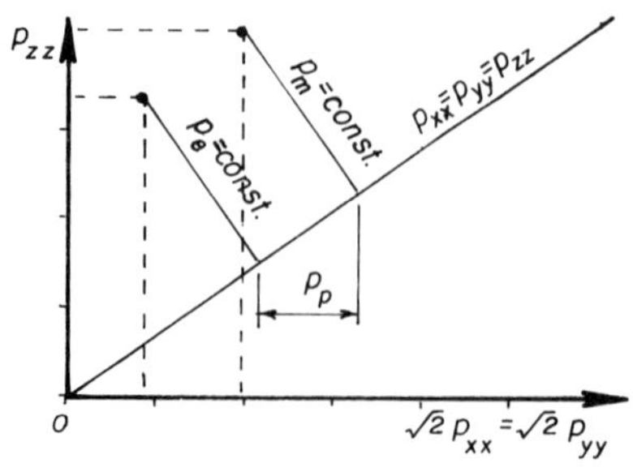

Fig 11.23.2. Octahedral representation of Fig. 11.23.1.

line to the left of the spherical pressure, at a horizontal distance equal to $p_{\mathfrak{w}}$. Also any stress p_{ij} with the mean spherical pressure equal to $p_{\mathfrak{w}}$ will be on a line normal to the hydrostatic directrix in the octahedral plane, and its corresponding effective stress $p_{\mathfrak{s}ij}$ will be on a parallel line at its left, at a horizontal distance equal to the pore-water pressure $p_{\mathfrak{w}}$.

In a dry soil, $p_{\mathfrak{w}}$ is a suction which is an internal stress acting in the water, and it is added to the mean total pressure p_{m}, Eq. (11.22.7). The mean effective pressure $p_{\mathfrak{e}}$ is given by

$$p(t)_{\mathfrak{e}} = \tfrac{1}{3}p_{\mathfrak{s}\alpha\alpha} = \tfrac{1}{3}p_{\alpha\alpha} - (-p_{\mathfrak{w}}) = \tfrac{1}{3}p_{\alpha\alpha} + p_{\mathfrak{w}} \qquad (11.23.2)$$

In the shear stresses versus normal stresses presentation, Fig. 11.23.3, the point of the effective pressure $p_{\mathfrak{e}}$ is to the right of the total pressure p_{m}, the distance between them being equal to the suction $p_{\mathfrak{w}}$. A Mohr circle of the effective stresses is shifted to the right by $p_{\mathfrak{w}}$. In the octahedral presentation in Fig. 11.23.4 the effective pressure $p_{\mathfrak{e}}$ of a pressure p_{m} is given by a point to the right of p_{m} on the hydrostatic directrix line, at a horizontal distance $p_{\mathfrak{w}}$. Similarly a stress p_{ij} with a mean spherical pressure p_{m} will be on a line normal to the hydrostatic directrix at a point corresponding to the spherical pressure p_{m}, and the effective stress $p_{\mathfrak{e}} = p_{\mathfrak{s}ij}$ will be on a line parallel to the line of the spherical pressure p_{m} and to its right, at a horizontal distance equal to the suction $p_{\mathfrak{w}}$.

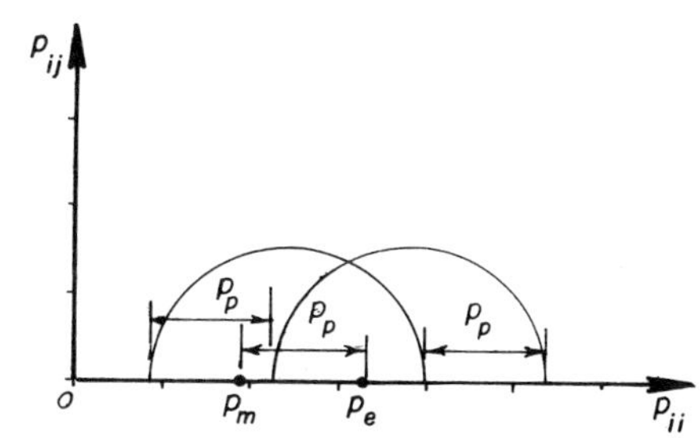

Fig 11.23.3. The effect of suction on the effective pressure in the Mohr circle representation.

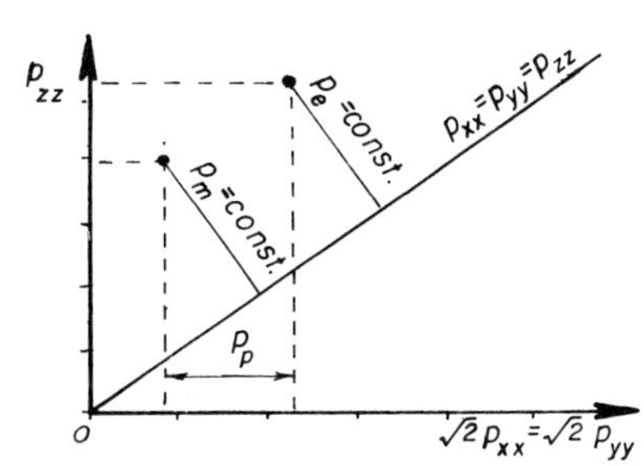

Fig 11.23.4. The effect of suction on the effective pressure in the octahedral representation.

11.24 The Internal Energy and Energy Rate of Spherical Phenomena

Following Sect. 7.8, the resolution of the internal energy rate into a dual equation, the *stored* internal energy rate or power and the *disbursed* internal energy rate, respectively, is

$$\begin{aligned}\dot{\mathfrak{I}}_{\mathfrak{s}\mathrm{i}} &= \Delta\dot{\mathfrak{I}}_{\mathfrak{s}\mathrm{i}} = \int_V \rho\dot{\epsilon}_{\mathfrak{s}\mathrm{i}}\,\mathrm{d}V = \int_V (p_{\mathrm{m}} - p_p)\dot{\varepsilon}_{\alpha\alpha}\,\mathrm{d}V \\ &\quad + \int_S \sum_{\mathrm{n}=1}^{\mathfrak{N}} h_{\mathrm{n}\alpha\mathfrak{d}} n_\alpha\,\mathrm{d}S + \int_V \sum_{\mathrm{n}=1}^{\mathfrak{N}} \rho_{\mathrm{n}} q_{\mathrm{ni}}\,\mathrm{d}V \\ &= \int_V p_{\mathrm{e}}\dot{\varepsilon}_{\alpha\alpha}\,\mathrm{d}V + \int_S \sum_{\mathrm{n}=1}^{\mathfrak{N}} h_{\mathrm{n}\alpha\mathrm{i}} n_\alpha\,\mathrm{d}S + \int_V \sum_{\mathrm{n}=1}^{\mathfrak{N}} \rho_{\mathrm{n}} q_{\mathrm{ni}}\,\mathrm{d}V \\ &= \int_V \sum_{\mathrm{n}=1}^{\mathfrak{N}} \rho_{\mathrm{n}} T\dot{\eta}_{\mathrm{ni}}\,\mathrm{d}V + \int \sum_{\mathrm{n}=1}^{\mathfrak{N}} \rho_{\mathrm{n}} \sum_{\mathrm{m}=1}^{\mathfrak{M}} \tau_{\mathrm{nmi}}\dot{v}_{\mathrm{nmi}}\,\mathrm{d}V \end{aligned} \tag{11.24.1}$$

$$\begin{aligned}\dot{\mathfrak{I}}_{\mathfrak{s}\mathfrak{d}}\,\mathrm{d}V &= \int_V \rho\dot{\epsilon}_{\mathfrak{s}\mathfrak{d}}\,\mathrm{d}V = \int_V s_{\alpha\beta}\dot{e}_{\alpha\beta}\,\mathrm{d}V + \int_S \sum_{\mathrm{n}=1}^{\mathfrak{N}} h_{\mathrm{n}\alpha\mathfrak{d}} n_\alpha\,\mathrm{d}S + \int_V \sum_{\mathrm{n}=1}^{\mathfrak{N}} \rho_{\mathrm{n}} q_{\mathrm{n}\mathfrak{d}}\,\mathrm{d}V \\ &= \int_V \sum_{\mathrm{n}=1}^{\mathfrak{N}} \rho_{\mathrm{n}} T\dot{\eta}_{\mathrm{ni}}\,\mathrm{d}V + \int_V \sum_{\mathrm{n}=1}^{\mathfrak{N}} \rho_{\mathrm{n}} \sum_{\mathrm{m}=1}^{\mathfrak{M}} \tau_{\mathrm{nm}\mathfrak{d}}\dot{v}_{\mathrm{nm}\mathfrak{d}}\,\mathrm{d}V \end{aligned} \tag{11.24.2}$$

where $\mathfrak{I}_{\mathfrak{s}\mathrm{i}}$ is the stored part of the internal energy of the solids, $\mathfrak{I}_{\mathfrak{s}\mathfrak{d}}$ is the disbursed part of the internal energy of the solids, $\epsilon_{\mathfrak{s}\mathrm{i}}$ is the *stored specific internal energy* of the solids, $\epsilon_{\mathfrak{s}\mathfrak{d}}$ is the *disbursed specific internal energy* or the deviatoric specific internal energy of the solids, $p_{\mathrm{m}} = \frac{1}{3}p_{\alpha\alpha}$ is the spherical pressure, $p_{\mathrm{e}} = \frac{1}{3}p_{\mathfrak{s}\alpha\alpha} = p_{\mathrm{m}} - p_{\mathrm{p}}$ is the *effective pressure*, $\dot{\varepsilon}_{\alpha\alpha}$ is the volumetric strain rate, s_{ij} is the deviatoric stress tensor, $\dot{e}_{ij}$ is the distortion rate tensor, $h_{\mathrm{n}\alpha\mathrm{i},\alpha}$ are the *spherical efflux gradients* of the respective constituents, $h_{\mathrm{n}\alpha\mathfrak{d},\alpha}$ are the deviatoric efflux gradients of the respective constitutents, $\rho_{\mathrm{n}} q_{\mathrm{ni}}$ are the *spherical supply rates* of the respective constituents, $\rho_{\mathrm{n}} q_{\mathrm{n}\mathfrak{d}}$ are the *deviatoric supply rates* of the respective constituents, τ_{nmi} and $\dot{v}_{\mathrm{nmi}}$ are the *spherical thermodynamic tensions* and the *spherical thermodynamic substates rate* of the respective constituents, and $\tau_{\mathrm{nm}\mathfrak{d}}$ and $\dot{v}_{\mathrm{nm}\mathfrak{d}}$ are the *deviatoric thermodynamic tensions* and the *deviatoric thermodynamic substates rate* of the respective constituents.

The stored internal energy of a material $\mathfrak{I}_{\mathfrak{s}\mathrm{i}}$ is made up of two parts, $\mathfrak{I}_{\mathfrak{s}\mathrm{i}0}$ is the *residing stored internal energy* and $\Delta\mathfrak{I}_{\mathfrak{s}\mathrm{i}}$ is the *excess stored internal energy* of the solids

$$\mathfrak{I}_{\mathfrak{s}\mathrm{i}} = \mathfrak{I}_{\mathfrak{s}\mathrm{i}0} + \Delta\mathfrak{I}_{\mathfrak{s}\mathrm{i}} \tag{11.24.3}$$

In terms of the specific internal energy components Eq. (11.24.3) can be written

$$\rho\epsilon_{\mathfrak{s}i} = \rho\epsilon_{\mathfrak{s}i0} + \Delta\rho\epsilon_{\mathfrak{s}i} \tag{11.24.4}$$

where $\rho\epsilon_{\mathfrak{s}i}$ is the stored specific internal energy of the solids, $\rho\epsilon_{\mathfrak{s}i0}$ is the residing stored specific internal energy of the solids and $\Delta\rho\epsilon_{\mathfrak{s}i}$ is the excess stored specific internal energy of the solids.

The residing stored internal energy is the internal energy inherent in the soil and is a function of its original density and the cohesive forces acting in it, thus it depends on its history of consolidation. During the test this stored internal energy remains unchanged and its time derivative vanishes, $\dot{\mathfrak{I}}_{\mathfrak{s}i0} = 0$, and does not appear in Eq. (11.24.1).

The excess stored free energy is the energy induced in the soil by the spherical pressure and consolidation in progress. After the soil has consolidated and has become stable its new stored energy state can be considered as its residing stored energy.

From Eq. (11.24.1) and Eq. (11.24.2) we obtain the following equations

$$\begin{aligned}\Delta\rho\epsilon_{\mathfrak{s}i} &= \int_0^{t'} (p_{\mathrm{m}} - p_{\mathrm{p}})\dot{\varepsilon}_{\alpha\alpha}\,\mathrm{d}t + \int_0^{t'} \sum_{\mathfrak{n}=1}^{\mathfrak{N}} h_{\mathfrak{n}\alpha,\alpha i}\,\mathrm{d}t + \int_0^{t'} \sum_{\mathfrak{n}=1}^{\mathfrak{N}} \rho_{\mathfrak{n}} q_{\mathfrak{n}i}\,\mathrm{d}t \\ &= \int_0^{t'} \sum_{\mathfrak{n}=1}^{\mathfrak{N}} \rho_{\mathfrak{n}} T\dot{\eta}_{\mathfrak{n}i}\,\mathrm{d}t + \int_0^{t'} \sum_{\mathfrak{n}=1}^{\mathfrak{N}} \rho_{\mathfrak{n}} \sum_{\mathfrak{m}=1}^{\mathfrak{M}} \tau_{\mathfrak{nm}i}\dot{v}_{\mathfrak{nm}i}\,\mathrm{d}t\end{aligned} \tag{11.24.5}$$

$$\begin{aligned}\rho\varepsilon_{\mathfrak{s}\mathfrak{d}} &= \int_0^{t''} s_{\alpha\beta}\dot{e}_{\alpha\beta}\,\mathrm{d}t + \int_0^{t''} \sum_{\mathfrak{n}=1}^{\mathfrak{N}} h_{\mathfrak{n}\alpha,\alpha\mathfrak{d}}\,\mathrm{d}t + \int_0^{t''} \sum_{\mathfrak{n}=1}^{\mathfrak{N}} \rho_{\mathfrak{n}} q_{\mathfrak{n}\mathfrak{d}}\,\mathrm{d}t \\ &= \int_0^{t''} \sum_{\mathfrak{n}=1}^{\mathfrak{N}} \rho_{\mathfrak{n}} T\dot{\eta}_{\mathfrak{n}i}\,\mathrm{d}t + \int_0^{t''} \sum_{\mathfrak{n}=1}^{\mathfrak{N}} \rho_{\mathfrak{n}} \sum_{\mathfrak{m}=1}^{\mathfrak{M}} \tau_{\mathfrak{nm}\mathfrak{d}}\dot{v}_{\mathfrak{nm}\mathfrak{d}}\,\mathrm{d}t\end{aligned} \tag{11.24.6}$$

which represent the stored specific internal energy and the disbursed specific internal energy, respectively.

In Eq. (5.11.1) we see that the internal energy rate is composed of a stress power and a flux and supply on one side of the equation and a mechanical and non-mechanical energy rate on the other side. Resolving Eq. (5.11.1) into the spherical and deviatoric components and considering also Eq. (6.9.3) for the spherical component we obtain

$$\begin{aligned}\Delta\rho\dot{\epsilon}_{\mathfrak{s}i} &= (p_{\mathrm{m}} - p_{\mathrm{p}})\dot{\varepsilon}_{\alpha\alpha} + \sum_{\mathfrak{n}=1}^{\mathfrak{N}} h_{\mathfrak{n}\alpha,\alpha i} + \sum_{\mathfrak{n}=1}^{\mathfrak{N}} \rho_{\mathfrak{n}} q_{\mathfrak{n}i} \\ &= \sum_{\mathfrak{n}=1}^{\mathfrak{N}} \rho_{\mathfrak{n}} T\dot{\eta}_{\mathfrak{n}i} + \sum_{\mathfrak{n}=1}^{\mathfrak{N}} \rho_{\mathfrak{n}} \sum_{\mathfrak{m}=1}^{\mathfrak{M}} \tau_{\mathfrak{nm}i}\dot{v}_{\mathfrak{nm}i}\end{aligned} \tag{11.24.7}$$

$$\rho\dot{\epsilon}_{\mathfrak{s}\mathfrak{d}} = s_{\alpha\beta}\dot{e}_{\alpha\beta} + \sum_{\mathfrak{n}=1}^{\mathfrak{N}} h_{\mathfrak{n}\alpha,\alpha\mathfrak{d}} + \sum_{\mathfrak{n}=1}^{\mathfrak{N}} \rho_{\mathfrak{n}} q_{\mathfrak{n}\mathfrak{d}} = \sum_{\mathfrak{n}=1}^{\mathfrak{N}} \rho_{\mathfrak{n}} T\dot{\eta}_{\mathfrak{n}i} + \sum_{\mathfrak{n}=1}^{\mathfrak{N}} \rho_{\mathfrak{n}} \sum_{\mathfrak{m}=1}^{\mathfrak{M}} \tau_{\mathfrak{nm}\mathfrak{d}}\dot{v}_{\mathfrak{nm}\mathfrak{d}} \tag{11.24.8}$$

where $\Delta\rho\epsilon_{\mathfrak{si}}$ is the excess spherical specific internal energy rate or the excess stored specific internal energy rate, $\rho\dot{\epsilon}_{\mathfrak{sd}}$ is the deviatoric specific internal energy rate or the disbursed specific internal energy rate, $p_{\mathrm{m}} - p_{\mathfrak{p}} = \frac{1}{3}p_{\mathfrak{s}\alpha\alpha}$ is the effective spherical pressure acting on the solids and

$$\dot{\varepsilon}_{\alpha\alpha} = \overset{\text{inf}}{\dot{\varepsilon}}_{\alpha\alpha}$$

is here the infinitestimal volumetric strain rate. Integrating the first part of Eq. (11.24.7) with time, the excess stored specific internal energy is obtained

$$\begin{aligned}\rho\epsilon_{\mathfrak{si}} &= \int_{-\infty}^{t'} \rho\dot{\epsilon}_{\mathfrak{si}}\,\mathrm{d}t = \int_{-\infty}^{0} \rho\dot{\epsilon}_{\mathfrak{si}}\,\mathrm{d}t + \int_{0}^{t'} \rho\dot{\epsilon}_{\mathfrak{si}}\,\mathrm{d}t = \rho\epsilon_{\mathfrak{si}0} + \rho\epsilon_{\mathfrak{si}}\Big|_{0}^{t'} \\ &= \rho\epsilon_{\mathfrak{si}0} + \Delta\rho\epsilon_{\mathfrak{si}} \qquad (11.24.9)\end{aligned}$$

where $\rho\epsilon_{\mathfrak{si}}$ is the specific stored internal energy, $\rho\epsilon_{\mathfrak{si}0}$ is the specific residing stored internal energy and $\Delta\rho\epsilon_{\mathfrak{si}}$ is the specific excess stored internal energy.

Eq. (11.24.5) is the spherical part of the dual specific stored internal energy equation and it can be obtained also by integrating Eq. (11.24.7). The deviatoric part of the dual equation is discussed in Sect. 12.24. The excess specific stored internal energy is an account of the energy invested in the soil by the effective pressure through any sort of consolidation of the soil and stored in it.

We will also see in Sect. 11.28 that the term

$$\int_{0}^{t'} (p_{\mathrm{m}} - p_{\mathfrak{p}})\dot{\varepsilon}_{\alpha\alpha}\mathrm{d}t$$

includes not only the volumetric strain $\varepsilon_{\alpha\alpha}$ resulting from the spherical stresses but also the volumetric strain $\varepsilon_{\beta\beta}$ resulting from the deviatoric stresses, when general isotropic relations are assumed. The time ranges of the two differ from one another. The time range of the spherical pressure is denoted by t' and the time range of the deviatoric stresses by t''.

11.25 The Excess Stored Specific Free Energy

Of all components involved in Eqs. (11.24.1)–(11.24.8) the non-mechanical energies are sometimes unknown and generally difficult to evaluate. Therefore, if the processes can be assumed to occur under an isentropic condition $\dot{\eta}_{\mathrm{ni}} = \dot{\eta}_{\mathrm{nd}} = 0$ and if no effluxes and supplies are available or if, at least, their sum vanishes, $h_{\mathfrak{s}\alpha,\alpha} + \rho_{\mathfrak{s}}q_{\mathfrak{s}} = 0$, then according to Eq. (5.11.3) it is convenient (and customary) to introduce the specific free energy rate rather than the internal energy equations, that is, to replace Eqs. (11.24.1) and (11.24.2) with equations void of entropy and heat effects

$$\Delta\rho\dot{\psi}_{\mathfrak{si}} = p_{\mathfrak{e}}\dot{\varepsilon}_{\alpha\alpha} = (p_{\mathrm{m}} - p_{\mathfrak{p}})\dot{\varepsilon}_{\alpha\alpha} = (p_{\mathrm{m}} - p_{\mathfrak{w}} - p_{\mathfrak{g}})\dot{\varepsilon}_{\alpha\alpha}$$

$$= \sum_{\mathrm{n}=1}^{\mathfrak{N}} \rho_{\mathrm{n}} \sum_{\mathrm{m}=1}^{\mathfrak{M}} \tau_{\mathfrak{s}\mathrm{mi}}\dot{v}_{\mathfrak{s}\mathrm{mi}} \tag{11.25.1}$$

$$\rho\dot{\psi}_{\mathfrak{sd}} = s_{\alpha\beta}\dot{e}_{\alpha\beta} = \sum_{\mathrm{n}=1}^{\mathfrak{N}} \rho_{\mathrm{n}} \sum_{\mathrm{m}=1}^{\mathfrak{M}} \tau_{\mathfrak{s}\mathrm{m}\mathfrak{d}}\dot{v}_{\mathfrak{s}\mathrm{m}\mathfrak{d}} \tag{11.25.2}$$

where $\dot{\psi}_{\mathfrak{si}}$ is the *spherical free energy rate* or the *excess stored specific free energy rate* of the solids and $\dot{\psi}_{\mathfrak{sd}}$ is the *deviatoric specific free energy rate* or the *disbursed specific free energy rate* of the solids. Both expressions are free of temperature and entropy rates, as well as of efflux and heat supply.

Integrating Eqs. (11.24.1) and (11.24.2) with time we obtain

$$\Delta\rho\psi_{\mathfrak{si}} = \int_0^{t'} \rho\dot{\psi}_{\mathfrak{si}}\,\mathrm{d}t = \int_0^{t'} \rho\,\mathrm{d}\psi_{\mathfrak{si}} = \rho\psi_{\mathfrak{si}}\Big|_0^{t'} = \int_0^{t'} (p_{\mathrm{m}} - p_{\mathfrak{p}})\,\mathrm{d}\varepsilon_{\alpha\alpha}$$

$$= \int_0^{t'} (p_{\mathrm{m}} - p_{\mathfrak{p}})\dot{\varepsilon}_{\alpha\alpha}\,\mathrm{d}t$$

$$= \int_0^{t'} \sum_{\mathrm{n}=1}^{\mathfrak{N}} \rho_{\mathrm{n}} \sum_{\mathrm{m}=1}^{\mathfrak{M}} \tau_{\mathrm{nmi}}\dot{v}_{\mathrm{nmi}}\,\mathrm{d}t = \int_0^{t'} \sum_{\mathrm{n}=1}^{\mathfrak{N}} \rho_{\mathrm{n}} \sum_{\mathrm{m}=1}^{\mathfrak{M}} \tau_{\mathrm{nmi}}\,\mathrm{d}v_{\mathrm{nmi}} \tag{11.25.3}$$

$$\rho\psi_{\mathfrak{sd}} = \int_0^{t''} \rho\dot{\psi}_{\mathfrak{sd}}\,\mathrm{d}t = \int_0^{t''} \rho\,\mathrm{d}\psi_{\mathfrak{sd}} = \rho\psi\,\big|_0^{t''} = \int_0^{t''} s_{\alpha\beta}\dot{e}_{\alpha\beta}\,\mathrm{d}t = \int_0^{e_{ij}} s_{\alpha\beta}de_{\alpha\beta}$$

$$= \int_0^{t''} \sum_{\mathrm{n}=1}^{\mathfrak{N}} \rho_{\mathrm{n}} \sum_{\mathrm{m}=1}^{\mathfrak{M}} \tau_{\mathrm{nm}\mathfrak{d}}\dot{v}_{\mathrm{nm}\mathfrak{d}}\,\mathrm{d}t = \int_0^{t''} \sum_{\mathrm{n}=1}^{\mathfrak{N}} \rho_{\mathrm{n}} \sum_{\mathrm{m}=1}^{\mathfrak{M}} \tau_{\mathrm{nm}\mathfrak{d}}\,\mathrm{d}v_{\mathrm{nm}\mathfrak{d}} \tag{11.25.4}$$

$\Delta\rho\psi_{\mathfrak{si}}$, or Eq. (11.25.3), represents the specific spherical free energy increment invested in the volume element through densification, stored to be available in full or in part whenever needed. On the other hand, $\rho\psi_{\mathfrak{sd}}$ or Eq. (11.25.4) represents the specific deviatoric free energy, a disbursed energy that thrives on the stored specific free energy increment and is consumed during the shearing process. The two equations together are known as the dual *specific free energy balance* equations. Almost all that was said about the dual specific internal energies is valid also with respect to the dual specific free energies, with the limitations of constant temperature and no heat exchange.

Thus, an equation analogous to (11.24.4) can be written for the specific internal energies

$$\rho\psi_{\mathfrak{si}} = \rho\psi_{\mathfrak{si}0} + \Delta\rho\psi_{\mathfrak{si}} = \rho\psi_{\mathfrak{si}0} + \int_0^{t'} p_{\mathfrak{e}}\,\mathrm{d}\varepsilon_{\alpha\alpha} = \rho\psi_{\mathfrak{si}0} + \int_0^{t'} (p_{\mathrm{m}} - p_{\mathfrak{p}})\,\mathrm{d}\varepsilon_{\alpha\alpha}$$

$$= \rho\psi_{\mathfrak{si}0} + \int_0^{t'} (p_{\mathrm{m}} - p_{\mathfrak{w}} - p_{\mathfrak{g}})\,\mathrm{d}\varepsilon_{\alpha\alpha} \tag{11.25.5}$$

indicating that Eq. (11.25.3) represents only the increment of the stored specific free energy, added to the residing specific free energy $\rho\psi_{\mathrm{\mathfrak{s}i0}}$ which has been stored in the soil before the present constraints and pressures were applied. After the application of the pressure and the resulting volume change the free energy reaches a new $\rho\psi_{\mathrm{\mathfrak{s}i}}$ level.

The residing specific free energy $\rho\psi_{\mathrm{\mathfrak{s}i0}} = \rho\epsilon_{\mathrm{\mathfrak{s}i0}}$ in a normally consolidated soil is related to the work done by the cohesion p_{m0}, defined

$$\rho\psi_{\mathrm{\mathfrak{s}i0}} = \rho\epsilon_{\mathrm{\mathfrak{s}i0}} = p_{\mathrm{m0}}\varepsilon_{\alpha\alpha0} \tag{11.25.6}$$

where $\varepsilon_{\alpha\alpha0}$ is an equivalent virtual volumetric strain related to the cohesion.

Assuming a linear functional relation between the cohesion p_{m0} and the virtual volumetric strain, such as, for instance

$$\varepsilon_{\alpha\alpha0} = \mathcal{R}p_{\mathrm{m0}} \tag{11.25.7}$$

where $\mathcal{R} = \mathcal{R}(p_{\mathrm{m0}}, \mathrm{t})$ could be a function of the cohesion and of the time, Eq. (11.25.6) can also be written

$$\rho\psi_{\mathrm{\mathfrak{s}i0}} = \rho\epsilon_{\mathrm{\mathfrak{s}i0}} = \mathcal{R}(p_{\mathrm{m0}}, t)p_{\mathrm{m0}}^2 \tag{11.25.8}$$

11.26 The Excess Stored Specific Free Energy of Solids in Linear Viscoelastic Media

Eq. (11.25.3) can now be applied to problems with specific loading conditions. Of particular interest to us is the application of a spherical pressure to a cylindrical soil sample, a problem common to any shear test, conventional or other, performed in a triaxial testing cell.

We shall discuss in this section the excess specific free energy in a linear viscoelastic medium and in the next section the excess specific free energy in an isotropic nonlinear medium.

We have seen in Sects. 11.18 and 11.19 that when a spherical pressure $p(t)_{\mathrm{m}} = \frac{1}{3}p(t)_{\alpha\alpha}$ is applied to a soil element an activating spherical pressure $p(t)_{\mathrm{m}}[1 - (p_0/p(t)_{\mathrm{m}})\,\vartheta(t)]$ produces the strains. The spherical pressure acting on the solids or the effective pressure is obtained with the help of Eq. (11.18.4), based on a four coefficient viscoelastic spherical constitutive equation

$$\begin{aligned}\tfrac{1}{3}p(t)_{\mathfrak{s}\alpha\alpha} &= \left[\tfrac{1}{3}p(t)_{\alpha\alpha}\left(1 - \frac{p_0}{p(t)\mu}\,\vartheta(t)\right) - p(t)_{\mathrm{p}}\right]\\ &= \left[p(t)_{\mathrm{m}}\left(1 - \frac{p_0}{p(t)_{\mathrm{m}}}\,\vartheta(t)\right) - p(t)_{\mathrm{p}}\right]\end{aligned}$$

$$= p(t)_{\mathrm{m}} \left(1 - \frac{p_0}{p(t)} \vartheta(t)\right)$$

$$+ \frac{1}{T_3} \exp \frac{-t}{T_3} \int_0^t p(\tau)_{\mathrm{m}} \left(1 - \frac{p_0}{p(\tau)_{\mathrm{m}}} \vartheta(\tau)\right) \exp \frac{\tau}{T_3} \, \mathrm{d}\tau \qquad (11.26.1)$$

where $\vartheta(t)$ is the plastic restraint function, p_0 is the coefficient of plastic restraint and $T_3 = \mu_3/\varkappa_3$ is the volumetric retardation time.

The volumetric strain, the volumetric strain rate and the pore pressure are, according to Eqs. (11.18.1), (11.18.2) and (11.18.4), respectively

$$(11.18.1) \qquad \varepsilon(t)_{\alpha\alpha} = \frac{p(t)_{\mathrm{m}}}{\varkappa_1} \left(1 - \frac{p_0}{p(t)_{\mathrm{m}}} \vartheta(t)\right)$$

$$+ \frac{1}{\mu_2} \int_0^t p(\tau)_{\mathrm{m}} \left(1 - \frac{p_0}{p(\tau)_{\mathrm{m}}} \vartheta(\tau)\right) \mathrm{d}\tau$$

$$+ \frac{1}{\mu_3} \exp \frac{-t}{T_3} \int_0^t p(\tau)_{\mathrm{m}} \left(1 - \frac{p_0}{p(\tau)_{\mathrm{m}}} \vartheta(\tau)\right) \exp \frac{\tau}{T_3} \, \mathrm{d}\tau$$

$$(11.18.2) \qquad \dot{\varepsilon}(t)_{\alpha\alpha} = \frac{1}{\varkappa_1} \left(\dot{p}(t)_{\mathrm{m}} - p_0 \dot{\vartheta}(t)\right) + \frac{p(t)_{\mathrm{m}}}{\mu_2} \left(1 - \frac{p_0}{p(t)} \vartheta(t)\right)$$

$$+ \frac{p(t)_{\mathrm{m}}}{\mu_3} \left(1 - \frac{p_0}{p(t)_{\mathrm{m}}} \vartheta(t)\right)$$

$$- \frac{1}{\mu_3 T_3} \exp \frac{-t}{T_3} \int_0^t p(\tau)_{\mathrm{m}} \left(1 - \frac{\dot{p}_0}{p(\tau)_{\mathrm{m}}} \vartheta(\tau)\right) \exp \frac{\tau}{T_3} \, \mathrm{d}\tau$$

$$(11.18.4) \qquad p_{\mathfrak{p}} = \mu_3 \dot{\varepsilon}(t)_{\mathfrak{p}} = - \frac{1}{T_3} \exp \frac{-t}{T_3} \int_0^t p(\tau)_{\mathrm{m}} \left(1 - \frac{p_0}{p(\tau)_{\mathrm{m}}} \vartheta(\tau)\right) \exp \frac{\tau}{T_3} \, \mathrm{d}\tau$$

We can now determine the excess stored specific free energy by substituting Eqs. (11.26.1), (11.18.2) and (11.18.4) into Eq. (11.25.3)

$$\Delta \rho \psi_{\mathfrak{si}} = \int_0^{t'} \tfrac{1}{3} p(\tau)_{\mathfrak{s}\alpha\alpha} \, \dot{\varepsilon}(\tau)_{\alpha\alpha} \, \mathrm{d}\tau = \int_0^{t'} p(\tau)_{\mathfrak{e}} \dot{\varepsilon}(\tau)_{\alpha\alpha} \, \mathrm{d}\tau$$

$$= \int_0^{t'} \Bigg[p(t)_{\mathrm{m}} \left(1 - \frac{p_0}{p(t)_{\mathrm{m}}} \vartheta(t)\right)$$

$$+ \frac{1}{T_3} \exp \frac{-t}{T_3} \int_0^t p(\tau)_{\mathrm{m}} \left(1 - \frac{p_0}{p(\tau)_{\mathrm{m}}} \vartheta(\tau)\right) \exp \frac{\tau}{T_3} \, \mathrm{d}\tau \Bigg]$$

$$\times \Bigg[\frac{1}{\varkappa_1} \left(\dot{p}(t)_{\mathrm{m}} - p_0 \dot{\vartheta}(t)\right) + \left(\frac{1}{\mu_2} + \frac{1}{\mu_3}\right) p(t)_{\mathrm{m}} \left(1 - \frac{p_0}{p(t)_{\mathrm{m}}} \vartheta(t)\right)$$

$$- \frac{1}{\mu_3 T_3} \exp \frac{-t}{T_3} \int_0^t p(\tau)_{\mathrm{m}} \left(1 - \frac{p_0}{p(\tau)} \vartheta(\tau)\right) \exp \frac{\tau}{T_3} \, \mathrm{d}\tau \Bigg] \mathrm{d}t \qquad (11.26.2)$$

and apply to it the particular conditions.

Specifically two loading conditions by which a required spherical pressure p_{m} is applied are explored:

1. *Single step loading*. If the spherical pressure is applied in a single step, $p_{\mathrm{m}} = \mathcal{P} = \text{const}$, the rate of pressure vanishes, $\dot{p}_{\mathrm{m}} = 0$. Introducing these values in Eq. (11.26.2) we obtain

$$
\begin{aligned}
\Delta\rho\psi_{\check{s}\mathrm{i}} &= \int_0^{t'} p(\tau)_{\mathrm{e}}\dot{\varepsilon}(\tau)_{\alpha\alpha}\,\mathrm{d}\tau = \int_0^{t'} \tfrac{1}{3}p(\tau)_{\check{s}\alpha\alpha}\dot{\varepsilon}(\tau)_{\alpha\alpha}\,\mathrm{d}\tau \\
&= \mathcal{P}^2 \int_0^{t'} \left[\left(1 - \frac{p_0}{\mathcal{P}}\,\vartheta(t)\right) + \frac{1}{T_3}\exp\frac{-t}{T_3}\int_0^t \left(1 - \frac{p_0}{\mathcal{P}}\,\vartheta(\tau)\right)\exp\frac{\tau}{T_3}\,\mathrm{d}\tau\right] \\
&\quad \times \left[-\frac{p_0}{\varkappa_1}\,\dot{\vartheta}(\tau) + \left(\frac{1}{\mu_2} + \frac{1}{\mu_3}\right)\left(1 - \frac{p_0}{\mathcal{P}}\,\vartheta(\tau)\right)\right. \\
&\quad \left. - \frac{1}{\mu_3 T_3}\exp\frac{-t}{T_3}\int_0^t \left(1 - \frac{p_0}{\mathcal{P}}\,\vartheta(\tau)\right)\exp\frac{\tau}{T_3}\,\mathrm{d}\tau\right]\mathrm{d}t
\end{aligned}
\tag{11.26.3}
$$

We may now substitute for the plastic restraint $\vartheta(t)$ one of the functions suggested in Sect. C.14 of Appendix C, say $\vartheta(t) = 1 - \exp(-\mathcal{B}T)$, where $\mathcal{B} = \text{const}$ with the dimensions of time reciprocal, $[\mathcal{B}] = [T^{-1}]$. Eq. (11.26.3) can now be integrated and rendered in a more closed form

$$
\begin{aligned}
\Delta\rho\psi_{\check{s}\mathrm{i}} &= \int_0^{t'} \tfrac{1}{3}p(\tau)_{\check{s}\alpha\alpha}\dot{\varepsilon}(\tau)_{\alpha\alpha}\,\mathrm{d}\tau \\
&= \mathcal{P}^2 \int_0^{t'} \left[\left(1 - \frac{p_0}{\mathcal{P}}\,[1 - \exp(-\mathcal{B}\tau)]\right)\right. \\
&\quad \left. + \frac{1}{T_3}\exp\frac{-t}{T_3}\int_0^t \left(1 - \frac{p_0}{\mathcal{P}}\,[1 - \exp(-\mathcal{B}\tau)]\right)\exp\frac{\tau}{T_3}\,\mathrm{d}\tau\right] \\
&\quad \times \left[\frac{\mathcal{B}p_0}{\varkappa_1\mathcal{P}}\exp(-\mathcal{B}t)\right) + \left(\frac{1}{\mu_2} + \frac{1}{\mu_3}\right)\left(1 - \frac{p_0}{\mathcal{P}}\,[1 - \exp(\mathcal{B}t)]\right) \\
&\quad \left. - \frac{1}{\mu_3 T_3}\exp\frac{-t}{T_3}\int_0^t \left(1 - \frac{p_0}{\mathcal{P}}\,[1 - \exp(-\mathcal{B}\tau)]\right)\exp\frac{\tau}{T_3}\,\mathrm{d}\tau\right]\mathrm{d}t \\
&= \mathcal{P}^2 T_3 \left[\left(\frac{\mathcal{P} + \mathcal{B}p_0}{\mathcal{P}}\right)\left(\frac{\mathcal{B}p_0}{\varkappa_1\mathcal{P}(1 + \mathcal{B}T_3)}\right)\exp\frac{-(1 + \mathcal{B}T_3)t'}{T_3}\right. \\
&\quad + \left(\frac{1}{\mu_2} - \frac{1}{\mu_3}\right)\left(\frac{\mathcal{B}p_0}{\mathcal{P}(1 + \mathcal{B}T_3)}\right)\left(\exp\frac{-(1 + \mathcal{B}T_3)t'}{T_3} - \exp\frac{-t'}{T_3}\right)
\end{aligned}
$$

$$+\left(\frac{\mathcal{B}p_0}{\mathcal{D}(1+\mathcal{B}T_3)}\right)\exp\frac{-2t'}{T_3}\bigg]$$

$$=\mathcal{D}^2\,\mathcal{F}(\varkappa_1,\mu_2,\mu_3,T_3,p_0,\mathcal{B},t')=\mathrm{I}_p^2\mathcal{F} \tag{11.26.4}$$

Eq. (11.26.4) gives the stored specific free energy at any time t of the consolidation process. The total stored specific free energy at time $t=t'$, at the end of consolidation or at any other stage when the consolidation is discontinued, is obtained when t' is substituted in the equation.

2. *Constant rate loading*. If the spherical pressure is applied at a constant rate $\dot{p}_{\mathrm{m}}=C$, the loading is linear $p(t)_{\mathrm{m}}=Ct$. When these values are inserted in Eq. (11.26.2) we obtain

$$\begin{aligned}\Delta\rho\psi_{\mathfrak{si}} = C^2\int_0^{t'}\Bigg[&t\left(1-\frac{p_0}{C\tau}\,\vartheta(t)\right)\\ &+\frac{1}{T_3}\exp\frac{-t}{T_3}\int_0^t\tau\left(1-\frac{p_0}{C\tau}\,\vartheta(\tau)\right)\exp\frac{\tau}{T_3}\,\mathrm{d}\tau\Bigg]\\ &\times\Bigg[-\frac{1}{\varkappa_1}\,p_0\dot{\vartheta}(t)+\left(\frac{t}{\mu_2}+\frac{t}{\mu_3}\right)\left(1-\frac{p_0}{C\tau}\,\vartheta(t)\right)\\ &-\frac{1}{\mu_3T_3}\exp\frac{-t}{T_3}\int_0^t\tau\left(1-\frac{p_0}{C\tau}\,\vartheta(\tau)\right)\exp\frac{\tau}{T_3}\,\mathrm{d}\tau\Bigg]\,\mathrm{d}t\end{aligned} \tag{11.26.5}$$

If $\vartheta(t)=1-\exp(-\mathcal{B}t)$ is substituted in Eq. (11.26.5) the integration can be performed and we obtain

$$\begin{aligned}\Delta\rho\psi_{\mathfrak{si}} = C^2\int_0^{t'}\Bigg[&t\left(1-\frac{p_0}{Ct}\,[1-\exp(-\mathcal{B}t)]\right)\\ &+\frac{1}{T_3}\exp\frac{-t}{T_3}\int_0^t\tau\left(1-\frac{p_0}{C\tau}\,[1-\exp(-\mathcal{B}\tau)]\right)\exp\frac{\tau}{T_3}\,\mathrm{d}\tau\Bigg]\\ &\times\Bigg[\frac{p_0}{\varkappa_1C}\,\mathcal{B}\exp(-\mathcal{B}t)+\left(\frac{t}{\mu_2}+\frac{t}{\mu_3}\right)\left(1-\frac{p_0}{Ct}\,[1-\exp(-\mathcal{B}t)]\right)\\ &-\frac{1}{\mu_3T_3}\exp\frac{-t}{T_3}\int_0^t\tau\left(1-\frac{p_0}{C\tau}\,[1-\exp(-\mathcal{B}\tau)]\right)\exp\frac{\tau}{T_3}\,\mathrm{d}\tau\Bigg]\,\mathrm{d}t\\ = C^2t'^2\Bigg[&\frac{1}{\varkappa_1}\left(\frac{1}{2}-\frac{T_3}{t'}-\frac{p_0}{Ct'}\right)\\ &+\frac{1}{\mu_2}\left(\frac{t'^2}{3}-\frac{T_3}{2}-\frac{p_0}{C}+\frac{p_0T_3}{Ct'^2}+\frac{p_0}{C^2t'}\right)\\ &-\frac{1}{\mu_3}\left(\frac{T_3}{2}-\frac{T_3^2}{t'}-\frac{p_0T_3}{Ct'}\right)\end{aligned}$$

$$+ \frac{1}{t'} \exp - \mathcal{B}t' \left[\frac{p_0}{\varkappa_1 C} - \frac{1}{\mu_2} \left(\frac{p_0}{C\mathcal{B}} + \frac{p_0}{C\mathcal{B}(1 - \mathcal{B}T_3)} \right) - \frac{p_0}{\mu_3 C\mathcal{B}} \right]$$

$$+ \frac{1}{t'^2} \exp - \mathcal{B}t' \left[\frac{1}{\varkappa_1} \left(\frac{p_0}{C\mathcal{B}} - \frac{p_0 T_3}{C} - \frac{p_0^2}{C^2} - \frac{p_0}{C\mathcal{B}(1 - \mathcal{B}T_3)} \right) \right.$$

$$+ \frac{1}{\mu_2} \left(\frac{p_0}{C\mathcal{B}^2} + \frac{p_0 T_3}{C\mathcal{B}} + \frac{p_0^2}{C^2} \right.$$

$$\left. + \frac{p_0^2}{C^2\mathcal{B}(1 - \mathcal{B}T_3)} + \frac{p_0}{C\mathcal{B}^2(1 - \mathcal{B}T_3)} \right)$$

$$- \frac{1}{\mu_3} \left(\frac{p_0}{C\mathcal{B}^2} + \frac{p_0 T_3}{C\mathcal{B}} + \frac{p_0^2}{C^2} - \frac{2 p_0 T_3}{C\mathcal{B}(1 - \mathcal{B}T_3)} \right.$$

$$\left. \left. - \frac{p_0^2}{C^2\mathcal{B}(1 - \mathcal{B}T_3)} - \frac{2 p_0}{C\mathcal{B}^2(1 - \mathcal{B}T_3)} \right) \right]$$

$$+ \frac{1}{t'^2} \exp - 2\mathcal{B}t \left[\frac{1}{\varkappa_1} \frac{p_0^2}{2C^2(1 - \mathcal{B}T_3)} - \frac{1}{\mu_2} \frac{p_0^2}{2C^2\mathcal{B}(1 - \mathcal{B}T_3)} \right.$$

$$\left. + \frac{1}{\mu_3} \frac{p_0^2}{C^2\mathcal{B}(1 - \mathcal{B}T_3)} \right]$$

$$= C^2 t'^2 \mathcal{F}(\varkappa_1, \mu_2, \mu_3, T_3, p_0, \mathcal{B}, t') = \mathrm{I}_p^2 \mathcal{F} \tag{11.26.6}$$

In both cases it is seen that the excess stored specific free energy is a function of the material coefficients, the time t and the spherical consolidation pressure. The spherical consolidation pressure is equal to the first invariant of the stress tensor and can replace it in the equation, and the excess stored specific free energy depends on its square.

11.27 Isotropic Functional Relationship Applied to the Excess Stored Specific Free Energy

If isotropic functional relationship is postulated between the strain, the stress and the stress rate, the free energy balance can be applied to Eqs. (7.9.6) and (7.9.7).

The excess stored specific free energy can then be analyzed only for total pressures. Since the current theories do not permit the consideration of pore pressures, isotropic relations have to be formulated in terms of total stresses rather than effective pressures. When a spherical consolidation pressure

$p_{\mathrm{m}} = \frac{1}{3} p_{\alpha\alpha}$ is applied to a soil element and isotropic stress–strain relationship is assumed, an equation of the form (7.9.6) can be written

$$\varepsilon_{\alpha\alpha} = \beth_1 p_{\mathrm{m}} + \beth_2 p_{\mathrm{m}}^2 + \beth_3 \dot{p}_{\mathrm{m}} + \beth_4 \dot{p}_{\mathrm{m}}^2 \tag{11.27.1}$$

where $\varepsilon_{\alpha\alpha}$ is the volumetric strain and $\dot{p}_{\mathrm{m}}$ is the rate at which the spherical pressure is applied.

The rate of volumetric strain is then

$$\dot{\varepsilon}_{\alpha\alpha} = \beth_1 \dot{p}_{\mathrm{m}} + 2\beth_2 p_{\mathrm{m}} \dot{p}_{\mathrm{m}} + \beth_3 \ddot{p}_{\mathrm{m}} + 2\beth_4 \dot{p}_{\mathrm{m}} \ddot{p}_{\mathrm{m}} \tag{11.27.2}$$

If the spherical consolidation pressure p_{m} is followed by a shear stress s_{ij} applied to the soil element, either by step loading or as a time dependent loading, we obtain, considering also Eq. (7.9.6)

$$\varepsilon_{ij} = \mathfrak{w}_1 s_{ij} + \mathfrak{w}_2 s_{ij}^2 + \mathfrak{w}_3 \dot{s}_{ij} + \mathfrak{w}_4 \dot{s}_{ij}^2 \tag{11.27.3}$$

where ε_{ij} is a function of the stress tensor s_{ij} and the stress-rate tensor $\dot{s}_{ij}$.

It is evident that although traceless shear stresses are applied to the soil element the strains obtained are not traceless, if isotropic stress–strain relations are postulated. Thus ε_{ij} appears in Eq. (11.27.3) and not e_{ij} like in linear stress–strain relationships and ε_{ij} can be resolved into the dual equations, one the trace of ε_{ij} and one the traceless part of ε_{ij}.

Let us assume that the shear stress s_{ij} is a pure deviatoric stress applied to a cylindrical soil sample and defined $s_{ij} = \mathcal{S}(t)\Xi_{ij}$ and consequently $\dot{s}_{ij} = \dot{\mathcal{S}}(t)\Xi_{ij}$, where $\mathcal{S}$ is an isotropic function of time. Substituting these values of the stress and the stress rate into Eq. (11.27.3) we obtain

$$\begin{aligned}
\varepsilon_{ij} &= \mathfrak{w}_1 \mathcal{S} \Xi_{ij} + \mathfrak{w}_2 \mathcal{S}^2 \Xi_{ij}^2 + \mathfrak{w}_3 \dot{\mathcal{S}} \Xi_{ij} + \mathfrak{w}_4 \dot{\mathcal{S}}^2 \Xi_{ij}^2 \\
&= \mathfrak{w}_1 \mathcal{S} \Xi_{ij} + \mathfrak{w}_2 \mathcal{S}^2 (2\delta_{ij} + \Xi_{ij}) + \mathfrak{w}_3 \dot{\mathcal{S}} \Xi_{ij} + \mathfrak{w}_4 \dot{\mathcal{S}}^2 (2\delta_{ij} + \Xi_{ij}) \\
&= 2(\mathfrak{w}_2 \mathcal{S}^2 + \mathfrak{w}_4 \dot{\mathcal{S}}^2)\delta_{ij} + (\mathfrak{w}_1 \mathcal{S} + 2\mathfrak{w}_2 \mathcal{S}^2 + \mathfrak{w}_3 \dot{\mathcal{S}} + 2\mathfrak{w}_4 \dot{\mathcal{S}}^2)\Xi_{ij} \\
&= \tfrac{1}{3} \varepsilon_{\beta\beta} \delta_{ij} + \mathcal{E} \Xi_{ij}
\end{aligned} \tag{11.27.4}$$

where $\mathcal{E}(t)\Xi_{ij}$ is the distortion tensor and $\varepsilon(t)_{\beta\beta}$ is a volumetric strain known as *dilatancy*, see Sects. 7.9 and 7.10. They are defined

$$\mathcal{E}\Xi_{ij} = (\mathfrak{w}_1 \mathcal{S} + 2\mathfrak{w}_2 \mathcal{S}^2 + \mathfrak{w}_3 \dot{\mathcal{S}} + 2\mathfrak{w}_4 \dot{\mathcal{S}}^2)\Xi_{ij} \tag{11.27.5}$$

$$\varepsilon_{\beta\beta} = 6(\mathfrak{w}_2 \mathcal{S}^2 + \mathfrak{w}_4 \dot{\mathcal{S}}^2) \tag{11.27.6}$$

We are at present interested in the volumetric strain $\varepsilon(t)_{\beta\beta}$ and its rate which is

$$\dot{\varepsilon}(t)_{\beta\beta} = 12(\mathfrak{w}_2 \mathcal{S}(t)\dot{\mathcal{S}}(t) + \mathfrak{w}_4 \dot{\mathcal{S}}(t)\ddot{\mathcal{S}}(t) \tag{11.27.7}$$

From here on the evaluation of the excess stored specific free energy is straightforward. Similarly to Eq. (11.25.3), the excess stored specific free energy can be defined, this time for the soil as a whole and not for the solids only

$$\Delta\rho\psi_i = \int_0^t \rho\dot{\psi}_i\,dt = \int_0^t \rho\,d\psi_i = \int_0^{t'} p_m\,d\varepsilon_{\alpha\alpha} + \int_0^{t''} p_m\,d\varepsilon_{\beta\beta}$$

$$= \int_0^{t'} p_m\dot{\varepsilon}_{\alpha\alpha}\,dt + \int_0^{t''} p_m\dot{\varepsilon}_{\beta\beta}\,dt \tag{11.27.8}$$

$\Delta\rho\psi_i$, or Eq. (11.27.8), represents the *excess stored specific free energy* invested in the volume element through densification, stored to be available in full or in part whenever needed. The volumetric strains $\varepsilon_{\alpha\alpha}$ and $\varepsilon_{\beta\beta}$ and the volumetric strain rates $\dot{\varepsilon}_{\alpha\alpha}$ and $\dot{\varepsilon}_{\beta\beta}$ are each isotropic functions of the stress and stress rate tensors. It is seen from Eqs. (11.27.2) and (11.27.6) that if isotropic relations are postulated, two volumetric strains and strain rates have to be considered when evaluating the excess stored specific free energy: the volumetric strain $\varepsilon_{\alpha\alpha}$ resulting from the spherical pressure that consolidates the soil prior to the shear stresses applied, and the volumetric strain $\varepsilon_{\beta\beta}$, which is a second-order effect, a result of the shear stresses applied thereafter. Both volume changes contribute to the strength of the soil, hence to the increase of the excess stored specific free energy. Since the two originate in different constraints that occur also at different times, the time limits of their integration differ: $\varepsilon_{\alpha\alpha}$ depends on the duration t' of the spherical pressure p_m, while $\varepsilon_{\beta\beta}$ depends on the duration t'' of the shear stresses s_{ij}. The coefficients $\beth_n$ and $ש_m$ are functions of the six stress and stress rate invariants.

11.28 The Excess Stored Specific Free Energy (Particular Cases)

Among the many possible modes of stress application the most frequent for the spherical consolidation pressure are the series of stepwise loadings, the single step steady loading and the constant rate of loading, and for the shear stresses the series of stepwise loadings and the constant rate of loading. The combination of the single step spherical pressure with stepwise shear stresses, and of the single step spherical pressure with the constant rate of shear stress will be explored here.

1. *Single step spherical pressure with stepwise shear stresses.* If the spherical pressure is applied to the soil element in a single step of constant pressure $p_m = \mathcal{P} = \text{const}$ and $\dot{p}_m = 0$, and if at the onset there was no other pressure on the soil $p(0)_m = 0$, then, according to Eq. (11.27.1) we may write the volumetric strain $\varepsilon_{\alpha\alpha}$

$$\varepsilon_{\alpha\alpha} = \beth_1 p_{\mathrm{m}} + \beth_2 p_{\mathrm{m}}^2 = \beth_1 \mathcal{P} + \beth_2 \mathcal{P}^2 \tag{11.28.1}$$

The part of the excess stored specific free energy resulting from the spherical pressure can be derived from Eq. (11.27.8)

$$\Delta\rho\psi_{\mathrm{i}}' = \int_0^{\varepsilon_{\alpha\alpha}} p_{\mathrm{m}}\, \mathrm{d}\varepsilon_{\alpha\alpha} = \int_0^{\mathcal{P}} p_{\mathrm{m}} \frac{\mathrm{d}\varepsilon_{\alpha\alpha}}{\mathrm{d}p_{\mathrm{m}}}\, \mathrm{d}p_{\mathrm{m}} = \int_0^{\mathcal{P}} (\beth_1 p_{\mathrm{m}} + 2\beth_2 p_{\mathrm{m}}^2)\, \mathrm{d}p_{\mathrm{m}}$$

$$= \left(\tfrac{1}{2}\beth_1 p_{\mathrm{m}}^2 + \tfrac{2}{3} p_{\mathrm{m}}^3\right)\Big|_0^{\mathcal{P}} = \tfrac{1}{2}\beth_1 \mathcal{P}^2 + \tfrac{2}{3}\beth_2 \mathcal{P}^3 = \tfrac{1}{2}\beth_1 p_{\mathrm{m}}^2 + \tfrac{2}{3}\beth_2 p_{\mathrm{m}}^3 \tag{11.28.2}$$

The coefficients $\beth_1$ and $\beth_2$ are evaluated from experimental data by curve fitting methods of the measured volumetric strain versus time, and then introduced in Eq. (11.28.2).

We assume that the shear stress is applied in a series of $\mathfrak{Q}$ steps $s_{ij\mathfrak{q}} = \mathcal{S}_{\mathfrak{q}}\Xi_{ij} = \text{const}$ and therefore $\dot{s}_{ij\mathfrak{q}} = 0$. It is possible to define the volumetric strain resulting from second-order effects in shear stresses for each of the steps, according to Eq. (11.27.6)

$$\varepsilon_{\beta\beta\mathfrak{q}} = 6\mathfrak{w}_2 s_{ij\mathfrak{q}}^2 = 6\mathfrak{w}_2 \mathcal{S}_{\mathfrak{q}}^2 \tag{11.28.3}$$

The total volumetric strain can now be evaluated

$$\varepsilon_{\beta\beta} = \sum_{\mathfrak{q}=1}^{\mathfrak{Q}} \varepsilon_{\beta\beta\mathfrak{q}} = 6\mathfrak{w}_2 \sum_{\mathfrak{q}=1}^{\mathfrak{Q}} s_{ij\mathfrak{q}}^2 = 6\mathfrak{w}_2 \sum_{\mathfrak{q}=1}^{\mathfrak{Q}} \mathcal{S}_{\mathfrak{q}}^2 \tag{11.28.4}$$

The excess stored specific free energy resulting from the shear stresses is obtained according to Eq. (11.27.8)

$$\Delta\rho\psi_{\mathrm{i}}'' = \int_0^{t''} \rho\dot{\psi}_{\mathrm{i}}''\mathrm{d}t = \int_0^{t''} \rho\,\mathrm{d}\psi_{\mathrm{i}}'' = \int_0^{t''} p_{\mathrm{m}}\, \mathrm{d}\varepsilon_{\beta\beta} = 6\mathfrak{w}_2 p_{\mathrm{m}} \sum_{\mathfrak{q}=1}^{\mathfrak{Q}} \mathcal{S}_{\mathfrak{q}}^2 \tag{11.28.5}$$

Since p_{m} is not a function of the volumetric strain $\varepsilon_{\beta\beta}$ it was possible to remove it from the integration. The sum of the two components of the excess stored energy, from Eqs. (11.28.2) and (11.28.5), is

$$\Delta\rho\psi_{\mathrm{i}} = \Delta\rho\psi_{\mathrm{i}}' + \Delta\rho\psi_{\mathrm{i}}'' = \int_0^{t'} p_{\mathrm{m}}\, \mathrm{d}\varepsilon_{\alpha\alpha} + \int_0^{t''} p_{\mathrm{m}}\, \mathrm{d}\varepsilon_{\beta\beta}$$

$$= \tfrac{1}{2}\beth_1 p_{\mathrm{m}}^2 + \tfrac{2}{3}\beth_2 p_{\mathrm{m}}^3 + 6\mathfrak{w}_2 p_{\mathrm{m}} \sum_{\mathfrak{q}=1}^{\mathfrak{Q}} \mathcal{S}_{\mathfrak{q}}^2$$

$$= \tfrac{1}{2}\beth_1 \mathrm{I}_p^2 + \tfrac{2}{3}\beth_2 \mathrm{I}_p^3 + 6\mathfrak{w}_2 \mathrm{I}_p \sum_{\mathfrak{q}=1}^{\mathfrak{Q}} \overline{\mathrm{II}}_{s\mathfrak{q}} \tag{11.28.6}$$

2. *Single step spherical pressure with constant rate of shear stresses.* The part of the excess stored specific free energy resulting from the spherical pressure

is similar to that of the stepwise loading of shear stresses, Eq. (11.28.2)

$$\text{(11.28.2)} \qquad \Delta\rho\psi_{\mathrm{i}}' = \int_0^{t''} p_{\mathrm{m}}\,\mathrm{d}\varepsilon_{\alpha\alpha} = \tfrac{1}{2}\beth_1 p_{\mathrm{m}}^2 + \tfrac{2}{3}\beth_2 p_{\mathrm{m}}^3$$

and the volumetric strain $\varepsilon_{\alpha\alpha}$ is equal to that in Eq. (11.28.1)

$$\text{(11.28.1)} \qquad \varepsilon_{\alpha\alpha} = \beth_1 p_{\mathrm{m}} + \beth_2 p_{\mathrm{m}}^2 = \beth_1 \mathcal{P} + \beth_2 \mathcal{P}^2$$

We assume that the shear stresses are applied at a constant rate, $\dot{s}_{ij} = \mathcal{S}\Xi_{ij} = \text{const}$, that is, they are linearly applied with time, $s_{ij} = \mathcal{S}\Xi_{ij}t$. The volumetric change $\varepsilon_{\beta\beta}$ resulting from second-order effects in shear stresses becomes, according to Eq. (11.27.6)

$$\varepsilon_{\beta\beta} = \mathfrak{w}_2 \mathcal{S}^2 t^2 + \mathfrak{w}_4 \mathcal{S}^2 \tag{11.28.7}$$

from which the strain rate is

$$\dot{\varepsilon}_{\beta\beta} = 2\mathfrak{w}_2 \mathcal{S}^2 t \tag{11.28.8}$$

The excess stored specific free energy resulting from the shear stresses can now be evaluated according to Eq. (11.27.8)

$$\Delta\rho\psi_{\mathrm{i}}'' = \int_0^{t''} p_{\mathrm{m}}\dot{\varepsilon}_{\beta\beta}\,\mathrm{d}t = 2\int_0^{t''} \mathfrak{w}_2 p_{\mathrm{m}} \mathcal{S}^2 t\,\mathrm{d}t = \mathfrak{w}_2 p_{\mathrm{m}} \mathcal{S}^2 t''^2 = \mathfrak{w}_2 p_{\mathrm{m}} s_{\alpha\beta}^2 \tag{11.28.9}$$

and the sum of the two components becomes

$$\Delta\rho\psi_{\mathrm{i}} = \int_0^{t'} p_{\mathrm{m}}\dot{\varepsilon}_{\alpha\alpha}\,\mathrm{d}t + \int_0^{t''} p_{\mathrm{m}}\dot{\varepsilon}_{\beta\beta}\,\mathrm{d}t = \tfrac{1}{2}\beth_1 p_{\mathrm{m}}^2 + \tfrac{2}{3}\beth_2 p_{\mathrm{m}}^3 + \mathfrak{w}_2 p_{\mathrm{m}} s_{\alpha\beta}^2$$

$$= \tfrac{1}{2}\beth_1 \mathrm{I}_p^2 + \tfrac{2}{3}\beth_2 \mathrm{I}_p^3 + \mathfrak{w}_2 \mathrm{I}_p \overline{\mathrm{II}}_s = \tag{11.28.10}$$

The two cases presented here, stepwise shear stresses and constant rate of shear stress, are based on isotropic stress–strain relations. It is interesting to point out that in both cases the excess stored specific free energy is a function of the first invariant of the stresses I_p, as in the case of linear viscoelasticity, and of the second moment invariant of the shear stresses $\overline{\mathrm{II}}_s$ as well.

The results of these findings will be further explored in Chap. 13.

12 Shear Stress–Strain Phenomena

12.1 Introduction

A fundamental message of this work is the advantage gained from the separation of volumetric and deviatoric phenomena in the study of the physical and mechanical behavior of soils. With respect to the treatment of the volumetric phenomenon, consolidation, there is almost no discrepancy between this and the triaxial consolidation tests as they are in fact performed: the treatment of volumetric phenomena is indeed separate from that of deviatoric phenomena. The fact that the shear test is usually preceded by consolidation of the soil, and that since early in the history of soil mechanics consolidation is regarded, at least *de facto*, as a volumetric phenomenon, is perhaps responsible for the fair compatibility that has been established between volumetric phenomena and theory. However, this is not so with respect to deviatoric phenomena. The study of the deviatoric phenomena and the course it has taken commanded a continuous search for a proper formulation of "theories", that caused lasting controversies and prompted revisions. A typical example is the theory of plasticity, "imported" from the mechanical behavior of metals. Two basic assumptions of this theory, lack of influence of hydrostatic pressure and incompressibility, are in direct contradiction to experimental facts, even in metals. But they are inconsistent even with other assumptions, for example, of softening and hardening of the material, which are certainly density related phenomena. In soils both assumptions, the lack of effect of hydrostatic pressure and the incompressibility, are definitely unacceptable.

The study of deviatoric stress–strain phenomena has two aspects:

1. The study of deviatoric stress–strain relationship through the application of deviatoric conditions, up to the point of failure.
2. The study of shear stress–strain conditions at failure.

In the forthcoming sections of this chapter the first aspect of deviatoric phenomena in soils will be discussed in a form which is different from the accepted concepts in classical soil mechanics, but not necessarily in contradiction with them. Discrepancies as well as agreements between the presented material and the accepted concepts will occasionally be pointed out. The difference in the presentation of the fundamental parameters, such as s_{ij} versus p_m adopted here instead of the accepted q versus p, will be stressed and critically evaluated as well. The second aspect of the deviatoric phenomena, the conditions at failure, is left for the next chapter, but some of the essential concepts common to both aspects of shear will be discussed here.

12.2 Density Effects on Shear Stresses

It is believed that, contrary to a variety of common materials like steel, copper, brass, concrete, marble, etc. the deviatoric stresses and the ultimate stress in soils are greatly affected by the density of the soil. Actually, the stresses and the ultimate stress of many other materials are affected by their density. The changes in density, however, under ordinary loads are very small and their effect is thus negligible. Fig. 12.2.1 shows that a decrease of approximately 20% in the density of concrete may cause a reduction of more than 80% of its strength. It should be noted that the densities in Fig. 12.2.1 were obtained on separate samples, by densifying the mortar previous to its curing and not by compressing the hardened concrete product; nevertheless, it is a fair indication that in other, more ductile materials, the change in density affects their strength as well. Density changes that could produce significant changes in strength would, however, require considerable compressive stresses, usually not encountered in daily practice.

In soils, as in other materials, the density change is a function of spherical stress–strain changes which are volumetric phenomena, in addition to initial packing conditions. The density can be specified by the porosity n, but more effectively, and therefore more often, by the void ratio e.

We may conclude that during densification an energy is stored in the soil, to be available when the soil is sheared; the stored energy increases with the increase of the density, and a commensurable energy is spent during shearing. Conversely, a distension of the soil dwindles its inventory of stored energy

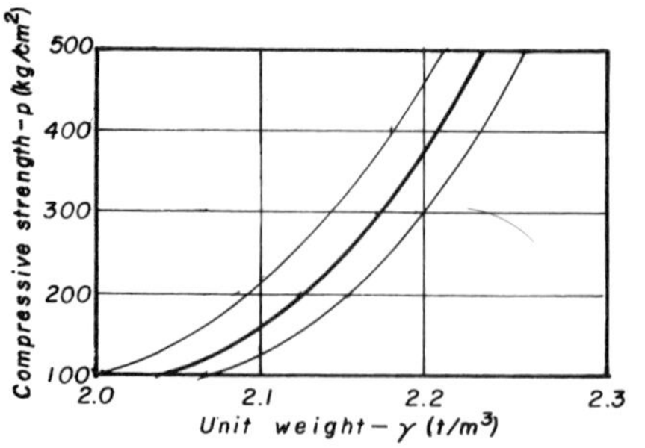

Fig. 12.2.1. Strength of concrete at different densities. [After O. Graf (1930), courtesy of Springer-Verlag.]

and spurs a reduction of its shear strength. Thus the shear test can best demonstrate the importance of the volumetric phenomena, i.e. that during consolidation a determined amount of energy is invested and stored in the soil, and during distention a corresponding amount of energy is lost.

12.3 The Shear Stress–Strain Relationship

From Eq. (9.4.11) it follows that if a pressure p_{ij} applied to a soil is decomposed into spherical and deviatoric stress components, then the deviatoric stress is carried by the solids

$$\begin{aligned} s_{ij} = s_{\mathfrak{s}ij} &= p_{ij} - \tfrac{1}{3}p_{\alpha\alpha}\delta_{ij} + \rho_{\mathfrak{s}}(\bar{v}_{\mathfrak{s}i}\bar{v}_{\mathfrak{s}j} - \tfrac{1}{3}\bar{v}_{\mathfrak{s}\alpha}\bar{v}_{\mathfrak{s}\alpha}\delta_{ij}) \\ &\quad + \rho_{\mathfrak{w}}(\bar{v}_{\mathfrak{w}i}\bar{v}_{\mathfrak{w}j} - \tfrac{1}{3}\bar{v}_{\mathfrak{w}\alpha}\bar{v}_{\mathfrak{w}\alpha}\delta_{ij}) + \rho_{\mathfrak{g}}(\bar{v}_{\mathfrak{g}i}\bar{v}_{\mathfrak{g}j} - \tfrac{1}{3}\bar{v}_{\mathfrak{g}\alpha}\bar{v}_{\mathfrak{g}\alpha}\delta_{ij}) \\ &= p_{\mathfrak{s}ij} - \tfrac{1}{3}p_{\mathfrak{s}\alpha\alpha}\delta_{ij} + \rho_{\mathfrak{s}}(\bar{v}_{\mathfrak{s}i}\bar{v}_{\mathfrak{s}j} - \tfrac{1}{3}\bar{v}_{\mathfrak{s}\alpha}\bar{v}_{\mathfrak{s}\alpha}\delta_{ij}) \\ &\quad + \rho_{\mathfrak{w}}(\bar{v}_{\mathfrak{w}i}\bar{v}_{\mathfrak{w}j} - \tfrac{1}{3}\bar{v}_{\mathfrak{w}\alpha}\bar{v}_{\mathfrak{w}\alpha}\delta_{ij}) + \rho_{\mathfrak{g}}(\bar{v}_{\mathfrak{g}i}\bar{v}_{\mathfrak{g}j} - \tfrac{1}{3}\bar{v}_{\mathfrak{g}\alpha}\bar{v}_{\mathfrak{g}\alpha}\delta_{ij}) \\ s_{\mathfrak{w}ij} &= s_{\mathfrak{g}ij} = 0 \end{aligned} \tag{12.3.1}$$

where s_{ij} is the total shear stress acting on the soil element, $s_{\mathfrak{s}ij}$ is the shear stress carried by the solids, $\frac{1}{3}p_{\mathfrak{s}ij}\delta_{ij}$ is the mean spherical pressure acting on the solids, considered already in the previous chapter, and $\rho_{\mathfrak{s}}(\bar{v}_{\mathfrak{s}i}\bar{v}_{\mathfrak{s}j} - \frac{1}{3}\bar{v}_{\mathfrak{s}\alpha}\bar{v}_{\mathfrak{s}\alpha}\delta_{ij})$, $\rho_{\mathfrak{w}}(\bar{v}_{\mathfrak{w}i}\bar{v}_{\mathfrak{w}j} - \frac{1}{3}\bar{v}_{\mathfrak{w}\alpha}\bar{v}_{\mathfrak{w}\alpha}\delta_{ij})$ and $\rho_{\mathfrak{g}}(\bar{v}_{\mathfrak{g}i}\bar{v}_{\mathfrak{g}j} - \frac{1}{3}\bar{v}_{\mathfrak{g}\alpha}\bar{v}_{\mathfrak{g}\alpha}\delta_{ij})$ are the dynamic shear stresses acting on the solids, water and air, respectively. The assumptions are that the water and the air cannot carry shear stresses.

If dynamic shear stresses do not exist or can be ignored due to their small size, Eq. (12.3.1) simplifies and reduces to

$$\begin{aligned} s_{ij} = s_{\mathfrak{s}ij} &= p_{ij} - \tfrac{1}{3}p_{\alpha\alpha}\delta_{ij} = p_{\mathfrak{s}ij} - \tfrac{1}{3}p_{\mathfrak{s}\alpha\alpha}\delta_{ij} \\ v_{\mathfrak{s}i}v_{\mathfrak{s}j} &= v_{\mathfrak{w}i}v_{\mathfrak{w}j} = v_{\mathfrak{g}i}v_{\mathfrak{g}j} \cong 0 \end{aligned} \tag{12.3.2}$$

Eqs. (12.3.1) and (12.3.2) imply that *the shear stress of soils s_{ij} is equal to the shear stress acting on the solids $s_{\mathfrak{s}ij}$ and is dependent only on the stresses acting on the solids*. This observation has already been made by Coleman (1961).

The traceless part of the strain tensor, obtained from Eq. (2.3.12) when the strain tensor is resolved into a spherical and a traceless tensor is

$$e_{ij} = \varepsilon_{ij} - \tfrac{1}{3}\varepsilon_{\alpha\alpha}\delta_{ij} \tag{2.3.12}$$

where, as we recall, e_{ij} is the traceless strain tensor, ε_{ij} is the total strain tensor and $\varepsilon_{\alpha\alpha}$ is the trace of tensor ε_{ij} forming the spherical strain tensor $\varepsilon_{\alpha\alpha}\delta_{ij}$. As was expounded in Sects. 7.5 and 11.1, only the infinitesimal strain or the finite Hencky strain can assure that the resolution of the strain tensor is not only mathematically possible but has also physical meaning. While the spherical strains for those strain measures are related to the volumetric strain, the traceless strain represents distortions of the volume element.

Under the circumstances just outlined, by which the strain tensor may be resolved into a volumetric strain and a distortional strain tensor, a constitutive relationship between the deviatoric stress $s_{ij} = s_{\mathfrak{s}ij}$ and the distortion e_{ij} can be postulated. If a linear relationship is postulated between the stress tensor p_{ij} and the strain tensor ε_{ij}, then Eq. (7.3.2) can be directly resolved into Eqs. (7.5.1) and (7.5.2) which are the dual constitutive equations, where Eq. (7.5.2) represents the relationship between the deviatoric stress $s_{ij} = s_{\mathfrak{s}ij}$, and the distortion e_{ij}. If, however, the relationship between the stress tensor p_{ij} and the strain tensor ε_{ij} is not linear, then isotropic relations can be postulated between the stress tensor p_{ij} and the strain tensor ε_{ij}, and the general stress–strain relationship, Eq. (7.3.1), can be resolved by a consecutive superposition of the traceless stress–strain relationship over the spherical stress–strain relationship. The physical meaning is still preserved, with some added provisions.

In the following sections of this chapter the deviatoric stress–strain relationship of soils is discussed.

12.4 Shear versus Volumetric Stress–Strain Relationship

Under laboratory conditions one may be able to approximate quite closely either volumetric or deviatoric stress conditions. In the field, however, when loads are applied, volumetric stresses come usually superimposed with deviatoric stresses. Unless pore pressures are present, the two act in opposing directions, the volumetric stresses tending to increase the density of the soil, concurrently increasing its resistance against deviatoric stresses and failure. Pore pressures, which are volumetric phenomena, as we have seen at the end of the previous chapter, act in opposition to volumetric stresses and alleviate their tendency to densify the soil, thus diminishing the amount of deviatoric stresses needed to cause distortion and failure. In practice, therefore, the nature of the deformation and failure of a structural element will depend on its geometry, the geometry of the forces involved and the relative intensity of the volumetric stresses versus the deviatoric stresses at any instant of the loading.

Both the volumetric and deviatoric stresses result in deformations, but there are great differences between the two. Volumetric stresses (pressure or tension) alone will never cause failure in the soil. They compress or decompress the soil, and if the structure is stiff enough and the forces evenly distributed they may cause settlements, sometimes excessive, but seldom detrimental to the soil. Moreover, the settlements decelerate with time and come to a final asymptotic value. Deviatoric stresses, on the other hand, increase with time and their rate accelerates, and ultimately the result is failure, *always accompanied by a tilt*, a change of angle. Therefore, failure by deviatoric stresses will occur in a relatively short time.

It is in general very easy to distinguish between deformations resulting mainly from volumetric stresses or mainly from deviatoric stresses. If the rate of deformation decreases with time the dominating stresses are volumetric, but if the rate increases with time the dominating stresses are deviatoric. A settlement dominated by volumetric stresses may transform into a settlement dominated by deviatoric stresses, due to changes in the distribution of the loads, caused by the settlement itself. However, a settlement dominated by deviatoric stresses can never revert to a settlement dominated by volumetric stresses, unless an outside intervention changes the geometrical circumstances.

Three classical examples can serve as an illustration for the above.

Mexico City (Carillo 1948; Cuevas 1936a; Rabe 1941) is located in a valley bounded on three sides by mountain ranges up to 4000 m high, some of volcanic origin, and open to the north, forming the elliptically shaped closed basin of the Mexican Plateau, approximately 2300 m above sea level. The basin, believed to have been a lake fed by many rivers and occasional floods, accumulated large deposits of water-transported sediments, Fig. 12.4.1. The

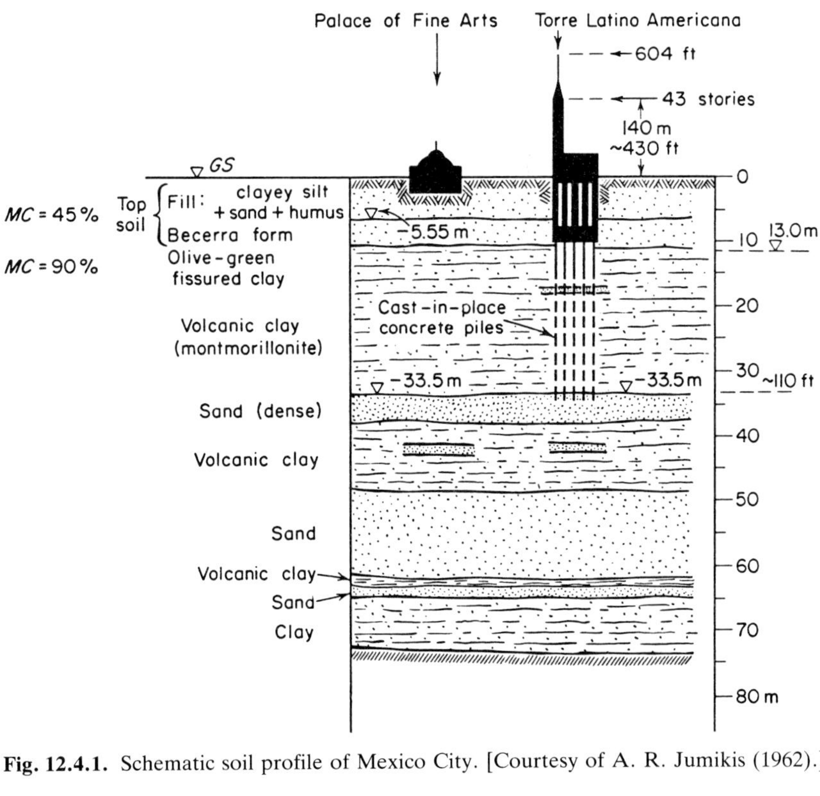

Fig. 12.4.1. Schematic soil profile of Mexico City. [Courtesy of A. R. Jumikis (1962).]

material accumulated formed a layered deposit of granular material and fines (gravel with sand and soft silts with clays), a total of several hundred meters. The mineralogical nature of the clay is 20% montmorillonite, 50% ashes, fossilized minute marine crustacea, 10% organic colloids and other substances. The natural water content of some of these clays is more than 400%, the void ratio up to 12 and the dry unit weight as low as 600 kg/m^3.

The soft deposits at Mexico City, and specifically two clay layers (13–33 m, 37–48 m), consolidate continuously. Subsidence measured in the downtown district of Mexico City during a period of 9 years (1939–1947) was some 140 cm (Carrillo 1948), Fig. 12.4.2. Subsidence due to intensive pumping from a large number of wells drilled to depths of 50–500 m, measured at 3 and 34 m depth and reported (Zeevaert 1953) for a period of $3\frac{1}{2}$ years (1949–1952), indicated 52 and 94 cm, respectively, Fig. 12.4.3. It is believed that since 1900 and until about 1957 Mexico City sunk close to 5 m and some buildings have settled more than 7.5 m.

Mexico City's soft soil structure with its high void ratio is a classical case where a volumetric phenomenon, consolidation, is the dominant factor and demands solutions accordingly. In order to achieve even settlement, buildings are constructed with stiff raft foundations, floating box foundations, weight lighter than the excavated soil (Cuevas 1936b) or cell foundations adjusted to

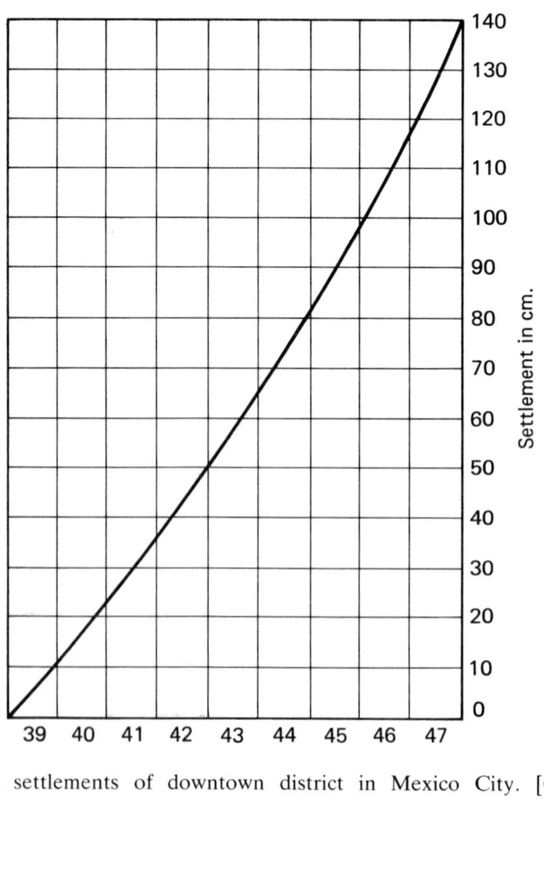

Fig. 12.4.2. Average settlements of downtown district in Mexico City. [Courtesy of Dr L. Zeevaert (1953).]

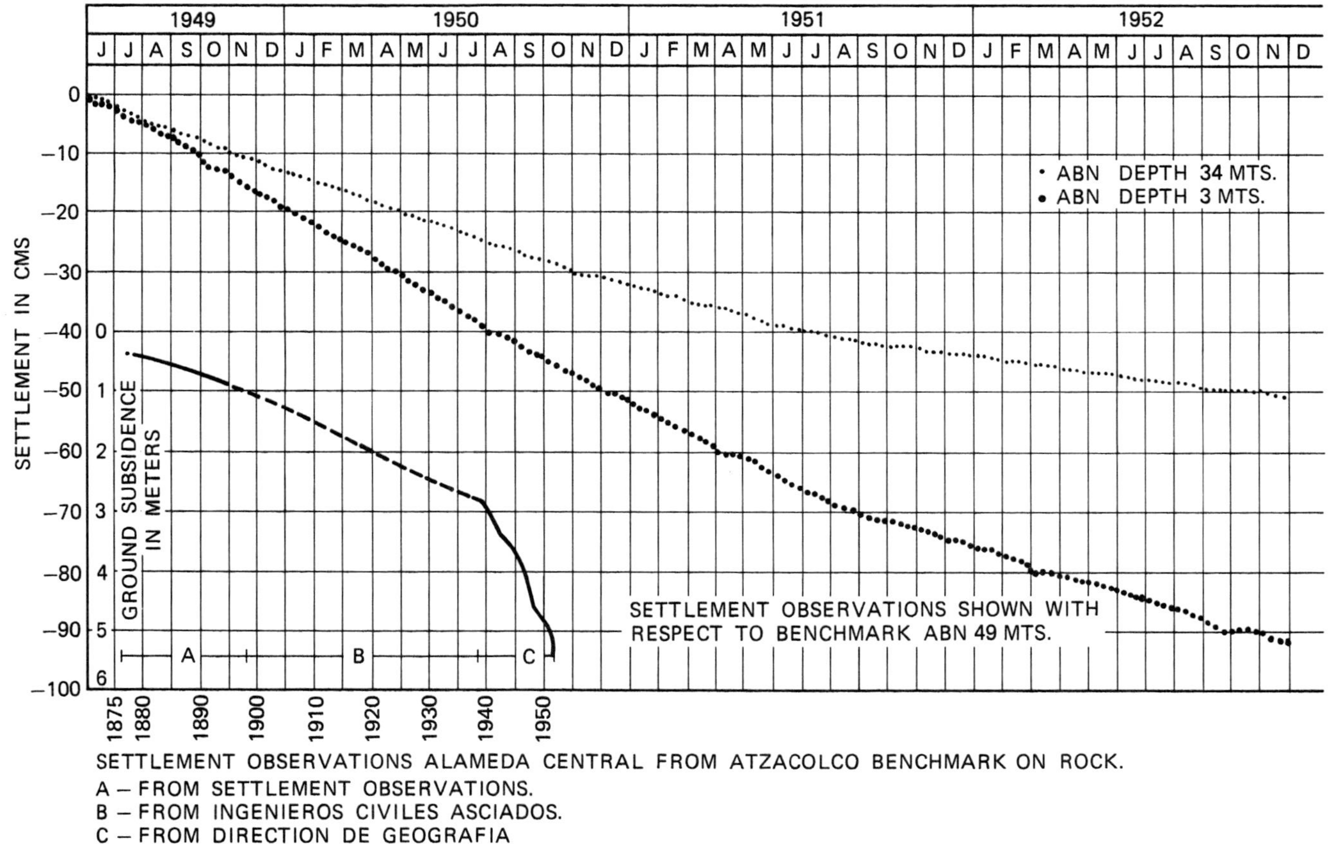

Fig. 12.4.3. Subsurface subsidence measured at 3 m and 34 m. [Courtesy of Dr L. Zeevaert (1953).]

settle evenly by pumping water into and out of the cells. Buildings on flexible foundations develop cracks due to *local* deviatoric stresses associated with differential settlements. Tall and slender structures on shallow foundations, susceptible to eccentricity even for small asymmetric loadings, develop excess deviatoric stresses, settle unevenly and go out of plumb. Therefore tall buildings are founded on point bearing piles driven to the sand layer, approximately 35 m (Zeevaert 1957).

The second example is the leaning tower of Pisa, Fig. 12.4.4, one of many similar structures, like fortification towers, castles, cathedrals and bell towers, all of them slender and tall. If not properly anchored into the soil, such structures settle due to consolidation, and develop eccentricity as a result of inevitable inhomogeneity of the soil, different soil profiles under the various parts of the structure, and uneven distribution of the loads.

As the settlement progresses, the eccentricity advances and becomes more and more pronounced. Contact pressures between the foundation and the soil, which at the beginning are evenly distributed normal pressures inducing

Fig. 12.4.4. The Leaning Tower of Pisa. [Courtesy of A. R. Jumikis (1962).]

consolidation, become unevenly distributed over the contact area and can even cause part of the foundation to detach from the soil, reducing the contact area. Such uneven distribution of contact pressures induces deviatoric stresses into the soil, ultimately causing failure.

The tower of Pisa was started in 1174 and began tilting from the very beginning of the construction. After several interruptions and slight changes in the plans it was finally completed in 1350. The tower, Fig. 12.4.4, comprising eight galleries, is 55.86 m high, outside diameter at the bottom 7.37 m, weighs 14500 tons and is founded on a circular ring foundation with an outside diameter of 19.48 m, inside diameter 4.47 m. The average contact pressure should be 5.14 kg/cm^2; due to eccentricity and wind pressure, the maximum contact pressure becomes 9.6 kg/cm^2.

The soil profile beneath the tower consists of 8 m of clayey sand or sand with clay lenses, underlaid by a plastic clay. The tilting of the tower was observed in 1186 when the masonry had reached some 12 m. In 1932 more than 1000 tons of high-strength cement was injected into the ground (Krynine 1947), through 361 holes of 2 in. diameter, in order to stabilize it. Terzaghi (1934) published a study of stress distribution underneath the tower, Fig. 12.4.5, based on the consolidation theory. He also published a time-settlement diagram for the tower, Fig. 12.4.6, indicating the differential settlements between the highest and lowest point of the base of the footing, and estimated the absolute settlement of the tower as 2.4 m. The center of gravity of the tower is 23.2 m from the bottom of the footing and is out of plumb by close to 5.5 m.

Unless stabilization measures to decelerate and arrest the settlements were taken, the tower would have reached, perhaps, a failure course similar to that of the bell tower of Venice, which collapsed in 1902.

Failure of the soil in the case of silos and other similar storage structures, with high live load–dead load ratio and high behavior and probability factor, is quite frequent. The picture in Fig. 12.4.7 shows a row of 10 circular silos supported by a mat foundation 49 ft by 225 ft, which collapsed eight months after their completion, due to failure of the soil.

The study of this failure (White 1940), resumed in 1949 (Tschebotarioff 1952), revealed that the foundation was at a depth of 3 ft from ground level on one side and 10.6 ft on the other side. The soil beneath the mat was a varved clay of glacial origin, of a brown-gray color; its unconfined compressive strength dropped from $p_u = 2.0$ ton/ft^2 at the depth of the foundation mat to $p_u = 1.0$ ton/ft^2 at 18 ft below the mat, a reflection of the corresponding increase of the water content from 34% to 46%, the water table varying in depth between 7 and 15 ft. The sensitivity of the soil increased with the depth, from 2.0 to 5.0 according to the definition by Terzaghi, and from 2.5 to 10 according to the definition by Tschebotarioff, see Sect. 8.23, indicating quite a sensitive clay.

The contact pressure between the foundation mat and the soil due to the dead load of the silos was approximately $p_c = 0.3p_{cmar}$ and produced a settlement of $\frac{1}{8}$ in. on one side and $\frac{1}{16}$ in. on the other side of the silo, where p_{cmar} represents the maximum contact pressure, which includes the dead load

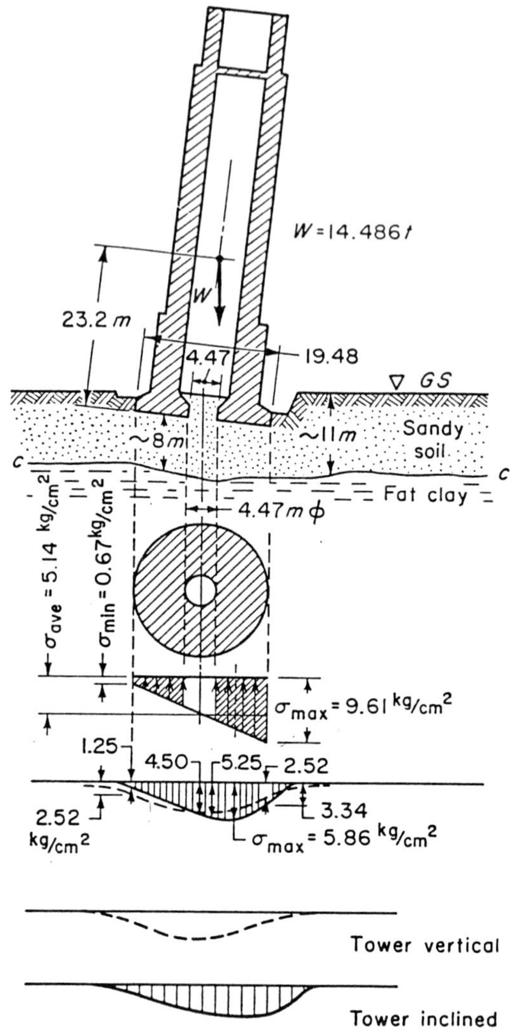

Fig. 12.4.5. Stress distribution under the Tower of Pisa according to Terzaghi. [Courtesy of A. R. Jumikis (1962).]

(the weight of the silos) and the live load (the contents of the silos). Within a month, owing to the filling of the silos the contact pressure reached $p_c = 0.82p_{cmar}$, and the tilt advanced and reversed, the settlements on each side reaching the values of 1.0 in. and $1\frac{5}{16}$ in., respectively. The silos remained partially filled for six more months, and the contact pressure varied within the limits of $p_c = 0.71\text{–}0.82p_{cmar}$, the settlement during this period increasing to the values of $8\frac{7}{8}$ in. and $10\frac{3}{4}$ in. and the rate of settlement increasing rapidly. The consolidation in these settlements was estimated as 4.8 in., thus more than half of each settlement was due to shear. No attention was paid to the rapid increase of the rate of settlement, and during the following month the continued filling of the silos increased the pressure to its highest final value,

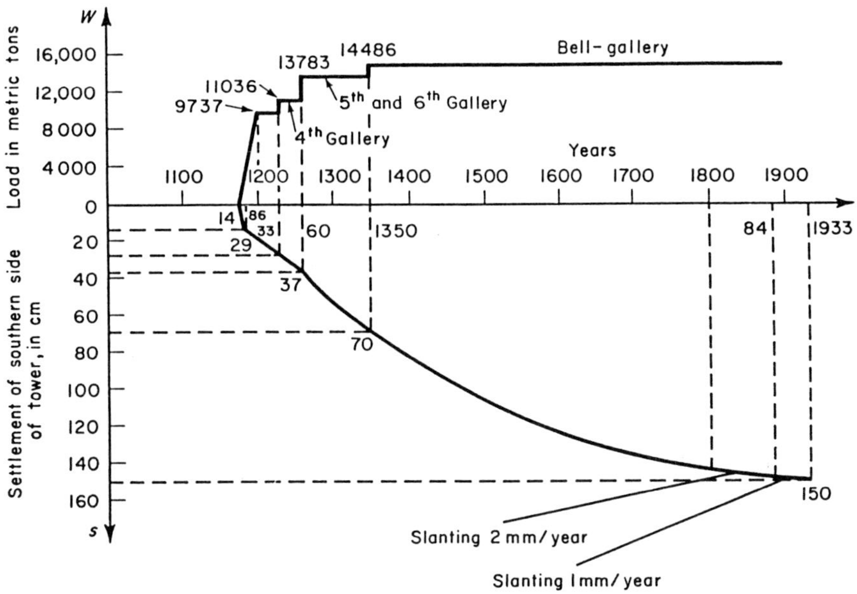

Fig. 12.4.6. Settlement with time of the Tower of Pisa. [Courtesy of A. R. Jumikis (1962).]

the rate of settlement increased, and prior to failure the settlement reached $11\frac{7}{16}$in. and $13\frac{7}{8}$in, respectively. Then suddenly the tilt again reversed direction, and within 2 minutes the silos overturned, as seen in Fig. 12.4.7, with an almost cylindrical sliding surface of a radius close to 60 ft, the rising edge seen in the picture.

Had the silos been filled at a slower pace, permitting slow and proportional consolidation of the clay layers, the failure could have been avoided. Also, if the foundation mat had been deeper, part of the silo filling would have compensated for the excavated soil, and the sliding surface would have somewhat lengthened.

Each of the three examples represents a specific pattern of the competing energies involved in straining soils. In Mexico City the dominating energies are due to spherical pressures, consequently the strains are spherical, namely there is consolidation, and however large the strains may be, failure is not involved. Existing distortions are mostly local and due to differential settlements of the foundation, and result in cracks. The settlement in consolidation is dissipative and its rate is decreasing for a constant load.

The second example, the Pisa tower and similar structures, shows the real competition between spherical pressures and deviatoric stresses; the spherical pressures result in settlements and an increase of the strength of the soil, due to consolidation, while the deviatoric stresses result in settlements and failure, due to distortion. In this manner, the energy stored by the spherical pressures is spent by the deviatoric stresses. The distinction between the two settlements is significant, and in the field they can be identified by establishing whether the rate of settlement increases or decreases.

a

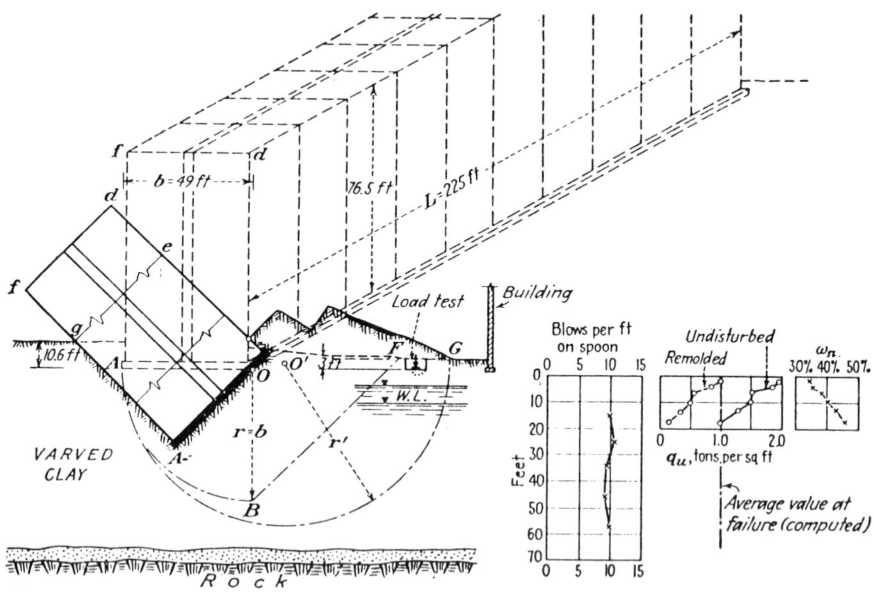

b

Fig. 12.4.7. Collapse of a row of heavy reinforced-concrete silos due to shear failure of the underlying varved clay. **a** View of the silos after failure. **b** Schematic drawing of the silos shown before and after failure. [From G. P. Tschebotarioff (1952), courtesy of McGraw-Hill Book Co. Inc.]

Finally, the collapse of the silo and similar stockhouse structures is a result of dominant deviatoric stresses, owing to distortion and failure of the soil articulated by settlements, the rate of which increases with time when no cumulative loads are added.

12.5 Deviatoric Tests

The immediate and distinct testing requirements for soils, as well as for other visco-elasto-plastic-compressible materials, as follows from the theory of the dual constitutive equation outlined in Sect. 7.5, are the separation between the volumetric stress–strain testing of soils and the deviatoric stress–strain testing. A deviatoric stress condition is characterized by the lack of spherical stresses, and one can now define a deviatoric loading as a state of stress in which the increment of the spherical tensor component and the increment of its highest-order time derivatives vanish

$$\Delta p_{aa} = 0; \quad \Delta \dot{p}_{aa} = 0; \quad \Delta \ddot{p}_{aa} = 0; \quad \ldots \tag{12.5.1}$$

Eq. (12.5.1), which is a necessary and sufficient condition for deviatoric loading, is the equation on which all deviatoric loading devices and tests, including pure deviatoric tests, are based.

In Sects. 11.10 and 11.11 we have seen that the triaxial testing device is a sophisticated equipment for the investigation of the spherical stress–strain phenomena of soils and its many facets. We will see that the same device is appropriate, if properly used, for the deviatoric testing of soils as well, and for the study of their deviatoric behavior. The fact that the same equipment can be used for both the spherical and the deviatoric study and testing of soils makes the triaxial testing device a powerful tool in soil science and engineering. The conventional triaxial shear test is reviewed in the next section, and the presentation of the direct shear and the simple shear tests and their merits follows. Deviatoric tests, torsion and pure deviatoric tests, are then introduced.

12.6 The Conventional Triaxial Shear Test

The triaxial device that performs triaxial shear tests, *drained, consolidated undrained* and *undrained*, and its even more diversified implementation (Bishop and Henkel 1962), has become the most versatile testing device for the investigation of shear in soils. From the start (Rendulic 1937) the use of the triaxial device has gained impetus, and in 1960 the ASCE Shear Strength Conference in Boulder, Colorado demonstrated and summarized the achievements arrived at through this device. The development and advance of the triaxial testing device has continued ever since.

The conventional triaxial shear test is performed on a cylindrical soil sample placed in the triaxial testing device described in Sect. 11.10. It generally succeeds a spherical consolidation, that is, a spherical pressure applied according to Eq. (11.11.1). After consolidation or equilibrium, an additional vertical pressure Δp_{33} is applied to the sample through the piston that penetrates the top plate. The load on the piston is produced by a loading device of some sort, mechanical, electrical, pneumatic, etc. and is monitored accordingly by a proving ring, strain gage, pressure gage, etc. The load may be applied stepwise, linearily increasing with time, or by any other loading procedure, as the specific need requires.

The triaxial shear test can be of the *drained* type, where there is either draining at both ends, or draining at one end and a possibility of measuring pore pressures at the other end, or of the *undrained* type, where there is no means for draining but pore pressures can be measured. The vertical strain of the sample is measured by a dial gage and the volume changes through the confining liquid burette, or, in the drained test, by the expelled water. Other optional measurements are possible, like the measurement of pore pressures in the sample at places other than the ends, through inserted probes if the size of the sample permits it, as well as measurements of the changes in the diameter of the sample at specific locations. Except for the control and the measurement of the vertical pressure applied, all other measurements are performed the same way as in the spherical consolidation.

The conventional triaxial shear test, irrespective of its type, does apply deviatoric stresses to the soil sample, but it has several features which prevent it from qualifying as a pure deviatoric test:

1. It induces, through the application of the vertical pressure Δp_{33}, not merely a deviatoric pressure but also an additional spherical pressure. If the applied pressure tensor Δp_{ij} is

$$\Delta p_{ij} \equiv \begin{vmatrix} 0 & 0 & 0 \\ 0 & 0 & 0 \\ 0 & 0 & \Delta p_{33} \end{vmatrix}$$

then it resolves into a spherical and a deviatoric stress tensor, Fig. 12.6.1

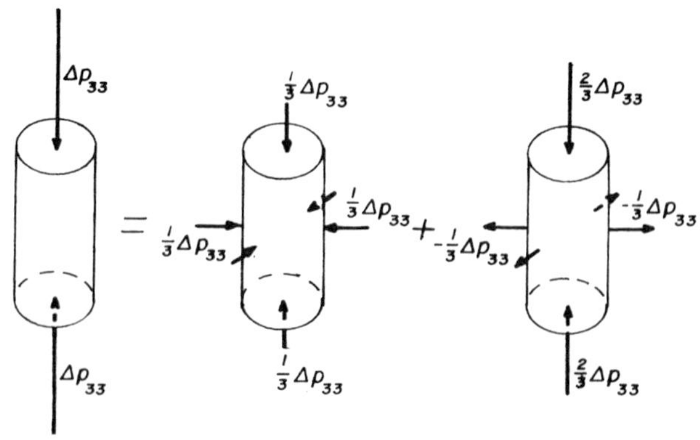

Fig. 12.6.1. Resolution of the pressure increment into spherical and deviatoric pressures in a triaxial loading.

$$\Delta p_{ij} \equiv \begin{vmatrix} 0 & 0 & 0 \\ 0 & 0 & 0 \\ 0 & 0 & \Delta p_{33} \end{vmatrix}$$

$$= \frac{\Delta p_{33}}{3} \begin{vmatrix} 1 & 0 & 0 \\ 0 & 1 & 0 \\ 0 & 0 & 1 \end{vmatrix} + \frac{\Delta p_{33}}{3} \begin{vmatrix} -1 & 0 & 0 \\ 0 & -1 & 0 \\ 0 & 0 & 2 \end{vmatrix}$$

$$\equiv \frac{\Delta p_{33}}{3} \delta_{ij} + \frac{\Delta p_{33}}{3} \Xi_{ij} \tag{12.6.1}$$

where $\frac{1}{3}\Delta p_{33}\delta_{ij}$ is the spherical stress tensor and $\frac{1}{3}\Delta p_{33}\Xi_{ij}$ is the deviatoric stress tensor. The pressure tensor that during consolidation was

$$p_{ij} \equiv \begin{vmatrix} p_c & 0 & 0 \\ 0 & p_c & 0 \\ 0 & 0 & p_c \end{vmatrix} \equiv p_c\delta_{ij} \tag{12.6.2}$$

becomes, after the application of pressure Δp_{33}, Fig. 12.6.2

$$p_{ij} \equiv \begin{vmatrix} p_c & 0 & 0 \\ 0 & p_c & 0 \\ 0 & 0 & p_c + \Delta p_{33} \end{vmatrix} = \begin{vmatrix} p_c & 0 & 0 \\ 0 & p_c & 0 \\ 0 & 0 & p_{33} \end{vmatrix}$$

$$\equiv \tfrac{1}{3}(p_{33} + 2p_c)\delta_{ij} + \tfrac{1}{3}(p_{33} - p_c)\Xi_{ij} = (p_c + \tfrac{1}{3}\Delta p_{33})\delta_{ij} + \tfrac{1}{3}\Delta p_{33}\Xi_{ij} \tag{12.6.3}$$

In terms of effective pressures Eq. (12.6.3) can be written

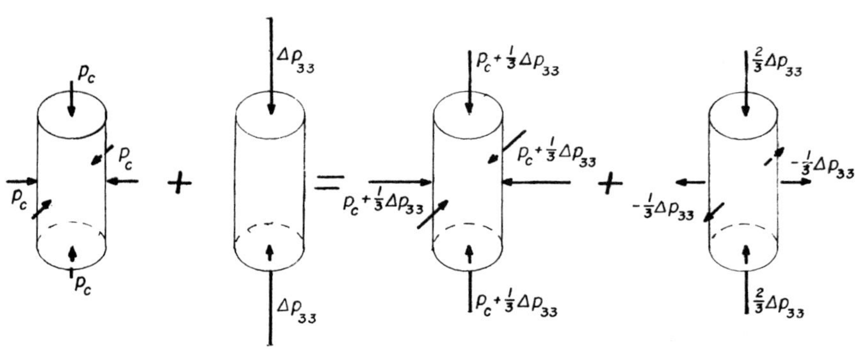

Fig. 12.6.2. Resolution of pressures of the triaxial test under constant vertical pressure.

$$p_{\mathfrak{s}ij} = \tfrac{1}{3}(p_{\mathfrak{s}33} + 2p_{\mathfrak{s}c})\,\delta_{ij} + \tfrac{1}{3}(p_{\mathfrak{s}33} - p_{\mathfrak{s}c})\,\Xi_{ij}$$

$$= (p_{\mathfrak{s}c} + \tfrac{1}{3}\Delta p_{\mathfrak{s}33})\delta_{ij} + \tfrac{1}{3}\Delta p_{\mathfrak{s}33}\,\Xi_{ij} \tag{12.6.4}$$

Consequently, the spherical presure $\frac{1}{3}\Delta p_{33}$ or the effective spherical pressure $\frac{1}{3}\Delta p_{\mathfrak{s}33}$ results in a volume change of the sample, compression. If the load applied is gradually increasing, the resulting gradual change in pressure $\frac{1}{3}\Delta p(t)_{33}$ will continuously change the density of the sample, thus the sample at the end of the test is different from the sample at the start.

2. The rate of deviatoric stress application affects the magnitude of the distortion and the other shear properties of the sample, and the rate of the added spherical pressure further affects the shear properties, due to the density changes and the pore pressures that develop in the sample. Thus, the resulting effect on the shear properties of the sample is a superposition of two rate effects, one resulting from the added spherical component and one resulting from the deviatoric component of the pressure applied.

3. The way the shear stresses are applied in a conventional triaxial shear test is partly the reason for the existence of volume changes or pore pressures that arise in the drained and consolidated-undrained tests. The added spherical pressure causes volume changes or pore pressures, while if the pressures applied were deviatoric pressures alone, no volume changes or pore pressures would develop, and so the results of the two types of tests would not differ.

4. According to definition, the *critical void ratio* is "the void ratio at which prevention of volume change leads to no strength change" (Taylor 1948), introduced earlier (US Engineers Corps 1938), in somewhat different terms. Dense sands generally reach greater strength when volume changes are prevented during shear, and loose sands lose strength. It was shown (Roscoe et al. 1958) that the critical void ratio concept is valid for clays as well, see also Sect. 12.22. The critical void ratio concept is a result of the way shear tests are performed, namely, that the stresses applied during the triaxial shear test are not pure deviatoric stresses, as was demonstrated. Assuming linear isotropic soil behavior, deviatoric stresses result in distortions alone, without volume changes. If, however, non-linear isotropic relations are postulated, the critical void ratio has to be formulated in a different way. Consequently, critical void ratio related "theories" are rooted in misconceived and misinterpreted tests.

It is noted, however, that the two parameters p and q, introduced to represent triaxial yield surfaces (Roscoe et al. 1958) and defined $\frac{1}{3}(\sigma_1' + 2\sigma_3')$ and $q = \sigma_1' - \sigma_3'$ (in the conventional notation), correspond to the spherical and deviatoric pressure components, respectively (in our notation), as expressed in Eq. (12.6.4), $\frac{1}{3}p_{\mathfrak{s}\alpha\alpha} = \frac{1}{3}(p_{\mathfrak{s}33} + 2p_{\mathfrak{s}c}) \equiv \frac{1}{3}(\sigma_1' + 2\sigma_3')$ and ab18$_3$. It must be remembered, however, that this representation is correct only for axially symmetric principal stresses and not, as often used, for general representation.

Usually the volume changes and the vertical deformations are measured during the triaxial shear test, then the volumetric strains $\varepsilon_{\alpha\alpha}$ and the vertical strains ε_{33} are evaluated and deviatoric stress–vertical strain curves and volumetric strain–vertical strain curves are drawn for the different spherical cell pressures p_c. From the volumetric strains and the vertical strains the average horizontal radial strains $\varepsilon_{11} = \varepsilon_{22}$ can be computed, irrespective of whether the strains are defined as infinitesimal or as Hencky strains

$$\varepsilon_{11} = \varepsilon_{22} = \tfrac{1}{2}(\varepsilon_{\alpha\alpha} - \varepsilon_{33}) \tag{12.6.5}$$

and the strain tensor ε_{ij} can be written and decomposed, considering Eq. (12.6.5)

$$\varepsilon_{ij} \equiv \begin{vmatrix} \tfrac{1}{2}(\varepsilon_{\alpha\alpha} - \varepsilon_{33}) & 0 & 0 \\ 0 & \tfrac{1}{2}(\varepsilon_{\alpha\alpha} - \varepsilon_{33}) & 0 \\ 0 & 0 & \varepsilon_{33} \end{vmatrix} \equiv \tfrac{1}{3}\varepsilon_{\alpha\alpha}\,\delta_{ij} + \tfrac{1}{6}(3\varepsilon_{33} - \varepsilon_{\alpha\alpha})\Xi_{ij} \tag{12.6.6}$$

where the right-hand side contains a spherical strain and a distortion.

Pursuing the stress–strain relationships, the two parts of Eq. (12.6.6) can be correlated with the two parts of either the total pressures, Eq. (12.6.3), or the effective pressures, Eq. (12.6.4), if pore pressures are also measured. Hence, the aforementioned representation can be replaced by a more meaningful display of the results, spherical pressures versus volumetric strain and deviatoric stresses versus distortions.

12.7 The Unconfined Compression Test

A particular case of the conventional triaxial shear test, where no initial spherical confining pressure is applied to the sample before the application of the vertical pressure p_{33}, that is, $p_{11} = p_{22} = 0$, is an *unconfined compression test.*

The unconfined compression test is performed on a cylindrical sample in a simple vertical loading device, the load applied mechanically or manually, and the only measurement performed is the vertical deformation. Fig. 12.7.1 shows an unconfined compression device, operated manually and activated by spring, which automatically records the load and the deformation. Unconfined compression tests are occasionally carried out in the triaxial testing device; in this case, although no confining pressure is applied, volume changes can be measured through the cell fluid, and pore pressures can be measured as well.

The pressure tensor p_{ij} decomposes

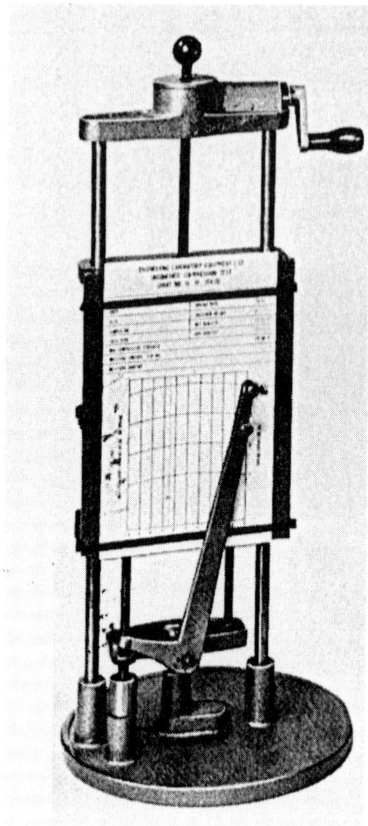

a

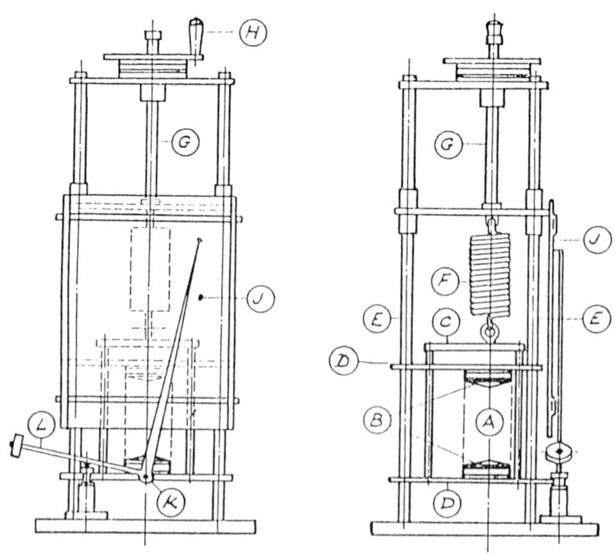

b

Fig. 12.7.1. Automatically recording unconfined compression device. **a** View of the device. **b** Schematic drawing of the device. [From P. L. Capper and W. F. Cassie (1953), courtesy of McGraw-Hill Book Co. Inc.]

$$p_{ij} \equiv \begin{vmatrix} 0 & 0 & 0 \\ 0 & 0 & 0 \\ 0 & 0 & p_{33} \end{vmatrix} = \frac{p_{33}}{3}\begin{vmatrix} 1 & 0 & 0 \\ 0 & 1 & 0 \\ 0 & 0 & 1 \end{vmatrix} + \frac{p_{33}}{3}\begin{vmatrix} -1 & 0 & 0 \\ 0 & -1 & 0 \\ 0 & 0 & 2 \end{vmatrix}$$

$$\equiv \frac{p_{33}}{3}(\delta_{ij} + \Xi_{ij}) \tag{12.7.1}$$

into a spherical and a deviatoric pressure tensor, proportional to the vertical pressure p_{33}. If pore pressures are measured, the effective pressures can be computed

$$p_{\bar{s}ii} = \tfrac{1}{3} p_{\bar{s}33}(\delta_{ij} + \Xi_{ij}) \tag{12.7.2}$$

The vertical strain is evaluated, then a stress–strain curve, Fig. 12.7.2, is

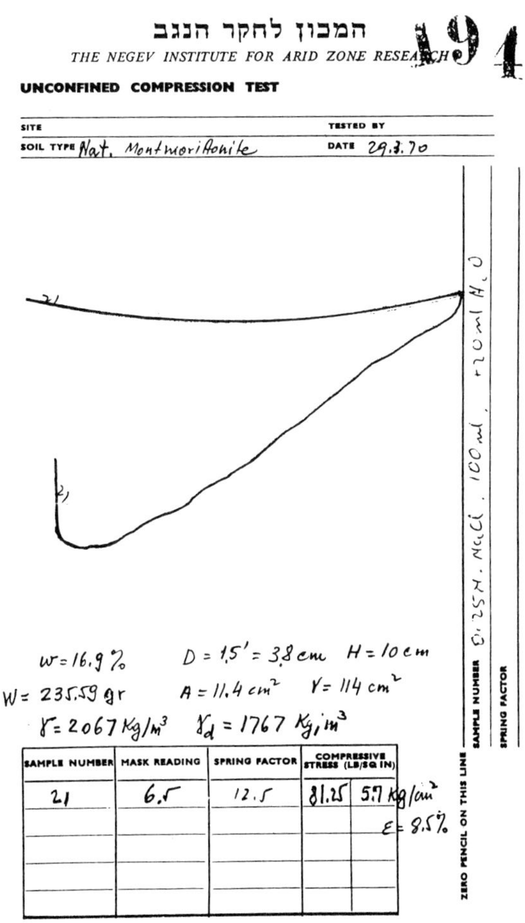

Fig. 12.7.2. Stress–strain curve of an unconfined compression test.

drawn. If, however, volume changes are also measured, the average horizontal radial strain $\varepsilon_{11} = \varepsilon_{22}$ can be computed as in Eq. (12.6.5) and the strain tensor ε_{ij} becomes the same as in Eq. (12.6.6).

12.8 The Simple Shear Test

The theoretical aspects of strains in a simple shear were discussed in Sect. 3.17. The simple shear strain is a plane strain, and according to Sect. 3.16 there exists a family of parallel planes which are identically preserved. In Fig. 3.17.1 all YZ planes normal to the X direction are identical and behave alike. Fig. 12.8.1 shows one of these typical planes, in which each line $Y = \text{const}$ representing a XZ plane rotates rigidly about the line of intersection between the XZ plane and the $Z = 0$ plane, in an angle α representing the angle of distortion. The $Z = \text{const}$ and $X = \text{const}$ planes remain unchanged.

The Green and Almansi strain tensors will be represented, respectively, by the matrices

$$\overset{G}{\varepsilon}_{ij} = \begin{vmatrix} 0 & 0 & 0 \\ 0 & 0 & \tfrac{1}{2}S \\ 0 & \tfrac{1}{2}S & \tfrac{1}{2}S^2 \end{vmatrix} ; \quad \overset{A}{\varepsilon}_{ij} = \begin{vmatrix} 0 & 0 & 0 \\ 0 & 0 & \tfrac{1}{2}S \\ 0 & \tfrac{1}{2}S & -\tfrac{1}{2}S^2 \end{vmatrix} \tag{12.8.1}$$

and the Hencky strain by

$$\overset{H}{\varepsilon}_{ij} = \begin{vmatrix} 0 & 0 & 0 \\ 0 & \tfrac{1}{2}\ln\left[1 + \tfrac{1}{2}S^2 + S\sqrt{(1 + \tfrac{1}{4}S^2)}\right] & 0 \\ 0 & 0 & \tfrac{1}{2}\ln\left[1 + \tfrac{1}{2}S^2 - S\sqrt{(1 + \tfrac{1}{4}S^2)}\right] \end{vmatrix} \tag{12.8.2}$$

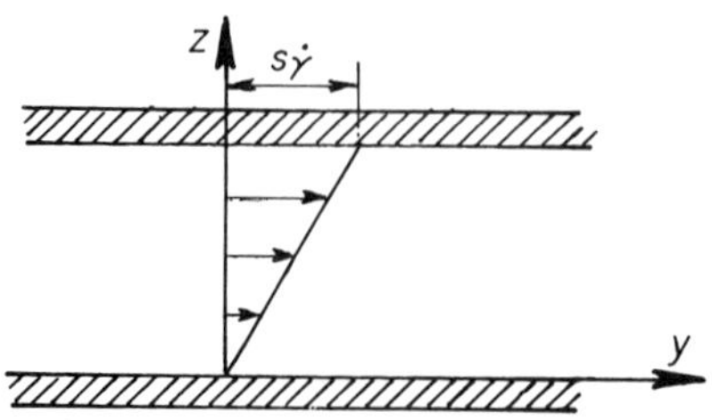

Fig. 12.8.1. Plain strain in simple shear flow.

The principal directions of the strain rotate and bisect the angle of the original coordinates Z and Y, making an angle θ with the xy plane, Fig. 12.8.1, so that

$$\tan\theta = \tfrac{1}{2}S + \sqrt{(\tfrac{1}{4}S^2)} \tag{12.8.3}$$

The velocity field is represented in the rectangular coordinates by the velocities

$$v_x = 0; \quad v_y = S\dot{Y}; \quad v_z = 0 \tag{12.8.4}$$

The construction of a simple shear test apparatus, however simple the test may seem, is quite a problem from the technical point of view. This is possibly the reason why the simple shear test did not become widely used. In the attempt to improve the apparatus, eight different models were developed at Cambridge (Roscoe 1970) and the various test results are found in many publications (Roscoe 1953; Bassett 1967; Cole 1967; Roscoe et al. 1967; Duncan and Dunlop 1969; Stroud 1971). The box housing the sample must be rectangular, Fig. 12.8.2. After the sample is placed in the shear box, and before the shear test is performed, a consolidation pressure is applied in order to be able to relate the shear stress to the density or void ratio.

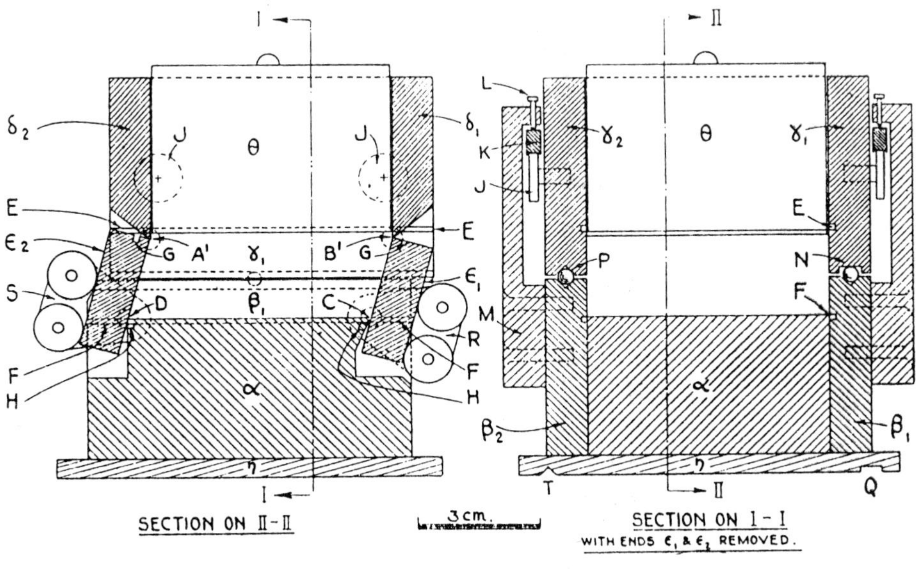

Fig. 12.8.2. Schematic drawing of a simple shear apparatus. [From K. H. Roscoe (1953).]

Aside from the technical problem that the settlement, due to consolidation, imposes on the design of the device, the test is not really a pure deviatoric test, especially under low initial consolidation pressures. Since there is no complete adherence of the sample to the vertical walls of the shear box, there are no complementary shear stresses along the walls, consequently the shear stresses, and the vertical pressures p_{zz} as well, are not evenly distributed over the end surfaces of the sample. As seen from Fig. 12.8.3, p_{zz} varies from infinite compression at the trailing edges of the end surfaces to infinite tension at the leading edges of the end surfaces. This tension results in a separation of the sample ends from the shear box, illustrated in Fig. 12.8.4 on a plasticine sample under 0.125 kg/cm^2 vertical pressure. Under higher vertical pressures, 5.0 kg/cm^2, such separation was not observed. Analytical studies indicate that the value of the shear stresses is maximal at the centers of the surfaces, falling off to zero towards the leading and trailing edges, Fig. 12.8.5.

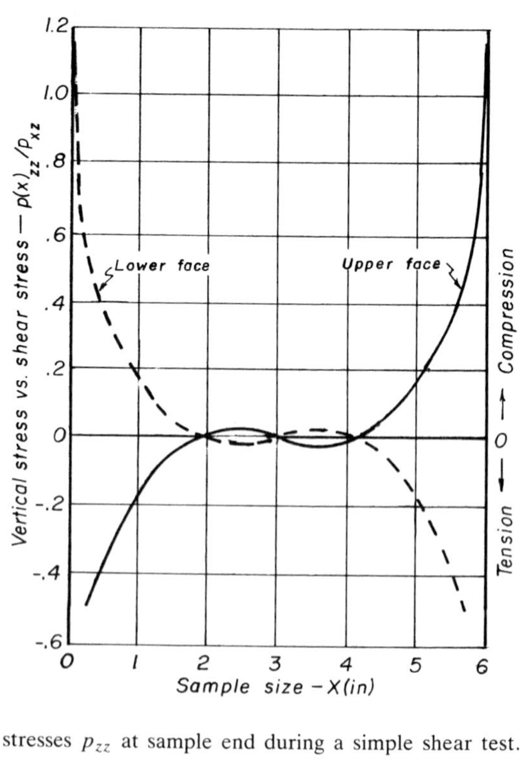

Fig. 12.8.3. Vertical stresses p_{zz} at sample end during a simple shear test. [From K. H. Roscoe (1953).]

Fig. 12.8.4. Sample at failure in a simple shear test under a vertical pressure of 0.125 kg/cm^2. [From K. H. Roscoe (1953).]

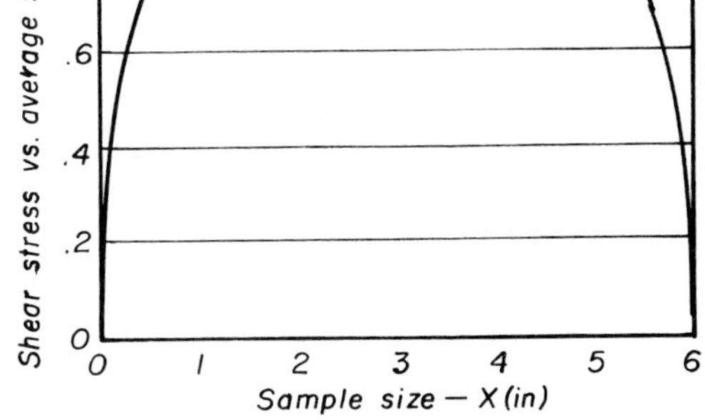

Fig. 12.8.5. Shear stresses on the sample ends under simple shear. [From K. H. Roscoe (1953).]

12.9 The Direct Shear Test

One of the oldest and simplest forms of testing the shearing resistance of soils is the *direct shear test*. Owing to its simplicity it is also the most common method for investigating the shear strength of all types of soils, granular or fine grained. Much before it was applied to soils, the direct shear test served for testing metals, wood, bricks, concrete, etc. In the Johnson type shear device, a rectangular section or a circular rod are sheared off, similarly to the way many structural elements are sheared.

The direct shear device for the measurement of shear strength of sand appears to have been used in about 1885 by Leygue in France. Its use was restored by Krey in Germany and Terzaghi and Casagrande in the USA. The first direct shear device allowing the measurement of vertical expansion and contraction, was built by A. Casagrande in 1930 as a two-part square box. From then on, many improvements were introduced, especially in the loading mechanism.

The direct shear apparatus has two parts: the shear box and the loading mechanism.

The *shear box*, Fig. 12.9.1, is a square or circular two part box, where one part can move horizontally with respect to the other, thus shearing the enclosed soil sample. The size of the box and the sample vary, from 5 cm by 5 cm square or 5 cm diameter to any larger size, accommodating samples that vary in thickness from 1 cm to several centimeters. The sample is fitted tightly in the box between two porous plates, a bottom porous plate connected to drainage provisions and a top porous plate. A solid top cap placed on the top porous plate serves to distribute the vertical load applied through it to the sample.

The shear box is usually located in a cup shaped solid container, maintaining it and the sample immersed in water.

The loading mechanism itself serves, in fact, two loading functions, vertical and horizontal, and therefore the device is fitted with two loading mechanisms:

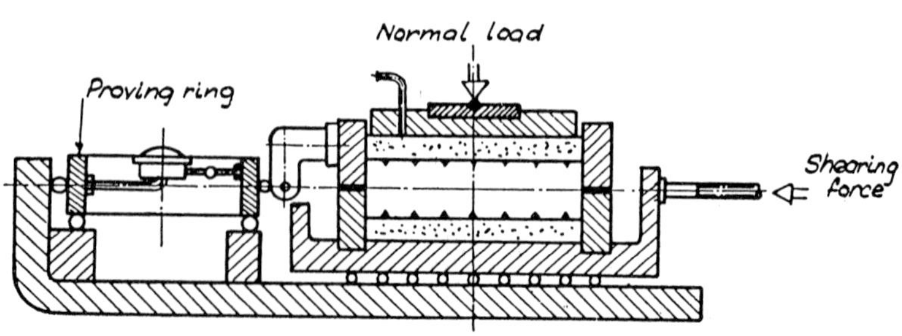

Fig. 12.9.1. Direct shear box. [From P. L. Capper and W. F. Cassie (1953), courtesy of McGraw-Hill Book Co. Inc.]

1. The *vertical loading mechanism* applies a vertical consolidating load on the sample through the top cap. The load can be of dead weights, or generated by hydraulic or pneumatic pressure, read on a pressure gage or a proving ring.

2. The *horizontal shear mechanism* applies a horizontal force to one half of the box, usually the upper half. Earlier horizontal loading was done by dead weights, a very cumbersome and not too efficient way of loading. Today electrically operated mechanical, hydraulic or pneumatic loadings are available and can provide stepwise, linear or any other form of controlled strain or controlled stress loadings, measured by a pressure gage or proving ring.

A vertical dial, attached usually to the solid container, serves as an independent reference, and measures the vertical deformations of the sample on the top plate. The dial indicates 10^{-3} of a centimeter or 10^{-4} of an inch and has a short travel range, of several millimeters.

A horizontal dial, attached to the solid container as well, measures the horizontal motion of the moving part of the shear box. The dial indicates 10^{-2} of a centimeter or 10^{-3} of an inch and has a travel range of several centimeters.

The direct shear test is performed on undisturbed samples trimmed to fit tightly into the shear box, its two parts clamped together, or on remolded samples compacted directly into the box, so that the plane passing between the two halves of the box passes through the soil sample.

Once the sample is in place, set on the bottom porous plate and covered with the top porous plate and the top cap, and the vertical and horizontal dials are installed, the assembly is ready for loading. Prior to the application of the horizontal load that provides the shear stress, a vertical load, normal to the sample surface, is applied. The vertical compression is measured as the consolidation proceeds and the sample compresses.

Like the conventional triaxial test, the direct shear test can be performed as a *drained*, *consolidated undrained* and *undrained* test, with some limitations. Since we have no control over the consolidation process and the pore pressure of the sample cannot be measured, the undrained test is meaningless, except in slowly consolidating fine grained soils, where it can be performed fast enough. It should be noted that the USDI Bureau of Reclamation has discontinued the direct shear test, because "drainage cannot be controlled and pore pressures cannot be measured in this type of test." Another reason for the discontinuance was the belief that "the triaxial shear test permits a far better simulation of the stresses produced within an actual structure or foundation." (USDI, BR Earth Manual, 1957). Yet they admit that the direct shear test "has received the most general popularity in the past because the test equipment is easier to construct, is easier to operate, and the analyses of the data appear to be more direct."

In the *drained* test, after consolidating the sample at a specific vertical pressure, the two parts of the shear box are unclamped and a shear stress is applied to the sample by a gradually increasing horizontal force, slow enough to allow the dissipation of positive or negative pore pressures that could develop during shear due to localized spherical pressures or tensions, respectively. The duration of a complete drained direct shear test may be weeks.

In the *consolidated undrained* test the procedure is similar to that of the drained test, except that the shear stresses should be applied as fast as possible, so that occasional spherical pressures or tensions will cause minimum drainage or sorption, respectively. Since the direct shear box is not as sophisticated as the triaxial shear device and it is not possible to close the drainage path of the sample, only slow draining soils with moderate to low coefficients of permeability are suitable for such a test. After consolidation, the shearing of the sample can be accomplished in less than 10 min.

An *undrained* test in direct shear has almost no meaning, as mentioned before, since the vertical load, and immediately thereafter the horizontal load, must be applied fast enough so that the soil sample does not consolidate while the test lasts. Only soils with low coefficients of permeability that can withstand tangible volume changes for 10–15 min., the duration of a complete test, qualify.

The direct shear test is, in a way, a degenerated simple shear test, where the region of the tilting walls of the device shrinks and reduces to a single plane, confined between the two halves of the direct shear box. The vertical load determines a vertical normal pressure p_{zz} or an effective vertical pressure $p_{\mathfrak{s}zz}$ acting on any area element of that plane, and the horizontal load determines a shear stress p_{zy} acting on the same element, Fig. 12.9.2. Both the vertical pressure and the shear stress form a resultant traction, p_z, acting on the plane normal to the direction Z and in an angle of obliquity β. The vertical pressure is constant while the shear stress increases gradually in order to overcome the frictional and cohesive resistance, and the obliquity angle increases as well. The effect of the shear stress does not become apparent until it equals the frictional resistance; at this time a motion sets in and the obliquity angle reaches its maximum value ϕ, which is the angle of internal friction. The sample fails and the following relation can be established

$$p_{zy} = c + p_{zz} \tan \phi \tag{12.9.1}$$

where c is the *cohesive intercept* and ϕ the *angle of internal friction* of the soil. In terms of effective pressure the above equation will be

$$p_{zy} = c_{\mathfrak{s}} + p_{\mathfrak{s}zz} \tan \phi_{\mathfrak{s}} \tag{12.9.2}$$

where $c_{\mathfrak{s}}$ is the *effective cohesive intercept* and $\phi_{\mathfrak{s}}$ is the *effective angle of internal friction*, both marked with the subscript $\mathfrak{s}$ indicating that the effective parameters relate to the constituent of solids. Eq. (12.9.1) and (12.9.2) will be discussed in detail in the next chapter.

Direction X of the sample is a principal stress direction, while directions Y and Z are not. To find the other two principal stress directions Y' and Z' is not a simple task, since the normal pressures in the X and Y directions are unknown. The pressure tensor p_{ij} of the consolidation prior to the application

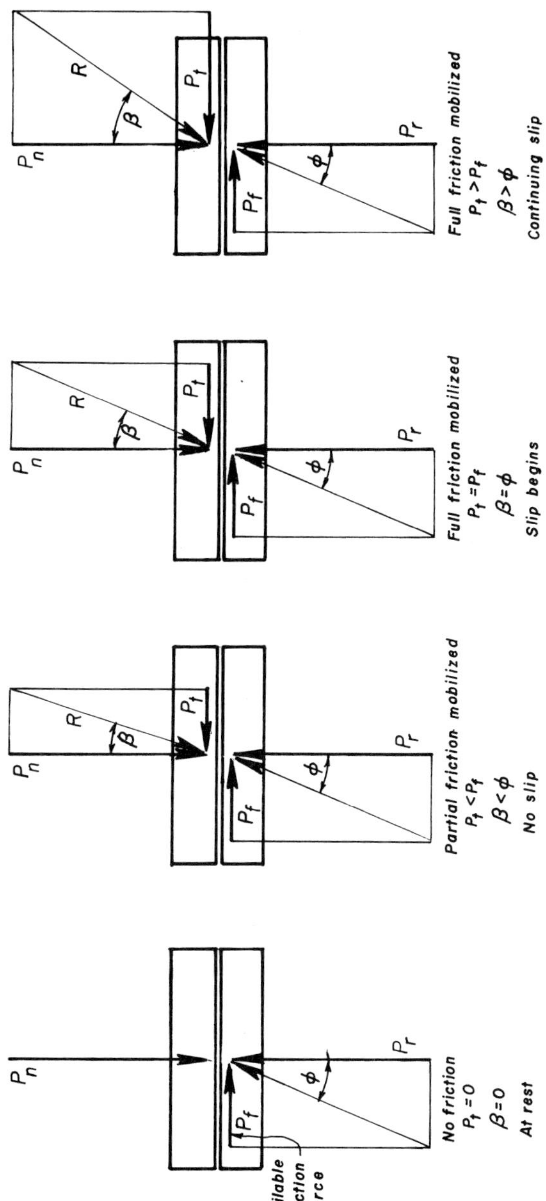

Fig. 12.9.2. Obliquity angle during a shear process.

of shear stresses is similar to the pressure tensor in the uniaxial consolidation test, Eqs. (11.9.1)–(11.9.3), and during the shearing of the sample the stress tensor becomes

$$
\begin{aligned}
p_{ij} &\equiv \begin{vmatrix} Kp_{zz} & 0 & 0 \\ 0 & Kp_{zz} & p_{zy} \\ 0 & p_{zy} & p_{zz} \end{vmatrix} \\
&= \frac{1+2K}{3} p_{zz} \begin{vmatrix} 1 & 0 & 0 \\ 0 & 1 & 0 \\ 0 & 0 & 1 \end{vmatrix} \\
&\quad + \begin{vmatrix} \frac{K-1}{3} p_{zz} & 0 & 0 \\ 0 & \frac{K-1}{3} p_{zz} & p_{zy} \\ 0 & p_{zy} & \frac{2-2K}{3} p_{zz} \end{vmatrix} \\
&= \frac{1+2K}{3} p_{zz} \begin{vmatrix} 1 & 0 & 0 \\ 0 & 1 & 0 \\ 0 & 0 & 1 \end{vmatrix} \\
&\quad + \frac{1-K}{3} p_{zz} \begin{vmatrix} -1 & 0 & 0 \\ 0 & -1 & 0 \\ 0 & 0 & 2 \end{vmatrix} \\
&\quad + p_{zy} \begin{vmatrix} 0 & 0 & 0 \\ 0 & 0 & 1 \\ 0 & 1 & 0 \end{vmatrix}
\end{aligned}
\tag{12.9.3}
$$

where the pressure tensor p_{ij} is a function of the vertical pressure p_{zz}, the shear stress p_{zy} and the coefficient of lateral earth pressure K, all functions of Z. According to the second part of Eq. (12.9.3) the pressure tensor decomposes into two tensors, a spherical pressure tensor and a deviatoric tensor, while in the last part of the equation the deviatoric tensor itself resolves into two tensors. Eq. (12.9.3) represents the stresses on the shear plane, i.e. the plane between the two parts of the shear box, and not the planes where the pressures are applied. Since the stress on the shear plane is applied via a normal pressure in the Y direction on the X-Y plane, its even distribution is not necessarily assured. $p(Y)_{zy}$ in the direction Y is uneven and a function of Y, consequently the normal pressure $p(Y, Z)_{zz}$ is also a function of Y. Since the stress condition along Y varies, the coefficient of lateral earth pressure varies as well, $K(Y)$. The only statements that can be made with respect to the normal and shear stresses acting on the shear plane are

$$F_z = \int_{A_z} p(Y)_{zz} \, dA; \quad F_y = \int_{A_z} p(Y)_{zy} \, dA \tag{12.9.4}$$

where F_z is the vertical force loading the sample and acting on the shear plane A_z, and F_y is the horizontal force applying the shear stress on the shear plane. The principal directions, Fig. 12.9.3, are at an angle θ which can be found from

$$\tan 2\theta = \frac{-2p_{zy}}{(1 - K)p_{zz}} \tag{12.9.5}$$

The principal stresses p_{22} and p_{33} are

$$p_{33} = \tfrac{1}{2}(1 + K)p_{zz} + \tfrac{1}{2}\sqrt{[(1 - K)^2 p_{zz}^2 + 4p_{zy}^2]}$$

$$p(Y)_{22} = \tfrac{1}{2}(1 + K(Y))p(Y)_{zz} - \tfrac{1}{2}\sqrt{((1 - K(Y))^2 p(Y)_{zz}^2 + 4p(Y)_{zy}^2)} \tag{12.9.6}$$

and the maximum shear stress $p_{zy\max}$ is

$$p_{zy\max} = \tfrac{1}{2}\sqrt{[(1 - K)^2 \, p_{zz}^2 + 4p_{zy}^2]} \tag{12.9.7}$$

In terms of the effective pressures, Eq. (12.9.3) becomes

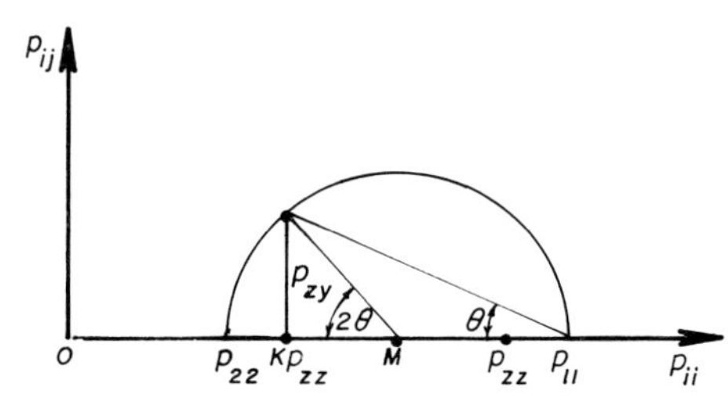

Fig. 12.9.3. Principal directions in the direct shear test.

$$p_{ij} = p_{\dot{s}ij} \equiv \begin{vmatrix} Kp_{\dot{s}zz} & 0 & 0 \\ 0 & Kp_{\dot{s}zz} & p_{zy} \\ 0 & p_{zy} & p_{\dot{s}zz} \end{vmatrix}$$

$$= \frac{1 + 2K}{3} p(Y)_{\dot{s}zz} \begin{vmatrix} 1 & 0 & 0 \\ 0 & 1 & 0 \\ 0 & 0 & 1 \end{vmatrix}$$

$$+ \begin{vmatrix} \frac{K-1}{3} p_{\dot{s}zz} & 0 & 0 \\ 0 & \frac{K-1}{3} p_{\dot{s}zz} & p_{zy} \\ 0 & p_{zy} & \frac{2-2K}{3} p_{\dot{s}zz} \end{vmatrix}$$

$$= \frac{1 + 2K}{3} p_{\dot{s}zz} \begin{vmatrix} 1 & 0 & 0 \\ 0 & 1 & 0 \\ 0 & 0 & 1 \end{vmatrix}$$

$$+ \frac{1 - K}{3} p_{\dot{s}zz} \begin{vmatrix} -1 & 0 & 0 \\ 0 & -1 & 0 \\ 0 & 0 & 2 \end{vmatrix}$$

$$+ p_{zy} \begin{vmatrix} 0 & 0 & 0 \\ 0 & 0 & 1 \\ 0 & 1 & 0 \end{vmatrix} \tag{12.9.8}$$

It is not at all simple to find the value of K, and in order to do this some assumptions must be made, therefore the possibility of evaluating the principal coordinates and their directions is somewhat doubtful. The evaluation of strains in a direct shear test is not straightforward either. Usually the relative displacement of the two parts of the shear box in the Y direction is measured and related to the dimension of the sample in that direction, and an equivalent of a "strain" is in this way determined.

Nevertheless, speculations have been made, and assumptions like perfect elasticity (Hansen 1961b) and deviation of the principal strain rate axes from the principal stress axes (De Josselin De Jong 1959) have been proposed. The different assumptions yielded, of course, debated results concerning topics like anisotropy, or direction of the principal axes of the stresses and strains and their deviations. Investigations in these topics (Roscoe 1953; Kisiel 1964;

Morgenstern and Tchalenko 1967a, b) were not at all conclusive. That the direct shear test is a particular case of the simple shear is inferred (Morgenstern and Tchalenko 1967b) in the fact that the shear plane, considered a plane without thickness, is actually a layer of finite thickness, within which the shear occurs. Such a finite layer is also seen in tests performed by Roscoe (1953) on plasticine, and is particularly noticeable in two samples, with low

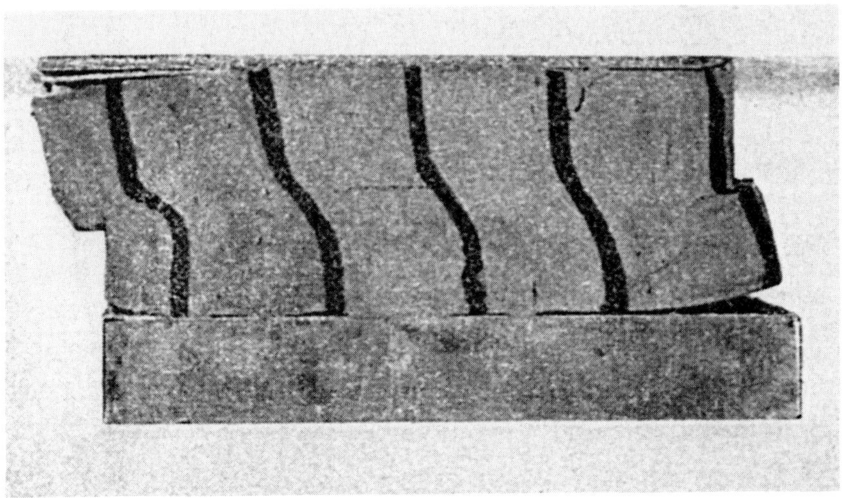

Fig. 12.9.4. Direct shear test on plasticine sample under 0.125 kg/cm^2 vertical pressure. [From K. H. Roscoe (1953).]

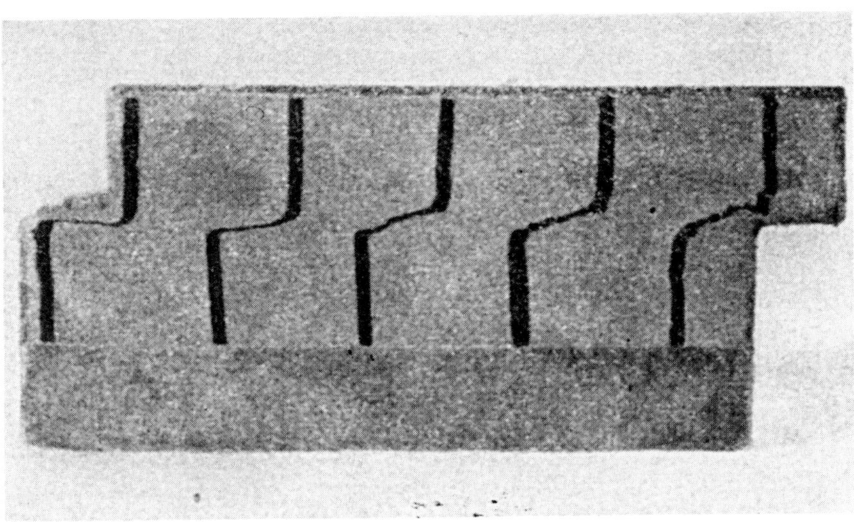

Fig. 12.9.5. Direct shear test on plasticine sample under 5.0 kg/cm^2 vertical pressure. [From K. H. Roscoe (1953).]

vertical pressure of 0.125 kg/cm^2 and high vertical pressure of 5.0 kg/cm^2, shown in Fig. 12.9.4 and Fig. 12.9.5, respectively. Similar to the simple shear test shown in Fig. 12.8.4, a separation of the leading end at the top of the sample and the trailing end at the bottom of the sample may be observed, owing to tensional stresses in the sample, an indication of uneven distribution of the pressure components $p(Y)_{xx}$, $p(Y)_{yy}$, $p(Y)_{zz}$, $p(Y)_{zy}$.

Direct shear devices, other than the one described here, are occasionally in use. In the *double shear plane* device (Proctor 1948), Fig. 12.9.6, the shear box is made of three parts where the center part is pulled out from between the two others so that the sample is sheared in two shear planes. Such a shear device is not more intricate than the single shear plane device, and has an advantage over it when soils with a low shearing resistance are tested: due to the two shear planes a better and more accurate result can be obtained. The *torsional direct shear* device (Hvorslev 1937, 1939), Fig. 12.9.7, is also a direct shear type, where a boxed ring shaped soil sample is sheared by torque. Another shear device is the *punch shear device*, Fig. 12.9.8, in which a soil sample is pierced by a cylindrical plunger and the shear stress is calculated over the area determined by the sample thickness and the plunger's perimeter.

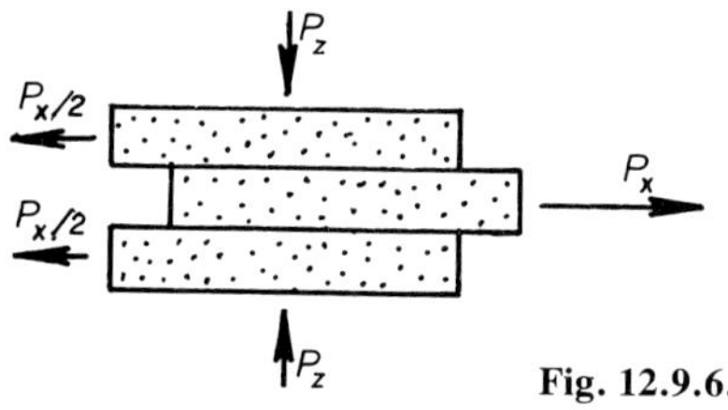

Fig. 12.9.6. Double shear device.

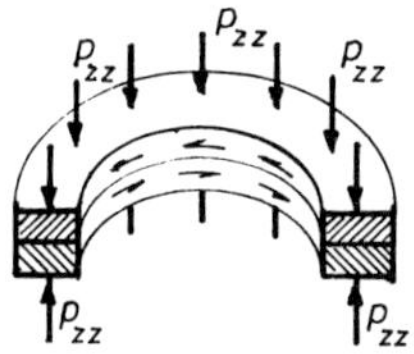

Fig. 12.9.7. Torsional direct shear test. [Courtesy of A. R. Jumikis (1962).]

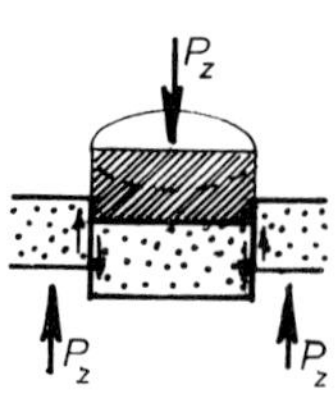

Fig. 12.9.8. Punch shear test.

12.10 The Torsion Test and Testing Device

The uniform twist of a circular cylinder along its length, as defined in Sect. 3.18, is a simple torsion. There is no doubt that a torsion, either on a solid or a hollow cylindrical sample, satisfies a deviatoric state of stresses, which, when combined with spherical pressure, is most appropriate for the investigation and testing of both spherical and deviatoric behavior of soils. Testing devices for such tests were constructed, but they are difficult to manipulate and pose technical problems, and their use is therefore limited to sporadic and localized research efforts.

Another aspect of the torsion is the possibility of investigating the conduct of soils where all three principal stresses vary, rather than only two as in the axially symmetric triaxial tests, in other words, the possibility of studying the stresses in planes other than the octahedral bisectrix plane.

Two types of torsional tests, differing in the nature of the samples tested, can be distinguished: torsion of solid cylindrical samples (Habib 1953, 1954; Badiey et al. 1988) and torsion of hollow cylindrical tubes (Geuze and Tan 1954: Tan 1957a–c; Haythornthwaite 1960a, b; Saada 1967, 1968; Biarez et al. 1969; Lomize et al. 1969; Fukushima and Tatsuoka 1982), each having its merits and drawbacks in the testing of soils as well as other materials. Except for minor technical deviations, the main features of all torsional testing devices are basically the same.

The torsional testing device, irrespective of the nature of the sample it tests, is very similar to the triaxial testing device. It consists of four main components:

1. The cell which contains the sample.
2. The auxiliary measuring and control unit.
3. The torsional loading device.
4. The vertical loading device.

The cell, Fig. 12.10.1 is similar to that of the triaxial cell and its main features are:

1. The base plate, made of solid metal, with a central pedestal to place the soil sample on it. The pedestal of the solid cylindrical sample is different from that of the cylindrical tube. For the former it is similar to that of the triaxial cell, except that it has built-in porous plates, sectorized by metal blades or pins to provide a slipless grip of the sample. The grip at the ends of the sample is one of the technical problems which must be solved. Occasionally, in solid cylindrical samples the pedestal is larger in diameter than the sample, Fig. 12.10.2, which is an alternative way to provide a good frictional grip to the sample ends.

For cylindrical tube samples the problem of grip is similar and similarly solved. Passages under the base plate sometimes provide the connection between the tubular hole of the sample and the cell, to permit the cell fluid to

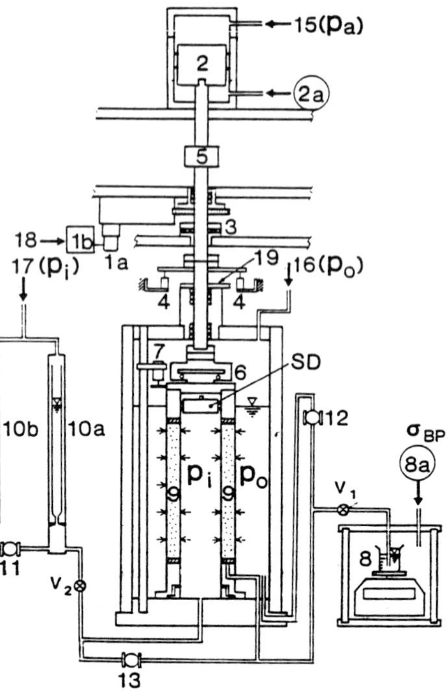

Fig. 12.10.1. Schematic drawing of a torsional shear device.
1a: Reversible motor
1b: Loading control unit
2 : Double-action bellofram cylinder
2a: Regulator for constant air pressure
3 : Clamp
4 : Axial displacement transducers
5 : Outer axial load cell (only for monitoring)
6 : Inner two-component load cell
7 : Potentiometer for rotational displacement measurement
8 : Electronic balance for volume change measurement
8a: Regulator for constant back air pressure
9 : Hollow cylindrical specimen
10a: Burette for inner volume change measurement
10b: Reference tube
11 : Low capacity differential pressure transducer (DPT) for volume change measurement of inner cylindrical space
12 : High capacity DPT for effective outer cell pressure (p_0-u) measurement
13 : High capacity DPT for effective inner cell pressure (p_i-u) measurement
14a: 16 Bit micro-computer (PC 9801)
14b: 12 bit A/D converter
14c: 12 bit D/A converter
14d: Floppy disk
not shown
15 : Controlled axial pressure (p_a)
16 : Controlled outer cell pressure (p_0)
17 : Controlled inner cell pressure (p_i)
18 : Controlled output voltage for motor control
[From F. Tatsuoka, T. B. S. Pradhan and H. Yoshi-Ie (1989).]

Fig. 12.10.2. Torsional sample with enlarged ends. [From P. Habib (1953).]

fill the hole. Through the base plate the cell is connected to the cell fluid supply tank and to the volume measuring facilities.

Pedestals for either type of sample can drain the water expelled from the sample through the base plate to the outside. Other connections leading out of the cell through the base plate are possible, such as pore pressure probes, strain gage wiring, etc.

2. The loading cap, placed on top of the sample and matching the pedestal in size and geometry, has features similar to the base plate pedestal, including built-in porous plates, grip blades and drainage connections. It is usually made of lightweight material such as Perspex, and, unlike the triaxial cap, it is connected rigidly to the loading piston, in order to transfer the torque to the sample.

3. The top plate, made of solid metal, is clamped to the base plate with a cylindrical Perspex tube between them, to form a pressure proof shear cell. The loading piston which penetrates through the top plate transfers not only the vertical pressure as in the triaxial cell, but has an additional purpose, that of transferring torque to the sample. Its construction is more intricate and

requires a highly frictionless bushing or ball bearing lining, which is at the same time leakproof.

The control panel and auxiliary measuring devices are similar to that of the triaxial shear test, see Sect. 11.10, including:

4. Pressure regulator and pressure gage, to control and monitor the spherical cell pressure.

5. Volume change burette, to measure volume changes in the shear cell.

6. Drainage burette, to measure water drained from the sample or absorbed by it.

7. Vertical strain dial, connected to the top plate and measuring vertical deformations of the soil sample on the loading piston.

8. Pore pressure measuring device, see Sect. 11.10.

9. Torque meter, which is of different types, from the simple long radial arm attached to the piston and moving around a graduated arc of ±0.0005 radian, to the more sophisticated mirror type troptometers that give a precision of ±0.00005 radian.

The torsional loading device can also be of different makes and degrees of sophistication, from the simple calibrated dead weight driving disc connected to the piston to most intricate, automatically recording pendulum type testing machines. The loading device can be operated mechanically, hydraulically or electronically, the torsional load being transferred to the sample, in all instances, through the loading piston.

The vertical loading device is optional; it is possible to apply a pure deviatoric load by a torque alone. If, however, a vertical load is used, it should be transferred by the piston connected to the torsional loading device, and impose minimum weight on it. All this must not prevent the piston from moving freely, following the vertical deformation of the sample.

Most torsion tests were performed on sands placed or compacted on the cell pedestal, in a form with a rubber membrane (for hollow cylindrical samples with an internal membrane as well). Until the top cap is in place and the cell assembled, clamped together and filled with cell fluid, the sample is held together by vacuum. If required, the sample is gradually saturated, and the vacuum is released. The dial gages and measuring instruments are attached and zeroed and the assembly is ready for the tests.

Cohesive soil samples, remolded or undisturbed, can also be tested for torsion; however, undisturbed hollow cylinders are difficult to prepare and keep erect, particularly if thin samples are used.

There are no rules as to the size of the sample or the height–average diameter ratio, $2H/D_{av}$. However, it is essential for the proper performance of the torsion test that there should be a difference between solid and hollow cylinders in the height–diameter ratio. In metals it is customary to require a height–average diameter ratio of 10 and a wall thickness–average diameter ratio of 0.08–0.1 (Davis et al. 1955). In soils, relatively thick samples with $d/D_{av} = 0.1$–0.5 and short samples with $2H/D_{av} = 1$–2 are used. The latter are unsatisfactory, since the width of a tubular sample, when sheared, is the circumference πD_{ao} and therefore a ratio of approximately $2H/D_{ao} = 5$–6 is required if the entire failure plane should be contained in the soil. For instance, a sample with an outside diameter of $D_0 = 55$ mm, inside diameter of $D_i = 25$ mm and height of $2H = 200$ mm is satisfactory.

12.11 Torsion of Solid Cylindrical Samples

The first to develop experimentally the relationship between the applied torque and the angle of twist of circular bars was Coulomb (Timoshenko 1953). He assumed that the cross-section of the bar remains a plane and rotates without any distortion during twist. This assumption, applied later by Navier (St Venant 1855), proved erroneous when it was applied to prismatic bars of constant non-circular cross-sections, but was soon rectified by St Venant (1864). He showed that the state of stresses and strains which can produce the correct solution is maintained by assuming forces and couples applied only at the ends of the bar. His treatise abounds in beautiful and instructive graphical illustrations of results on warping cross-sections, and he used the results to prove the well-known principle that carries his name.

As was mentioned in the previous section, the torsion test is performed in a cell similar to that of the triaxial cell, enabling us to apply a spherical pressure to the sample prior to torsion. This spherical pressure p_m is similar to the pressure defined by Eqs. (11.11.1) and (11.11.2), and is followed by a torque T_z applied at the ends of the sample, which in cylindrical coordinates is defined

$$T_z = \int_A e_{\alpha\beta z} R p_{\alpha\beta} \, dA = 2\pi \int_0^R e_{\alpha\beta z} p_{\alpha\beta} R^2 \, dR \tag{12.11.1}$$

where T_z is a torque acting in the plane normal to the direction Z, R are the radii of the area elements dA normal to the direction z and $p_{\theta z} = p(R, Z)_{\theta z}$ is the shear stress, a function of the radius and the direction Z, Fig. 12.11.1. It is assumed that, in accordance with the principle of St Venant, a short distance from the ends the shear stress p ceases to be a function of Z and practically remains a function of R alone, $p_{\theta z} = p(R)_{\theta z}$.

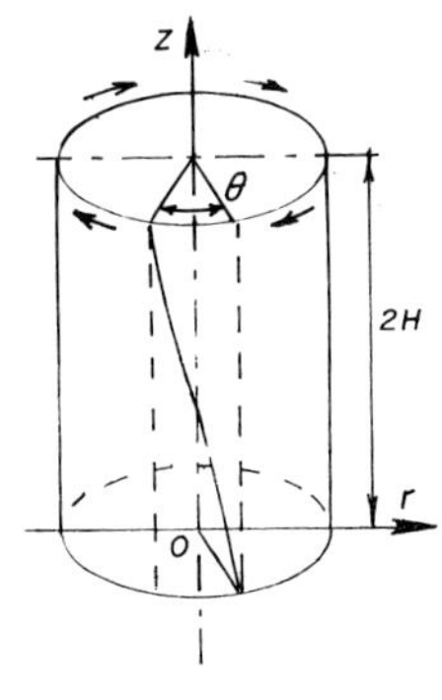

Fig. 12.11.1. Torsion of a solid cylinder.

The deviatoric stress tensor s_{ij} can be written

$$s_{ij} \equiv \begin{vmatrix} 0 & 0 & 0 \\ 0 & 0 & p_{\theta z} \\ 0 & p_{\theta z} & 0 \end{vmatrix} \tag{12.11.2}$$

The strains can be evaluated from the general strain equations given in cylindrical coordinates, Appendix B Eqs. (B.8.1)–(B.8.3), where the following assumptions are made:

1. Owing to axial symmetry, all derivatives with respect to θ will vanish. This introduces changes in $\varepsilon_{\theta\theta}$, $\varepsilon_{r\theta}$ and $\varepsilon_{\theta z}$.
2. The displacement in the direction R is laminar, therefore $u_r = 0$. All expressions containing u_r and ∂u_r vanish.
3. For simple torsion as defined by Eq. $(3.18.1)_2$ the tangential deformation u_θ is proportional to the twist per unit length B and to the angle of twist Ω

$$u_\theta = BZ = \frac{\Omega}{2H} Z \tag{12.11.3}$$

where $B = \Omega/2H$, the length of the sample being $2H$, and the initial angle of twist vanishes, $\theta = 0$. The strain $\varepsilon_{r\theta}$ vanishes and ε_{zz}, $\varepsilon_{\theta z}$ and ε_{zr} change.

These assumptions reduce the strains to

$$\begin{aligned}
\varepsilon_{rr} &= -\frac{1}{2}\left[R^2\left(\frac{\partial u_\theta}{\partial R}\right)^2 + \left(\frac{\partial u_z}{\partial R}\right)^2\right] \\
\varepsilon_{\theta\theta} &= 0 \\
\theta_{zz} &= \frac{\partial u_z}{\partial Z} - \frac{1}{2}\left[R^2 B^2 + \left(\frac{\partial u_z}{\partial Z}\right)^2\right] \\
\varepsilon_{r\theta} &= 0 \\
\varepsilon_{\theta z} &= \tfrac{1}{2} BR \\
\varepsilon_{zr} &= \frac{1}{2}\frac{\partial u_z}{\partial R} - \frac{1}{2}\left[R^2 \frac{\partial u_\theta}{\partial R} B + \frac{\partial u_z}{\partial R}\frac{\partial u_z}{\partial Z}\right]
\end{aligned} \tag{12.11.4}$$

For infinitesimal strains, when second-order terms vanish, we get from Eqs. (12.11.4) the strains

$$\begin{aligned}
\overset{\text{inf}}{\varepsilon}_{rr} &= \overset{\text{inf}}{\varepsilon}_{\theta\theta} = \overset{\text{inf}}{\varepsilon}_{zz} = \overset{\text{inf}}{\varepsilon}_{r\theta} = \overset{\text{inf}}{\varepsilon}_{zr} = 0 \\
\overset{\text{inf}}{\varepsilon}_{\theta z} &= \tfrac{1}{2} BR
\end{aligned} \tag{12.11.5}$$

where the deviatoric strain tensor $\overset{\text{inf}}{e}_{ij}$ can now be defined

$$\overset{\text{inf}}{e}_{ij} \equiv \begin{vmatrix} 0 & 0 & \frac{1}{2}\frac{\partial u_z}{\partial Z} \\ 0 & 0 & \frac{BR}{2} \\ \frac{1}{2}\frac{\partial u_z}{\partial Z} & \frac{BR}{2} & \frac{\partial u_z}{\partial Z} - \frac{B^2R^2}{2} \end{vmatrix} \tag{12.11.6}$$

If a linear stress–strain relationship is postulated, the volumetric strain of the deviatoric strain tensor and the distortion $\overset{\text{inf}}{\varepsilon}_{zr}$ necessarily vanish, $\overset{\text{inf}}{e}_{\alpha\alpha} = 0$ and $\overset{\text{inf}}{\varepsilon}_{zr} = 0$, from which the deviatoric strain tensor is obtained

$$\overset{\text{inf}}{e}_{ij} = \begin{vmatrix} 0 & 0 & 0 \\ 0 & 0 & \frac{BR}{2} \\ 0 & \frac{BR}{2} & 0 \end{vmatrix} \tag{12.11.7}$$

If a non-linear stress–strain relationship is postulated, as in Eq. (7.9.2), the deviatoric strain, Eq. (12.11.6), remains unchanged.

The assumption generally made with respect to other materials, specifically metals, that the deformation in the direction of Z vanishes, $u_z = 0$, and consequently its rate in the direction Z vanishes as well, $\partial u_z/\partial Z = 0$, cannot be applied to soils. With respect to other materials it is indeed a fair assumption, but for the soil it is erroneous, particularly its second part. The volume changes in soil are a result of consolidation, thus the deformation changes along the drainage path are not constant and therefore $\partial u_z/\partial Z \neq 0$.

If second-order terms are not ignored and the strains are as defined by Eqs. (12.11.4), then only non-linear stress–strain relationships like the one in Eq. (7.9.2) can be applied. None of these relationships, linear or non-linear, has been worked out so far for soils, nor was the Hencky strain measure applied to torsion of soils. A general exposure of torsion of materials was presented by Reiner in his work *Rheology* (1958), Sects. 29 and 30.

12.12 Torsion of Hollow Cylindrical Samples

The technical difficulties encountered by the torsion test in general were noted in Sect. 12.10, and the testing of tubular cylindrical samples augments those difficulties. The grip of a thin hollow sample, its ability to stand up, the clamping of the double membranes, the sensitive measurements required and the technical problems concerning the more complicated loading devices render the use of the torsion test of hollow cylindrical samples quite infrequent. Yet, some investigators have found it essential to counter the hardships and to perform such tests, and rightly so.

There are two main advantages to the torsional test of hollow cylindrical samples:

1. By testing a thin hollow cylinder for torsion, the shear stress $p_{\theta z}$ applied at the end of the sample by a couple force is not a function of the coordinates R and Z any more, as in Eq. (12.11.1), but is evenly distributed over its cross-section

$$T_z = 2\pi R_\alpha d p_{\alpha z} = \pi(R_{o\alpha}^2 - R_{i\alpha}^2) p_{\alpha z} \tag{12.12.1}$$

where R_{oi} is the outside radius of the sample, R_{ii} is the inside radius of the sample, $R_i = \frac{1}{2}(R_{oi} + R_{ii})$ is the mean radius of the sample, d is the thickness of the cylinder and T_z is the torque applied to the cylinder, Fig. 12.12.1. As the sample becomes thinner, the distribution of the shear stress over the ends of the sample is more uniform, but its preparation and support becomes more and more problematic, leading to a poor compromise, mostly a thicker sample.

2. The cylindrical hollow sample usually placed in a triaxial cell can be subjected to a spherical pressure p_m, to a vertical pressure Δp_{zz} which is actually a pressure increment, or to a torque T_z resulting in a torsional stress $p_{\theta z}$, applied simultaneously or separately. Thus there are many possible pressure combinations, which makes the test suitable for the investigation of the behavior of soils. This applies especially to the investigation of anisotropy and failure conditions, the latter being determined by all three principal stress combinations rather than by two as in most other soil tests (Haythornthwaite 1960a,b; Saada 1967, 1968; Biarez et al. 1969; Lomize et al. 1969; Tatsuoka et al. 1982; Fukushima and Tatsuoka 1982).

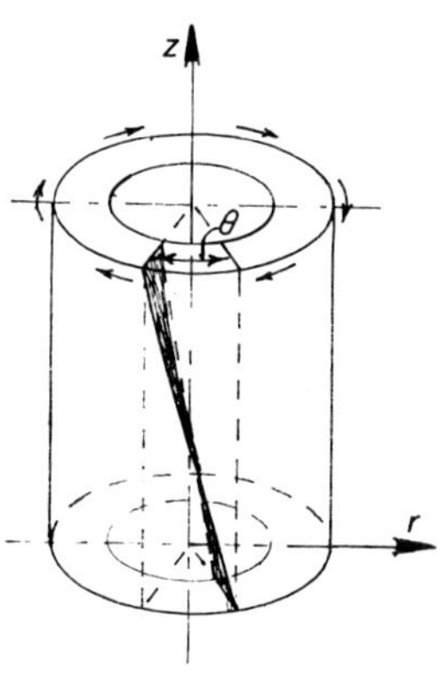

Fig. 12.12.1. Torsion of a hollow cylinder.

The following pressure combinations are possible:

1. *Spherical pressure* p_{m} *and vertical pressure* Δp_{zz}. If the spherical pressure p_{m} and a vertical pressure Δp_{zz} are applied, the principal directions remain R, θ and Z and the pressure tensor p_{ij} is

$$p_{ij} \equiv \begin{vmatrix} p_{rr} & 0 & 0 \\ 0 & p_{\theta\theta} & 0 \\ 0 & 0 & p_{zz} \end{vmatrix} = \begin{vmatrix} p_{11} & 0 & 0 \\ 0 & p_{22} & 0 \\ 0 & 0 & p_{33} \end{vmatrix}$$

$$= \begin{vmatrix} p_{\mathrm{m}} & 0 & 0 \\ 0 & p_{\mathrm{m}} & 0 \\ 0 & 0 & p_{\mathrm{m}} + \Delta p_{zz} \end{vmatrix}$$

$$\equiv (p_{\mathrm{m}} + \tfrac{1}{3}\Delta p_{zz})\delta_{ij} + \tfrac{1}{3}\Delta p_{zz}\Xi_{ij} \tag{12.12.2}$$

where it is seen that the total pressure tensor may be decomposed into a spherical pressure $\frac{1}{3}p_{\alpha\alpha} = p_{\mathrm{m}} + \frac{1}{3}\Delta p_{zz}$ and a deviatoric stress tensor $\frac{1}{3}\Delta p_{zz}\Xi_{ij}$. The Mohr circle representation shows a single circle, Fig. 12.12.2, while two of the other circles coincide and reduce to a point since $p_{rr} = p_{\theta\theta} = p_{\mathrm{m}}$. On any element of the sample the principal stresses act in the directions of the coordinates R, θ and Z, Fig. 12.12.3.

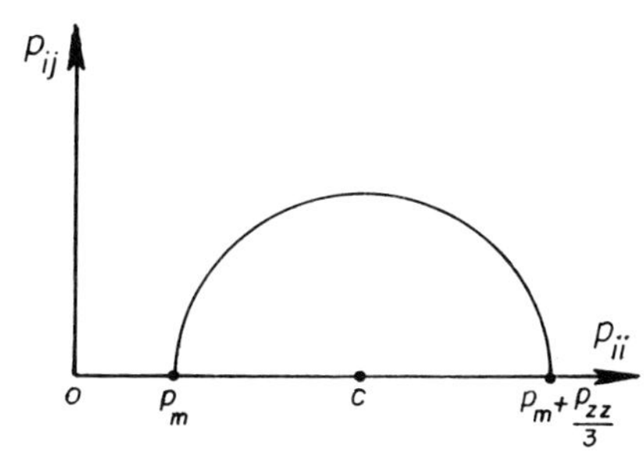

Fig. 12.12.2. Mohr circle of $p_{\mathrm{m}} + \Delta p_{zz}$.

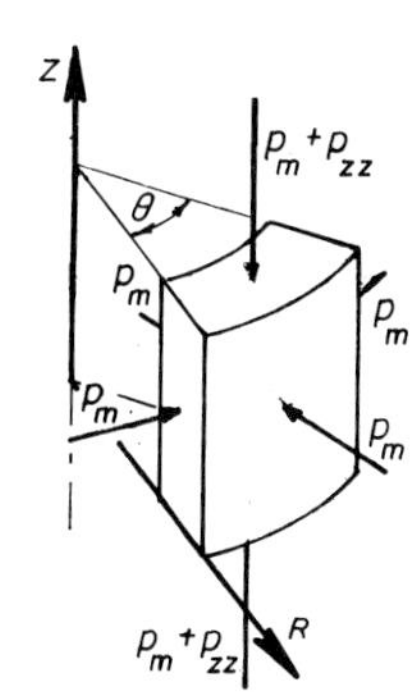

Fig. 12.12.3. Stresses acting on a volume element in $R\theta Z$ coordinates.

2. *Spherical pressure p_{m} and torsional stress $p_{\theta z}$*. When a spherical pressure p_{m} and a torsional stress $p_{\theta z}$ are applied to the sample, the pressure tensor p_{ij} is

$$p_{ij} \equiv \begin{vmatrix} p_{rr} & 0 & 0 \\ 0 & p_{\theta\theta} & p_{\theta z} \\ 0 & p_{\theta z} & p_{zz} \end{vmatrix} = \begin{vmatrix} p_{\mathrm{m}} & 0 & 0 \\ 0 & p_{\mathrm{m}} & p_{\theta z} \\ 0 & p_{\theta z} & p_{\mathrm{m}} \end{vmatrix}$$

$$\equiv p_{\mathrm{m}} \begin{vmatrix} 1 & 0 & 0 \\ 0 & 1 & 0 \\ 0 & 0 & 1 \end{vmatrix} + p_{\theta z} \begin{vmatrix} 0 & 0 & 0 \\ 0 & 0 & 1 \\ 0 & 1 & 0 \end{vmatrix} \tag{12.12.3}$$

where $p_{rr} = p_{\theta\theta} = p_{zz} = p_{\mathrm{m}}$, and the pressure tensor p_{ij} decomposes into a spherical stress and a simple shear in a state of plane stress. Clearly any element of the sample is subjected to a simple shear state of stresses, Fig. 12.12.5. From the Mohr circle display of the pressures, given in Fig. 12.12.4, it is seen that two of the principal stresses are obtained by rotating the coordinates in the $\theta - Z$ plane by 45°, while the coordinate in the R direction and the pressure p_{rr} in that direction remain intact. Thus, the pressure tensor $\bar{p}_{ij}$ in its principal directions becomes

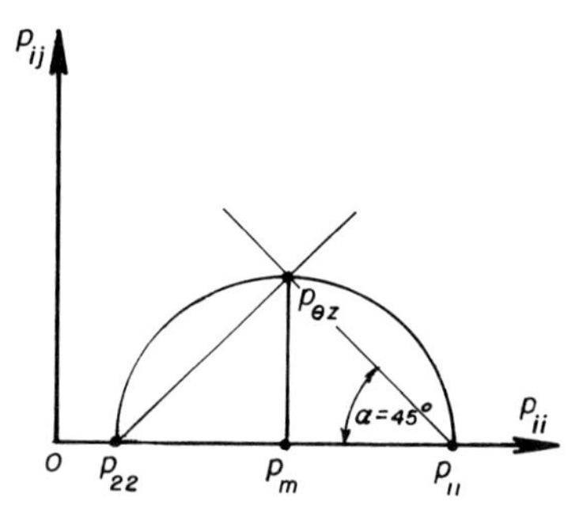

Fig. 12.12.4. Mohr circle of $p_{\mathrm{m}} + p_{\theta z}$.

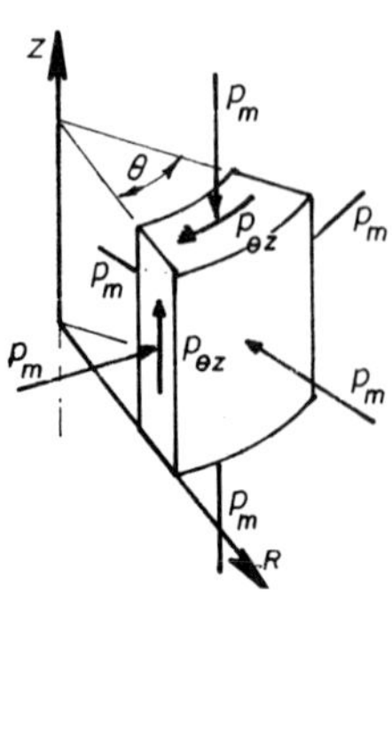

Fig. 12.12.5. Stresses acting on a volume element in cylindrical coordinates.

$$\bar{p}_{ij} \equiv \begin{vmatrix} p_{11} & 0 & 0 \\ 0 & p_{22} & 0 \\ 0 & 0 & p_{33} \end{vmatrix} = \begin{vmatrix} p_{\rm m} & 0 & 0 \\ 0 & p_{\rm m} + p_{\theta z} & 0 \\ 0 & 0 & p_{\rm m} - p_{\theta z} \end{vmatrix}$$

$$= p_{\rm m} \begin{vmatrix} 1 & 0 & 0 \\ 0 & 1 & 0 \\ 0 & 0 & 1 \end{vmatrix} + p_{\theta z} \begin{vmatrix} 0 & 0 & 0 \\ 0 & 1 & 0 \\ 0 & 0 & -1 \end{vmatrix} \qquad (12.12.4)$$

Here again, when Eq. (12.12.4) is decomposed into spherical and deviatoric pressure tensors, the deviatoric tensor indicates a simple shear, rotated at an angle of 45° to the θ and Z directions.

3. *Vertical pressure* p_{zz} *and torsional stress* $p_{\theta z}$. If a vertical pressure p_{zz} and a torsional stress are applied to the sample, then the pressure tensor p_{ij} is in a plane stress state without any pressure in the R direction, and, as seen, the pressure tensor is not in the principal directions

$$p_{ij} \equiv \begin{vmatrix} 0 & 0 & 0 \\ 0 & 0 & p_{\theta z} \\ 0 & p_{\theta z} & p_{zz} \end{vmatrix} = \tfrac{1}{3} p_{zz} \begin{vmatrix} 1 & 0 & 0 \\ 0 & 1 & 0 \\ 0 & 0 & 1 \end{vmatrix}$$

$$+ \tfrac{1}{3} p_{zz} \begin{vmatrix} -1 & 0 & 0 \\ 0 & -1 & 0 \\ 0 & 0 & 2 \end{vmatrix}$$

$$+ p_{\theta z} \begin{vmatrix} 0 & 0 & 0 \\ 0 & 0 & 1 \\ 0 & 1 & 0 \end{vmatrix} \qquad (12.12.5)$$

There are only two principal stresses and therefore in the Mohr circle display, Fig. 12.12.6, we have only one circle, and the principal stress tensor $\bar{p}_{ij}$ is

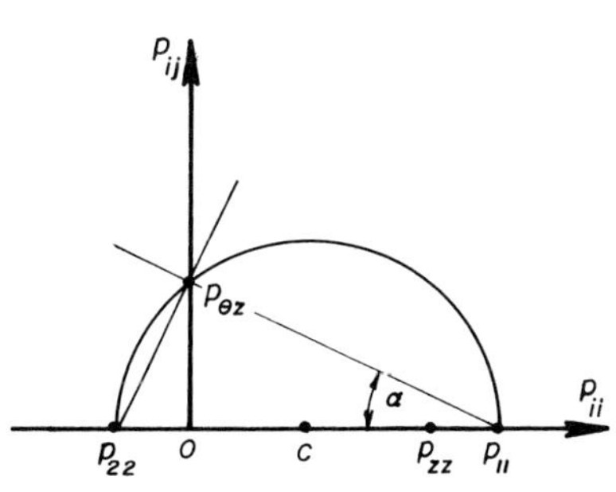

Fig. 12.12.6. Mohr circle of $p_{zz} + p_{\theta z}$.

$$\bar{p}_{ij} = \begin{vmatrix} 0 & 0 & 0 \\ 0 & p_{22} & 0 \\ 0 & 0 & p_{33} \end{vmatrix} \tag{12.12.6}$$

where the principal stresses are evaluated according to Eq. (A.20.24) Appendix A

$$p_{22};\ p_{33} = \tfrac{1}{2}p_{zz} \pm \surd(\tfrac{1}{2}p_{zz}^2 + p_{\theta z}^2) \tag{12.12.7}$$

and the angle α of the principal axis, according to Eq. (A.20.23) Appendix A, is

$$\tan 2\alpha = \frac{2p_{\theta z}}{p_{zz}} \tag{12.12.8}$$

Fig. 12.12.7 shows the stresses acting on any element of the sample.

4. *Spherical pressure p_m, vertical pressure p_{zz} and torsional stress $p_{\theta z}$.* If all three pressures, the spherical pressure p_m, the vertical pressure Δp_{zz} and the torsional stress $p_{\theta z}$ are applied then the pressure tensor p_{ij} becomes

$$p_{ij} \equiv \begin{vmatrix} p_m & 0 & 0 \\ 0 & p_m & p_{\theta z} \\ 0 & p_{\theta z} & p_m + \Delta p_{zz} \end{vmatrix}$$

$$= (p_m + \tfrac{1}{3}\Delta p_{zz}) \begin{vmatrix} 1 & 0 & 0 \\ 0 & 1 & 0 \\ 0 & 0 & 1 \end{vmatrix}$$

$$+ \tfrac{1}{3}\Delta p_{zz} \begin{vmatrix} -1 & 0 & 0 \\ 0 & -1 & 0 \\ 0 & 0 & 2 \end{vmatrix} + p_{\theta z} \begin{vmatrix} 0 & 0 & 0 \\ 0 & 0 & 1 \\ 0 & 1 & 0 \end{vmatrix} \tag{12.12.9}$$

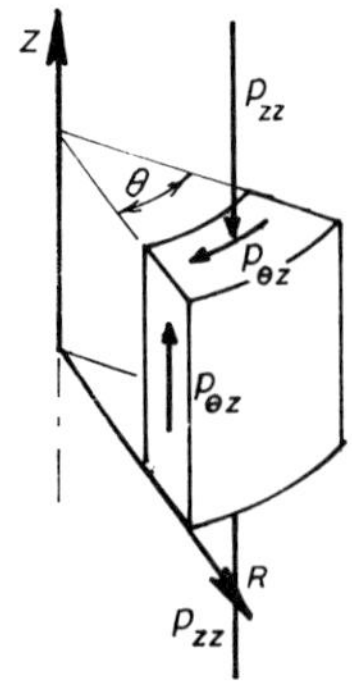

Fig. 12.12.7. Stresses acting on a volume element in cylindrical coordinates given in Fig. 12.12.6.

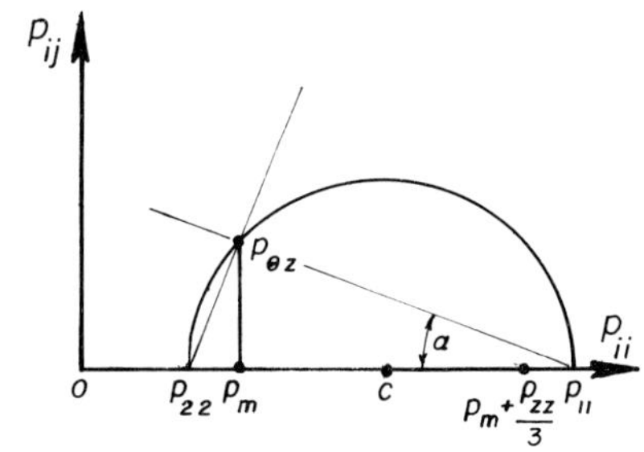

Fig. 12.12.8. Mohr circles for stresses p_{m}, Δp_{zz} and $p_{\theta z}$ acting on a soil element.

The Mohr circle representation, Fig. 12.12.8, indicates that there are three circles and evidently there are three principal pressures. The first remains the same as that in the direction R while the second and third are obtained from Eq. (A.20.24) Appendix A

$$p_{11} = p_{rr} = p_{\mathrm{m}}; \quad p_{22}; p_{33} = \tfrac{1}{2}(2p_{\mathrm{m}} + \Delta p_{zz}) \pm \sqrt{\{(\tfrac{1}{2}\Delta p_{zz})^2 + p_{\theta z}^2\}} \tag{12.12.10}$$

and the directions p_{22} and p_{33} are obtained from Eq. (A.20.23) Appendix A

$$\tan 2\alpha = \frac{2p_{\theta z}}{\Delta p_{zz}} \tag{12.12.11}$$

Fig. 12.12.9 shows the pressures acting on any element of the sample, and as can be observed it is a combination of cases 1 and 2, 2 and 3 or 1 and 3.

The strain tensor ε_{ij} for a hollow cylinder can be construed from Eq. (12.11.4) by adding the assumption $\partial u_\theta / \partial R = 0$ to the assumptions made in Sect. 12.11. Since the pressure change across the thickness d of the tube is zero, namely, the pressure on both sides of the tube is the same, either zero as in case 3 or p_{m} as in cases 1, 2 and 4, and since the tubular sample is

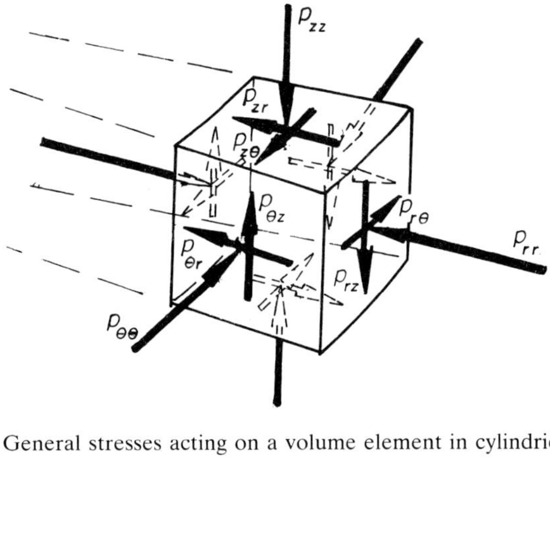

Fig. 12.12.9. General stresses acting on a volume element in cylindrical coordinates.

located around a mean radius R_i, there cannot be a change in the circumference of the sample, which remains $2\pi R_i$. Thus the pressure tensor ε_{ij} will be

$$\varepsilon_{ij} \equiv \begin{vmatrix} \varepsilon_{rr} & 0 & 0 \\ 0 & 0 & \varepsilon_{\theta z} \\ \varepsilon_{11} & \varepsilon_{\theta z} & \varepsilon_{zz} \end{vmatrix} = \begin{vmatrix} \varepsilon_{11} & 0 & 0 \\ 0 & \varepsilon_{22} & 0 \\ 0 & 0 & \varepsilon_{33} \end{vmatrix}$$

$$= \tfrac{1}{3}(\varepsilon_{rr} + \varepsilon_{zz}) \begin{vmatrix} 1 & 0 & 0 \\ 0 & 1 & 0 \\ 0 & 0 & 1 \end{vmatrix} + \begin{vmatrix} e_{11} & 0 & 0 \\ 0 & e_{22} & 0 \\ 0 & 0 & e_{33} \end{vmatrix} \tag{12.12.12}$$

where

$$\varepsilon_{rr} + \varepsilon_{zz} = \varepsilon_{11} + \varepsilon_{22} + \varepsilon_{33} \tag{12.12.13}$$

$$e_{11} = \varepsilon_{11} - \tfrac{1}{3}(\varepsilon_{rr} + \varepsilon_{zz})\tfrac{1}{6}(\varepsilon_{zz} - 2\varepsilon_{rr}) - \sqrt{[(\tfrac{1}{2}\varepsilon_{zz})^2 + \varepsilon_{z\theta}^2]}$$

$$e_{22} = \varepsilon_{22} - \tfrac{1}{3}(\varepsilon_{rr} + \varepsilon_{zz}) = \tfrac{2}{3}(\varepsilon_{rr} - \varepsilon_{zz});$$

$$e_{33} = \varepsilon_{33} - \tfrac{1}{3}(\varepsilon_{rr} + \varepsilon_{zz}) = \tfrac{1}{6}(\varepsilon_{zz} - 2\varepsilon_{rr}) + \sqrt{[(\tfrac{1}{2}\varepsilon_{zz})^2 + \varepsilon_{z\theta}^2]} \tag{12.12.14}$$

and the individual strains are

$$\varepsilon_{rr} = -\frac{1}{2}\left[\frac{\partial u_z}{\partial R}\right]^2$$

$$\varepsilon_{\theta\theta} = 0$$

$$\varepsilon_{zz} = \frac{\partial u_z}{\partial Z} - \frac{1}{2}\left[R_i^2 B^2 + \left(\frac{\partial u_z}{\partial Z}\right)^2\right]$$

$$\varepsilon_{r\theta} = 0$$

$$\varepsilon_{\theta z} = \tfrac{1}{2}BR_i$$

$$\varepsilon_{zr} = \frac{1}{2}\frac{\partial u_z}{\partial R} - \frac{1}{2}\left[\frac{\partial u_z}{\partial R}\frac{\partial u_z}{\partial Z}\right] \tag{12.12.15}$$

If the strains are considered infinitesimal, the higher-order terms of Eq. (12.12.15) vanish and Eqs. (12.12.12)–(12.12.15) become

$$\overset{\text{inf}}{\varepsilon}_{ij} \equiv \begin{vmatrix} 0 & 0 & \overset{\text{inf}}{\varepsilon}_{zr} \\ 0 & 0 & \overset{\text{inf}}{\varepsilon}_{\theta z} \\ \overset{\text{inf}}{\varepsilon}_{zr} & \overset{\text{inf}}{\varepsilon}_{\theta z} & \overset{\text{inf}}{\varepsilon}_{zz} \end{vmatrix} = \begin{vmatrix} \overset{\text{inf}}{\varepsilon}_{11} & 0 & 0 \\ 0 & \overset{\text{inf}}{\varepsilon}_{22} & 0 \\ 0 & 0 & \overset{\text{inf}}{\varepsilon}_{33} \end{vmatrix}$$

$$= \frac{\overset{\text{inf}}{\varepsilon}_{zz}}{3}\begin{vmatrix} 1 & 0 & 0 \\ 0 & 1 & 0 \\ 0 & 0 & 1 \end{vmatrix} + \begin{vmatrix} \overset{\text{inf}}{e}_{11} & 0 & 0 \\ 0 & \overset{\text{inf}}{e}_{22} & 0 \\ 0 & 0 & \overset{\text{inf}}{e}_{33} \end{vmatrix} \tag{12.12.16}$$

where, similarly to Eqs. (12.12.13) and (12.12.14), we have

$$\overset{\text{inf}}{\varepsilon}_{zz} = \overset{\text{inf}}{\varepsilon}_{11} + \overset{\text{inf}}{\varepsilon}_{22} + \overset{\text{inf}}{\varepsilon}_{33} \tag{12.12.17}$$

$$\overset{\text{inf}}{e}_{11} = -\frac{\overset{\text{inf}}{\varepsilon}_{zz}}{3}; \quad \overset{\text{inf}}{e}_{22} = -\frac{\overset{\text{inf}}{\varepsilon}_{zz}}{3}; \quad \overset{\text{inf}}{e}_{33} = \overset{\text{inf}}{\varepsilon}_{33} - \frac{\overset{\text{inf}}{\varepsilon}_{zz}}{3} \tag{12.12.18}$$

and the individual infinitesimal strains are

$$\overset{\text{inf}}{\varepsilon}_{rr} = \overset{\text{inf}}{\varepsilon}_{\theta\theta} = \overset{\text{inf}}{\varepsilon}_{r\theta} = 0;$$

$$\overset{\text{inf}}{\varepsilon}_{zz} = \frac{\partial u_z}{\partial Z}; \quad \overset{\text{inf}}{\varepsilon}_{\theta z} = \tfrac{1}{2}\mathrm{B}R_i; \quad \overset{\text{inf}}{\varepsilon}_{zr} = \frac{1}{2}\frac{\partial u_z}{\partial R} \tag{12.12.19}$$

For linear stress–strain relationships, either of the pressure tensors, Eqs. (12.12.2)–(12.12.6) and (12.12.9), are correlated with the infinitesimal strain tensor, Eq. (12.12.16), while for non-linear stress–strain relationships the same pressure tensor equations are correlated with the strain tensor, Eq. (12.12.12).

12.13 Introduction to the Pure Deviatoric Test

The pure deviatoric loading of soils, a technique known for some time and applied occasionally by several investigators (Bishop and Henkel 1962; Geuze 1960; Henkel 1960) was revived at Princeton University by Schmid during the years 1958–1961, and several studies have since appeared in print (Klausner 1960, 1964, 1967, 1969b, 1970c; Saada 1961, 1967; Klausner and Rooney 1967; Zur 1969). Except for the fact that it allows a deviatoric stress to be applied to a soil sample, its main advantage is that it uses the conventional triaxial testing equipment and techniques.

The test is performed on a conventional cylindrical sample and is preceded, as in the conventional triaxial shear test, by consolidation, achieved by applying a spherical pressure p_{m} to the sample. After this pressure is applied and the sample reaches volumetric equilibrium, the pure deviatoric test actually begins. The principal stresses applied to the sample from here on are so controlled as to keep the spherical pressure unchanged

$$p(t)_{aa} = 3p_{\mathrm{m}} = \text{const}; \quad \Delta p(t)_{\alpha\alpha} = 0 \tag{12.13.1}$$

This is the condition for pure deviatoric stresses, therefore the pressure tensor $p(t)_{ij}$ at time t is

$$p(t)_{ij} \equiv \begin{vmatrix} p(t)_{11} & 0 & 0 \\ 0 & p(t)_{22} & 0 \\ 0 & 0 & p(t)_{33} \end{vmatrix}$$

$$= \begin{vmatrix} p(t)_c & 0 & 0 \\ 0 & p(t)_c & 0 \\ 0 & 0 & p(t)_c + \delta p(t)_{33} \end{vmatrix};$$

$$p(t)_{11} = p(t)_{22} \tag{12.13.2}$$

where $p(t)_{11}$, $p(t)_{22}$, $p(t)_{33}$ are the principal pressures, $p(t)_c$ is the existing hydrostatic cell pressure and $\delta p(t)_{33}$ is the excess vertical pressure at time t.* The pressure tensor $p(t)'_{ij}$ at time $t' > t$ can also be written

$$p(t)'_{ij} \equiv \begin{vmatrix} p(t)'_{11} & 0 & 0 \\ 0 & p(t)'_{22} & 0 \\ 0 & 0 & p(t)'_{33} \end{vmatrix}$$

$$= \begin{vmatrix} p(t)'_c & 0 & 0 \\ 0 & p(t)'_c & 0 \\ 0 & 0 & p'_c + \delta p(t)'_{33} \end{vmatrix}$$

$$= \begin{vmatrix} p(t)_{11} + \Delta p(t)_{11} & 0 & 0 \\ 0 & p(t)_{22} + \Delta p(t)_{22} & 0 \\ 0 & 0 & p(t)_{33} + \Delta p(t)_{33} \end{vmatrix} \tag{12.13.3}$$

$$p(t)'_{11} = p(t)'_{22}; \quad \Delta p(t)_{11} = \Delta p(t)_{22}$$

where $p(t)'_{11}$, $p(t)'_{22}$, $p(t)'_{33}$ are the principal pressures, $p(t)'_c$ is the hydrostatic cell pressure and $\delta p(t)_{33}$ is the excess vertical pressure, all at time t'. The first parts of Eqs. (12.13.1) and (12.13.3) imply

$$\begin{aligned} \Delta p(t)_{\alpha\alpha} &= p(t)'_{11} + p(t)'_{22} + p(t)'_{33} - p(t)_{11} - p(t)_{22} - p(t)_{33} \\ &= \Delta p(t)_{11} + \Delta p(t)_{22} + \Delta p(t)_{33} = 2\Delta p(t)_{11} + \Delta p(t)_{33} = 0 \end{aligned} \tag{12.13.4}$$

* The excess vertical pressure is either a compression, or an extension, in which case it is negative, $-\delta p(t)_{33}$. Extensional pure deviatoric tests were performed by Zur (1969).

or

$$\Delta p(t)_{11} = \Delta p(t)_{22} = -\tfrac{1}{2}\,\Delta p(t)_{33} \tag{12.13.5}$$

Therefore, in an axially symmetric system a pure deviatoric stress loading can be obtained by decreasing the lateral pressure increments by half the amount of the vertical pressure increment, Fig. 12.13.1.

Since the pure deviatoric test is preceded by a spherical pressure p_m which during the shear test is maintained constant, the pressure tensor $p(t)'_{ij}$ according to Eqs. (12.13.3) and (12.13.5) is as follows

$$p(t)'_{ij} \equiv \begin{vmatrix} p_m - \frac{1}{2}\Delta p(t)_{33} & 0 & 0 \\ 0 & p_m - \frac{1}{2}\Delta p(t)_{33} & 0 \\ 0 & 0 & p_m + \Delta p(t)_{33} \end{vmatrix}$$

$$= p_m \begin{vmatrix} 1 & 0 & 0 \\ 0 & 1 & 0 \\ 0 & 0 & 1 \end{vmatrix} + \tfrac{1}{2}\Delta p(t)_{33} \begin{vmatrix} -1 & 0 & 0 \\ 0 & -1 & 0 \\ 0 & 0 & 2 \end{vmatrix}$$

$$\equiv p_m\,\delta_{ij} + \tfrac{1}{2}\Delta p(t)_{33}\Xi_{ij} \tag{12.13.6}$$

where $\Delta p(t)_{33}$ is a vertical pressure applied through the piston in addition to the spherical pressure p_m. The last part of the equation indicates that the pure deviatoric loading test is a superposition of a deviatoric test on a spherical test, and that as the vertical pressure increment is applied to the sample a corresponding reduction in the lateral pressure must take place.

Since the hydrostatic cell pressure in the triaxial cell acts on all sides of the sample and in all directions including the vertical direction at the ends of the

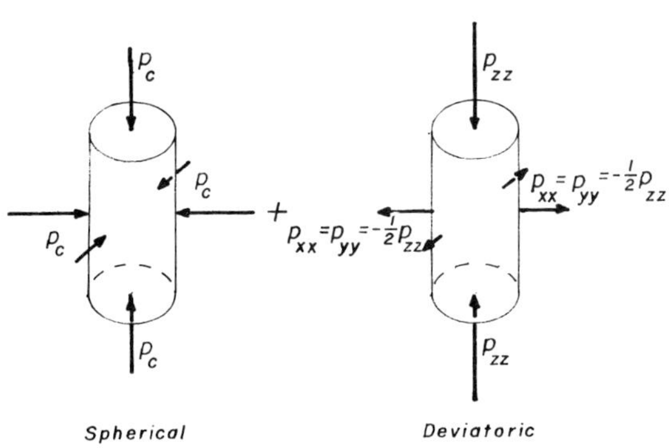

Fig. 12.13.1. Stresses in a pure deviatoric test.

sample, Eq. (12.13.6) expressed in terms of the hydrostatic cell pressure $p(t)'_c$ will be

$$p'_{ij} \equiv \begin{vmatrix} p(t)'_c & 0 & 0 \\ 0 & p(t)'_c & 0 \\ 0 & 0 & p(t)'_c + \frac{3}{2}\Delta p(t)_{33} \end{vmatrix}$$

$$= p'_c \begin{vmatrix} 1 & 0 & 0 \\ 0 & 1 & 0 \\ 0 & 0 & 1 \end{vmatrix} + \tfrac{3}{2}\Delta p(t)_{33} \begin{vmatrix} 0 & 0 & 0 \\ 0 & 0 & 0 \\ 0 & 0 & 1 \end{vmatrix} \tag{12.13.7}$$

where $p'_c = p_c - \frac{1}{2}\Delta p(t)_{33} = p_m - \frac{1}{2}\Delta p(t)_{33}$.

Since the spherical pressure p_m is maintained constant throughout the pure deviatoric test, the following may be written, from Eqs. (12.13.2) and (12.13.3)

$$\begin{aligned} p(t)_{\alpha\alpha} - p(t)'_{\alpha\alpha} &= 3p_c + \delta p(t)_{33} - 3p'_c - \delta p(t)'_{33} \\ &= 3(p_c - p'_c) + [\delta p(t)_{33} - \delta_p(t)'_{33}] \\ &= 3\Delta p_c + \Delta p(t)_{33} = 0 \end{aligned} \tag{12.13.8}$$

where $\Delta p_c = p_c - p'_c$ and $\Delta p(t)_{33} = \delta p(t)'_{33} - \delta p(t)_{33}$, while $p_c \geq p'_c$ and $\delta p(t)'_{33} \geq \delta p(t)_{33}$, and from which we obtain

$$\Delta p(t)_c = \tfrac{1}{3}\Delta p(t)_{33} \tag{12.13.9}$$

Hence,* in order to produce a pure deviatoric state of stresses in a cylindrical sample in a triaxial cell, the cell pressure $\Delta p(t)_c$ is decreased at a ratio of one third of the excess vertical pressure $\Delta p(t)_{33}$. If before the deviatoric loading the sample was consolidated in the cell by an initial spherical pressure $p_m = p_{c0}$, then at any time t of the deviatoric loading the cell pressure will be

$$p(t)_c = p_{c0} - \tfrac{1}{3}\Delta p(t)_{33} \tag{12.13.10}$$

If $p_c = p_m$, in other words if the vertical excess pressure is applied at time $t = 0$ directly after the sample is consolidated under the spherical pressure p_m, then $\delta p(t)_{33} = 0$ and

$$\Delta p(t)_{33} = \delta p(t)'_{33} \tag{12.13.11}$$

* In triaxial cells, where the diameter of the loading piston is equal to the diameter of the sample, Eqs. (12.13.8)–(12.13.10) do not apply.

12.14 The Pure Deviatoric Test and its Equipment

It was already mentioned that one of the advantages of the pure deviatoric test is that the triaxial testing equipment and test techniques can be readily used, and only the loading device need be supplemented.

After applying the spherical pressure in the conventional way $p_{c0} = p_m$, see Sect. 11.11, and after the consolidation is completed, the deviatoric stress is applied according to Eq. (12.13.6).

The shear stresses applied thereafter are functions of the vertical pressure $\Delta p(t)_{33}$, according to Eq. (12.13.6) which specifies the relation between the principal pressures in an axially symmetric system, specifically the change in the lateral pressures $\Delta p(t)_{11} = \Delta p(t)_{22}$. Similarly to the conventional triaxial shear test, Eq. (12.13.6) of the pure deviatoric test does not specify the form of $\Delta p(t)_{33}$ and how it is applied. It can be a stepwise function, a linear function or any other preferred function, see Sect. 12.15.

For stepwise loading any conventional triaxial loading device is satisfactory. Constant vertical load increments are applied and the corresponding reductions in the hydrostatic cell pressures are manually adjusted. In the early stage of the exploration of the pure deviatoric loading (Saada 1961; Klausner 1964, 1987b), the tests were performed with stepwise loadings using a platform scale type loading device, Fig. 12.14.1. Zur (1969) performed pure deviatoric tests of stepwise loadings with a conventional triaxial loading device.

Two more sophisticated loading devices were designed to facilitate the application of a continuous constant rate of loadings (Klausner 1964, 1970b; Klausner and Rooney 1967) both with pneumatic control. Since all signals in the triaxial cell were pneumatic, and the loading jack in both cases was also pneumatically operated, a pneumatically controlled loading device seemed the simplest and most straightforward. The pneumatic flow diagram, Fig. 12.14.2, and the pneumatic component circuitry, Fig. 12.14.3, were identical in the design of both loading devices. The difference between the two designs was that the first, designed and constructed during 1962–64 and reported in detail by Klausner and Rooney (1967), was installed individually with each triaxial cell, while the second, designed later during 1964–66 and reported by Klausner (1970b), consisted of a single mobile deviatoric loading console, which could be connected with each of the triaxial cells, Fig. 12.14.4.

The latter model of the pure deviatoric loading device consists of five major components:

1. The triaxial cell.
2. The control panel.
3. The pneumatic loading jack.
4. The pneumatic control unit.
5. The linear load rate-control generator unit.

The first three components, the triaxial cell, the control panel and the

Fig. 12.14.1. Scale platform type loading device.

pneumatic loading jack constitute a single cell unit, as in the conventional triaxial test. They are described in Sects. 11.10 and 12.6. The pneumatic control unit and the load rate control generator unit are incorporated in the deviatoric loading console, a mobile unit that can be attached to each of the cells in order to perform the deviatoric shear, after consolidation has been reached for a given spherical pressure. Since the consolidation of a sample takes sometimes weeks, one loading console serves several cells.

The pneumatic control receives several inputs and provides two outputs. The inputs are:

1. The spherical pressure $p_m = p_{c0}$, initially applied to consolidate the sample.

2. The specific rate of vertical pressure, regulated by the linear load rate generator unit.

3. The cross-section areas A_1 of the loading jack, A_2 of the soil sample and A_3 of the triaxial cell piston, introduced as pressure signals.

4. The force F', corresponding to the weight of the triaxial cell with the sample and the loading jack friction, in general constant values throughout the test. The force is also introduced as a pressure signal.

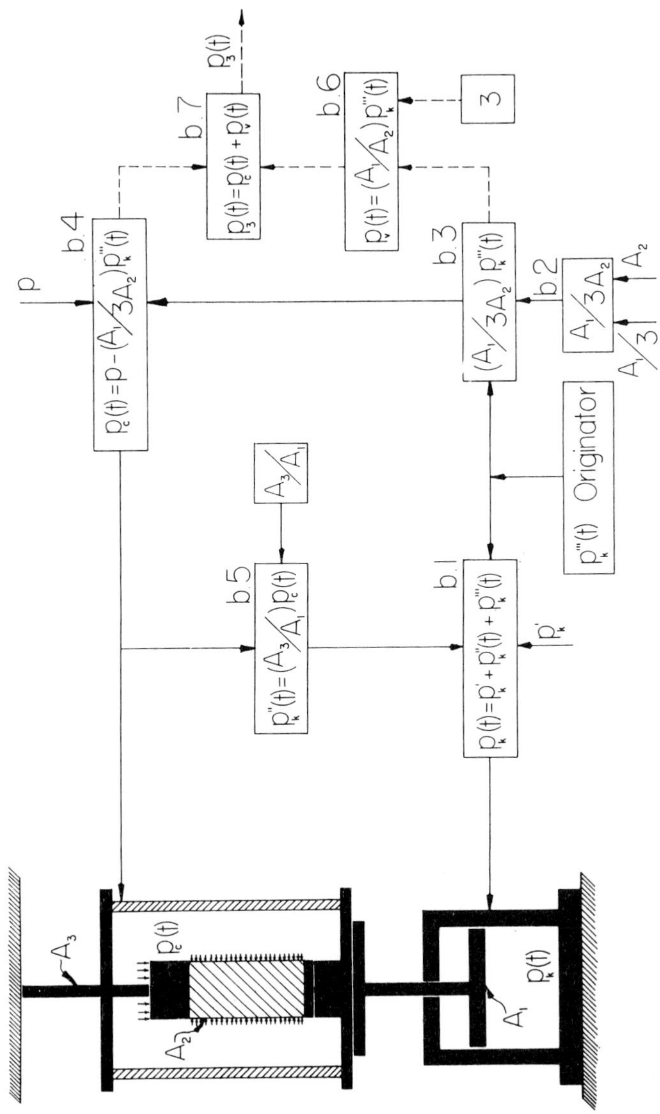

Fig. 12.14.2. Flow diagram of pure deviatoric loading device.

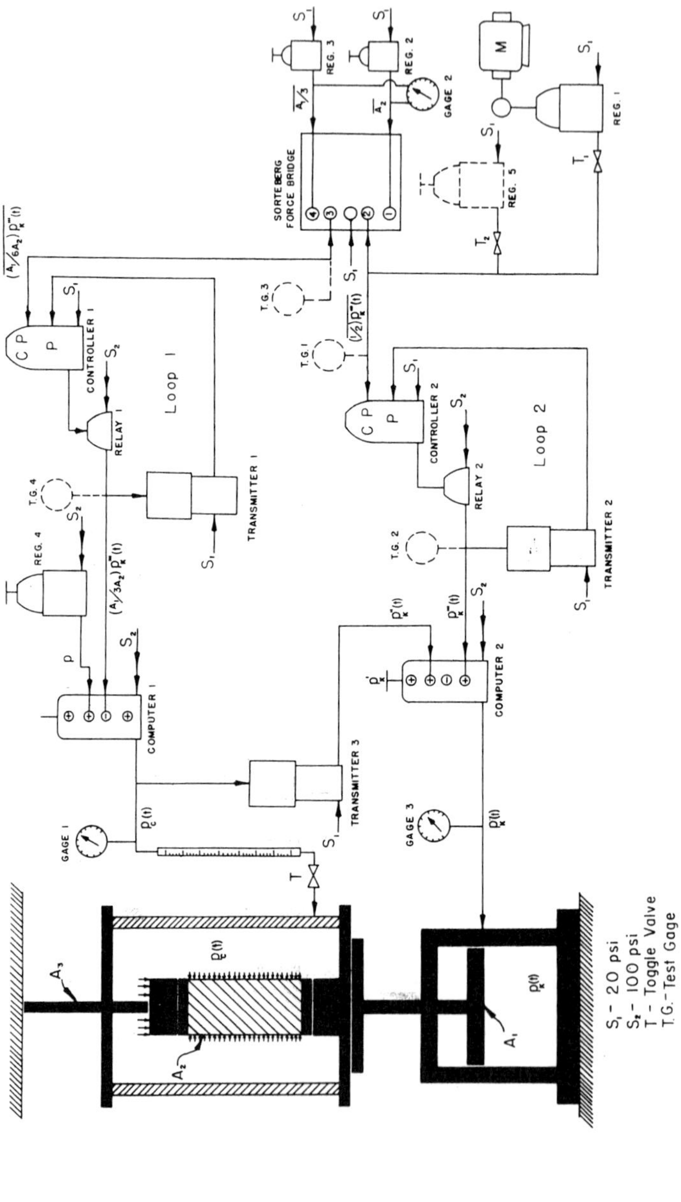

Fig. 12.14.3. Pneumatic circuit of pure deviatoric loading device.

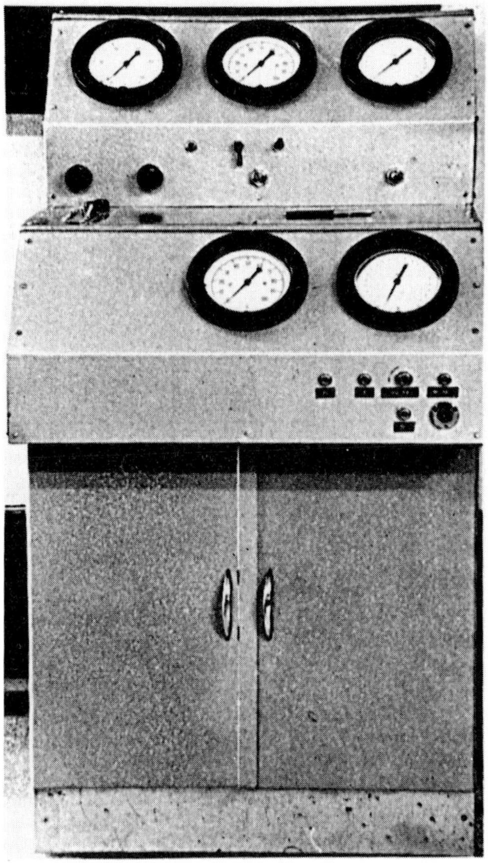

Fig. 12.14.4. Mobile deviatoric loading console.

The two outputs that control the pure deviatoric test are:

1. The triaxial cell pressure p_c.

2. The pressure of the pneumatic loading jack p_k that applies the excess vertical pressure $\Delta p(t)_{33}$.

The control unit considers, corrects and compensates for the following:

1. Considers the weight of the triaxial cell and the frictional resistance of the loading jack and compensates for it with a loading jack pressure, $p_k' = F'/A_1$.

2. Considers the force that the triaxial cell pressure exerts as an uplift pressure on the piston and makes the corresponding correction for it in the loading jack pressure, $p(t)_k'' = (A_3/A_1)p(t)_c$.

3. Allows the testing of samples of any diameter, adjusting for it by considering the ratio of the sample diameter to the diameter of the loading jack piston, $p(t)_k''' = (A_2/A_1)\Delta p(t)_{33}$.

The flow diagram, Fig. 12.14.2, shows that once the values p_m, F', A_1, A_2 and A_3 have been provided as inputs and the rate of vertical loading has been adjusted, the triaxial cell pressure $p(t)_c$ and the loading jack pressure $p(t)_k$ are generated and controlled while the proper corrections are made.

It is worth noting that a pure deviatoric loading device can be built with electronic or fluidic control as well, using reliable and accurate transducers at the interfaces. An electronic control unit would probably be the most promising, since it provides all computing facilities and is readily connected to a plotter of stress–strain curves.

In the pure deviatoric test pore pressures can be measured as well, either in an undrained test (Zur 1969) or at one end of the sample while the sample drains at the other end.

12.15 Stresses and Strains in the Pure Deviatoric Test

Pure deviatoric loadings are by nature control stress tests: the lateral pressures are adjusted according to the vertical pressure, therefore a control stress unit, specifically in an automatic control unit, is most reasonable.

Two of the familiar methods of pure deviatoric loadings are discussed here.

1. *Constant stepwise vertical pressure application*, $\Delta p_{33} = \text{const}$. Some of the early pure deviatoric tests were performed with stepwise increments (Saada 1961; Klausner 1964, 1987b), and the method is still applied, since the conventional triaxial testing device permits such loading.

After application of the spherical pressure $p_c = p_m$ and consolidation of the soil sample, vertical constant pressure increments $\Delta p_{33} = \text{const}$ are applied while at the same time the triaxial cell pressure is decreased, according to Eq. (12.13.9), by $\frac{1}{3}\Delta p_{33}$. The pressure tensor for each loading step becomes

$$p_{ij} \equiv \begin{vmatrix} p_{11} & 0 & 0 \\ 0 & p_{22} & 0 \\ 0 & 0 & p_{33} \end{vmatrix}$$

$$= \begin{vmatrix} p_m - \frac{1}{2}\Delta p_{33} & 0 & 0 \\ 0 & p_m - \frac{1}{2}\Delta p_{33} & 0 \\ 0 & 0 & p_m + \Delta p_{33} \end{vmatrix}$$

$$= \begin{vmatrix} p_c & 0 & 0 \\ 0 & p_c & 0 \\ 0 & 0 & p_c + \frac{3}{2}\Delta p_{33} \end{vmatrix} \equiv p_m\delta_{ij} + \tfrac{1}{2}\Delta p_{33}\Xi_{ij} = p_m\delta_{ij} + s_{ij} \qquad (12.15.1)$$

where $p_c = p_m - \frac{1}{2}\Delta p_{33}$ is the triaxial cell pressure, Ξ_{ij} the dimensionless deviatoric tensor and s_{ij} the deviatoric stress tensor.

It is clear from Eq. (12.15.1) that the pure deviatoric test is a deviatoric shear test, since the spherical pressure is maintained constant throughout the whole test, $p_{\alpha\alpha} = 3p_{\mathrm{m}} = \text{const}$.

The deformations measured are the same as in the triaxial test discussed in Sect. 12.6, that is, they are vertical deformations and volume changes, and from them the vertical strains $\varepsilon_{33} = e_{33}$ and the volumetric strains $\varepsilon_{\alpha\alpha}$ are calculated, respectively. The deviatoric stress–deviatoric strain curves, Fig. 12.15.1a, and the deviatoric stress–volumetric strain curves, Fig. 12.15.1b, are shown for the different spherical pressures p_{m} of a specific series of tests. The deviatoric strain tensor that corresponds to the deviatoric stress tensor can be written

$$e_{ij} \equiv \begin{vmatrix} e(t)_{11} & 0 & 0 \\ 0 & e(t)_{22} & 0 \\ 0 & 0 & e(t)_{33} \end{vmatrix}; \qquad e(t)_{11} = e(t)_{22} \tag{12.15.2}$$

The average horizontal radial strains, $\varepsilon_{11} = \varepsilon_{22} = e_{11} = e_{22}$, can be computed by assuming that strain due to deviatoric stresses will be only deviatoric. Thus the volumetric strain of the deviatoric test vanishes, $e_{\beta\beta} = 0$, and the deviatoric shear strain, Eq. (12.15.2), may be written explicitly

$$e_{ij} \equiv \begin{vmatrix} -\frac{1}{2}e(t)_{33} & 0 & 0 \\ 0 & -\frac{1}{2}e(t)_{33} & 0 \\ 0 & 0 & e(t)_{33} \end{vmatrix} \equiv \tfrac{1}{2}e(t)_{33}\Xi_{ij} \tag{12.15.3}$$

where we see that although the loadings are constant step loadings the strains are time dependent. In Fig. 12.15.1a the changes with time of the deviatoric strain are seen, marked in dotted lines.

There is a discrepancy between the assumption of $e_{\beta\beta} = 0$ and the experimental fact that volumetric strain is measured. This cannot be reconciled by a linear stress–strain relationship, since according to the linear stress–strain relationship, Sect. 7.5, volumetric strains cannot be present during shear stresses. Only an anisotropic stress–strain relationship can accommodate volumetric strains during deviatoric stresses and provide a reasonable explanation.

2. *Constant rate of vertical pressure application*, $\overline{\Delta \dot{p}(t)_{33}} = \text{const}$. A constant rate of vertical pressure application is a controlled stress loading where the vertical pressure is increased linearly with time

$$\Delta p(t)_{33} = \mathscr{C}t; \tag{12.15.4}$$

where $e = \overline{\Delta \dot{p}(t)_{33}} = \text{const} > 0$.

The loading was made possible by the pure deviatoric control unit described in the previous section. Pure deviatoric tests with a constant rate of loading were performed on different types of soil, and the effect of the rate of loading was investigated (Klausner 1967, 1970c).

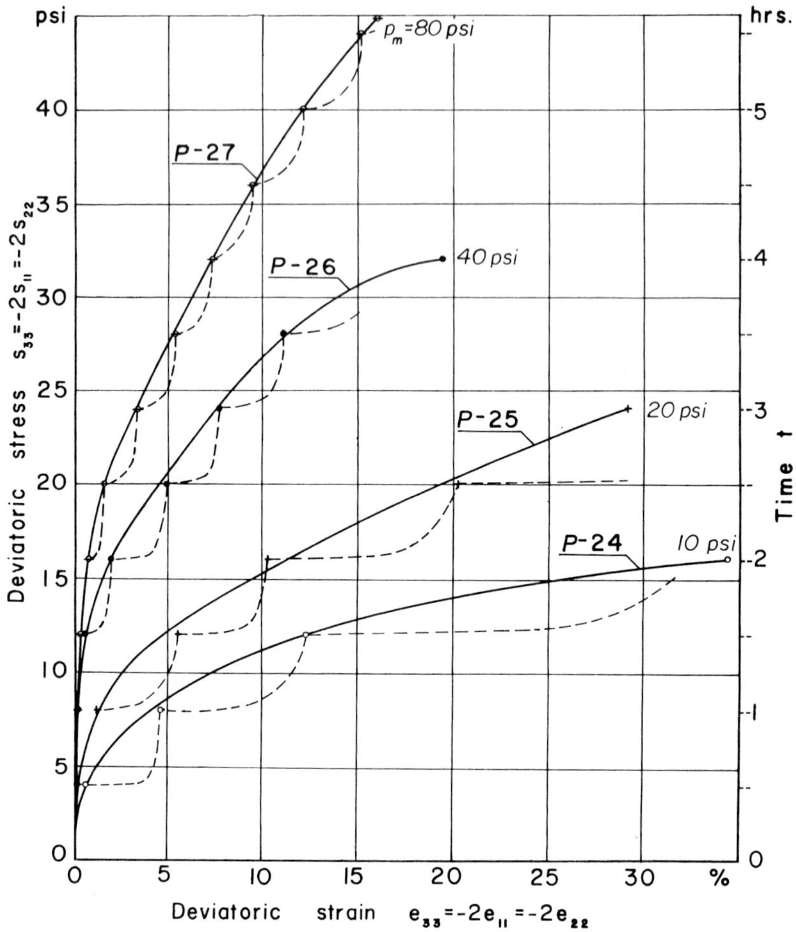

a

After the spherical pressure $p_m = p_c$ is applied on the sample and the consolidation process ends, the pure deviatoric loading with a constant rate of vertical load is applied and the pressure tensor p_{ij} becomes

$$p_{ij} \equiv \begin{vmatrix} p_{11} & 0 & 0 \\ 0 & p_{22} & 0 \\ 0 & 0 & p_{33} \end{vmatrix}$$

$$= \begin{vmatrix} p_m - \frac{1}{2}\Delta p(t)_{33} & 0 & 0 \\ 0 & p_m - \frac{1}{2}\Delta p(t)_{33} & 0 \\ 0 & 0 & p_m + \Delta p(t)_{33} \end{vmatrix}$$

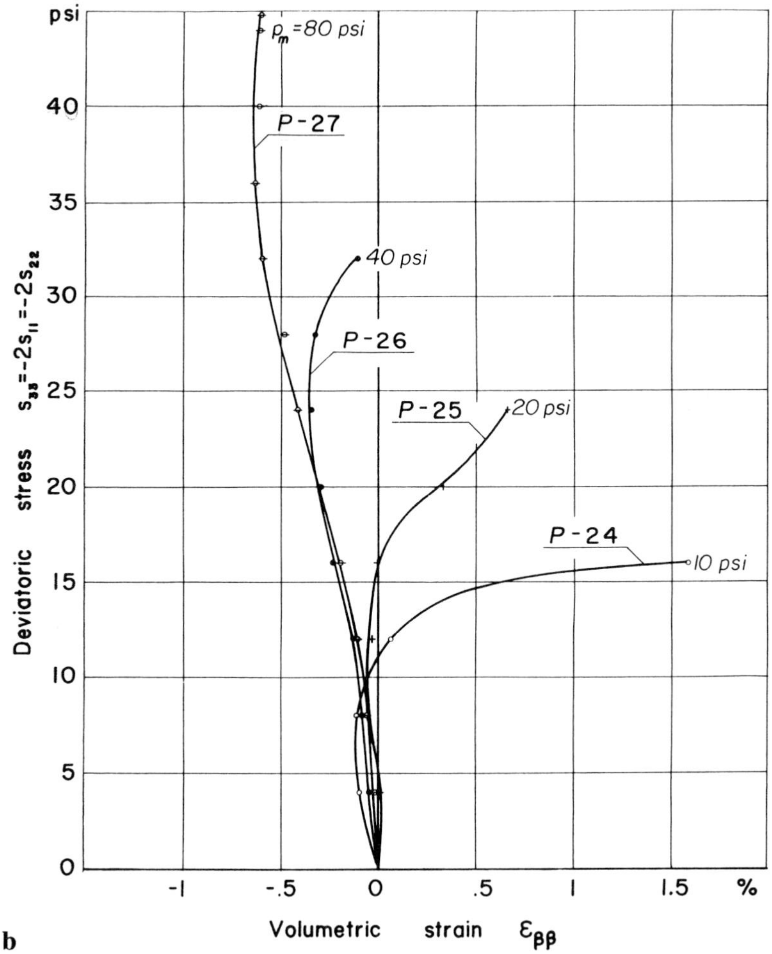

b

Fig. 12.15.1. Typical pure deviatoric shear test with stepwise loading. **a** Deviatoric stress versus deviatoric strain. **b** Deviatoric stress versus volumetric strain.

$$= \begin{vmatrix} p_m - \frac{1}{2}\mathscr{C}t & 0 & 0 \\ 0 & p_m - \frac{1}{2}\mathscr{C}t & 0 \\ 0 & 0 & p_m + \mathscr{C}t \end{vmatrix}$$

$$= \begin{vmatrix} p_c & 0 & 0 \\ 0 & p_c & 0 \\ 0 & 0 & p_c + \frac{3}{2}\mathscr{C}t \end{vmatrix}$$

$$\equiv p_m \delta_{ij} + \tfrac{1}{2}\mathscr{C}t\, \Xi_{ij} = p_m \delta_{ij} + s_{ij} \qquad (12.15.5)$$

In the case of constant rate of deviatoric loading we may consider the rate

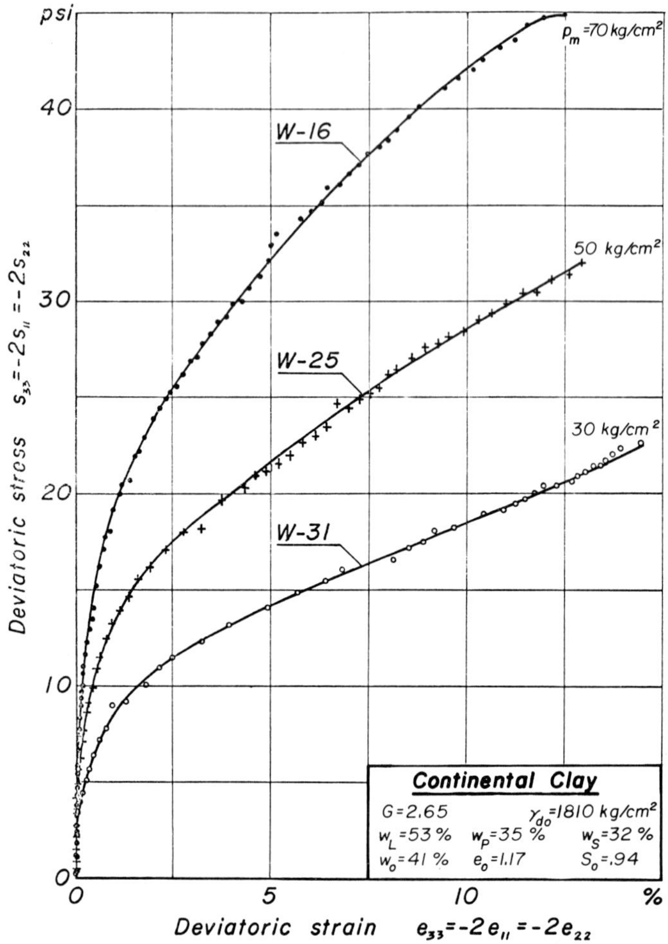

a

of the pressure tensor $\dot{p}_{ij}$ as follows

$$\dot{p}_{ij} \equiv \begin{vmatrix} \dot{p}_{11} & 0 & 0 \\ 0 & \dot{p}_{22} & 0 \\ 0 & 0 & \dot{p}_{33} \end{vmatrix}$$

$$= \begin{vmatrix} -\frac{1}{2}\Delta\dot{p}(t)_{33} & 0 & 0 \\ 0 & -\frac{1}{2}\Delta\dot{p}(t)_{33} & 0 \\ 0 & 0 & \Delta\dot{p}(t)_{33} \end{vmatrix}$$

$$= \begin{vmatrix} -\frac{1}{2}\mathscr{C} & 0 & 0 \\ 0 & -\frac{1}{2}\mathscr{C} & 0 \\ 0 & 0 & \mathscr{C} \end{vmatrix} \equiv \tfrac{1}{2}\mathscr{C}\Xi_{ij} = \dot{s}_{ij} \tag{12.15.6}$$

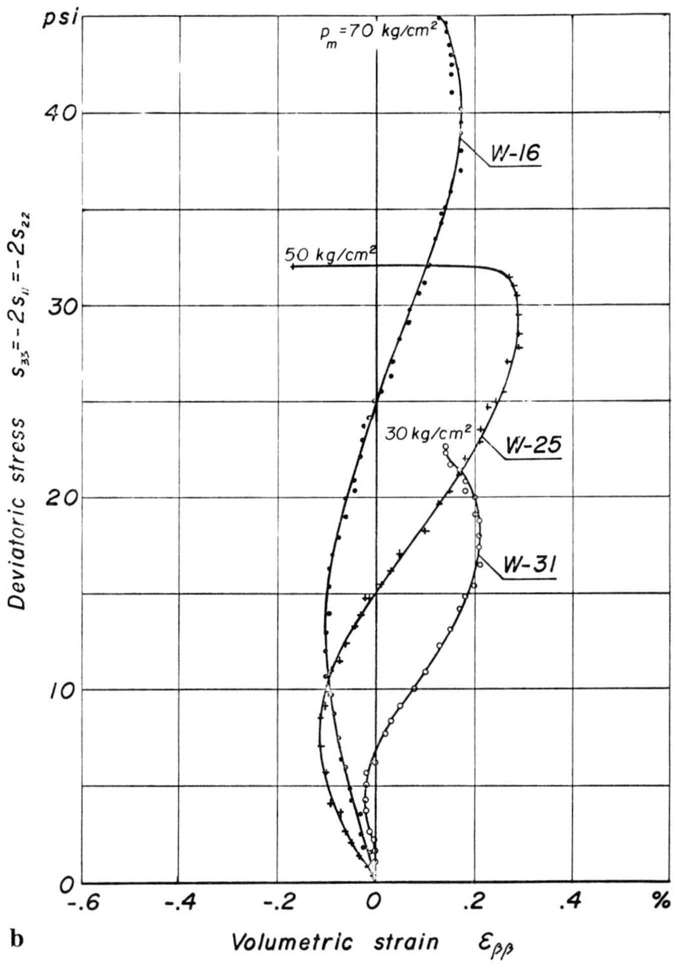

b

Fig. 12.15.2. Typical pure deviatoric test with constant rate of loading. **a** Deviatoric stress versus deviatoric strain. **b** Deviatoric stress versus volumetric strain.

It is seen that the rate of total pressure tensor $\dot{p}_{ij}$ is equal to the rate of the deviatoric stress tensor $\dot{s}_{ij}$.

The deformations from which the volumetric strains $\varepsilon_{\beta\beta}$ and the vertical strains $\varepsilon_{33} = e_{33}$ are calculated are the volume changes and the vertical deformation, respectively, the same as in step loading. Deviatoric stress–deviatoric strain curves, Fig. 12.15.2a, and deviatoric stress–volumetric strain curves, Fig. 12.15.2b, are shown for the different spherical pressures p_m of a specific series of tests. The deviatoric strain tensor is the same as that shown in Eqs. (12.15.2) and (12.15.3)

$$e_{ij} \equiv \begin{vmatrix} e(t)_{11} & 0 & 0 \\ 0 & e(t)_{22} & 0 \\ 0 & 0 & e(t)_{33} \end{vmatrix}$$

$$= \begin{vmatrix} -\frac{1}{2}e(t)_{33} & 0 & 0 \\ 0 & -\frac{1}{2}e(t)_{33} & 0 \\ 0 & 0 & e(t)_{33} \end{vmatrix} \equiv \tfrac{1}{2}e(t)_{33}\Xi_{ij}$$

$$e(t)_{11} = e(t)_{22} \tag{12.15.7}$$

and the same remarks with respect to the discrepancy between the assumed and measured volume strains are valid here as well.

12.16 Linear Deviatoric Stress–Strain Relationship

Resolving the stress and strain tensors into volumetric and deviatoric components, the general linear shear stress–strain relationship can be determined according to Eq. (7.6.7), represented by a general Kelvin model, Fig. 12.6.1

$$\text{(7.6.7)} \qquad e(t'')_{ij} = \frac{1}{2}\left[\frac{s(t'')_{ij}}{G_1} + \frac{1}{\eta_2}\int_0^{t''} s(t)_{ij}\,dt + \right.$$

$$\left. + \sum_{i=3}^{\mathfrak{I}} \frac{1}{\eta_i} \exp\frac{-t''}{T'_{\text{reti}}} \int_0^{t''} s(t)_{ij} \exp\frac{t}{T'_{\text{reti}}}\,dt\right]$$

where the total distortion e_{ij} is the sum of an instantaneous elastic distortion $s(t'')_{ij}/2G_1$, a viscous distortion $1/2\eta_2 \int_0^{t''} s(t)_{ij}\,dt$ and a series of viscoelastic distortions

$$\sum_{i=3}^{\mathfrak{I}} \frac{1}{\eta_i} \exp\frac{-t''}{T'_{\text{reti}}} \int_0^{t''} s(t)_{ij} \exp(t/T'_{\text{reti}})\,dt.$$

The formulation of the deviatoric constitutive equations cannot be deduced from the general balance equations, but has to be based on the data obtained from the existing deviatoric tests of the specific material. It is evident from that data that there are differences between the requirements for the formulation of constitutive equations for the volumetric behavior of soils, as outlined in Sect. 11.2, and their formulation for the deviatoric behavior. The requirements that follow are based on deviatoric test results (Geuze 1953; Geuze and Tan 1954; Tan 1954; Klausner 1970c), the objective being to attain a reliable expression of the deviatoric stress–strain relationship. The deviatoric test itself will be described in a later section.

1. The total distortion should contain an instantaneously recoverable dis-

tortion. This is a reasonable assumption, and it would be particularly apparent at low ranges of deviatoric stresses.

2. The total distortion must contain a time-dependent unrecoverable viscous distortion.

3. The total distortion should, probably, contain a time dependent recoverable distortion.

4. The total distortion will be dependent not only on the deviatoric stress but also on the rate at which that stress is applied, and perhaps even on its higher order rates.

The support for the above requirements has to come from tests and the analytical proof from thermodynamic considerations.

12.17 The Linear Deviatoric Constitutive Equation for Soils

Applying the above requirements to Eq. (7.6.7), the formulation of a deviatoric constitutive equation for soils is satisfied by a four parameter equation, $\mathfrak{i} \equiv \mathfrak{I} = 3$, as follows

$$e(t'')_{ij} = \frac{1}{2}\left[\frac{s(t'')_{ij}}{G_1} + \frac{1}{\eta_2}\int_0^{t''} s(t)_{ij}\, dt + \right.$$

$$\left. + \frac{1}{\eta_3}\exp\frac{-t''}{T'_{\text{ret}3}}\int_0^{t''} s(t)_{ij}\exp\frac{t}{T'_{\text{ret}3}}\, dt\right] \tag{12.17.1}$$

where $e(t'')_{ij}$ is the distortion, $s(t'')_{ij}$ is the deviatoric stress, G_1 and G_3 are the *shear moduli*, η_2 and η_3 are *shear viscous coefficients* and $T'_{\text{ret}3} = \eta_3/G_3$ is the *viscoelastic retardation time*.

Since in a pure deviatoric loading $e(t'')_{ij} = \frac{1}{2}\mathscr{E}(t'')\Xi_{ij}$ and $s(t'')_{ij} = \frac{1}{2}\mathscr{S}(t'')\Xi_{ij}$, Eq. (12.17.1) becomes a scalar equation, which may be written

$$\mathscr{E}(t'') = \frac{1}{2}\left[\frac{\mathscr{S}(t'')}{G_1} + \frac{1}{\eta_2}\int_0^{t''} \mathscr{S}(t)\, dt + \frac{1}{\eta_3}\exp\frac{-t''}{T'_{\text{ret}3}}\int_0^{t''} \mathscr{S}(t)\exp\frac{t}{T'_{\text{ret}3}}\, dt\right] \tag{12.17.2}$$

From (12.17.2) we obtain the equation of the distortion rate $\dot{\mathscr{E}}(t'')$, as follows

$$\dot{\mathscr{E}}(t'') = \frac{1}{2}\left[\frac{\dot{\mathscr{S}}(t'')}{G_1} + \frac{\mathscr{S}(t'')}{\eta_2} + \left(\mathscr{S}(t'') - \frac{G_3}{\eta_3}\exp\frac{-t''}{T'_{\text{ret}3}}\int_0^{t''} \mathscr{S}(t)\exp\frac{t}{T'_{\text{ret}3}}\, dt\right)\right] \tag{12.17.3}$$

Further conditions can be introduced in Eq. (12.17.3) concerning the deviatoric stress, the distortion, the rate of the deviatoric stress and the rate of distortion. Some of the conditions common in testing of soils follow:

1. *Step loading of deviatoric stress*. If the deviatoric loading is applied in Ω constant steps until failure is reached, $\mathcal{S}(t)_q = \mathcal{S}_q = \text{const}$, then Eq. (12.17.2) for the individual step becomes

$$\mathcal{E}(t'')_q = \frac{\mathcal{S}_q}{2}\left[\frac{1}{G_1} + \frac{t''}{\eta_2} + \frac{1}{G_3}\left(1 - \exp\frac{-t''}{T'_{\text{ret3}}}\right)\right] \tag{12.17.4}$$

and the rate of distortion, Eq. (12.17.3), becomes

$$\dot{\mathcal{E}}(t'')_q = \frac{\mathcal{S}_q}{2}\left[\frac{1}{\eta_2} + \frac{1}{\eta_3}\exp\frac{-t''}{T'_{\text{ret3}}}\right] \tag{12.17.5}$$

Fig. 12.17.1 shows that according to Eq. (12.17.4) the distortion for a constant deviatoric step loading is the sum of the elastic, viscous and viscoelastic distortions. The rate of distortion depends, according to Eq. (12.17.5), on the viscous and viscoelastic deviatoric stresses, Fig. 12.17.2.

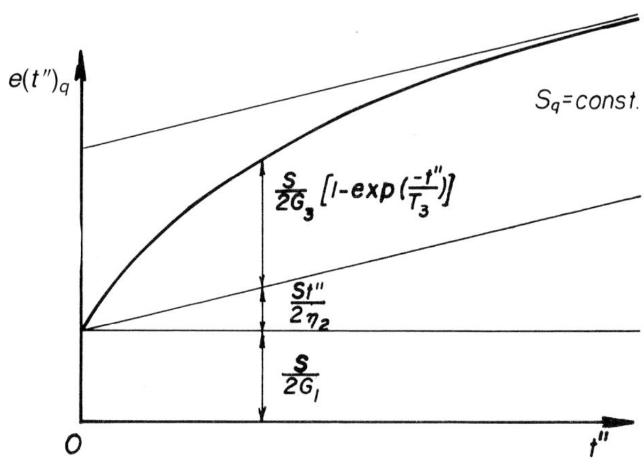

Fig. 12.17.1. Distortion versus time for stepwise deviatoric loading.

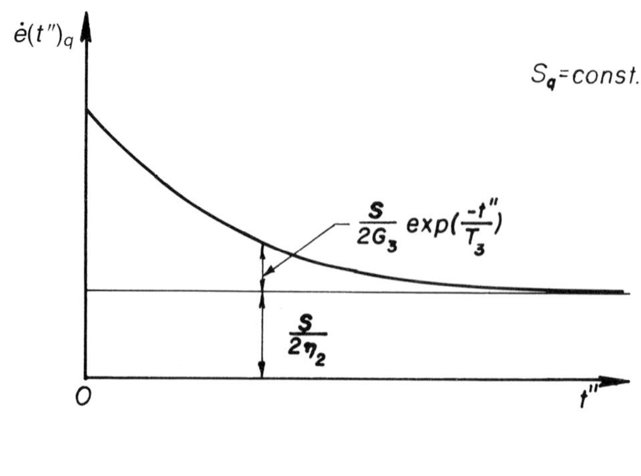

Fig. 12.17.2. Distortion rate versus time for stepwise deviatoric loading.

The total distortion is the sum of the individual $\mathfrak{Q}$ step distortions

$$\mathcal{E}(t'') = \sum_{q=1}^{\mathfrak{Q}} \mathcal{E}(t)_q = \frac{1}{2}\left[\frac{1}{G_1} + \frac{t''}{\eta_2} + \frac{1}{G_3}\left(1 - \exp\frac{-t''}{T'_{\mathrm{ret}3}}\right)\right] \sum_{q=1}^{\mathfrak{Q}} \mathcal{S}_q \tag{12.17.6}$$

assuming the coefficients G_1, G_3, η_2 and $T'_{\mathrm{ret}3}$ remain unchanged throughout the deviatoric steps. Consequently the distortion rate is

$$\dot{\mathcal{E}}(t'') = \sum_{q=1}^{\mathfrak{Q}} \dot{\mathcal{E}}(t'')_q = \frac{1}{2}\left[\frac{1}{\eta_2} + \frac{1}{\eta_3}\exp\frac{-t''}{T'_{\mathrm{ret}3}}\right] \sum_{q=1}^{\mathfrak{Q}} \mathcal{S}_q \tag{12.17.7}$$

2. *Constant rate of deviatoric stress*. If the deviatoric stress is applied at a constant rate, $\mathcal{S}(t) = \dot{\mathcal{S}}t$, the distortion will be

$$\mathcal{E}(t'') = \frac{\dot{\mathcal{S}}}{2}\left[\frac{t''}{G_1} + \frac{t''^2}{\eta_2} + \frac{1}{G_3}\left\{t'' + \frac{T'_{\mathrm{ret}3}}{t''}\left(\exp\frac{-t''}{T'_{\mathrm{ret}3}} - 1\right)\right\}\right] \tag{12.17.8}$$

and the rate of distortion, Eq. (12.17.3), becomes

$$\dot{\mathcal{E}}(t'') = \frac{\dot{\mathcal{S}}}{2}\left[\frac{1}{G_1} + \frac{t''}{\eta_2} + \frac{1}{G_3}\left(1 - \exp\frac{-t''}{T'_{\mathrm{ret}3}}\right)\right] \tag{12.17.9}$$

It is interesting to note that the rate of distortion in the case of constant rate of deviatoric stress application, Eq. (12.17.9), Fig. 12.17.4, is equal to the distortion in the case of constant deviatoric stress application, Eq. (12.17.4), Fig. 12.17.2. The distortion, Fig. 12.17.3, is equal to the area below the curve in Fig. 12.17.4 and can be obtained by integration.

3. *Constant distortion*. Differentiating Eq. (12.17.3) and eliminating the last term, a result similar to the one obtained in Appendix C Eq. (C.9.4) is obtained

$$\dot{\mathcal{E}}(t) + \frac{\eta_3}{G_3}\ddot{\mathcal{E}}(t) = \frac{\mathcal{S}(t)}{2\eta_2} + \frac{\eta_2 G_3 + \eta_2 G_1 + \eta_3 G_1}{2\eta_2 G_1 G_3}\dot{\mathcal{S}}(t) + \frac{\eta_3}{2G_1G_3}\ddot{\mathcal{S}}(t) \tag{12.17.10}$$

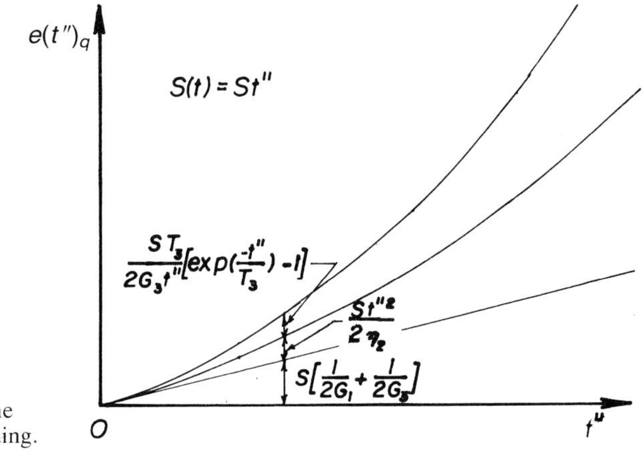

Fig. 12.17.3. Distortion versus time for constant rate of deviatoric loading.

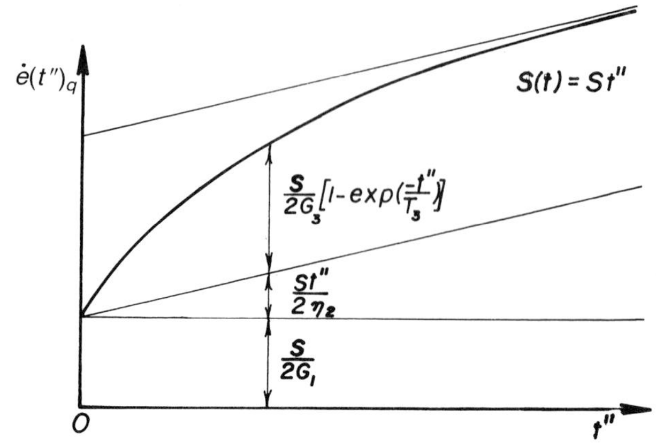

Fig. 12.17.4. Distortion rate versus time for constant rate of deviatoric loading.

If a constant distortion $\mathcal{E}(t) = \mathcal{E} = \text{const}$ is imposed on a soil sample, then the first and second derivatives of the distortion vanish, $\dot{\mathcal{E}} = \ddot{\mathcal{E}} = 0$, and Eq. (12.17.10) yields

$$\mathcal{S}(t) + \frac{\eta_2 G_3 + \eta_2 G_1 + \eta_3 G_1}{G_1 G_3} \dot{\mathcal{S}}(t) + \frac{\eta_2 \eta_3}{G_1 G_3} \ddot{\mathcal{S}}(t) = 0 \tag{12.17.11}$$

The solution of this homogeneous linear equation will be (Reiner 1958)

$$\mathcal{S}(t) = C_1 \exp\frac{-t}{T_1} + C_2 \exp\frac{-t}{T_2} \tag{12.17.12}$$

where T_1 and T_2 are two relaxation times given by

$$\frac{1}{T_{1,2}} = \frac{\eta_2 G_3 + \eta_2 G_1 + \eta_3 G_1 \pm \sqrt{\Delta}}{2 G_1 G_3} \tag{12.17.13}$$

with

$$\Delta = (\eta_2 G_3 + \eta_2 G_1 + \eta_3 G_1)^2 - 4\eta_2 \eta_3 G_1 G_3 \tag{12.17.14}$$

Since at time $t = 0$ the stress is equal to $2G_1\mathcal{E}$, we have from Eq. (12.17.12)

$$\mathcal{S}(0) = 2G_1\mathcal{E} = C_1 + C_2 \tag{12.17.15}$$

It is also seen from Eq. (12.17.12) that for a constant distortion the stress dissipates from the initial value $\mathcal{S}(0) = 2G_1\mathcal{E}$ to $\mathcal{S}(\infty) = 0$.

4. *Constant rate of distortion.* If the constant rate of distortion is $\mathcal{E}(t) = \mathcal{E}t$, then the first derivative of the distortion is constant $\dot{\mathcal{E}}(t) = \mathcal{E} = \text{const}$ and the second derivative vanishes $\ddot{\mathcal{E}} = 0$, and Eq. (12.17.10) yields

$$\mathcal{S}(t) + \frac{\eta_2 G_2 + \eta_2 G_1 + \eta_3 G_1}{G_1 G_3} \dot{\mathcal{S}}(t) + \frac{\eta_2 \eta_3}{G_1 G_3} \ddot{\mathcal{S}}(t) = 2\eta_2\mathcal{E} \tag{12.17.16}$$

The solution of this non-homogeneous equation consists of the solution of the homogeneous equation (12.17.12) and a particular solution

$$\mathcal{S}(t) = C_1 \exp\frac{-t}{T_1} + C_2 \exp\frac{-t}{T_2} + 2\eta_2\mathcal{E} \tag{12.17.17}$$

where T_1 and T_2 are the two relaxation times identical to the relaxation times given in Eq. (12.17.13) and (12.17.14).

Since the initial stress is zero, $\mathcal{S}(0) = 0$, Eq. (12.17.17) yields

$$C_1 + C_2 = -2\eta_2\mathcal{E} \tag{12.17.18}$$

while the stress curve reaches an asymptotic value, $\mathcal{S}(\infty) = 2\eta_2\mathcal{E}$, as time tends to infinity.

12.18 Isotropic Strain Functions of Shear Constitutive Equations

The isotropic stress–strain relationship (Reiner 1945; Rivlin and Ericksen 1955) allows a great deal of latitude, in that when defining the relationship between the stresses and strains and their time derivatives of a higher order, it admits the expression of any one of them in terms of any other two or more.

The general isotropic stress–strain relationship introducing tensorial non-linearity was discussed in Sect. 7.9, and one set of equations relevant to deviatoric tests are Eqs. (7.9.6) and (7.9.7), which give the strain tensor $\varepsilon(t)_{ij}$ as a function of the stress tensor $p(t)_{ij} = s(t)_{ij}$ and the stress rate tensor $\dot{p}(t)_{ij} = \dot{s}(t)_{ij}$. Equivalent sets of equations are also possible, each providing a dual constitutive equation. Let us pursue the constitutive equation of four coefficients introduced in Sect. 11.27

$$(11.27.3) \qquad \varepsilon_{ij} = \mathfrak{w}_1 s_{ij} + \mathfrak{w}_2 s_{ij}^2 + \mathfrak{w}_3 \dot{s}_{ij} + \mathfrak{w}_4 \dot{s}_{ij}^2$$

It was indicated that the strain ε_{ij} is composed of a volumetric strain, a dilatancy $\varepsilon_{\beta\beta}$ and a distortion $\mathcal{E}\Xi_{ij}$

$$(11.27.6) \qquad \varepsilon_{\beta\beta} = 6(\mathfrak{w}_2\mathcal{S}^2 + \mathfrak{w}_4\dot{\mathcal{S}}^2)$$

$$(11.27.5) \qquad \mathcal{E}\Xi_{ij} = (\mathfrak{w}_1\mathcal{S} + 2\mathfrak{w}_2\mathcal{S}^2 + \mathfrak{w}_3\dot{\mathcal{S}} + 2\mathfrak{w}_4\dot{\mathcal{S}}^2)\Xi_{ij}$$

Eq. (11.27.6) was considered in Chap. 11 for the excess stored specific free energy, and now we shall concentrate on Eq. (11.27.5).

In the axially symmetric loading only two independent strain and stress components exist, therefore only two coefficients can be determined and they will be chosen accordingly. Additional assumptions can be formulated if the shape of the strain curves is known. The assumptions are:

1. The principal strains in the radial direction are deviatoric at any time and derivable from the measured vertical strain, $\varepsilon(t)_{33} = \mathscr{E}(t)$, thus $\varepsilon(t)_{11} = \varepsilon(t)_{22} = -\frac{1}{2}\varepsilon(t)_{33} = -\frac{1}{2}\mathscr{E}(t)$.

2. The applied stresses are deviatoric, $p(t)_{33} = \mathscr{S}(t)$, $p(t)_{11} = p(t)_{22} = -\frac{1}{2}p(t)_{33} = -\frac{1}{2}\mathscr{S}(t)$.

3. The measured volumetric strain $\varepsilon_{\beta\beta}$ is a consequence of *second-order effects* resulting from the assumption of isotropic stress–strain relations (Reiner 1945; Rivlin and Ericksen 1955).

The above conditions determine a pure deviatoric state of stresses.
Two cases of interest will be presented here:

1. *Stepwise deviatoric stress increments*. If after the application of a constant spherical pressure p_{m} the shear stress is applied in $\mathfrak{Q}$ increments up to failure, then the axially symmetric shear stress tensor of the $\mathfrak{q}$th increment will be $s_{ij\mathfrak{q}} = \frac{1}{2}\mathscr{S}_{\mathfrak{q}}\Xi_{ij}$, where $S_{\mathfrak{q}} = \mathrm{const}$ and thus $\dot{s}_{ij\mathfrak{q}} = 0$ and $\dot{\mathscr{S}}_{\mathfrak{q}} = 0$. We shall obtain the strain tensor of the $\mathfrak{q}$th increment according to Eq. (7.9.5)

$$\begin{aligned}\varepsilon_{ij\mathfrak{q}} &= \mathfrak{w}_1 s_{ij\mathfrak{q}} + \mathfrak{w}_2 s^2_{ij\mathfrak{q}} = \tfrac{1}{2}\mathfrak{w}_1\mathscr{S}\Xi_{ij} + \tfrac{1}{4}\mathfrak{w}_2\mathscr{S}^2(2\delta_{ij} + \Xi_{ij})\\ &= \tfrac{1}{2}\mathfrak{w}_2\mathscr{S}^2_{\mathfrak{q}}\delta_{ij} + (\tfrac{1}{2}\mathfrak{w}_1\mathscr{S}_{\mathfrak{q}} + \tfrac{1}{4}\mathfrak{w}_2\mathscr{S}^2_{\mathfrak{q}})\Xi_{ij} = \tfrac{1}{3}\varepsilon_{\beta\beta\mathfrak{q}}\delta_{ij} + \tfrac{1}{2}\mathscr{E}\Xi_{ij}\end{aligned} \tag{12.18.1}$$

where the strain $\varepsilon_{ij\mathfrak{q}}$ can be resolved into a volumetric strain $\varepsilon_{\beta\beta\mathfrak{q}}$ and a distortion $\frac{1}{2}\mathscr{E}\Xi_{ij\mathfrak{q}}$

$$\varepsilon_{\beta\beta\mathfrak{q}} = \tfrac{3}{2}\mathfrak{w}_2\mathscr{S}^2_{\mathfrak{q}} \tag{12.18.2}$$

$$e_{ij\mathfrak{q}} = \tfrac{1}{2}\mathscr{E}_{\mathfrak{q}}\Xi_{ij} = (\tfrac{1}{2}\mathfrak{w}_1\mathscr{S}_{\mathfrak{q}} + \tfrac{1}{4}\mathfrak{w}_2\mathscr{S}^2_{\mathfrak{q}})\Xi_{ij} \tag{12.18.3}$$

The components of the distortion tensor $e_{ij\mathfrak{q}}$ can be written from Eq. (12.18.3)

$$e_{11\mathfrak{q}} = e_{22\mathfrak{q}} = -\,(\tfrac{1}{2}\mathfrak{w}_1\mathscr{S}_{\mathfrak{q}} + \tfrac{1}{4}\mathfrak{w}_2\mathscr{S}^2_{\mathfrak{q}});\quad e_{33\mathfrak{q}} = (\mathfrak{w}_1\mathscr{S}_{\mathfrak{q}} + \tfrac{1}{2}\mathfrak{w}_2\mathscr{S}^2_{\mathfrak{q}}) \tag{12.18.4}$$

It was pointed out in Sect. 11.28 that the volumetric strain $\varepsilon_{\beta\beta\mathfrak{q}}$ contributes to the excess stored specific free energy. Here we are concerned with the distortions.

The distortions of the individual step loads determine the total distortion $\frac{1}{2}\mathscr{E}\Xi_{ij}$, which is the sum of the $\mathfrak{Q}$ distortions of the individual steps

$$\mathscr{E}\Xi_{ij} = \sum_{\mathfrak{q}=1}^{\mathfrak{Q}} \mathscr{E}_{\mathfrak{q}}\Xi_{ij} = \sum_{\mathfrak{q}=1}^{\mathfrak{Q}} (\mathfrak{w}_1\mathscr{S}_{\mathfrak{q}} + \tfrac{1}{2}\mathfrak{w}_2\mathscr{S}^2_{\mathfrak{q}})\Xi_{ij} \tag{12.18.5}$$

Fig. 12.15.1 shows the distortions obtained as well as the volumetric strain measured during the stepwise pure deviatoric shear stress application on sand samples.

In a pure deviatoric shear test we know s_{33q} while the volumetric strain $\varepsilon_{\beta\beta}$ and the vertical strain e_{33} are measured. On the other hand we have two equations, (12.18.2) and (12.18.4), with two coefficients $\mathfrak{w}_1$ and $\mathfrak{w}_2$ to be evaluated. We get

$$\mathfrak{w}_1 = \frac{3e_{33q} - \varepsilon_{\beta\beta q}}{3\mathcal{S}}; \quad \mathfrak{w}_2 = \frac{2\varepsilon_{\beta\beta q}}{3\mathcal{S}^2} \tag{12.18.6}$$

The contribution of the distortions to the disbursed specific free energy will be discussed in the next section.

2. *Constant rate of shear stress*. If the stress is applied at a constant rate defined $s(\mathrm{t})_{ij} = \frac{1}{2}\mathcal{S}\mathrm{t}\Xi_{ij}$ and $\dot{s}(\mathrm{t})_{ij} = \frac{1}{2}\mathcal{S}\Xi_{ij}$ where $\mathcal{S} = \text{const}$ and $\dot{\mathcal{S}} = 0$, then when these terms are substituted into Eq. (11.27.3) the corresponding strain ε_{ij} is obtained as a function of the stress alone

$$\begin{aligned}
\varepsilon_{ij} &= \tfrac{1}{2}\mathfrak{w}_1\mathcal{S}t\Xi_{ij} + \tfrac{1}{4}\mathfrak{w}_2\mathcal{S}^2t^2\Xi_{ij}^2 + \tfrac{1}{2}\mathfrak{w}_3\mathcal{S}\Xi_{ij} + \tfrac{1}{4}\mathfrak{w}_4\mathcal{S}^2\Xi_{ij}^2 \\
&= \tfrac{1}{2}(\mathfrak{w}_1\mathcal{S}t + \mathfrak{w}_3\mathcal{S})\Xi_{ij} + \tfrac{1}{4}(\mathfrak{w}_2\mathcal{S}^2t^2 + \mathfrak{w}_4\mathcal{S}^2)(2\delta_{ij}\,\Xi_{ij}) \\
&= \tfrac{1}{2}(\mathfrak{w}_2\mathcal{S}^2t^2 + \mathfrak{w}_4\mathcal{S}^2)\delta_{ij} + (\tfrac{1}{2}\mathfrak{w}_1\mathcal{S}t + \tfrac{1}{4}\mathfrak{w}_2\mathcal{S}^2t^2 + \tfrac{1}{2}\mathfrak{w}_3\mathcal{S} + \tfrac{1}{4}\mathfrak{w}_4\mathcal{S}^2)\Xi_{ij} \\
&= \tfrac{1}{3}\varepsilon_{\beta\beta}\delta_{ij} + \tfrac{1}{2}\mathcal{E}\Xi_{ij} = \tfrac{1}{3}\varepsilon_{\beta\beta}\delta_{ij} + e_{ij}
\end{aligned} \tag{12.18.7}$$

where it is seen that the strain ε_{ij} resolves into a volumetric strain $\varepsilon_{\beta\beta}$ and a distortion $\frac{1}{2}\mathcal{E}\Xi_{ij}$, as follows

$$\varepsilon_{\beta\beta} = \tfrac{3}{2}(\mathfrak{w}_2\mathcal{S}^2t^2 + \mathfrak{w}_4\mathcal{S}^2) \tag{12.18.8}$$

$$e_{ij} = \tfrac{1}{2}\mathcal{E}\Xi_{ij} = (\tfrac{1}{2}\mathfrak{w}_1\mathcal{S}t + \tfrac{1}{4}\mathfrak{w}_2\mathcal{S}^2t^2 + \tfrac{1}{2}\mathfrak{w}_3\mathcal{S} + \tfrac{1}{4}\mathfrak{w}_4\mathcal{S}^2)\Xi_{ij} \tag{12.18.9}$$

from which the components of the distortion tensor e_{ij} are obtained

$$\begin{aligned}
e_{11} &= e_{22} = -(\tfrac{1}{2}\mathfrak{w}_1\mathcal{S}t + \tfrac{1}{4}\mathfrak{w}_2\mathcal{S}^2t^2 + \tfrac{1}{2}\mathfrak{w}_3\mathcal{S} + \tfrac{1}{4}\mathfrak{w}_4\mathcal{S}^2) \\
e_{33} &= (\mathfrak{w}_1\mathcal{S}t + \tfrac{1}{2}\mathfrak{w}_2\mathcal{S}^2t^2 + \mathfrak{w}_3\mathcal{S} + \tfrac{1}{2}\mathfrak{w}_4\mathcal{S}^2)
\end{aligned} \tag{12.18.10}$$

In a pure deviatoric test the coefficients $\mathfrak{w}_1$, $\mathfrak{w}_2$, $\mathfrak{w}_3$ and $\mathfrak{w}_4$ are evaluated under the additional conditions: $e_{\beta\beta} = 0$ for $t = 0$, $e(t)_{ij} = 0$ for $i = j$, and e_{33} is continuous and rising with $\dot{e}_{33} > 0$ and $\ddot{e}_{33} \leq 0$. Solving for all these conditions we obtain

$$\mathfrak{w}_1 = \frac{3\varepsilon_{33} - \varepsilon_{\beta\beta}}{3\mathcal{S}t}; \quad \mathfrak{w}_2 = \frac{\varepsilon_{\beta\beta}}{6\mathcal{S}^2t^2}; \quad \mathfrak{w}_3 = \mathfrak{w}_4 = 0 \tag{12.18.11}$$

Instead of Eq. (11.27.3) many other three or four coefficient equations could have been chosen. For instance, $\varepsilon_{ij} = \mathfrak{w}_1 s_{ij} + \mathfrak{w}_2 s_{ij}^2 + \mathfrak{w}_4 \dot{s}_{ij}^2 + \mathfrak{w}_5 (s_{ij}\dot{s}_{ij} + \dot{s}_{ij}s_{ij})$, $\varepsilon_{ij} = \mathfrak{w}_1 s_{ij} + \mathfrak{w}_2 s_{ij}^2 + \mathfrak{w}_7 (s_{ij}\dot{s}_{ij}^2 + \dot{s}_{ij}^2 s_{ij})$, $\varepsilon_{ij} = \mathfrak{w}_1 s_{ij} + \mathfrak{w}_2 s_{ij}^2 + \mathfrak{w}_5 (s_{ij}\dot{s}_{ij} + \dot{s}_{ij}s_{ij}) + \mathfrak{w}_8 (s_{ij}^2 \dot{s}_{ij}^2 + \dot{s}_{ij}^2 s_{ij}^2)$, $\varepsilon_{ij} = \mathfrak{w}_1 s_{ij} + \mathfrak{w}_2 s_{ij}^2 + \mathfrak{w}_3 \dot{s}_{ij} + \mathfrak{w}_6 (s_{ij}^2 \dot{s}_{ij} + \dot{s}_{ij} s_{ij}^2)$, and many other combinations. The coefficients, then, would have different values in the various combinations, but an equivalence between them can be established.

12.19 Isotropic Stress Functions of Shear Constitutive Equations

Isotropic shear stress–strain relationships can be developed around the stresses as well, and for soils it has special meaning. We shall follow Eq. (7.9.1), where the stress tensor $p(t)_{ij}$ is defined as a function of the strain $e(t)_{ij}$ and the strain-rate $\dot{e}(t)_{ij}$. The strain and the strain-rate tensor are $e(t)_{ij} = \frac{1}{2}\mathscr{E}(t)\Xi_{ij}$ and $\dot{e}(t)_{ij} = \frac{1}{2}\dot{\mathscr{E}}(t)\Xi_{ij}$, respectively.

The assumptions underlying our treatment are:

1. The vertical strains $2\mathscr{E}t$ are as measured in the axially symmetric shear tests.

2. The radial strains are evaluated from the measured volumetric strain $\varepsilon(t)$ and the vertical strain $e(t) = \mathscr{E}(t)$, $e(t)_{11} = e(t)_{22} = -\frac{1}{2}\mathscr{E}(t) = \frac{1}{2}[\varepsilon(t)_{\beta\beta} - e(t)_{33}]$.

3. The total stress $p(t)_{ij}$ is equal to the sum of a shear stress $s(t)_{ij}$ and a spherical stress $p(t)_\flat$, which is a result of second-order effects and is derived from the isotropic stress–strain relationship.

For simplicity we shall use a two-coefficient function, and the stress tensor $p(t)_{ij}$ for the axially symmetric deviatoric loading will be

$$p(t)_{ij} = \beth_1 e(t)_{ij} + \beth_4 \dot{e}(t)_{ij}^2 = \tfrac{1}{2}\beth_1 \mathscr{E}(t)\Xi_{ij} + \tfrac{1}{4}\beth_4 \dot{\mathscr{E}}(t)^2 \Xi_{ij}^2$$

$$= \tfrac{1}{2}\beth_1 \mathscr{E}(t)\Xi_{ij} + \tfrac{1}{4}\beth_4 \dot{\mathscr{E}}(t)^2 (2\delta_{ij} + \Xi_{ij}) \tag{12.19.1}$$

which can be resolved into a spherical pressure $p(t)_\flat = \frac{1}{3}p(t)_{\beta\beta}$ and a deviatoric stress tensor $s(t)_{ij}$

$$p(t)_\flat = \tfrac{1}{3}p(t)_{\beta\beta} = \tfrac{1}{2}\beth_4 \dot{\mathscr{E}}(t)^2 \tag{12.19.2}$$

$$s(t)_{ij} = [\tfrac{1}{2}\beth_1 \mathscr{E}(t) + \tfrac{1}{4}\beth_4 \dot{\mathscr{E}}(t)^2]\Xi_{ij} \tag{12.19.3}$$

When $s(t)_{ij} = \frac{1}{2}\mathscr{S}(t)\Xi_{ij}$ is substituted in Eq. (12.19.3), the dimensionless octahedral tensor Ξ_{ij} can be removed and the following is obtained

$$s(t)_{11} = s(t)_{22} = -[\tfrac{1}{2}\beth_1 \mathscr{E}(t) + \tfrac{1}{4}\beth_4 \dot{\mathscr{E}}(t)^2]; \quad s(t)_{33} = \beth_1 \mathscr{E}(t) + \tfrac{1}{2}\beth_4 \dot{\mathscr{E}}(t)^2 \tag{12.19.4}$$

From Eqs. (12.19.2) and (12.19.4) $\beth_1$ and $\beth_4$ can now be evaluated

$$\beth_1 = \frac{\mathcal{S}(t) - 2p(t)_{\mathfrak{b}}}{\mathcal{E}(t)}; \quad \beth_4 = \frac{2p(t)_{\mathfrak{b}}}{\dot{\mathcal{E}}(t)^2} \tag{12.19.5}$$

where $\mathcal{S}(t) = s(t)_{33}$ is the applied vertical deviatoric pressure, $\mathcal{E}(t) = e(t)_{33}$ is the vertical strain calculated form the measured vertical deformation, $\dot{\mathcal{E}}(t)$ is the rate of vertical strain calculated from the vertical strain and $p(t)_{\mathfrak{b}}$ is the pore pressure that can be measured in an undrained soil sample.

12.20 Spherical Components in the Pure Deviatoric Test

According to Eq. (12.5.1), a necessary and sufficient condition that an applied stress increment be a deviatoric stress is that the spherical pressure remains constant during the application of the stress increment, $p(t)_{\alpha\alpha} = \text{const}$ or $\Delta p(t)_{\alpha\alpha} = 0$.

In the following we will try to show some of the specific features of a deviatoric test, its properties and its advantages and disadvantages as they are revealed by experiments and theory. We shall also show how it compares with the pure deviatoric test.

It was pointed out that the torsional test, especially the torsion of cylindrical tubes, perhaps most closely satisfies the required conditions of pure deviatoric shear. However, technical impediments prevent its extensive use. Added to this is the fact that the torsional test is limited to certain types of soils, owing to the difficulty of maintaining an erect undisturbed or remolded sample, particularly of hollow cylindrical shape.

The pure deviatoric test discussed in Sects. 12.13–12.15 is the next best deviatoric test. Its deficiency, namely that the shear stresses are applied to the sample by means of manipulating the principal normal stresses, is minor and inconsequential to the fact that it can be performed in the same way and by the same means as the conventional triaxial test.

From Eq. (12.6.1) we see that in the conventional triaxial shear test, when the vertical pressure Δp_{33} is applied following the consolidation due to the spherical pressure, a spherical stress $\frac{1}{3}\Delta p_{33}\delta_{ij}$ is also added to the deviatoric stress $\frac{1}{3}\Delta p_{33}\Xi_{ij}$. Thus in a conventional triaxial test the spherical properties of the sample are altered continuously. In the pure deviatoric test, however, when the vertical pressure Δp_{33} is applied, only deviatoric stresses $\frac{1}{2}\Delta p_{33}\Xi_{ij}$ are exerted on the sample, according to Eq. (12.13.6).

Undoubtedly the torsional test is the most appropriate deviatoric test from the point of view of the spherical components, since the deviatoric stresses are obtained by direct application of shear stress components, while in the pure deviatoric test the deviatoric effect is reached through manipulating the normal stress components. This has many consequences which will be discussed in the following sections, and unless further advances are achieved to overcome the technical difficulties of the torsional test, the pure deviatoric test will remain the most appropriate test for soils, and it should replace the conventional triaxial test.

One cannot but marvel at the fact that the conventional triaxial shear test became the leading test in soil mechanics, even though it very soon became obvious that it does not uniquely determine the shear properties of the soil.

In a pure deviatoric test the vertical stress is increased and the cell pressure is reduced, and although deviatoric stress conditions are attained and $\Delta p(t)_{\alpha\alpha} = 0$, the volumetric compression $\frac{1}{3}\Delta p_{33}$ corresponding to the vertical stress $\Delta p(t)_{33}$ is not equal to the volumetric decompression $-\frac{1}{3}\Delta p_{33}$ corresponding to the reduction of the cell pressure $\Delta p_c = -\frac{1}{3}\Delta p_{33}$. This difference is due to the hysteresis phenomenon known to us from the consolidation process.* It can be said with certainty that the compression due to the vertical stress $\Delta p(t)_{33}$ exceeds the decompression due to the cell pressure $\Delta p(t)_c$ and the net result is a constriction, a volume decrease and a density increase. Even when linear stress–strain relations are postulated for a pure deviatoric test, a volumetric decrease due to this discrepancy in the compressive and decompressive pressures can be expected, if the theory of hysteresis is applied. However, this topic has not been thoroughly investigated.

As was shown in Sects. 12.18 and 12.19, when an isotropic stress–strain relationship is postulated and deviatoric stresses are applied second-order effects arise and volume changes, *dilatancy***, occurs. According to the theory, these volume changes should always be positive, namely an increase in volume and a decrease in density, since they evolve from the second power of the terms involved. However, it is not certain that even soil samples tested in torsion will always indicate a volume increase. In pure deviatoric tests certainly not all volume changes are volume increases, which complicates even further the study of volume changes. When the dilatancy, a volume increase expected from the second-order effects is superimposed on the constriction, a volume decrease resulting from the conflicting actions of the pressures, the net sum will be either an increase or a decrease, depending on the relative magnitude of the two effects. This phenomenon is associated with the magnitude of the existing density of the soil relative to its *critical density* or *critical void ratio*, a parameter of the soil defined in Sect. 12.23.

In the pure deviatoric test, different soils exhibit different volumetric responses, but generally the strains measured were quite small, 2% and only in marginal cases of Beer-Sheva Loess 3%. The discrepancy between them and the volumetric strains of −16% to +20%, and perhaps even more, measured in conventional triaxial shear tests (Gibbs et al. 1960; Hvorslev 1960; Shockley and Ahlvin 1960)† should be attributed to the type of the test.

---

* The term *pure deviatoric* is probably not the proper name for the test because it really is not a pure deviatoric test. A pure deviatoric test is a torsion. A more proper term would be *neo-deviatoric* or *quasi-deviatoric*.

** Very often any volumetric changes during conventional triaxial shear tests are termed dilatancy, disregarding the formal definition accepted and recognized in mechanics.

† Except for a few studies early in the development of soil mechanics, it is difficult to find data on volumetric strains measured during conventional triaxial tests.

12.21 Pore Pressures in the Pure Deviatoric Test

We are aware of the fact that spherical pressures induce pore pressures and volumetric strains are indications of dissipating pore pressures. We have also seen that according to linear viscoelasticity deviatoric stresses cannot cause volumetric strains and distortions cannot be the result of spherical stresses. Only isotropic viscoelasticity provides a clue for the occurrence of volumetric strain under deviatoric stresses (Sect. 12.18), and of spherical stresses due to distortions (Sect. 12.19). These are two aspects of the same phenomenon resulting in dilatancy, a decrease of the density accompanied by a negative pore pressure.

In a conventional triaxial shear test, when the rate of loading is high the pore pressures due to the added spherical pressure cannot dissipate fast enough and the angle of internal friction decreases. If the preconsolidation pressure p_0 is higher than the spherical pressure p_m the cohesive intercept remains constant. If the test is run slow enough so that the pore pressures could readily dissipate, followed by an increase in the density of the sample, the angle of internal friction will reach its maximum value while the cohesive intercept will reach, usually but not necessarily, its lowest value. Consequently, *the conventional triaxial test does not define uniquely the deviatoric properties, the angle of internal friction ϕ and the cohesive intercept c, of the soil.*

To complicate matters, second-order effects in the conventional triaxial test result in a negative pore pressure and a decrease in density. The relative part of the two pore pressure effects and their distribution was never investigated, and only their superimposed total result is known.

In the pure deviatoric test the combined effects of the dilatancy resulting in a decrease in pore pressures and of the constriction resulting in an increase in pore pressures act sometimes in the same direction and sometimes in opposite directions. If it is an extension type test the effect of the constriction is a decrease in density of the sample and a decrease in the pore pressure, and this is added to the dilatancy which is always a decrease in the density and a decrease in pore pressure.

As stated before, when the dilatancy and the constriction are superimposed, the relative magnitude of the two determines whether the end result is a decrease or increase in pore pressure. In a pure deviatoric test in compression the effect of the constriction and the related increase in pore pressure generally, but not always, dominates in the early stages of the test. Negative pore pressures were measured by Zur (1969) in the early stages of pure deviatoric tests performed in compression on spherically consolidated undrained samples of a Mishmar Hanegev loess, Fig. 12.21.1. Fig. 12.21.2 shows pore pressure measurements by Zur on the same soil in tests performed in extension. There are indications that in granular soils the dilatancy tends to override the constriction effect throughout the shear test, the result being a volume increase and density and pore pressure decrease. In fine grained soils the constriction surpasses the dilatancy in the early stages of the shear test, while in the later stages it is exceeded by the dilatancy.

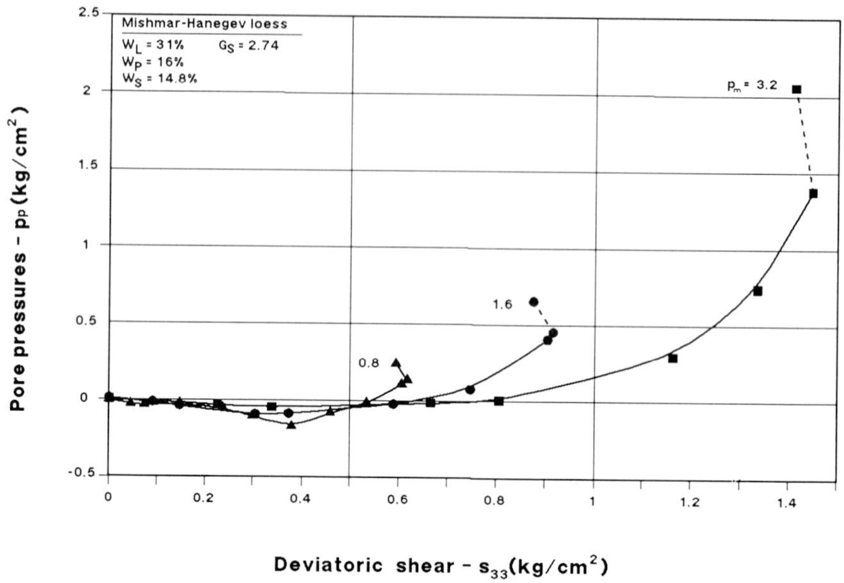

Fig. 12.21.1. Pore pressure measurements in undrained pure deviatoric tests in compression.

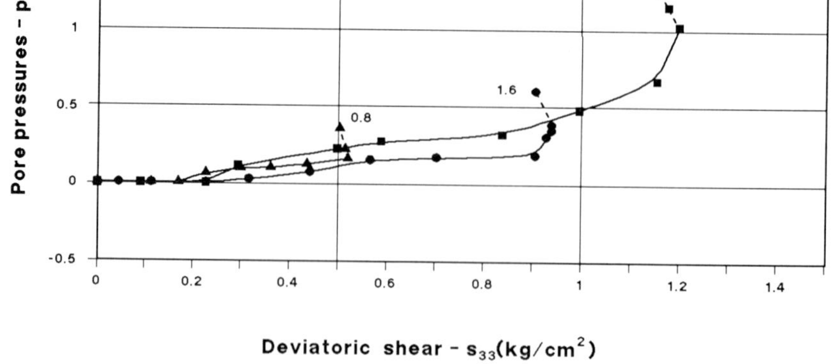

Fig. 12.21.2. Pore pressure measurements in undrained pure deviatoric tests in extension.

12.22 The Effect of the Rate of Loading in the Pure Deviatoric Test

Since during a deviatoric test no spherical pressures are added to the sample, pore pressures do not develop and changes in the density of the sample do not occur, except as second-order effects and as constrictions. If the effect of the constrictions is minimal, the changes in density occur only as second-order effects, and there should be no difference in the test results whether the test is run fast or slow; the results are independent of the rate of loading. In linear viscoelasticity the changes stem only from the constrictions, since linear viscoelasticity does not consider second-order effects.

Experimental studies of pure deviatoric tests were performed in three laboratories (Klausner 1987b–d) on remolded samples of different types of soils. The results which are summarized in Table 12.22.1 (Klausner 1967, 1970c) indeed show the almost minimal effect of the rate of loading on the deviatoric properties of the various soils. The differences in the angles of internal friction ϕ, up to 5° at the most, can be attributed, as explained, to second-order and constrictional effects. While the cohesive intercept c and the cohesion p_0 generally varied only slightly, their variation in the Grantham clay was more marked. Further studies are required on the effect of the rate of loading in the pure deviatoric test and on the influence of second-order and constrictional effects.

Fig. 12.22.1–12.22.3 present typical test results, showing the effect of the rate of loading on a Continental clay of an average initial void ratio of 1.17, tested at average deviatoric rates of loadings of 2.72, 1.72, 0.2 p.s.i/h. Table 12.22.1 presents the results of several other soils tested at different rates of loading.

The effect of the rate of loading in the conventional triaxial shear test is most dramatic in comparison to that of the pure deviatoric test. Pore pressures due mainly to the excess spherical pressures induced by the loading, Eq. (12.6.1), necessitated the introduction of slow, quick, drained and undrained tests, and the correlation between the shear properties that all the combinations of these tests yielded became essential. While this is an adequate and workable method, it is not the most logical and straightforward or the simplest, since the test is performed under continuously varying spherical pressure conditions.

12.23 Critical Void Ratio and Pure Deviatoric Loading

The concept of critical void ratio was introduced early in the development of classical soil mechanics through a study on compacted granular soils (US Engineers Corps 1938). It was found that in the shear process of a conventional traixial drained test for a specific soil and a specific loading, the

Table 12.22.1. Effect of deviatoric stress–rate on the strength of soils

Initial conditions						Conditions at failure						
Test No.	Soil type	Density dry γ_{d0} (kg/m^3)	Void ratio e_0	Moisture content w_0 (%)	Hydrostatic content $p_{\alpha\alpha}$ (psi)	Principal stresses: Lateral $p_{11} = p_{22}$ (psi)	Principal stresses: Vertical p_{33} (psi)	Vertical pressure p_v (psi)	Deviatoric stress s_{33} (psi)	Rate of deviatoric stress $\dot{s}_{33}$ (psi/h)	Angle of internal friction ϕ	Cohesion p_0 (psi)
P-24		1313	1.05	40.10	10.00	2.00	26.00	24.00	16.00	8.00		
P-25		1272	1.12	43.70	20.00	8.00	44.00	36.00	24.00	8.00		
P-26		1278	1.11	44.00	40.00	14.00	72.00	58.00	32.00	8.00	22°45′	15.00
P-27		1275	1.11	44.40	60.00	37.60	104.80	67.20	44.80	8.00		
	Grantham clay											
P-28		1264	1.14	40.80	10.00	4.00	24.00	20.00	12.00	0.167		
P-29		1277	1.12	44.00	20.00	10.00	40.00	30.00	20.00	0.167		
P-30		1260	1.14	44.40	40.00	23.00	74.00	51.00	34.00	0.167	27°10′	6.50
P-31		1280	1.11	44.40	60.00	36.00	108.00	72.00	48.00	0.167		
W-29		1240	1.14	38.40	30.00	20.00	50.40	30.40	20.20	303.0		
W-23		1218	1.18	41.00	50.00	37.50	79.50	42.00	28.00	304.0	16°55′	14.50
W-9		1217	1.18	40.80	70.00	54.20	110.50	56.30	37.50	342.0		
W-28		1242	1.13	40.80	30.00	21.50	51.60	30.10	20.10	186.0		
W-22	Continental clay	1228	1.16	40.70	50.00	37.50	83.20	45.50	30.20	204.0	17°37′	14.00
W-11		1240	1.14	41.20	70.00	54.00	112.20	58.20	38.80	210.0		
W-31		1227	1.16	41.00	30.00	20.00	53.80	33.80	22.50	21.0		
W-25		1218	1.18	40.60	50.00	36.40	84.30	47.90	31.90	22.0	21°11′	9.00
W-16		1186	1.22	42.60	70.00	51.00	118.20	67.20	44.70	28.0		

Table 12.22.1. *Cont.*

Initial conditions						Conditions at failure						
Test No.	Soil type	Density dry	Void ratio	Moisture content	Hydrostatic content	Principal stresses: Lateral	Principal stresses: Vertical	Vertical pressure	Deviatoric stress	Rate of deviatoric stress	Angle of internal friction	Cohesion
		γ_{d0} (kg/m^3)	e_0	w_0 (%)	$p_{\alpha\alpha}$ (psi)	$p_{11} = p_{22}$ (psi)	p_{33} (psi)	p_v (psi)	s_{33} (psi)	$\dot{s}_{33}$ (psi/h)	ϕ	p_0 (psi)
N-9		1808	0.47	18.20	1.00	0.10	3.21	3.31	2.21	27.1		
N-12		1792	0.48	18.40	2.00	0.65	4.70	4.05	2.70	54.0		
N-20		1795	0.48	13.60	4.00	1.62	8.75	7.13	4.75	28.5	35°35′	0.90
N-11		1765	0.50	18.70	6.00	3.15	11.71	8.56	5.71	25.6		
N-5		1785	0.48	18.20	1.00	0.08	2.85	2.77	1.85	11.7		
N-6	Beer Sheva	1802	0.47	18.65	2.00	0.42	5.17	4.75	3.17	14.6		
N-7	loess	1800	0.47	18.65	4.00	1.42	9.17	7.75	5.17	15.5	39°20′	0.90
N-8		1785	0.48	18.70	6.00	2.25	13.50	11.25	7.50	15.5		
N-1		1792	0.48	18.85	1.00	0.10	2.82	2.72	1.82	3.12		
N-2		1800	0.47	18.45	2.00	0.55	4.90	4.35	2.90	3.34		
N-3		1800	0.47	18.60	4.00	1.73	8.53	6.79	4.53	3.62	35°45′	0.90
N-4		1780	0.49	18.70	6.00	2.66	12.68	10.02	6.68	3.78		

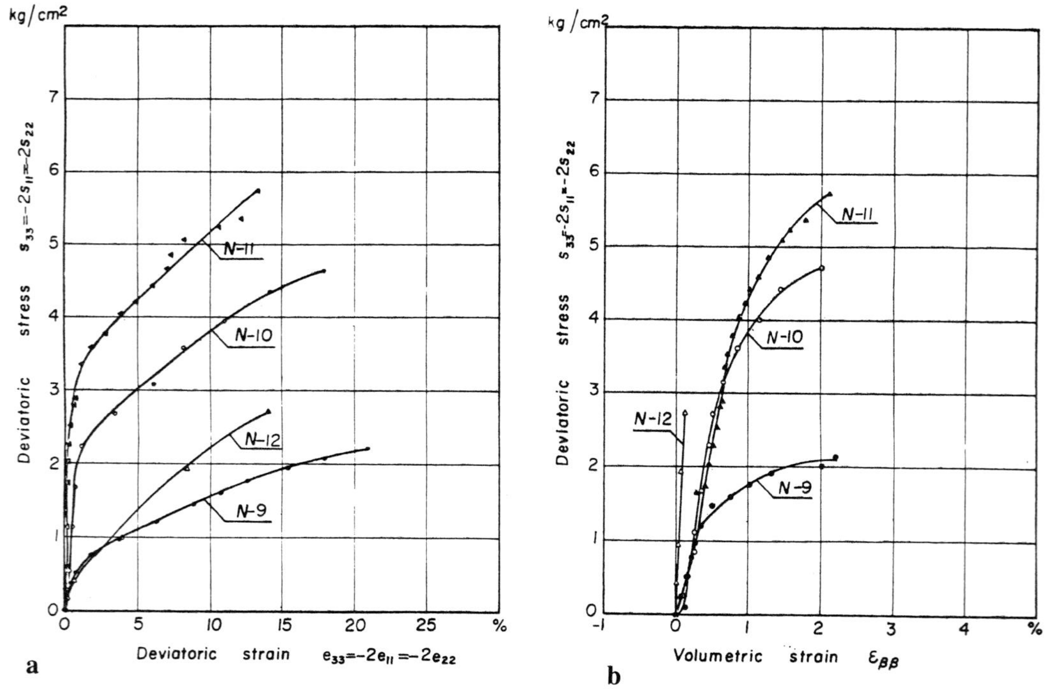

Fig. 12.22.1. Pure deviatoric test results at a high rate of loading. **a** Deviatoric stress s_{33} versus deviatoric strain e_{33}. **b** Deviatoric stress s_{33} versus volumetric strain $\varepsilon_{\beta\beta}$.

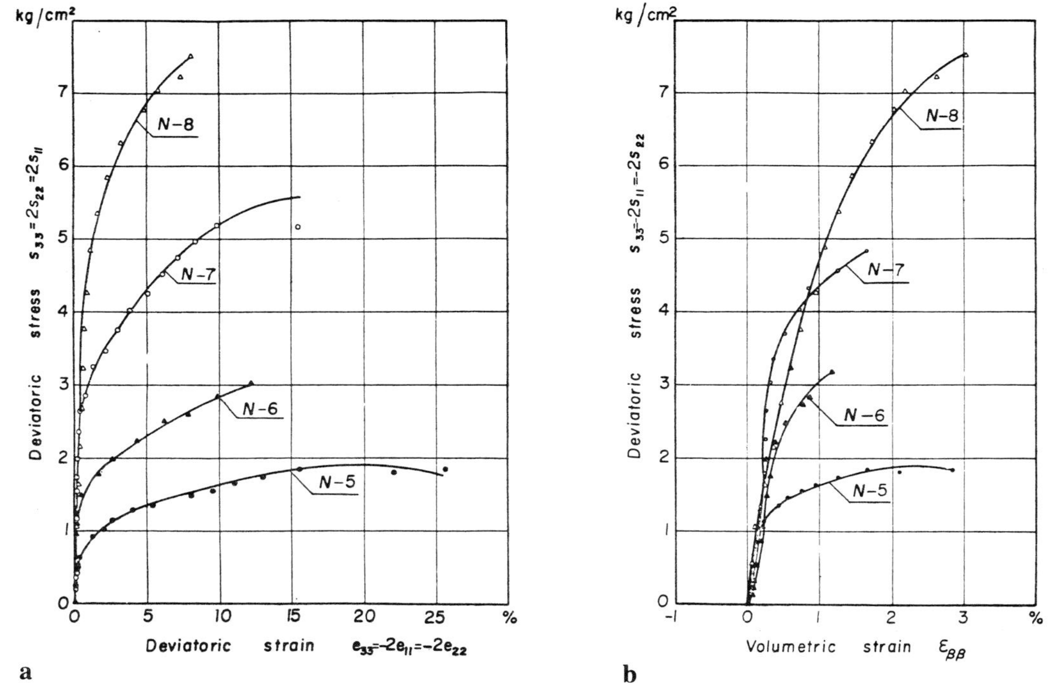

Fig. 12.22.2. Pure deviatoric test results at a medium rate of loading. **a** Deviatoric stress s_{33} versus deviatoric strain e_{33}. **b** Deviatoric stress s_{33} versus volumetric strain $\varepsilon_{\beta\beta}$.

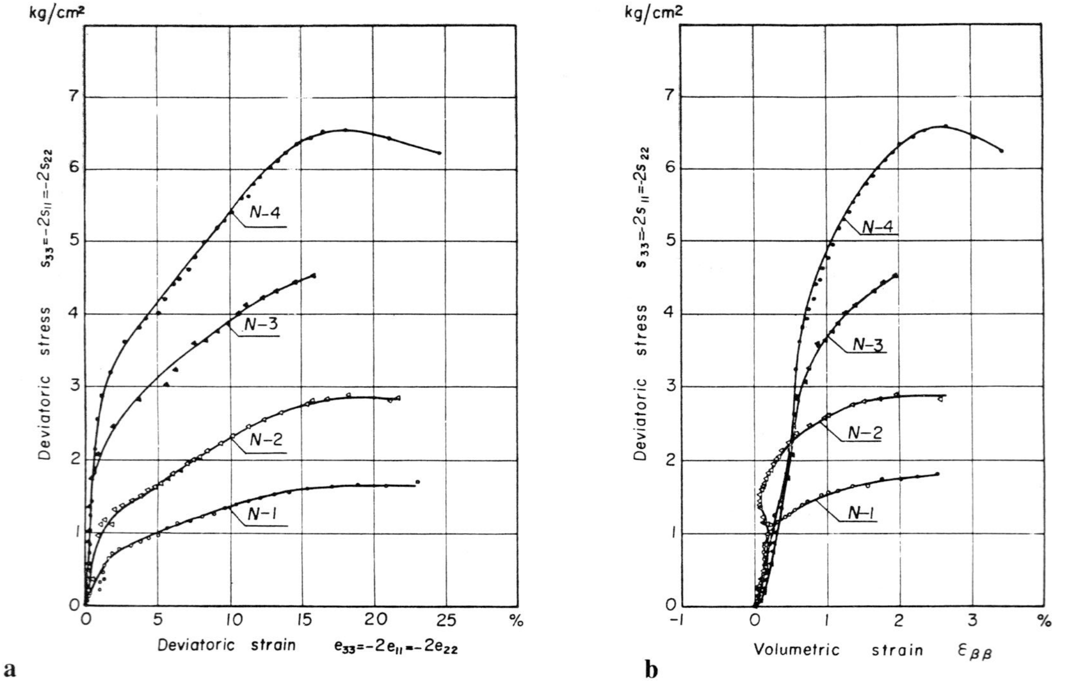

Fig. 12.22.3. Pure deviatoric test results at a low rate of loading. **a** Deviatoric stress s_{33} versus deviatoric strain e_{33}. **b** Deviatoric stress s_{33} versus volumetric strain $\varepsilon_{\beta\beta}$.

change in volume can be either a decrease or an increase, depending on the density or initial void ratio. This means that there must be a specific density or void ratio defined as the *critical void ratio* where no change in volume will occur. Taylor (1948) introduced another definition, by which the critical void ratio is the density or void ratio in which the effective pressure remains unchanged during a conventional triaxial undrained shear test. The concept was generalized and extended to include fine grained soils.

An attempt to check whether the two definitions are compatible (Roscoe et al. 1958) proved futile, especially since in the first definition the critical void ratio is related to the void ratio at the state of failure, while in the second definition it is related to the initial void ratio of the sample. The discrepancy in the results is due mainly to the fact that the conventional triaxial test contributes a spherical component during the shear loading and in a drained test these results are greatly affected by the rate of loading. In a deviatoric loading where spherical stress components do not exist at all, changes in density or in void ratio cannot be considered if linear viscoelasticity is assumed. In a pure deviatoric test, even if non-linear isotropic viscoelasticity is assumed only small changes in density or in void ratio can occur, as dilatancy or as constriction. When the dilatancy and the constriction do not act in the same direction, the volumetric effects are further reduced.

A critical void ratio can nevertheless be defined, but the changes in void ratio due to dilatancy and constriction are secondary in magnitude and importance in the pure deviatoric test compared to the conventional triaxial test. Thus, *the critical void ratio is the void ratio for which the transfer from an increase in density to a decrease in density occurs during shear stresses*. This topic will be further explored in the forthcoming sections.

Since there is a unique single-valued relation between the void ratio and the dry density of the soil, some investigators prefer the term *critical density* to the term critical void ratio. The higher the critical density of a soil the lower its critical void ratio.

12.24 The Free Energy and Energy Rate of Deviatoric Phenomena

In Sect. 7.8 the internal energy was resolved into a dual equation, a stored internal energy and a disbursed internal energy. In Sect. 11.24 the dual internal energy equation was derived for the solids constituent as follows:

$$(11.24.1) \qquad \dot{\mathfrak{I}}_{\mathfrak{si}} = \Delta\dot{\mathfrak{I}}_{\mathfrak{isi}} = \int_V \rho\dot{\epsilon}_{\mathfrak{si}} \mathrm{d}V$$

$$= \int_V (\mathrm{p}_{\mathfrak{m}} - \mathrm{p}_{\mathfrak{p}})\dot{\varepsilon}_{\alpha\alpha} \mathrm{d}V + \int_S h_{\mathfrak{s}\alpha\mathfrak{i}} n_\alpha \mathrm{d}S + \int_V \rho_{\mathfrak{s}} q_{\mathfrak{si}} \mathrm{d}V$$

$$(11.24.2) \qquad \mathfrak{I}_{\mathfrak{si}} = \int_V \rho\dot{\epsilon}_{\mathfrak{sd}} = \int_V s_{\alpha\beta}\dot{e}_{\alpha\beta}\mathrm{d}t$$

where $\mathfrak{I}_{\mathfrak{si}}$ is the stored internal energy, $\mathfrak{I}_{\mathfrak{sd}}$ is the disbursed internal energy, $\epsilon_{\mathfrak{si}}$ is the stored specific internal energy or the volumetric specific internal energy, $\epsilon_{\mathfrak{sd}}$ is the disbursed specific internal energy or the deviatoric specific internal energy, $p_{\mathfrak{m}}$ is the spherical consolidating pressure, $p_{\mathfrak{p}}$ is the pore pressure, $\dot{\varepsilon}_{\alpha\alpha}$ is the spherical strain rate, s_{ij} is the deviatoric stress tensor and $\dot{e}_{ij}$ is the distortion rate tensor.

In Eq. (5.11.1) we see that the internal energy rate is composed of a stress power and a flux and supply on one side of the equation and a mechanical and non-mechanical energy rate on the other side. When resolved into the spherical and deviatoric components, Eqs. (5.11.1) and (6.9.2) yield for the deviatoric component the following

$$\begin{aligned}\rho\dot{\epsilon}_{\mathfrak{sd}} &= s_{\mathfrak{s}\alpha\beta}(f_{\mathfrak{s}\alpha\beta} - \tfrac{1}{3}f_{\mathfrak{s}\gamma\gamma}\delta_{\alpha\beta}) = s_{\mathfrak{s}\alpha\beta}\overset{\text{inf}}{\dot{e}}_{\mathfrak{s}\alpha\beta} - \tfrac{1}{3}s_{\mathfrak{s}\alpha\beta}\overset{\text{inf}}{\dot{e}}_{\mathfrak{s}\gamma\gamma}\delta_{\alpha\beta}\\ &= \rho_{\mathfrak{s}}T\dot{\eta}_{\mathfrak{sd}} + \rho_{\mathfrak{s}}\sum_{\mathfrak{m}=\mathfrak{i}}^{\mathfrak{M}}\tau_{\mathfrak{smd}}\dot{v}_{\mathfrak{smd}} - h_{\mathfrak{s}\alpha,\alpha} - \rho_{\mathfrak{s}}q_{\mathfrak{s}} - \rho H_{\mathfrak{s}} \qquad (12.24.1)\end{aligned}$$

where $\dot{\epsilon}_{\mathfrak{isd}}$ is the deviatoric specific internal energy rate, $f_{\mathfrak{s}ij} = \overset{\text{inf}}{\dot{e}}_{\mathfrak{s}ij}$ is the deviatoric flow tensor, $\dot{\eta}_{\mathfrak{sd}}$ is the deviatoric rate of the specific entropy of the solids, $\tau_{\mathfrak{smd}}$ and $\dot{v}_{\mathfrak{smd}}$ are the deviatoric thermodynamic tensions and substates rates of the solids, respectively, $h_{\mathfrak{s}i}$ is the heat efflux of the solids, $q_{\mathfrak{s}}$ is the heat supply to the mass of the solids and $H_{\mathfrak{s}}$ is the excess interacting entropy of the solids. The thermodynamic tensions $\tau_{\mathfrak{smd}}$ could be of any tensorial order, and the corresponding rates of thermodynamic substates $\dot{v}_{\mathfrak{smd}}$ will have a matching order.

Integrating Eq. (12.24.1) with time, the disbursed specific internal energy is obtained

$$\begin{aligned}\rho\epsilon_{\mathfrak{sd}} &= \int_0^{t''}\rho\dot{\epsilon}_{\mathfrak{sd}}\,\mathrm{d}t = \int_0^{t''}\rho\,\mathrm{d}\epsilon_{\mathfrak{sd}} = \rho\mathrm{d}\epsilon_{\mathfrak{sd}}\Big|_0^{t''} = \int_0^{t''} s_{\mathfrak{s}\alpha\beta}\overset{\text{inf}}{\dot{e}}_{\mathfrak{s}\alpha\beta}\,\mathrm{d}t\\ &= \int_0^{t''} s_{\mathfrak{s}\alpha\beta}\,\mathrm{d}\overset{\text{inf}}{e}_{\mathfrak{s}\alpha\beta} = \int_0^{t''}\rho_{\mathfrak{s}}T\dot{\eta}_{\mathfrak{sd}}\mathrm{d}t + \int_0^{t''}\rho_{\mathfrak{s}}\sum_{\mathfrak{m}=1}^{\mathfrak{M}}\tau_{\mathfrak{smd}}\dot{v}_{\mathfrak{smd}}\mathrm{d}t\\ &= \int_0^{t''}\rho_{\mathfrak{s}}T\mathrm{d}\eta_{\mathfrak{sd}} + \int_0^{t''}\rho\sum_{\mathfrak{m}=1}^{\mathfrak{M}}\tau_{\mathfrak{smd}}\mathrm{d}v_{\mathfrak{smd}} \qquad (12.24.2)\end{aligned}$$

Eq. (12.24.2) is the deviatoric part of the dual specific internal energy equation.

We maintain that the amount of stored specific energy invested in the volume element is manifested through the strength of the element, as discussed in the previous chapter. Ultimately, the strength of the element is a manifestation of the disbursed specific internal energy which, as seen, is an account of the internal energy expended through the shear stresses. Indeed

Eq. (12.24.2) expresses such an energy. The stored specific internal energy and the disbursed specific internal energy have the dimensions of a scalar, which allows us to compare $\rho\epsilon_{\mathfrak{s}\mathfrak{d}}$ with $\rho\epsilon_{\mathfrak{s}\mathfrak{i}}$ and correlate them.

12.25 The Disbursed Specific Free Energy

In Eq. (12.24.1) and (12.24.2) the non-mechanical energies are unknown and difficult to evaluate. Therefore, if the processes can be assumed to occur at constant temperatures, it is convenient and customary to introduce the specific free energy rate, Eq. (5.11.3) rather than using the internal energy equations, that is, to replace Eq. (12.24.1) with an equation void of temperature and heat effects

$$\rho\dot{\psi}_{\mathfrak{s}\mathfrak{d}} = s_{\alpha\beta}(f_{\alpha\beta} - \tfrac{1}{3}f_{\gamma\gamma}\delta_{\alpha\beta}) = s_{\mathfrak{s}\alpha\beta}\dot{e}_{\alpha\beta} = \rho_{\mathfrak{s}} \sum_{\mathfrak{m}=1}^{\mathfrak{M}} \tau_{\mathfrak{s}\mathfrak{m}\mathfrak{d}}\dot{v}_{\mathfrak{s}\mathfrak{m}\mathfrak{d}} \tag{12.25.1}$$

where $\dot{\psi}_{\mathfrak{s}\mathfrak{d}}$ is the deviatoric specific free energy rate of the solids.
Integrating Eq. (12.25.1) with time we obtain

$$\begin{aligned}\rho\psi_{\mathfrak{s}\mathfrak{d}} &= \int_0^{t''} \rho\dot{\psi}_{\mathfrak{s}\mathfrak{d}}\mathrm{d}t = \int_0^{t''} \rho\,\mathrm{d}\psi_{\mathfrak{s}\mathfrak{d}} = \rho\,\psi_{\mathfrak{s}\mathfrak{d}}\Big|_0^{t''} \\ &= \int_0^{t''} s_{\mathfrak{s}\alpha\beta}\,\dot{\overset{\mathrm{inf}}{e}}_{\alpha\beta}\mathrm{d}t = \int_0^{t''} s_{\alpha\beta}\,\mathrm{d}\overset{\mathrm{inf}}{e}_{\alpha\beta} \\ &= \int_0^{t''} \rho \sum_{\mathfrak{m}=1}^{\mathfrak{R}} \tau_{\mathfrak{s}\mathfrak{m}\mathfrak{d}}\dot{v}_{\mathfrak{s}\mathfrak{m}\mathfrak{d}}\,\mathrm{d}t = \int_0^{t''} \rho \sum_{\mathfrak{m}=1}^{\mathfrak{R}} \tau_{\mathfrak{s}\mathfrak{m}\mathfrak{d}}\mathrm{d}v_{\mathfrak{s}\mathfrak{m}\mathfrak{d}}\end{aligned} \tag{12.25.2}$$

$\rho\psi_{\mathfrak{s}\mathfrak{d}}$ or Eq. (12.25.2) represents the specific deviatoric free energy, a disbursed energy that thrives on the stored specific spherical free energy increment and is consumed during the shearing process. Almost all that was said about the dual specific internal energies is also valid with respect to the dual specific free energies, with the limitations of constant temperature and no heat exchange.

12.26 The Disbursed Free Energy Applied to Linear Stress–Strain Relations

If we apply the specific free energy balance equation (12.25.2) to the equation of the general linear viscoelastic medium (7.6.7), or to the equation of a specific viscoelastic medium such as (7.6.9), we obtain the disbursed free energy. Here we shall apply Eq. (12.25.2) to Eq. (7.6.9)

$$\rho\psi_{\mathfrak{s}\mathfrak{d}} = \int_0^{t''} s_{\mathfrak{s}\alpha\beta}\overset{\text{inf}}{\dot{e}}_{\alpha\beta}\mathrm{d}t = \int_0^{t''}\left\{s(t)_{\alpha\beta}\left[\frac{1}{2G}\,\dot{s}(t)_{\alpha\beta} + \frac{1}{2\eta}\,s(t)_{\alpha\beta}\right]\right\}\mathrm{d}t \quad (12.26.1)$$

Since linear viscoelastic relations were assumed, volumetric strains cannot occur during shear, thus $\varepsilon_{\beta\beta} = 0$.

Eq. (12.26.1) appears to be somewhat intricate, yet less so than the equation of the stored specific free energy (11.26.1) and also less than the four-coefficient equation of the disbursed specific free energy.

12.27 The Disbursed Free Energy in the Pure Deviatoric Test of Linear Viscoelastic Soils

Eq. (12.26.1) can now be applied to the pure deviatoric test or to any of the other conventional tests. We shall discuss the pure deviatoric test.

The assumption of pure deviatoric loading prescribes axial symmetry of the stresses at all times and $s_{\alpha\alpha} = 0$ during the shearing. Thus we have $s_{ij} = \frac{1}{2}p(t)_{33}\Xi_{ij} = \frac{1}{2}\mathcal{S}(t)\Xi_{ij}$ for the deviatoric part of the test, as was shown in Sect. 12.17. We shall now consider the same two cases for deviatoric loadings.

1. *Step loadings of deviatoric stresses*. The first part of the test, the consolidation, is performed under the spherical pressure $p_{\mathrm{m}} = \frac{1}{3}p_{\alpha\alpha} = \text{const.}$ It is followed by shear stresses of $\mathfrak{Q}$ steploads $s_{ij\mathfrak{q}} = \frac{1}{2}\mathcal{S}_{\mathfrak{q}}\Xi_{ij}$, until the soil sample reaches failure. Inserting these loadings into the equation of the disbursed free energy, Eq. (12.26.1), and recalling also Eq. (A.21.19) from Appendix A, we have

$$\begin{aligned}\rho\psi_{\mathfrak{s}\mathfrak{d}} &= \sum_{\mathfrak{q}=1}^{\mathfrak{Q}}\int_0^{t''}\left\{s_{\alpha\beta\mathfrak{q}}\left[\frac{1}{2G}\,\dot{s}_{\alpha\beta\mathfrak{q}} + \frac{1}{2\eta}\,s_{\alpha\beta\mathfrak{q}}\right]\right\}\mathrm{d}t \\ &= \sum_{\mathfrak{q}=1}^{\mathfrak{Q}}\frac{3\mathcal{S}_{\mathfrak{q}}^2}{4\eta}\,t''_{\mathfrak{q}} = \sum_{\mathfrak{q}=1}^{\mathfrak{Q}}\frac{t''_{\mathfrak{q}}}{2\eta}\,s_{\alpha\beta\mathfrak{q}}\,s_{\alpha\beta\mathfrak{q}} \\ &= \sum_{\mathfrak{q}=1}^{\mathfrak{Q}}\frac{t''_{\mathfrak{q}}}{\eta}\,\overline{\mathrm{II}}_{\mathfrak{q}s} = \sum_{\mathfrak{q}=1}^{\mathfrak{Q}}\frac{4t''_{\mathfrak{q}}}{3\eta}\,\Pi_{\mathfrak{q}s}\end{aligned} \quad (12.27.1)$$

Fig. 12.27.1 shows the pure deviatoric test represented by the Mohr circles. Each of the $\mathfrak{Q}$ circles indicates one step load around the spherical pressure p_{m} which is located at $\frac{1}{3}$ of the diameter and divides the diameters of all circles at a 1 : 2 ratio. The same test in the axially symmetric bisectrix plane representa-

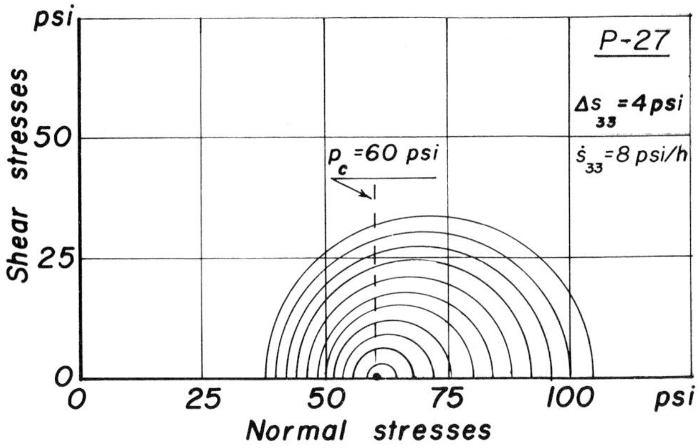

Fig. 12.27.1. Mohr circles of pure deviatoric test of step loadings.

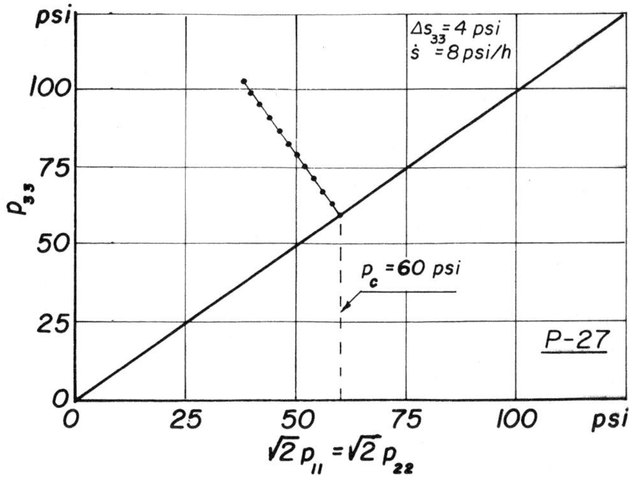

Fig. 12.27.2. Axially symmetric bisectrix plane of step loadings.

tion, Fig. 12.27.2, is given by Ω points indicating the step loads, all of them on a line normal to the hydrostatic directrix.

Eq. (12.27.1) is a scalar and depends at any time *t* on the Ω octahedral invariants of the shear stress Π_{qs}. The octahedral shear stress is a function of the time *t* and of the consolidation pressure p_{m} as well.

2. *Constant stress–rate loading*. For a constant rate of deviatoric loading we shall assume $s_{ij} = \frac{1}{2}p_{33}\Xi_{ij} = \frac{1}{2}\mathcal{S}t\,\Xi_{ij}$ from which we have $\dot{s}_{ij} = \frac{1}{2}\mathcal{S}\Xi_{ij}$, and substitute it in Eq. (12.26.1). The consolidation preceding the pure deviatoric loading is similar to that preceding the stepwise loading, and therefore the stored specific free energy is equal to that given in Eq. (11.26.3). The disbursed specific free energy, however, is equal to

$$\rho\psi_{\mathfrak{sd}} = \int_0^{t''} \left\{ s_{\alpha\beta} \left[\frac{1}{2G} \dot{s}_{\alpha\beta} + \frac{1}{2\eta} s_{\alpha\beta} \right] \right\} \mathrm{d}t = \left\{ \frac{3}{8G} \mathcal{S}^2 t''^2 + \frac{1}{4\eta} \mathcal{S}^2 t''^3 \right\}$$

$$= \left(\frac{1}{4G} t''^2 + \frac{1}{6\eta} t''^3 \right) \dot{s}_{\alpha\beta} \dot{s}_{\alpha\beta} = \left(\frac{1}{4G} + \frac{1}{6\eta} t'' \right) s_{\alpha\beta} s_{\alpha\beta}$$

$$= \mathcal{S}^2 \left(\frac{1}{16G} t''^2 + \frac{1}{24G} t''^3 \right) \Xi_{\alpha\beta} \Xi_{\alpha\beta}$$

$$= \left(\frac{1}{2G} t''^2 + \frac{1}{3\eta} t''^3 \right) \overline{\Pi}_{\dot{s}} = \left(\frac{2}{3G} t''^2 + \frac{4}{9\eta} t''^3 \right) \Pi_{\dot{s}}$$

$$= \left(\frac{2}{3G} + \frac{4}{9\eta} t'' \right) \Pi_s \tag{12.27.2}$$

As in the previous case, we see that Eq. (12.27.2) is a scalar and depends at any time t not only on the octahedral invariant of the shear stress Π_s, but also on that of the shear stress rate $\Pi_{\dot{s}}$.

Fig. 12.27.3 shows the Mohr circles for a pure deviatoric test with constant stress-rate loading. We see that a single circle represents the stress condition of the sample at failure.

Fig. 12.27.4 shows the axially symmetric bisectrix plane representation of the same. The result is indicated by a point reached at failure, the test having proceeded along the line perpendicular to the octahedral directrix. A line perpendicular to the octahedral directrix indicates a constant spherical pressure and in fact is a projection of a plane normal to the bisectrix plane. The point obtained here is located on the line indicating the pressure p_{m}.

This analysis is based on the particular case of the constitutive equation (7.6.9). If more intricate equations were chosen, for instance a four-coefficient model instead of the model of Eq. (7.6.9), somewhat different

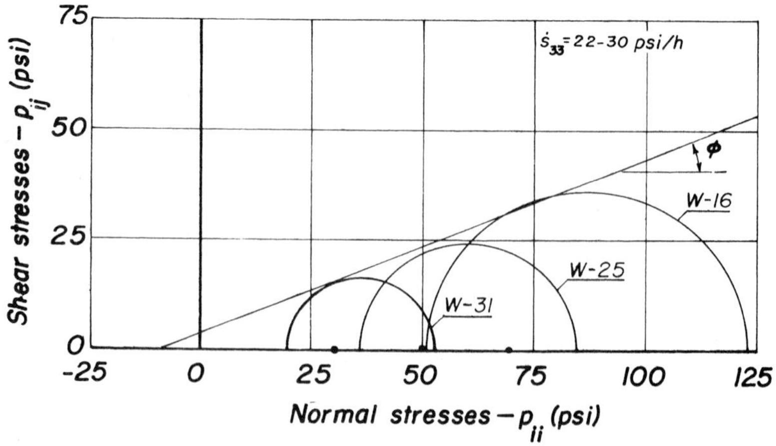

Fig. 12.27.3. Mohr circles of pure deviatoric test with constant stress–rate of loadings.

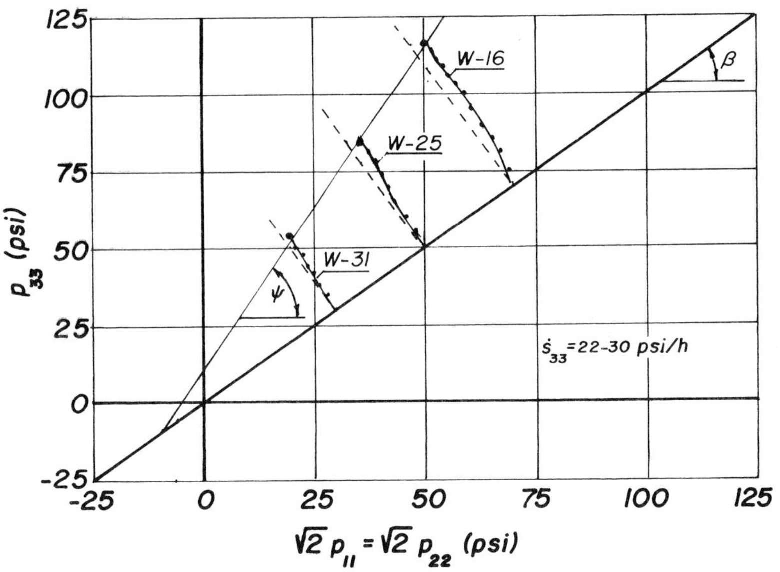

Fig. 12.27.4. Axially symmetric bisectrix plane of step loadings.

results would have been obtained. Still the general pattern would have been preserved.

12.28 The Disbursed Free Energy in Pure Deviatoric Tests for Isotropic Stress–Strain Relations of Soils

It was shown that when an isotropic stress–strain relationship is postulated, then volumetric strains $\varepsilon_{\beta\beta}$ are obtained even during deviatoric stresses, in addition to the deviatoric strains e_{ij}. These volumetric strains increase or decrease the stored energy.

The same loading cases discussed for linear stress–strain relationships will be discussed here:

1. *Step loadings of deviatoric stresses*. After the first part of the test when the soil sample consolidates under a spherical pressure $p_{\mathrm{m}} = \frac{1}{3}p_{\alpha\alpha} = \text{const}$, $\mathfrak{Q}$ steploads $s_{ij\mathfrak{q}} = \frac{1}{2}\mathcal{S}_{\mathfrak{q}}\Xi_{ij\mathfrak{q}}$ are applied until the sample reaches failure. If Eqs. (12.18.2) and (12.18.3) are substituted into Eq. (12.24.1) and (12.24.2), respectively, and we recall also Eqs. (A.21.18) and (A.21.19) from Appendix A, we obtain the excess stored free energy originating from the dilatancy of the soil and the disbursed free energy, respectively

$$\Delta\rho\psi_{\mathfrak{s}\mathrm{i}} = 3p_{\mathrm{m}}\mathfrak{W}_2 \sum_{\mathfrak{q}=1}^{\mathfrak{Q}} (s)^2_{\mathfrak{q}\alpha\beta\mathrm{i}}\, \mathrm{t}'' + 12p_{\mathrm{m}} \sum_{\mathfrak{q}=1}^{\mathfrak{Q}} [\mathfrak{W}_5 s_{\mathfrak{q}\alpha\beta\mathrm{i}} + \mathfrak{W}_6 (s^2)_{\alpha\beta\mathfrak{q}\mathrm{i}}] \int_0^{t''_{\mathfrak{q}}} e(t)_{\alpha\beta\mathrm{i}}\, \mathrm{d}t$$

$$= \tfrac{3}{4}\,\mathfrak{w}_2 p_m \sum_{q=1}^{\Omega} \mathcal{S}_q^2\, t_q'' + 3p_m \sum_{q=1}^{\Omega} \Bigg[2\mathfrak{w}_5\, \mathcal{S}_q \int_0^{t_q''} e(t)_q\, dt$$

$$+ \mathfrak{w}_6 \mathcal{S}_q^2 \int_0^{t_q''} \left(\frac{\varepsilon(t)_{q\beta\beta} + e(t)_q}{3}\right) dt$$

$$= p_m \sum_{q=1}^{\Omega} \left[2\mathfrak{w}_5 \sqrt{(6s^2_{\alpha\beta q})} \int_0^{t_q''} e(t)_q dt + 2\mathfrak{w}_6 s^2_{\alpha\beta q} \int_0^{t_q''} \left(\frac{\varepsilon(t)_{q\beta\beta} + e(t)_q}{3}\right) dt\right]$$

$$+ \mathfrak{w}_2 p_m \sum_{q=1}^{\Omega} s^2_{\alpha\beta q}\, t_q''$$

$$= p_m \sum_{q=1}^{\Omega} \left[2\mathfrak{w}_5 \sqrt{(6\overline{\mathrm{II}}_{qs})} \int_0^{t_q''} e(t)_q\, dt + 2\mathfrak{w}_6\, \overline{\mathrm{II}}_{qs} \int_0^{t_q''} \left(\frac{\varepsilon(t_{q\beta\beta} + e(t)_q}{3}\right) dt\right]$$

$$+ \mathfrak{w}_2 p_m \sum_{q=1}^{\Omega} \overline{\mathrm{II}}_{qs} t_q'' \qquad (12.28.1)$$

$$\rho\psi_{\mathfrak{s}\mathfrak{d}} = 6 \sum_{q=1}^{\Omega} \int_0^{t_q''} s_{\alpha\beta q} \Bigg[\mathfrak{w}_1 s_{\alpha\beta q} + \mathfrak{w}_2 (s)^2_{\alpha\beta\mathfrak{d}q} + \mathfrak{w}_3 e_{\alpha\beta q} + \mathfrak{w}_5 (se + es)_{\alpha\beta\mathfrak{d}q}$$

$$+ \mathfrak{w}_6 (s^2 e + es^2)_{\alpha\beta\mathfrak{d}q}\Bigg] dt$$

$$= \sum_{q=1}^{\Omega} [\tfrac{3}{2}\mathfrak{w}_1 \mathcal{S}_q^2 + \tfrac{3}{4}\mathfrak{w}_2 \mathcal{S}_q^3] t_q'' + \sum_{q=1}^{\Omega} \Bigg[3\mathfrak{w}_3 \mathcal{S}_q \int_0^{t_q''} e(t)_q dt$$

$$+ 3\mathfrak{w}_5 \mathcal{S}_q^2 \int_0^{t_q''} \frac{\varepsilon(t)_{q\beta\beta} + e(t)_q}{3}\, dt + \tfrac{3}{4}\mathfrak{w}_6 \mathcal{S}_q^3 \int_0^{t_q''} \frac{\varepsilon(t)_{q\beta\beta} + 2e(t)_q}{3}\, dt\Bigg]$$

$$= \sum_{q=1}^{\Omega} [\mathfrak{w}_1 s^2_{\alpha\beta q} + \mathfrak{w}_2 \sqrt{(\tfrac{1}{6} s^2_{\alpha\beta q})} s^2_{\alpha\beta q}]\, t_q'' + \sum_{q=1}^{\Omega} \Bigg[\mathfrak{w}_3 \sqrt{(6s^2_{\alpha\beta q})} \int_0^{t_q''} e(t)_q\, dt$$

$$+ 2\mathfrak{w}_5 s^2_{\alpha\beta q} \int_0^{t_q''} \frac{\varepsilon(t)_{q\beta\beta} + e(t)_q}{3}\, dt$$

$$+ \mathfrak{w}_6 \sqrt{(\tfrac{1}{6} s^2_{\alpha\beta q})} s^2_{\alpha\beta q} \int_0^{t_q''} \frac{\varepsilon(t)_{q\beta\beta} + 2e(t)_q}{3}\, dt$$

$$= \sum_{q=1}^{\Omega} \left[\mathfrak{w}_1 \overline{\mathrm{II}}_{sq} + \frac{1}{18}\mathfrak{w}_2 \overline{\mathrm{III}}_{sq}\right] t_q'' + \sum_{q=1}^{\Omega} \mathfrak{w}_3 \sqrt{(6\overline{\mathrm{II}}_{sq})} \int_0^{t_q''} e(t)_q dt$$

$$+ 2\mathfrak{w}_5 \sum_{q=1}^{\Omega} \overline{\mathrm{II}}_{sq} \int_0^{t_q''} \frac{\varepsilon(t)_{q\beta\beta} + e(t)_q}{3}\, dt$$

$$+\frac{1}{18}\mathfrak{w}_6\sum_{q=1}^{\Omega}\overline{\mathrm{III}}_{sq}\int_0^{t''_q}\frac{\varepsilon(t)_{q\beta\beta}+2e(t)_q}{3}\,dt \tag{12.28.2}$$

where $\rho\Delta\psi'_{\mathrm{\check{s}i}}$ is the excess stored free energy due to the dilatancy of the soil and $\varepsilon(t)_{\beta\beta}$ is the volumetric strain or the dilatancy during the deviatoric stresses. While the excess stored specific free energy $\Delta\rho\psi'_{\mathrm{\check{s}i}}$ in Eq. (12.28.1) depends on the same spherical pressure p_{m} as the excess stored specific free energy $\Delta\rho\psi_{\mathrm{\check{s}i}}$ in Eq. (11.28.2), the time limit of integration of the two equations is different since the duration of the two is different, $t' \neq t''$. The total strain tensor is

$$\varepsilon(t)_{ij} = \tfrac{1}{3}\varepsilon(t)_{\beta\beta}\delta_{ij} + \tfrac{1}{3}e(t)\Xi_{ij} \tag{12.28.3}$$

where $e(t) = e_{33} - e_{11}$ is the *deviatoric factor* of the deviatoric strain tensor e_{ij}. It is seen that the excess stored free energy and the disbursed free energy can be expressed as functions of the second moment invariant $\overline{\mathrm{II}}_{sq}$ of the shear stress s_{ijq} at the different step loadings, and the latter also as a function of the third moment invariant $\overline{\mathrm{III}}_{sq}$.

Fig. 12.28.1 shows the excess stored free energy of the tests in Fig. 12.15.1 as functions of the vertical strain due to dilatancy.

2. *Constant rate of deviatoric stress*. If the deviatoric stress is applied at a continuous and constant rate, $s_{ij} = \frac{1}{2}\mathcal{S}t\,\Xi_{ij}$, an excess stored energy and a disbursed energy is obtained when we substitute Eqs. (12.18.1) and (12.18.2)

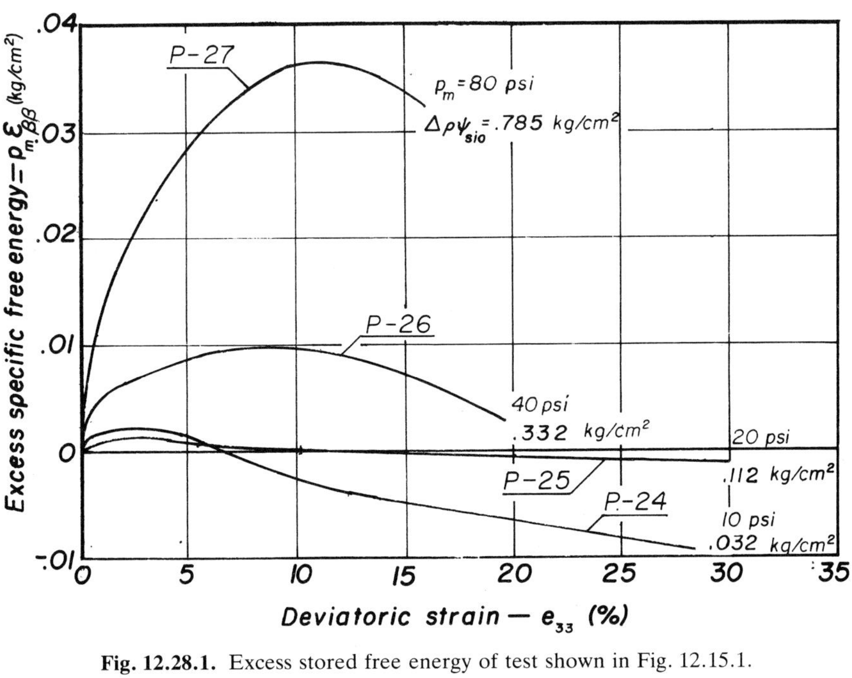

Fig. 12.28.1. Excess stored free energy of test shown in Fig. 12.15.1.

into Eqs. (12.24.1) and (12.24.2), respectively

$$\rho\Delta\psi'_{\mathfrak{si}} = 3p_{\mathrm{m}}\mathfrak{W}_2(s)^2_{\alpha\alpha i}t'' + 12p_{\mathrm{m}}[\mathfrak{W}_5 s_{\alpha\beta i} + \mathfrak{W}_6(s^2)_{\alpha\beta i}] \int_0^{t''} e(t)_{\alpha\beta i}\,dt$$

$$= \tfrac{1}{2}\mathfrak{W}_2 p_{\mathrm{m}}\mathcal{S}^2 t''^3 + 3p_{\mathrm{m}}\Bigg[2\mathfrak{W}_5\mathcal{S}^3 \int_0^{t''} te(t)\,dt + \mathfrak{W}_6\mathcal{S}^2 \int_0^{t''} \frac{\varepsilon(t)_{\beta\beta} + e(t)}{3} t^2\,dt\Bigg]$$

$$= \tfrac{2}{3}p_{\mathrm{m}}\mathfrak{W}_2\mathrm{II}_s + 2p_{\mathrm{m}}\Bigg[\mathfrak{W}_5 \frac{1}{t''} \sqrt{(12\overline{\mathrm{II}}_s)} \int_0^{t''} e(t)t\,dt + 2\mathfrak{W}_6\frac{\overline{\mathrm{II}}_s}{t''^2} \int_0^{t''} \frac{\varepsilon(t)_{\beta\beta} + e(t)}{3} t^2 dt\Bigg] \tag{12.28.4}$$

$$\rho\psi_{\mathfrak{s}\mathfrak{d}} = 6\int_0^{t''} s_{\alpha\beta}\,[\mathfrak{W}_1 s_{\alpha\beta} + \mathfrak{W}_2(s)^2_{\alpha\beta\mathfrak{d}} + \mathfrak{W}_3 e_{\alpha\beta} + \mathfrak{W}_5(se + es)_{\alpha\beta\mathfrak{d}} + \mathfrak{W}_6(s^2e + es^2)_{\alpha\beta\mathfrak{d}}]\,dt$$

$$= [\tfrac{1}{2}\mathfrak{W}_1\mathcal{S}^2 t''^3 + \tfrac{3}{16}\mathfrak{W}_2\mathcal{S}^3 t''^4] + \mathfrak{W}_3\mathcal{S} \int_0^{t''} te(t)\,dt + \tfrac{3}{4}\mathfrak{W}_5\mathcal{S}^2 \int_0^{t''} \frac{\varepsilon(t)_{\beta\beta} + e(t)}{3} t^2\,dt + \tfrac{1}{16}\mathfrak{W}_6\mathcal{S}^3 \int_0^{t''} \frac{\varepsilon(t)_{\beta\beta} + 2e(t)}{3} t^3\,dt$$

$$= [\tfrac{2}{3}\mathfrak{W}_1 s^2_{\alpha\beta} + \tfrac{1}{4}\mathfrak{W}_2 \sqrt{(\tfrac{4}{3}s^2_{\alpha\beta})}s^2_{\alpha\beta}\,t''] + \mathfrak{W}_3 \frac{1}{t''} \sqrt{(\tfrac{4}{3}s^2_{\alpha\beta})} \int_0^{t''} te(t)\,dt + \mathfrak{W}_5 \frac{s^2_{\alpha\beta}}{t''^2} \int_0^{t''} \frac{\varepsilon(t)_{\beta\beta} + e(t)}{3} t^2\,dt + \tfrac{1}{12}\mathfrak{W}_6 \sqrt{(\tfrac{4}{3}s^2_{\alpha\beta})}\frac{s^2_{\alpha\beta}}{t''^3} \int_0^{t''} \frac{\varepsilon(t)_{\beta\beta} + 2e(t)}{3} t^4\,dt$$

$$= [\tfrac{2}{3}\mathfrak{W}_1\overline{\mathrm{II}}_s + \tfrac{1}{12}\mathfrak{W}_2\sqrt{\tfrac{2}{9}}\,\overline{\mathrm{III}}_s t''] + \mathfrak{W}_3 \frac{1}{t''}\sqrt{\tfrac{4}{3}\overline{\mathrm{II}}_s} \int_0^{t''} te(t)\,dt + \mathfrak{W}_5 \frac{\overline{\mathrm{II}}_s}{t''^2} \int_0^{t''} \frac{\varepsilon(t)_{\beta\beta} + e(t)}{3} t^2\,dt + \frac{1}{12t''^3}\mathfrak{W}_6 \sqrt{\tfrac{2}{9}}\,\overline{\mathrm{III}}_s \int_0^{t''} \frac{\varepsilon(t)_{\beta\beta} + 2e(t)}{3} t^4\,dt \tag{12.28.5}$$

As remarked in Sect. 7.9, the coefficients $\mathfrak{W}_{\mathrm{n}}$ can be defined as functions of the invariants of the respective stresses. At present, however, this is beyond our scope. Fig. 12.28.2 shows excess stored free energy due to dilatation for the tests shown in Fig. 12.15.2.

12.29 Note on General Non-linear Stress–Strain Relations

As anticipated from the theoretical presentation in Sect. 7.9, the isotropic stress–strain relationship introduces non-linearity in Eqs. (12.28.1), (12.28.2) and (12.28.4), (12.28.5).

When general isotropic relations are postulated, the theoretical presentation predicts that volume changes will occur even when shear stresses alone are applied, and that spherical pressures appear when distortions are imposed on the material. This is not a quality of linear stress–strain relations. In linear

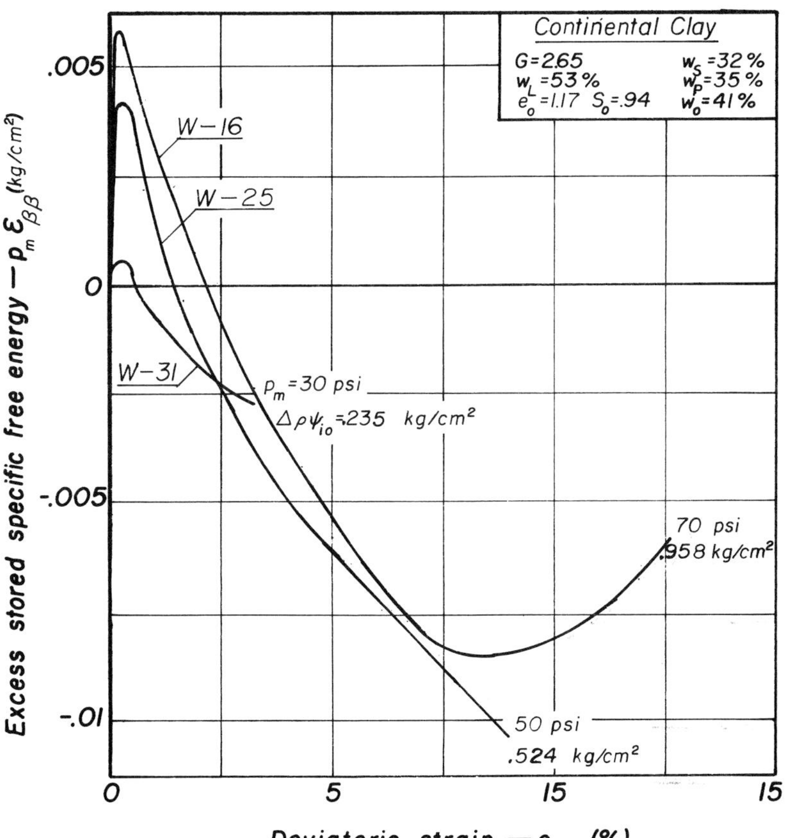

Fig. 12.28.2. Excess stored free energy at constant stress-rate of test shown in Fig. 12.15.2.

viscoelaticity volume changes are produced only by spherical pressures, while distortions are produced only by shear stresses.

A general theory of non-linear viscoelasticity is beyond our purpose and has to be deferred to a later time, together with other important topics such as polar forces, couple stresses, anisotropy, discontinuities, etc. Of the general theory of viscoelasticity only isotropic viscoelastic relations were discussed here, because of their simplicity and capability to impart reliably second-order phenomena or second-order effects, linked to other important phenomena in soils. Anisotropy of materials, known since the nineteenth century (Thomson and Tait 1867; Warburg 1879; Finger 1894) received systematic formal treatment in mechanics only years later (Ericksen and Rivlin 1954; Ericksen 1960b, 1961). The introduction of a directional vector acting at each point of the material permits linearization of certain problems.

12.30 Closing Remarks on Shear Stresses

The general isotropic relations presented in Sects. 7.9 and 7.10 and their extension to soils, Sects. 12.18 and 12.19, indicate that when shear stresses are applied, an increase of volumetric strains occurs, regardless of the direction and sign of the stresses. When the stored free energy of this volume increase is evaluated by multiplying it with the respective spherical stress, Sect. 12.28, terms of negative stored energies are obtained as well. Since the coefficients $\mathfrak{W}_n$ are always positive, the sign of the stored free energy depends on the sign of the volumetric strain $\varepsilon_{\beta\beta}$ in Eqs. (12.28.1) and (12.28.4).

In the particular case presented, the terms of the excess stored free energy with the coefficients $\mathfrak{W}_2$ and $\mathfrak{W}_6$ are analytically always negative while the term with the coefficients $\mathfrak{W}_5$ could be either negative or positive. Practically, however, even the former could become positive if the volumetric strains involved are positive, that is, if there is a decrease in the volume. This can occur in a pure deviatoric loading and perhaps not in a torsional shear test. In a torsional test the deviatoric shear is a simple shear and the volume changes are mostly a result of the second-order effect, thus a volume increase. In a pure deviatoric test the deviatoric shear is applied via normal pressures, by which the vertical stress is increased (decreased) as the horizontal stresses are decreased (increased). Whether the volume changes are increases or decreases depends on the state of the density of the soil with respect to its critical void ratio. If the void ratio of the soil is above the critical void ratio, the volume decrease due to the vertical stress can be greater than the combined volume increase of the rebound from the horizontal stresses and of the second-order effect. If, however, the void ratio of the soil is below the critical void ratio, the combined volume increase from the rebound of the volume due to the horizontal pressures and the second-order effect is larger than the compression due to the vertical pressure. In this case the net result is a decompression of the soil.

Comparing the disbursed free energy between a soil assumed to be linear, Eqs. (12.27.1) and (12.27.2), and a soil where isotropic stress–strain relations are assumed, Eqs. (12.28.2) and (12.28.5), we see that in both cases it depends on the second moment invariant II_s which is always positive, and on the third moment invariant $\overline{\mathrm{III}}_s$ which can be positive or negative. The strength of the soil will also depend on the sign of $\varepsilon_{\alpha\alpha\tau}$, $\varepsilon_{\alpha\alpha}$, $e(t)_q$ and $e(t)$. Theoretically the last terms of Eqs. (12.28.2) and (12.28.5) should always yield a decrease in the disbursed free energy. When experimental data of $\varepsilon_{\alpha\alpha\tau}$, $\varepsilon_{\alpha\alpha}$, $e(t)_q$ and $e(t)$ are introduced in these equations the effect of the constriction is also obtained, although this was not explicitly considered.

13 Failure

13.1 Brief Review of Failure Theories

The two previous chapters deal at length with the two dominant features of soils and other compressible materials. They are, on one hand, the relationship between compression–decompression and volumetric deformation or strain, Chap. 11, and on the other hand, the relationship between shear and distortional deformation or strain, Chap. 12. This permitted the development of two time dependent stress–strain relationships, the two constitutive equations. However, neither of these relationships can, by itself, provide any hint as to the strength of the material. In the present chapter we will show that the strength of the material depends on the interaction between the two relationships mentioned. We will also show that some of the existing failure theories evolve as particular cases of this interaction.

Many theories known in mechanics of materials as failure theories or yield criteria such as Rankine's maximum stress theory (Rankine 1857), St Venant's maximum elastic strain theory (St Venant 1870), Tresca's maximum shear theory (Tresca 1868, 1869; Unwin 1888; Guest 1900) have not actually been effective as failure criteria, although each holds some veracity. Even the more advanced failure theories like the Huber–Hencky elastic strain energy of distortion theory (Huber 1904; Hencky 1924) do not provide concepts general enough to confirm the empirical data of real materials.

The combination of the von Mises distortion energy theory (von Mises 1913), an extension of the Huber–Hencky theory, with the Coulomb–Mohr friction theory (Coulomb 1821; Mohr 1914) comes closest to providing an appropriate solution to the problem of strength in general. These two approaches are empirical, attempting to provide a sensible failure theory based on experimental results.

With the development of soil mechanics, experimental evidence indicated that these failure theories, related to linear behavior of materials, could not explain all phenomena observed in soils, such as kinematic effects (rate of shear), excess volumetric strains, geometric non-linearities of strains, volumetric changes during shear stresses (observed earlier by Reynolds 1885;

Poynting 1905). Many investigators of soils and mechanics considered these phenomena and sought to improve the existing theories (Drucker and Prager 1952; Prager 1953, 1959; Drucker et al. 1955; Green 1956; Roscoe et al. 1958; Henkel 1960; Iwan 1967; Schofield and Wroth 1968; Scott and Ko 1969; Gudehus 1973; Drescher and Vardoulakis 1982; Bardet 1983; Chen and Baladi 1985; Bolton 1986). In addition, many studies on the topics of coaxiality of the strain and stress tensors during the shear process and on the degree of inherent and induced anisotropy in soils are found in literature (Oda 1981; Hight et al. 1983; Wood 1984; Wong and Arthur 1985; Boehler and Kabbaj 1985; Hayashi et al. 1985; Mayne 1985). Elasto-plasticity, limit analysis, critical state, the cam-clay soil, work hardening and clay softening are some of the theories that evolved from these studies and offered partial explanations of the observed phenomena.

Without extending the discussion to the details of the theories themselves, which may be found in many textbooks, we shall outline in the next section the requirements of a failure theory, from which the shortcomings of the available theories will be evident.

In the following sections a strength theory based on the specific free energy of the material will be presented. A study by Schmid reported to the ONR Department of the Navy (Schmid et al. 1960) may be considered as the forerunner of the theory presented here. A comparable study was presented to the 5th INCOSOMEFE (De Wet 1961).

13.2 Failure Criteria

It was pointed out briefly in Sect. 7.7 that the mode of failure of a material is represented by a failure envelope, similar to that shown in Fig. 7.7.2, or an envelope somewhat more involved as, for instance, that shown in Fig. 13.2.1.

A viable failure theory should satisfy three requirements. Two of the requirements are usually well recognized, and pertain to the failure envelopes, while the last exists as a hidden feature in many of the theories, and we shall explicitly formulate it here.

1. *Dependence of the failure on the spherical pressure.* The theory should reflect the effect of spherical pressure on the failure of the material, represented by the failure envelope in the octahedral bisectrix plane, Fig. 13.2.2. While this may be obvious to engineers trained in soil mechanics, who are acquainted with the Mohr–Coulomb theory, Fig. 13.2.3, it seems not to have carried much weight in mechanics of materials,

The experimental evidence by von Kármán on marble cylinders subjected to high hydrostatic pressures dates back to 1912 and was followed by works of many others mostly on solids, rocks, concrete and metals (Böker 1915; Richart et al. 1928; Bridgman 1931, 1939, 1949, 1952; Taylor and Quinney 1931; Griggs 1936; Lubahn 1948; Balmer 1949; Handin 1953; Handin and

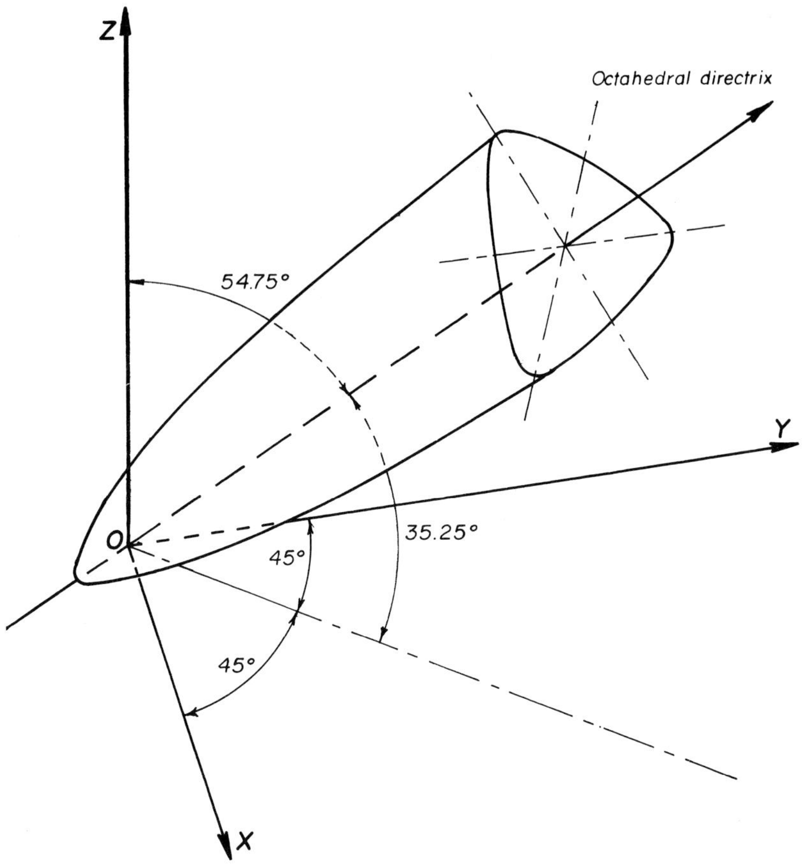

Fig. 13.2.1. General failure envelope.

Hager 1957; Bredthauer 1957; Crossland and Dearden 1958; Ilyushin and Ogibalov 1959; Serata 1961; Serdengecti and Boozer 1961; Akroyd 1961; McClintock and Walsh 1962; Schwartz 1964; Chinn and Zimmermann 1965; Niwa et al. 1967; Hannant and Frederick 1968). Many of these works indicate a strong influence of the hydrostatic pressure on the strength of the materials, and its inclusion in the failure conditions was suggested as early as the sixties (Hu 1960; Hull 1961; Hu and Pae 1963).

It was found that when the Mohr–Coulomb theory, originally conceived as a method for representing stresses acting on a volume element (Mohr 1900, 1914), is combined with the earlier studies on earth pressure by Coulomb (1776), it provides a failure envelope.

2. *Determination of the failure envelope in the octahedral plane*. The failure theory should give a description, as complete as possible, of the failure envelope in any octahedral plane normal to the octahedral directrix. Most failure theories mentioned satisfy this requirement. Fig. 13.2.4 shows the failure envelope determined by the *von Mises theory* and defined by the

equation

$$\tfrac{1}{2}\overline{\mathrm{II}}_s = \tfrac{1}{2}[\overline{\mathrm{II}}_p - \tfrac{1}{3}\mathrm{I}_p^2] = \tfrac{1}{6}[(p_{11} - p_{22})^2 + (p_{22} - p_{33})^2 + (p_{33} - p_{11})^2]$$

$$= \tfrac{3}{2}\tau_{\mathrm{oct}}^2 = k^2 \tag{13.2.1}$$

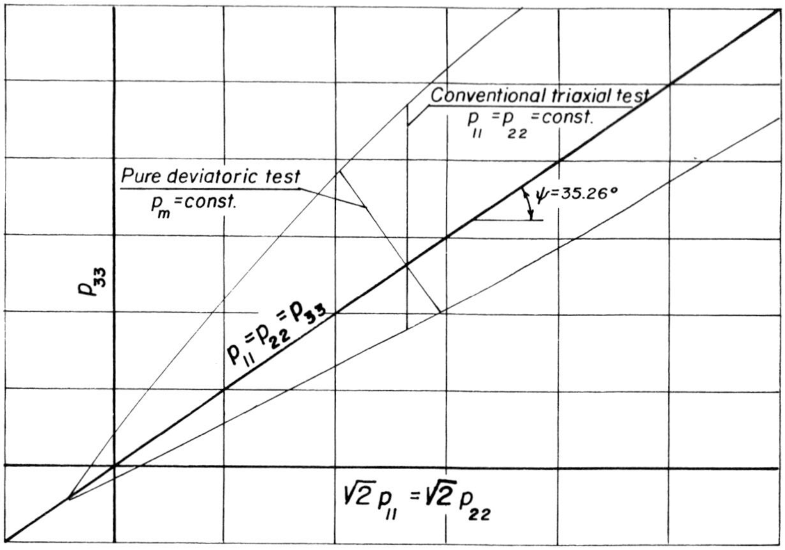

Fig. 13.2.2. Failure envelope along the octahedral directrix.

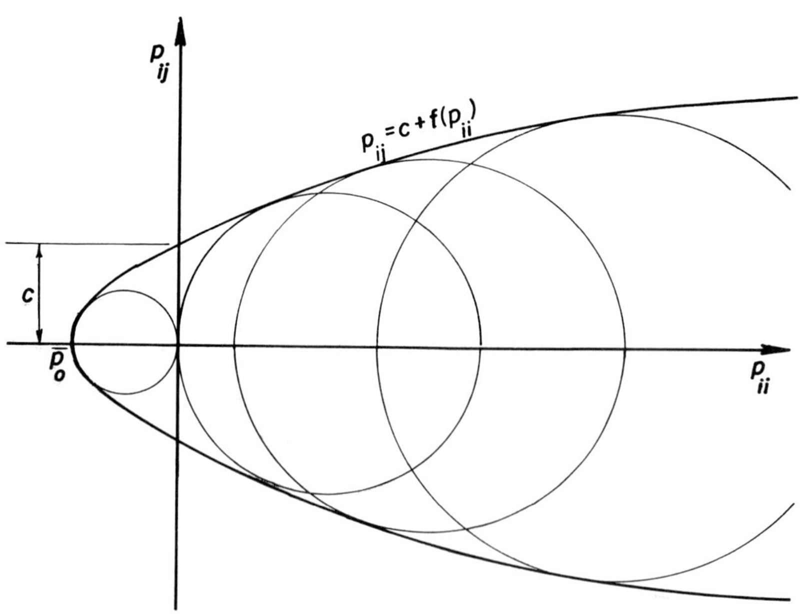

Fig. 13.2.3. Mohr–Coulomb linear failure curve.

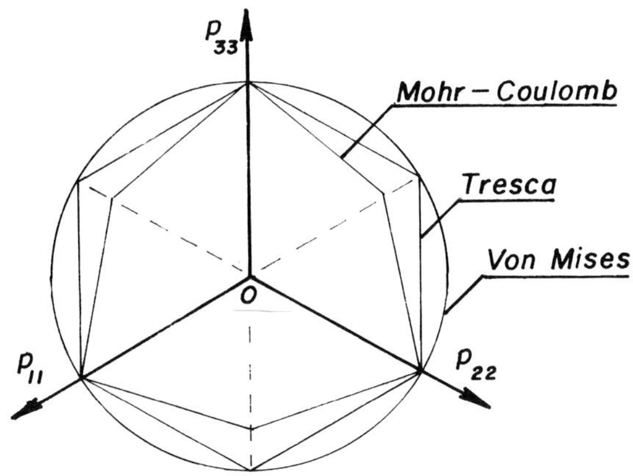

Fig. 13.2.4. Failure envelope in the octahedral plane.

where k is a constant, and $\sqrt{\mathrm{II}_s} = \sqrt{2}k$ is the diameter of the circle that defines the envelope of the von Mises theory in the octahedral plane.

Tresca's maximum shear stress criterion asserts that yield will occur when any one of the following six conditions are satisfied

$$p_{11} - p_{22} = \pm 2k; \quad p_{22} - p_{33} = \pm 2k; \quad p_{33} - p_{11} = \pm 2k \tag{13.2.2}$$

which defines a regular hexagon inscribed in the von Mises circle, Fig. 13.2.4, and from which we see that the octahedral shear stress is equal to the maximum shear stress, $\tau_{\text{oct}} = \tau_{\max} = k$.

When the *Mohr–Coulomb theory* is applied to triaxial tests of soils, the results in the principal stress directions define an irregular hexagonal failure envelope in the octahedral plane. In tension the maximum shear stress (identical to the octahedral shear stress) is smaller than in compression, Fig. 13.2.4. The formulation is similar to that in Eq. (13.2.2)

$$\begin{aligned} &p_{11} - p_{22} = 2k; \quad p_{22} - p_{33} = 2k; \quad p_{33} - p_{11} = 2k; \\ &p_{11} - p_{22} = -2k'; \quad p_{22} - p_{33} = -2k'; \quad p_{33} - p_{11} = -2k'; \quad k > k' \end{aligned} \tag{13.2.3}$$

Since the Mohr–Coulomb theory distinguishes between compression and tension, it is the closest suitable basis for the description of failure in compressible materials such as soils.

Experimental studies on hollow cylindrical samples tested in torsion or in principal stress combinations (Habib 1953; Kirkpatrick 1957; Haythornthwaite 1960a, b; Wu et al. 1978) and on cuboid samples tested in compression and extension (Bishop 1966; Lorenz et al. 1965; Green and Bishop 1969; Lade 1972; Matsuoka and Nakai 1974) indicate a slight departure from the irregular hexagonal failure envelope of the Mohr–Coulomb theory: in these studies the failure envelope in the octahedral plane is a continuous curve-shaped envelope with rounded tips, similar to that shown in Fig. 13.2.5.

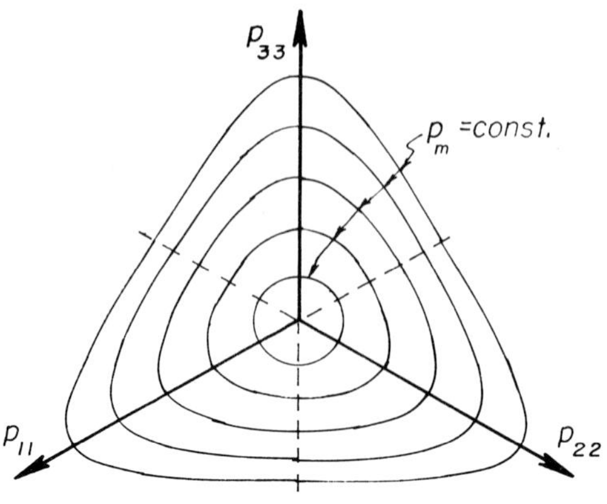

Fig. 13.2.5. Curvilinear failure envelope in the octahedral plane.

3. *Determination of the yield criterion on the basis of the internal energies of the material*. This is perhaps the most important requirement. It will be evident upon implementing it that in fact it could be the only requirement needed, since the two others evolve directly from it.

The theory based on the specific internal energy and presented in the forthcoming sections will show that the constant k is not just another constant obtained from tests, but is a physical entity that can be evaluated quantitatively.

13.3 The Dual Specific Internal Energy

In Eq. (5.4.6) the balance or conservation of internal energy was formulated, and the state equation was given in Eq. (5.9.9). We may say that one represents the energy input and the other the energy output of the material, and Eq. (5.11.1), the production rate of internal energy, asserts the equivalence of the two. Eq. (5.11.1) indicates that the internal energy rate is composed of a stress-power and a flux or a supply on the input side, and a mechanical and non-mechanical energy rate on the output side.

In Sect. 11.24 it was shown that Eq. (5.11.1) can be resolved into two components, a spherical component and a deviatoric component, as follows

$$(11.24.7)\qquad \rho\dot{\epsilon}_{\mathfrak{s}\mathfrak{i}} = -\tfrac{1}{3} p_{\mathfrak{s}\alpha\alpha} f_{\mathfrak{s}\alpha\alpha} + \sum_{\mathfrak{n}=1}^{\mathfrak{N}} h_{\mathfrak{s}\alpha,\alpha} + \sum_{\mathfrak{n}=1}^{\mathfrak{N}} \rho_{\mathfrak{s}} q_{\mathfrak{s}}$$

$$= (p_{\mathfrak{m}} - p_{\mathfrak{p}})(\dot{\varepsilon}_{\mathfrak{s}\alpha\alpha} + \dot{\varepsilon}_{\mathfrak{s}\beta\beta}) = (p_{\mathfrak{m}} - p_{\mathfrak{w}} - p_{\mathfrak{g}})(\dot{\varepsilon}_{\alpha\alpha} + \dot{\varepsilon}_{\beta\beta})$$

for
$$\sum_{\mathfrak{n}=1}^{\mathfrak{N}} h_{\mathfrak{s}\alpha,\alpha\mathfrak{i}} = \sum_{\mathfrak{n}=1}^{\mathfrak{N}} h_{\mathfrak{s}\alpha,\alpha}; \qquad \sum_{\mathfrak{n}=1}^{\mathfrak{N}} \rho_{\mathfrak{s}} q_{\mathfrak{s}\mathfrak{i}} = \sum_{\mathfrak{n}=1}^{\mathfrak{N}} \rho_{\mathfrak{s}} q_{\mathfrak{s}};$$

$$\sum_{\mathfrak{n}=1}^{\mathfrak{N}} h_{\mathfrak{s}\alpha,\alpha\mathfrak{d}} \equiv \sum_{\mathfrak{n}=1}^{\mathfrak{N}} \rho_{\mathfrak{s}} q_{\mathfrak{s}\mathfrak{d}} = 0$$

$$(11.24.8) \qquad \rho\dot{\varepsilon}_{\mathfrak{s}\mathfrak{d}} = s_{\mathfrak{s}\alpha\beta}\dot{e}_{\mathfrak{s}\alpha\beta}(\dot{\varepsilon}_{\mathfrak{s}\alpha\beta} - \tfrac{1}{3}\dot{\varepsilon}_{\mathfrak{s}\gamma\gamma}\delta_{\alpha\beta}) = s_{\mathfrak{s}\alpha\beta}(f_{\mathfrak{s}\alpha\beta} - \tfrac{1}{3}f_{\mathfrak{s}\gamma\gamma}\delta_{\alpha\beta})$$

where $\dot{\varepsilon}_{\mathfrak{s}\mathfrak{i}}$ and $\dot{\varepsilon}_{\mathfrak{s}\mathfrak{d}}$ are the spherical and deviatoric specific internal energy rates defined by Eqs. (11.24.7) and (11.24.8), respectively, $\frac{1}{3}p_{\mathfrak{s}\alpha\alpha} = (\frac{1}{3}p_{\alpha\alpha} - p_{\mathfrak{w}} - p_{\mathfrak{g}}) = (p_{\mathrm{m}} - p_{\mathfrak{w}} - p_{\mathfrak{g}}) = (p_{\mathrm{m}} - p_{\mathfrak{p}})$ is the effective pressure defined by Eq. (9.4.8), and $f_{\mathfrak{s}ij} = \dot{\varepsilon}_{\mathfrak{s}ij}$ is the *spherical flow tensor*, which is the sum of the infinitesimal volumetric strain rates of the solids, $\dot{\varepsilon}_{\mathfrak{s}\alpha\alpha}$ measured during the spherical stresses and $\dot{\varepsilon}_{\mathfrak{s}\beta\beta}$ measured during the deviatoric stresses. All other terms were defined in Sect. 11.24.

When Eqs. (11.24.7) and (11.24.8) are integrated with time, the stored specific internal energy increment and the disbursed specific internal energy, respectively, were obtained

$$(11.24.5) \qquad \rho\epsilon_{\mathfrak{s}\mathfrak{i}} = \int_{-\infty}^{t'} (p_{\mathrm{m}} - p_{\mathfrak{p}})\dot{\varepsilon}_{\alpha\alpha}\,\mathrm{d}t$$

$$= \int_{-\infty}^{t'} (p_{\mathrm{m}} - p_{\mathfrak{w}} - p_{\mathfrak{g}})\,\mathrm{d}(\varepsilon_{\alpha\alpha} + \varepsilon_{\beta\beta})$$

$$\text{for} \int_{-\infty}^{t'} \sum_{\mathfrak{n}=1}^{\mathfrak{N}} h_{\mathfrak{n}\alpha,\alpha\mathfrak{i}}\,\mathrm{d}t = \int_{-\infty}^{t'} \sum_{\mathfrak{n}=1}^{\mathfrak{N}} \rho_{\mathfrak{n}}q_{\mathfrak{n}\mathfrak{i}}\,\mathrm{d}t = 0$$

$$(11.24.6) \qquad \rho\epsilon_{\mathfrak{s}\mathfrak{d}} = \int_{-\infty}^{t''} \rho\dot{\epsilon}_{\mathfrak{s}\mathfrak{d}}\,\mathrm{d}t = \int_{-\infty}^{t''} \rho\,\mathrm{d}\epsilon_{\mathfrak{s}\mathfrak{d}} = \int_{\infty}^{t''} s_{\mathfrak{s}\alpha\beta}\dot{e}_{\mathfrak{s}\alpha\beta}\,\mathrm{d}t$$

$$\text{for} \int_{-\infty}^{t''} \sum_{\mathfrak{n}=1}^{\mathfrak{N}} h_{\mathfrak{n}\alpha,\alpha\mathfrak{d}}\,\mathrm{d}t = \int_{-\infty}^{t''} \sum_{\mathfrak{n}=1}^{\mathfrak{N}} \rho_{\mathfrak{n}}q_{\mathfrak{n}\mathfrak{d}}\,\mathrm{d}t$$

Eqs. (11.24.5) and (11.24.6) are the *dual specific internal energy* equations. Note the fact that the spherical internal energy is a measure of the stored energy existing in materials and in soils as well. It was shown in Eq. (11.24.4) that prior to any mechanical, thermal or chemical stresses an *initial stored specific internal energy* $\rho\epsilon_{\mathfrak{s}\mathfrak{i}0}$, called the *residing stored specific internal energy*, exists in the volume element. This energy is a function of the internal forces acting on the solids and holding them together or of the densification of the solids. It is a scalar and has the dimensions of a spherical pressure $[FL^{-2}]$. Eq. (11.24.5) represents an increment of the stored specific internal energy, resulting from an added stress p_{m}. This stress increases the density of the soil and the total stored specific internal energy to $\rho\epsilon_{\mathfrak{s}\mathfrak{i}}$, as derived in equation (11.24.4)

$$(11.24.4) \qquad \rho\epsilon_{\mathfrak{s}\mathfrak{i}} = \rho\epsilon_{\mathfrak{s}\mathfrak{i}0} + \Delta\rho\epsilon_{\mathfrak{s}\mathfrak{i}}$$

The amount of the stored specific internal energy invested in the volume element is manifested through the strength of the element. Along with the increase of the density or the internal forces holding the particles together, the shearing resistance of the element increases. This must be evidenced through an invested energy commensurate with the stored internal energy.

Indeed Eq. (11.24.6) provides us with such an energy, the disbursed specific internal energy that, as seen, is an account of the internal energy expended through shear stresses and distortions. The disbursed specific internal energy is also a scalar with the dimensions of a spherical pressure, which allows us to compare $\rho\epsilon_{\mathfrak{si}}$ with $\rho\epsilon_{\mathfrak{sd}}$ and correlate them. Thus we may formulate a general failure criterion

$$\epsilon_{\mathfrak{sd}} = \epsilon_{\mathfrak{si}} = \epsilon_{\mathfrak{si}0} \pm \Delta\epsilon_{\mathfrak{si}} \tag{13.3.1}$$

relating the disbursed internal energy to the stored internal energy per unit weight.

Proposition. The functional relationship between the spherical part of the internal energy (the stored specific internal energy) and its deviatoric part (the disbursed specific internal energy) defines a criterion necessary and sufficient for failure.

In the following sections the failure and shear strength of soils based on the above definition will be analyzed.

13.4 Stored and Disbursed Specific Free Energy

In Sect. 11.25 it was shown that if isentropic conditions $\dot{\eta}_{\mathrm{ni}} = \dot{\eta}_{\mathrm{nd}} = 0$ can be assumed, or if no effluxes or supplies exist or their sum vanishes $h_{\mathfrak{s}\alpha,\alpha} + \rho_{\mathfrak{s}} q_{\mathfrak{s}} = 0$, then the specific free energy concept can be introduced instead of that of the internal energy, and Eqs. (11.24.7) and (11.24.8) can be replaced by Eqs. (11.25.3) and (11.25.4), respectively, which are free of entropy and heat effects

$$\text{(11.25.3)} \qquad \Delta\rho\psi_{\mathfrak{si}} = \int_0^{t'} \rho\dot{\psi}_{\mathfrak{si}}\,\mathrm{d}t = \int_0^{t'} \rho\,\mathrm{d}\psi_{\mathfrak{si}} = \rho\psi_{\mathfrak{si}}\Big|_0^{t'} = \int_0^{t'} (p_{\mathrm{m}} - p_{\mathrm{p}})\,\mathrm{d}\varepsilon_{\mathfrak{s}\alpha\alpha}$$

$$\text{(11.25.4)} \qquad \rho\psi_{\mathfrak{sd}} = \int_0^{t''} \rho\dot{\psi}_{\mathfrak{sd}}\,\mathrm{d}t = \int_0^{t''} \rho\,\mathrm{d}\psi_{\mathfrak{sd}} = \rho\psi_{\mathfrak{sd}}\Big|_0^{t''} = \int_0^{t''} s_{\alpha\beta}\,\mathrm{d}e_{\mathfrak{s}\alpha\beta}$$

Eqs. (11.25.3) and (11.25.4) were obtained by integration from Eqs. (11.25.1) and (11.25.2)

$$\rho\dot{\psi}_{\mathfrak{si}} = \tfrac{1}{3}p_{\mathfrak{s}\alpha\alpha}f_{\alpha\alpha} = (p_{\mathrm{m}} - p_{\mathrm{w}} - p_{\mathrm{g}})(\dot{\varepsilon}_{\mathfrak{s}\alpha\alpha} + \dot{\varepsilon}_{\mathfrak{s}\beta\beta})$$

$$\text{(11.25.1)} \qquad (p_{\mathrm{m}} - p_{\mathrm{p}})(\dot{\varepsilon}_{\mathfrak{s}\alpha\alpha} + \dot{\varepsilon}_{\mathfrak{s}\beta\beta})$$

$$\text{(11.25.2)} \qquad \rho\dot{\psi}_{\mathfrak{sd}} = s_{\alpha\beta}(f_{\alpha\beta} - \tfrac{1}{3}f_{\beta\beta}) = s_{\mathfrak{s}\alpha\beta}\dot{e}_{\mathfrak{s}\alpha\beta}$$

where $\dot{\psi}_{\mathfrak{s}\mathrm{i}}$ is the spherical specific free energy rate of the solids or the stored specific free energy rate and $\dot{\psi}_{\mathfrak{s}\mathfrak{d}}$ is the deviatoric specific free energy rate of the solids or the disbursed specific free energy rate. Both expressions are free of temperature and entropy rates, as well as of efflux and heat supply.

The excess stored specific free energy, $\Delta\rho\psi_{\mathfrak{s}\mathrm{i}}$, indicates the amount of free energy *stored* in the soil by densification, compaction or compression, in addition to the already *residing* free energy in the volume element of the soil $\psi_{\mathfrak{s}\mathrm{i}0}$. The sum of the two resists the *disbursed* free energy $\psi_{\mathfrak{s}\mathfrak{d}}$, which tends to disrupt the soil by shear stresses and distortions. Almost all that was said about the dual specific internal energy is valid also with respect to the dual specific free energy, with the limitations of constant temperature and no heat exchange.

Thus, as for the specific internal energy, an equation analogous to (11.24.7) can be written

$$\rho\psi_{\mathfrak{s}\mathrm{i}} = \rho\psi_{\mathfrak{s}\mathrm{i}0} + \Delta\rho\psi_{\mathfrak{s}\mathrm{i}} = \rho\psi_{\mathfrak{s}\mathrm{i}0} + \int_{-\infty}^{t'} (p_{\mathrm{m}} - p_{\mathfrak{w}} - p_{\mathfrak{g}})\mathrm{d}(\varepsilon_{\mathfrak{s}\alpha\alpha} + \varepsilon_{\mathfrak{s}\beta\beta})$$

$$= \rho\psi_{\mathfrak{s}\mathrm{i}0} + \int_{-\infty}^{t'} (p_{\mathrm{m}} - p_{\mathrm{p}})\mathrm{d}(\varepsilon_{\mathfrak{s}\alpha\alpha} + \varepsilon_{\mathfrak{s}\beta\beta}) \tag{13.4.1}$$

This equation shows that Eq. (11.25.3) represents only the excess stored specific free energy, which is the energy added to the residing specific free energy $\rho\psi_{\mathfrak{s}\mathrm{i}0}$ existing in the soil before pressure was applied. After the application of the pressure and the resulting volume change the free energy reaches a new level $\rho\psi_{\mathfrak{s}\mathrm{i}}$.

Eq. (13.3.1) also has a counterpart for the specific free energy. When we combine it with Eq. (13.4.1) we obtain the balance of the dual specific free energies

$$\rho\psi_{\mathfrak{s}\mathfrak{d}} = \rho\psi_{\mathfrak{s}\mathrm{i}} = \rho\psi_{\mathfrak{s}\mathrm{i}0} + \Delta\rho\psi_{\mathfrak{s}\mathrm{i}}$$

$$= \rho\psi_{\mathfrak{s}\mathrm{i}0} \int_0^{t''} s_{\alpha\beta}\,\mathrm{d}e_{\alpha\beta} = \rho\psi_{\mathfrak{s}\mathrm{i}0} + \int_0^{t''} (p_{\mathrm{m}} - p_{\mathfrak{w}} - p_{\mathrm{g}})\mathrm{d}(\varepsilon_{\alpha\alpha} + \varepsilon_{\beta\beta})$$

$$= \rho\psi_{\mathfrak{s}\mathrm{i}0} + \int_0^{t} (p_{\mathrm{m}} - p_{\mathrm{p}})\mathrm{d}(\varepsilon_{\alpha\alpha} + \varepsilon_{\beta\beta}) \tag{13.4.2}$$

indicating a linear relationship and an equivalence between the disbursed specific free energy and the stored specific free energy and an interdependence of the two, that can be described by the equation

$$\psi_{\mathfrak{s}\mathfrak{d}} = A\psi_{\mathfrak{s}\mathrm{i}} = A\psi_{\mathfrak{s}\mathrm{i}0} + A\Delta\psi_{\mathfrak{s}\mathrm{i}} \tag{13.4.3}$$

where $A \leq 1$ and is not necessarily constant throughout the range of the free energy changes. The angle α of the line shown in Fig. 13.4.1 may also vary, $\alpha \leq 45°$. If no dissipation is assumed in Eq. (13.4.2), then $A = 1$ and the line

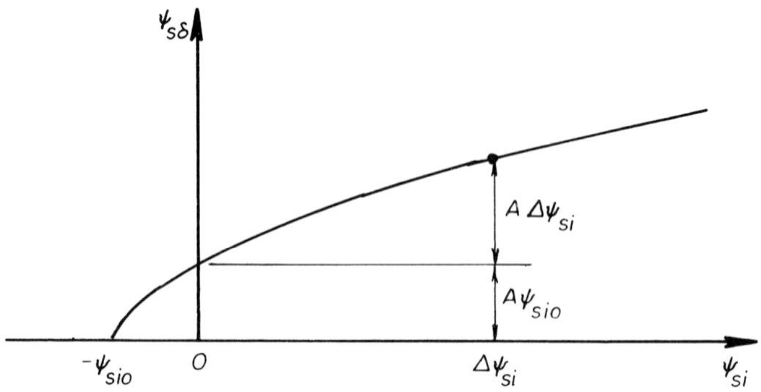

Fig. 13.4.1. Relation between the disbursed and stored free energy.

shown in Fig. 13.4.1 is at $\alpha = 45°$. However, if the energy dissipates as in any real material, then $A < 1$ whether Eq. (13.4.2) is linear or not, and the angle becomes smaller than 45°, $\alpha < 45°$. If the dissipation is proportional to the stored specific free energy, then $A \sim 1/\psi$ and the curve becomes non-linear in $\psi_{\mathfrak{si}}$, as seen in Fig. 13.4.1.

13.5 Specific Free Energy Balance of the Elastic Medium (A Particular Case)

We shall consider a homogeneous isotropic elastic medium where $d\varepsilon_{\alpha\alpha} = (1/\varkappa)\,dp_{\mathrm{m}}$ and $de_{ij} = (1/2G)\,ds_{ij}$, and $\varkappa$ and G are the bulk and the shear moduli, respectively. We insert these values into Eq. (13.4.2) and apply Eq. (4.10.11). The integrals are not functions of time t any more, since the stresses and strains are not functions of t. Thus we obtain

$$\begin{aligned}\int_0^{e_{ij}} s_{\alpha\beta}\,de_{\alpha\beta} &= \int_0^{s_{ij}} \frac{1}{2G}\, s_{\alpha\beta}\,ds_{\alpha\beta} = \frac{1}{4G}\, s_{\alpha\beta}^2 = \frac{1}{3G}\,\Pi_s = \frac{1}{3G}\,\Pi_p\\ &= \frac{1}{4G}\,\overline{\Pi}_s = \frac{3}{4G}\,\tau^2\\ &= \rho\psi_{\mathfrak{si}0} + \int_0^{\varepsilon_{\alpha\alpha}} \frac{1}{\varkappa}\, p_{\mathrm{m}}\,d\varepsilon_{\alpha\alpha} = \rho\psi_{\mathfrak{si}0} + \int_0^{p_{\mathrm{m}}} \frac{1}{\varkappa}\, p_{\mathrm{m}}\,dp_{\mathrm{m}}\\ &= \rho\psi_{\mathfrak{si}0} + \frac{1}{2\varkappa}\, p_{\mathrm{m}}^2 \qquad (13.5.1)\end{aligned}$$

The pore pressures $p_{\mathfrak{w}}$ and $p_{\mathfrak{g}}$ have been omitted and the limits of integration replaced, since they are irrelevant to the elastic case. After rearrangement of Eq. (13.5.1) we have

$$\tau = \sqrt{(\tfrac{1}{3}s_{\alpha\beta}^2)} = \sqrt{\left(\tfrac{4}{3}\rho G\psi_{\check{s}i0} + \frac{2G}{3\varkappa}\, p_{\mathrm{m}}^2\right)}$$

$$= \sqrt{\left(\frac{2G}{3\varkappa}\right)}\,(-p_{\mathrm{m}} \pm \sqrt{(p_{\mathrm{m}}^2 + 2\rho\varkappa\psi_{\check{s}i0})}) + \sqrt{\left(\frac{G}{3\varkappa}\right)}\, p_{\mathrm{m}}$$

$$= c + p_{\mathrm{m}}\mu = c + p_{\mathrm{m}}\tan\phi = (p_{\mathrm{m}0} + p_{\mathrm{m}})\tan\phi \qquad (13.5.2)$$

where τ is the *octahedral shear stress*, p_{m} is the spherical pressure or the *octahedral normal stress* denoted also by σ, $c = \sqrt{(2G/3\varkappa)}[-p_{\mathrm{m}} \pm \sqrt{(p_{\mathrm{m}}^2 + 2\rho\varkappa\psi_{\check{s}io})}]$ is the *cohesive shear intercept* while $p_{\mathrm{mo}} = c/\tan\phi = [-p_{\mathrm{m}} \pm \sqrt{(p_{\mathrm{m}}^2 + 2\rho\varkappa\psi_{\check{s}io})}]$ is the *cohesion*, and μ is the *coefficient of friction* which is defined $\mu = \tan\phi = \sqrt{(G/3\varkappa)}$ and is a function of ϕ, the *angle of internal friction*.

Eq. (13.5.2) represents the Mohr–Coulomb failure theory for the elastic medium and Fig. 13.5.1 is its familiar graphical display revealing the Mohr–Coulomb failure envelope. It is evident from Eq. (13.5.2) that the cohesion $p_{\mathrm{m}0}$ is an inherent spherical pressure related to the initially stored specific free energy which, as already seen, depends on the existing internal forces that hold the particles together. Thus cohesion depends on those

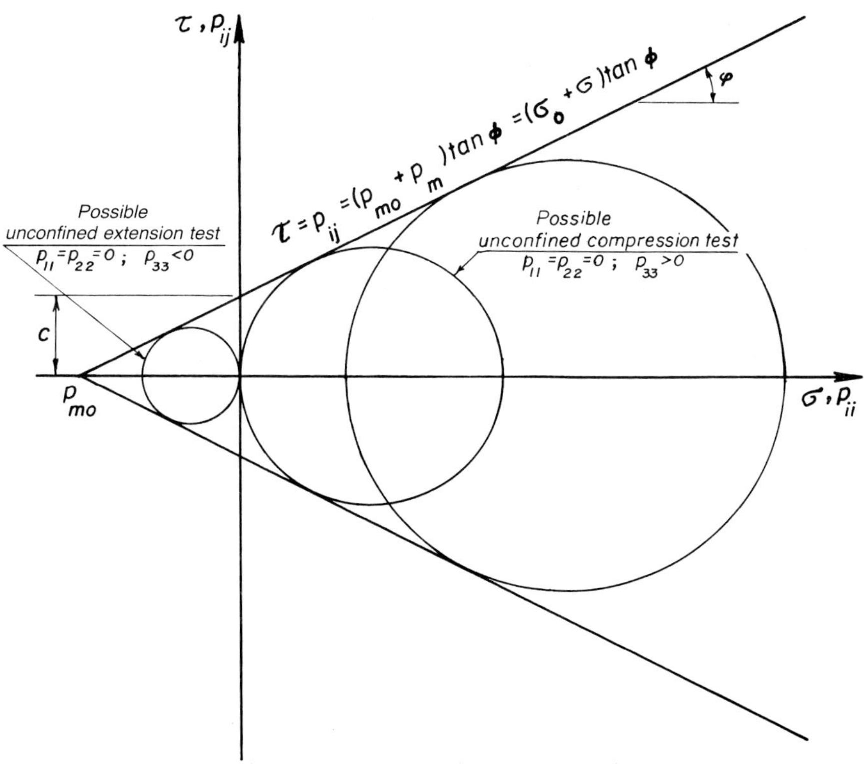

Fig. 13.5.1. Mohr–Coulomb failure envelope for the elastic medium.

internal forces. If such internal forces do not exist, the material is cohesionless, and its shear strength depends on its induced densification. And while it seems logical that the coefficient of friction and the angle of internal friction should depend on the ratio of the bulk modulus to the shear elastic modulus, this was never verified experimentally. The cohesion also depends on the spherical pressure p_{m} and is, therefore, not necessarily constant in the various pressures.

Another expression derived from Eq. (13.5.1) is

$$\overline{\mathrm{II}}_s = \tfrac{4}{3}\Pi_s = \tfrac{4}{3}\Pi_p = s_{\alpha\beta}^2 = 3\tau^2 = 4G\rho\psi_{\mathfrak{si}0} + 2\,\frac{G}{\varkappa}\,p_{\mathrm{m}}^2$$

$$= 4G\rho\psi_{\mathfrak{si}0} + \frac{2G}{9\varkappa}\,\mathrm{I}_p^2 \tag{13.5.3}$$

Eq. (13.5.3) is known as the von Mises yield condition (von Mises 1913). One can note that this condition is closely related to the octahedral shear stress and to the octahedral invariant, and depends on the spherical pressure p_{m} or on the first invariant of the pressure tensor p_{ij}. However, it is independent of the third invariant of the pressure tensor.

In the preceding discussion it was shown how the failure theories of an elastic medium evolve from energy considerations and from the balance between the disbursed free energy and the stored free energy. It can be similarly shown how these energies balance for a general linear viscoelastic medium as well, although it might not be that forthright.

13.6 Specific Free Energy Balance of Linear Viscoelastic Media

Applying the specific free energy balance equations (11.25.3) and (11.25.4) to the equations of the general linear viscoelastic medium (7.6.6) and (7.6.7), or to the specific viscoelastic equations such as (7.6.8) and (7.6.9), we obtain the excess stored free energy and the disbursed free energy, respectively. Here we shall apply Eqs. (11.25.3) and (11.25.4) to the particular case, Eqs. (7.6.8) and (7.6.9)

$$\Delta\rho\psi_{\mathfrak{si}} = \int_0^{t'} p_{\mathrm{e}}\,\mathrm{d}\varepsilon_{\alpha\alpha} = \int_0^{t'} (p_{\mathrm{m}} - p_{\mathrm{p}})\,\mathrm{d}\varepsilon_{\alpha\alpha} = \int_0^{t'} (p_{\mathrm{m}} - p_{\mathfrak{w}} - p_{\mathfrak{g}})\dot{\varepsilon}_{\alpha\alpha}\,\mathrm{d}t$$

$$= \frac{1}{3}\int_0^{t'} \left[(p(t)_{\mathrm{m}} - p_0\vartheta(t)) + \frac{1}{T_3}\exp\frac{-t'}{T_3}\int_0^{t'} (p(t)_{\mathrm{m}} - p_0\vartheta(t))\exp\frac{t}{T_3}\,\mathrm{d}t\right]$$

$$\times \left[\frac{1}{\varkappa_1}(\dot{p}(t')_{\mathrm{m}} - p_0\dot{\vartheta}(t')) + \frac{1}{\mu_2}\int_0^{t'}[p(t)_{\mathrm{m}} - p_0\vartheta(t)]\,\mathrm{d}t\right.$$

$$\left. + \frac{1}{\mu_3}\exp\frac{-t}{T_3}\int_0^{t'}\left\{[p(t)_{\mathrm{m}} - p_0\vartheta(t)]\exp\frac{t}{T_3}\,\mathrm{d}t\right\}\mathrm{d}t\right] \tag{13.6.1}$$

$$\rho\psi_{\mathfrak{s}\mathfrak{d}} = \int_0^{t''} s_{\mathfrak{s}\alpha\beta}\dot{e}_{\alpha\beta}\,\mathrm{d}t = \int_0^{t''}\left\{s(t)_{\alpha\beta}\left[\frac{1}{2G}\dot{s}(t)_{\alpha\beta} + \frac{1}{2\eta}s(t)_{\alpha\beta}\right]\right\}\mathrm{d}t \tag{13.6.2}$$

Since linear viscoelastic relations were assumed, volumetric strains cannot occur during shear, thus $\varepsilon_{\beta\beta} = 0$.

Eqs. (13.6.1) and (13.6.2) appear to be intricate, particularly the former. If, however, the shape of $\vartheta(t)$ is assumed and substituted and the nature of the spherical pressure $p(t)_{\mathrm{m}}$ is known and inserted, Eq. (13.6.1), however lengthy, is easily solvable.

13.7 The Pure Deviatoric Test of a Linear Viscoelastic Medium

Eqs. (13.6.1) and (13.6.2) can now be applied to the spherical and pure deviatoric tests or to any of the other conventional tests.

The assumption of pure deviatoric loading prescribes axial symmetry and $s_{\alpha\alpha} = 0$. Thus, for the consolidation part of the test we have $p_{\mathrm{m}} = \frac{1}{3}p_{\alpha\alpha} = p_{11} = p_{22} = p_{33}$ and the plastic restraint $\vartheta(t)$ according to Eq. (C.14.9) in Appendix C, and for the deviatoric part of the test we have $s_{ij} = \frac{1}{2}\Delta p_{33}\Xi_{ij}$ and $\dot{s}_{ij} = 0$. We shall consider two cases of deviatoric loadings:

1. *Constant stepwise deviatoric loading*. For the stored and disbursed specific free energies we insert

$$p_{\mathrm{m}} = \tfrac{1}{3}p_{\alpha\alpha} = \mathrm{const}$$

and

$$\sum_{\mathfrak{q}=1}^{\mathfrak{Q}} s_{\mathfrak{q}ij} = \sum_{\mathfrak{q}=1}^{\mathfrak{Q}} \tfrac{1}{2}\mathcal{P}_{\mathfrak{q}}\Xi_{ij}$$

in Eqs. (13.6.1) and (13.6.2), and recalling also Eqs. (A.21.18) and (A.21.19) from Appendix A, we have

$$\Delta\rho\psi_{\mathfrak{s}\mathrm{i}} = \frac{1}{3}\int_0^{t'}\left\{[p(t)_{\mathrm{m}} - p_0\vartheta(t)] + \frac{1}{T_3}\exp\frac{-t'}{T_3}\int_0^{t'}[p(t)_{\mathrm{m}} - p_0\vartheta(t)]\exp\frac{t}{T_3}\,\mathrm{d}t\right\}$$

$$\left\{\frac{1}{\varkappa_1}[\dot{p}(t)_{\mathrm{m}} - p_0\dot{\vartheta}(t)] + \frac{1}{\mu_2}[p(t)_{\mathrm{m}} - p_0\vartheta(t)]\right.$$

$$+\frac{1}{\mu_3}\exp\frac{-t}{T_3}\times\int_0^{t'}[p(t)_{\mathrm{m}}-p_0\vartheta(t)]\exp\frac{t}{T_3}\,\mathrm{d}t\Big\}\,\mathrm{d}t$$

$$=\frac{1}{3}\int_0^{t'}\Big[\Big(p_{\mathrm{m}}-p_0[1-\exp-Ct]\Big)$$

$$-\frac{1}{T_3}\left(\exp\frac{-t'}{T_3}\right)\times\int_0^{t'}(p_{\mathrm{m}}-p_0[1-\exp-Ct])\exp\frac{t}{T_3}\,\mathrm{d}t\Big]$$

$$\times\Big[\frac{1}{\mu_2}(p_m-p_0)[1-\exp-Ct]-\frac{1}{\mu_3T_3}\left(\exp\frac{-1}{T_3}\right)$$

$$\times\int_0^{t'}(p_{\mathrm{m}}-p_0[1-\exp-Ct])\exp\frac{t}{T_3}\,\mathrm{d}t\Big]\,\mathrm{d}t$$

$$=\tfrac{1}{3}p_{\mathrm{m}}^2\int_0^{t'}\Big[(1-B[1-\exp-Ct])-\frac{1}{T_3}\left(\exp\frac{-t'}{T_3}\right)$$

$$\times\int_0^{t'}(1-B[1-\exp-Ct])\exp\frac{t}{T_3}\,\mathrm{d}t\Big]$$

$$\times\Big[\frac{1}{\mu_2}(1-B[1-\exp-Ct])-\frac{1}{\mu_3T_3}\left(\exp\frac{-1}{T_3}\right)$$

$$\times\int_0^{t'}(1-B[1-\exp-Ct])\exp\frac{t}{T_3}\,\mathrm{d}t\Big]\,\mathrm{d}t$$

$$=p_{\mathrm{m}}^2\mathfrak{F}(t')=\tfrac{1}{9}p_{\alpha\alpha}^2\mathfrak{F}(t')\tag{13.7.1}$$

$$\rho\psi_{\mathfrak{sd}}=\sum_{\mathfrak{q}=1}^{\mathfrak{Q}}\int_0^{t''}\left\{s_{\mathfrak{q}\alpha\beta}\left[\frac{1}{2G}\dot{s}_{\mathfrak{q}\alpha\beta}+\frac{1}{2\eta}s_{\mathfrak{q}\alpha\beta}\right]\right\}\mathrm{d}t$$

$$=\sum_{\mathfrak{q}=1}^{\mathfrak{Q}}\frac{3\mathcal{P}_{\mathfrak{q}}^2}{4\eta}t''_{\mathfrak{q}}=\sum_{\mathfrak{q}=1}^{\mathfrak{Q}}\frac{t'_{\mathfrak{q}}}{2\eta}s_{\alpha\beta}s_{\alpha\beta}$$

$$=\sum_{\mathfrak{q}=1}^{\mathfrak{Q}}\frac{t''_{\mathfrak{q}}}{\eta}\overline{\mathrm{II}}_s=\sum_{\mathfrak{q}=1}^{\mathfrak{Q}}\frac{4t''_{\mathfrak{q}}}{3\eta}\Pi_s=\sum_{\mathfrak{q}=1}^{\mathfrak{Q}}\frac{3t''_{\mathfrak{q}}}{2\eta}\tau_{\mathfrak{q}}^2\tag{13.7.2}$$

where $p_0=Bp_{\mathrm{m}}$ assuming $B<1$ and $p_0=\mathrm{const}<p_{\mathrm{m}}$.

Eq. (13.7.2) can be written as follows

$$\rho\psi_{\mathfrak{sd}} = \sum_{\mathfrak{q}=1}^{\mathfrak{Q}} \frac{3t''_{\mathfrak{q}}}{2\eta}\,\tau_{\mathfrak{q}}^2 = \sum_{\mathfrak{q}=1}^{\mathfrak{Q}-1} \frac{3t''_{\mathfrak{q}}}{2\eta}\,\tau_{\mathfrak{q}}^2 + \frac{3t''_{\mathfrak{Q}}}{2\eta}\,\tau_{\mathfrak{Q}}^2 = \mathfrak{D}(t'')_{\mathfrak{Q}-1} + \frac{3t''_{\mathfrak{Q}}}{2\eta}\,\tau_{\mathfrak{Q}}^2 \qquad (13.7.3)$$

where it is resolved into two parts, one containing the first $\mathfrak{Q}-1$ step loadings and the other the last $\mathfrak{Q}$th step loading which corresponds to the maximum octahedral shear stress, $\tau_{\mathfrak{Q}} = \tau$.

When we now equate the stored specific free energy with the disbursed specific free energy according to Eq. (13.4.3) we obtain

$$\mathfrak{D}(t'')_{\mathfrak{Q}-1} + \frac{3t''_{\mathfrak{Q}}}{2\eta}\,\tau^2 = \rho\psi_{\mathfrak{si}0} + \mathfrak{S}(t')p_{\mathrm{m}}^2 \qquad (13.7.4)$$

where

$$\mathfrak{D}(t'')_{\mathfrak{Q}-1} = \sum_{\mathfrak{q}=1}^{\mathfrak{Q}-1} \frac{3t''_{\mathfrak{q}}}{2\eta}\,\tau_{\mathfrak{q}}^2$$

from which we obtain, according to Eq. (13.5.2), the octahedral shear stress or in short the shear stress

$$\begin{aligned}
\tau = \sqrt{(\tfrac{1}{3}s_{\alpha\beta}^2)} &= \sqrt{\left(\frac{2\eta\rho\psi_{\mathfrak{si}0}}{3t''_{\mathfrak{Q}}\mathfrak{D}(t'')} + \frac{2\eta\mathfrak{S}(t')}{3t''_{\mathfrak{Q}}\mathfrak{D}(t'')}\,p_{\mathrm{m}}^2\right)} \\
&= \sqrt{\left(\frac{2\eta\mathfrak{S}(t')}{3t''_{\mathfrak{Q}}\mathfrak{D}(t'')}\right)}\left[-p_{\mathrm{m}} \pm \sqrt{\left(p_{\mathrm{m}}^2 + \frac{2\eta\rho\psi_{\mathfrak{si}0}}{3t''_{\mathfrak{Q}}\mathfrak{S}(t')}\right)}\right] + \sqrt{\left(\frac{2\eta\mathfrak{S}(t')}{3t''_{\mathfrak{Q}}\mathfrak{D}(t'')}\right)}\,p_{\mathrm{m}} \\
&= c + \mu p_{\mathrm{m}} = c + p_{\mathrm{m}}\tan\phi = (p_{\mathrm{m}0} + p_{\mathrm{m}})\tan\phi \qquad (13.7.5)
\end{aligned}$$

where the cohesive shear intercept is

$$c = \sqrt{\frac{2\eta\mathfrak{S}(t')}{3t''_{\mathfrak{Q}}\mathfrak{D}(t'')}}\left[-p_{\mathrm{m}} \pm \sqrt{\left(p_{\mathrm{m}}^2 + \frac{2\eta\rho\psi_{\mathfrak{si}0}}{3t''_{\mathfrak{Q}}\mathfrak{S}(t')}\right)}\right]$$

the cohesion is

$$p_{\mathrm{m}0} = \frac{c}{\tan\phi} = \left[-p_{\mathrm{m}} \pm \sqrt{\left(p_{\mathrm{m}}^2 + \frac{2\eta\rho\psi_{\mathfrak{si}0}}{3t''_{\mathfrak{Q}}\mathfrak{S}(t')}\right)}\right]$$

and

$$\mu = \tan\phi = \sqrt{\left(\frac{2\eta\mathfrak{S}(t')}{3t''_{\mathfrak{Q}}\mathfrak{D}(t'')}\right)}.$$

It was shown here that, for the pure deviatoric loading based on linear viscoelasticity, as in the case of an elastic medium, Sect. 13.5, the balance of the stored and disbursed specific free energies leads to the von Mises failure criterion in the octahedral plane and can be traced to the Mohr–Coulomb failure theory. Eq. (13.7.1) is a scalar, the magnitude of which depends on the spherical pressure which is proportional to the first invariant of the stress tensor, $p_{\mathrm{m}} = \frac{1}{3}p_{\alpha\alpha} = \frac{1}{3}\mathrm{I}_p$. Eq. (13.7.2), a scalar as well, depends at any time t on the octahedral invariant of the shear stress Π_s. The octahedral shear stress, the cohesion, the cohesive intercept and the angle of internal friction are all functions of the time t, and except for the last they are functions of the consolidation pressure p_{m} as well.

2. *Constant rate of loading*. For a constant rate of deviatoric loading we shall assume $s_{ij} = \frac{1}{2}s_{zz}\Xi_{ij} = \frac{1}{2}\mathcal{S}t\Xi_{ij}$ from which we have $\dot{s}_{ij} = \frac{1}{2}\mathcal{S}\Xi_{ij}$, and substitute it into Eq. (13.6.2). The consolidation preceding the pure deviatoric loading is similar to that in the previous case of stepwise loading, and therefore the stored specific free energy is equal to Eq. (13.7.1). The disbursed specific free energy, however, is equal to

$$\begin{aligned}
\rho\psi_{\mathfrak{sd}} &= \int_0^{t''}\left\{s_{\alpha\beta}\left[\frac{1}{2G}\dot{s}_{\alpha\beta} + \frac{1}{2\eta}s_{\alpha\beta}\right]\right\}dt = \left\{\frac{3}{8G}\mathcal{S}^2t''^2 + \frac{1}{4\eta}\mathcal{S}^2t''^3\right\} \\
&= \left(\frac{1}{4G}t''^2 + \frac{1}{6\eta}t''^3\right)\dot{s}_{\alpha\beta}\dot{s}_{\alpha\beta} = \left(\frac{1}{4G} + \frac{1}{6\eta}t''\right)s_{\alpha\beta}s_{\alpha\beta} \\
&= \mathcal{S}^2\left(\frac{1}{16G}t''^2 + \frac{1}{24\eta}t''^3\right)\Xi_{\alpha\beta}\Xi_{\alpha\beta} \\
&= \left(\frac{1}{2G}t''^2 + \frac{1}{3\eta}t''^3\right)\overline{\Pi}_{\dot{s}} = \left(\frac{2}{3G}t''^2 + \frac{4}{9\eta}t''^3\right)\Pi_{\dot{s}} \\
&= \left(\frac{2}{3G} + \frac{4}{9\eta}t''\right)\Pi_s = \left(\frac{3}{2G} + \frac{1}{\eta}t''\right)\tau^2 \\
&= \left(\frac{3\eta + 2Gt''}{2G\eta}\right)\tau^2 = \mathfrak{D}(t'')\tau^2
\end{aligned} \tag{13.7.6}$$

As in the previous case, we see that Eq. (13.7.6) is a scalar and depends at any time t not only on the octahedral invariant of the shear stress Π_s but also on that of the shear stress rate $\Pi_{\dot{s}}$.

When the stored free energy is equated with the disbursed free energy, one of the possible results is again

$$\mathfrak{D}(t'')\tau^2 = \rho\psi_{\mathfrak{si}0} + \mathfrak{S}(t')p_{\mathrm{m}}^2 \tag{13.7.7}$$

where $\mathfrak{D}(t'') = (3\eta + 2Gt'')/(2G\eta)$. As in Sect. 13.5, we may write similarly to Eq. (13.5.2)

$$\tau = \sqrt{(\tfrac{1}{3} s_{\alpha\beta}^2)} = \sqrt{\left(\frac{\rho\psi_{\mathfrak{si}0}}{\mathfrak{D}(t'')} + \frac{\mathfrak{S}(t')}{\mathfrak{D}(t'')}\, p_{\mathrm{m}}^2\right)}$$

$$= \sqrt{\left(\frac{\mathfrak{S}(t')}{\mathfrak{D}(t'')}\right)}\left[-p_{\mathrm{m}} \pm \sqrt{\left(p_{\mathrm{m}}^2 + \frac{\rho\psi_{\mathfrak{si}0}}{\mathfrak{S}(t')}\right)}\right] + \sqrt{\left(\frac{\mathfrak{S}(t')}{\mathfrak{D}(t'')}\, p_{\mathrm{m}}\right)}$$

$$= c + \mu p_{\mathrm{m}} = c + p_{\mathrm{m}} \tan\phi = (p_{\mathrm{m}0} + p_{\mathrm{m}}) \tan\phi \qquad (13.7.8)$$

where the cohesive shear intercept is

$$c = \sqrt{\left(\frac{\mathfrak{S}(t')}{\mathfrak{D}(t'')}\right)}\left[-p_{\mathrm{m}} \pm \sqrt{\left(p_{\mathrm{m}}^2 + \frac{\rho\psi_{\mathfrak{si}0}}{\mathfrak{S}(t')}\right)}\right]$$

the cohesion

$$p_{\mathrm{m}0} = \frac{c}{\tan\phi} = \left[-p_{\mathrm{m}} \pm \sqrt{\left(p_{\mathrm{m}}^2 + \frac{\rho\psi_{\mathfrak{si}0}}{\mathfrak{S}(t')}\right)}\right]$$

and $\mu = \tan\phi = \sqrt{(\mathfrak{S}(t')/[\mathfrak{D}(t'')]}$.

It should be remarked that as before the balance of the stored and disbursed specific free energy leads to the von Mises failure criterion and to the Mohr–Coulomb failure theory.

This analysis is based on the particular case of the constitutive equations given by (7.6.8) and (7.6.9). If more intricate equations were chosen, for instance a four-coefficient model instead of the model of Eq. (7.6.9), somewhat different results would have been obtained. Still the general pattern would have been preserved.

13.8 Free Energy Balance with Non-linear Isotropic Stress–Strain Relations

If isotropic relations between strains and stresses and their time derivatives are postulated, the free energy balance can be applied to Eqs. (7.9.6) and (7.9.7).

Isotropic relations were formulated in terms of total pressure and they do not lend themselves to direct application to effective pressures in a purely analytical form. Nevertheless we shall attempt to show how to apply them indirectly when pore pressure measurements are available. Considering pore pressures, the equations derived in Sects. 11.27 and 11.28 can be adapted to failure conditions of soils that did not reach full consolidation.

When a soil is consolidated and subjected to shear stresses and pore pressures are measured, and Eqs. (13.4.2) and (11.25.4), the excess stored free energy and the disbursed free energy, respectively, are applied to the isotropic stress–strain relationships, Eqs. (7.9.6) and (7.9.7), we obtain

$$\Delta\rho\psi_{\mathfrak{si}} = \int_0^{t'} [p_{\mathrm{m}} - p(t)_{\mathrm{p}}]\dot{\varepsilon}_{\alpha\alpha}\,\mathrm{d}t + \int_0^{t''} [p_{\mathrm{m}} - p(t)_{\mathrm{p}}]\dot{\varepsilon}_{\beta\beta}\,\mathrm{d}t$$

$$= \int_0^{t'} [p_{\mathrm{m}} - p(t)_{\mathrm{p}}]\{\beth_0 + \beth_1[p_{\mathrm{m}} - p(t)_{\mathrm{p}}] + \beth_2[p_{\mathrm{m}}' - p(t)_{\mathrm{p}}]^2$$

$$+ \beth_3\varepsilon_{\alpha\alpha} + \beth_5[p_{\mathrm{m}} - p(t)_{\mathrm{p}}]\varepsilon_{\alpha\alpha} + \beth_6[p_{\mathrm{m}} - p(t)_{\mathrm{p}}]^2\varepsilon_{\alpha\alpha}$$

$$+ \mathfrak{w}_2(s)_{\mathrm{i}}^2 + \mathfrak{w}_5(es + se)_{\mathrm{i}} + \mathfrak{w}_6(es^2 + s^2e)_{\mathrm{i}}\}\,\mathrm{d}t \tag{13.8.1}$$

$$\rho\psi_{\mathfrak{sd}} = \int_0^{t''} s_{\alpha\beta}\dot{e}_{\alpha\beta}\,\mathrm{d}t' = \int_0^{t''} s_{\alpha\beta}\,[\mathfrak{w}_1 s_{\alpha\beta} + \mathfrak{w}_2(s)_{\alpha\beta\mathfrak{d}}^2 + \mathfrak{w}_3 e_{\alpha\beta}$$

$$+ \mathfrak{w}_5(se + es)_{\alpha\beta\mathfrak{d}} + \mathfrak{w}_6(s^2e + es^2)_{\mathfrak{d}}]\,\mathrm{d}t \tag{13.8.2}$$

where $\Delta\rho\psi_{\mathfrak{si}}$ and $\rho\psi_{\mathfrak{sd}}$ are the excess stored free energy and the disbursed free energy of the total stresses, respectively, the strain rate tensor is an isotropic function of the stress and strain tensors, $\dot{\varepsilon}_{ij} = \dot{\varepsilon}(p, \varepsilon)_{ij}$, and the coefficients $\beth_{\mathrm{n}}$ and $\mathfrak{w}_{\mathrm{n}}$ are functions of six stress and strain invariants. The pore pressure $p(t)_{\mathrm{p}}$ is a time-dependent function even when the applied spherical pressure p_{m} is steady. The times t' and t'' are not necessarily identical.

13.9 The Pure Deviatoric Test with Isotropic Relations

The application of the specific free energy balance equations (Sect. 13.8) to the pure deviatoric loading entails the assumption of isotropic stress–strain relationships.

The assumptions in pure deviatoric loading are, among others, axial symmetry and $s_{\alpha\alpha} = 0$. Thus, we have $p_{\mathrm{m}} = \frac{1}{3}p_{\alpha\alpha} = p_{11} = p_{22} = p_{33}$ for the consolidation and $s_{ij} = \frac{1}{2}\Delta p_{33}\Xi_{ij}$ for the deviatoric stresses. For the deviatoric shear stress in Eq. (13.8.2) we shall consider the same two cases as for linear viscoelasticity:

1. *Constant stepwise deviatoric loading*. For the stored and disbursed specific free energies we insert in Eqs. (13.8.1) and (13.8.2), respectively, $p_{\mathrm{m}} = \frac{1}{3}p_{\alpha\alpha} = \mathrm{const}$ and

$$\sum_{\mathfrak{q}=1}^{\mathfrak{Q}} s_{\mathfrak{q}ij} = \sum_{\mathfrak{q}=1}^{\mathfrak{Q}} \tfrac{1}{2}\mathcal{S}_{\mathfrak{q}}\Xi_{ij}$$

and, recalling also Eqs. (A.21.18) and (A.21.19) from Appendix A we get, after considerable labor

$$
\begin{aligned}
\Delta\rho\psi_{\mathfrak{s}i} = & \int_0^{t'} \{\beth_0[p_{\mathrm{m}} - p(t)_{\mathfrak{p}}] + \beth_1[p_{\mathrm{m}} - p(t)_{\mathfrak{p}}]^2 + \beth_2[p_{\mathrm{m}} - p(t)_{\mathfrak{p}}]^3\}\, dt \\
& + 3\int_0^{t'} \{\beth_3[p_{\mathrm{m}} - p(t)_{\mathfrak{p}}] + \beth_5[p_{\mathrm{m}} - p(t)_{\mathfrak{p}}]^2 \\
& + \beth_6[p_{\mathrm{m}} - p(t)_{\mathfrak{p}}]^3\}\varepsilon(t)_{\alpha\alpha}\, dt + 3\mathfrak{w}_2 \sum_{\mathfrak{q}=1}^{\mathfrak{Q}} (s)^2_{\mathfrak{q}\alpha\alpha i} \int_0^{t''_{\mathfrak{q}}} [p_{\mathrm{m}} - p(t)_{\mathfrak{p}}]\, dt \\
& + 12 \sum_{\mathfrak{q}=1}^{\mathfrak{Q}} [\mathfrak{w}_5 s_{\mathfrak{q}\alpha\beta i} + \mathfrak{w}_6 (s^2)_{\mathfrak{q}\alpha\beta i}] \int_0^{t''_{\mathfrak{q}}} [p_{\mathrm{m}} - p(t)_{\mathfrak{p}}] e_{\alpha\beta i}\, dt \\
= & \int_0^{t'} \{\beth_0[p_{\mathrm{m}} - p(t)_{\mathfrak{p}}] + \beth_1[p_{\mathrm{m}} - p(t)_{\mathfrak{p}}]^2 + \beth_2[p_{\mathrm{m}} - p(t)_{\mathfrak{p}}]^3\}\, dt \\
& + 3\int_0^{t'} \{\beth_3[p_{\mathrm{m}} - p(t)_{\mathfrak{p}}] + \beth_5[p_{\mathrm{m}} - p(t)_{\mathfrak{p}}]^2 \\
& + \beth_6[p_{\mathrm{m}} - p(t)_{\mathfrak{p}}]^3\}\varepsilon(t)_{\alpha\alpha}\, dt + \tfrac{3}{4}\mathfrak{w}_2 \sum_{\mathfrak{q}=1}^{\mathfrak{Q}} \mathcal{S}^2_{\mathfrak{q}} \int_0^{t''_{\mathfrak{q}}} [p_{\mathrm{m}} - p(t)_{\mathfrak{p}}]\, dt \\
& + 3 \sum_{\mathfrak{q}=1}^{\mathfrak{Q}} \Bigg[2\mathfrak{w}_5 \mathcal{S}_{\mathfrak{q}} \int_0^{t''_{\mathfrak{q}}} [p_{\mathrm{m}} - p(t)_{\mathfrak{p}}] e(t)_{\mathfrak{q}}\, dt \\
& + \mathfrak{w}_6 \mathcal{S}^2_{\mathfrak{q}} \int_0^{t''_{\mathfrak{q}}} [p_{\mathrm{m}} - p(t)_{\mathfrak{p}}] \left(\frac{\varepsilon(t)_{\mathfrak{q}\beta\beta} + e(t)_{\mathfrak{q}}}{3} \right) \Bigg]\, dt \\
= & \int_0^{t'} \{\beth_0[p_{\mathrm{m}} - p(t)_{\mathfrak{p}}] + \beth_1[p_{\mathrm{m}} - p(t)_{\mathfrak{p}}]^2 + \beth_2[p_{\mathrm{m}} - p(t)_{\mathfrak{p}}]^3\}\, dt \\
& + 3\int_0^{t'} \{\beth_3[p_{\mathrm{m}} - p(t)_{\mathfrak{p}}] + \beth_5[p_{\mathrm{m}} - p(t)_{\mathfrak{p}}]^2 \\
& + \beth_6[p_{\mathrm{m}} - p(t)_{\mathfrak{p}}]^3\}\varepsilon(t)_{\alpha\alpha}\, dt + \mathfrak{w}_2 \sum_{\mathfrak{q}=1}^{\mathfrak{Q}} s^2_{\mathfrak{q}\alpha\beta} \int_0^{t''_{\mathfrak{q}}} [p_{\mathrm{m}} - p(t)_{\mathfrak{p}}]\, dt \\
& + 2\mathfrak{w}_5 \sum_{\mathfrak{q}=1}^{\mathfrak{Q}} \surd(6 s^2_{\mathfrak{q}\alpha\beta}) \int_0^{t''_{\mathfrak{q}}} [p_{\mathrm{m}} - p(t)_{\mathfrak{p}}] e(t)_{\mathfrak{q}}\, dt \\
& + 2\mathfrak{w}_6 \sum_{\mathfrak{q}=1}^{\mathfrak{Q}} s^2_{\mathfrak{q}\alpha\beta} \int_0^{t''_{\mathfrak{q}}} [p_{\mathrm{m}} - p(t)_{\mathfrak{p}}] \left(\frac{\varepsilon(t)_{\mathfrak{q}\beta\beta} + e(t)_{\mathfrak{q}}}{3} \right) dt
\end{aligned}
$$

$$= \int_0^{t'} \{\beth_0[p_\mathrm{m} - p(t)_\mathrm{p}] + \beth_1[p_\mathrm{m} - p(t)_\mathrm{p}]^2 + \beth_2[p_\mathrm{m} - p(t)_\mathrm{p}]^3\}\,\mathrm{d}t$$

$$+ 3\int_0^{t'} \{\beth_3[p_\mathrm{m} - p(t)_\mathrm{p}] + \beth_5[p_\mathrm{m} - p(t)_\mathrm{p}]^2$$

$$+ \beth_6[p_\mathrm{m} - p(t)_\mathrm{p}]^3\}\varepsilon(t)_{\alpha\alpha}\,\mathrm{d}t$$

$$+ 2\mathfrak{w}_5 \sum_{\mathfrak{q}=1}^{\mathfrak{Q}} \surd(6\mathrm{II}_{s\mathfrak{q}}) \int_0^{t''_\mathfrak{q}} [p_\mathrm{m} - p(t)_\mathrm{p}]e(t)_\mathfrak{q}\,\mathrm{d}t$$

$$+ \mathfrak{w}_2 \sum_{\mathfrak{q}=1}^{\mathfrak{Q}} \overline{\mathrm{II}}_{s\mathfrak{q}} \int_0^{t''_\mathfrak{q}} [p_\mathrm{m} - p(t)_\mathrm{p}]\,\mathrm{d}t$$

$$+ 2\mathfrak{w}_6 \sum_{\mathfrak{q}=1}^{\mathfrak{Q}} \overline{\mathrm{II}}_{s\mathfrak{q}} \int_0^{t''_\mathfrak{q}} [p_\mathrm{m} - p(t)_\mathrm{p}]\left(\frac{\varepsilon(t)_{\mathfrak{q}\beta\beta} + e(t)_\mathfrak{q}}{3}\right)\mathrm{d}t \qquad (13.9.1)$$

$$\rho\psi_{\mathfrak{sd}} = 6\sum_{\mathfrak{q}=1}^{\mathfrak{Q}} \int_0^{t''_\mathfrak{q}} s_{\mathfrak{q}\alpha\beta}\,[\mathfrak{w}_1 s_{\mathfrak{q}\alpha\beta} + \mathfrak{w}_2(s)^2_{\mathfrak{q}\alpha\beta\mathfrak{d}} + \mathfrak{w}_3 e_{\mathfrak{q}\alpha\beta} + \mathfrak{w}_5(se + es)_{\mathfrak{q}\alpha\beta\mathfrak{d}}$$

$$+ \mathfrak{w}_6(s^2e + es^2)_{\mathfrak{q}\alpha\beta\mathfrak{d}}]\,\mathrm{d}t$$

$$= \sum_{\mathfrak{q}=1}^{\mathfrak{Q}} [\tfrac{3}{2}\mathfrak{w}_1\mathcal{S}^2_\mathfrak{q} + \tfrac{3}{4}\mathfrak{w}_2\mathcal{S}^3_\mathfrak{q}]t''_\mathfrak{q} + \sum_{\mathfrak{q}=1}^{\mathfrak{Q}} \Bigg[3\mathfrak{w}_3\mathcal{S}_\mathfrak{q} \int_0^{t''_\mathfrak{q}} e(t)_\mathfrak{q}\,\mathrm{d}t$$

$$+ 3\mathfrak{w}_5\mathcal{S}^2_\mathfrak{q} \int_0^{t''_\mathfrak{q}} \frac{\varepsilon(t)_{\mathfrak{q}\beta\beta} + e(t)_\mathfrak{q}}{3}\,\mathrm{d}t + \tfrac{3}{4}\mathfrak{w}_6\mathcal{S}^3_\mathfrak{q} \int_0^{t''_\mathfrak{q}} \frac{\varepsilon(t)_{\mathfrak{q}\beta\beta} + 2e(t)_\mathfrak{q}}{3}\,\mathrm{d}t\Bigg]$$

$$= \sum_{\mathfrak{q}=1}^{\mathfrak{Q}} [\mathfrak{w}_1 s^2_{\mathfrak{q}\alpha\beta} + \mathfrak{w}_2\surd(\tfrac{1}{6}s^2_{\mathfrak{q}\alpha\beta})s^2_{\mathfrak{q}\alpha\beta}]t''_\mathfrak{q} + \sum_{\mathfrak{q}=1}^{\mathfrak{Q}} \Bigg[\mathfrak{w}_3\surd(6s^2_{\mathfrak{q}\alpha\beta}) \int_0^{t''_\mathfrak{q}} e(t)_\mathfrak{q}\,\mathrm{d}t$$

$$+ 2\mathfrak{w}_5 s^2_{\mathfrak{q}\alpha\beta} \int_0^{t''_\mathfrak{q}} \frac{\varepsilon(t)_{\mathfrak{q}\beta\beta} + e(t)_\mathfrak{q}}{3}\,\mathrm{d}t$$

$$+ \mathfrak{w}_6\surd(\tfrac{1}{6}s^2_{\mathfrak{q}\alpha\beta})s^2_{\mathfrak{q}\alpha\beta} \int_0^{t''_\mathfrak{q}} \frac{\varepsilon(t)_{\mathfrak{q}\beta\beta} + 2e(t)_\mathfrak{q}}{3}\,\mathrm{d}t\Bigg]$$

$$= \sum_{\mathfrak{q}=1}^{\mathfrak{Q}} [\mathfrak{w}_1\overline{\mathrm{II}}_{s\mathfrak{q}} + \tfrac{1}{18}\mathfrak{w}_2\overline{\mathrm{III}}_{s\mathfrak{q}}]t''_\mathfrak{q} + \sum_{\mathfrak{q}=1}^{\mathfrak{Q}} \mathfrak{w}_3\surd(6\overline{\mathrm{II}}_{s\mathfrak{q}}) \int_0^{t''_\mathfrak{q}} e(t)_\mathfrak{q}\,\mathrm{d}t$$

$$+ 2\mathfrak{w}_5 \sum_{\mathfrak{q}=1}^{\mathfrak{Q}} \overline{\mathrm{II}}_{s\mathfrak{q}} \int_0^{t''_\mathfrak{q}} \frac{\varepsilon(t)_{\mathfrak{q}\beta\beta} + e(t)_\mathfrak{q}}{3}\,\mathrm{d}t$$

$$+ \tfrac{1}{18}\mathfrak{w}_6 \sum_{\mathfrak{q}=1}^{\mathfrak{Q}} \overline{\mathrm{III}}_{s\mathfrak{q}} \int_0^{t''_\mathfrak{q}} \frac{\varepsilon(t)_{\mathfrak{q}\beta\beta} + 2e(t)_\mathfrak{q}}{3}\,\mathrm{d}t \qquad (13.9.2)$$

where $\varepsilon(t)_{\alpha\alpha}$ is the volumetric strain during the consolidation and $\varepsilon(t)_{\beta\beta}$ is the volumetric strain during the deviatoric stresses, and the total strain tensor computed is

$$\varepsilon(t)_{ij} = \tfrac{1}{3}\varepsilon(t)_{\beta\beta}\delta_{ij} + \tfrac{1}{3}e(t)\Xi_{ij} \tag{13.9.3}$$

where $e(t) = e_{33} - e_{11}$ is the deviatoric strain factor of the deviatoric strain tensor e_{ij}.

By balancing the stored specific free energy with the disbursed specific free energy we can write

$$\begin{aligned}
&\sum_{q=1}^{\Omega} [\mathfrak{w}_1\overline{\mathrm{II}}_{sq} + \tfrac{1}{18}\mathfrak{w}_2\overline{\mathrm{III}}_{sq}]t''_q + \sum_{q=1}^{\Omega} \mathfrak{w}_3\sqrt{(6\overline{\mathrm{II}}_{sq})} \int_0^{t''_q} e(t)_q\, dt \\
&\quad + 2\mathfrak{w}_5 \sum_{q=1}^{\Omega} \overline{\mathrm{II}}_{sq} \int_0^{t''_q} \frac{\varepsilon(t)_{q\beta\beta} + e(t)_q}{3}\, dt \\
&\quad + \tfrac{1}{18}\mathfrak{w}_6 \sum_{q=1}^{\Omega} \overline{\mathrm{III}}_{sq} \int_0^{t''_q} \frac{\varepsilon(t)_{q\beta\beta} + 2e(t)_q}{3}\, dt \\
&= \rho\psi_{\bar{s}0} + \int_0^{t'} \{\beth_0[p_m - p(t)_p] + \beth_1[p_m - p(t)_p]^2 + \beth_2[p_m - p(t)_p]^3\}\, dt \\
&\quad + 3\int_0^{t'} \{\beth_3[p_m - p(t)_p] \\
&\quad + \beth_5[p_m - p(t)_p]^2 + \beth_6[p_m - p(t)_p]^3\}\varepsilon(t)_{\alpha\alpha}\, dt \\
&\quad + 2\mathfrak{w}_5 \sum_{q=1}^{\Omega} \sqrt{(6\mathrm{II}_{sq})} \int_0^{t''_q} [p_m - p(t)_p]e(t)_q\, dt \\
&\quad + \mathfrak{w}_2 \sum_{q=1}^{\Omega} \overline{\mathrm{II}}_{sq} \int_0^{t''_q} [p_m - p(t)_p]\, dt \\
&\quad + 2\mathfrak{w}_6 \sum_{q=1}^{\Omega} \overline{\mathrm{II}}_{sq} \int_0^{t''_q} [p_m - p(t)_p]\left(\frac{\varepsilon(t)_{q\beta\beta} + e(t)_q}{3}\right) dt
\end{aligned} \tag{13.9.4}$$

2. *Constant rate of loading*. The consolidation of the soil is similar to the previous case where a consolidation pressure $p_m = p_{11} = p_{22} = p_{33} = p_{\alpha\alpha}/3$ is applied. The pure deviatoric loading, however, is continuous and linear with time $s_{ij} = \tfrac{1}{2}p_{zz}\Xi_{ij} = \tfrac{1}{2}\mathcal{S}t\Xi_{ij}$. It is assumed that throughout the loadings vertical and volumetric strains are available. The specific stored and disbursed free energies according to Eqs. (13.8.1) and (13.8.2) can now be computed as follows

$$\Delta\rho\psi_{\mathfrak{si}} = \int_0^{t'} \{\beth_0[p_{\mathfrak{m}} - p(t)_{\mathfrak{p}}] + \beth_1[p_{\mathfrak{m}} - p(t)_{\mathfrak{p}}]^2 + \beth_2[p_{\mathfrak{m}} - p(t)_{\mathfrak{p}}]^3\}\,\mathrm{d}t$$

$$+ 3\int_0^{t'} \{\beth_3[p_{\mathfrak{m}} - p(t)_{\mathfrak{p}}] + \beth_5[p_{\mathfrak{m}} - p(t)_{\mathfrak{p}}]^2$$

$$+ \beth_6[p_{\mathfrak{m}} - p(t)_{\mathfrak{p}}]^3\}\varepsilon(t)_{\alpha\alpha}\,\mathrm{d}t + 3\mathfrak{w}_2(s)^2_{\alpha\alpha\mathrm{i}} \int_0^{t''} [p_{\mathfrak{m}} - p(t)_{\mathfrak{p}}]\,\mathrm{d}t$$

$$+ 12[\mathfrak{w}_5 s_{\alpha\beta\mathrm{i}} + \mathfrak{w}_6(s^2)_{\alpha\beta\mathrm{i}}] \int_0^{t''} [p_{\mathfrak{m}} - p(t)_{\mathfrak{p}}]e_{\alpha\beta\mathrm{i}}\,\mathrm{d}t$$

$$= \int_0^{t'} \{\beth_0[p_{\mathfrak{m}} - p(t)_{\mathfrak{p}}] + \beth_1[p_{\mathfrak{m}} - p(t)_{\mathfrak{p}}]^2 + \beth_2[p_{\mathfrak{m}} - p(t)_{\mathfrak{p}}]^3\}\,\mathrm{d}t$$

$$+ 3\int_0^{t'} \{\beth_3[p_{\mathfrak{m}} - p(t)_{\mathfrak{p}}] + \beth_5[p_{\mathfrak{m}} - p(t)_{\mathfrak{p}}]^2$$

$$+ \beth_6[p_{\mathfrak{m}} - p(t)_{\mathfrak{p}}]^3\}\varepsilon(t)_{\alpha\alpha}\,\mathrm{d}t$$

$$+ \tfrac{3}{4}\mathfrak{w}_2\mathcal{S}^2 \int_0^{t''} [p_{\mathfrak{m}} - p(t)_{\mathfrak{p}}]t\,\mathrm{d}t + 6\mathfrak{w}_5\mathcal{S} \int_0^{t''} [p_{\mathfrak{m}} - p(t)_{\mathfrak{p}}]te(t)\,\mathrm{d}t$$

$$+ 3\mathfrak{w}_6\mathcal{S}^2 \int_0^{t''} [p_{\mathfrak{m}} - p(t)_{\mathfrak{p}}]\,\frac{\varepsilon(t)_{\beta\beta} + e(t)}{3}\,t^2\,\mathrm{d}t$$

$$= \int_0^{t'} \{\beth_0[p_{\mathfrak{m}} - p(t)_{\mathfrak{p}}] + \beth_1[p_{\mathfrak{m}} - p(t)_{\mathfrak{p}}]^2 + \beth_2[p_{\mathfrak{m}} - p(t)_{\mathfrak{p}}]^3\}\,\mathrm{d}t$$

$$+ 3\int_0^{t'} \{\beth_3[p_{\mathfrak{m}} - p(t)_{\mathfrak{p}}] + \beth_5[p_{\mathfrak{m}} - p(t)_{\mathfrak{p}}]^2$$

$$+ \beth_6[p_{\mathfrak{m}} - p(t)_{\mathfrak{p}}]^3\}\varepsilon(t)_{\alpha\alpha}\,\mathrm{d}t + \tfrac{4}{3}\mathfrak{w}_2 s^2_{\alpha\beta} \int_0^{t''} [p_{\mathfrak{m}} - p(t)_{\mathfrak{p}}]t\,\mathrm{d}t$$

$$+ 2\mathfrak{w}_5\,\frac{1}{t''}\,\surd(12 s^2_{\alpha\beta}) \int_0^{t''} [p_{\mathfrak{m}} - p(t)_{\mathfrak{p}}]te(t)\,\mathrm{d}t$$

$$+ 2\mathfrak{w}_6\,\frac{s^2_{\alpha\beta}}{t''^2} \int_0^{t''} [p_{\mathfrak{m}} - p(t)_{\mathfrak{p}}]\,\frac{\varepsilon(t)_{\beta\beta} + e(t)}{3}\,\mathrm{d}t$$

$$= \int_0^{t'} \{\beth_0[p_{\mathfrak{m}} - p(t)_{\mathfrak{p}}] + \beth_1[p_{\mathfrak{m}} - p(t)_{\mathfrak{p}}]^2 + \beth_2[p_{\mathfrak{m}} - p(t)_{\mathfrak{p}}]^3\}\,\mathrm{d}t$$

$$+ 3\int_0^{t'} \{\beth_3[p_{\mathfrak{m}} - p(t)_{\mathfrak{p}}] + \beth_5[p_{\mathfrak{m}} - p(t)_{\mathfrak{p}}]^2$$

$$+ \beth_6[p_{\mathfrak{m}} - p(t)_{\mathfrak{p}}]^3\}\varepsilon(t)_{\alpha\alpha}\,\mathrm{d}t + \tfrac{4}{3}\mathfrak{w}_2\overline{\mathrm{II}}_s \int_0^{t''} [p_{\mathfrak{m}} - p(t)_{\mathfrak{p}}]t\,\mathrm{d}t$$

$$+ 2\mathfrak{w}_5 \frac{1}{t''} \sqrt{(12\overline{\mathrm{II}}_s)} \int_0^{t''} [p_{\mathfrak{m}} - p(t)_{\mathfrak{p}}]te(t)\,\mathrm{d}t$$

$$+ 2\mathfrak{w}_6 \frac{\overline{\mathrm{II}}_s}{t''^2} \int_0^{t''} [p_{\mathfrak{m}} - p(t)_{\mathfrak{p}}] \frac{\varepsilon(t)_{\beta\beta} + e(t)}{3}\,\mathrm{d}t \qquad (13.9.5)$$

$$\rho\psi_{\mathfrak{sd}} = 6\int_0^{t''} s_{\alpha\beta}\,[\mathfrak{w}_1 s_{\alpha\beta} + \mathfrak{w}_2(s)^2_{\alpha\beta\mathfrak{d}} + \mathfrak{w}_3 e_{\alpha\beta} + \mathfrak{w}_5(se + es)_{\alpha\beta\mathfrak{d}}$$

$$+ \mathfrak{w}_6(s^2e + es^2)_{\alpha\beta\mathfrak{d}}]\,\mathrm{d}t$$

$$= [\tfrac{1}{2}\mathfrak{w}_1\mathcal{S}^2t''^2 + \tfrac{3}{16}\mathfrak{w}_2\mathcal{S}^3t''^4] + \mathfrak{w}_3\mathcal{S}\int_0^{t''} te(t)\,\mathrm{d}t$$

$$+ \tfrac{3}{4}\mathfrak{w}_5\mathcal{S}^2 \int_0^{t''} \frac{\varepsilon(t)_{\beta\beta} + e(t)}{3}\,t^2\,\mathrm{d}t + \tfrac{1}{16}\mathfrak{w}_6\mathcal{S}^3 \int_0^{t''} \frac{\varepsilon(t)_{\beta\beta} + 2e(t)}{3}\,t^4\,\mathrm{d}t$$

$$= [\tfrac{2}{3}\mathfrak{w}_1 s^2_{\alpha\beta} + \tfrac{1}{4}\mathfrak{w}_2\sqrt{(\tfrac{4}{3}s^2_{\alpha\beta})}s^2_{\alpha\beta}t''] + \mathfrak{w}_3 \frac{1}{t''}\sqrt{(\tfrac{4}{3}s^2_{\alpha\beta})}\int_0^{t''} te(t)\,\mathrm{d}t$$

$$+ \mathfrak{w}_5 \frac{s^2_{\alpha\beta}}{t''^2} \int_0^{t''} \frac{\varepsilon(t)_{\beta\beta} + e(t)}{3}\,t^2\,\mathrm{d}t$$

$$+ \tfrac{1}{12}\mathfrak{w}_6\sqrt{(\tfrac{4}{3}s^2_{\alpha\beta})}\,\frac{s^2_{\alpha\beta}}{t''^3} \int_0^{t''} \frac{\varepsilon(t)_{\beta\beta} + 2e(t)}{3}\,t^4\,\mathrm{d}t$$

$$= [\tfrac{2}{3}\mathfrak{w}_1\overline{\mathrm{II}}_s + \tfrac{1}{12}\mathfrak{w}_2\sqrt{\tfrac{2}{9}\overline{\mathrm{III}}_s}\,t''] + \mathfrak{w}_3 \frac{1}{t''}\sqrt{(\tfrac{4}{3}\overline{\mathrm{II}}_s)}\int_0^{t''} te(t)\,\mathrm{d}t$$

$$+ \mathfrak{w}_5 \frac{\overline{\mathrm{II}}_s}{t''^2} \int_0^{t''} \frac{\varepsilon(t)_{\beta\beta} + e(t)}{3}\,t^2\,\mathrm{d}t$$

$$+ \frac{1}{12t''^3}\,\mathfrak{w}_6\sqrt{\tfrac{2}{9}\overline{\mathrm{III}}_s} \int_0^{t''} \frac{\varepsilon(t)_{\beta\beta} + 2e(t)}{3}\,t^4\,\mathrm{d}t \qquad (13.9.6)$$

Balancing the disbursed specific free energy with the stored specific free energy we obtain

$$
\begin{aligned}
&[\tfrac{2}{3}\mathfrak{w}_1\overline{\mathrm{II}}_s + \tfrac{1}{12}\mathfrak{w}_2\sqrt{\tfrac{2}{9}\overline{\mathrm{III}}_s}\,t''] + \mathfrak{w}_3\,\frac{1}{t''}\,\sqrt{(\tfrac{4}{3}\overline{\mathrm{II}}_s)}\int_0^{t''} te(t)\,\mathrm{d}t \\
&\quad + \mathfrak{w}_5\,\frac{\overline{\mathrm{II}}_s}{t''^2}\int_0^{t''}\frac{\varepsilon(t)_{\beta\beta} + e(t)}{3}\,t^2\,\mathrm{d}t \\
&\quad + \frac{1}{12t''^3}\,\mathfrak{w}_6\sqrt{\tfrac{2}{9}\overline{\mathrm{III}}_s}\int_0^{t''}\frac{\varepsilon(t)_{\beta\beta} + 2e(t)}{3}\,t^4\,\mathrm{d}t \\
&= \rho\psi_{s0} + \int_0^{t'}\{\beth_0[p_{\mathrm{m}} - p(t)_{\mathrm{p}}] + \beth_1[p_{\mathrm{m}} - p(t)_{\mathrm{p}}]^2 + \beth_2[p_{\mathrm{m}} - p(t)_{\mathrm{p}}]^3\}\,\mathrm{d}t \\
&\quad + 3\int_0^{t'}\{\beth_3[p_{\mathrm{m}} - p(t)_{\mathrm{p}}] + \beth_5[p_{\mathrm{m}} - p(t)_{\mathrm{p}}]^2 \\
&\quad + \beth_6[p_{\mathrm{m}} - p(t)_{\mathrm{p}}]^3\}\varepsilon(t)_{\alpha\alpha}\,\mathrm{d}t + \tfrac{4}{3}\mathfrak{w}_2\overline{\mathrm{II}}_s\int_0^{t''}[p_{\mathrm{m}} - p(t)_{\mathrm{p}}]t\,\mathrm{d}t \\
&\quad + 2\mathfrak{w}_5\,\frac{1}{t''}\,\sqrt{(12\overline{\mathrm{II}}_s)}\int_0^{t''}[p_{\mathrm{m}} - p(t)_{\mathrm{p}}]te(t)\,\mathrm{d}t \\
&\quad + 2\mathfrak{w}_6\,\frac{\overline{\mathrm{II}}_s}{t''^2}\int_0^{t''}[p_{\mathrm{m}} - p(t)_{\mathrm{p}}]\,\frac{\varepsilon(t)_{\beta\beta} + e(t)}{3}\,\mathrm{d}t
\end{aligned}
\tag{13.9.7}
$$

This equation constitutes a failure criterion.

13.10 Appraisal of the Presented Failure Criteria

Eqs. (13.7.4) and (13.7.7), which represent failure criteria for the step and the constant stress rate loadings, respectively, of a linear viscoelastic medium, are fairly complicated; it may be shown, however, that they lead to the Mohr–Coulomb theory. For the non-linear isotropic relations the failure criteria for step and constant stress rate loadings, Eqs. (13.9.4) and (13.9.7) respectively, are far more involved, and seeking an analytical linkage between them and the Mohr–Coulomb theory is of little consequence.

Even the linkage between Eqs. (13.7.4) and (13.7.7) and the Mohr–Coulomb theory is not as simple as generally assumed in soil mechanics. These equations conform with the requirements for failure criteria outlined in Sect.

13.2, that is, they are functions of the spherical pressure p_m, they determine a failure envelope in the octahedral plane and they are derived on the basis of internal energies, and they also depend on the deformations and strains, functions of the time t. Consequently, since the cohesion p_{m0}, the cohesive intercept c and the angle of internal friction ϕ depend on time dependent functions like $\mathcal{F}(t)$ and $\mathcal{G}(t)$, it is not certain whether they remain steady throughout the variations of p_m and result in a linear Mohr–Coulomb envelope.

There are indications in the literature of the possibility of curved failure envelopes, specifically with respect to rocks and stiff materials (Jáky 1948a,b; Newman and Newman 1969). These envelopes are shown here in the octahedral bisectrix plane, Fig. 13.10.1. Further investigation is required in order to prove that in the pressure ranges in which soil engineers are working the assumption of linear failure envelopes is sufficient.

In the equations of balance between the stored and disbursed free energies, (13.9.4)–(13.9.7), it can be observed that the failure envelope according to the first criterion outlined in Sect. 13.2 is controlled by the stored free energy, while the failure envelope according to the second criterion is controlled by the disbursed free energy. It can be further observed that the stored free energy is a function of the first invariant of the stress tensor p_{ij}, $\mathrm{I}_p = \bar{\mathrm{I}}_p = p_m$ or of its higher power up to the third power, and that the disbursed free energy is a function of the second moment invariant of the shear stress tensor s_{ij}, $\overline{\mathrm{II}}_s = s_{\alpha\beta}s_{\alpha\beta} = s^2_{\alpha\beta}$, if linear viscoelasticity is postulated. If isotropic non-linear stress–strain relations are postulated the disbursed free energy is a function of the third moment invariant of the shear stress tensor as well, $\overline{\mathrm{III}}_s = s_{\alpha\beta}s_{\beta\gamma}s_{\gamma\alpha} = \frac{1}{3}\sqrt{(\frac{1}{6}s^2_{\alpha\beta})}s^2_{\alpha\beta}$. This concurs with the studies of other investigators (Lade 1972; Matsuoka and Nakai 1974), among them those who for some time tried to advance the inclusion of the invariants,

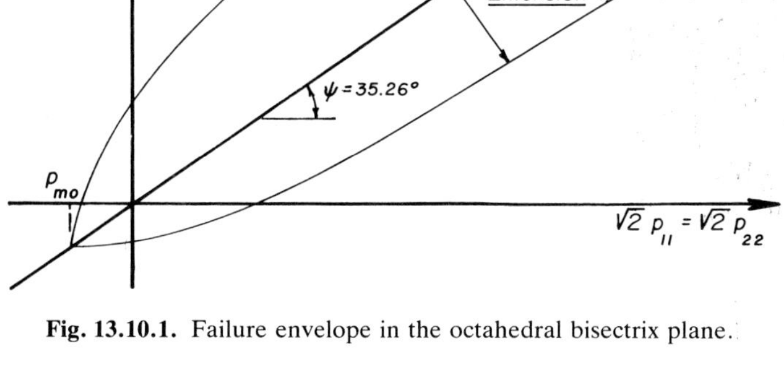

Fig. 13.10.1. Failure envelope in the octahedral bisectrix plane.

particularly the first invariant, in the failure theory (Hu 1960; Hu and Pae 1963; Satake 1981).

Studies of energy based on work input have been occasionally presented by investigators both in mechanics and in soils (Roscoe et al. 1963; Rowe et al. 1964; Green and Rivlin 1964; Olszak and Perzyna 1964; Besseling 1966; Reiner 1966; Partom 1966; Palmer 1967; Sternstein and Lederle 1968; Keinonen 1973; Morgan and Gerrard 1973; Houlsby 1979). However, the adverse effects of the stored free energy and the disbursed free energy were not fully appreciated. While these energies can be either kinetic or potential and act on the same material element, their effects are different. One affects the density of the material, the other indicates its strength. In the studies mentioned the stored free energy and the disbursed free energy are usually added in order to account for the work performed on the element. However, as was shown in the proposed failure criterion, Eq. (13.3.1), at failure the disbursed free energy balances the stored free energy, which is the sum of the residing stored free energy and the excess stored free energy. The excess stored free energy can be either positive, an increase of the stored internal energy, or negative, a decrease of the stored internal energy. A positive increment is achieved, according to Eq. (11.25.3), if the effective spherical stress and the spherical strain have the same sign, and the increment is negative if they are of opposite signs.

13.11 The Study of Failure through the Pure Deviatoric Test

In the pure deviatoric test, as in the conventional triaxial shear test discussed in Sect, 12.6, the vertical deformations and the volume changes of the sample are measured. The vertical strains $\varepsilon_{zz} = \varepsilon_{33}$ are evaluated from the vertical deformations and the volumetric strains $\varepsilon_{\alpha\alpha}$ and $\varepsilon_{\beta\beta}$ from the volume changes, $\varepsilon_{\alpha\alpha}$ measured during the consolidation and $\varepsilon_{\beta\beta}$ during the shearing of the sample. The measurement of the pore pressure $p_{\mathfrak{p}}$ is required as well, either at one end of the sample but preferably at mid-height. Unless the rate of the applied shear stress is slow enough to let pore pressures dissipate instantaneously without allowing their build-up and instantaneous consolidation of the sample, the measurement of pore pressures is essential. Additional measurements, such as the diameter of the sample measured at one or several elevations, the pore pressures at several elevations, the volume of the pore-water drained from the sample and the volume of the pore-air drained from the sample, are always useful.

Based on experimental data, the excess stored energy and the disbursed energy are

$$\Delta\rho\psi_{\mathfrak{si}} = \int_0^t \rho\dot{\psi}_{\mathfrak{si}}\,\mathrm{d}t = \int_{\psi_1}^{\psi_2} \rho\,\mathrm{d}\psi_{\mathfrak{si}} = \tfrac{1}{3}\int_0^{t'} p_{\mathfrak{s}\alpha\alpha}\dot{\varepsilon}_{\alpha\alpha}\,\mathrm{d}t + \tfrac{1}{3}\int_0^{t''} p_{\mathfrak{s}\alpha\alpha}\dot{\varepsilon}_{\beta\beta}\,\mathrm{d}t$$

$$= \int_0^{t'} (p_{\mathrm{m}} - p'_{\mathrm{w}} - p'_{\mathrm{g}})\dot{\varepsilon}_{\alpha\alpha}\,\mathrm{d}t + \int_0^{t''} (p_{\mathrm{m}} - p''_{\mathrm{w}} - p''_{\mathrm{g}})\dot{\varepsilon}_{\beta\beta}\,\mathrm{d}t$$

$$= \int_0^{t'} (p_{\mathrm{m}} - p'_{\mathrm{p}})\dot{\varepsilon}_{\alpha\alpha}\,\mathrm{d}t + \int_0^{t''} (p_{\mathrm{m}} - p''_{\mathrm{p}})\dot{\varepsilon}_{\beta\beta}\,\mathrm{d}t$$

$$= p_{\mathrm{m}}\varepsilon_{\alpha\alpha}\Big|_0^{t'} + p_{\mathrm{m}}\varepsilon_{\beta\beta}\Big|_0^{t''} - \int_0^{t'} p'_{\mathrm{p}}\dot{\varepsilon}_{\alpha\alpha}\,\mathrm{d}t - \int_0^{t''} p''_{\mathrm{p}}\dot{\varepsilon}_{\beta\beta}\,\mathrm{d}t \qquad (13.11.1)$$

where p_{m} is the applied spherical pressure, $\varepsilon_{\alpha\alpha}$ is the volumetric strain measured during the spherical pressure, $\varepsilon_{\beta\beta}$ is the volumetric strain measured during the deviatoric shear, p'_{w} is the pore-water pressure and p'_{g} the pore-air pressure measured during the spherical pressure, and p''_{w} is the pore-water and p''_{g} the pore-air pressure measured during the deviatoric shear. The sums of the pore-water and pore-air pressures are $p'_{\mathrm{p}} = p'_{\mathrm{w}} + p'_{\mathrm{g}}$ and $p''_{\mathrm{p}} = p''_{\mathrm{w}} + p''_{\mathrm{g}}$, respectively. Note that at time $t' = 0$ the volumetric strains vanish, $\varepsilon(0)_{\alpha\alpha} = 0$ and $\varepsilon(0)_{\beta\beta} = 0$, and that t' is the time at the end of the spherical consolidation and t'' the time at the end of the deviatoric shear, t' and t'' not necessarily equal.

Whether the spherical effective pressure of the consolidation process in the first part of the test, $p_{\mathrm{e}} = p_{\mathrm{m}} - p'_{\mathrm{p}}$, is a compression or a distention, its multiplication by the respective decrement or increment of volume strain $\varepsilon_{\alpha\alpha}$ results always in a positive excess of stored free energy, since the sign of the volume strain $\varepsilon_{\alpha\alpha}$ is always identical to that of the spherical effective pressure $p_{\mathrm{e}} = p_{\mathrm{m}} - p_{\mathrm{p}}$. The change in the stored free energy in the shearing process due to the product of the spherical pressure p_{e} and the volume strain $\varepsilon_{\beta\beta}$ is, however, not always an increment, since the volume strain $\varepsilon_{\beta\beta}$ could be either positive or negative depending on whether the density of the soil is below or above the critical void ratio. The sign of $\varepsilon_{\beta\beta}$ in this case does not necessarily coincide with the sign of the effective spherical pressure.

If the test is a conventional shear test, undrained or drained, the total spherical pressure p_{m} is not maintained constant throughout the shearing part of the test but increases by a spherical pressure increment $\Delta p_{\mathrm{m}} = \frac{1}{3}[p(t)_{33} - p_{\mathrm{m}}]$. Since this pressure increment is a function of time, Eq. (13.11.1) will also be a function of time. Thus, for a conventional shear test another term will have to be added to Eq. (13.11.1)

$$\Delta\rho\psi_{\mathrm{si}} = \int_0^{t} \rho\dot{\psi}_{\mathrm{si}}\,\mathrm{d}t = \int_{\psi_1}^{\psi_2} \rho\,\mathrm{d}\psi_{\mathrm{si}}$$

$$= \int_0^{t'} (p_{\mathrm{m}} - p'_{\mathrm{p}})\dot{\varepsilon}_{\alpha\alpha}\,\mathrm{d}t + \int_0^{t''} (p_{\mathrm{m}} - p''_{\mathrm{p}})\dot{\varepsilon}_{\beta\beta}\,\mathrm{d}t + \int_0^{t''} \Delta p_{\mathrm{m}}\,\mathrm{d}\varepsilon_{\beta\beta}$$

$$= p_{\mathrm{m}}\varepsilon_{\alpha\alpha}\Big|_0^{t'} + p_{\mathrm{m}}\varepsilon_{\beta\beta}\Big|_0^{t''} - \int_0^{t'} p'_{\mathrm{p}}\dot{\varepsilon}_{\alpha\alpha}\,\mathrm{d}t - \int_0^{t''} p''_{\mathrm{p}}\dot{\varepsilon}_{\beta\beta}\,\mathrm{d}t + \int_0^{t''} \Delta p_{\mathrm{m}}\,\mathrm{d}\varepsilon_{\beta\beta}$$

$$= p_{\mathrm{m}}\varepsilon_{\alpha\alpha}\Big|_0^{t'} + p_{\mathrm{m}}\varepsilon_{\beta\beta}\Big|_0^{t''} - \int_0^{t''} p'_{\mathrm{p}}\dot{\varepsilon}_{\alpha\alpha}\,\mathrm{d}t - \int_0^{t''} p''_{\mathrm{p}}\dot{\varepsilon}_{\beta\beta}\,\mathrm{d}t$$

$$+ \tfrac{1}{3}\int_0^{t''} [p(t)_{33} - p_{\mathrm{m}}]\,\mathrm{d}\varepsilon_{\beta\beta} \qquad (13.11.2)$$

The values of $p''_{\mathfrak{w}}$ and $p''_{\mathfrak{g}}$ versus $\varepsilon_{\beta\beta}$ in the conventional shear test are different from the values obtained in the pure deviatoric shear test. The excess stored specific free energy of the conventional triaxial test can be evaluated according to Eq. (13.11.2).

We shall concentrate on the pure deviatoric shear test and try to evaluate the excess stored free energy and the disbursed free energy from the results of the pure deviatoric test, presented in Fig. 13.11.1a,b and Fig. 13.11.2a,b, respectively. For the conventional shear test the corresponding terms are evaluated from results similar to those presented in Fig. 13.11.1a,b and from the data of the test itself. The first part, the spherical consolidation, is identical in both tests.

Excess stored and disbursed specific free energies can be evaluated from test results either graphically or by simple computational methods. For the excess stored specific free energy we have Eq. (13.11.2) and for the disbursed specific free energy we have Eq. (11.25.4).

Of the many possible loading combinations, the four most frequent will be considered here.

1. *Stepwise spherical pressure with stepwise deviatoric loading*. If the spherical pressure is applied in $\mathfrak{R}$ steps $p_{\mathrm{m}\mathfrak{r}} = \mathcal{P}_{\mathfrak{r}} = \mathrm{const}$ we may say that the total spherical pressure $\mathcal{P}$ applied to the soil sample through the $\mathfrak{R}$ steps is equal to the sum of the $\mathfrak{R}$ step loads

$$\mathcal{P} = \sum_{\mathfrak{r}=1}^{\mathfrak{R}} \mathcal{P}_{\mathfrak{r}} \qquad (13.11.3)$$

where each $\mathcal{P}_{\mathfrak{r}}$ represents a pressure increment.

The volumetric strain increment of the $\mathfrak{r}$th step is $3\varepsilon_{\mathrm{m}\mathfrak{r}} = \varepsilon_{\alpha\alpha\mathfrak{r}}$ while the total volumetric strain of the $\mathfrak{R}$ steps is the sum of the $\mathfrak{R}$ volumetric strains

$$3\varepsilon'_{\mathrm{m}} = 3\sum_{\mathfrak{r}=1}^{\mathfrak{R}} \varepsilon_{\mathrm{m}\mathfrak{r}} = \sum_{\mathfrak{r}=1}^{\mathfrak{R}} \varepsilon_{\alpha\alpha\mathfrak{r}} \qquad (13.11.4)$$

During the spherical pressure application pore pressures $p(t)_{\mathrm{p}\mathfrak{r}}$ are measured for each step loading $\mathfrak{r}$. The pore pressures are functions of time since they dissipate during the course of each loading, particularly in a drained test. Even in an undrained test there is a time lag during which the pore pressures come to an equilibrium. Thus, the effective pressure at each loading becomes

$$p(t)_{\mathrm{e}\mathfrak{r}} = \tfrac{1}{3}p(t)_{\mathfrak{s}\alpha\alpha\mathfrak{r}} = p_{\mathrm{m}\mathfrak{r}} - p(t)_{\mathrm{p}\mathfrak{r}} = \mathcal{P}_{\mathfrak{r}} - p(t)_{\mathrm{p}\mathfrak{r}} \qquad (13.11.5)$$

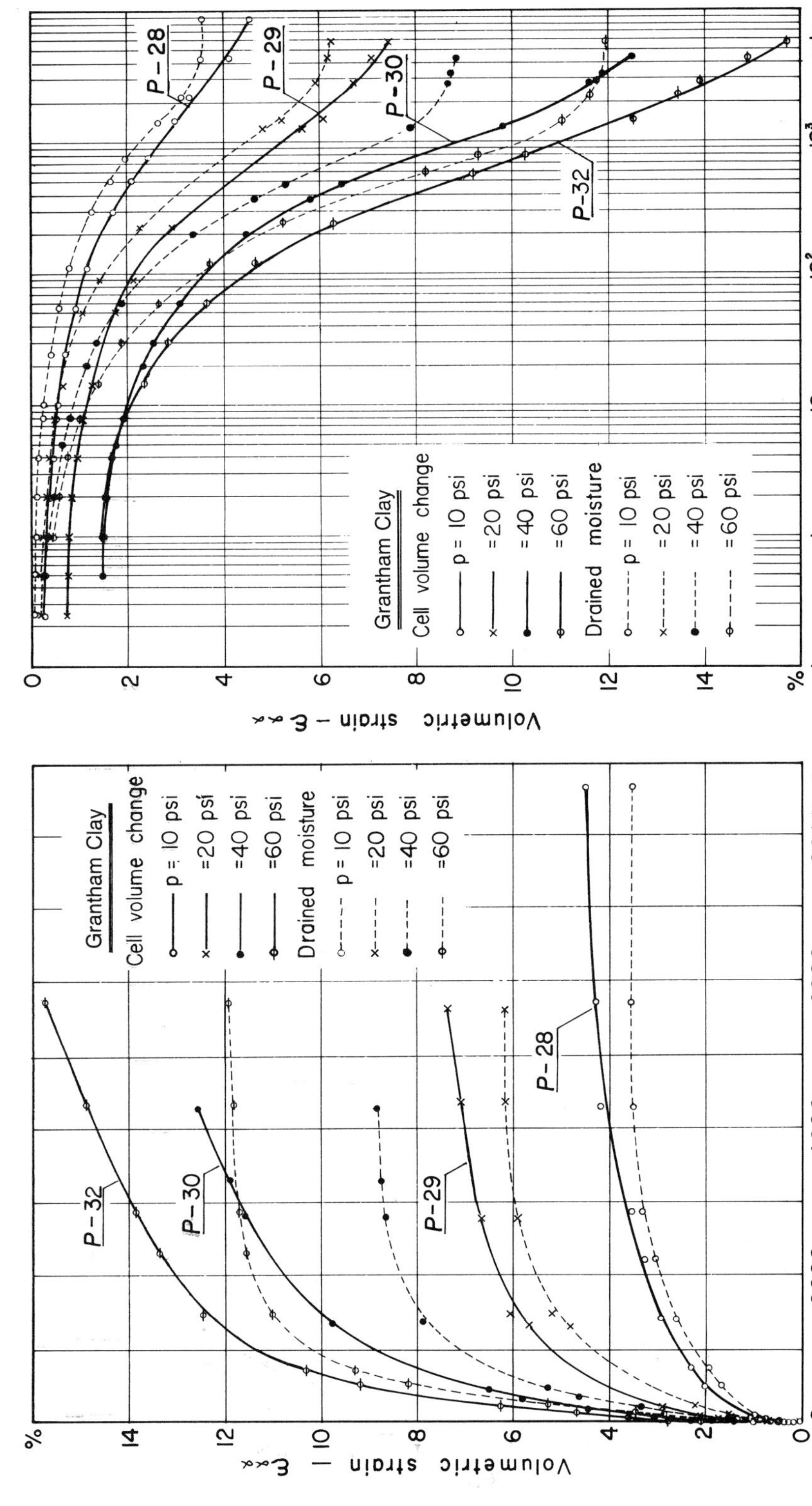

Fig. 13.11.1. Volumetric strain under spherical pressure versus time. **a** Volumetric strain $\varepsilon_{\alpha\alpha}$ versus time t. **b** Volumetric strain $\varepsilon_{\alpha\alpha}$ versus $\log t$.

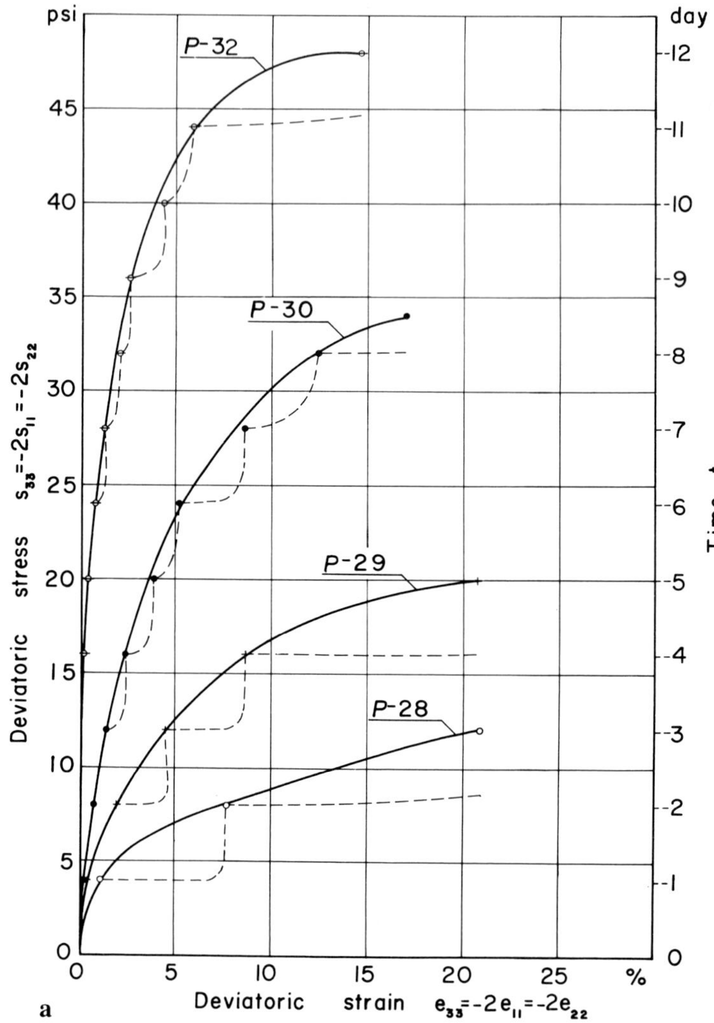

a

The pore pressures $p(t)_{\mathfrak{r}}$ in the above equation are not pressure increments, but actually measured pore pressures and they should not be added up.

The shear stress is also applied in a series of $\mathfrak{Q}$ steps $s_{ij\mathfrak{q}} = \frac{1}{2}\mathcal{S}_{33\mathfrak{q}}\Xi_{ij} = \text{const}$, where the $\mathfrak{q}$th rate of shear stress vanishes, $\dot{s}_{ij\mathfrak{q}} = 0$. Failure is reached at the $\mathfrak{Q}$th increment when the shear stress s_{ij} is the sum of the $\mathfrak{Q}$ increments

$$s_{ij} = \sum_{\mathfrak{q}=1}^{\mathfrak{Q}} s_{ij\mathfrak{q}} = \frac{1}{2}\sum_{\mathfrak{q}=1}^{\mathfrak{Q}} \mathcal{S}_{33\mathfrak{q}}\Xi_{ij} = \tfrac{1}{2}\mathcal{S}_{33}\Xi_{ij} \tag{13.11.6}$$

where $\mathcal{S}_{33}$ is the vertical stress applied to the soil sample when it reaches failure.

Each shear stress increment results in a strain increment $\varepsilon_{ij\mathfrak{q}}$ which can be

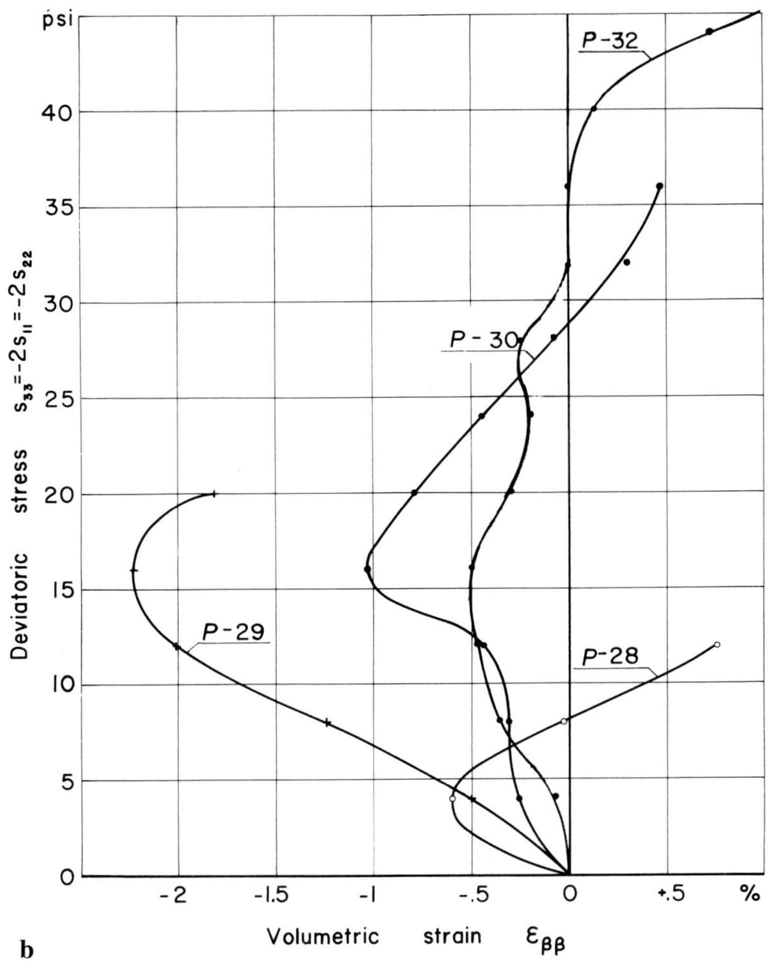

b

Fig. 13.11.2. Distortions and volumetric strains under pure deviatoric stresses. **a** Deviatoric stress s_{33} versus deviatoric strain $\varepsilon_{\beta\beta}$. **b** Deviatoric stress s_{33} versus volumetric strain e_{33}.

resolved into a volumetric strain increment $\varepsilon_{\beta\beta\mathfrak{q}}$ and a distortion increment $e_{ij\mathfrak{q}} = \frac{1}{2}\mathscr{E}_{33\mathfrak{q}}\Xi_{ij}$

$$\varepsilon_{ij\mathfrak{q}} = \tfrac{1}{3}\varepsilon_{\beta\beta\mathfrak{q}}\delta_{ij} + e_{ij\mathfrak{q}} = \tfrac{1}{3}\varepsilon_{\beta\beta\mathfrak{q}}\delta_{ij} + \tfrac{1}{2}\mathscr{E}_{33\mathfrak{q}}\Xi_{ij} \tag{13.11.7}$$

while the total strain ε_{ij} at failure is the sum of the $\mathfrak{Q}$ strain increments

$$\varepsilon_{ij} = \sum_{\mathfrak{q}=1}^{\mathfrak{Q}} \varepsilon_{ij\mathfrak{q}} = \frac{1}{3}\sum_{\mathfrak{q}=1}^{\mathfrak{Q}} \varepsilon_{\beta\beta\mathfrak{q}}\delta_{ij} + \frac{1}{2}\sum_{\mathfrak{q}=1}^{\mathfrak{Q}} \mathscr{E}_{33\mathfrak{q}}\Xi_{ij} = \tfrac{1}{3}\varepsilon_{\beta\beta}\Xi_{ij} + \tfrac{1}{2}\mathscr{E}_{33}\Xi_{ij} \tag{13.11.8}$$

Since volumetric strains are expected, pore pressures $p_{\mathfrak{pq}}$ can be measured as well for each shear stress increment, and the corresponding effective pressure can be evaluated

$$p_{\text{eq}} = \tfrac{1}{3}p(t)_{\mathfrak{s}\beta\beta\mathfrak{q}} = p_{\text{mq}} - p(t)_{\text{pq}} = \mathcal{P} - p(t)_{\text{pq}} \tag{13.11.9}$$

As in the case of spherical pressures, the pore pressures are not to be added up, since they are not pressure increments but actually measured pressures.

The partial excess stored specific free energy of the $\mathfrak{r}$th spherical loading $\Delta\rho\psi'_{\mathfrak{sir}}$ can be evaluated when the $\mathfrak{r}$th step load of Eq. (13.11.5) and the $\mathfrak{r}$th volumetric strain are substituted in Eq. (11.25.5)

$$\Delta\rho\psi'_{\mathfrak{sir}} = \int_0^{\varepsilon_{\alpha\alpha}} p_{\text{er}}\, d\varepsilon_{\alpha\alpha\mathfrak{r}} = \int_0^{\varepsilon_{\alpha\alpha}} (\mathcal{P}_{\mathfrak{r}} - p(t)_{\text{pr}})\, d\varepsilon_{\alpha\alpha\mathfrak{r}} = \mathcal{P}_{\mathfrak{r}}\varepsilon_{\alpha\alpha\mathfrak{r}} - p(t')_{\text{pr}}\varepsilon_{\alpha\alpha\mathfrak{r}} \tag{13.11.10}$$

When summed over $\mathfrak{r}$, Eq. (13.11.10) yields the excess stored specific free energy resulting from the spherical step loading

$$\Delta\rho\psi'_{\mathfrak{si}} = \sum_{\mathfrak{r}=1}^{\mathfrak{R}} \Delta\rho\psi'_{\mathfrak{sir}} = \sum_{\mathfrak{r}=1}^{\mathfrak{R}} \mathcal{P}_{\mathfrak{r}}\varepsilon_{\alpha\alpha\mathfrak{r}} - \sum_{\mathfrak{r}=1}^{\mathfrak{R}} p(t')_{\text{pr}}\varepsilon_{\alpha\alpha\mathfrak{r}} \tag{13.11.11}$$

The partial excess stored specific free energy resulting from the deviatoric shear stress loading can be computed from Eq. (13.11.7)

$$\Delta\rho\psi'_{\mathfrak{siq}} = \int_0^{\varepsilon_{\beta\beta}} (\mathcal{P} - p(t)_{\text{pq}})\, d\varepsilon_{\beta\beta\mathfrak{q}} = \mathcal{P}\varepsilon_{\beta\beta\mathfrak{q}} - p(t'')_{\text{pq}}\varepsilon_{\beta\beta\mathfrak{q}} \tag{13.11.12}$$

where $\mathcal{P}$ is the maximum spherical pressure which is applied at the $\mathfrak{R}$th step load and acts throughout the entire deviatoric loading, and $p(t'')_{\text{pq}}$ is the pore pressure measured during the $\mathfrak{q}$th deviatoric step load. The sum of the partial excess stored specific free energies resulting from the deviatoric stresses is

$$\Delta\rho\psi''_{\mathfrak{si}} = \sum_{\mathfrak{q}=1}^{\mathfrak{Q}} \Delta\rho\psi''_{\mathfrak{iq}} = \sum_{\mathfrak{q}=1}^{\mathfrak{Q}} \mathcal{P}\varepsilon_{\beta\beta\mathfrak{q}} - \sum_{\mathfrak{q}=1}^{\mathfrak{Q}} p(t'')_{\text{pq}}\varepsilon_{\beta\beta\mathfrak{q}} = \mathcal{P}\varepsilon_{\beta\beta} - \sum_{\mathfrak{q}=1}^{\mathfrak{Q}} p(t'')_{\text{pq}}\varepsilon_{\beta\beta\mathfrak{q}} \tag{13.11.13}$$

The partial disbursed specific free energy can be calculated according to Eq. (11.25.3), when Eq. (13.11.6) and the distortion from Eq. (13.11.7) are substituted

$$\rho\psi''_{\mathfrak{sdq}} = \int_0^{e_{ij}} s_{\alpha\beta\mathfrak{q}}\, de_{\alpha\beta\mathfrak{q}} = s_{\alpha\beta\mathfrak{q}}e_{\alpha\beta\mathfrak{q}} = \tfrac{3}{2}\mathcal{S}_{33\mathfrak{q}}\mathcal{E}_{33\mathfrak{q}} \tag{13.11.14}$$

Eq. (13.11.14) may be summed over $\mathfrak{q}$ and we get the total disbursed specific free energy

$$\rho\psi''_{\mathfrak{sd}} = \sum_{\mathfrak{q}=1}^{\mathfrak{Q}} \rho\psi''_{\mathfrak{sdq}} = \sum_{\mathfrak{q}=1}^{\mathfrak{Q}} s_{\alpha\beta\mathfrak{q}}e_{\alpha\beta\mathfrak{q}} = \frac{3}{2}\sum_{\mathfrak{q}=1}^{\mathfrak{Q}} \mathcal{S}_{33\mathfrak{q}}\mathcal{E}_{33\mathfrak{q}} \tag{13.11.15}$$

We can now write the failure condition

$$\sqrt{(\rho\psi''_{\mathfrak{s}\mathfrak{d}})} = \sqrt{(\rho\psi_{\mathfrak{s}i0} + \Delta\rho\psi'_{\mathfrak{s}i} + \Delta\rho\psi''_{\mathfrak{s}i})} = \sqrt{\left(\frac{3}{2}\sum_{\mathfrak{q}=1}^{\mathfrak{Q}} \mathcal{S}_{33\mathfrak{q}}\mathcal{E}_{33\mathfrak{q}}\right)}$$

$$= \sqrt{\left(\rho\psi_{\mathfrak{s}i0} + \sum_{\mathfrak{r}=1}^{\mathfrak{R}} \mathcal{P}_{\mathfrak{r}}\varepsilon_{\alpha\alpha\mathfrak{r}} - \sum_{\mathfrak{r}=1}^{\mathfrak{R}} p(t')_{\mathfrak{p}\mathfrak{r}}\varepsilon_{\alpha\alpha\mathfrak{r}} + \mathcal{P}\varepsilon_{\beta\beta} - \sum_{\mathfrak{q}=1}^{\mathfrak{Q}} p(t'')_{\mathfrak{p}\mathfrak{q}}\varepsilon_{\beta\beta\mathfrak{q}}\right)}$$

$$= \sqrt{\left(\rho\psi_{\mathfrak{s}i0} + \sum_{\mathfrak{r}=1}^{\mathfrak{R}} p_{\mathfrak{m}\mathfrak{r}}\varepsilon_{\alpha\alpha\mathfrak{r}} - \sum_{\mathfrak{r}=1}^{\mathfrak{R}} p(t')_{\mathfrak{p}\mathfrak{r}}\varepsilon_{\alpha\alpha\mathfrak{r}} + p_{\mathfrak{m}}\varepsilon_{\beta\beta} - \sum_{\mathfrak{q}=1}^{\mathfrak{Q}} p(t'')_{\mathfrak{p}\mathfrak{q}}\varepsilon_{\beta\beta\mathfrak{q}}\right)} \tag{13.11.16}$$

$e(t'')_{ij\mathfrak{q}}$, $\varepsilon(t')_{\alpha\alpha\mathfrak{r}}$, $\varepsilon(t'')_{\beta\beta}$, $\varepsilon(t'')_{\beta\beta\mathfrak{q}}$, $\mathcal{E}(t'')_{33}$ and $\mathcal{E}(t'')_{33\mathfrak{q}}$ are all functions of the respective times t' and t'', although it is not specifically so stated. Their values as well as the values of $\mathcal{P}_{\mathfrak{r}}$, $p(t')_{\mathfrak{p}\mathfrak{r}}$ and $p(t'')_{\mathfrak{p}\mathfrak{q}}$ which appear in the equations, are the final values at the end of the respective processes.

Considering Eq. (13.11.6), the octahedral shear stress denoted τ, which is proportional to the square root of the second moment invariant, is also equal to

$$\sqrt{3}\tau = \sqrt{\mathrm{II}_s} = \sqrt{(s_{\alpha\beta}s_{\alpha\beta})} = \sqrt{\left(\sum_{\mathfrak{q}=1}^{\mathfrak{Q}} s_{\alpha\beta\mathfrak{q}}s_{\alpha\beta\mathfrak{q}}\right)} = \sqrt{\tfrac{3}{2}}\mathcal{S}_{33} \tag{13.11.17}$$

When the square root of the disbursed specific free energy, Eq. (13.11.16), is compared to the octahedral shear stress, Eq. (13.11.17) by calculating their ratio, we get

$$\frac{\sqrt{(\rho\psi_{\mathfrak{s}\mathfrak{d}})}}{\tau} = \frac{\sqrt{(\frac{3}{2}\Sigma_{\mathfrak{q}=1}^{\mathfrak{Q}} \mathcal{S}_{33\mathfrak{q}}\mathcal{E}_{33\mathfrak{q}})}}{\sqrt{\frac{1}{2}}\mathcal{S}_{33}} = \frac{\sqrt{(3\Sigma_{\mathfrak{q}=1}^{\mathfrak{Q}} \mathcal{S}_{33\mathfrak{q}}\mathcal{E}_{33\mathfrak{q}})}}{\mathcal{S}_{33}} \tag{13.11.18}$$

If Eq. (13.11.16) is normalized we obtain

$$\sqrt{3}\tau = \sqrt{(\mathrm{II}_s)} = \sqrt{\tfrac{3}{2}}\mathcal{S}_{33}$$

$$= \mathcal{S}_{33}\frac{\sqrt{[\rho\psi_{\mathfrak{s}i0} + \Sigma_{\mathfrak{r}=1}^{\mathfrak{R}} p_{\mathfrak{m}\mathfrak{r}}\varepsilon_{\alpha\alpha\mathfrak{r}} - \Sigma_{\mathfrak{r}=1}^{\mathfrak{R}} p(t')_{\mathfrak{p}\mathfrak{r}}\varepsilon_{\alpha\alpha\mathfrak{r}} + p_{\mathfrak{m}}\varepsilon_{\beta\beta} - \Sigma_{\mathfrak{q}=1}^{\mathfrak{Q}} p(t'')_{\mathfrak{p}\mathfrak{q}}\varepsilon_{\beta\beta\mathfrak{q}}]}}{\sqrt{(\Sigma_{\mathfrak{q}=1}^{\mathfrak{Q}} \mathcal{S}_{33\mathfrak{q}}\mathcal{E}_{33\mathfrak{q}})}} \tag{13.11.19}$$

Note that the quotient of the square roots on the right-hand side of the equation is equal to $\sqrt{\frac{3}{2}}$

$$\frac{\sqrt{[\rho\psi_{\mathfrak{s}i0} + \Sigma_{\mathfrak{r}=1}^{\mathfrak{R}} p_{\mathfrak{m}\mathfrak{r}}\varepsilon_{\alpha\alpha\mathfrak{r}} - \Sigma_{\mathfrak{r}=1}^{\mathfrak{R}} p(t')_{\mathfrak{p}\mathfrak{r}}\varepsilon_{\alpha\alpha\mathfrak{r}} + p_{\mathfrak{m}}\varepsilon_{\beta\beta} - \Sigma_{\mathfrak{q}=1}^{\mathfrak{Q}} p(t'')_{\mathfrak{p}\mathfrak{q}}\varepsilon_{\beta\beta\mathfrak{q}}]}}{\sqrt{(\Sigma_{\mathfrak{q}=1}^{\mathfrak{Q}} \mathcal{S}_{33\mathfrak{q}}\mathcal{E}_{33\mathfrak{q}})}} = \sqrt{\tfrac{3}{2}} \tag{13.11.20}$$

2. *Stepwise spherical pressure with constant rate of deviatoric loading.* The first part of the test, the application of the spherical pressure, is similar to the previous case and consequently Eqs. (13.11.3)–(13.11.5), (13.11.10) and (13.11.11), relevant to this part, remain the same.

The deviatoric shear stress is applied at a constant rate, $s(t)_{ij} = \frac{1}{2}\mathcal{S}_{33}t\Xi_{ij}$ and $\dot{s}(t)_{ij} = \frac{1}{2}\mathcal{S}_{33}\Xi_{ij} = 0$, where $\mathcal{S}_{33} = \text{const}$ and $\dot{\mathcal{S}}_{33} = 0$. The resulting strain $\varepsilon(t)_{ij}$ can be resolved into a volumetric strain $\varepsilon_{\beta\beta}$ and a distortion e_{ij}

$$\varepsilon_{ij} = \tfrac{1}{3}\varepsilon_{\beta\beta}\Xi_{ij} + e_{ij} = \tfrac{1}{3}\varepsilon_{\beta\beta}\Xi_{ij} + \tfrac{1}{2}\mathcal{E}(t)_{33}\Xi_{ij} \tag{13.11.21}$$

where $e_{ij} = \frac{1}{2}\mathcal{E}(t)_{33}\Xi_{ij}$ and $\mathcal{E}(t)_{33}$ is the vertical strain measured during the deviatoric loading.

The rate of strain $\dot{\varepsilon}_{ij}$ can then be computed

$$\dot{\varepsilon}_{ij} = \tfrac{1}{3}\dot{\varepsilon}_{\beta\beta}\Xi_{ij} + \tfrac{1}{2}\dot{\mathcal{E}}(t)_{33}\Xi_{ij} \tag{13.11.22}$$

If volumetric strains can be expected, then pore pressures $p_{\mathfrak{p}}$ can also be expected and measured and the corresponding effective pressure is

$$p_{\mathfrak{e}} = \tfrac{1}{3}p(t)_{\mathfrak{s}\beta\beta} = p_{\mathrm{m}} - p(t)_{\mathfrak{p}} = \mathcal{P} - p(t)_{\mathfrak{p}} \tag{13.11.23}$$

from which the excess stored specific free energy resulting from shear stresses can be written

$$\Delta\rho\psi''_{\mathfrak{si}} = \int_0^{t''} (\mathcal{P} - p(t)_{\mathfrak{p}})\dot{\varepsilon}_{\beta\beta}\,\mathrm{d}t = \mathcal{P}\varepsilon_{\beta\beta} - \int_0^{t''} p(t)_{\mathfrak{p}}\dot{\varepsilon}_{\beta\beta}\,\mathrm{d}t \tag{13.11.24}$$

where $\mathcal{P}$ is the maximum spherical pressure that is applied at the $\mathfrak{N}$th step load and acts throughout the deviatoric loading, as was shown also in the previous case, and $p(t)_{\mathfrak{p}}$ is the pore pressure measured during the deviatoric shear stresses.

The stored specific free energy is the sum of the partial stored specific free energies, Eqs. (13.11.11) and (13.11.24)

$$\Delta\rho\psi_{\mathfrak{si}} = \Delta\rho\psi'_{\mathfrak{si}} + \Delta\rho\psi''_{\mathfrak{si}} = \sum_{\mathfrak{r}=1}^{\mathfrak{N}} \mathcal{P}_{\mathfrak{r}}\varepsilon_{\alpha\alpha\mathfrak{r}} - \sum_{\mathfrak{r}=1}^{\mathfrak{N}} p(t')_{\mathfrak{pr}}\varepsilon_{\alpha\alpha\mathfrak{r}} + \mathcal{P}\varepsilon_{\beta\beta} - \int_0^{t''} p(t)_{\mathfrak{p}}\dot{\varepsilon}_{\beta\beta}\,\mathrm{d}t \tag{13.11.25}$$

The disbursed specific free energy can be calculated when the shear stress and the distortion rate from Eq. (13.11.22) are substituted in Eq. (11.25.3)

$$\rho\psi''_{\mathfrak{sd}} = \int_0^{t''} s_{\alpha\beta}\dot{e}_{\alpha\beta}\,\mathrm{d}t = \tfrac{3}{2}\mathcal{S}_{33}\int_0^{t''} \dot{\mathcal{E}}(t)_{33}\,\mathrm{d}t = \tfrac{3}{2}\mathcal{S}_{33}\mathcal{E}(t'')_{33} \tag{13.11.26}$$

The failure condition is determined according to Eqs. (13.11.25) and (13.11.26)

$$\sqrt{(\rho\psi''_{\mathfrak{s}\mathfrak{b}})} = \sqrt{(\rho\psi_{\mathfrak{s}\mathfrak{i}0} + \Delta\rho\psi'_{\mathfrak{s}\mathfrak{i}} + \Delta\rho\psi''_{\mathfrak{s}\mathfrak{i}})} = \sqrt{[\tfrac{3}{2}\mathcal{S}_{33}\mathcal{E}(t'')_{33}]}$$

$$= \sqrt{\left(\rho\psi_{\mathfrak{s}\mathfrak{i}0} + \sum_{\mathfrak{r}=1}^{\mathfrak{R}} \mathcal{P}_{\mathfrak{r}}\varepsilon_{\alpha\alpha\mathfrak{r}} - \sum_{\mathfrak{r}=1}^{\mathfrak{R}} p(t')_{\mathfrak{p}\mathfrak{r}}\varepsilon_{\alpha\alpha\mathfrak{r}} + \mathcal{P}\varepsilon_{\beta\beta} - \int_0^{t''} p(t)_{\mathfrak{p}}\dot{\varepsilon}_{\beta\beta}\,\mathrm{d}t\right)}$$

$$= \sqrt{\left(\rho\psi_{\mathfrak{s}\mathfrak{i}0} + \sum_{\mathfrak{r}=1}^{\mathfrak{R}} p_{\mathrm{m}\mathfrak{r}}\varepsilon_{\alpha\alpha\mathfrak{r}} - \sum_{\mathfrak{r}=1}^{\mathfrak{R}} p(t')_{\mathfrak{p}\mathfrak{r}}\varepsilon_{\alpha\alpha\mathfrak{r}} + p_{\mathrm{m}}\varepsilon_{\beta\beta} - \int_0^{t''} p(t)_{\mathfrak{p}}\dot{\varepsilon}_{\beta\beta}\,\mathrm{d}t\right)} \tag{13.11.27}$$

The octahedral shear stress in this case is

$$\sqrt{3}\tau = \sqrt{(\overline{\mathrm{II}}_s)} = \sqrt{\tfrac{3}{2}}\mathcal{S}_{33}$$

$$= \mathcal{S}_{33}\frac{\sqrt{[\rho\psi_{\mathfrak{s}\mathfrak{i}0} + \Sigma_{\mathfrak{r}=1}^{\mathfrak{R}} p_{\mathrm{m}\mathfrak{r}}\varepsilon_{\alpha\alpha\mathfrak{r}} - \Sigma_{\mathfrak{r}=1}^{\mathfrak{R}} p(t')_{\mathfrak{p}\mathfrak{r}}\varepsilon_{\alpha\alpha\mathfrak{r}} + p_{\mathrm{m}}\varepsilon_{\beta\beta} - \int_0^{t''} p(t)_{\mathfrak{p}\mathfrak{q}}\dot{\varepsilon}_{\beta\beta}\,\mathrm{d}t]}}{\sqrt{(\mathcal{S}_{33}\mathcal{E}_{33})}} \tag{13.11.28}$$

3. *Constant spherical pressure with stepwise deviatoric loading.* The spherical pressure is applied as a constant pressure in a single step $p_{\mathrm{m}} = \mathcal{P} = \text{const}$, and a volumetric strain $\varepsilon_{\alpha\alpha}$ as well as a pore pressure $p(t)_{\mathfrak{p}}$ are measured. The volumetric strain is a function of time and the rate of volumetric strain $\dot{\varepsilon}_{\alpha\alpha}$ can be computed. The excess stored specific free energy $\Delta\rho\psi'_{\mathfrak{s}\mathfrak{i}}$ resulting from the spherical pressure is then

$$\Delta\rho\psi'_{\mathfrak{s}\mathfrak{i}} = \int_0^{\varepsilon_{\alpha\alpha}} p(t)_{\mathfrak{e}}\,\mathrm{d}\varepsilon_{\alpha\alpha}$$

$$= \int_0^{\varepsilon_{\alpha\alpha}} [\mathcal{P} - p(t)_{\mathfrak{p}}]\,\mathrm{d}\varepsilon_{\alpha\alpha} = \mathcal{P}\varepsilon(t')_{\alpha\alpha} - \int_0^{t'} p(t)_{\mathfrak{p}}\dot{\varepsilon}(t)_{\alpha\alpha}\,\mathrm{d}t \tag{13.11.29}$$

The deviatoric shear stress is applied similarly to that in case 1 and therefore the respective equations (13.11.6)–(13.11.9) and (13.11.12)–(13.11.15) are valid.

The excess stored specific free energy is the sum of the stored specific free energies resulting from the spherical pressure and from the deviatoric shear stress, Eqs. (13.11.29) and (13.11.13)

$$\Delta\rho\psi_{\mathfrak{s}\mathfrak{i}} = \Delta\rho\psi'_{\mathfrak{s}\mathfrak{i}} + \Delta\rho\psi''_{\mathfrak{s}\mathfrak{i}} = \mathcal{P}\varepsilon(t')_{\alpha\alpha} - \int_0^{t'} p(t)_{\mathfrak{p}}\dot{\varepsilon}(t)_{\alpha\alpha}\,\mathrm{d}t + \mathcal{P}\varepsilon_{\beta\beta} - \sum_{\mathfrak{q}=1}^{\mathfrak{Q}} p(t'')_{\mathfrak{p}\mathfrak{q}}\varepsilon_{\beta\beta\mathfrak{q}} \tag{13.11.30}$$

The disbursed specific free energy is the same as in Eq. (13.11.15).

The failure condition can be formulated from Eqs. (13.11.15) and (13.11.30)

$$\sqrt{(\rho\psi''_{\mathfrak{sb}})} = \sqrt{(\rho\psi_{\mathfrak{si}0} + \Delta\rho\psi'_{\mathfrak{si}} + \Delta\rho\psi''_{\mathfrak{si}})} = \sqrt{\left(\frac{3}{2}\sum_{q=1}^{\Omega} \mathcal{S}_{33q}\mathcal{E}(t'')_{33q}\right)}$$

$$= \sqrt{\left(\rho\psi_{\mathfrak{si}0} + \mathcal{P}\varepsilon(t')_{\alpha\alpha} - \int_0^{t'} p(t)_{\mathrm{p}}\dot{\varepsilon}(t)_{\alpha\alpha}\,\mathrm{d}t\right.}$$

$$\left. + \mathcal{P}\varepsilon_{\beta\beta} - \sum_{q=1}^{\Omega} p(t'')_{\mathrm{pq}}\varepsilon_{\beta\beta q}\right)$$

$$= \sqrt{\left(\rho\psi_{\mathfrak{si}0} + p_{\mathrm{m}}\varepsilon(t')_{\alpha\alpha} - \int_0^{t'} p(t)_{\mathrm{p}}\dot{\varepsilon}(t)_{\alpha\alpha}\,\mathrm{d}t\right.}$$

$$\left. + p_{\mathrm{m}}\varepsilon_{\beta\beta} - \sum_{q=1}^{\Omega} p(t'')_{\mathrm{pq}}\varepsilon_{\beta\beta q}\right) \tag{13.11.31}$$

In this case the octahedral shear stress is given

$$\sqrt{3}\tau = \sqrt{(\mathrm{II}_s)} = \sqrt{\tfrac{3}{2}}\mathcal{S}_{33}$$

$$= \mathcal{S}_{33}\frac{\sqrt{[\rho\psi_{\mathfrak{si}0} + p_{\mathrm{m}}\varepsilon_{\alpha\alpha} - \int_0^{t''} p(t)_{\mathrm{p}}\dot{\varepsilon}_{\alpha\alpha}\,\mathrm{d}t + p_{\mathrm{m}}\varepsilon_{\beta\beta} - \Sigma_{q=1}^{\Omega}\, p(t'')_{\mathrm{pq}}\varepsilon_{\beta\beta q}]}}{\sqrt{(\Sigma_{q=1}^{\Omega}\,\mathcal{S}_{33q}\mathcal{E}_{33q})}} \tag{13.11.32}$$

4. *Constant spherical pressure with constant rate of deviatoric loading*. The spherical pressure in this case is the same as in case 3 and Eq. (13.11.29) is valid. The deviatoric shear, however, is the same as in case 2 and Eqs. (13.11.21)–(13.11.24) and (13.11.26) hold for it. Consequently the excess stored specific free energy is the sum of the excess specific free energies resulting from the spherical pressure and the deviatoric shear stress, Eqs. (13.11.29) and (13.11.24), respectively

$$\Delta\rho\psi_{\mathfrak{si}} = \Delta\rho\psi'_{\mathfrak{si}} + \Delta\rho\psi''_{\mathfrak{si}} = \mathcal{P}\varepsilon(t')_{\alpha\alpha} - \int_0^{t'} p(t)_{\mathrm{p}}\dot{\varepsilon}(t)_{\alpha\alpha}\,\mathrm{d}t$$

$$+ \mathcal{P}\varepsilon_{\beta\beta} - \int_0^{t''} p(t)_{\mathrm{p}}\dot{\varepsilon}_{\beta\beta}\,\mathrm{d}t \tag{13.11.33}$$

and the disbursed specific free energy is the same as in Eq. (13.11.14).

The failure condition in this case is determined by Eqs. (13.11.14) and (13.11.33)

$$\sqrt{(\rho\psi''_{\mathrm{sb}})} = \sqrt{[\rho\psi_{\mathrm{si0}} + \Delta\rho\psi'_{\mathrm{si}} + \Delta\rho\psi''_{\mathrm{si}}]} = \sqrt{[\tfrac{3}{2}\mathcal{S}_{33}\mathcal{E}(t'')_{33}]}$$

$$= \sqrt{\Big(\rho\psi_{\mathrm{si0}} + \mathcal{P}\varepsilon(t')_{\alpha\alpha} - \int_0^{t'} p(t)_{\mathrm{p}}\dot{\varepsilon}(t)_{\alpha\alpha}\,\mathrm{d}t + \mathcal{P}\varepsilon_{\beta\beta} - \int_0^{t''} p(t)_{\mathrm{p}}\dot{\varepsilon}_{\beta\beta}\,\mathrm{d}t\Big)}$$

$$= \sqrt{\Big(\rho\psi_{\mathrm{si0}} + p_{\mathrm{m}}\varepsilon(t')_{\alpha\alpha} - \int_0^{t'} p(t)_{\mathrm{p}}\dot{\varepsilon}(t)_{\alpha\alpha}\,\mathrm{d}t + p_{\mathrm{m}}\varepsilon_{\beta\beta} - \int_0^{t''} p(t)_{\mathrm{p}}\dot{\varepsilon}_{\beta\beta}\,\mathrm{d}t\Big)} \tag{13.11.34}$$

The octahedral shear stress is

$$\sqrt{3}\tau = \sqrt{(\mathrm{II}_s)} = \sqrt{\tfrac{3}{2}}\mathcal{S}_{33}$$

$$= \mathcal{S}_{33}\,\frac{\sqrt{[\rho\psi_{\mathrm{si0}} + p_{\mathrm{m}}\varepsilon_{\alpha\alpha} - \int_0^{t''} p(t)_{\mathrm{p}}\dot{\varepsilon}_{\alpha\alpha}\,\mathrm{d}t + p_{\mathrm{m}}\varepsilon_{\beta\beta} - \int_0^{t''} p(t)_{\mathrm{p}}\dot{\varepsilon}_{\beta\beta}\,\mathrm{d}t]}}{\sqrt{[\mathcal{S}_{33}\mathcal{E}(t'')_{33}]}} \tag{13.11.35}$$

The left-hand sides of the equations of failure conditions in the four cases discussed, Eqs. (13.11.16), (13.11.27), (13.11.31) and (13.11.34) depend on the work of the shear stresses, while the right-hand sides depend on the following terms constituting the stored specific free energy:

1. The residing specific free energy, $\rho\psi_{\mathrm{si0}}$.
2. The excess stored specific free energy, that is, the work of the applied spherical pressure p_{m} or p_{mr}, positive for compression and negative for tension.
3. The dissipative work of the pore pressure $p(t)_{\mathrm{p}}$ or $p(t)_{\mathrm{pr}}$, that decreases the excess stored specific free energy. The energy is negative when measured as a pressure and positive when measured as a suction.
4. The second-order effect measured during deviatoric stresses in terms of a volumetric strain $\varepsilon_{\beta\beta}$ or $\varepsilon_{\beta\beta\mathrm{q}}$, positive when the density of the soil is below the critical density and negative when it is above the critical density. The positive value is known as *work hardening* or *strain hardening* and the negative value is known as *work softening* or *strain softening*.
5. The dissipative work of the pore pressure $p(t)_{\mathrm{p}}$ or $p(t)_{\mathrm{pq}}$, due to the second-order effect, negative when measured as a pressure and positive when measured as a suction.

The combined effect of the stress work of the spherical pressure (2) and of the pore pressure (3) is the work of the *effective* pressure, and the combined effect of the strain hardening (4) and the pore pressure during the strain hardening (5) is the work of the *effective* strain hardening. All factors appearing in the equation, except p_{m} and p_{mr}, are time dependent.

13.12 Drained Pure Deviatoric Shear Test

It is instructive to see how the strength of the soil sample is affected by the drainage conditions. The three main types of tests discussed in Sects. 12.6 and 12.9 are examined here.

If the test is run under full drainage conditions and the deviatoric shear is applied so slowly that the pore pressures dissipate before having a chance to build up, $p(t)_{\mathrm{p}} = p(t)_{\mathrm{pr}} = 0$, then the failure, Eqs. (13.11.16), (13.11.27), (13.11.31) and (13.11.34), depends only on four components: the residing specific free energy, the work of the spherical pressure, the strain hardening and the disbursed specific free energy. None of them are functions of the pore pressure.

The strength of the soil in a drained test in compression is the highest, since all components except the strain hardening are always positive. The strain hardening can be either positive or negative, but its magnitude in either case is minute compared to the other components.

The octahedral shear stresses, Eqs. (13.11.20), (13.11.28), (13.11.32) and (13.11.35), corresponding to the four cases of differing loading conditions discussed in the previous section, simplify greatly. For cases (1) and (2) we have, respectively

$$\sqrt{3}\tau_{\mathfrak{d}} = \sqrt{\overline{\mathrm{II}}_s} = \sqrt{\tfrac{3}{2}}\mathcal{S}_{33} = \mathcal{S}_{33}\,\frac{\sqrt{[\rho\psi_{\mathfrak{s}i0} + \Sigma_{\mathfrak{q}=1}^{\Omega}\, p_{\mathrm{mr}}\varepsilon(t')_{\alpha\alpha\mathrm{r}} + p_{\mathrm{m}}\varepsilon(t'')_{\beta\beta}]}}{\sqrt{[\Sigma_{\mathfrak{k}=1}^{\mathfrak{K}}\, \mathcal{S}_{33\mathfrak{q}}\mathcal{E}(t'')_{33\mathfrak{q}}]}} \tag{13.12.1}$$

$$\sqrt{3}\tau_{\mathfrak{d}} = \sqrt{(\overline{\mathrm{II}}_s)} = \sqrt{\tfrac{3}{2}}\mathcal{S}_{33}$$

$$= \mathcal{S}_{33}\,\frac{\sqrt{[\rho\psi_{\mathfrak{s}i0} + \Sigma_{\mathfrak{k}=1}^{\mathfrak{K}}\, p_{\mathrm{mr}}\varepsilon(t')_{\alpha\alpha\mathrm{r}} + p_{\mathrm{m}}\varepsilon(t'')_{\beta\beta}]}}{\sqrt{[\mathcal{S}_{33}\mathcal{E}(t'')_{33}]}} \tag{13.12.2}$$

The two differ only in their denominators, the disbursed specific free energy. For cases (3) and (4) we have

$$\sqrt{3}\tau_{\mathfrak{d}} = \sqrt{(\overline{\mathrm{II}}_s)} = \sqrt{\tfrac{3}{2}}\mathcal{S}_{33} = \mathcal{S}_{33}\,\frac{\sqrt{[\rho\psi_{\mathfrak{s}i0} + p_{\mathrm{m}}\varepsilon(t')_{\alpha\alpha} + p_{\mathrm{m}}\varepsilon(t'')_{\beta\beta}]}}{\sqrt{[\Sigma_{\mathfrak{q}=1}^{\Omega}\, \mathcal{S}_{33\mathfrak{q}}\mathcal{E}(t'')_{33\mathfrak{q}}]}} \tag{13.12.3}$$

$$\sqrt{3}\tau_{\mathfrak{d}} = \sqrt{(\overline{\mathrm{II}}_s)} = \sqrt{\tfrac{3}{2}}\mathcal{S}_{33} = \mathcal{S}_{33}\,\frac{\sqrt{[\rho\psi_{\mathfrak{s}i0} + p_{\mathrm{m}}\varepsilon(t')_{\alpha\alpha} + p_{\mathrm{m}}\varepsilon(t'')_{\beta\beta}]}}{\sqrt{[\mathcal{S}_{33\mathfrak{q}}\mathcal{E}(t'')_{33\mathfrak{q}}]}} \tag{13.12.4}$$

Eqs. (13.12.3) and (13.12.4) also differ in their denominators.

As seen in Eqs. (13.12.1)–(13.12.4), the measured quantities are the volumetric strain $\varepsilon(t')_{\alpha\alpha}$ or $\varepsilon(t')_{\alpha\alpha\mathfrak{r}}$ measured during the spherical pressure, and the volumetric strain $\varepsilon(t'')_{\beta\beta}$ and the vertical strain $\mathscr{E}(t'')_{33}$ or $\mathscr{E}(t'')_{33\mathfrak{q}}$ measured during the deviatoric shear. p_{m} or $p_{\mathrm{m}\mathfrak{r}}$ and $\mathscr{S}_{33}$ are the applied spherical pressure and deviatoric stresses, respectively. The residing specific free energy $\rho\psi_{\mathfrak{s}i0}$ is the only unknown quantity and it can be evaluated from the tests. It represents a free energy present in the soil prior to the tests, from previous densification of the soil. The residing specific free energy is closely related to the virtual work of the cohesion, and in a normally consolidated soil this relationship is defined by Eq. (11.25.6). If the soil is overconsolidated the residing specific free energy is larger than the product of the cohesion and the virtual volumetric strain, $\rho\psi_{\mathfrak{s}i0} > p_{\mathrm{m}0}\varepsilon_{\alpha\alpha0}$. The right-hand sides of Eqs. (13.12.1)–(13.12.4) and the shear stress at failure $\tau_{\mathfrak{b}}$ on the left-hand sides become larger. Fig. 13.12.1 and Fig. 13.12.2 are the Coulomb failure envelopes of a normally consolidated and of an overconsolidated soil, respectively.

A dry soil in which suction can be measured is equivalent to a soil subjected to a negative pressure, a spherical tension. The volumetric strain anticipated from such a tension is a decompression, which is a negative volumetric strain. The implicit stored specific free energy will be positive because it is the product of two negative quantities, and it should be added to the total stored specific free energy.* Since the suction in a dry soil is occasionally quite high, the strength of dry soils is sometimes excessive.

13.13 Consolidated Undrained Pure Deviatoric Shear Tests

The spherical pressure is applied under fully drained conditions while the deviatoric stresses are applied under undrained conditions. Consequently, pore pressures do not develop under the spherical pressure, but during the deviatoric shear stresses they come to full expression and reduce somewhat the strain work. For the cases (1), (2), (3) and (4) we have, respectively

$$\sqrt{3}\tau_{\mathrm{cu}} = \sqrt{\mathrm{II}_s} = \sqrt{\tfrac{3}{2}}\mathscr{S}_{33}$$
$$= \mathscr{S}_{33}\,\frac{\sqrt{[\rho\psi_{\mathfrak{s}i0} + \Sigma_{\mathfrak{k}=1}^{\mathfrak{K}} p_{\mathrm{m}\mathfrak{r}}\varepsilon_{\alpha\alpha\mathfrak{r}} + p_{\mathrm{m}}\varepsilon_{\beta\beta} - \Sigma_{\mathfrak{q}=1}^{\mathfrak{Q}} p(t'')_{\mathfrak{p}\mathfrak{q}}\varepsilon_{\beta\beta\mathfrak{q}}]}}{\sqrt{[\Sigma_{\mathfrak{q}=1}^{\mathfrak{Q}} \mathscr{S}_{33\mathfrak{q}}\mathscr{E}(t'')_{33\mathfrak{q}}]}} \tag{13.13.1}$$

* Another way to explain this is to point out that a suction is a negative pore pressure and if it were measured during the test it would be recorded as such. When it is introduced in Eqs. (13.12.10), (13.12.28), (13.12.32) or (13.12.35) as a negative pore pressure it becomes positive on account of the negative sign preceding it.

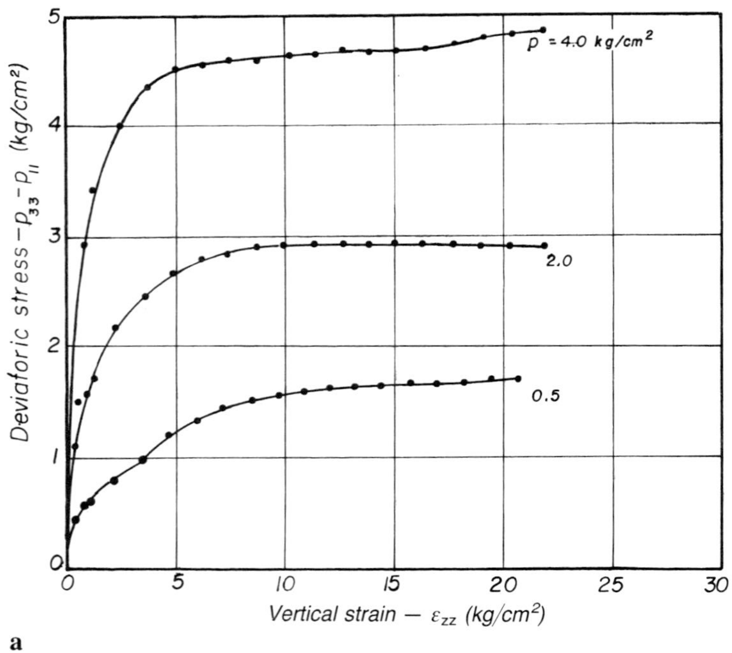

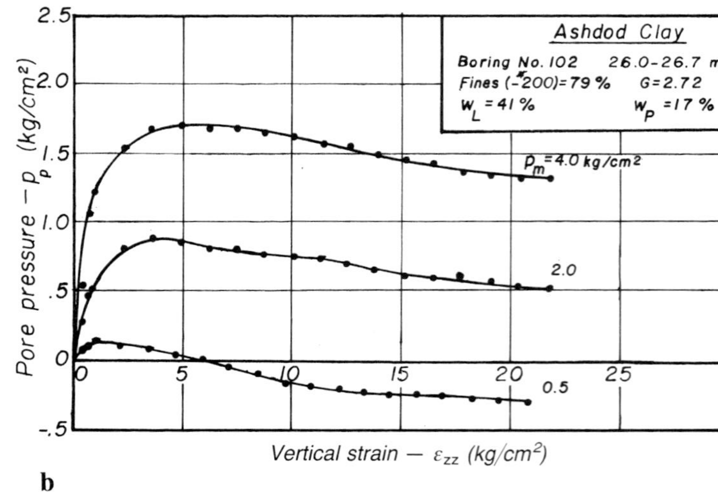

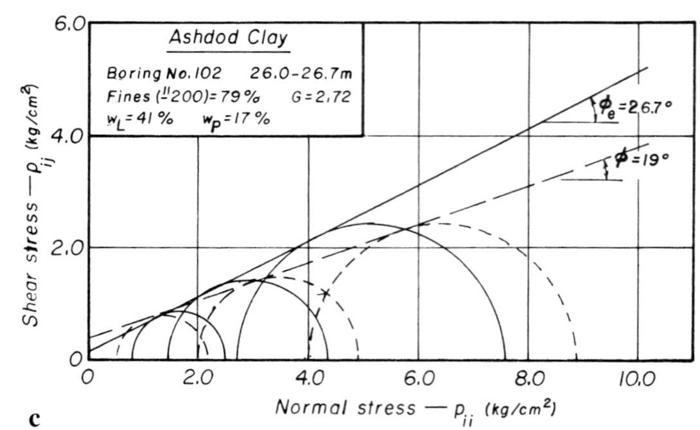

Fig. 13.12.1. Consolidated undrained tests of normally consolidated soils. **a** Deviatoric stress $p_{33} - p_{11}$ versus vertical strain ε_{zz}. **b** Pore pressure p_p versus vertical strain ε_{zz}. **c** Shear stress p_{ij} versus normal stress p_{ii}.

$$\sqrt{3}\tau_{cu} = \sqrt{\mathrm{II}_s} = \sqrt{\tfrac{3}{2}}\mathcal{S}_{33}$$

$$= \mathcal{S}_{33} \frac{\sqrt{[\rho\psi_{\mathfrak{s}i0} + \Sigma_{\mathfrak{k}=1}^{\mathfrak{K}} p_{\mathrm{mr}}\varepsilon_{\alpha\alpha\mathfrak{r}} + p_{\mathrm{m}}\varepsilon_{\beta\beta} - \int_0^{t''} p(t)_{\mathrm{pq}}\dot{\varepsilon}_{\beta\beta}\,\mathrm{d}t]}}{\sqrt{[\mathcal{S}_{33}\mathcal{E}(t'')_{33}]}} \quad (13.13.2)$$

$$\sqrt{3}\tau_{cu} = \sqrt{\mathrm{II}_s} = \sqrt{\tfrac{3}{2}}\mathcal{S}_{33}$$

$$= \mathcal{S}_{33} \frac{\sqrt{[\rho\psi_{\mathfrak{s}i0} + p_{\mathrm{m}}\varepsilon_{\alpha\alpha} + p_{\mathrm{m}}\varepsilon_{\beta\beta} - \Sigma_{\mathfrak{q}=1}^{\mathfrak{Q}} p(t'')_{\mathrm{pq}}\varepsilon_{\beta\beta\mathfrak{q}}]}}{\sqrt{[\Sigma_{\mathfrak{q}=1}^{\mathfrak{Q}} \mathcal{S}_{33\mathfrak{q}}\mathcal{E}(t'')_{33\mathfrak{q}}]}} \quad (13.13.3)$$

$$\sqrt{3}\tau_{cu} = \sqrt{\mathrm{II}_s} = \sqrt{\tfrac{3}{2}}\mathcal{S}_{33}$$

$$= \mathcal{S}_{33} \frac{\sqrt{[\rho\psi_{\mathfrak{s}i0} + p_{\mathrm{m}}\varepsilon_{\alpha\alpha} + p_{\mathrm{m}}\varepsilon_{\beta\beta} - \int_0^{t''} p(t)_{\mathrm{p}}\dot{\varepsilon}_{\beta\beta}\,\mathrm{d}t]}}{\sqrt{[\mathcal{S}_{33}\mathcal{E}(t'')_{33}]}} \quad (13.13.4)$$

In the above equations the first three terms in the numerator are always positive while the last term of the pore pressure is generally negative, except for the case of suction discussed in the previous section. Since the pore pressure is measured during the deviatoric stresses and is due to the second-order effects, it is small and the strength of the sample is slightly reduced. Note that the results of the conventional triaxial undrained test are lower than the results of the pure deviatoric test.

The remarks made in the previous section with respect to overconsolidated soils and suction hold for consolidated undrained soils as well.

13.14 Undrained Pure Deviatoric Shear Tests

We shall consider a pure deviatoric test in which drainage is prevented throughout the entire test. Pore pressures $p(t')_{\mathrm{p}}$, $p(t')_{\mathrm{pr}}$, $p(t'')_{\mathrm{p}}$ and $p(t'')_{\mathrm{pq}}$ and volumetric strains $\varepsilon(t')_{\alpha\alpha}$, $\varepsilon(t')_{\alpha\alpha\mathfrak{r}}$, $\varepsilon(t'')_{\beta\beta}$ and $\varepsilon(t'')_{\beta\beta\mathfrak{q}}$ are measured, and vertical strains $\mathcal{E}(t'')_{33}$ and $\mathcal{E}(t'')_{33\mathfrak{q}}$, Eqs. (13.11.19), (13.11.28), (13.11.32) and (13.11.35) are applied in full. For cases (1), (2), (3) and (4) in Sect. 3.11 we have, respectively

$$\sqrt{3}\tau_{u} = \sqrt{\mathrm{II}_s} = \sqrt{\tfrac{3}{2}}\mathcal{S}_{33} = \mathcal{S}_{33}$$

$$= \frac{\sqrt{[\rho\psi_{\mathfrak{s}i0} + \Sigma_{\mathfrak{k}=1}^{\mathfrak{K}} p_{\mathrm{mr}}\varepsilon_{\alpha\alpha\mathfrak{r}} - \Sigma_{\mathfrak{k}=1}^{\mathfrak{K}} p(t')_{\mathrm{pr}}\varepsilon_{\alpha\alpha\mathfrak{r}} + p_{\mathrm{m}}\varepsilon_{\beta\beta} - \Sigma_{\mathfrak{q}=1}^{\mathfrak{Q}} p(t'')_{\mathrm{pq}}\varepsilon_{\beta\beta\mathfrak{q}}]}}{\sqrt{[\Sigma_{\mathfrak{q}=1}^{\mathfrak{Q}} \mathcal{S}_{33\mathfrak{q}}\mathcal{E}(t'')_{33\mathfrak{q}}]}} \quad (13.14.1)$$

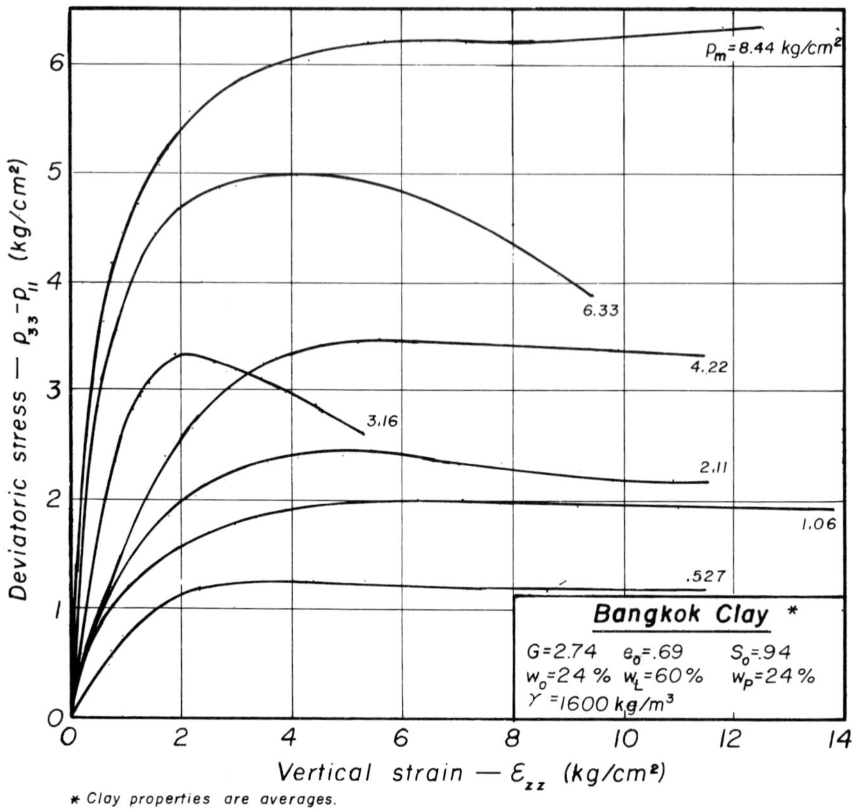
Deviatoric stress — $p_{33} - p_{11}$ (kg/cm²)
Vertical strain — ε_{zz} (kg/cm²)
p_m=8.44 kg/cm²
6.33
4.22
3.16
2.11
1.06
.527
Bangkok Clay *
G=2.74 e_0=.69 S_0=.94
w_0=24 % w_L=60 % w_p=24 %
γ =1600 kg/m³
* Clay properties are averages.

a

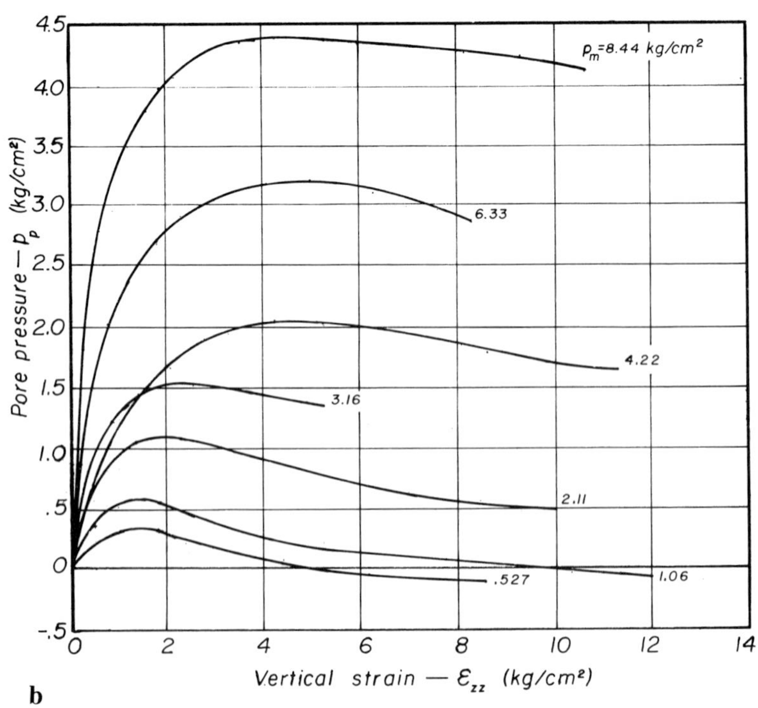
Pore pressure — p_p (kg/cm²)
Vertical strain — ε_{zz} (kg/cm²)
p_m=8.44 kg/cm²
6.33
4.22
3.16
2.11
1.06
.527

b

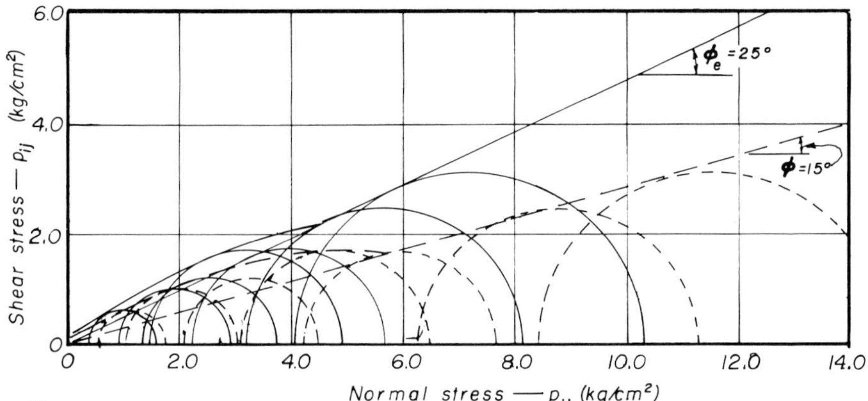

c

Fig. 13.12.2. Consolidated undrained tests of overconsolidated soils. **a** Deviatoric stress $p_{33} - p_{11}$ versus vertical strain ε_{zz}. **b** Pore pressure p_{p} versus vertical strain ε_{zz}. **c** Shear stress p_{ij} versus normal stress p_{ii}.

$$\sqrt{3}\tau_{\mathrm{u}} = \sqrt{\mathrm{II}_s} = \sqrt{\tfrac{3}{2}}\mathcal{S}_{33} = \mathcal{S}_{33}$$

$$= \frac{\sqrt{[\rho\psi_{\check{s}i0} + \Sigma_{\mathfrak{k}=1}^{\mathfrak{K}} p_{\mathrm{m}\mathfrak{r}}\varepsilon_{\alpha\alpha\mathfrak{r}} - \Sigma_{\mathfrak{k}=1}^{\mathfrak{K}} p(t')_{\mathrm{p}\mathfrak{r}}\varepsilon_{\alpha\alpha\mathfrak{r}} + p_{\mathrm{m}}\varepsilon_{\beta\beta} - \int_0^{t''} p(t)_{\mathrm{p}\mathfrak{q}}\dot{\varepsilon}_{\beta\beta}\,\mathrm{d}t]}}{\sqrt{[\mathcal{S}_{33}\mathcal{E}(t'')_{33}]}} \tag{13.14.2}$$

$$\sqrt{3}\tau_{\mathrm{u}} = \sqrt{\mathrm{II}_s} = \sqrt{\tfrac{3}{2}}\mathcal{S}_{33}$$

$$= \mathcal{S}_{33}\frac{\sqrt{[\rho\psi_{\check{s}io} + p_{\mathrm{m}}\varepsilon_{\alpha\alpha} - \int_0^{t'} p(t)_{\mathrm{p}}\dot{\varepsilon}_{\alpha\alpha}\,\mathrm{d}t + p_{\mathrm{m}}\varepsilon_{\beta\beta} - \Sigma_{\mathfrak{q}=1}^{\mathfrak{Q}} p(t'')_{\mathrm{p}\mathfrak{q}}\varepsilon_{\beta\beta\mathfrak{q}}]}}{\sqrt{[\Sigma_{\mathfrak{q}=1}^{\mathfrak{Q}}\mathcal{S}_{33\mathfrak{q}}\mathcal{E}(t'')_{33\mathfrak{q}}]}} \tag{13.14.3}$$

$$\sqrt{3}\tau_{\mathrm{u}} = \sqrt{\mathrm{II}_s} = \sqrt{\tfrac{3}{2}}\mathcal{S}_{33}$$

$$= \mathcal{S}_{33}\frac{\sqrt{[\rho\psi_{\check{s}io} + \rho_{\mathrm{m}}\varepsilon_{\alpha\alpha} - \int_0^{t'} p(t)_{\mathrm{p}}\dot{\varepsilon}_{\alpha\alpha}\,\mathrm{d}t + p_{\mathrm{m}}\varepsilon_{\beta\beta} - \int_0^{t''} p(t)_{\mathrm{p}}\dot{\varepsilon}_{\beta\beta}\,\mathrm{d}t]}}{\sqrt{[\mathcal{S}_{33}\mathcal{E}(t'')_{33}]}} \tag{13.14.4}$$

The difference between the undrained and the consolidated undrained tests is that in the undrained test pore pressures develop and are measured during the application of the spherical pressures as well. In a soil close to saturation the pore pressure corresponds to the degree of saturation, and as the soil approaches saturation the pore pressure approaches the value of the applied spherical pressure $p_{\alpha\alpha}$ or $p_{\alpha\alpha\mathrm{r}}$. The difference approaches zero and ultimately may vanish. The strength of the soil in an undrained pure deviatoric test is lowest, and except for the residing specific free energy, the total of all other components in the numerator of the equations approaches zero as the soil approaches saturation. At full saturation the residing specific free energy remains the sole component, and Eqs. (13.14.1)–(13.14.4) become constants for whatever spherical pressure is applied. Fig. 13.14.1 illustrates the strength of a soil in an undrained pure deviatoric test close to saturation. Note the small angle of internal friction. Fig. 13.14.2 shows that in a saturated soil the angle of internal friction is equal to zero.

13.15 Slip Surfaces

Two aspects of failure in soils should be considered. The first aspect, strength, representing the conditions that contribute to the occurrence of failure, was studied in the previous sections. The second aspect considers how and where failure takes place. It was observed that failure occurs along surfaces, not necessarily planar, known as *slip surfaces* or *failure planes*. It will be shown that slip surfaces have predetermined inclinations and predetermined locations. The presentation by Tschebotarioff (1952) is followed here, because of its simplicity and generality.

Let us consider an infinitesimal soil element of the dimensions dx_1, dx_2 and dx_3 oriented in the direction of the principal stresses and in balance with the deviatoric stresses acting on the element,* Fig. 13.15.1. It is also assumed that the principal stresses p_{11}, p_{22} and p_{33} act in the direction of the principal coordinates and that $p_{33} > p_{11}$ and $p_{33} > p_{22}$. We aim to find the inclination of the slip plane along which failure will occur. The slip surface divides the soil into two wedge shaped sections that slide along it in opposite directions. Along this plane the resistance to shearing is lowest, and its inclination is called the critical angle of failure and denoted θ_{cr}.

From the balance of the forces acting on any inclined plane which forms the angle θ with the horizontal, Fig. 13.15.1, the forces normal and tangential to the plane are obtained

$$F_{\mathrm{n}} = F_3 \cos\theta + F_1 \sin\theta \tag{13.15.1}$$

* The presentation by Tschebotarioff assumes a plane stress condition, along which an element of unit length is considered. There is no loss of generality by using either approach, element or plane stress, both being general enough to be valid for cylindrical elements or any other shape of elements.

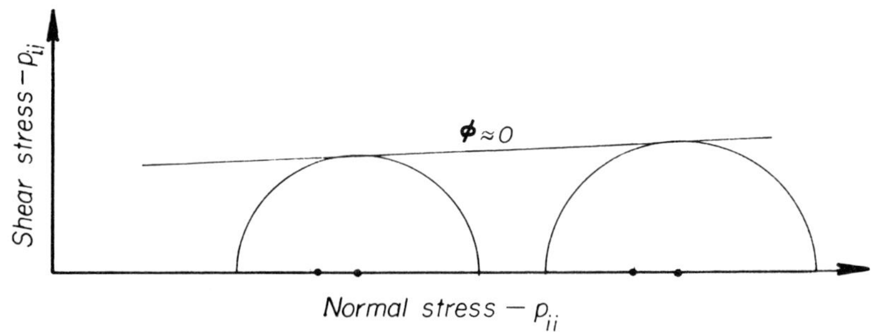

Fig. 13.14.1. Undrained test of soil close to saturation.

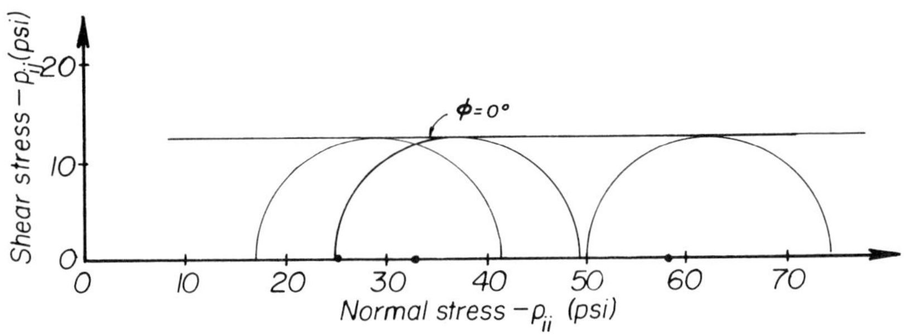

Fig. 13.14.2. Undrained test of a saturated sand.

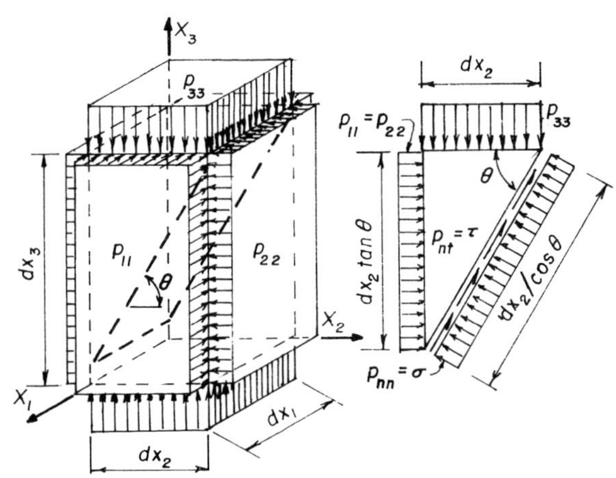

Fig. 13.15.1. Volume element in equilibrium at failure.

$$F_t = F_3 \sin\theta - F_1 \cos\theta \tag{13.15.2}$$

In terms of stresses these balance equations are

$$\sigma \frac{dx_1 \, dx_2}{\cos\theta} = p_{nn} \frac{dx_1 \, dx_2}{\cos\theta} = p_{33} \, dx_1 \, dx_2 \cos\theta + p_{11} \, dx_2 \, dx_1 \tan\theta \sin\theta \tag{13.15.3}$$

$$\tau \frac{dx_1\,dx_2}{\cos\theta} = p_{nt}\frac{dx_1\,dx_2}{\cos\theta} = p_{33}\,dx_2\,dx_1 \sin\theta - p_{11}\,dx_2\,dx_1 \tan\theta\cos\theta \tag{13.15.4}$$

where $\sigma = p_{nn}$ is the stress normal to the inclined plane and $\tau = p_{nt}$ is the stress tangential to the inclined plane. Since $dx_1\,dx_2$ is common to all terms of the foregoing equations, they can be simplified

$$\sigma = p_{11}\sin^2\theta + p_{33}\cos^2\theta = p_{11} + (p_{33} - p_{11})\cos^2\theta \tag{13.15.5}$$

$$\tau = p_{33}\sin\theta\cos\theta - p_{11}\sin\theta\cos\theta = \tfrac{1}{2}(p_{33} - p_{11})\sin 2\theta \tag{13.15.6}$$

Assuming Coulomb's linear relationship between the octahedral shear stress and the octahedral normal stress, Eq. (13.5.2), and substituting the values of σ and τ from Eqs. (13.15.5) and (13.15.6) into that equation we get

$$\begin{aligned}\tau &= p_{33}\sin\theta\cos\theta - p_{11}\sin\theta\cos\theta = c + \sigma\tan\phi \\ &= c + p_{11}\tan\phi + p_{33}\cos^2\theta\tan\phi - p_{11}\cos^2\theta\tan\phi \end{aligned} \tag{13.15.7}$$

where c is the cohesive intercept and ϕ is the angle of internal friction.
Solving for p_{33} we have

$$p_{33} = p_{11} + \frac{c + p_{11}\tan\phi}{\sin\theta\cos\theta - \cos^2\theta\tan\phi} \tag{13.15.8}$$

The plane with the least resistance to shearing is obtained when p_{33} is minimum, that is, when the denominator in the second part of Eq. (13.15.8) is maximum

$$\frac{d}{d\theta}(\sin\theta\cos\theta - \cos^2\theta\tan\phi) = 0 \tag{13.15.9}$$

Performing the differentiation on Eq. (13.15.9) we obtain

$$\cos^2\theta_{cr} - \sin^2\theta_{cr} + 2\cos\theta_{cr}\sin\theta_{cr}\tan\phi = 0 \tag{13.15.10}$$

leading to

$$\cos 2\theta_{cr} = -\tan\phi\sin 2\theta_{cr} \tag{13.15.11}$$

or

$$-\tan\phi = \cot 2\theta_{cr} = \cot(90° + \phi); \quad \tan\phi = \cot(90° - \phi) \tag{13.15.12}$$

from which the value of the critical angle of failure is obtained

$$\theta_{cr} = 45° + \tfrac{1}{2}\phi \qquad (13.15.13)$$

θ_{cr} is the angle of the slip plane with the horizontal line. It is seen that the inclination of the slip surface at failure, θ_{cr}, depends only on the angle of internal friction ϕ of the soil and is independent of any other property. However, properties like the density ρ, the pore pressure p_p, the cohesive intercept c and the applied normal pressure p_{nn} affect the angle of internal friction and indirectly the inclination of the slip surface.

In a cylindrical sample* Eq. (13.15.13) of the inclination of the slip surface is valid, but because of axial symmetry any plane of inclination θ_{cr} around the cylindrical sample is a potential slip surface. Moreover, there are two symmetric sets of potential slip surfaces, in both directions around the sample.

Although a sample is generally considered homogeneous and symmetric, minute induced asymmetries or frail spots in the sample will cause a slip surface to develop in a specific location, Fig. 13.15.2. Sometimes more than one slip surface develops, Fig. 13.15.3, and occasionally symmetric slip surfaces in opposite directions evolve, Fig. 13.15.4.

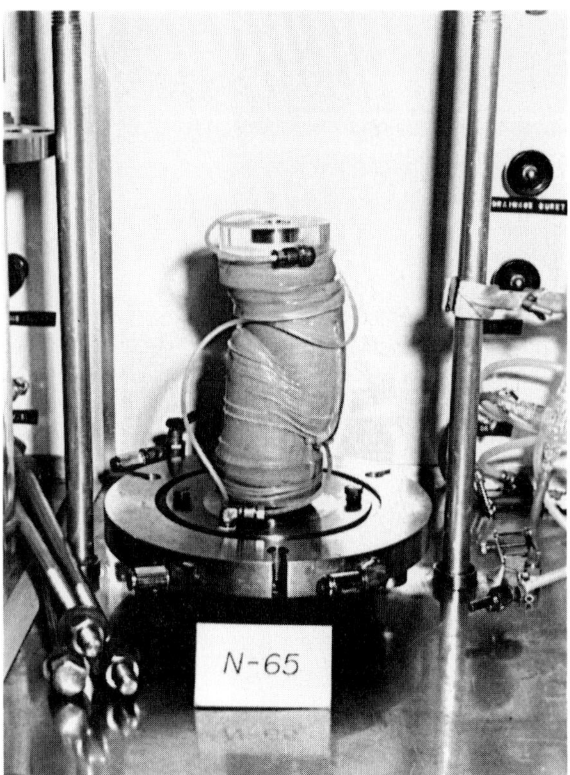

Fig. 13.15.2. Slip surface in soil sample.

* See footnote on p. 480.

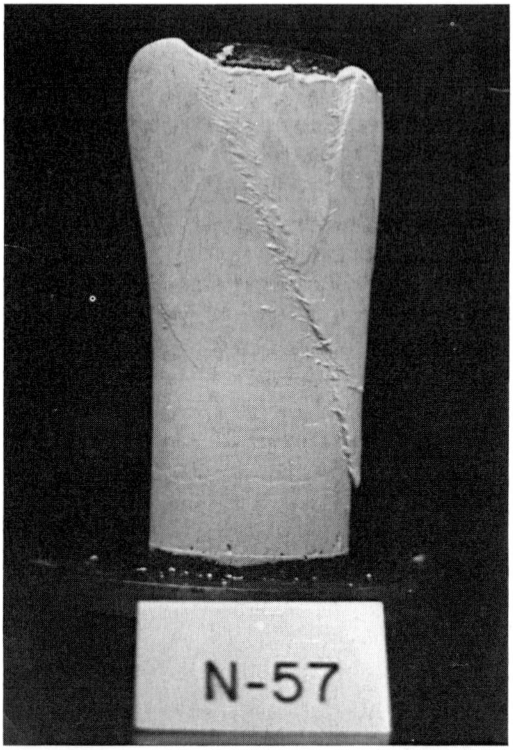

Fig. 13.15.3. Multiple slip surfaces in soils.

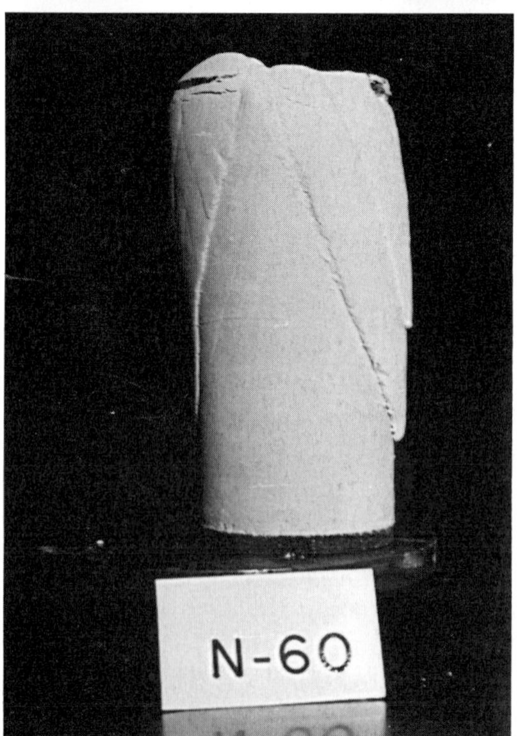

Fig. 13.15.4. Two directional slip surfaces.

Fig. 13.15.5. Bulging failure in an undrained sample.

The tendency to develop several slip surfaces is shared by soft soil samples and samples which develop pore pressures in undrained tests. In these samples the angle of internal friction is zero or close to it, and the angle of inclination of the critical slip surface is down to 45°. Because of the softness of the sample, an array of an infinite number of slip surfaces develops in both directions. This type of failure is attested by a bulging of the sample, particularly at mid height where pore pressures are usually higher, Fig. 13.15.5. Because of the numerous slip surfaces around the bulge, the failure planes are not always discernible and only the sphere-like shape of the bulge is an indication of its existence.

13.16 Lateral Earth Pressure

Substituting the value of θ_{cr} in Eq. (13.15.8) and considering also Eq. $(13.15.12)_2$, the relation between the principal stresses at failure is determined in two ways

$$p_{33} = p_{11}\tan^2(45^\circ + \tfrac{1}{2}\phi) + 2c\tan(45^\circ + \tfrac{1}{2}\phi) = p_{11}k_p + 2c\sqrt{k_p}$$

$$= p_{11}K_p = p_{11}\frac{1}{K_a} \tag{13.16.1}$$

$$p_{11} = p_{33}\tan^2(45^\circ - \tfrac{1}{2}\phi) - 2c\tan(45^\circ - \tfrac{1}{2}\phi) = p_{33}k_a - 2c\sqrt{k_a}$$

$$= p_{33}K_a = p_{33}\frac{1}{K_p} \tag{13.16.2}$$

where K_p is the *passive coefficient of lateral earth pressure*, K_a is the *active coefficient of lateral earth pressure*, k_p is the *restricted passive coefficient of lateral earth pressure* and k_a is the *restricted active coefficient of lateral earth pressure*.* It was further assumed here that both the frictional resistance and the cohesive resistance reach their maximum values simultaneously and after the same amount of deformation. This assumption was never proved correct or incorrect.

Eqs. (13.16.1) and (13.16.2) are the general equations of passive and active lateral earth pressure, respectively, for soils possessing cohesive properties. Coulomb (1776) was the first to present results similar to those in Eqs. (13.16.1) and (13.16.2). For non-cohesive soils where the cohesive intercept vanishes, $c = 0$, Eqs. (13.16.1) and (13.16.2) become

$$p_{33} = p_{11}\tan^2(45^\circ + \tfrac{1}{2}\phi) = p_{11}k_p = p_{11}K_p = p_{11}\frac{1}{K_a}; \quad c = 0 \tag{13.16.3}$$

$$p_{11} = p_{33}\tan^2(45^\circ - \tfrac{1}{2}\phi) = p_{33}k_a = p_{33}K_a = p_{33}\frac{1}{K_p}; \quad c = 0 \tag{13.16.4}$$

It is seen that in a cohesionless soil the coefficients of lateral earth pressure K_p and K_a are reduced to the values of the restricted coefficients of lateral earth pressure k_p and k_a. The latter can be written in several forms

$$k_p = \tan^2(45^\circ + \tfrac{1}{2}\phi) = \frac{\sqrt{(\tan^2\phi + 1)} + 1}{\sqrt{(\tan^2\phi + 1)} - 1} = \frac{1 + \sin\phi}{1 - \sin\phi}$$

$$= \frac{1}{(1 - \sin\phi)/(1 + \sin\phi)} = \frac{1}{\tan^2(45^\circ - \tfrac{1}{2}\phi)} = \frac{1}{k_a} \tag{13.16.5}$$

In this particular derivation where it was assumed that $p_{33} > p_{11}$ and $p_{33} > p_{22}$ the vertical pressure p_{33} is named the *passive earth pressure* and the

* The term restricted is derived from the fact that k_p and k_a are restricted to cohesionless soils. In the literature there is great confusion between the coefficients K_p and K_a and the restricted coefficients k_p and k_a. All coefficients are functions of the angle of internal friction ϕ, while K_p and K_a are also functions of the cohesive intercept c and of the normal pressure p.

horizontal pressures p_{11} and p_{22} are named the *active earth pressure*. To distinguish between the two, the general rule is that *the slide along the slip surface is always from the direction of the passive pressure towards the direction of the active pressure*.

The assumption made at the beginning of the derivation was that a slip surface exists and that the soil has passed the stage of failure due to a deviatoric stress $s_{ij} = \frac{1}{3}(p_{33} - p_{11})\Xi_{ij}$, under a spherical stress $p_m = \frac{1}{3}(p_{33} + 2p_{11})$. The active and passive coefficients of lateral earth pressure were all determined for that failure condition and without reference to deformations.

Let us suppose that the soil element is subjected as before to a vertical pressure p_{33} but the horizontal pressure vanishes, $p_{11} = 0$. From Eq. (13.16.1) we see that the vertical pressure can be supported only if the soil is cohesive. In this case p_{33} is limited to twice the product of the cohesive intercept and the square root of the restricted coefficient of passive earth pressure

$$p_{33} = 2c \tan(45° + \tfrac{1}{2}\phi) = 2c\sqrt{k_p} \tag{13.16.6}$$

Consequently a cohesive soil can carry a vertical pressure p_{33} up to the limit described by Eq. (13.16.6) even without the support of a horizontal pressure $p_{11} = p_{22}$, while a cohesionless soil requires the support of a horizontal pressure.

For a frictionless soil where the angle of internal friction is zero, $\phi = 0$, the restricted passive earth pressure coefficient becomes equal to unity, $k_p = 1$, and we get from Eq. (13.16.1)

$$p_{33} = p_{11} + 2c = p_{11}(1 + 2c/p_{11}); \quad k_p = k_a = 1 \tag{13.16.7}$$

We see that the horizontal pressure varies hydrostatically with the vertical pressure. This is more emphasized in a cohesionless soil, $c = 0$, where we get

$$p_{33} = p_{11} \quad k_p = k_a = 1 \tag{13.16.8}$$

and the soil behaves like water. A similar condition can occur in an undrained saturated soil when the angle of internal friction ϕ vanishes.

13.17 The Coefficient of Lateral Earth Pressure

Let us assume an infinitesimal volume element of a natural soil in situ at a depth h in equilibrium with its surroundings. On the element acts a vertical pressure $p_{33} = \gamma h + \Delta p_{33}$, which is the sum of the weight of the soil column above it and a pressure Δp_{33} from whatever overburden load is present, the weight of the column being the product of the unit weight of the soil γ and the depth h of the element. The horizontal pressures $p_{11} = p_{22}$ acting on the element are proportional functions of the vertical pressure p_{33}, introduced in Sect. 11.9

$$K_0 = \frac{p_{11}}{p_{33}} = \frac{p_{22}}{p_{33}} \quad \text{for } \varepsilon_{ij} = 0 \tag{13.17.1}$$

where K_0 is the coefficient of proportionality called the *coefficient of lateral earth pressure at rest*.

If the soil in situ is not in equilibrium for some reason, for instance if it is in the process of consolidation due to an added overburden load, a *coefficient of lateral earth pressure* K may still be defined

$$K = \frac{p_{11}}{p_{33}} = \frac{p_{22}}{p_{33}} \tag{13.17.2}$$

where $K = K(\phi, c, \varepsilon_{ij}) \equiv K(\phi, c, t)$ is a function of the properties of the soil, the angle of internal friction ϕ and the cohesive intercept c, as well as of the strain in process, thus a function of time. If a saturated soil in situ is in equilibrium and is subjected to a consolidating process, its equilibrium is disrupted and so is its coefficient of lateral earth pressure K_0 that has been so far at rest. Owing to a high pore pressure at the initial stages of the consolidation process, the coefficient of lateral earth pressure K assumes a value close to unity, then it decreases and finally returns to its value at rest K_0, Fig. 13.17.1. We see that the coefficient of lateral earth pressure is dependent on and related to the spherical stress–strain behavior of soils and that $1 \geq K \geq K_0$.

If, however, the equilibrium of the soil in situ is upset by a deviatoric stress

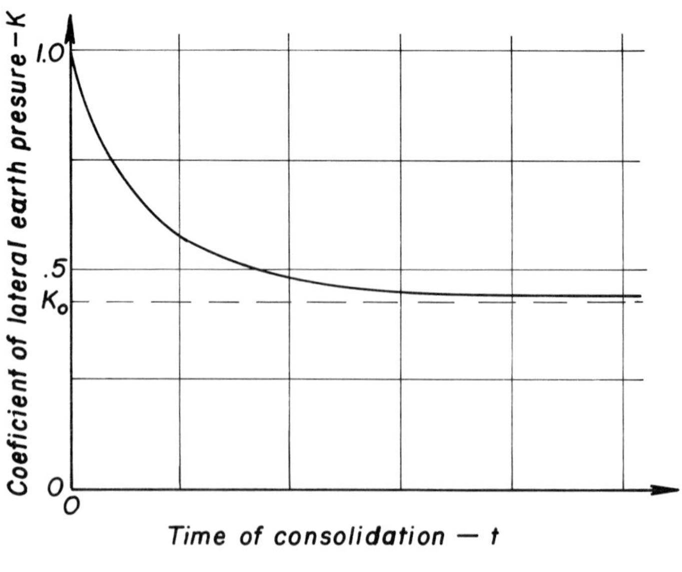

Fig. 13.17.1. Coefficient of lateral earth pressure change during the process of consolidation.

which ultimately produces failure, the coefficient of lateral earth pressure reaches either its highest, passive value K_p or its lowest, active value K_a, as was shown in Sects. 13.15 and 13.16. To answer the question *what happens to the coefficient of lateral earth pressure between the at rest state and these extreme failure states*, we shall consider again the infinitesimal element at rest in situ.

Since the soil in situ is in equilibrium, all its volume elements, at any depth, are at rest. Let us assume that we insert an imaginary vertical partition without disturbing the soil, and excavate the soil on one side keeping the partition from moving, Fig. 13.17.2. To maintain the partition stable, a horizontal pressure equal to the product of the vertical pressure and the coefficient of lateral earth pressure at rest, $p_{11} = p_{22} = K_0 p_{33}$, will have to be applied. If the imaginary partition is now displaced by moving it in parallel fashion towards the soil, the horizontal pressure $p_{11} = p_{22}$ has to be increased in order to overcome the resistance of the soil, until it attains a maximum value $p_{11max} = p_{22max}$ and the soil reaches failure, displaying a slip surface. Since the vertical pressure remains unchanged, the coefficient of lateral earth pressure K must increase, until, at failure, it reaches asymptotically its passive value of K_p. The right-hand branch of the curve in Fig. 13.17.3 shows the increase in the coefficient of lateral earth pressure K in a positive displacement, that is, a displacement moving towards the soil.

If the imaginary partition is displaced in the opposite direction, that is, away from the soil, the horizontal pressure $p_{11} = p_{22}$ decreases continuously with the displacement, as less and less soil is "leaning" on the partition, until

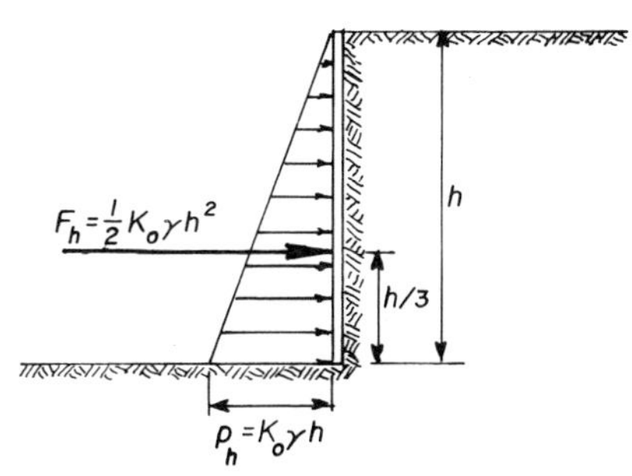

Fig. 13.17.2. Soil in situ with imaginary partition.

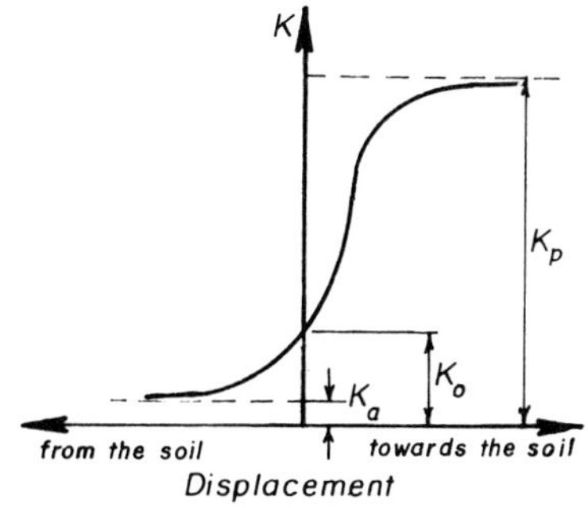

Fig. 13.17.3. Variation of lateral earth pressure coefficient with distortion.

it reaches a minimal value $p_{11\min} = p_{22\min}$. The coefficient of lateral earth pressure K decreases correspondingly and reaches asymptotically its minimal active value K_a. The left-hand branch of the curve in Fig. 13.17.3 represents the coefficient of lateral earth pressure K for a negative displacement, a displacement moving away from the soil.

The coefficient of lateral earth pressure at rest K_0 varies from soil to soil, from $K_0 = 1$ for a saturated soil to any value $K_0 < 1$. A stiff rock at rest is considered to have a coefficient of lateral earth pressure approaching zero. Fig. 13.17.4 illustrates K values for a clay at different consistencies.

In the absence of a more comprehensive theory, the development of slip surfaces presented here is based on Coulomb's classical failure theory. No attempt has been made to develop a slip surface theory from the failure theory based on the internal energy balance as presented in this chapter, although there is no doubt that it can be done.

The theory of *discontinuities* introduced by Helmholtz (1858) and the *jump* functions introduced by Christoffel (1877) and developed further (Hadamard 1901, 1903; Emde 1915; Spielrein 1916; Kotchine 1926; Thomas 1957) are also not presented here. These theories, developed from the study of continuum mechanics and extended to slip surfaces, dislocations and shock waves in fluids are not yet applied to soils.

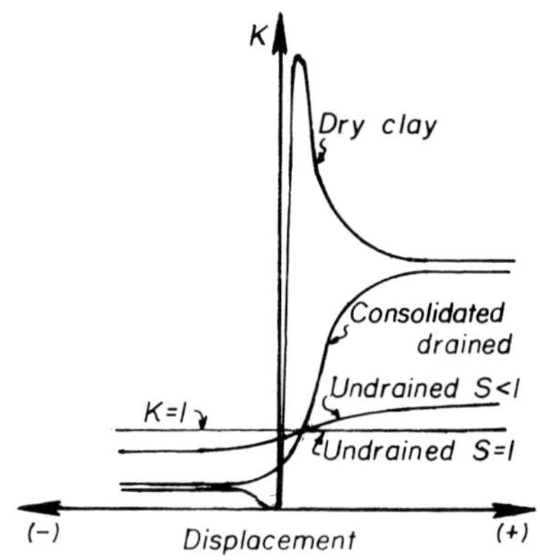

Fig. 13.17.4. K values for soils with various K_0.

Appendix A
Tensor Mathematics

A.1 Introduction

In physics we are accustomed to relating quantities to other quantities by means of mathematical expressions. The quantities that represent physical properties of a point or of an infinitesimal volume in space may be of a single numerical value without direction, such as temperature, mass, speed, distance, specific gravity, energy, etc.; they are defined by a single *magnitude* and we call them *scalars*. Other physical quantities have direction as well and their magnitude is represented by an array of numerical values, their number corresponding to the number of dimensions of the space, three numerical values in a three-dimensional space; such quantities are called *vectors*, for example velocity, acceleration, force, etc., and the three numerical values are the *components* of the vector in the direction of the coordinates of the space. Still other physical quantities such as moment of inertia, stress, strain, permeability coefficients, electric flux, electromagnetic flux, etc., are represented in a three-dimensional space by nine numerical quantities, called components, and are known as *tensors*.

We will introduce a slightly different definition. We shall call the scalars tensors of order zero, vectors – tensors of order one and tensors with nine components – tensors of order two. There are, of course, in physics and mathematics, tensors of higher order. Tensors of order three have an array of 27 components and tensors of order four have 81 components and so on. Couple stresses that arise in materials with polar forces are examples of tensors of order three, and the Riemann curvature tensor that appears in geodesics is an example of a tensor of order four. The coefficients of an elastic crystal also form a tensor of the fourth order with 81 components, but owing to isotropy and to symmetry in geometry and in energies they are soon reduced to the two known coefficients. In general we can say that a tensor of order n in a three-dimensional space will have 3^n components.

Why tensor mathematics? For two reasons. First, a shorthand presentation of the equations is possible, since one equation identically replaces three or more equations in any other mathematical notation, thus affording a great simplification in the mathematics. Second, any mathematical equation presented is valid for any coordinate system, the rules of transformation from one coordinate system to another being predetermined.

This appendix contains not only the basics of tensor analysis, which is necessary to follow the exposition of this book, but also additional material to enable the reader to continue his studies, perhaps in other directions. For further insight and more detailed derivations a great selection of publications is available, among them, Synge and Schild (1949), Lass (1950), Schouten (1965), Spain (1953), Ericksen (1960a), Borisenko and Tarapov (1968).

A.2 The Indicial Notation

If we are given a set of N independent variables, say, coordinates x, y, z, . . ., N, we find it more convenient to denote them by the same letter, distinguishing between them by means of indices. Thus we shall write the N variables x_1, x_2, x_3, . . ., x_N or x_i, x_j, x_k, . . ., x_N, or it may be written more compactly x_r, where r takes in turn the values of 1, 2, 3, . . ., or of i, j, k, . . ., up to N. In the same way that we write the index r as a subscript, we can use superscripts instead, by writing x^r. The italic characters used for superscripts and subscripts distinguish them from power exponents, which are roman characters.

Non-tensorial indices, denoting generic groups of expressions or numericals of summation, will be marked by gothic indices, while the use of greek indices is reserved for tensorial summations, as will be explained later.

If we have a three variable system or a three manifold coordinate system or *triad*, ($N = 1, 2, 3$ or $\mathsf{N} = x, y, z$), then x^r, x_i, X^j, X_q represent notations for tensors of the first order, that is, vectors, in a tridimensional triad, where x and X are the respective values of the components of the vectors in the directions r, i, j, q, which may take in turn the values of 1, 2, 3 or x, y, z, or of any other triad.

A.3 Transformation of Coordinates

Let us consider a point in an N dimensional space, defined by the coordinates of its position vector, X^1, X^2, X^3, . . ., X^N. The N equations

$$x^i = x^i(X^1, X^2, \ldots, X^N) \quad (i = 1, 2, \ldots, N) \tag{A.3.1}$$

where X^i are single-valued continuous differentiable functions of the coordinates, define a *transformation of coordinates* into a new coordinate system x^i. The necessary and sufficient condition that the N equations (A.3.1) be independent is satisfied if the determinant formed from the partial derivatives $\partial x^i/\partial X^j$, named *Jacobian*, does not vanish

$$J \equiv \begin{vmatrix} \dfrac{\partial x^1}{\partial X^1} & \dfrac{\partial x^1}{\partial X^2} & \cdots & \dfrac{\partial x^1}{\partial X^N} \\ \dfrac{\partial x^2}{\partial X^1} & \dfrac{\partial x^2}{\partial X^2} & \cdots & \dfrac{\partial x^2}{\partial X^N} \\ \dfrac{\partial x^N}{\partial X^1} & \dfrac{\partial x^N}{\partial X^2} & \cdots & \dfrac{\partial x^N}{\partial X^N} \end{vmatrix} \equiv \frac{\partial(x^1, x^2, x^3, \ldots, x^N)}{\partial(X^1, X^2, X^3, \ldots, X^N)} \neq 0 \tag{A.3.2}$$

Conversely, Eq. (A.3.1) may be solved for the X^i as a function of x^i

$$X^i = X^i(x^1, x^2, \ldots, x^N) \tag{A.3.3}$$

In a tridimensional coordinate system, where $N = 3$, Eq. (A.3.2) becomes

$$J \equiv \begin{vmatrix} \dfrac{\partial x^1}{\partial X^1} & \dfrac{\partial x^1}{\partial X^2} & \dfrac{\partial x^1}{\partial X^3} \\ \dfrac{\partial x^2}{\partial X^1} & \dfrac{\partial x^2}{\partial X^2} & \dfrac{\partial x^2}{\partial X^3} \\ \dfrac{\partial x^3}{\partial X^1} & \dfrac{\partial x^3}{\partial X^2} & \dfrac{\partial x^3}{\partial X^3} \end{vmatrix} \equiv \frac{\partial(x^1, x^2, x^3)}{\partial(X^1, X^2, X^3)} \neq 0 \tag{A.3.4}$$

As our study is concerned with tridimensional systems, we shall restrict ourselves, from here on, to tridimensional coordinates, unless otherwise stated.

A.4 The Summation Convention

An *italic* index appearing twice in a term implies a summation. As a convention, we shall transpose such repeated indices to *Greek* indices in order to stress the fact that the indices are no longer tensorial indices but "dummy" indices. For instance, the expression

$$S = A_1x^1 + A_2x^2 + \ldots + A_Nx^N = \sum_{n=1}^{N} A_ix^i = A_\alpha x^\alpha \tag{A.4.1}$$

and the total differential of x^i from Eq. (A.3.1)

$$\mathrm{d}x^i = \frac{\partial x^i}{\partial X^1}\,\mathrm{d}X^1 + \frac{\partial x^i}{\partial X^2}\,\mathrm{d}X^2 + \frac{\partial x^i}{\partial X^3}\,\mathrm{d}X^3 = \sum \frac{\partial x^i}{\partial X^j}\,\mathrm{d}X^j = \frac{\partial x^i}{\partial X^\alpha}\,\mathrm{d}X^\alpha \tag{A.4.2}$$

contain summations over the Greek indices.

The expression $A_{\alpha\alpha}$ represents a summation of the identically indexed components of a second-order tensor A_{rs}

$$A_{\alpha\alpha} = A_{11} + A_{22} + A_{33} \tag{A.4.3}$$

We shall introduce two conventions:

Range convention. A free unrepeated italic index will have the range of values 1, 2, 3.

Summation convention. A repeated Greek index is a dummy index, i.e. a result of transposed repeated italic indices, and is to be summed from 1 *to* 3.

A summation reduces by two the tensorial order of the term in which it appears.

A.5 The Kronecker Delta

The *Kronecker delta* δ^i_j, also known as the *unit tensor*, is defined so that its components equal zero whenever $i \neq j$, and 1 if $i = j$. Its matricial form is

$$|\delta^i_j| = \begin{vmatrix} 1 & 0 & 0 \\ 0 & 1 & 0 \\ 0 & 0 & 1 \end{vmatrix} \tag{A.5.1}$$

The obvious property of the unit tensor is that when it is multiplied by a tensor of any order it maintains the tensor intact, $\delta^i_\alpha A^\alpha = A^i$. Also $\partial X^k/\partial X^j = \delta^k_j$ and $\delta^\alpha_\alpha = 3$.

A.6 Contravariant and Covariant Tensors

The components of a vector A^i, a first-order tensor, are said to be components of a *contravariant* tensor if by changing coordinates from X^i to x^i they transform according to equation

$$a^i = \frac{\partial x^i}{\partial X^\alpha} A^\alpha \tag{A.6.1}$$

Conversely, by multiplying Eq. (A.6.1) by $\partial X^k/\partial x^i$ and summing over i, we obtain

$$\frac{\partial X^k}{\partial x^\alpha} a^\alpha = \frac{\partial X^k}{\partial x^\alpha} \frac{\partial x^\alpha}{\partial X^\beta} A^\beta = \frac{\partial X^k}{\partial X^\alpha} A^\alpha = \delta^k_\alpha A^\alpha = A^k \tag{A.6.2}$$

From Eq. (A.6.2) follows the transformation

$$x^i = \frac{\partial x^i}{\partial X^\alpha} A_\alpha \tag{A.6.3}$$

The components of a vector A_i, are said to be components of a *covariant* tensor if they transform at the change of the coordinate X^i to the coordinate x^i, according to equation

$$a_i = \frac{\partial X^\alpha}{\partial x^i} A_\alpha \tag{A.6.4}$$

Similarly to Eq. (A.6.2), by multiplying Eq. (A.6.4) by $\partial x^i/\partial X^k$ and summing over the index i from 1 to 3, we obtain

$$\frac{\partial x^\alpha}{\partial X^k} a_\alpha = \frac{\partial x^\alpha}{\partial X^k} \frac{\partial X^\beta}{\partial x^\alpha} A_\beta = \frac{\partial X^\alpha}{\partial x^k} A_\alpha = A_k \tag{A.6.5}$$

The term $\partial f/\partial X^i$ which forms a first-order tensor from a scalar function f which is a zero-order tensor, will, in any other coordinate system, have the components

$$\frac{\partial f}{\partial X^\alpha} \frac{\partial X^\alpha}{\partial x^i} = \frac{\partial f}{\partial x^i} \tag{A.6.6}$$

Such a covariant tensor is called the *gradient* of f.

A second-order contravariant tensor A^{ij} transforms according to Eq. (A.6.1)

$$\alpha^{ij} = \frac{\partial x^i}{\partial X^\alpha} \frac{\partial x^j}{\partial X^\beta} A^{\alpha\beta} \tag{A.6.7}$$

Similarly, a covariant second-order tensor A_{ij} will transform, according to Eq. (A.6.4), as follows

$$a_{ij} = \frac{\partial X^{\alpha}}{\partial x^{i}} \frac{\partial X^{\beta}}{\partial x^{j}} A_{\alpha\beta} \tag{A.6.8}$$

A second-order tensor whose components are A^{i}_{j}, will transform

$$a^{i}_{j} = \frac{\partial x^{i}}{\partial X^{\alpha}} \frac{\partial X^{\beta}}{\partial x^{j}} A^{\alpha}_{\beta} \tag{A.6.9}$$

A^{i}_{j} are the components of a *mixed* tensor.

Higher-order tensors transform according to the above rules as well. The components of a fifth-order mixed tensor A^{ij}_{klm}, for instance, will transform as follows

$$a^{ij}_{klm} = \frac{\partial x^{i}}{\partial X^{\alpha}} \frac{\partial x^{j}}{\partial X^{\beta}} \frac{\partial X^{\gamma}}{\partial x^{k}} \frac{\partial X^{\rho}}{\partial x^{l}} \frac{\partial X^{\lambda}}{\partial x^{m}} A^{\alpha\beta}_{\gamma\rho\lambda} \tag{A.6.10}$$

The order of transformation is important.

Finally, it should be noted that the coordinates X^{i}, x^{i} do not form components of a contravariant tensor, although they seem to suggest it by appearance.

A most important deduction from Eq. (A.6.10) is that if all components of a tensor are zero in one coordinate system, they are zero in every other coordinate system as well.

A.7 Symmetric and Skew-symmetric Tensors

The order of indices in a tensor is meaningful. The tensor A^{ij} is not necessarily the same as the tensor A^{ji}. Tensor A^{ij} may be called the *transpose* of tensor A^{ji}, a name borrowed from the mathematics of matrices. If the indices of a twice covariant or twice contravariant tensor can be changed without altering the tensor, we have a *symmetric* tensor, $A^{ij} = A^{ji}$. If, however, by interchanging the indices of a twice contravariant or twice covariant tensor the tensor changes its sign, the tensor is then *skew-symmetric*, $A^{ij} = -A^{ji}$.

Since in a second-order symmetric tensor $A_{ij} = A_{ji}$, only six out of the nine components are independent, the tensor being symmetric with respect to its diagonal.

A skew-symmetric tensor has, at the most, three independent components. Its diagonal components are all zero, while the components symmetric with respect to the diagonal are either zero or differ merely in sign.

Since the order of transformation is important, symmetry cannot properly be defined with respect to two indices of which one is covariant and the other contravariant.

A.8 Addition, Subtraction and Multiplication

Two tensors of the same order and type (contravariant, covariant, mixed), may be added or subtracted. It is clear also that $A^{ij} + B^i$ or $A^{ij} + B^i_j$ do not have any meaning because A^{ij}, B^i_j, B^i transform differently. The same holds for subtractions. Therefore we have only

$$A^i + B^i = C^i$$

$$A_{ij} - B_{ij} = C_{ij} \tag{A.8.1}$$

$$A^i_{jk} - B^i_{jk} = C^i_{jk}$$

We may even have $\lambda A^{ij}_k - \mu B^{ij}_k = C^{ij}_{ki}$ provided λ and μ are scalar quantities.

For any second-order tensor A_{ij}, we can write

$$A_{ij} = \tfrac{1}{2}(A_{ij} + A_{ji}) + \tfrac{1}{2}(A_{ij} - A_{ji}) \tag{A.8.2}$$

where $A_{ij} + A_{ji}$ is symmetric while $A_{ij} - A_{ji}$ is skew-symmetric. Thus a covariant, and similarly a contravariant second-order tensor is always a sum of a symmetric and a skew-symmetric tensor.

In generalized coordinates the *symmetrization* or *skew-symmetrization* may be performed on tensors of higher order as well, but only with respect to two indices of the same type, either covariant or contravariant. It should also be remembered that the fundamental tensors g_{ij}, g^{ij}, G_{ij} and G^{ij}, to be encountered later in equation (A.10.3) and elsewhere, are always symmetric.

From (A.6.10) it follows also that a product of two tensors, one contravariant of order $\mathfrak{s}$ and covariant of order $\mathfrak{p}$, the other contravariant of order $\mathfrak{t}$ and covariant of order $\mathfrak{q}$, will be contravariant of order $\mathfrak{s} + \mathfrak{t}$ and covariant of order $\mathfrak{p} + \mathfrak{q}$. For instance

$$A^{ij}_k B^l_{mnt} = C^{ijl}_{kmnt} \tag{A.8.3}$$

This operation is called an *outer product*.

The division of one tensor by another tensor is not defined. A tensor may be divided only by a scalar quantity.

A.9 Contraction

The process of contraction is an extension of the summation convention discussed in Sect. A.4. Let us take a mixed tensor of any order, say A^{ij}_{klm}, which transforms

$$a^{sf}_{pqr} = \frac{\partial x^s}{\partial X^i}\frac{\partial x^f}{\partial X^j}\frac{\partial X^l}{\partial x^p}\frac{\partial X^m}{\partial x^q}\frac{\partial X^n}{\partial x^r}A^{ij}_{lmn} \tag{A.9.1}$$

If we now set f equal to r, we get the tensor A^{sr}_{pqr}, and since r is now a repeated index it will be transposed to, say, α, and summed over from 1 to 3

$$a^{sr}_{pqr} = \frac{\partial x^s}{\partial X^i}\frac{\partial x^r}{\partial X^j}\frac{\partial X^l}{\partial x^p}\frac{\partial X^m}{\partial x^q}\frac{\partial X^n}{\partial x^r}A^{ij}_{lmn} = \frac{\partial x^s}{\partial X^i}\frac{\partial X^l}{\partial x^p}\frac{\partial X^m}{\partial x^q}\delta^n_j A^{ij}_{lmn}$$

$$= \frac{\partial x^s}{\partial X^i}\frac{\partial X^l}{\partial x^p}\frac{\partial X^m}{\partial x^q}A^{i\alpha}_{lm\alpha} \tag{A.9.2}$$

We observe now that the mixed tensor $A^{i\alpha}_{lm\alpha}$ is a third-order tensor, once contravariant and twice covariant. This process, called *contraction*, reduces the tensorial order by two, as we have seen in Sect. A.4. A contraction may occur in a single tensorial quantity, such as that in Eq. (A.4.3) or it may succeed a multiplication of two tensors of any order, in which case the operation is called an *inner product*. The component *ilm* of the contracted tensor A^{ij}_{klm} is $A^{i\alpha}_{lm\alpha} = A^{i1}_{lm1} + A^{i2}_{lm2} + A^{i3}_{lm3}$. Tensor $A^{i\alpha}_{lm\alpha}$ may be contracted further, for example $A^{\beta\alpha}_{l\beta\alpha}$ or $A^{\beta\alpha}_{\beta l\alpha}$ as the tensor is reduced again to the first order, i.e. a vector.

The contraction operates on two indices which are not of the same type, one is a superscript and the other a subscript, otherwise the resulting sum is not necessarily a tensor.

A.10 The Line Element

Let X_i be the components of a first-order tensor, a position vector of a point, defined in a rectangular Cartesian triad. A point neighboring to X_i will have the components $X_i + \mathrm{d}X_i$. The distance $\mathrm{d}S$ between the two points is given by the relation

$$\mathrm{d}S^2 = \mathrm{d}X_1^2 + \mathrm{d}X_2^2 + \mathrm{d}X_3^2 = \delta_{\alpha\beta}\,\mathrm{d}X^\alpha\,\mathrm{d}X^\beta \tag{A.10.1}$$

If we introduce a curvilinear coordinate system x^i into which the points X^L and $X^L + \mathrm{d}X^L$ are transformed, then, instead of Eq. (A.10.1) we get

$$\mathrm{d}S^2 = \delta_{\alpha\beta} \frac{\partial x^\alpha}{\partial X^\gamma} \frac{\partial x^\beta}{\partial X^\rho} \mathrm{d}X^\gamma \mathrm{d}X^\rho = G_{\alpha\beta} \mathrm{d}X^\alpha \mathrm{d}X^\beta \tag{A.10.2}$$

where G_{LM} is a covariant tensor of the second order, known as the *fundamental tensor* or the *Riemann metric* tensor, defined

$$G_{LM} = \delta_{\alpha\beta} \frac{\partial x^\alpha}{\partial X^L} \frac{\partial x^\beta}{\partial X^M} \tag{A.10.3}$$

The tensor G_{LM} is symmetric, since, according to Eq. (A.8.1) the diagonals of the term $(G_{LM} - G_{ML})$ are all zero and do not contribute to the sum $\mathrm{d}S^2$. Similarly, in the x^i coordinates we get

$$\mathrm{d}s^2 = g_{\alpha\beta} \mathrm{d}x^\alpha \mathrm{d}x^\beta \tag{A.10.4}$$

Thus, the most general form of the line element $\mathrm{d}S^2$ or $\mathrm{d}s^2$, in Euclidean curvilinear space, is the quadratic form, Eqs. (A.10.2) or (A.10.4), respectively.

The right-hand side of Eq. (A.10.4) is invariant since the length $\mathrm{d}s$ is the same, irrespective of the coordinate system.

Similarly, the length a of a vector a^i in a Riemann space is defined

$$a^2 = g_{\alpha\beta} a^\alpha a^\beta \tag{A.10.5}$$

On the other hand, the product of the metric tensor g_{ij} with the vector a^i evolves from Eq. (A.10.5)

$$g_{i\alpha} a^\alpha = a_i \tag{A.10.6}$$

where a^i is the *conjugate* or *inverse* vector of a_i, so that

$$a^i a_j = \delta^i_j \tag{A.10.7}$$

Similarly we may define the contravariant metric tensor g^{lm}, the conjugate of g_{lm}, and G^{LM} the conjugate of G_{LM}

$$g^{lm} = \delta^{\alpha\beta} \frac{\partial x^l}{\partial X^\alpha} \frac{\partial x^m}{\partial X^\beta}; \quad G^{LM} = \delta^{\alpha\beta} \frac{\partial X^L}{\partial x^\alpha} \frac{\partial X^M}{\partial x^\beta} \tag{A.10.8}$$

from which it follows

$$g_{l\alpha} g^{m\alpha} = \delta^m_l; \quad G_{L\alpha} G^{M\alpha} = \delta^M_L \tag{A.10.9}$$

The fundamental covariant and contravariant tensors G_{LM} and G^{LM}, respectively, transform according to equations

$$G_{LM} = \frac{\partial x^\alpha}{\partial X^L} \frac{\partial x^\beta}{\partial X^M} g_{\alpha\beta} \tag{A.10.10}$$

$$G^{LM} = g^{\alpha\beta} \frac{\partial X^L}{\partial x^\alpha} \frac{\partial X^M}{\partial x^\beta} \tag{A.10.11}$$

Eq. (A.10.10), when differentiated with respect to X^K, results in

$$\frac{\partial G_{LM}}{\partial X^K} = \frac{\partial x^\alpha}{\partial X^L} \frac{\partial x^\beta}{\partial X^M} \frac{\partial g_{\alpha\beta}}{\partial x^\gamma} \frac{\partial x^\gamma}{\partial X^K} + \frac{\partial^2 x^\alpha}{\partial X^L \partial X^K} \frac{\partial x^\beta}{\partial X^M} g_{\alpha\beta} + \frac{\partial x^\alpha}{\partial X^L} \frac{\partial^2 x^\beta}{\partial X^M \partial X^K} g_{\alpha\beta} \tag{A.10.12}$$

In orthogonal curvilinear coordinates $g_{lm} = 0$ and $G_{LM} = 0$ for all $i \neq j$ and, validating Eq. (A.10.9) by $g_{lm} = 1/g^{lm}$ and $G_{LM} = 1/G^{LM}$ for all $i = j$.

From Eq. (A.10.5) the *physical components* of a first order tensor may be obtained. These are the components that retain their physical significance and dimensions, rather than their geometrical meaning and may be identified by the added overbar

$$\overline{a}_i = \sqrt{(g_{\alpha i})} a^\alpha = \sqrt{(g_{\alpha i})} g^{\alpha\beta} a_\beta \tag{A.10.13}$$

where $\overline{a}_i$ is the physical component of tensor a_i.

Similarly the physical components of a second-order tensor a_{ij} may be derived

$$\overline{a}_{ij} = \sqrt{(g_{\alpha i} g_{\beta j})} a^{\alpha\beta} = \sqrt{(g_{\alpha i} g_{\alpha j})} g^{\alpha\gamma} g^{\beta\rho} a_{\gamma\rho} \tag{A.10.14}$$

As is seen, the physical tensor is neither contravariant nor covariant, therefore its subscript does not indicate covariance, but tensorial order.

A.11 The Angle between Vectors

The angle between two unit vectors a^i and b^i is defined by the cosine of that angle

$$\cos\theta = g^{\alpha\beta} a_\alpha b_\beta = g_{\alpha\beta} a^\alpha b^\beta = a^\alpha b^\alpha \tag{A.11.1}$$

If the vectors a^i and b^i are not unit vectors, the angle is given

$$\cos\theta = \frac{g_{\alpha\beta}a^{\alpha}b^{\beta}}{(g_{\gamma\delta}a^{\gamma}a^{\delta})^{1/2}(g_{\rho\mu}b^{\rho}b^{\mu})^{1/2}} \tag{A.11.2}$$

The vectors are *orthogonal* if $\cos\theta = 0$, that is, if

$$g_{\alpha\beta}a^{\alpha}b^{\beta} = 0; \quad \therefore\ \theta = 90^{\circ} \tag{A.11.3}$$

A.12 Lowering and Raising Indices

A contravariant vector A^L will lower its index to become a covariant vector as follows

$$A^L = G_{L\alpha}A^{\alpha} \tag{A.12.1}$$

which is the associate vector to A^L. Vice versa, a covariant vector A_L raises its index to become a contravariant vector and associate to A_L

$$A_L = G^{L\alpha}A_{\alpha} \tag{A.12.2}$$

This process is often referred to as lowering the superscript or raising the subscript, respectively. Associate second-order tensors of any type may be generated by raising subscripts or lowering superscripts, as follows

$$A^{L}_{M} = G_{M\alpha}A^{L\alpha} = G^{L\alpha}A_{\alpha M} \tag{A.12.3}$$

$$A^{M}_{L} = G_{L\alpha}A^{\alpha M} = G^{M\alpha}A_{L\alpha} \tag{A.12.4}$$

$$A_{LM} = G_{L\alpha}A^{\alpha}_{M} = G_{\alpha M}A^{\alpha}_{L} \tag{A.12.5}$$

$$A^{LM} = G^{L\alpha}A^{M}_{\alpha} = G^{\alpha M}A^{L}_{\alpha} \tag{A.12.6}$$

Associate tensors of any order may be generated by changing their type, as subscripts are raised and superscripts are lowered, like, for instance, $A_{Q}^{LMRP} = G_{\alpha Q}G^{\beta R}G^{\gamma P}A^{L\alpha M}_{\alpha\gamma}$ or $A^{LM}_{RPQ} = G_{R\alpha}A^{\alpha LM}_{PQ}$.

A.13 The Christoffel Symbols

Our aim is now to define how tensors of various order differentiate. In order to do this, we have to investigate first two functions, called the *Christoffel symbols* of the *first kind* and of the *second kind*, respectively, formed from the fundamental tensor

$$\Gamma_{k,ij} = \frac{1}{2}\left(\frac{\partial g_{ik}}{\partial x^j} + \frac{\partial g_{jk}}{\partial x^i} - \frac{\partial g^{ij}}{\partial x^k}\right) \tag{A.13.1}$$

$$\Gamma^k_{ij} = \tfrac{1}{2} g^{k\alpha}\Gamma_{\alpha,ij} \tag{A.13.2}$$

The number of components possible in each kind is $\frac{1}{2}\mathfrak{N}^2(\mathfrak{N}+1)$ and in a three-dimensional coordinate system it is $\frac{1}{2}3^2(3+1) = 18$.

These symbols are not tensors, yet the notation with respect to the summation and contraction convention seems to apply to the Christoffel symbol of the second kind. According to Eq. (A.10.9) we obtain

$$g_{\beta m}\Gamma^\beta_{ij} = g_{\beta m} g^{\beta\alpha}\Gamma_{\alpha,ij} = \delta^\alpha_m \Gamma_{\alpha,ij}\Gamma_{m,ij} \tag{A.13.3}$$

From Eq. (A.13.1) the following equation may be derived

$$\Gamma_{k,ij} + \Gamma_{i,kj} = \frac{\partial g_{ij}}{\partial x^j} \tag{A.13.4}$$

By differentiating Eq. (A.10.9) with respect to x^i we obtain $(\partial g_{\alpha k}/\partial x^l)g^{\alpha j} + (\partial g_{\alpha j}/\partial x^l)g^{\alpha k} = 0$ and, by inner multiplication with g^{im} we have $\partial g^{mk}/\partial x^l = \; g^{\alpha m} g^{\beta k}(\partial g_{\alpha\beta}/\partial x^l) = 0$. In view of Eqs. (A.13.2) and (A.13.4), we obtain

$$\frac{\partial g^{mk}}{\partial x^l} = -g^{m\alpha}\Gamma^k_{\alpha l} - g^{k\alpha}\Gamma^m_{\alpha l} \tag{A.13.5}$$

By applying Eq. (A.10.11) to Eq. (A.13.1), the transformation of the Christoffel symbol of the first kind is obtained, after appropriate manipulations of the equations

$$\Gamma_{N,ML} = \Gamma_{k,ij}\frac{\partial x^i}{\partial X^L}\frac{\partial x^j}{\partial X^M}\frac{\partial x^k}{\partial X^N} + g_{ij}\frac{\partial x^i}{\partial X^N}\frac{\partial^2 x^j}{\partial X^L \partial X^N} \tag{A.13.6}$$

where capital indices indicate that the Christoffel symbol is given in the X^L coordinates. Multiplying Eq. (A.13.6) by (A.10.11) results in the transformation of the Christoffel symbol of the second kind

$$\Gamma^P_{LM} = \Gamma^\alpha_{\beta\gamma}\frac{\partial X^P}{\partial x^\alpha}\frac{\partial x^\beta}{\partial X^L}\frac{\partial x^\gamma}{\partial X^M} + \frac{\partial X^P}{\partial x^\gamma}\frac{\partial^2 x^\gamma}{\partial X^L \partial X^M} \tag{A.13.7}$$

From the second derivatives of the last term of Eqs. (A.13.6) and (A.13.7) it is seen that the Christoffel symbols do not transform as tensors and are therefore not considered such. We may, however, isolate the squared expression from the last equation, (A.13.7), to get

$$\frac{\partial^2 x^r}{\partial X^L \partial X^M} = \Gamma^P_{LM}\frac{\partial x^r}{\partial X^P} - \Gamma^r_{ij}\frac{\partial x^i}{\partial X^L}\frac{\partial x^j}{\partial X^M} \tag{A.13.8}$$

an important equation expressing second partial derivatives in terms of first derivatives and of Christoffel symbols of the second kind.

A.14 Covariant Differentiation of Tensors

Eq. (A.6.1) presents us with the transformation of a contravariant vector a^i. We may now differentiate it with respect to x^i and obtain

$$\frac{\partial A^L}{\partial X^M} = \frac{\partial a^\alpha}{\partial x^\beta}\frac{\partial x^\beta}{\partial X^M}\frac{\partial X^L}{\partial x^\alpha} + a^\alpha \frac{\partial^2 X^L}{\partial x^\alpha \partial x^\beta}\frac{\partial x^\beta}{\partial X^M}$$

By eliminating the second partial derivative in the last term with the help of Eq. (A.13.8), and after certain changes, we obtain

$$\frac{\partial A^L}{\partial X^M} + \Gamma^L_{\alpha M}A^\alpha = \left[\frac{\partial A^\alpha}{\partial X^\beta} + \Gamma^\alpha_{\gamma\beta}A^\gamma\right]\frac{\partial x^\beta}{\partial X^M}\frac{\partial X^M}{\partial x^\alpha}$$

where we see that the expression representing the covariant derivative of a vector transforms as a tensor. We also introduce, by convention, the *comma*, followed by the covariant index denoting the derivative

$$A^L_{,M} \equiv \frac{\partial A^L}{\partial X^M} + \Gamma^L_{\alpha M}A^\alpha = A^\alpha_{,\beta}\frac{\partial x^\beta}{\partial X^M}\frac{\partial X^L}{\partial x^\alpha} \tag{A.14.1}$$

Thus $A^L_{,M}$ is a mixed tensor of the second order, named the *covariant derivative* of A^L with respect to X^L. To get the covariant derivative of a covariant vector α^i the same routine is followed and the final result obtained is

$$A_{L,M} = \frac{\partial A_L}{\partial X^M} - \Gamma^\alpha_{LM}A_\alpha = a_{\alpha,\beta}\frac{\partial x^\alpha}{\partial X^L}\frac{\partial x^\beta}{\partial X^M} \tag{A.14.2}$$

The covariant derivative raises the order of the tensor by one degree.

The covariant differentiation of a second-order tensor is defined in a similar way, i.e. Eqs. (A.6.7), (A.6.8), (A.6.9)

$$A^{LM}_{,R} \equiv \frac{\partial A^{LM}}{\partial X^R} + \Gamma^L_{R\alpha}A^{\alpha L} + \Gamma^M_{R\alpha}A^{L\alpha} \tag{A.14.3}$$

$$A^L_{M,R} \equiv \frac{\partial A_M}{\partial X^R} - \Gamma^\alpha_{MR}A^L_\alpha + \Gamma^L_{R\alpha}A^\alpha_M \tag{A.14.4}$$

$$A_{LM,R} \equiv \Gamma^\alpha_{LR}A_{\alpha M} - \Gamma^\alpha_{MR}A_{L\alpha} \tag{A.14.5}$$

As before in the case of the vector, the covariant derivative raises the order of the tensor by one degree.

Applying Eqs. (A.14.3)–(A.14.5) to the fundamental tensors G_{LM} or G^{LM}, g_{lm} or g^{lm}, we obtain

$$G_{LM,R} = G^{LM}_{,R} = 0; \quad g_{lm,r} = g^{lm}_{,r} = 0 \tag{A.14.6}$$

and in general, a tensor of any order, say $A^{NPS}_{,LM}$, differentiates

$$A^{NPS}_{LM,R} \equiv \frac{\partial A^{NPS}_{LM}}{\partial X^R} + \Gamma^N_{R\alpha} A^{\alpha PS}_{LM} + \Gamma^P_{R\alpha} A^{N\alpha S}_{LM} + \Gamma^S_{R\alpha} A^{NP\alpha}_{LM} - \Gamma^\alpha_{LR} A^{NPS}_{\alpha M} - \Gamma^\alpha_{MR} A^{NPS}_{L\alpha} \tag{A.14.7}$$

The contraction of a covariant derivative is also possible, so we may write

$$A^\alpha_{,\alpha} \equiv \frac{\partial A}{\partial X^\alpha} + \Gamma^\alpha_{\beta\alpha} A^\beta = \frac{\partial A}{\partial X^\alpha} + A^\beta \frac{\partial[\log \sqrt{A}]}{\partial A^\beta} \equiv \operatorname{div} \mathbf{A} \tag{A.14.8}$$

which is called the *divergence* of the contravariant vector A^L.

As the differentiation raises the order of the vector by one degree and the contraction reduces the tensorial order by two degrees, there is a net reduction in a contracted differentiation by one degree, and a vector reduces to a scalar. This is compatible with what we know from vector analysis, that the divergence of a vector is a scalar.

In Cartesian rectangular coordinates, where $g_{ij} = \delta_{ij}$, the Christoffel symbols vanish and the covariant derivatives reduce to the plain partial derivatives, and there is therefore no distinction between the covariant derivatives of a contravariant vector or that of a covariant vector.

A scalar function B, when differentiated, becomes a vector by increasing its tensorial order by one degree. We may write

$$B_{,M} = \frac{\partial B}{\partial X^M} \equiv \operatorname{grad} B \tag{A.14.9}$$

and we have a covariant vector that we recognize as the *gradient* of the scalar function B.

If an additional contracted differentiation is applied on the gradient, we get

$$B_{,\alpha\alpha} = (B_{,\alpha})_{,\alpha} = \frac{\partial^2 B}{\partial X^\alpha \partial X^\alpha} - \Gamma^\beta_{\alpha\alpha} \frac{\partial B}{\partial X^\beta} = \nabla^2 B \tag{A.14.10}$$

known as the *Laplacian* of B, and obtained by differentiating twice a scalar function and by contracting it.

Lastly, let us consider the covariant derivative $A_{L,R}$ of the covariant vector A_L and thus for the vector C^L

$$C^L = \sqrt{(G)} e^{L\alpha\beta} A_{\alpha,\beta} \equiv \operatorname{curl} \mathbf{A} \tag{A.14.11}$$

where G is the numerical value of the determinant $|G_{LM}|$, and e^{LMN} is the *permutation symbol*, defined to have the following three distinct values:

$e^{LMN} = 0$ if any two of the indices are identical.
$e^{LMN} = 1$ if the order of the indices form a right-hand manifold.
$e^{LMN} = -1$ if the order of the indices form a left-hand manifold,

and the vector C^L is called the *curl*, or *rotation* of vector A^L.

The definitions show that e^{LMN} is a skew-symmetric system with the same values as those of e_{LMN}.

A.15 Principal Directions of Second-order Tensors

Let A_{LM} be a second-order symmetric tensor, represented by the matrix $|A_{LM}|$ of nine components, six of which are independent, and let B_L be any vector. We may then form an inner product $A_{\alpha M}B_\alpha = C_M$ and obtain a new vector C_M. Vector C_M will differ from vector B_L both in direction and in size. The operation $A_{\alpha M}B_\alpha$ rotates the vector B_L and changes its length. We may look for the vectors B_L that do not rotate but change only their length, such that

$$A_{\alpha M}B_\alpha = \lambda B_M \tag{A.15.1}$$

where λ is a scalar. If such vectors exist they are called *characteristic vectors* of tensor A_{LM} and their directions are the *principal directions* of tensor A_{LM}, where the axes determined by the principal directions are the *principal axes* or *principal triad* of A_{LM}. The problem now is to find the tensor A_{LM} in its principal axes.

In the generalized coordinates there are several ways to reduce the tensor to its principal axes. We may choose, for instance, covariant components and apply Eq. (A.15.1). Since $B_L = G_{L\alpha}B^\alpha$, Eq. (A.15.1) implies

$$(A_{L\alpha} - \lambda G_{L\alpha})B^\alpha = 0 \tag{A.15.2}$$

We may also obtain this equation in its mixed tensorial components

$$(A^L_\alpha - \lambda G^L_\alpha)B^\alpha = 0 \tag{A.15.3}$$

From Eq. (A.15.2), and analogously from Eq. (A.15.3), we may conclude that the determinant formed by the matrix $|A_{LM} - G_{LM}|$ vanishes identically

$$|A_{LM} - G_{LM}| = \begin{vmatrix} A_{XX} - G_{XX} & A_{XY} & A_{XZ} \\ A_{XY} & A_{YY} - G_{YY} & A_{YZ} \\ A_{XZ} & A_{YZ} & A_{ZZ} - G_{ZZ} \end{vmatrix} = 0 \tag{A.15.4}$$

Eq. (A.15.4) when expanded results in a cubic polynomial equation

$$\lambda^3 - \mathrm{I}\lambda^2 + \mathrm{II}\lambda - \mathrm{III} = 0 \tag{A.15.5}$$

called the *characteristic equation*, which has three *latent roots* $\lambda_{\mathfrak{K}}$ also known as *proper numbers* of the equation. For the time being we will just say that the coefficients of the polynomial, I, II and III, are invariants, functions of the components of the tensor A_{LM} (Hanin and Reiner 1956), discussed in Sect. A.17. Since tensor A_{LM} is symmetric, $A_{LM} = A_{ML}$, all the roots of the characteristic equation are real. The $\mathfrak{N} = 3$ values of $\lambda_{\mathfrak{K}}$ are not uniquely determined, but depend on the vector B^L and vice versa. We may thus say that the values of $\lambda_{\mathfrak{K}}$ determine the values of vector B^L, and in general we may write

$$(A_{\alpha M} - \lambda_{\mathfrak{K}} G_{\alpha M}) B^{\alpha}_{\mathfrak{K}} = 0 \tag{A.15.6}$$

By changing the coordinates from X_L to x_i in Eq. (A.15.6) we go through the transformation

$$(a_{\alpha\beta} - \lambda_{\mathfrak{K}} g_{\alpha\beta}) \frac{\partial x^{\alpha}}{\partial X^{\gamma}} \frac{\partial x^{\beta}}{\partial X^{M}} B^{\gamma}_{\mathfrak{K}} = 0 \tag{A.15.7}$$

which, by inner multiplication with $\partial X^L/\partial x^i$ yields

$$(a_{\alpha j} - \lambda_{\mathfrak{K}} g_{\alpha j}) \frac{\partial x^{\alpha}}{\partial X^{\beta}} B^{\beta}_{\mathfrak{K}} = 0 \tag{A.15.8}$$

The $\mathfrak{N} = 3$ quantities $(\partial x^i/\partial x^{\beta}) B^{\beta}_{\mathfrak{K}}$ are determined by the values of $\lambda_{\mathfrak{K}}$.

According to Eq. (A.10.2) we shall define the vectors $B^L_{\mathfrak{K}}$ so as to determine a unit vector

$$\mathrm{d}S^2 = G_{\alpha\beta} B^{\alpha}_{\mathfrak{K}} B^{\beta}_{\mathfrak{K}} = I_{\mathfrak{K}} \tag{A.15.9}$$

The direction of the unit length vector $I_{\mathfrak{K}}$ is dependent on the vectors $B^L_{\mathfrak{K}}$. Accordingly it may be shown that to any proper number $\lambda_{\mathfrak{M}}$ corresponds a unit vector $B^L_{\mathfrak{M}}$ satisfying Eqs. (A.15.6) and (A.15.9)

$$(A_{\alpha M} - \lambda_{\mathfrak{M}} G_{\alpha M}) B^{\alpha}_{\mathfrak{M}} = 0 \tag{A.15.10}$$

$$G_{\alpha\beta} B^{\alpha}_{\mathfrak{M}} B^{\beta}_{\mathfrak{M}} = I_{\mathfrak{M}} \tag{A.15.11}$$

Let us choose two of the roots $B^L_{\mathfrak{K}}$ and $B^L_{\mathfrak{M}}$. Since $\lambda_{\mathfrak{K}} \neq \lambda_{\mathfrak{M}}$, the inner product of Eq. (A.15.6) by $B^L_{\mathfrak{M}}$ and Eq. (A.15.10) by $B^L_{\mathfrak{K}}$ results, after subtraction

$$(\lambda_{\mathfrak{M}} - \lambda_{\mathfrak{K}})G_{\alpha\beta}B^{\alpha}_{\mathfrak{K}}B^{\beta}_{\mathfrak{M}} = 0 \tag{A.15.12}$$

and since $\lambda_{\mathfrak{K}} \neq \lambda_{\mathfrak{M}}$, as stated before, we get necessarily

$$G_{\alpha\beta}B^{\alpha}_{\mathfrak{K}}B^{\beta}_{\mathfrak{M}} = 0 \tag{A.15.13}$$

meaning that, according to Eq. (A.11.3), the unit vectors $B^L_{\mathfrak{K}}$ and $B^L_{\mathfrak{M}}$ are orthogonal. Thus Eq. (A.15.4) represents three mutually orthogonal vectors at a point, the principal triad of tensor A_{LM}.

The roots λ resulting from Eq. (A.15.5) are real numbers, if $G_{\alpha\beta}\,\mathrm{d}X^{\alpha}\,\mathrm{d}X^{\beta}$ are positive-definite, and are invariants.

By integrating Eq. (A.15.6) according to $B^L_{\mathfrak{K}}$ we get the general equation $A_{\alpha\beta}B^{\alpha}B^{\beta} = \lambda G_{\alpha\beta}B^{\alpha}B^{\beta}$, from which

$$\lambda = \frac{A_{\alpha\beta}B^{\alpha}B^{\beta}}{G_{\alpha\beta}B^{\alpha}B^{\beta}} \tag{A.15.14}$$

To find the maximum and minimum values of λ we differentiate twice with respect to B^L, setting both results equal to zero, and so we get Eq. (A.15.5) for the first differentiation and Eq. (A.15.4) for the second differentiation. Thus the maximum and minimum values of λ are those that correspond to the principal directions.

A.16 Differential Operators

The differential operators, the *gradient*, the *divergence* and the *curl* have analogous forms in tensor mathematics. It will be shown how these operators are applied to tensors of order zero, order one and any higher order, which correspond to a scalar φ, a vector $\mathbf{a} \equiv a^i$ and a tensor, say, $A^{\,k}_{lm}$, of any order, respectively. For that purpose we shall define the vector $g^i \equiv \mathbf{g}$, formed from the fundamental tensor g^{ij}

$$\mathbf{g} = g^i\mathbf{i} + g^j\mathbf{j} + g^k\mathbf{k} \equiv g^i \tag{A.16.1}$$

The *gradient* of a scalar φ is defined

$$\operatorname{grad}\varphi = \boldsymbol{\nabla}\varphi \equiv g^i\,\frac{\partial\varphi}{\partial x^i} \equiv \varphi_{,i}g^i\mathbf{i} + \varphi_{,j}g^j\mathbf{j} + \varphi_{,k}g^k\mathbf{k} \equiv \varphi_{,i}g^i \tag{A.16.2}$$

where

$$\nabla \equiv g^i \frac{\partial}{\partial x^i} \tag{A.16.3}$$

Similarly, the gradient of a vector or of a tensor **T** of any order can be defined

$$\operatorname{grad}(T^{ij}_{kl}) \equiv g_i g_j g^k g^l g^m \frac{\partial T^{ij}_{kl}}{\partial x^m} = T^{ij}_{kl,m}\, g_i g_j g^k g^l g^m \tag{A.16.4}$$

The *divergence* of a tensor **T** of any order is defined as the tensor obtained by contracting any one contravariant index with the covariant differentiation index

$$\operatorname{div}(T^{im}_{jkl}) = \nabla_m(T^{im}_{jkl}) \equiv T^{i\alpha}_{jkl,\alpha}\, g_i g^j g^k g^l \tag{A.16.5}$$

The *curl* of a vector **T** which is a covariant tensor of the first order, is defined

$$\begin{aligned}\operatorname{curl}\mathbf{T} &= \nabla \times \mathbf{T} \equiv \frac{\partial}{\partial x^i} \times T_i = \left(\frac{\partial T_k}{\partial x^j} - \frac{\partial T_j}{\partial x^k}\right)\mathbf{i} \\ &\quad + \left(\frac{\partial T_i}{\partial x^k} - \frac{\partial T_k}{\partial x^i}\right)\mathbf{j} + \left(\frac{\partial T_j}{\partial x^i} - \frac{\partial T_i}{\partial x^j}\right)\mathbf{k} \\ &\equiv \overline{T_i} \equiv e^{i\alpha\beta} T_{\beta,\alpha} g_i \end{aligned} \tag{A.16.6}$$

where $\overline{T_i}$ is a vector normal to the plane formed by the vectors ∇ and T_i and is positive when these vectors form a right-hand manifold and negative if they form a left-hand manifold.

A.17 Orthogonal and Cartesian Coordinates

The Riemann space contains, as a subgroup, spaces of orthogonal coordinates, with perpendicular tangents at the points of intersection of the coordinates. Orthogonal coordinates, in turn, contain the Cartesian space, in which the coordinates are rectangular. The transformations involved in all those subspaces are from one orthogonal triad to another. In the general case, such a transformation consists of a rotation about the origin plus a translation.

It can be shown that in curvilinear orthogonal coordinates the non-diagonal components of all fundamental tensors vanish, that is, $G^{LM} = G_{LM} = g^{ij} = g_{ij} = 0$ if $L \neq M$ and $i \neq j$, respectively, and hence $G^{LM} = 1/G_{LM}$ and $g^{ij} = 1/g_{ij}$.

In Cartesian coordinates, as is seen further, all fundamental tensors are reduced to the unit tensor, and the distinction between covariant and contravariant tensors is also abrogated, thus all indices in Cartesian coordinates are subscripts, on the condition that we allow only transformations into Cartesian coordinates.

Since most of our derivations are worked out in Cartesian coordinates and some in curvilinear orthogonal, our attention to those spaces is apportioned accordingly.

The line element for the Cartesian coordinates has already been given in Eq. (A.10.1)

$$ds^2 = dy_\alpha dy_\alpha = \delta_{\alpha\beta} dy_\alpha dy_\beta \tag{A.17.1}$$

where δ_{ij} is the Kronecker unit tensor already mentioned. The immediate meaning of this equation is that the fundamental tensor g_{ij} reduces to a unit tensor δ_{ij}

$$g_{ij} \equiv \delta_{ij}; \quad g_{ij} = 0 \quad \text{for } i \neq j \quad \text{and} \quad g_{ij} = 1 \quad \text{for } i = j \tag{A.17.2}$$

The vector x_i, defined in one Cartesian triad transforms into another Cartesian triad, say y_i, by the linear equations

$$y_i = a_{i\alpha} x_\alpha + b_i \tag{A.17.3}$$

where the first part of the right-hand side represents the rotation about the origin, and the second part represents the translation of the origin. The necessary and sufficient condition that y_i form a set of rectangular coordinates is, according to Eq. (A.17.1)

$$ds^2 = dy_\alpha dy_\alpha = a_{\alpha\beta} a_{\alpha\gamma} dx_\beta dx_\gamma = \delta_{\alpha\beta} dx_\alpha dx_\beta \tag{A.17.4}$$

from which

$$(a_{\alpha i} a_{\beta j} - \delta_{ij}) dx_\alpha dx_\beta = 0 \tag{A.17.5}$$

for all values of dx_i, or

$$a_{\alpha i} a_{\alpha j} = \delta_{ij} \tag{A.17.6}$$

from which it follows that the determinant $|a_{ij}|$ is either $+1$ or -1, depending on whether it is a right-handed or a left-handed orthogonal triad, respectively. By inner multiplication of Eq. (A.17.2) with a_{ij} we obtain

$$x_i = a_{\alpha i} y_\alpha - a_{\alpha i} b_\alpha \tag{A.17.7}$$

from which follows

$$\frac{\partial y^i}{\partial x^j} = \frac{\partial x^j}{\partial y^i} = a_{ij} \tag{A.17.8}$$

The meaning of Eq. (A.17.8) is that in a Cartesian triad the distinction between contravariance and covariance is meaningless. Consequently, as already noted in the beginning of this section and in Sect. A.14, all indices in Cartesian coordinates will be subscripts.

A tridimensional Cartesian tensor of the $\mathfrak{N}$th order transforms according to Eqs. (A.6.10) and (A.17.8)

$$c_{1,2\ldots i\ldots\mathfrak{N}} = a_{\alpha_1 1} a_{\alpha_2 2} \ldots a_{\alpha_{\mathfrak{N}}\mathfrak{N}} b_{\alpha_1} b_{\alpha_2} \ldots b_{\alpha_{\mathfrak{N}}} \tag{A.17.9}$$

We may further indicate that since in Cartesian coordinates the fundamental tensor reduces to a constant, namely the unit tensor, both Christoffel symbols become zero, in which case any differentiation of a tensor reduces to the partial derivative of the corresponding coordinates. Consequently, the gradient and the Laplacian of a scalar function φ and the divergence and the curl of a vector A_i in Cartesian coordinates may be derived

$$\operatorname{grad} \varphi \equiv \varphi_{,i} \equiv \frac{\partial \varphi}{\partial x}\mathbf{i} + \frac{\partial \varphi}{\partial y}\mathbf{j} + \frac{\partial \varphi}{\partial z}\mathbf{k} \tag{A.17.10}$$

$$\operatorname{div} \mathbf{A} \equiv A_{\alpha,\alpha} \equiv \frac{\partial A_x}{\partial x} + \frac{\partial A_y}{\partial y} + \frac{\partial A_z}{\partial z} \tag{A.17.11}$$

$$\nabla^2 \varphi \equiv \varphi_{,\alpha\alpha} \equiv \frac{\partial^2 \varphi}{\partial x^2} + \frac{\partial^2 \varphi}{\partial y^2} + \frac{\partial^2 \varphi}{\partial z^2} \tag{A.17.12}$$

$$\operatorname{curl} A_i = \left[\frac{\partial A_k}{\partial x_j} - \frac{\partial A_j}{\partial x_k}\right]\mathbf{i} + \left[\frac{\partial A_i}{\partial x_k} - \frac{\partial A_k}{\partial x_i}\right]\mathbf{j} + \left[\frac{\partial A_j}{\partial x_i} - \frac{\partial A_i}{\partial x_j}\right]\mathbf{k} \equiv e_{i\alpha\beta} A_{\alpha,\beta} \tag{A.17.13}$$

Eqs. (A.17.10)–(A.17.13) can be derived from Eqs. (A.16.2) to (A.16.6) since the vectors formed from the Riemann metric become unity in Cartesian coordinates, $g^i \equiv g_i = 1$.

Further vector field expressions in orthogonal Cartesian coordinates and their tensorial transcripts are

$$\mathbf{a} \cdot \mathbf{b} \equiv a_\alpha b_\alpha \tag{A.17.14}$$

$$\mathbf{a} \times \mathbf{b} = \mathbf{c} \equiv c_i = d_{jk} - d_{kj} \tag{A.17.15}$$

$$\nabla \cdot f\mathbf{a} = \operatorname{div} f\mathbf{a} = f(\nabla \cdot \mathbf{a}) + (\nabla f) \cdot \mathbf{a} = f \operatorname{div} \mathbf{a} + \operatorname{grad} f \cdot \mathbf{a} \equiv (f a_\alpha)_{,\alpha} = f_{,\alpha} a_\alpha + f a_{\alpha,\alpha} \tag{A.17.16}$$

$$\nabla \times f\mathbf{a} = \text{curl}\, f\mathbf{a} = f(\nabla \times \mathbf{a}) + (\nabla f) \times \mathbf{a} = f\,\text{curl}\,\mathbf{a} + \text{grad}\, f \times \mathbf{a}$$

$$\equiv e_{i\alpha\beta}(fa_\alpha)_{,\beta} \tag{A.17.17}$$

$$\text{div}\,(\mathbf{a} \times \mathbf{b}) = \mathbf{b} \cdot \text{curl}\,\mathbf{a} - \mathbf{a} \cdot \text{curl}\,\mathbf{b} = \mathbf{b} \cdot (\nabla \times \mathbf{a}) - \mathbf{a} \cdot (\nabla \times \mathbf{b})$$

$$\equiv c_{\alpha,\alpha} = b_\alpha e_{\alpha\beta\gamma} a_{\beta,\gamma} + a_\alpha e_{\alpha\beta\gamma} b_{\beta,\gamma} \tag{A.17.18}$$

$$\text{curl}\,(\mathbf{a} \times \mathbf{b}) = \nabla \times (\mathbf{a} \times \mathbf{b}) = \mathbf{a}\,\text{div}\,\mathbf{b} + (\mathbf{b} \cdot \nabla)\mathbf{a} - \mathbf{b}\,\text{div}\,\mathbf{a} - (\mathbf{a} \cdot \nabla)\mathbf{b}$$

$$= \mathbf{a}(\nabla \cdot \mathbf{b}) + (\mathbf{b} \cdot \nabla)\mathbf{a} - \mathbf{b}(\nabla \cdot \mathbf{a}) - (\mathbf{a} \cdot \nabla)\mathbf{b}$$

$$\equiv e_{i\alpha\beta} c_{\alpha,\beta} = a_i b_{\alpha,\alpha} + b_{i,\alpha} a_\alpha - b_i a_{\alpha,\alpha} - a_{i,\alpha} b_a \tag{A.17.19}$$

$$\text{grad}\,(\mathbf{a} \cdot \mathbf{b}) = (\mathbf{a} \cdot \nabla)\mathbf{a} + (\mathbf{a} \cdot \nabla)\mathbf{b} + \mathbf{b} \times (\nabla \times \mathbf{a})$$

$$+ \mathbf{a} \times (\nabla \times \mathbf{b})$$

$$\equiv (a_\alpha b_\alpha)_{,i} = a_{\alpha,i} b_\alpha + a_\alpha b_{\alpha,i} \tag{A.17.20}$$

$$\text{curl}\,\text{grad}\,\varphi = \nabla \times \nabla\varphi \equiv \varphi_{,jk} - \varphi_{,kj} = 0 \tag{A.17.21}$$

$$\text{div}\,\text{curl}\,\mathbf{a} = \nabla \cdot (\nabla \times \mathbf{a}) \equiv e_{\alpha\beta\gamma} a_{\beta,\gamma\alpha} = 0 \tag{A.17.22}$$

$$\text{curl}\,\text{curl}\,\mathbf{a} = \nabla \times (\nabla \times \mathbf{a}) = \text{grad}\,\text{div}\,\mathbf{a} - (\nabla \cdot \nabla)\mathbf{a} = \nabla(\nabla \cdot \mathbf{a}) - \nabla^2\mathbf{a}$$

$$\equiv a_{\alpha,\alpha i} - a_{i,\alpha\alpha} \tag{A.17.23}$$

$$\text{div}\,\text{grad}\,\varphi = \nabla \cdot \nabla\varphi \equiv \varphi_{,\alpha\alpha} \tag{A.17.24}$$

$$(\text{div}\,\text{grad})\mathbf{a} = \nabla^2\mathbf{a} \equiv a_{i,\alpha\alpha} \tag{A.17.25}$$

$$\text{grad}\,\text{div}\,\mathbf{a} = \nabla(\nabla \cdot \mathbf{a}) \equiv a_{\alpha,\alpha i} \tag{A.17.26}$$

A.18 Invariants

If a tensor originated quantity transforms in such a way that it maintains its value, we call it an *invariant*. A scalar is definitely an invariant, and an inner product of two tensors of order one, vectors, say $a_\alpha x^\alpha$, is also an invariant since, according to Eqs. (A.6.3) and (A.6.4), $a_\alpha x^\alpha = (\partial x^\gamma/\partial X^\alpha)(\partial x^\alpha/\partial X^\rho)A_\gamma X^\rho = A_\alpha X^\alpha$. Similarly it may be shown that a magnitude such as $g^{\alpha\beta}a_{\alpha\beta}$ is also invariant when transformed from one coordinate to another. We are particularly interested in invariants resulting from second-order symmetric tensors, and shall confine ourselves to the discussion of such invariants.

Let A_{LM} be a second-order symmetric tensor represented by the matrix $|A_{LM}|$ of nine components, six of which are independent. Since we are dealing with a tridimensional space, Eq. (A.15.1) has three latent roots, so that the determinant formed by the matrix $|A_{LM} - G_{LM}|$ vanishes identically, and we obtain Eq. (A.15.4). In Cartesian coordinates Eq. (A.15.4) will have the following form

$$|A_{LM} - \delta_{LM}| = \begin{vmatrix} A_{XX} - \lambda\delta_{XX} & A_{XY} & A_{XZ} \\ A_{XY} & A_{YY} - \lambda\delta_{YY} & A_{YZ} \\ A_{XZ} & A_{YZ} & A_{ZZ} - \lambda\delta_{ZZ} \end{vmatrix} = 0 \tag{A.18.1}$$

When expanded, Eq. (A.18.1) results in an equation analogous to (A.15.5)

$$|A_{LM} - \delta_{LM}| \equiv \lambda^3 - \mathrm{I}\lambda^2 + \mathrm{II}\lambda - \mathrm{III} = 0 \tag{A.18.2}$$

where I, II and III are functions of the components of A_{LM}, as follows

$$\mathrm{I_A} = \delta_{\alpha\beta}A_{\alpha\beta} = A_{\alpha\alpha} = A_{XX} + A_{YY} + A_{ZZ} = A_{11} + A_{22} + A_{33} \tag{A.18.3}$$

$$\begin{aligned}\mathrm{II_A} &= \tfrac{1}{2!}\,\delta_{\alpha\beta}\delta_{\gamma\rho}A_{\alpha\beta}A_{\gamma\rho} = |A_{\alpha\alpha}| = A_{XX}A_{YY} + A_{YY}A_{ZZ} + A_{ZZ}A_{XX} \\ &\quad - A_{XY}^2 - A_{YZ}^2 - A_{ZX}^2 = A_{11}A_{22} + A_{22}A_{33} + A_{33}A_{11}\end{aligned} \tag{A.18.4}$$

$$\begin{aligned}\mathrm{III_A} &= \tfrac{1}{3!}\,e_{\alpha\beta\gamma}e_{\rho\eta\mu}A_{\alpha\beta}A_{\gamma\rho}A_{\eta\mu} = |A_{LM}| \\ &= A_{XX}A_{YY}A_{ZZ} - A_{XX}A_{YZ}^2 - A_{YY}A_{XZ}^2 \\ &\quad - A_{ZZ}A_{XY}^2 + 2A_{XY}A_{YZ}A_{XZ} = A_{11}A_{22}A_{33}\end{aligned} \tag{A.18.5}$$

where A_{11}, A_{22} and A_{33} are the components of tensor A_{LM} in its principal directions.

When the tensor A_{LM} is transformed into another Cartesian orthogonal coordinate, it will become, for instance, α_{ij}, and if it is inserted into Eq. (A.18.1) it will yield the same results as Eqs. (A.18.3)–(A.18.5). Thus Eqs. (A.18.3)–(A.18.5) represent invariants that do not change with the changes in the coordinates, and are known as the *principal invariants* of tensor A_{LM}.

It can be shown that any function of the principal invariants is also an invariant (Gurevich 1948). Most invariants have geometrical or physical meaning. Such are the *moment invariants*, which are functions of the principal invariants and are defined as the sum of the powers of the proper numbers of tensor A_{LM}

$$\bar{\mathrm{I}}_{\mathrm{A}} = A_{\alpha\alpha} = A_{XX} + A_{YY} + A_{ZZ} = A_{11} + A_{22} + A_{33} = \mathrm{I_A} \tag{A.18.6}$$

$$\begin{aligned}\overline{\mathrm{II}}_{\mathrm{A}} &= A_{\alpha\beta}A_{\alpha\beta} = A_{XX}^2 + A_{YY}^2 + A_{ZZ}^2 - 2A_{XY}^2 - 2A_{YZ}^2 - 2A_{XZ}^2 \\ &= A_{11}^2 + A_{22}^2 + A_{33}^2 = \mathrm{I_A^2} - 2\mathrm{II_A}\end{aligned} \tag{A.18.7}$$

$$\overline{\mathrm{III}}_{\mathbf{A}} = A_{\alpha\beta}A_{\beta\gamma}A_{\gamma\alpha} = A_{XX}^3 + A_{YY}^3 + A_{ZZ}^3$$

$$+ 3A_{XY}^2(A_{XX} + A_{YY}) + 3A_{YZ}^2(A_{YY} + A_{ZZ})$$

$$+ 3A_{XZ}^2(A_{XX} + A_{ZZ}) + 6A_{XY}A_{YZ}A_{XZ} = A_{11}^3 + A_{22}^3 + A_{33}^3$$

$$= \mathrm{I}_{\mathbf{A}}^3 - 3\mathrm{I}_{\mathbf{A}}\mathrm{II}_{\mathbf{A}} + 3\mathrm{III}_{\mathbf{A}} \tag{A.18.8}$$

Conversely, we have

$$\mathrm{II}_{\mathbf{A}} = \tfrac{1}{2}(\mathrm{I}_{\mathbf{A}}^2 - \overline{\mathrm{II}}_{\mathbf{A}}) \tag{A.18.9}$$

$$\mathrm{III}_{\mathbf{A}} = \tfrac{1}{6}\mathrm{I}_{\mathbf{A}}^3 - \tfrac{1}{2}\mathrm{I}_{\mathbf{A}}\overline{\mathrm{II}}_{\mathbf{A}} + \tfrac{1}{3}\overline{\mathrm{III}}_{\mathbf{A}} \tag{A.18.10}$$

Another relation between the principal invariants may be derived from Eq. (A.18.2) as the proper number is set to one, $\lambda = 1$

$$\mathrm{I}_{\mathbf{A}} - \mathrm{II}_{\mathbf{A}} + \mathrm{III}_{\mathbf{A}} = 1 \tag{A.18.11}$$

The octahedral invariant Π is defined

$$\Pi_{\mathbf{A}} = \tfrac{1}{2}(\overline{\mathrm{II}}_{\mathbf{A}} - \mathrm{II}_{\mathbf{A}}) = \tfrac{1}{2}(\mathrm{I}_{\mathbf{A}}^2 - 3\mathrm{II}_{\mathbf{A}}) = \tfrac{1}{4}(3\overline{\mathrm{II}}_{\mathbf{A}} - \mathrm{I}_{\mathbf{A}}^2)$$

$$= \tfrac{1}{4}[(A_{XX} - A_{YY})^2 + (A_{YY} - A_{ZZ})^2 + (A_{ZZ} - A_{XX})^2$$

$$+ 6(A_{XY}^2 + A_{YZ}^2 + A_{XZ}^2)]$$

$$= \tfrac{1}{4}[(A_{11} - A_{22})^2 + (A_{22} - A_{33})^2 + (A_{33} - A_{11})^2] \tag{A.18.12}$$

In general a *Kth moment invariant* $\bar{\mathrm{I}}_A^{(K)}$ of a tensor $\mathbf{A}$ is the Kth power of the proper numbers of $\mathbf{A}$ and is defined (Ericksen 1960a)

$$\bar{\mathrm{I}}_A^{(K)} = A_{m1m2}A_{m2m3} \ldots A_{mKm(K+1)} \tag{A.18.13}$$

For higher-order invariants in general use see Gurevich (1948); Goldenblat (1962).

A.19 Integrals of Tensor Fields

Let S be a closed surface bounding a volume V, and ϕ a uniform continuous and integrable tensorial field, then $\int \phi_{\alpha\ldots j}\, \mathrm{d}X_\alpha$ is the line integral, called the *flow* of ϕ along line L. If the line L is a closed contour C, then the integral $\oint_C \phi_{\alpha\ldots j}\, \mathrm{d}X_\alpha$ is called the *circulation* along C.

The surface integral $\int_S \phi_{\alpha\ldots j} n_\alpha \, \mathrm{d}S$ is called the *flux* of ϕ across the surface S in the direction n_i, which is a unit vector normal to the surface S.

The volume integral $\int_V \phi_{i\ldots j} \, \mathrm{d}V$ is the *total* of ϕ in the volume V.

We shall now describe three key theorems of mathematical analysis.

Green–Gauss theorem. Let a tensor field $\phi_{ij\ldots k}$ be defined throughout the volume V and bounded by a closed surface S, and let $\phi_{\alpha ij\ldots k,\alpha}$ be its derivatives, both continuous in V and S, then

$$\int_V \phi_{\alpha j\ldots k,\alpha} \, \mathrm{d}V = \int_S \phi_{\alpha j\ldots k} n_\alpha \, \mathrm{d}S \tag{A.19.1}$$

where n_i is the unit vector normal to S, Fig. A.19.1. This theorem was presented by Gauss (1813) and further developed by Green (1828), and holds for volumes with "holes" bounded by closed surfaces. Also known as the *divergence theorem*, it describes the equality of the quantitative change of the field in a given volume, as expressed on the left-hand side of Eq. (A.19.1), and the flux across the surface as given on the right-hand side of the equation.

The components of the normal unit tensor n_i are the direction cosines of the angles between the unit tensor and the respective coordinates

$$n_i = \frac{\mathrm{d}r^i}{\mathrm{d}r} \equiv \cos(r^i, \mathrm{d}r) \tag{A.19.2}$$

and they satisfy, therefore, the equation

$$n_\alpha n_\alpha = 1 \tag{A.19.3}$$

Stokes' theorem. Given a surface S bounded by a closed contour C, through which a tensor field $\phi_{i\ldots j}$ and its derivatives $\phi_{\alpha\ldots j,\alpha}$, are defined, we have

$$\int_S e_{\gamma\beta\alpha} \phi_{\alpha\ldots j,\beta} n_\gamma \, \mathrm{d}S = \oint_C \phi_{\alpha\ldots j} \, \mathrm{d}X_\alpha \tag{A.19.4}$$

where n_i is the unit tensor normal to the surface S, Fig. A.19.2. As in Gauss' theorem, this theorem named after Stokes (1854), also known as the *curl* or *circulation* theorem, assumes that the surface S is continuous or at least piecewise continuous and if the surface S is pierced by "holes", they are bounded.

Green's second theorem. Let a tensor be defined as a product of two tensors, $\psi_{rq\ldots s}$ and $\phi_{ij\ldots k,l}$, then according to Eq. (A.19.1) we obtain

$$\int_V \psi_{rq\ldots s,\alpha} \phi_{ij\ldots k,\alpha} \, \mathrm{d}V + \int_V \psi_{rq\ldots s} \phi_{ij\ldots k,\alpha\alpha} \, \mathrm{d}V$$

$$= \int_S \psi_{rq\ldots s} \phi_{ij\ldots k,\alpha} n_\alpha \, \mathrm{d}S = \int_S \psi_{rq\ldots s} \phi_{ij\ldots k,\alpha} \, \mathrm{d}S_\alpha \tag{A.19.5}$$

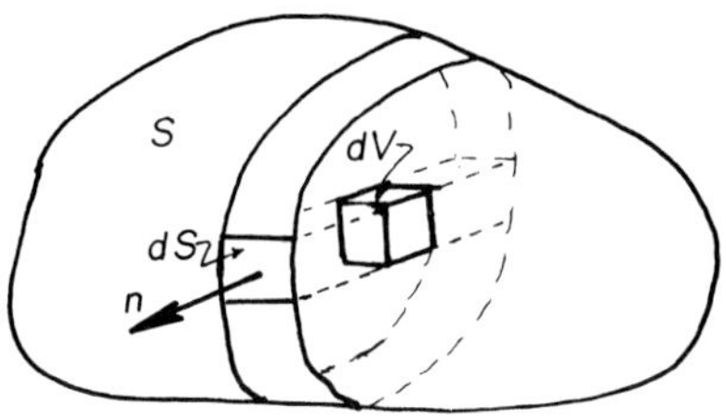

Fig. A.19.1. Green–Gauss theorem.

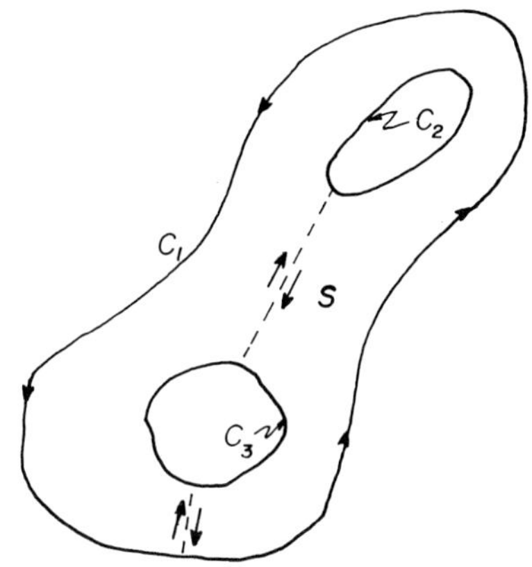

Fig. A.19.2. Stokes theorem.

If the tensors ψ and ϕ are interchanged in Eq. (A.19.5) and the equation obtained is subtracted from Eq. (A.19.5), Green's second theorem also known as the *uniqueness theorem* follows

$$\int_V (\psi_{rq\ldots s}\phi_{ij\ldots k,\alpha\alpha} - \phi_{ij\ldots k}\psi_{rq\ldots s,\alpha\alpha})\,\mathrm{d}V$$

$$= \int_S (\psi_{rq\ldots s}\phi_{ij\ldots k,\alpha} - \phi_{ij\ldots k}\psi_{rq\ldots s,\alpha})\,\mathrm{d}S_\alpha \tag{A.19.6}$$

Eqs. (A.19.5) and (A.19.6) are of extreme importance in hydrodynamics and electromagnetics.

The Stokes theorem, also called the Gauss or Kelvin theorem, is a recursion equation by which volume integrals are reduced to surface integrals and surface integrals can further be reduced to closed line integrals.

Eqs. (A.19.1) and (A.19.4) can now be further developed. A direct conclusion of Eq. (A.19.1) is that if the tensor ϕ is of order zero, that is, a scalar, or of order one, a vector, we can write, respectively

$$\int_V \operatorname{grad}\phi\,\mathrm{d}V \equiv \int_V \phi_{,i}\,\mathrm{d}V = \int_S \phi n_i\,\mathrm{d}S \tag{A.19.7}$$

$$\int_V \operatorname{div}\phi_i\,\mathrm{d}V \equiv \int_V \phi_{\alpha,\alpha}\,\mathrm{d}V = \int_S \phi_\alpha n_\alpha\,\mathrm{d}S \tag{A.19.8}$$

and from Eq. (A.19.4) we deduce

$$\int_S \operatorname{curl}\phi_\alpha n_\alpha\,\mathrm{d}S \equiv \int_S e_{\alpha\beta\gamma}\phi_{\alpha,\beta}n_\gamma\,\mathrm{d}S = \oint_C \phi_\alpha\,\mathrm{d}X_\alpha \tag{A.19.9}$$

A.20 Geometrical Representation of Second-order Tensors

We have learned that a triad X^L may be rotated so that a second-order symmetric tensor A_{LM}, defined in that triad by its six independent components, changes its components according to the rules of transformation. Eventually it may be rotated into its principal axes where only its diagonal components are retained, all other non-diagonal components vanishing identically, $A_{LM} = 0$ for $L \neq M$. A second-order symmetric tensor can be visualized as a set of values acting at a point in the corresponding coordinate directions. For the sake of demonstration, the point in question will be considered an infinitesimal cubic volume with its mutually perpendicular edges oriented in the direction of the coordinates, Fig. A.20.1. Any rotation of the coordinates changes the components. Let us cut through the element, close to the origin, with an oblique plane normal to a vector n^L, so as to form a tetrahedron, Fig. A.20.2. The components of vector n^L will then represent the direction cosines of the vector with the coordinates. Thus, they have to satisfy

$$\delta_{\alpha\beta} n_\alpha n_\beta = 1 \tag{A.20.1}$$

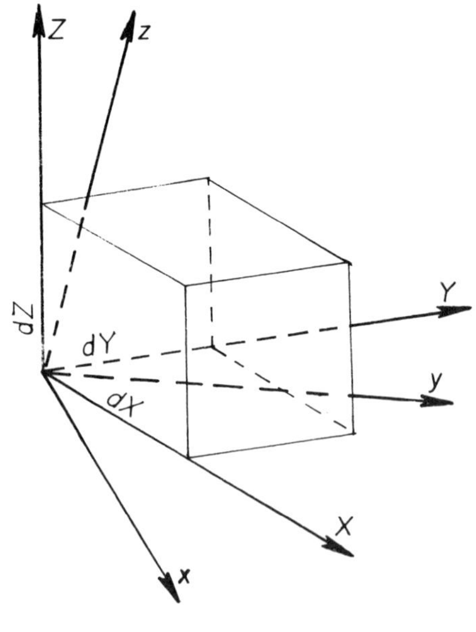

Fig. A.20.1. Infinitesimal cubic volume.

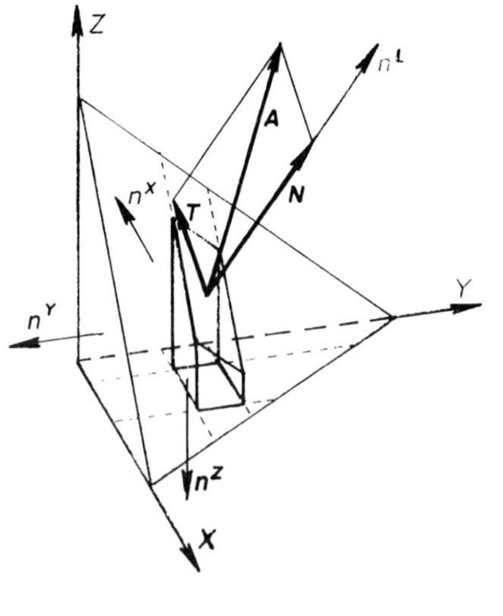

Fig. A.20.2. Principal triad directions cut by an oblique plane perpendicular to unit vector n^L.

It is interesting to note that if the oblique plane is chosen so that its area is equal to unity, then the areas of the other three orthogonal faces of the octahedron are equal to those cosine values.

By inner multiplication of tensor A_{LM} with n^L a vector A_M is obtained

$$A_L = A_{\lambda L} n^\alpha \tag{A.20.2}$$

The components of vector A_L are in the oblique plane and in the directions of the coordinates. For orthogonal coordinates the length A of vector A_L is

$$A^2 = \delta^{\alpha\beta} A_\alpha A_\beta = \delta^{\alpha\beta} A_{\beta\rho} A_{\alpha\gamma} n^\rho n^\gamma \tag{A.20.3}$$

In addition to its components in the direction of the coordinates, vector A_L may be resolved into two other components, one *normal* to the oblique plane, say **N**, and one in the oblique plane and *tangential* to **N**, **T**. The length of the normal component may be found by inner multiplication of vector A_L with the directions cosine n^L

$$N = A_\alpha n^\alpha = A_{\alpha\beta} n^\alpha n^\beta \tag{A.20.4}$$

while the length of the tangential component T is found by combining Eqs. (A.20.3) and (A.20.4)

$$\begin{aligned} T^2 &= A^2 - N^2 = \delta^{\alpha\beta} A_\alpha A_\beta - \delta^{\alpha\beta} \delta^{\beta\gamma} A_\alpha n^\gamma \\ &= A_\alpha A_\beta (\delta^{\alpha\beta} - n^\alpha n^\beta) = A_{\alpha\beta} (\delta^{\alpha\gamma} A_{\gamma\rho} n^\rho n^\beta - \delta^{\alpha\beta} A_{\rho\rho}) \end{aligned} \tag{A.20.5}$$

Fig. A.20.2 shows vectors **A**, **N**, **T** acting on the tetrahedron.

Inserting the values of A_L from Eq. (A.20.2) into Eq. (A.20.4) we obtain

$$N = A_{\alpha\beta} n^\alpha n^\beta \tag{A.20.6}$$

which is the equation of a conical section, and in the present specific case an *ellipsoid*. This may be seen in a much simpler way when tensor A_{ij} is expressed in its principal axes, and then the form of Eq. (A.20.6) will be

$$N = A_{11} n_1^2 + A_{22} n_2^2 + A_{33} n_3^2 \tag{A.20.7}$$

or, by dividing it by N, we obtain

$$\frac{A_{11}}{N} n_1^2 + \frac{A_{22}}{N} n_2^2 + \frac{A_{33}}{N} n_3^2 = 1 \tag{A.20.8}$$

an equation of an ellipsoid in the variables n_1, n_2, n_3. Also, by taking tensor

A_{LM} from Eq. (A.20.2) in its principal directions and inserting it in Eq. (A.20.1), we get in explicit form

$$\frac{A_1^2}{A_{11}^2} + \frac{A_2^2}{A_{22}^2} + \frac{A_3^2}{A_{33}^2} = 1 \tag{A.20.9}$$

expressing again the equation of an ellipsoid in the variables A_1, A_2, A_3. This is the ellipsoid of tensor A_{LM}, with the three orthogonal radii equal to A_{11}, A_{22} and A_{33}, and it represents the location of the end points of the resultant vector A_L in the oblique plane.

If the radii are numerically equal we obtain a sphere, thus the tensor A_{LM} is *isotropic* and has only diagonal components, which are equal and may be written

$$A_{LM} = A_{\alpha\alpha}\delta_{LM}; \quad A_{LL} = A_{MM} = A_{NN}; \quad A_{LM} = 0 \text{ for } L = M \tag{A.20.10}$$

If one of the principal components is zero, the tensor ellipsoid reduces to an area of an ellipse and we have a two-dimensional or a plane system.

If two of the principal components are equal, say $A_{11} = A_{22} \neq A_{33}$, we have an eggoid and the tensor forms an *axially symmetric* system.

The explicit form of Eq. (A.20.5) in principal coordinates is

$$T^2 = A^2 - N^2 = A_{11}^2 n_1^2 + A_{22}^2 n_2^2 + A_{33}^2 n_3^2 - [A_{11} n_1^2 + A_{22} n_2^2 + A_{33} n_3^2]^2 \tag{A.20.11}$$

We now eliminate one of the direction cosines, say n_3, by applying Eq. (A.20.1) to Eq. (A.20.11) and differentiating once with respect to n_1 and once with respect to n_2. Two equations are obtained, which are then set equal to zero in order to get the maximum value of **T**

$$n_1[(A_{11} - A_{33})n_1^2 + (A_{22} - A_{33})n_2^2 - \tfrac{1}{2}(A_{11} - A_{33})] = 0 \tag{A.20.12}$$

$$n_2[(A_{11} - A_{33})n_1^2 + (A_{22} - A_{33})n_2^2 - \tfrac{1}{2}(A_{22} - A_{33})] = 0 \tag{A.20.13}$$

One solution of these equations is $n_1 = n_2 = 0$, then $n_3 = \pm 1$. Other solutions are $n_1 = 0$, $n_2 = \pm\sqrt{\tfrac{1}{2}}$, $n_3 = \pm\sqrt{\tfrac{1}{2}}$ and $n_1 = \pm\sqrt{\tfrac{1}{2}}$, $n_2 = 0$, $n_3 = \pm\sqrt{\tfrac{1}{2}}$. The calculation is repeated by eliminating from Eq. (A.20.11) n_2 and then n_1. Finally two types of solutions are available, 0, 0, ± 1 and 0, $\pm\sqrt{\tfrac{1}{2}}$, $\pm\sqrt{\tfrac{1}{2}}$, which are rotated for the three direction cosines, altogether six solutions, for which the tangential components are maximum. The first solution gives the plane of the coordinates coinciding with the principal axes as we assumed, the second solution results in planes bisecting the angles between the two principal axes. Substituting these values into Eq. (A.20.11), the following is obtained for the tangential component

$$T' = \pm\tfrac{1}{2}(A_{11} - A_{22}); \quad T'' = \pm\tfrac{1}{2}(A_{11} - A_{33}); \quad T''' = \pm\tfrac{1}{2}(A_{22} - A_{33}) \tag{A.20.14}$$

showing that the maximum tangential component is equal to half the difference between these two principal components.

Eqs. (A.20.1), (A.20.4) and (A.20.5) may be solved for values n^L, the explicit form of which, for the case of principal axes, is

$$\begin{aligned} n_1^2 &= \frac{(A_{22} - N)(A_{33} - N) + T^2}{(A_{22} - A_{11})(A_{33} - A_{11})} \\ n_2^2 &= \frac{(A_{33} - N)(A_{11} - N) + T^2}{(A_{33} - A_{22})(A_{11} - A_{22})} \\ n_3^2 &= \frac{(A_{11} - N)(A_{22} - N) + T^2}{(A_{11} - A_{33})(A_{22} - A_{33})} \end{aligned} \tag{A.20.15}$$

which, after some rearrangement, becomes

$$\begin{aligned} &[N - \tfrac{1}{2}(A_{22} + A_{33})]^2 + T^2 \\ &\quad = n_1^2(A_{22} - A_{11})(A_{33} - A_{11}) + \tfrac{1}{4}(A_{22} - A_{33})^2 \\ &[N - \tfrac{1}{2}(A_{33} + A_{11})]^2 + T^2 \\ &\quad = n_2^2(A_{33} - A_{22})(A_{11} - A_{22}) + \tfrac{1}{4}(A_{33} - A_{11})^2 \\ &[N - \tfrac{1}{2}(A_{11} + A_{22})]^2 + T^2 \\ &\quad = n_3^2(A_{11} - A_{33})(A_{22} - A_{33}) + \tfrac{1}{4}(A_{11} - A_{22})^2 \end{aligned} \tag{A.20.16}$$

If we hold $n^L = \text{const}$ in these equations, a set of concentric circles in the variables N and T, with centers at $\tfrac{1}{2}(A_{22} - A_{33})$, $\tfrac{1}{2}(A_{33} - A_{11})$, $\tfrac{1}{2}(A_{11} - A_{22})$, respectively, is obtained for each of the equations, Fig. A.20.3. For $n^L = 0$, that is, an angle of 90°, the equations yield

$$\begin{aligned} &[N - \tfrac{1}{2}(A_{22} + A_{33})]^2 + T^2 = \tfrac{1}{4}(A_{22} - A_{33})^2 \qquad n^L = 0 \\ &[N - \tfrac{1}{2}(A_{33} + A_{11})]^2 + T^2 = \tfrac{1}{4}(A_{33} - A_{11})^2 \\ &[N - \tfrac{1}{2}(A_{11} + A_{22})]^2 + T^2 = \tfrac{1}{4}(A_{11} - A_{22})^2 \end{aligned} \tag{A.20.17}$$

which are three specific circles, maintaining a plane state of components in X^1, X^2 and X^3, with radii equal to $\tfrac{1}{2}(A_{22} + A_{33})$, $\tfrac{1}{2}(A_{33} + A_{11})$ and $\tfrac{1}{2}(A_{11} + A_{22})$, respectively.

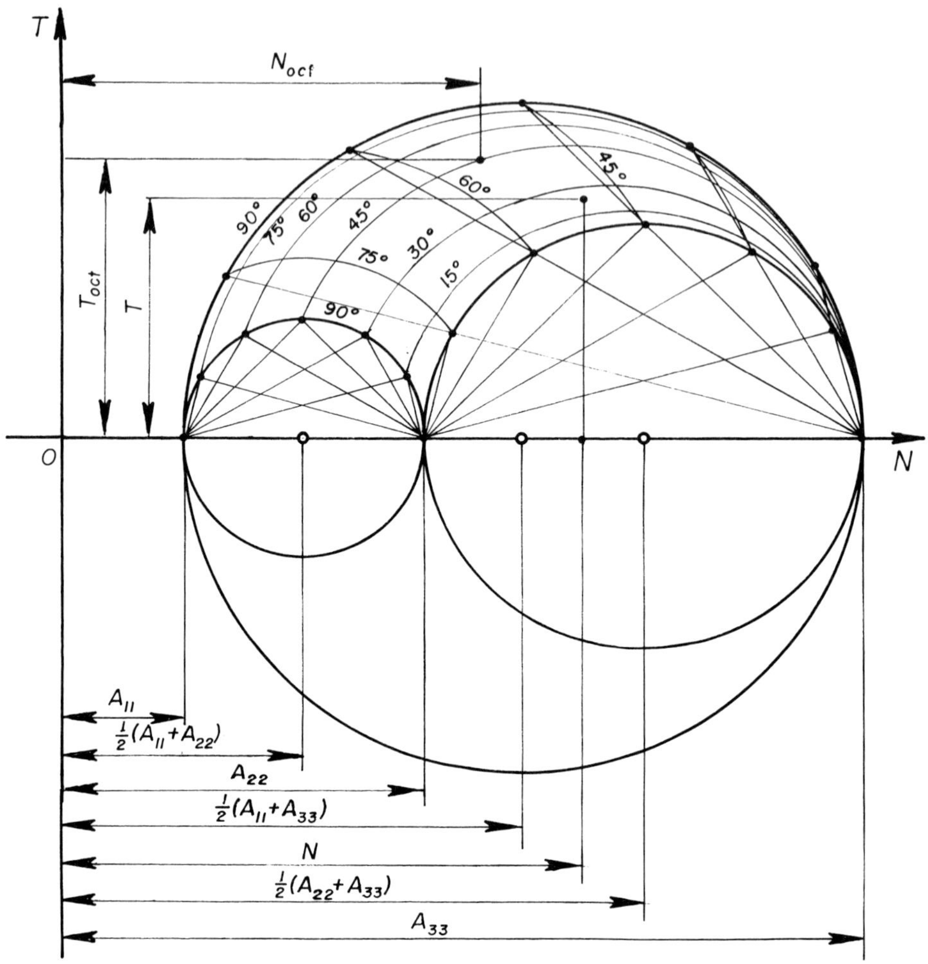

Fig. A.20.3. Mohr circles representing a second order tensor of three dimensions.

The graphical representation of tensor A_{ij} in Fig. A.20.3 has been presented by Mohr (1914) and is named after him, the *Mohr circles*.

In a two-dimensional coordinate system, say X^L and X^M, tensor A_{LM} has the form

$$|A_{LM}| = \begin{vmatrix} A_{XX} & A_{XY} \\ A_{XY} & A_{YY} \end{vmatrix} \tag{A.20.18}$$

and therefore Eqs. (A.20.4) and (A.20.5), after rearrangement, receive the form

$$\begin{aligned} N &= A_{XX}\cos^2\alpha + 2A_{XY}\cos\alpha\sin\alpha + A_{YY}\sin^2\alpha \\ &= \tfrac{1}{2}(A_{XX} + A_{YY}) + \tfrac{1}{2}(A_{XX} + A_{YY})\cos 2\alpha + A_{XY}\sin 2\alpha \end{aligned} \tag{A.20.19}$$

$$T^2 = A^2 - N^2 = (A_{XX}\cos\alpha + A_{XY}\sin\alpha)(1 - \cos^2\alpha)$$

$$-2(A_{XX}\cos\alpha + A_{XY}\sin\alpha)(A_{XX}\cos\alpha + A_{YY}\sin\alpha)\sin\alpha\cos\alpha$$

$$+ (A_{XY}\cos\alpha + A_{YY}\sin\alpha)(1 - \sin^2\alpha)$$

$$= [-\tfrac{1}{2}(A_{XX} + A_{YY})\sin 2\alpha + A_{XY}\cos 2\alpha]^2 \tag{A.20.20}$$

After some further rearrangement we have

$$[N - \tfrac{1}{2}(A_{XX} + A_{YY})]^2 + T^2 = \tfrac{1}{2}(A_{XX} - A_{YY})^2 + A_{XY}^2 \tag{A.20.21}$$

in which N is given as a function of T and represents the equation of a circle in the T–N coordinates, Fig. A.20.4, with its center at $\frac{1}{2}(A_{XX} + A_{YY})$ and its radius equal to $\sqrt{[\frac{1}{2}(A_{XX} - A_{YY})^2 + A_{XY}^2]}$. As we have already mentioned, N

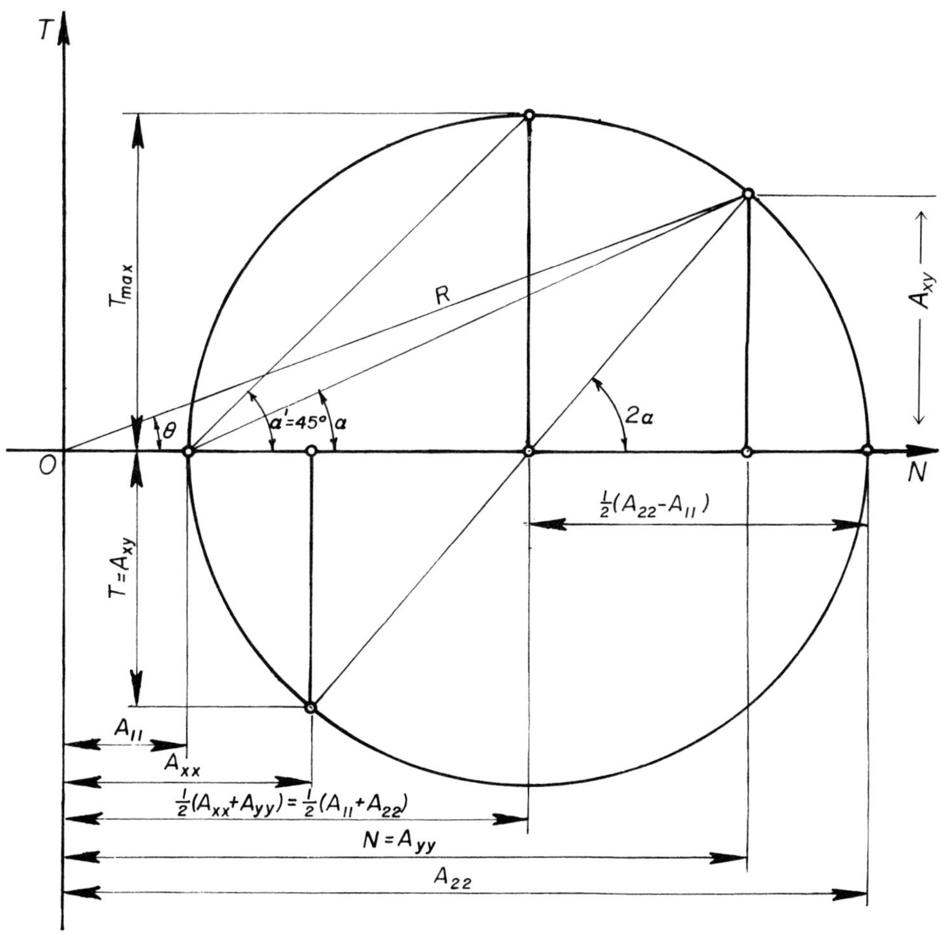

Fig. A.20.4. Mohr circle in a two-dimensional coordinate system.

is the component normal to the oblique plane of vector A_L and T is the tangential component to that plane, Fig. A.20.5.

T reaches its maximum value in Eq. (A.20.20) when $dT/d\alpha = 0$ and angle α' is obtained

$$\tan 2\alpha' = -\frac{A_{XX} - A_{YY}}{2A_{XY}} \tag{A.20.22}$$

The extreme values of N may be found from Eq. (A.20.19), when $dN/d\alpha = 0$. This value is obtained when

$$\tan 2\alpha'' = \frac{2A_{XY}}{A_{XX} - A_{YY}} \tag{A.20.23}$$

and is equal to A_{11} and A_{22}, the principal components of tensor A_{LM}.

From Eqs. (A.20.22) and (A.20.23) we have $\tan 2\alpha'$ and $\tan 2\alpha'' = -1$, indicating that the difference between the angles $2\alpha'$ and $2\alpha''$ is a right angle and that the maximum value of T is at 45° to the principal axes. The principal components, A_{11} and A_{22}, are obtained from Eq. (A.20.21) by setting $T = 0$

$$A_{11}, A_{22} = \tfrac{1}{2}(A_{XX} + A_{YY}) \pm \sqrt{[\tfrac{1}{2}(A_{XX} - A_{YY})^2 + A_{XY}^2]} \tag{A.20.24}$$

Also, the largest value of the tangential component is

$$T_{\max} = \sqrt{[\tfrac{1}{2}(A_{XX} - A_{YY})^2 + A_{XY}^2]} = \tfrac{1}{2}(A_{11} - A_{22}) \tag{A.20.25}$$

Eqs. (A.20.19) and (A.20.20), in terms of the principal components, are then

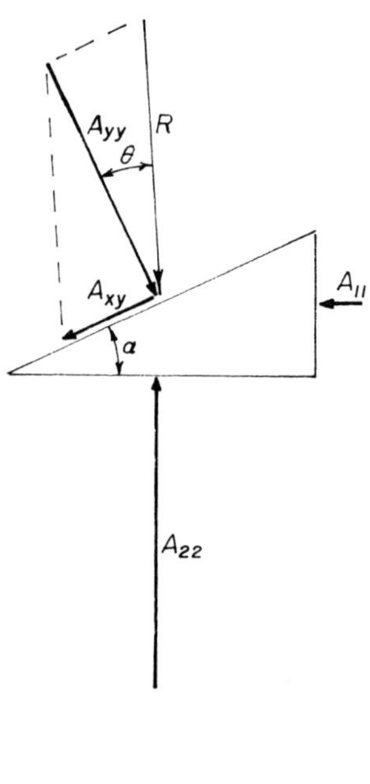

Fig. A.20.5. Oblique plane of an element on which the tensorial components shown in Fig. A.20.4. are acting.

$$N = \tfrac{1}{2}(A_{11} + A_{22}) + \tfrac{1}{2}(A_{11} - A_{22})\cos 2\alpha \tag{A.20.26}$$

$$T = -\tfrac{1}{2}(A_{11} - A_{22})\sin 2\alpha \tag{A.20.27}$$

Conversely, by knowing the principal components A_{11}, A_{22} and the angle α, the components A_{XX}, A_{YY} and A_{XY} can be readily derived

$$A_{XX}, A_{YY} = \tfrac{1}{2}(A_{11} + A_{22}) + \tfrac{1}{2}(A_{11} - A_{22})\cos 2\alpha \tag{A.20.28}$$

$$A_{XY} = \tfrac{1}{2}(A_{11} - A_{22})\sin 2\alpha \tag{A.20.29}$$

A.21 Axially Symmetric Second-order Tensors

A second-order tensor is axially symmetric, see Sect. A.20, if two of its principal components are equal, say $A_{11} = A_{22} \neq A_{33}$. The axially symmetric tensor has, therefore, two coinciding Mohr circles, and they coincide also with the third circle.

Suppose our triad is in the principal axes of tensor A_{LM}, which is axially symmetric, Fig. A.21.1, and we cut through the volume element with an oblique plane that intersects the axes at equal distances. Given an axially symmetric tensor A_{LM} in its principal components, one can always resolve it, according to Eq. (A.8.2), into two tensors

$$A_{LM} = \tfrac{1}{3}H_{\alpha\alpha}\delta_{LM} + S_{LM} \tag{A.21.1}$$

where $\frac{1}{3}H\delta_{LM}$ is a *spherical tensor* and S_{LM} a *deviatoric tensor*, Fig. A.21.2. The deviatoric tensor in the axially symmetric case has the following form

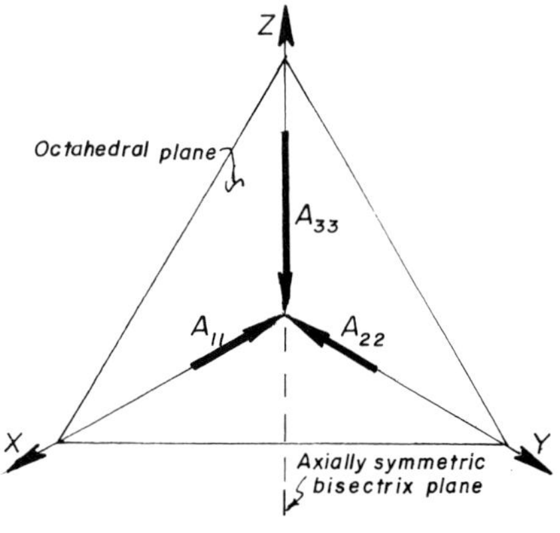

Fig. A.21.1. Axially symmetric triad viewed in the octahedral plane.

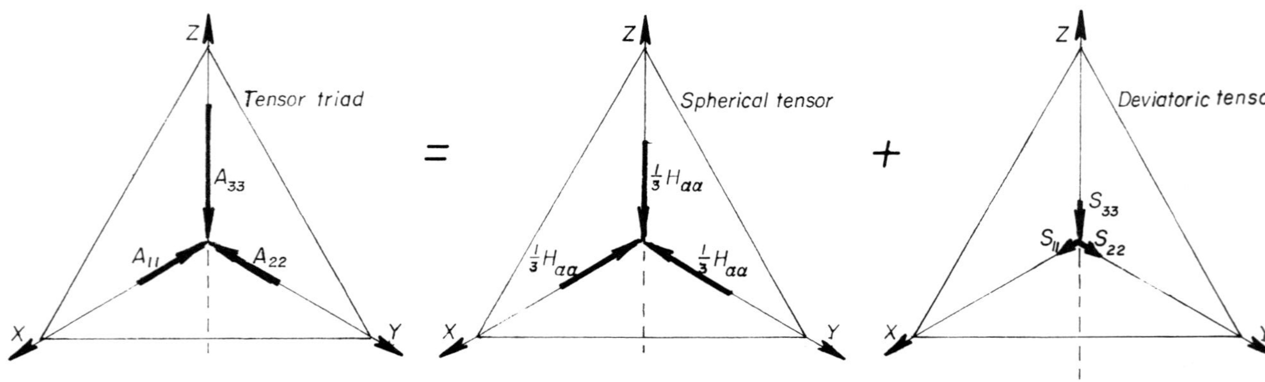

Fig. A.21.2. Decomposition of a triad.

$$|S_{LM}| = \begin{vmatrix} \frac{1}{3}(A_{11} - A_{33}) & 0 & 0 \\ 0 & \frac{1}{3}(A_{11} - A_{33}) & 0 \\ 0 & 0 & -\frac{2}{3}(A_{11} - A_{33}) \end{vmatrix}$$

$$= \tfrac{1}{3}(A_{33} - A_{11}) \begin{vmatrix} -1 & 0 & 0 \\ 0 & -1 & 0 \\ 0 & 0 & 2 \end{vmatrix} = \tfrac{1}{3}(A_{33} - A_{11})|\Xi_{LM}| \quad \text{(A.21.2)}$$

where Ξ_{LM} is the *dimensionless deviatoric tensor*

$$|\Xi_{LM}| = \begin{vmatrix} -1 & 0 & 0 \\ 0 & -1 & 0 \\ 0 & 0 & 2 \end{vmatrix} \quad \text{(A.21.3)}$$

The principal invariants, the moment invariants and the octahedral invariant of the dimensionless deviatoric tensor are

$$\mathrm{I}_\Xi = 0; \quad \mathrm{II}_\Xi = -3; \quad \mathrm{III}_\Xi = 2 \quad \text{(A.21.4)}$$

$$\bar{\mathrm{I}}_\Xi = 0; \quad \overline{\mathrm{II}}_\Xi = 6; \quad \overline{\mathrm{III}}_\Xi = 6 \quad \text{(A.21.5)}$$

$$\Pi_\Xi = \tfrac{9}{2} \quad \text{(A.21.6)}$$

We shall now look for a tensor B_{LM} that transforms the tensor Ξ_{LM} into a symmetric tensor Ξ'_{LM} with vanishing diagonals, having the form

$$|\Xi'_{LM}| = \begin{vmatrix} 0 & \Xi'_{XY} & \Xi'_{XZ} \\ \Xi'_{XY} & 0 & \Xi'_{YZ} \\ \Xi'_{XZ} & \Xi'_{YZ} & 0 \end{vmatrix} \quad \text{(A.21.7)}$$

We demand that the transformation matrix B_{LM} be also orthogonal (Goldstein 1953), thus

$$B_{L\alpha}C_{M\alpha} = \delta_{LM} \quad \text{(A.21.8)}$$

where C_{LM} is the reciprocal tensor of B_{LM}, obtained by interchanging the columns with the rows

$$C_{LM} = B_{LM} \quad \text{(A.21.9)}$$

and the required transformation can be written as

$$\Xi'_{LM} = B_{L\alpha}\Xi_{\alpha\beta}B_{M\beta} \quad \text{(A.21.10)}$$

To find B_{LM} we must calculate first the tensor Ξ'_{LM}, and this can be done by requiring the invariants I_{Ξ}, II_{Ξ} and III_{Ξ}, Eqs. (A.21.4), to remain the same for tensor Ξ'_{LM}. As Ξ'_{LM} is a traceless tensor, Eq. (A.21.7), its first invariant, will obviously vanish, as in Eq. (A.21.4). For the second and third invariants we require

$$II_{\Xi'} = -(\Xi'^{2}_{XY} + \Xi'^{2}_{YZ} + \Xi'^{2}_{XZ}) = -3 \tag{A.21.11}$$

$$III_{\Xi'} = 2\Xi'_{XY}\Xi'_{YZ}\Xi'_{XZ} \tag{A.21.12}$$

A real solution of Eq. (A.21.10), also requiring $\Xi'_{XZ} = \Xi'_{YZ}$ because of axial symmetry, is

$$\Xi'_{XY} = \Xi'_{YZ} = \Xi'_{XZ} = 1 \tag{A.21.13}$$

and therefore tensor Ξ'_{LM}, Eq. (A.21.7), has the form

$$|\Xi'_{LM}| = \begin{vmatrix} 0 & 1 & 1 \\ 1 & 0 & 1 \\ 1 & 1 & 0 \end{vmatrix} \tag{A.21.14}$$

In order to find the transformation tensor B_{LM} we multiply both sides of Eq. (A.21.10) by B_{LM} and we get

$$\Xi'_{\alpha M} B_{L\alpha} = B_{L\alpha}\Xi_{\alpha M} \tag{A.21.15}$$

Adding to it the orthogonality condition, Eq. (A.21.8), the solution for the components of B_{LM} is evident. In its matricial form B_{LM} is

$$|B_{LM}| = \begin{vmatrix} \frac{1}{\sqrt{2}} & \frac{1}{\sqrt{6}} & \frac{1}{\sqrt{3}} \\ -\frac{1}{\sqrt{2}} & \frac{1}{\sqrt{6}} & \frac{1}{\sqrt{3}} \\ 0 & -\frac{\sqrt{2}}{3} & \frac{1}{\sqrt{3}} \end{vmatrix} \tag{A.21.16}$$

representing a compound rotation with the *Euler angles*, $\alpha = 0$, $\beta = 54.75°$, $\gamma = 45°$. The angles β and γ determine, respectively, the inclination of the octahedral plane, an oblique plane cutting the coordinates at equal distances, and the direction of the *symmetric bisectrix plane*, the plane that bisects the X–Y plane in the middle, passing through the Z axis, Fig. A.21.1.

Observing the symmetric bisectrix plane, Fig. A.21.3, it is seen that the octahedral plane, which is represented here by a line, forms an angle $\beta = 54.75°$ with the Z coordinate. The line piercing the octahedral plane at its

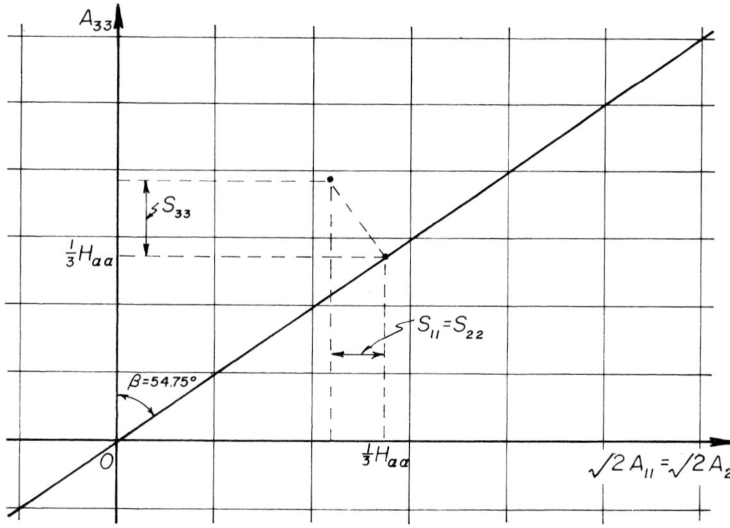

Fig. A.21.3. Axially symmetric triad shown in the octahedral bisectrix plane.

center and at equal distance from the coordinates is the isotropic or spherical directrix, Fig. A.21.1. In Fig. A.21.3 it is seen at a 54.75° angle with the Z coordinate. It should be noted that the abscissa in Fig. A.21.3 is distorted for coordinates X and Y, and represents actually $\sqrt{2}X$, $\sqrt{2}Y$.

We are also interested in finding the values of the higher powers of the dimensionless deviatoric tensor. In general, the dimensionless deviatoric tensor raised to the power n is

$$\Xi_{ij}^{n} = A_{n}\delta_{ij} + B_{n}\Xi_{ij} \tag{A.21.17}$$

where $A_{n} = \frac{1}{3}[2^{n} + 2(-1)^{n}]$ and $B_{n} = \frac{1}{3}[2^{n} - (-1)^{n}]$. Consequently for the second and third powers we get

$$\Xi_{ij}^{2} = 2\delta_{ij} + \Xi_{ij} \tag{A.21.18}$$

$$\Xi_{ij}^{3} = 2\delta_{ij} + 3\Xi_{ij} \tag{A.21.19}$$

Appendix B
Cylindrical Coordinates

B.1 Introduction

While many of the calculations in applied mathematics can be satisfied by using Cartesian coordinates, many others can be simplified by choosing instead another coordinate system, that takes advantage of the geometry, the symmetry, the nature of the boundary conditions and the form of the mathematical expressions involved in the problem under consideration.

In mechanics of soils, our wide interest in cylindrical coordinates springs from the extensive use of cylindrical samples in the experimental studies of soils. The topic of cylindrical coordinates discussed here is considered an extension of Appendix A, and treated as such. It presents a short version of a study by a graduate student* and is adapted to the notation employed in the present work.

B.2 Definition of the Cylindrical Coordinate System

A three-dimensional Euclidean space consists of:

1. Three families of independent surfaces, determined by the coordinates.
2. Three families of curves defined by the intersections of the coordinate surfaces.
3. The intersection of the coordinate lines determining the points in a three-dimensional space.

The coordinates in a cylindrical space are r, θ and z, Fig. B.2.1. Accordingly, $r = \text{const}$ defines circular cylinders about axis z, $\theta = \text{const}$ defines half

* Kouskoulas, V. 1962 Transformation of the Basic Tensorial Equations of Mechanics into Cylindrical Coordinates. Wayne State University, Detroit.

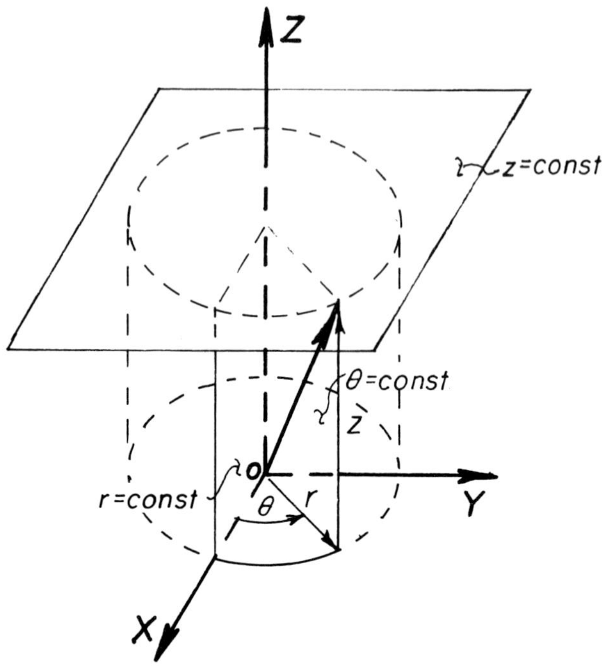

Fig. B.2.1. Cylindrical coordinates related to the Cartesian coordinates.

planes through axis z and $z = \text{const}$ defines planes perpendicular to axis z. The intersection of the surfaces thus determined yields a point or a position vector in the cylindrical coordinates, $r(r, \theta, z)^i$. In vectorial notation we have

$$r(r, \theta, z)^i \equiv re_r + \theta e_\theta + ze_z \tag{B.2.1}$$

where e_r, e_θ and e_z are the *base vectors* in the directions r, θ and z, respectively, defined

$$\begin{aligned} e_r &= 1_x \cos\theta + 1_y \sin\theta \\ e_\theta &= -1_x r \sin\theta + 1_y r \cos\theta \\ e_z &= 1_z \end{aligned} \tag{B.2.2}$$

From the definition of the cylindrical coordinates and by simple geometry we deduce the coordinate transformation from coordinates $x^i = x(x, y, z)^i$ to $r^i = r(r, \theta, z)^i$ and we get

$$x^i = x = r\cos\theta$$

$$x^j = y = r\sin\theta \tag{B.2.3}$$

$$x^k = z = z$$

and inversely

$$\begin{aligned} r^i &= r = (x^2 + y^2)^{1/2} \\ r^j &= \theta = \tan^{-1}(x/y) \\ r^k &= z = z \end{aligned} \tag{B.2.4}$$

since a one to one correspondence exists between the coordinates.

The Cartesian coordinates transform according to the covariant law, Eq. (A.6.4)

$$\mathrm{d}x_i = \frac{\partial x^i}{\partial r^\alpha}\,\mathrm{d}r_\alpha \tag{B.2.5}$$

which defines the Jacobian, Eq. (A.3.4)

$$J \equiv \begin{vmatrix} \dfrac{\partial x^i}{\partial r^i} & \dfrac{\partial x^i}{\partial r^j} & \dfrac{\partial x^i}{\partial r^k} \\[2ex] \dfrac{\partial x^j}{\partial r^i} & \dfrac{\partial x^j}{\partial r^j} & \dfrac{\partial x^j}{\partial r^k} \\[2ex] \dfrac{\partial x^k}{\partial r^i} & \dfrac{\partial x^k}{\partial r^j} & \dfrac{\partial x^k}{\partial r^k} \end{vmatrix} = \begin{vmatrix} \cos\theta & -r\sin\theta & 0 \\ \sin\theta & r\cos\theta & 0 \\ 0 & 0 & 1 \end{vmatrix} \tag{B.2.6}$$

The inverse transformation and its Jacobian are

$$\mathrm{d}r_i = \frac{\partial r^i}{\partial x^\alpha}\,\mathrm{d}x_\alpha \tag{B.2.7}$$

$$J' = \frac{1}{J} \equiv \begin{vmatrix} \dfrac{\partial r^i}{\partial x^i} & \dfrac{\partial r^i}{\partial x^j} & \dfrac{\partial r^i}{\partial x^k} \\[2ex] \dfrac{\partial r^j}{\partial x^i} & \dfrac{\partial r^j}{\partial x^j} & \dfrac{\partial r^j}{\partial x^k} \\[2ex] \dfrac{\partial r^k}{\partial x^i} & \dfrac{\partial r^k}{\partial x^j} & \dfrac{\partial r^k}{\partial x^k} \end{vmatrix}$$

$$= \begin{vmatrix} \cos\theta & \sin\theta & 0 \\[1ex] \dfrac{-\sin\theta}{r} & \dfrac{\cos\theta}{r} & 0 \\[2ex] 0 & 0 & 1 \end{vmatrix} \tag{B.2.8}$$

Finally, by dividing the base vectors by their length, we obtain the *unit vector* in the cylindrical coordinates

$$1_r = \frac{1_x \cos\theta + 1_y \sin\theta}{(\cos^2\theta + \sin^2\theta)^{1/2}} = 1_x \cos\theta + 1_y \sin\theta = e_r$$

$$1_\theta = \frac{-1_x r\sin\theta + 1_y r\cos\theta}{(r^2 \sin\theta^2 + r^2 \cos^2\theta)^{1/2}} = -1_x \sin\theta + 1_y \cos\theta = \frac{e_\theta}{r} \qquad \text{(B.2.9)}$$

$$1_z = 1_z = e_z$$

From Eq. (B.2.9) the unit vectors of the Cartesian coordinates may be evaluated

$$1_x = 1_r \cos\theta - 1_\theta \sin\theta$$

$$1_y = 1_r \sin\theta + 1_\theta \cos\theta \qquad \text{(B.2.10)}$$

$$1_z = 1_z$$

It should be noted that the base and unit vectors in the Cartesian coordinates are identical, while in the cylindrical coordinates they differ. Note also that the unit vectors in Cartesian coordinates are constant while the cylindrical unit vectors vary.

That the cylindrical coordinates are orthogonal coordinates may be seen by multiplying the unit vectors

$$1_r \cdot 1_\theta = (1_x \cos\theta + 1_y \sin\theta)\cdot(-1_x \sin\theta + 1_y \cos\theta) = 0$$

$$1_r \cdot 1_k = (1_x \cos\theta + 1_y \sin\theta)\cdot 1_k = 0 \qquad \text{(B.2.11)}$$

$$1_\theta \cdot 1_k = (-1_x \sin\theta + 1_y \cos\theta)\cdot 1_k = 0$$

B.3 The Fundamental Tensor

In any Riemannian space the fundamental tensor is defined according to one of Eqs. (A.10.8)–(A.10.11), say

$$g_{ij} = \begin{vmatrix} g_{11} & g_{12} & \cdots & g_{1n} \\ g_{21} & g_{22} & \cdots & g_{2n} \\ \vdots & \vdots & & \vdots \\ g_{n1} & g_{n2} & \cdots & g_{nn} \end{vmatrix} \qquad \text{(B.3.1)}$$

where n is the dimension of the space and any component of the fundamental tensor is

$$g_{ij} = \frac{\partial X^\alpha \partial X^\alpha}{\partial x^i \partial x^j} \tag{B.3.2}$$

In cylindrical coordinates $n = 3$, $X^i \equiv X^i(r, \theta, z)^i$, $x^i \equiv x^i(r, \theta, z)^i$, as defined in Eq. (B.2.1). Therefore the fundamental tensor obtained from Eqs. (B.2.2), (B.3.1) and (B.3.2) is

$$g_{ij} = \begin{vmatrix} 1 & 0 & 0 \\ 0 & r^2 & 0 \\ 0 & 0 & 1 \end{vmatrix} \qquad \begin{aligned} g_{ij} &= 0 \text{ for } i \neq j \\ g_{ij} &= e_i e_j \text{ for } i = j \end{aligned} \tag{B.3.3}*$$

By definition, Eq. (A.10.8), g^{ij} is the conjugate of g_{ij} or a conjugate fundamental tensor, if and only if Eqs. (A.10.9) are satisfied. From this very definition and Eq. (B.3.3) we get

$$g^{ij} = \begin{vmatrix} 1 & 0 & 0 \\ 0 & 1/r^2 & 0 \\ 0 & 0 & 1 \end{vmatrix} \tag{B.3.4}$$

B.4 The Christoffel Symbols

From Sects. A.13 and B.3 the Christoffel symbols for cylindrical coordinates are straightforward. Out of the 18 possible components only three subsisting components of the first kind remain,** and they are

$$\begin{aligned}
\Gamma_{r,\theta\theta} &= \frac{1}{2}\left(\frac{\partial g_{r\theta}}{\partial \theta} + \frac{\partial g_{r\theta}}{\partial r} - \frac{\partial g_{\theta\theta}}{\partial r}\right) = \frac{1}{2}\left(0 + 0 + \frac{\partial r^2}{\partial r}\right) = -r \\
\Gamma_{\theta,\theta r} &= \frac{1}{2}\left(\frac{\partial g_{\theta r}}{\partial \theta} + \frac{\partial g_{\theta\theta}}{\partial r} - \frac{\partial g_{\theta r}}{\partial \theta}\right) = \frac{1}{2}\left(0 + \frac{\partial r^2}{\partial r} - 0\right) = r \\
\Gamma_{\theta,r\theta} &= \frac{1}{2}\left(\frac{\partial g_{\theta\theta}}{\partial r} + \frac{\partial g_{\theta r}}{\partial \theta} - \frac{\partial g_{r\theta}}{\partial \theta}\right) = \frac{1}{2}\left(\frac{\partial r^2}{\partial r} + 0 - 0\right) = r
\end{aligned} \tag{B.4.1}$$

Likewise, the Christoffel symbols of the second kind are all zero, except for

---

* $g_{ij} = 0$ for $i \neq j$ proves the orthogonality once more.

** Since $g_{ij} = 0$ for $i \neq j$ and $g_{ij} \neq 0$ for $i = j$, as indicated by Eq. (B.3.3), the 18 components are reduced to 9. But from Eq. (B.3.3) where $g_{rr} = g_{zz} = 1$ and from Eq. (A.13.1) where $\Gamma_{k,ij} = f(\partial g_{ki}/\partial x^j)$ we see that $\Gamma_{k,ij} \neq 0$ for $k = i = \theta$ and $j = r$, since $\partial g_{ki}/\partial x^j = 0$ and $\Gamma_{k,ij} = f(0) = 0$ for $k = i \neq 0$ and $j = r, \theta, z$. Therefore there are only three remaining non-zero Christoffel symbols.

three. By means of the conjugate fundamental tensor, Eq. (B.3.4), and the first kind of the Christoffel symbol, Eq. (B.4.1), we get the three Christoffel symbols of the second kind in cylindrical coordinates

$$
\begin{aligned}
\Gamma^{r}_{\theta\theta} &= g^{rr}\Gamma_{r,\theta\theta} = (1)(-r) = -r \\
\Gamma^{\theta}_{\theta r} &= g^{\theta\theta}\Gamma_{\theta,\theta r} = (1/r^2)(r) = 1/r \\
\Gamma^{\theta}_{r\theta} &= g^{\theta\theta}\Gamma_{\theta,r\theta} = (1/r^2)(r) = 1/r
\end{aligned}
\tag{B.4.2}
$$

As explained in Sect. A.13, Appendix A, the Christoffel symbols are not tensors but behave like tensors with respect to mathematical operations.

B.5 Covariant Derivatives

In Sect. A.14, Appendix A, the covariant derivative of tensors in coordinates other than Cartesian was discussed in detail. Let $A_i = A(r, \theta, z)_i$ be a tensor of order one (a vector) given in cylindrical coordinates. Such a tensor has nine covariant derivatives

$$
\begin{aligned}
&A_{r,r} = \frac{\partial A_r}{\partial r}; \quad A_{r,\theta} = \frac{\partial A_r}{\partial \theta} + \frac{A_\theta}{r^2}; \quad A_{r,z} = \frac{\partial A_r}{\partial z} \\
&A_{\theta,r} = \frac{1}{r}\frac{\partial A_\theta}{\partial r} + \frac{A_\theta}{r^2}; \quad A_{\theta,\theta} = r\frac{\partial A_\theta}{\partial \theta} - rA_r; \quad A_{\theta,z} = r\frac{\partial A_\theta}{\partial z} \\
&A_{z,r} = \frac{\partial A_z}{\partial r}; \quad A_{z,\theta} = \frac{\partial A_z}{\partial \theta}; \quad A_{z,z} = \frac{\partial A_z}{\partial z}
\end{aligned}
\tag{B.5.1}^*
$$

Likewise the covariant derivative of a contravariant tensor (vector) in general curvilinear coordinates is defined by a set of nine functions according to Eq. (A.14.1), and just like in Eq. (B.5.1) we obtain the derivative $A^i_{,j}$ in cylindrical coordinates

$$
\begin{aligned}
&A^r_{,r} = \frac{\partial A^r}{\partial r}; \quad A^r_{,\theta} = \frac{\partial Az^r}{\partial \theta} + \frac{A^\theta}{r^2}; \quad A^r_{,z} = \frac{\partial A^r}{\partial z} \\
&A^\theta_{,r} = \frac{1}{r}\frac{\partial A^\theta}{\partial r} + \frac{A^\theta}{r^2}; \quad A^\theta_{,\theta} = r\frac{\partial A^\theta}{\partial \theta} - rA^r; \quad A^\theta_{,z} = \frac{1}{r}\frac{\partial A^\theta}{\partial z} \\
&A^z_{,r} = \frac{\partial A^z}{\partial r}; \quad A^z_{,\theta} = \frac{\partial A^z}{\partial \theta}; \quad A^z_{,z} = \frac{\partial A^z}{\partial z}
\end{aligned}
\tag{B.5.2}
$$

---

* $$A_{\theta,\theta} = \frac{\partial(A_\theta)/r}{\partial \theta} - \Gamma^{\alpha}_{\theta\theta}A_\alpha = \frac{1}{r}\frac{\partial A_\theta}{\partial \theta} + \Gamma^{\alpha}_{\theta\theta}A_\alpha = \frac{1}{r}\frac{\partial A_\theta}{\partial \theta} - rA_r$$

B.6 Basic Operations of First-order Tensors in Cylindrical Coordinates

First-order tensors correspond to vectors and we realize that certain mathematical operations are valid with respect to vectorial quantities.

Dot Product of Two Vectors (Inner Product)

Let A_i and B_i be covariant tensors, then their inner product becomes

$$\mathbf{A} \cdot \mathbf{B} \equiv g_{ij} A_i B_j = g^{rr} A_r B_r + g^{\theta\theta} A_\theta B_\theta + g^{zz} A_z B_z$$

$$= A_r B_r + \frac{1}{r^2} A_\theta B_\theta + A_z B_z \tag{B.6.1}$$

If the tensors are contravariant, then

$$\mathbf{A} \cdot \mathbf{B} \equiv A^r B^r + r^2 A^\theta B^\theta + A^z B^z \tag{B.6.2}$$

From the definition $\mathbf{A} \cdot \mathbf{B} \equiv AB \cos \gamma$, the angle γ between the vectors $\mathbf{A}$ and $\mathbf{B}$ is found

$$\cos \gamma = \frac{g_{\alpha\beta} A^\alpha B^\beta}{(g_{\delta\eta} A^\delta A^\eta)^{1/2} (g_{\mu\nu} B^\mu B^\nu)^{1/2}} \tag{B.6.3}$$

Cross Product of Two Vectors (Outer Product)

The outer product of two first-order tensors A_i and B_i results in a first-order tensor C_i normal to the plane formed by the tensors A_i and B_i, Fig. B.6.1. This corresponds to the cross product of vectors $\mathbf{A}$ and $\mathbf{B}$ resulting in a vector $\mathbf{C}$, so directed that the vectors $\mathbf{A}$, $\mathbf{B}$ and $\mathbf{C}$ form a right-handed manifold.

$$|g|^{-1/2} e^{k\alpha\beta} (A_\alpha B_\beta - A_\beta B_\alpha) = C^k \equiv \mathbf{A} \times \mathbf{B} = \mathbf{C}$$

$$= |g|^{-1/2} \begin{vmatrix} 1_r & r\,1_\theta & 1z \\ A_r & rA_\theta & A_z \\ B_r & rB_\theta & B_z \end{vmatrix} = \frac{1}{r} [(rA_\theta B_z - A_z r B_\theta) 1_r$$

$$+ (A_r B_z - A_z B_r) r 1_\theta + (rA_\theta B_r - A_r r B_\theta) 1_z]$$

$$= (A_\theta B_z - A_z B_\theta) 1_r + (A_r B_z - A_z B_r) 1_\theta + (A_\theta B_r - A_r B_\theta) 1_z \tag{B.6.4}$$

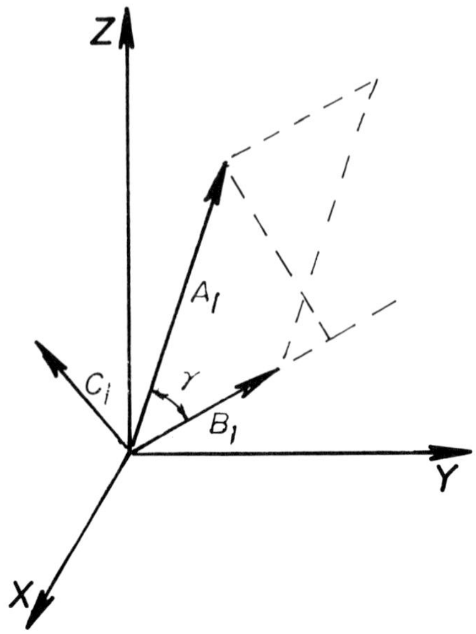

Fig. B.6.1. Outer product between two first-order tensors.

where e^{ijk} is the permutation unit tensor defined in Sect. A.14, and 1_r, 1_θ and 1_z are the unit vectors discussed in Sect. B.2.

Note that there is a difference between the outer product of A_i by B_i and that of B_i by A_i. The results are tensors equal in magnitude but opposite in direction

$$|g|^{-1/2}e^{\kappa\alpha\beta}(A_\alpha B_\beta - A_\beta B_\alpha) = -|g|^{-1/2}e^{\kappa\alpha\beta}(B_\alpha A_\beta - B_\beta A_\alpha) \qquad \text{(B.6.5)}$$

The Gradient

As defined in Sect. A.14, the gradient is a covariant derivative of a scalar function B. Since B is an invariant, it is easily transformed to any coordinate system. Thus in cylindrical coordinates we have

$$\operatorname{grad} B = \nabla B \equiv B_{,i} = |e_i| \frac{\partial B}{\partial X^i} 1_i = \frac{\partial B}{\partial r} \sum_{1_r} + \frac{1}{r}\frac{\partial B}{\partial \theta} \sum_{1_\theta} + \frac{\partial B}{\partial z} \sum_{1_z} \qquad \text{(B.6.6)}$$

Divergence

Let A^i be a contravariant tensor of the first order (a vector) given in general curvilinear coordinates. The divergence is then given by

$$\begin{aligned}
\operatorname{div}\mathbf{A} = \nabla\cdot\mathbf{A} \equiv A^i_{,i} = A^{\alpha}_{,\alpha} &= \frac{\partial A^\alpha}{\partial X^\alpha} + \Gamma^{\beta}_{\beta\alpha}A^\alpha \\
&= \frac{\partial A^\alpha}{\partial X^\alpha} + \left(\frac{\partial}{\partial X^\alpha}\ln\sqrt{g}\right)A^\alpha \\
&= g^{-1/2}\frac{\partial}{\partial X^\alpha}(g^{1/2}A^\alpha) = \frac{1}{r}\left(\frac{\partial(rA^r)}{\partial r} + \frac{\partial A^\theta}{\partial\theta} + \frac{\partial(rA^z)}{\partial z}\right)
\end{aligned} \tag{B.6.7}$$

From Eq. (B.6.7) it is obvious that the divergence of a first-order tensor A^i results in a contracted covariant derivative of the tensor A^i, a scalar.

The Laplacian

The Laplacian of a scalar function $B = B(X^i)$ is defined as the divergence of its gradient. Thus

$$\begin{aligned}
\nabla^2 B = \nabla\cdot\nabla B &\equiv \frac{1}{r}\left(\frac{\partial}{\partial r}\left(r\frac{\partial B}{\partial r}\right) + \frac{\partial}{\partial\theta}\left(r\frac{1}{r^2}\frac{\partial B}{\partial\theta}\right) + \frac{\partial}{\partial z}\left(r\frac{\partial B}{\partial z}\right)\right) \\
&= \frac{1}{r}\frac{\partial}{\partial r}\left(r\frac{\partial B}{\partial r}\right) + \frac{1}{r^2}\frac{\partial^2 B}{\partial^2\theta} + \frac{\partial^2 B}{\partial^2 z}
\end{aligned} \tag{B.6.8}$$

The Curl

The curl is defined in Sect. A.14 with respect to a vector. Let $A_i = A(X^i)$ be a covariant vector in general curvilinear coordinates. Then its curl is given by

$$\begin{aligned}
\nabla\times\mathbf{A} &\equiv g^{-1/2}e^{\kappa\alpha\beta}A_{\alpha,\beta} = g^{-1/2}(A_{\alpha,\beta} - A_{\beta,\alpha}) \\
&= \frac{1}{r}\begin{vmatrix} 1_r & r1_\theta & 1_z \\ \dfrac{\partial}{\partial r} & \dfrac{\partial}{\partial\theta} & \dfrac{\partial}{\partial z} \\ A_r & rA_\theta & A_z \end{vmatrix} \\
&= \frac{1}{r}\left(\frac{\partial A_r}{\partial\theta} - \frac{\partial A_\theta}{\partial r}\right)_r + \frac{1}{r}\left(\frac{\partial A_r}{\partial z} - \frac{r\partial A_z}{\partial r}\right)_\theta + \left(\frac{\partial A_\theta}{\partial r} - \frac{\partial A_r}{\partial\theta}\right)_z
\end{aligned} \tag{B.6.9}$$

B.7 Elements of Differential Geometry

Several elements of differential geometry, discussed in general in Appendix A, will be appraised in view of cylindrical coordinates and formulated accordingly.

The Line Element

Let dx^i represent an infinitesimal displacement, from a point u^i determined by the position vector $u(r, \theta, z)^i$ to another point $u'(r', \theta', z')^i = u'(r + dr, \theta + d\theta, z + dz)^i$. The distance between the two points determines a line element ds, whose square is defined by Eqs. (A.10.2) or (A.10.4) and which in cylindrical coordinates becomes

$$ds^2 = dx^\alpha\, dx^\alpha = g_{\alpha\beta}\, du^\alpha\, du^\beta = g_{11}\, dr\, dr + g_{22}\, d\theta\, d\theta + g_{33}\, dz\, dz$$

$$= dr\, dr + r^2\, d\theta\, d\theta + dz\, dz = (dr)^2 + (r\, d\theta)^2 + (dz)^2 \qquad \text{(B.7.1)}$$

Area Element

Let ds_1 correspond to an infinitesimal displacement at a point p^i, then

$$ds_1 = \sqrt{(g_{11})}\, du_1 \qquad \text{(B.7.2)}$$

and similarly

$$ds_2 = \sqrt{(g_{22})}\, du_2; \quad ds_3 = \sqrt{(g_{33})}\, du_3 \qquad \text{(B.7.3)}$$

Consider the area betwen vectors ds_2 and ds_3, Fig. B.7.1, equal to the vector dS_1 in the direction u_1 and normal to the plane between vectors ds_2 and ds_3, as follows

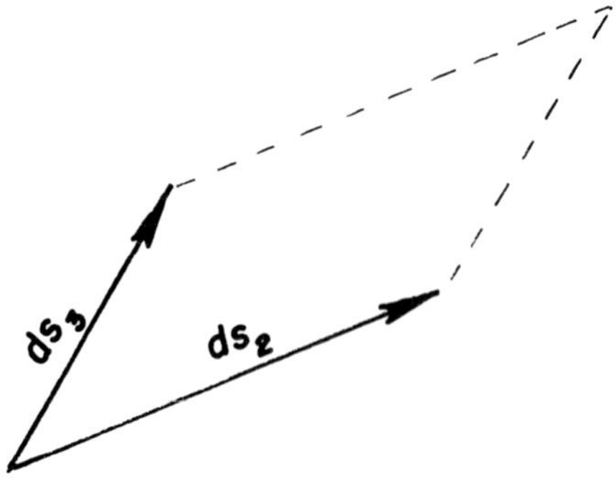

Fig. B.7.1. Area between two vectors.

$$dS_1 = ds_2 \times ds_3 \equiv \sqrt{(g_{22}g_{33})}\, du_2\, du_3 = r\, d\theta\, dz \tag{B.7.4}$$

where dS_1 defines a surface area on the cylinder of radius r, normal to that radius. Similarly we can derive the areas dS_2 and dS_3

$$dS_2 = ds_1 \times ds_3 \equiv \sqrt{(g_{11}g_{33})}\, du_1\, du_3 = dr\, dz$$

$$dS_3 = ds_1 \times ds_2 \equiv \sqrt{(g_{11}g_{22})}\, du_1\, du_2 = r\, dr\, d\theta \tag{B.7.5}$$

which define, respectively, a surface on the plane passing through the z axis and normal to $d\theta$, and a surface on the plane normal to the coordinate z.

Volume Element

The volume element produced by the displacements ds_1, ds_2 and ds_3, functions of r, θ and z, is given by

$$\begin{aligned} dV = |d\mathbf{s}_1\, d\mathbf{s}_2\, d\mathbf{s}_3| &= ds_1 \cdot (ds_2 \times ds_3) \\ &\equiv \sqrt{g_{11}} \cdot (\sqrt{g_{22}} \times \sqrt{g_{33}})\, du_1\, du_2\, du_3 \\ &= \sqrt{(g_{11}g_{22}g_{33})}\, du_1\, du_2\, du_3 = r\, dr\, d\theta\, dz \end{aligned} \tag{B.7.6}$$

The Distance Between Two Points

The distance s between two points l_1 and l_2 on a curve $x^i = x(l)^i$ is given by the integral equation

$$s = \int_{l_1}^{l_2} \sqrt{g_{\alpha\beta} \frac{dx^\alpha\, dx^\beta}{dl\, dl}}\, dl \tag{B.7.7}$$

In cylindrical coordinates, where $x^i = r, \theta, z$ and $g_{ij} = 0$ for $i \neq j$, $g_{rr} = 1$, $g_{\theta\theta} = r^2$, $g_{zz} = 1$, we have

$$\begin{aligned} g_{11} \frac{dx^1\, dx^1}{dl\, dl} &= g_{rr} \frac{dr\, dr}{dl\, dl} = \left(\frac{dr}{dl}\right)^2 \\ g_{22} \frac{dx^2\, dx^2}{dl\, dl} &= g_{\theta\theta} \frac{d\theta\, d\theta}{dl\, dl} = r^2 \left(\frac{d\theta}{dl}\right)^2 \\ g_{33} \frac{dx^3\, dx^3}{dl\, dl} &= g_{zz} \frac{dz\, dz}{dl\, dl} = \left(\frac{dz}{dl}\right)^2 \end{aligned} \tag{B.7.8}$$

Thus Eq. (B.7.7) becomes in cylindrical coordinates

$$s = \int_{l_1}^{l_2} \left[\left(\frac{dr}{dl}\right)^2 + r^2\left(\frac{d\theta}{dl}\right)^2 + \left(\frac{dz}{dl}\right)^2 \right]^{1/2} dl$$

$$= \int_{l_1}^{l_2} [(\dot{r})^2 + (r\dot{\theta})^2 + (\dot{z})^2]^{1/2}\, dl \tag{B.7.9}$$

B.8 Equations of Kinematics

Let $x^i = x(R, t)^i$ be the coordinates of a particle P describing a certain curve in space, where R^i represents the coordinates r, θ, and z. By definition, Eq. (3.2.2), the *velocity* v^i is given by

$$v^i = \frac{dx^i}{dt} \tag{B.8.1}$$

The velocity of the particle in cylindrical coordinates is obtained by the simple substitution

$$v^i = \sqrt{g_{11}}\,\frac{dr}{dt} + \sqrt{g_{22}}\,\frac{d\theta}{dt} + \sqrt{g_{33}}\,\frac{dz}{dt} = \frac{dr}{dt} + r\,\frac{d\theta}{dt} + \frac{dz}{dt}$$

$$= \dot{r} + r\dot{\theta} + \dot{z} \tag{B.8.2}$$

The contravariant tensor a^i representing the *acceleration* of a particle, with a velocity v^i whose coordinates are the same as those of the acceleration, was defined by Eq. (3.3.1)

$$a^i = \frac{dv^i}{dt} = \frac{d^2x^i}{dt^2} \tag{B.8.3}$$

For curvilinear coordinates the expression of the acceleration becomes

$$a^i = \frac{dv^i}{dt} = \frac{d^2x^i}{dt^2} + \Gamma^i_{\alpha\beta}\,\frac{dx^\alpha\, dx^\beta}{dt\,dt} \tag{B.8.4}$$

We have seen, Eq. (B.4.2), that in cylindrical coordinates only three components of the Christoffel symbol of the second kind remain. Consequently, substituting these values we get for the components of the acceleration

$$a^r = \frac{dv^r}{dt} = \frac{d^2r}{dt^2} + \Gamma^r_{\theta\theta}\frac{d\theta}{dt}\frac{d\theta}{dt} = \ddot{r} - r\dot{\theta}^2$$

$$a^\theta = \frac{dv^\theta}{dt} = \frac{d^2\theta}{dt^2} + (\Gamma^\theta_{\theta r} + \Gamma^\theta_{r\theta})\frac{dr}{dt}\frac{d\theta}{dt} = \ddot{\theta} + \frac{2}{r}\dot{r}\dot{\theta} \tag{B.8.5}$$

$$a^z = \frac{dv^z}{dt} = \frac{d^2z}{dt^2} = \ddot{z}$$

Finally, the physical components of the acceleration have the form

$$a \equiv \sqrt{g_{11}}a^r 1_r + \sqrt{g_{22}}a^\theta 1_\theta + \sqrt{g_{33}}a^z 1_z$$

$$= (\ddot{r} - \dot{\theta}^2)1_r + (r\ddot{\theta} + 2\dot{r}\dot{\theta})1_\theta + \ddot{z}1_z \tag{B.8.6}$$

On the basis of the velocity and acceleration of a particle in cylindrical coordinates, other formulas of kinematics of particles can easily be presented in cylindrical coordinates.

A *force* F^i is given by

$$F^i = ma^i \equiv m[(\ddot{r} - \dot{\theta}^2)1_r + (r\ddot{\theta} + 2\dot{r}\dot{\theta})1_\theta + \ddot{z}1_z] \tag{B.8.7}$$

where m is the *mass* of the particles and a^i is the acceleration derived in Eq. (B.8.6). Likewise the *kinetic energy* $\mathfrak{K}$ defined in Eq. (5.2.2) is expressed in cylindrical coordinates

$$\mathfrak{K} = \tfrac{1}{2}mv^\alpha\beta^\alpha = \tfrac{1}{2}m[(\dot{r})^2 + (r\dot{\theta})^2 + (\dot{z})^2] \tag{B.8.8}$$

B.9 The Strain Tensor

From Eq. (2.2.1) the *Lagrangian* and the *Eulerian strain tensors* are obtained. The Eulerian strain, Eq. (2.2.2), in its simple covariant form is

$$\overset{E}{\varepsilon}_{ij} = \tfrac{1}{2}[\delta_{i\gamma}u_{\gamma,j} + \delta_{\rho j}u_{i,\rho} - \delta_{\gamma\rho}u_{\gamma,i}u_{\rho,j}] \tag{B.9.1}$$

If u_r, u_θ and u_z are the projections of the displacement vector on the coordinates r, θ and z, then the physical components of the *Almansi* strain measure $\overset{A}{\varepsilon}_{ij}$, Eq. (2.3.9), can be written explicitly in cylindrical coordinates, following Reiner (1958)

$$\overset{A}{\varepsilon}_{rr} = u_{r,r} + ru_r(u_{\theta,r})^2 - \tfrac{1}{2}[u_{r,r}^2 + r^2(u_{\theta,r})^2 + (u_{z,r})^2 + u_r^2(u_{\theta,r})^2]$$

$$\overset{A}{\varepsilon}_{\theta\theta} = u_{\theta,\theta} - \tfrac{1}{2}u_{\theta,\theta}^2 + \frac{1}{r}[u_r - u_r u_{\theta,\theta}^2 - 2u_r u_{\theta,\theta}]$$

$$+ \frac{1}{2r^2}[-u_r^2 - u_{r,\theta}^2 - u_{z,\theta}^2 + 2u_r^2 u_{\theta,\theta} + u_r^2 u_{\theta,\theta}^2]$$

$$\overset{A}{\varepsilon}_{zz} = u_{z,z} + ru_r u_{\theta,z}^2 - \tfrac{1}{2}[u_{r,z}^2 + r^2 u_{\theta,z}^2 + u_{z,z}^2 + u_r^2 u_{\theta,z}^2]$$

$$\overset{A}{\varepsilon}_{zr} = ru_r u_{\theta,z} u_{\theta,r}$$

$$+ \tfrac{1}{2}[u_{r,z}u_{r,r} + u_r^2 u_{\theta,z}u_{\theta,r} + u_{z,z}u_{z,r} + u_r^2 u_{\theta,z}u_{\theta,z} - u_{r,z} - u_{z,r}]$$

$$\overset{A}{\varepsilon}_{r\theta} = -u_r u_{\theta,r} + u_r u_{\theta,r} u_{\theta,\theta} + \tfrac{1}{2}\Big[ru_{\theta,r} - u_{\theta,r}u_{\theta,\theta}$$

$$+ \frac{1}{r}(u_{r,\theta} - u_{r,r}u_{r,\theta} + u_r^2 u_{\theta,r} - u_r^2 u_{\theta,r}u_{\theta,\theta}) - \frac{1}{r^2}u_{z,r}u_{z,\theta}\Big]$$

$$\overset{A}{\varepsilon}_{\theta z} = -u_r u_{\theta,z} + u_r u_{\theta,\theta}u_{\theta,z} + \tfrac{1}{2}\Big[ru_{\theta,z} - ru_{\theta,\theta}u_{\theta,z}$$

$$+ \frac{1}{r}(u_{z,\theta} - u_{r,\theta}u_{r,z} - u_{z,\theta}u_{z,z} + u_r^2 u_{\theta,z} - u_r^2 u_{\theta,\theta}u_{\theta,z})\Big] \tag{B.9.2}$$

The expressions $\overset{G}{\varepsilon}_{ij}$ for the *Green* measure of strains are obtained by replacing r, θ, z by r_0, θ_0, z_0 and changing all negative signs to positive ones.

The *Hencky* measure of strains is obtained from the Almansi measure as follows

$$\overset{H}{\varepsilon}_{ij} = -\tfrac{1}{2}\log(1 - 2\overset{A}{\varepsilon}_{ij}) \tag{B.9.3}$$

In the case of axial symmetry all derivatives with respect to θ vanish. This reduces the expressions $\overset{A}{\varepsilon}_{\theta\theta}$, $\overset{A}{\varepsilon}_{\theta z}$, $\overset{A}{\varepsilon}_{r\theta}$ to Eqs. (B.9.2)

$$\overset{A}{\varepsilon}_{\theta\theta} = \frac{u_r}{r} - \left(\frac{u_r}{r}\right)^2$$

$$\overset{A}{\varepsilon}_{\theta z} = -u_r u_{\theta,z} + \tfrac{1}{2}[ru_{\theta,z} + (1/r)u_r^2 u_{\theta,z}]$$

$$\overset{A}{\varepsilon}_{r\theta} = -u_r u_{\theta,r} + \tfrac{1}{2}[ru_{\theta,r} + (1/r) + u_r^2\, u_{\theta,r}] \tag{B.9.4}$$

When terms of second order are neglected, the infinitesimal strain measure is obtained.

Except for the infinitesimal strain, the resolution of the strain measures into spherical and deviatoric strains according to Eqs. (2.3.13) and (2.3.14) can be very laborious.

B.10 The Balance Equations

The density balance and the balance of momentum will be explored here. They are the most consequential among the balance equations.

The *density* of a material particle is given as a function of its location in space and a function of time, $\rho(x^i, t)$ and the particle moves with a velocity $v(x^i, t)^i$. If there are no sources or sinks, the *continuity equation* or the *density balance* equation is formulated in Eq. (4.3.1)

$$\nabla \cdot (\rho \boldsymbol{v}) + \frac{\partial \rho}{\partial t} \equiv g^{-1/2} \frac{\partial}{\partial x^{\alpha}} (g^{1/2} \rho v^{\alpha}) + \frac{\partial \rho}{\partial t} = 0 \tag{B.10.1}$$

If v^i is a contravariant tensor of velocity in cylindrical coordinates, $x^i = r, \theta, z$, the density balance equation in cylindrical coordinates is obtained

$$g^{-1/2} \frac{\partial}{\partial x^{\alpha}} (g^{1/2} \rho v^{\alpha}) + \frac{\partial \rho}{\partial t}$$

$$= \frac{\partial}{\partial r} (\rho v^r) + \frac{\partial}{\partial \theta} (\rho v^{\theta}) + \frac{\partial}{\partial z} (\rho v^z) + \frac{\rho v^r}{r} + \frac{\partial \rho}{\partial t}$$

$$= \frac{\partial}{\partial r} (\rho \dot{r}) + \frac{\partial}{\partial \theta} (\rho \dot{\theta}) + \frac{\partial}{\partial z} (\rho \dot{z}) + \frac{\rho \dot{r}}{r} + \frac{\partial \rho}{\partial t} = 0 \tag{B.10.2}$$

The *balance of momentum*, also known as the *equation of motion*, formulated in Eq. (4.6.6), is given in general curvilinear coordinates as

$$\rho \dot{v}^i - t^{i\alpha}_{,\alpha} - \rho \sigma^i = p^{i\alpha}_{,\alpha} + \rho(\dot{v}^i - \sigma^i) = 0 \tag{B.10.3}$$

where t^{ij} is the *stress tensor*, p^{ij} the *pressure tensor* and σ^i the *body force*, while Eq. (4.6.6) represents in fact three equations.

Let the motion of the particle be uniformly accelerated, then $\dot{v}^i$ is given, according to Eq. (3.2.5)

$$\dot{v}^i = a^i = \frac{\partial v^i}{\partial t} + v^i_{,\alpha} v^{\alpha} \tag{B.10.4}$$

If a^r, a^{θ}, a^z and v^r, v^{θ}, v^z are the physical components of the acceleration and velocity, respectively, then in cylindrical coordinates we have

$$a^r = \frac{\partial v^r}{\partial t} + v^r \frac{\partial v^r}{\partial r} + r v^\theta \frac{\partial v^r}{\partial \theta} + v^z \frac{\partial v^z}{\partial z} - r(v^\theta)^2$$

$$= \frac{\partial \dot{r}}{\partial t} + \dot{r} \frac{\partial \dot{r}}{\partial r} + r\left(\dot{\theta} \frac{\partial \dot{r}}{\partial \theta} - (\dot{\theta})^2\right) + \dot{z} \frac{\partial \dot{z}}{\partial z}$$

$$a^\theta = \frac{\partial v^\theta}{\partial t} + \frac{1}{r}\left(v^r \frac{\partial (r v^\theta)}{\partial r} + v^\theta \frac{\partial v}{\partial \theta}\right) + v^z \frac{\partial v^\theta}{\partial z}$$

$$= \frac{\partial \dot{\theta}}{\partial t} + \frac{1}{r}\left(\dot{r}\dot{\theta} + \dot{\theta} \frac{\partial \dot{\theta}}{\partial \theta}\right) + \dot{r} \frac{\partial \dot{\theta}}{\partial r} + \dot{z} \frac{\partial \dot{\theta}}{\partial z}$$

$$a^z = \frac{\partial v^z}{\partial t} + v^r \frac{\partial v^z}{\partial r} + \frac{(v^\theta)^2 \partial v^z}{r \partial \theta} + v^z \frac{\partial v^z}{\partial z}$$

$$= \frac{\partial \dot{z}}{\partial t} + \dot{r} \frac{\partial \dot{z}}{\partial r} \sum + \frac{(\dot{\theta})^2 \partial \dot{z}}{r \partial \theta} + \dot{z} \frac{\partial \dot{z}}{\partial z} \tag{B.10.5}$$

Note that the only problem in deriving Eqs. (B.10.5) is the proper substitution of v^i and v_i in terms of the physical components, that is, $v_\theta = \sqrt{(g_{\theta\theta})} v_i$ and $v_\theta = [1/\sqrt{(g_{\theta\theta})}] v_i$, from which follow $v^i = v_\theta / r$ and $v_i = r v_\theta$, respectively.

In terms of their physical components, the pressure tensor gradient $p^{i\alpha}_{,\alpha}$, in the second part of Eq. (B.10.3) may be written

$$g_{ii}^{-1/2} p^{i\alpha}_{,\alpha} = g_{ii}^{-1/2}\left(g^{1/2} \frac{\partial (g^{1/2} p^{i\alpha})}{\partial r} + \Gamma^\alpha_{\alpha\beta} p^{i\beta}\right) \tag{B.10.6}$$

The three equations represented by (B.10.3) in cylindrical coordinates can now been written from Eqs. (B.10.5) and (B.10.6)

$$\rho\left(\frac{\partial \dot{r}}{\partial t} + \dot{r} \frac{\partial \dot{r}}{\partial r} + \frac{1}{r}\left(\dot{\theta} \frac{\partial \dot{r}}{\partial \theta} - (\dot{\theta})^2\right) + \dot{z} \frac{\partial \dot{z}}{\partial z} - \sigma^r\right)$$

$$+ \frac{\partial p^{rr}}{\partial r} + \frac{\partial p^{r\theta}}{\partial \theta} + \frac{1}{r}\left(\frac{\partial p^{rz}}{\partial z} + p^{rr} - p^{\theta\theta}\right) = 0$$

$$\rho\left(\frac{\partial \dot{\theta}}{\partial t} + \frac{1}{r}\left(\dot{r}\dot{\theta} + \dot{\theta} \frac{\partial \dot{\theta}}{\partial \theta}\right) + \dot{r} \frac{\partial \dot{\theta}}{\partial r} + \dot{z} \frac{\partial \dot{\theta}}{\partial z} - \sigma^\theta\right)$$

$$+ \frac{\partial p^{r\theta}}{\partial r} + \frac{\partial p^{\theta z}}{\partial z} + \frac{1}{r}\left(\frac{\partial p^{\theta\theta}}{\partial \theta} + 2p^{r\theta}\right) = 0$$

$$\rho\left(\frac{\partial \dot{z}}{\partial t} + \dot{r}\,\frac{\partial \dot{z}}{\partial r}\sum + \frac{(\dot{\theta})^2\partial \dot{z}}{r\partial\theta} + \dot{z}\,\frac{\partial \dot{z}}{\partial z} - \sigma^z\right)$$

$$+ \frac{\partial p^{rz}}{\partial r} + \frac{\partial p^{zz}}{\partial z} + \frac{1}{r}\left(\frac{\partial p^{\theta z}}{\partial \theta} + p^{rz}\right) = 0 \tag{B.10.7}$$

Further extension of the topic of cylindrical coordinates and their transformations is left for the learned investigator. Here the basis has been presented, hopefully to assist such further studies.

Appendix C
Rheological Modeling

C.1 Introduction

The mathematical relationship between *cause* and *effect* in a physical realm, is, in general, a relationship between the external variables that control the causes and the internal variables that constitute the effects, and is represented in the form of either *linear* or *non-linear* equations.

The similarity between the mathematical relationships of different physical systems, called *analogy*, permits the borrowing of images from a more elaborate and familiar system to a less developed one, in order to enhance the knowledge of the latter.

Linear systems are characterized by two properties (Naslin 1965):

1. Proportionality of causes and effects. If the excitations acting on a linear system are multiplied by a constant factor, the corresponding responses are multiplied by the same factor.

2. Superposition of causes and effects. If several excitations are simultaneously applied to a linear system, their total effect is the sum of the effects of each excitation acting separately.

The stability of non-linear systems depends essentially on the initially applied conditions. A valid approach for the solution of problems in non-linear systems is often their linearization, by *approximation* to linear systems, by *piecewise linearization*, or by *incremental variables*.

Electric networks, mechanical systems, electromechanical systems are *true linear analogue systems* or *models*, while non-linear friction, saturated electric machines, hydraulic systems, pneumatic systems and thermal systems are *approximate linear analogues*.

Linear stress–strain relationships of materials under isothermal conditions, known in mechanics as *constitutive equations*, are the subject of *rheology*, and the study of the interaction between the stresses and strains, aided by analogue models, is *rheological modeling*. Modeling is subject to accepted rules.

The rheological modeling presented here is based on the analogy between the mechanical behavior of a spring, a dashpot and a friction block, on one hand, and elastic, viscous and plastic phenomena, on the other hand (Reiner 1949, 1958). This behavior is described by a linear differential equation with constant coefficients, of the form

$$a_0 \frac{d^n F}{dt^n} + a_1 \frac{d^{n-1} F}{dt^{n-1}} + \ldots + a_n F = b_0 \frac{d^m E}{dt^m} + b_1 \frac{d^{m-1} E}{dt^{m-1}} + \ldots + b_m E \tag{C.1.1}$$

where F is an *excitation* in the form of a force, E is a *response* in the form of deformation, and $a_0, \ldots, a_n, b_0, \ldots, b_n$ are the constant coefficients.

If equation (C.1.1) contains additional constant terms A and B, they can easily be eliminated by substituting new variables for $(a_n F + A)$ and $(b_m E + B)$.

We also notice that if we substitute for d/dt the symbol $\mathcal{D}$, then the superposition and proportionality property can be written as follows

$$\mathcal{D}^n (F_1 + F_2) = \mathcal{D}^n F_1 + \mathcal{D}^n F_2 \tag{C.1.2}$$

$$\mathcal{D}^n (kF) = k\mathcal{D}^n F \tag{C.1.3}$$

and we may also write

$$\mathcal{D}^m \mathcal{D}^n F = \mathcal{D}^{m+n} F \tag{C.1.4}$$

If we denote by $\mathcal{R}(\mathcal{D})$ and $\mathcal{Q}(\mathcal{D})$ the polynomials on the left-hand side and right-hand side of equation (C.1.1), respectively, we can write

$$\mathcal{R}(\mathcal{D}) F = \mathcal{Q}(\mathcal{D}) E \tag{C.1.5}$$

from which it follows that

$$\frac{E}{F} = \frac{\mathcal{R}(\mathcal{D})}{\mathcal{Q}(\mathcal{D})} = \mathcal{T}(\mathcal{D}) \tag{C.1.6}$$

where $\mathcal{T}(\mathcal{D})$ is the *transmittance*, relating the response E to the corresponding excitation F.

There are three elements which serve as the building blocks for all mechanical models and rheological bodies.* In the following sections these elements and some of the familiar and important bodies and models are discussed.

* The effects of inertia and of mass acceleration are not considered; thus a fourth element, the *mass*, is not introduced.

It should be stressed again that the rheological model, while it is a convenient tool to analyze the response to an excitation of any nature, is only a duplicate of a mathematical equation.

C.2 The Hookean Elastic Element

A response which depends on the excitation alone and is always proportional to it is an *elastic* response, known also as the *Hookean elastic element*, represented in Fig. C.2.1 by a *spring* and defined

$$F = k_1 E \tag{C.2.1}$$

where k_1 is the coefficient of proportionality, known as the *elastic modulus*. The constants of the corresponding polynomial, Eq. (C.1.1), are $a_0 = 1$, $b_0 = k_1$, where $n = m = 0$, while all other constants vanish.

C.3 The Newtonian Viscous Element

A rheological element with a response rate proportional to the excitation and represented in Fig. C.3.1 by a *dashpot*, is a *Newtonian viscous* element, defined

$$F = q_2 \dot{E} \tag{C.3.1}$$

where q_2 is the *viscosity coefficient* and represents the angle of the tangent to the F–E curve, Fig. C.3.1b. The corresponding constants of the polynomial,

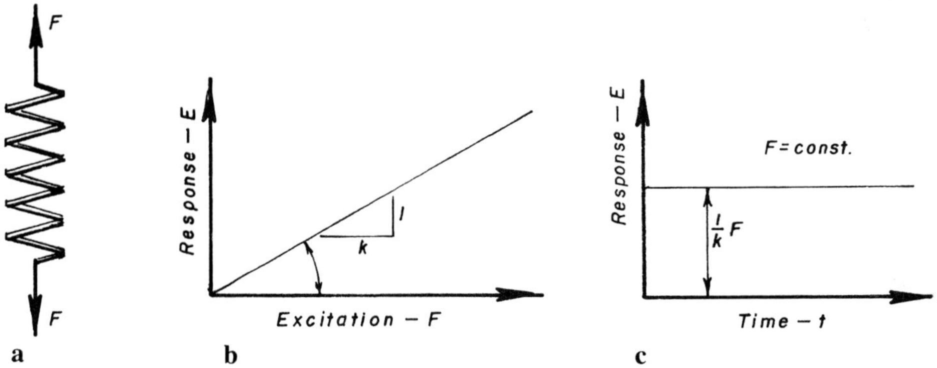

Fig. C.2.1. Hookean element. **a** Elastic spring-mechanical model. **b** Response versus excitation. **c** Response versus time.

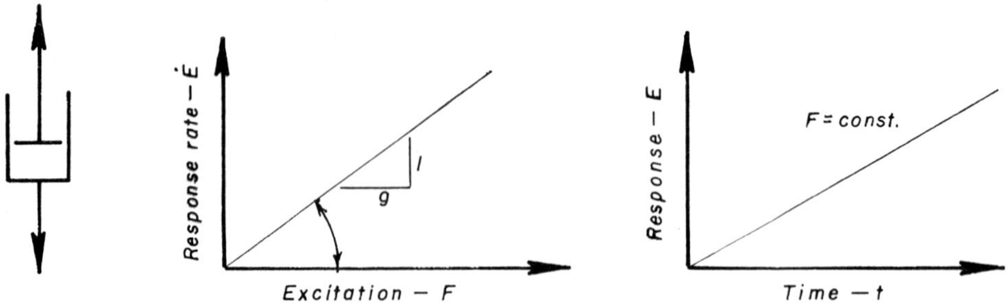

a b c

Fig. C.3.1. Newtonian element. **a** Dashpot-mechanical model. **b** Response rate versus excitation. **c** Response versus time.

Eq. (C.1.1), are $a_0 = 1$, $b_0 = q_2$, where $n = 0$, $m = 1$, and all other constants vanish.

C.4 Coupling of Rheological Elements

The three rheological elements, the elastic spring, the viscous dashpot and the plastic friction block, especially the two former, can be coupled into rheological *bodies* in two ways: in series and in parallel.

When coupled in series, an excitation acting on the body is transferred to each of the elements and the response of the body is the sum of the individual responses of the elements.

$$F = F_1 = F_2; \quad E = E_1 + E_2 \tag{C.4.1}$$

When coupled in parallel, the excitation acting on the body is divided between the individual elements, so that their responses are equal.

$$F = F' + F''; \quad E = E' = E'' \tag{C.4.2}$$

The above holds also for two elements of the same kind, two elastic springs or two viscous dashpots, coupled in series or in parallel. For two elastic springs with elastic moduli k_1 and k_2 coupled in series we can write

$$F = F_1 = F_2; \quad E_1 = \frac{F_1}{k_1} = \frac{F}{k_1}; \quad E_2 = \frac{F_2}{k_2} = \frac{F}{k_2} \tag{C.4.3}$$

$$E = E_1 + E_2 = F\left(\frac{1}{k_1} + \frac{1}{k_2}\right) = \frac{F}{k} \tag{C.4.4}$$

from which it follows

$$\frac{1}{k} = \frac{1}{k_1} + \frac{1}{k_2} \tag{C.4.5}$$

which means that the reciprocal of the combined modulus of elasticity of two or more elastic bodies coupled in series is the sum of their individual reciprocals.

A similar result is obtained for viscous dashpots coupled in series

$$\frac{1}{q} = \frac{1}{q_1} + \frac{1}{q_2} \tag{C.4.6}$$

On the other hand, for two elastic springs with moduli k_1 and k_2 coupled in parallel we can write

$$E = E' = E''; \quad F' = k_1 E' = k_1 E; \quad F'' = k_2 E'' = k_2 E; \tag{C.4.7}$$

$$F = F' + F'' = k_1 E + k_2 E = (k_1 + k_2)E = kE \tag{C.4.8}$$

from which it follows

$$k = k_1 + k_2 \tag{C.4.9}$$

which means that the combined modulus of elasticity of two elastic springs coupled in parallel is equal to the sum of their moduli of elasticity.

A similar equation is obtained from two viscous dashpots coupled in parallel

$$q = q_1 + q_2 \tag{C.4.10}$$

C.5 St Venant's Element of Plastic Restraint

The element represented by the *plastic restraint* $F_{\mathfrak{S}} = \mathfrak{S}_0 \vartheta(t)$, which dissipates part of the energy invested by the excitation as a result of an internal restraint ϑ that may change with time, is known as the *St Venant plastic restraint*.

$\mathfrak{S}_0$ is a constant characteristic of the material and it controls the magnitude of the plastic restraint $\vartheta(t)$, which is a monotonically increasing function equal to zero at time $t = 0$, $\vartheta(0) = 0$, and equal to unity at time $t = \infty$, $\vartheta(\infty) = 1$. The effect of the internal excitation $\mathfrak{S}_0$ is always negative and is subtracted from the total externally applied excitation.

This element exists *only* if it is aroused by an external excitation. Fig. C.5.1 shows the element, represented by a *friction block* or rather a series of friction blocks, which are activated gradually. The mathematical form of the

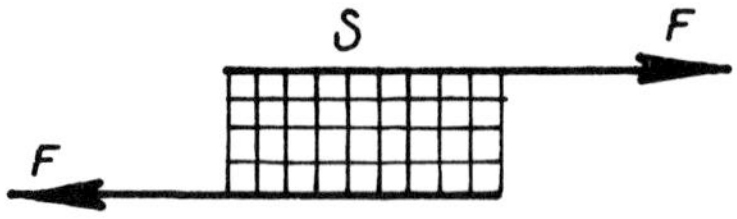

Fig. C.5.1. St Venant plastic element.

plastic restraint has not been fully investigated, but it is known to be a function of time, even under a constant excitation, see Fig. C.6.2.

C.6 The Prandtl Body

It was remarked in the previous section that the St Venant restraint does not sustain itself, but is aroused only by an external excitation. We define an *activating* excitation $\overline{F}$ as follows

$$\overline{F(t)} = F(t) - \mathfrak{F}_0 \vartheta(t) \tag{C.6.1}$$

where $\mathfrak{F}_0$ has the dimensions of a stress or an elastic modulus $[\mathfrak{F}_0] = [FL^{-2}]$.

Therefore the St Venant element comes always coupled in series with an elastic spring, which measures the magnitude of the excitation, Fig. C.6.1, and forms, with the St Venant restraint, a rheological body, the *Prandtl body*.

We see in Fig. C.6.2 that under constant excitation, $F = \text{const}$, the activating excitation $\overline{F}$ is nevertheless a function of time

$$\overline{F(t)} = F - \mathfrak{F}_0 \vartheta(t); \quad F = \mathcal{F} = \text{const} \tag{C.6.2}$$

The Prandtl body is also known as the *elasto-plastic body*.

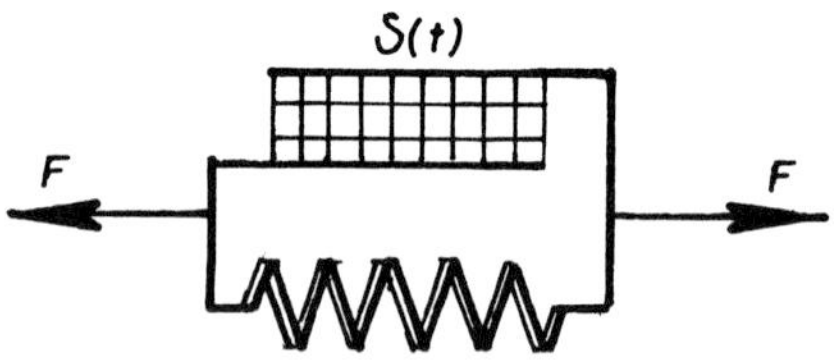

Fig. C.6.1. Prandtl body.

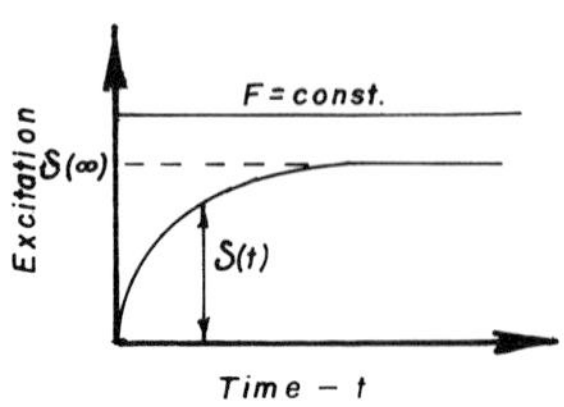

Fig. C.6.2. Prandtl body and response for constant excitation.

C.7 The Maxwell Body

An elastic spring and a viscous dashpot, coupled in series, Fig. C.7.1, form a *Maxwell body*.

If an excitation F is applied on the body, the responses of the spring and the dashpot will be, respectively

$$E_1 = \frac{F}{k_1}; \quad \dot{E}_2 = \frac{F}{g_2} \tag{C.7.1}$$

By differentiating Eq. (C.7.1)$_1$ with respect to time and summing up the two responses, it can be shown that

$$\dot{E} = \dot{E}_1 + \dot{E}_2 = \frac{\dot{F}}{k_1} + \frac{F}{g_2} \tag{C.7.2}$$

The constant coefficients of the polynomial equation (C.1.1), corresponding to Eq. (C.7.2), are $a_0 = 1$, $a_1 = 1/T_{\text{ref}}$ and $b_0 = k_1$, where $n = m = 1$.

By integrating Eq. (C.7.2) we get

$$\begin{aligned} F &= \left(\exp\frac{-k_1 t}{g_2}\right)\left(F_0 + k_1 \int_0^t \dot{E} \exp\frac{k_1 t}{g_2}\,\mathrm{d}t\right) \\ &= \left(\exp\frac{-t}{T_{\text{ref}}}\right)\left(F_0 + k_1 \int_0^t \dot{E} \exp\frac{t}{T_{\text{ref}}}\,\mathrm{d}t\right) \end{aligned} \tag{C.7.3}$$

where F_0 is the initial excitation acting on the body and $T_{\text{ref}} = g_2/k_1$ is the *relaxation time*.

Several observations should be made with respect to Eq. (C.7.3):

1. From Eqs. (C.7.2) we see that if a constant excitation is applied, $F = \mathcal{F} = \text{const}$, and Eq. (C.7.2) is integrated, we have

$$E = E_1 + E_2 = \frac{\mathcal{F}}{k_1} + \frac{\mathcal{F}t}{g_2} = \mathcal{F}\left(\frac{1}{k_1} + \frac{t}{g_2}\right) \tag{C.7.4}$$

Fig. C.7.1. Maxwell body–spring and dashpot coupled in series.

containing an instantaneous elastic response $\mathcal{F}/k_1$ and a time-dependent viscous response, Fig. C.7.2.

2. As the response rate increases the excitation increases.

3. For constant rates of response a series of excitation–time curves are obtained, Fig. C.7.3, if $\dot{E} = \mathcal{E} =$ const is substituted into Eq. (C.7.3)

$$F = F_0 \exp \frac{-t}{T_{\mathrm{rel}}} + g_2\mathcal{E}; \quad \dot{E} = \mathcal{E} = \mathrm{const} \tag{C.7.5}$$

and the various curves are asymptotic to $g_2\mathcal{E}$.

4. When the response is kept constant, $E =$ const, the rate of response vanishes, $\dot{E} = 0$, and Eq. (C.7.3) becomes

$$F = F_0\left(\exp \frac{-t}{T_{\mathrm{rel}}}\right) \tag{C.7.6}$$

The curve $E =$ const is the lowest curve in Fig. C.7.4, bounding the $\dot{E} = \mathcal{E} =$ const curves.

5. If T_{rel} is substituted for t in Eq. (C.7.5), the value of the excitation for the various curves with constant response rates is obtained

$$F = \frac{F_0}{e} + g_2\mathcal{E} \tag{C.7.7}$$

where, for the curve with constant response, $\mathcal{E} = 0$, the second part of the

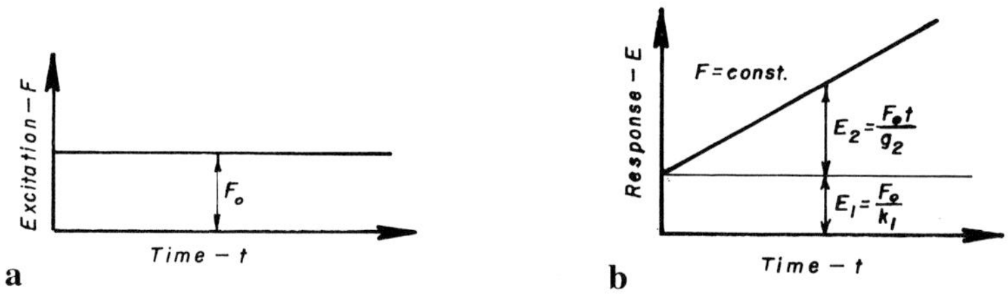

Fig. C.7.2. Response of a Maxwell body for constant excitation. **a** Excitation versus time. **b** Response versus time.

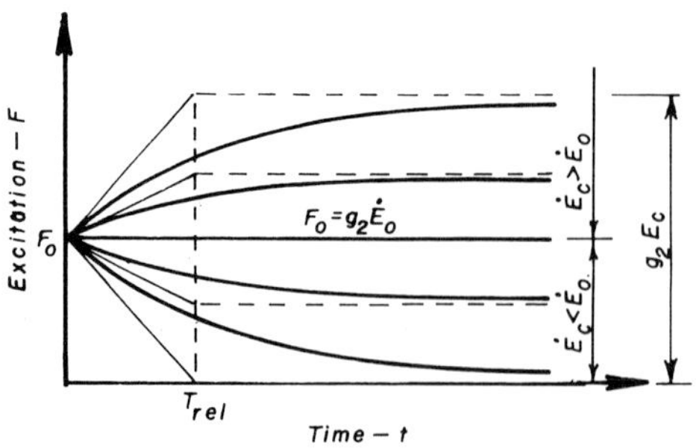

Fig. C.7.3. Excitation of a Maxwell body for a constant rate of response.

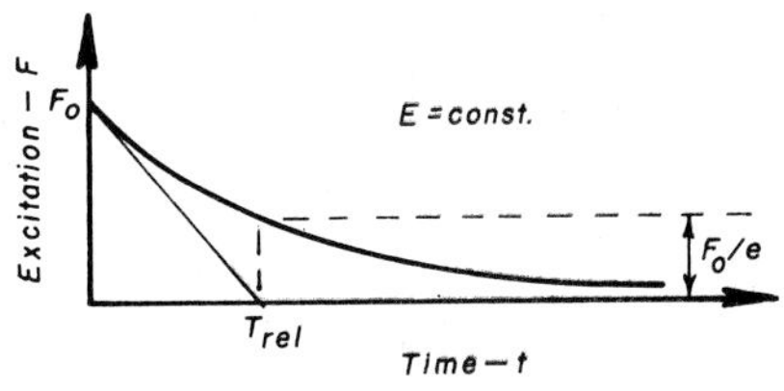

Fig. C.7.4. Excitation of a Maxwell body for a constant response.

equation vanishes

$$F = \frac{F_0}{e} \tag{C.7.8}$$

and $e = 2.71828$ is a constant, the base of the natural logarithm.

Eq. (C.7.2) can be written in the following two forms

$$E = \frac{1}{k_1}\left(F + \frac{1}{T_{\text{rel}}}\int F\,\mathrm{d}t\right) \tag{C.7.9}$$

$$F = q_2\dot{E} - \dot{F}T_{\text{rel}} \tag{C.7.10}$$

where the differential equation representing the Maxwell body is a function of the relaxation time T_{rel}, which is an intrinsic time scale. Eqs. (C.7.9) and (C.7.10) indicate that for a relatively short time, $t \ll T_{\text{rel}}$, the response is close to that of a Hookean element, $E \cong F/k$, while if the time is long compared to the relaxation time, $t \gg T_{\text{rel}}$, the response would be close to a Newtonian response, $F = q_2\dot{E}$.

C.8 The Kelvin Body

An elastic spring and a viscous dashpot coupled in parallel, Fig. C.8.1, form a *Kelvin body*, defined by the equations

$$E = E' = E'' \tag{C.8.1}$$

$$F = F' + F'' = k_3E' + q_3\dot{E}'' = k_3E + q_3\dot{E} \tag{C.8.2}$$

where F', F'' and E', E'' are the excitations and responses of the elastic spring and viscous dashpot, respectively.

The response of the body is equal to the responses of each of the participating elements, and the excitation applied to the body is the sum of the excitations of the individual elements. The constants corresponding to the polynomial equation (C.1.1) are $a_0 = 1$, $b_0 = q_3$ and $b_1 = k_3$, where $n = 0$ and $m = 1$.

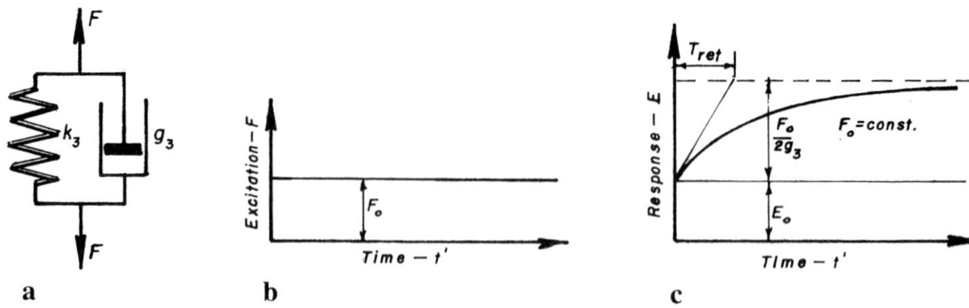

Fig. C.8.1. Kelvin body and response diagram. **a** Spring and dashpot parallelly coupled-creep element. **b** Excitation versus time. **c** Response versus time.

This is a non-relaxing elastic body, and for a constant response $E = \mathcal{E} =$ const and $\dot{E} = 0$ we have

$$F = k_3 \mathcal{E} \tag{C.8.3}$$

If, after the body has reached a response E_∞ the excitation is removed, $F = 0$, the response does not vanish at once as in an elastic body, and we obtain

$$k_3 E + g_3 \dot{E} = 0 \tag{C.8.4}$$

Solving the homogeneous differential equation (C.8.4), we get

$$E = E_\infty \exp \frac{-t}{T_{\text{ret}}} \tag{C.8.5}$$

where E_∞ is the final stage of the applied excitation and it serves as an initial response for the stage of the removal of the excitation, and $T_{\text{ret}} = g_3/k_3$ is the *retardation time*, a time scale of the response.

To obtain the response E we integrate Eq. (C.8.2)

$$E = \left(\exp \frac{-t}{T_{\text{ret}}}\right)\left(E_0 + \frac{1}{g_3}\int_0^t F \exp \frac{t'}{T_{\text{ret}}}\, dt'\right) \tag{C.8.6}$$

where E_0 is an initial response already present in the body.

Two remarks will be made here:

1. Let the excitation be constant, $F = \mathcal{F} =$ const, then Eq. (C.8.6) yields

$$E = \frac{\mathcal{F}}{k} + \left(E_0 - \frac{\mathcal{F}}{k_3}\right) \exp \frac{-t}{T_{\text{ret}}} \tag{C.8.7}$$

Eq. (C.8.7) gives a series of excitation–response curves, as shown in Fig. C.8.2.

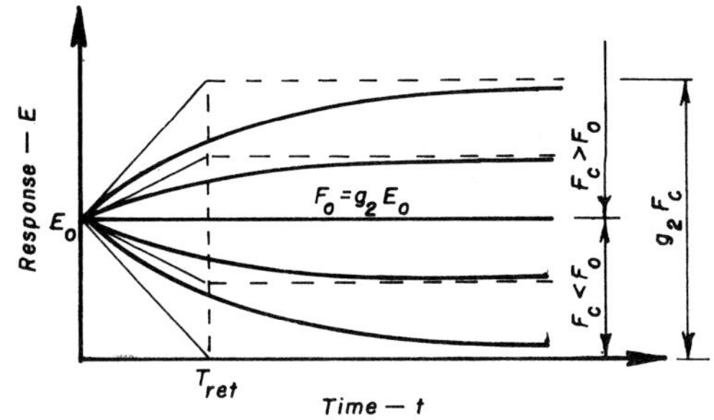

Fig. C.8.2. Response curves in a Kelvin body for constant excitation.

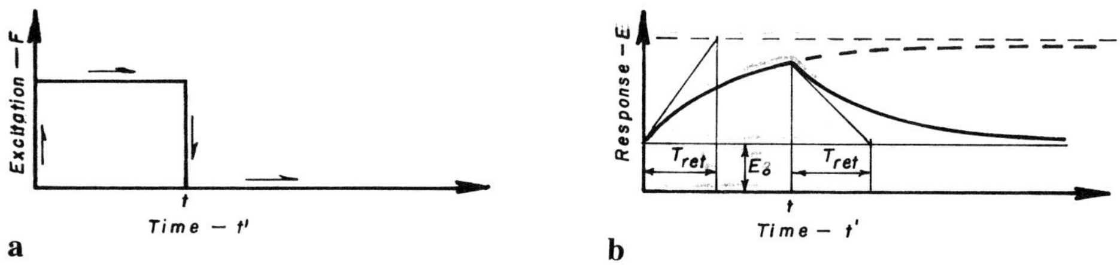

Fig. C.8.3. Loading–unloading excitation sequence in a Kelvin body. **a** Excitation versus time. **b** Response versus time.

2. When $\mathcal{F}$ is applied to the body, the response does not occur instantaneously, but is delayed in an elastic *fore-effect* or *creep*, with T_{ret} as the time of retardation, and comes gradually to an equilibrium as it approaches $t = \infty$. At equilibrium and with no initial response, the Kelvin body behaves as a Hookean elastic spring

$$E = \frac{\mathcal{F}}{k_3} = E_\infty \qquad \text{(C.8.8)}$$

When the excitation is removed, $F = 0$, the response E_∞ recovers in an *elastic after-effect* or *creep recovery* at time $t = \infty$. If, however, T_{ret} is not too large, it recovers in a finite time. The elastic fore- and after-effects constitute a *delayed* or *retarded elasticity* (Reiner 1958). Fig. C.8.3 shows the response-time curves for the applied and removed excitation.

C.9 The Burgers Body

A Maxwell body and a Kelvin body coupled together in series and exhibiting instantaneous elasticity, viscous flow and also delayed elasticity, is a *Burgers body*, Fig. C.9.1. The requirements with respect to the excitations and responses are

$$F = F_1 = F_2 = F_3 = F_3' + F_3'' \qquad \text{(C.9.1)}$$

$$E = E_1 + E_2 + E_3 = E_1 + E_2 + E_3' = E_1 + E_2 + E_3'' \qquad \text{(C.9.2)}$$

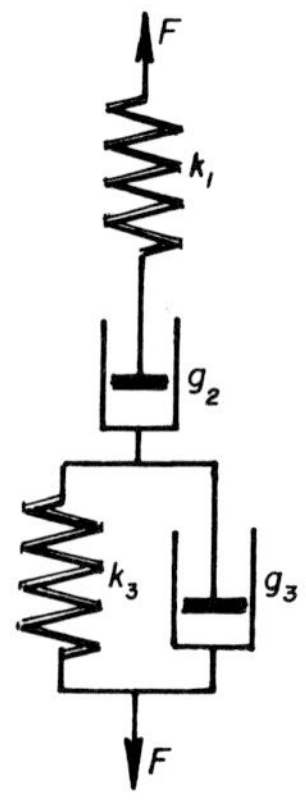

Fig. C.9.1. Burgers body.

where the indices 1, 2, 3 indicate the Hookean elastic, the Newtonian viscous and the viscoelastic Kelvin bodies, respectively, while the prime and double prime distinguish between the elastic and viscous elements in the Kelvin body. From Eq. (C.9.2) one can write

$$\dot{E} = \dot{E}_1 + \dot{E}_2 + \dot{E}_3 = \dot{E}_1 + \dot{E}_2 + \dot{E}'_3 = \dot{E}_1 + \dot{E}_2 + \dot{E}''_3 \tag{C.9.3}$$

The values of $\dot{E}_1$, $\dot{E}_2$ and $\dot{E}_3$ are eliminated by using equations (C.3.1), (C.7.3), (C.8.6) and (C.9.1)–(C.9.3), and we obtain

$$\ddot{F} + \left(\frac{k_1}{g_2} + \frac{k_1}{g_3} + \frac{k_3}{g_3}\right)\dot{F} + \frac{k_1 k_3}{g_2 g_3} F = k_1 \ddot{E} + \frac{k_1 k_3}{g_3} \dot{E} \tag{C.9.4}$$

which is also the equivalent of the polynomial equation (C.1.1) with constants

$$a_0 = 1, \quad a_1 = \frac{k_1}{g_2} + \frac{k_1}{g_2} + \frac{k_3}{g_3}, \quad a_2 = \frac{k_1 k_3}{g_2 g_3}, \quad b_0 = k_1, \quad b_1 = \frac{k_1 k_3}{g_3}$$

where $n = 2$ and $m = 1$.

The solution of Eq. (C.9.4) will be

$$E(t) = \frac{F(t)}{k_1} + \frac{1}{g_2}\int F(t)\,dt + \left(\exp\frac{-t}{T_{ret}}\right)\left(E_{30} + \frac{1}{g_3}\int F(t)\exp\frac{t}{T_{ret}}\,dt\right) \tag{C.9.5}$$

where E_{30} is an initial response of the Kelvin body.

Differentiating Eq. (C.9.5) and eliminating the two integrals, we have

$$\dot{E}(t) = \frac{\dot{F}(t)}{k_1} + \frac{F(t)}{g_2} + \left(E_{30} + \frac{F(t)}{g_3}\right)\exp\frac{-t}{T_{ret}} \tag{C.9.6}$$

Several remarks should be made here:

1. On application of an excitation to the Burgers body, the responses are instantaneous elasticity, delayed elasticity and viscous flow; on removal of the excitation, the instantaneous and delayed elasticity are recovered, and an unrecovered viscous flow is evident, Fig. C.9.2.

The Burgers body exhibits both a relaxation time originating from the Maxwell body, and a retardation time originating from the Kelvin body.

2. Applying a constant and steady initial excitation, $F = \text{const} = \mathcal{F}$ and assuming $E_{30} = 0$, we obtain, from Eqs. (C.9.5) and (C.9.6)

$$E(t) = \left[\frac{1}{k_1}\left(1 - \frac{t}{T_{\text{rel}}}\right) + \frac{1}{k_3}\left(1 - \exp\frac{-t}{T_{\text{ret}}}\right)\right]\mathcal{F} \tag{C.9.7}$$

$$\dot{E}(t) = \left(\frac{1}{g_2} + \frac{1}{g_3}\exp\frac{-t}{T_{\text{ret}}}\right)\mathcal{F} \tag{C.9.8}$$

from which it is seen that the response increases from $E = E_0 = \mathcal{F}/k_1$ to $E(t')$ at time t' and thus the response rate decreases from $\dot{E}_0 = \mathcal{F}(1/g_2) + (1/g_3)$ at $t = 0$ to $\dot{E}(\infty) = \mathcal{F}/g_2$ at $t = \infty$.

3. If the load is removed after a time t', we obtain the final response $E'_c = E(t')$ from Eq. (C.9.7), substituting $\mathcal{F} = 0$

$$\begin{aligned} E &= E'_c - \left[\frac{1}{k_1} + \frac{1}{k_3}\left(1 - \exp\frac{-t}{T_{\text{ret}}}\right)\right]\mathcal{F} \\ &= \left[\frac{1}{k_1}\left(1 - \frac{t'}{T_{\text{rel}}}\right) + \frac{1}{k_3}\left(1 - \exp\frac{-t'}{T_{\text{ret}}}\right) - \frac{1}{k_1} - \frac{1}{k_3}\left(1 - \exp\frac{-t}{T_{\text{ret}}}\right)\right]\mathcal{F} \\ &= \frac{1}{k_1}\frac{t'}{T_{\text{rel}}}\mathcal{F} = \frac{t'}{g_2}\mathcal{F} \end{aligned} \tag{C.9.9}$$

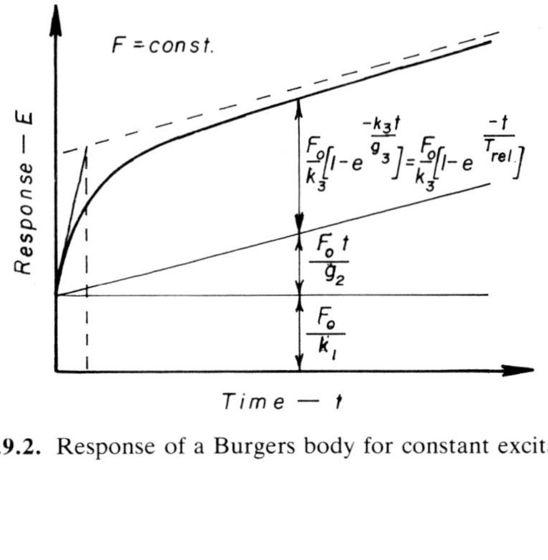

Fig. C.9.2. Response of a Burgers body for constant excitation.

where

$$E_c' = \frac{F_0}{k_1} + \frac{F_0}{g_2} + \left(1 - \exp\frac{-t'}{T_{\text{ret}}}\right)\left(\frac{F_0}{k_3} - E_{30}\right)$$

If an initial response does not exist in the first place, $E_{30} = 0$, then it is obvious that a constant excitation results in an elastic response, a viscous response and a delayed elastic response. At $t = 0$, right after the removal of the excitation, the elastic response vanishes and only the viscous and delayed elastic responses are retained. The viscous response attained at $t = t'$, the time of removal of the excitation, remains unchanged, while the delayed elasticity is recovered completely.

4. If the response is kept constant, $E = \text{const} = \mathscr{E}$, so that $\dot{E} = \ddot{E} = 0$, Eq. (C.9.4) yields

$$\ddot{F} + \left(\frac{k_1}{g_2} + \frac{k_1}{g_3} + \frac{k_3}{g_3}\right)\dot{F} + \frac{k_1 k_3}{g_2 g_3}F = 0 \tag{C.9.10}$$

from which

$$F = C_1 \exp\frac{-t}{T_1} + C_2 \exp\frac{-t}{T_2} \tag{C.9.11}$$

where T_1 and T_2 are the relaxation times, given by

$$\frac{1}{T_{1,2}} = \frac{k_3 g_2 + k_1 g_2 + k_1 g_3 \pm \sqrt{\Delta}}{2k_1 k_2}; \tag{C.9.12}$$

$$\Delta = (k_3 g_2 + k_1 g_2 + k_1 g_3)^2 - 4k_1 k_3 g_2 g_3$$

If, however, when the response reaches E_0 the excitation is removed, $F = \dot{F} = \ddot{F} = 0$, then Eq. (C.9.6) yields

$$\frac{g_3}{k_3}\ddot{E} + \dot{E} = 0 \tag{C.9.13}$$

from which, with the help of Eq. (C.9.4), we see that the recovery of the response is equal to

$$E = \left(\exp\frac{-t}{T_{\text{ret}}} - 1\right)E_{30} + E_{10} \tag{C.9.14}$$

The response at time $t = 0$ is $E_0 = E_{10}$, while at time $t = \infty$ it reduces to $E_{10} - E_{30}$.

C.10 The Relations between Excitation and Response

The Burgers body will be used here as a basic model, therefore we will spend some time in the investigation of its behavior under various excitation and response conditions. We shall see that the application of a constant excitation results in a response proportional to the excitation, and vice versa, when a constant response is imposed, a proportional excitation is aroused. Sinusoidal excitations result in sinusoidal responses.

1. Constant Excitation

It was shown, Eq. (C.9.5), that if we take a three-component body such as the Burgers body and apply to it an assumed constant excitation $F = \mathcal{F} = \text{const}$, we get a combination of three responses, Fig. C.9.1: an instantaneous, a viscous and a delayed elastic response. The total response of a general Kelvin body, Fig. C.12.1, which is in fact an extended Burgers body, will then be

$$E = E_1 + E_2 + \sum_{i=1}^{\Im} E_3 = \frac{\mathcal{F}}{k_1} + \frac{\mathcal{F}}{q_2}t + \sum_{i=3}^{\Im} \frac{\mathcal{F}}{k_3}\left(1 - \exp\frac{-t}{T_{\text{ret}}}\right)$$

$$= \left(\frac{1}{k_1} + \frac{1}{q_2}t + \psi(t)\right)\mathcal{F} \qquad \mathcal{F} = \text{const} \tag{C.10.1}$$

where $\psi(t)$ represents the *creep function* of the delayed elastic response.

2. Constant Response

Application of a constant response, $E = \mathcal{E} = \text{const}$, produces, according to Eq. (C.9.11)

$$F = k_1'\mathcal{E} + q_2'\delta(t)\mathcal{E} + \chi(t)\mathcal{E} = [k_1' + q_2'\delta(t) + \chi(t)]\mathcal{E} \tag{C.10.2}$$

where $\chi(t)$ is the relaxation function with $\chi(\infty) = 0$, Fig. C.10.1, and $\delta(t)$ is the *Dirac delta function*, defined

$$\int_{-\epsilon}^{\epsilon} \delta(t)\,\mathrm{d}t = 1 \quad \text{for } \epsilon > 0; \quad \delta(t) = 0 \quad \text{for } t \neq 0 \tag{C.10.3}$$

and k_1' and q_2' are elastic and viscous coefficients of a general Maxwell body, functions of k_1, k_i, q_2, q_i, see Sect. C.12.

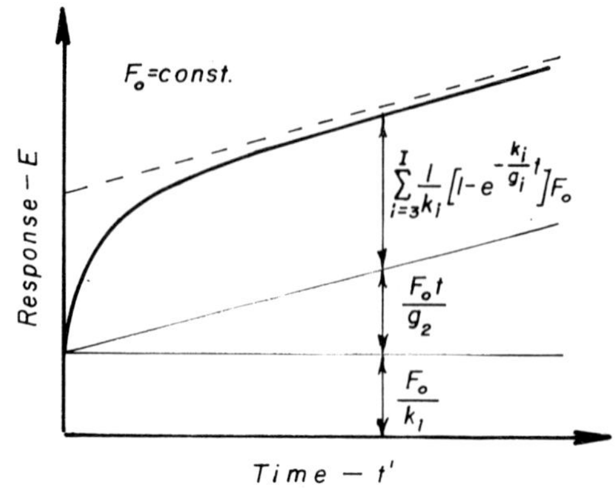

Fig. C.10.1. Creep of a general Burgers body for constant excitation.

3. Alternating Excitation

If an alternating excitation $F = F_0 \exp(i\omega t)$ is applied, it produces, in a steady state, a sinusoidal response, partly in phase and partly in quadrature with the excitation. Thus

$$E = \mathfrak{F}(i\omega)^* F = \mathfrak{F}(i\omega)^* F_0 \exp(i\omega t); \quad F = F_0 \exp(i\omega t) \tag{C.10.4}$$

where $\mathfrak{F}(i\omega)^*$ is the *complex elastic compliance*, ω is the *frequency* and i is an *imaginary unit*. Decomposing the complex elastic compliance into the real and imaginary parts we get

$$\mathfrak{F}(i\omega)^* = \left(\frac{1}{k_1} + \mathfrak{F}(\omega)_1\right) + i\left(\frac{-1}{\omega q_2} + \mathfrak{F}(\omega)_2\right) \tag{C.10.5}$$

where k_1 is the *coefficient of elasticity* of the instantaneous elastic response, $i\omega q_2$ is the *coefficient of viscosity* of the viscous response and $\mathfrak{F}_1$ and $\mathfrak{F}_2$ are, respectively, the real and imaginary parts of the *compliance*, related to the creep phenomenon. It should be noted that

$$\mathfrak{F}(i\omega) = \mathfrak{F}(\omega)_1 + i\mathfrak{F}(\omega)_2 \tag{C.10.6}$$

4. Alternating Response

An alternating response is satisfied by a sinusoidal exciation, which is part in phase and part in quadrature with the response. Consequently

$$F = \mathfrak{E}(i\omega)^* E = \mathfrak{E}(i\omega)^* \mathfrak{E}_0 \exp(i\omega t); \quad E = \mathfrak{E}_0 \exp(i\omega t) \tag{C.10.7}$$

where $\mathfrak{E}(i\omega)^*$ is the *complex elastic modulus*. Decomposing the complex elastic modulus into the real and imaginary parts we get

$$\mathfrak{E}(i\omega)^* = (\mathfrak{E}_0 + \mathfrak{E}(\omega)_1) + i\mathfrak{E}(\omega)_2 \tag{C.10.8}$$

where $\mathfrak{E}_0$ is the *static elastic modulus*, $\mathfrak{E}(\omega)_1$ is the *dynamic modulus*, $\mathfrak{E}(\omega)_2$ is the *dynamic friction* and $\mathfrak{E}(\omega)_1 + i\mathfrak{E}(\omega)_2$ is the part of the *complex modulus* associated with the phenomenon of relaxation.

From Eqs. (C.10.7) and (C.10.8) we obtain, by differentiation

$$F = \left[\frac{\mathfrak{E}(t)_2}{\omega} - \frac{i}{\omega}(\mathfrak{E}_0 - \mathfrak{E}_2)\right]\frac{dE}{dt} = \left[g(\omega) - \frac{i}{\omega}(\mathfrak{E}_0 - \mathfrak{E}_2)\right]\frac{dE}{dt} \tag{C.10.9}$$

where $g(\omega) = \mathfrak{E}(\omega)_2/\omega$ is the *dynamic viscosity*, analogous to the viscous flow in Eq. (C.3.1).

5. Generalization

All the foregoing equations lead us to the general expression of the relations between excitation, response and time

$$\begin{aligned} E(t) &= \int_{-\infty}^{t} \frac{dF(t')}{dt'}\left(\frac{1}{k_1} + \frac{t-t'}{g_2} + \psi(t-t')\right)dt' \\ &= \frac{1}{k_1}F(t) + \frac{1}{g_2}\int_{-\infty}^{t} F(t')\,dt' + \int_{-\infty}^{t} \psi(t-t')\,dt' \end{aligned} \tag{C.10.10}$$

$$\begin{aligned} F(t) &= \int_{-\infty}^{t} \frac{dE(t')}{dt'}(k_1' + g_2'\delta(t-t') + \chi(t-t'))\,dt' \\ &= k_1'E(t) + g_2'\dot{E}(t) + \int_{-\infty}^{t} \chi(t-t')\,dt' \end{aligned} \tag{C.10.11}$$

where $\delta(t)$ is the Dirac function, and $\psi(t)$ and $\chi(t)$ are the *creep* and *relaxation functions*, respectively, all defined earlier.

Eqs. (C.10.10) and (C.10.11) are dependent equations, meaning that from each of the equations the other can be derived. We will elaborate on Eq. (C.10.11).

C.11 The Relaxation and Creep Functions

The relaxation function $\chi(t)$ is a continuous function of t, which decreases monotonically to zero and can be represented in the integral form

$$\chi(t) = \int_0^{\infty} \beta B(t')\exp\frac{-t}{t'}\,dt' = \int_0^{\infty} N(s)\exp(-ts)\,ds \tag{C.11.1}$$

where β is a *normalization factor*, defined

$$\beta = \chi(0) \tag{C.11.2}$$

$B(t')\,dt'$ is the *distribution function* of the relaxation times or the *relaxation spectrum*, a continuous function defined

$$\int_0 B(t')\,dt' = 1 \tag{C.11.3}$$

and $N(s)\,ds$ is the *frequency function* introduced by the transformation $s = 1/t'$, defined

$$N(s) = \beta B(1/s)/s^2; \quad s = \frac{1}{t'} = \frac{1}{T_{rel}} \tag{C.11.4}$$

The relaxation function $\chi(t)$ is expressed

$$\chi(t) = \int_0^\infty N(s) \exp(-ts)\,ds \tag{C.11.5}$$

$B(t')$ may also be expressed as a discrete spectrum, a *line spectrum*

$$B(t') = \sum_{i=3}^{\mathfrak{J}} B_i \delta(t - \bar{t}') \tag{C.11.6}$$

and Eq. (C.11.1) degenerates into

$$\chi(t) = \sum_{i=3}^{\mathfrak{J}} \beta B_i \exp(-t/T_{reli}) \tag{C.11.7}$$

where T_{reli} is the *relaxation time* of the ith element.

Similarly, the creep function $\psi(t)$ is represented by an integral function

$$\psi(t) = \int \alpha A(t')(1 - \exp(-t/t'))\,dt' \tag{C.11.8}$$

where $A(t)\,dt'$ is the *distribution function* of the retardation times or the *retardation spectrum*, provided the normalization factor is determined by

$$\int_0^\infty A(t')\,dt' = 1 \tag{C.11.9}$$

and

$$\alpha = \psi(\infty) \tag{C.11.10}$$

The spectrum may be continuous or discontinuous. In the latter case we have

$$A(t') = \sum_{i=3}^{\Im} A_i \delta(t - t'_i) \tag{C.11.11}$$

For a discrete system, Eq. (C.11.9) degenerates into

$$\psi(t) = \sum_{i=3}^{\Im} \alpha A_i \left(1 - \exp \frac{-t}{T_{\text{reti}}}\right) \tag{C.11.12}$$

where T_{reti} is the *retardation time* of the ith element.

Introducing the rate of creep function

$$\frac{d\psi(t)}{dt} = \int_0^\infty \frac{\alpha A(t')}{t'} \exp \frac{-t}{t'} \, dt' \tag{C.11.13}$$

and a frequency distribution

$$M(s) = \alpha A(1/s)s^2 \tag{C.11.14}$$

Eq. (C.11.13) is transformed into

$$\frac{d\psi(t)}{dt} = \int_0^\infty sM(s) \exp(-ts) \, ds \tag{C.11.15}$$

Eqs. (C.11.5) and (C.11.15) are Laplace integrals and their treatment is similar.

The complex alternating functions are not discussed here. A complete treatment of the topic is presented by Gross (1953) and Bland (1960).

C.12 The General Rheological Models

Rheological modeling is the step by step assembly of rheological elements, resulting in models that fit linear mathematical equations and illustrate the time dependent effects of viscoelastic behavior, dominated by the principle of superposition. For some time this has captured the attention of scientists who aim to generalize the theory of viscoelastic modeling.

Within the framework of linearity, if inertia effects are neglected, any viscoelastic theory of modeling must lead to the simple exponential laws of excitation decay and response relaxation. Comparison with experimental results, however, indicates a behavior more complex than that represented by the simple models, suggesting the need for additional elements; mathematically this means an additional number of exponential functions, each labeled by its amplitude and time characteristics. When even this representation was not satisfactory, a further assumption was introduced, of a continuous set of exponential functions with amplitudes and time characteristics distributed continuously over a finite or infinite interval. In fact, two such functions were introduced: one referring to creep effects under given excitations, and one to relaxation effects for given responses.

Viscoelastic modeling has benefited from the earlier mathematical developments of dielectric phenomena and electrical networks. Only the elementary part of this development is presented here. For further study the reader is referred to the classical literature on the subject (Alfrey 1948; Aseltine 1958; Bland 1960; Brown 1961; Gross 1953; Naslin 1965).

It was shown that for the viscoelastic behavior described by two types of parameters (elasticity and viscosity), *two* fundamental systems of models can be considered. Each system assumes one of four simple forms, see Table C.12.1 (Roscoe 1950).

Let us assume that the distribution function is a line spectrum, then the creep and relaxation functions are sums of exponentials. The total response produced by the application of a constant excitation $\mathcal{F}$ is given by

$$E = \left(\frac{1}{k_1} + \frac{t}{q_2} + \alpha \sum_{i=3}^{\mathfrak{Z}} A_i\left(1 - \exp \frac{-t}{T_{\mathrm{ret}i}}\right)\right)\mathcal{F} \tag{C.12.1}$$

and the total excitation under constant response $\mathcal{E}$ is given by

$$F = \mathcal{E}\beta \sum_{i=1}^{\mathfrak{Z}} B_i \exp \frac{-t}{T_{\mathrm{rel}i}} \tag{C.12.2}$$

In Eq. (C.12.1), the first term in the brackets represents an elastic spring response; the second term a viscous dashpot response; and each of the following terms under the summation corresponds to a viscoelastic Kelvin body response. Since the total response is the sum of all the individual responses, all these single elements must be connected in series. A more general form of response can be deduced from Eq. (C.9.5)

$$E(t) = \frac{F(t)}{k_1} + \frac{1}{q_2}\int_0^t F(t')\,\mathrm{d}t' + \sum_{i=3}^{\mathfrak{Z}} \frac{1}{q_i}\left(\exp \frac{-t}{T_{\mathrm{ret}i}}\right)\int_0^t F(t')\exp \frac{t'}{T_{\mathrm{ret}i}}\,\mathrm{d}t' \tag{C.12.3}$$

Table C.12.1. Four forms of viscoelastic models

	Elastic	Viscous	Kelvin model		Maxwell model		Simple models
1	+	−	$1/k_1 \neq 0$	$1/q_2 = 0$	$\bar{k}_1 \neq 0$	$\bar{q}_2 = 0$	Poynting–Thomson
2	+	+	$1/k_1 \neq 0$	$1/q_2 \neq 0$	$\bar{k}_1 = 0$	$\bar{q}_2 = 0$	Burgers
3	−	−	$1/k_1 = 0$	$1/q_2 = 0$	$\bar{k}_1 = 0$	$\bar{q}_2 = 0$	Lethersich
4	−	+	$1/k_1 = 0$	$1/q_2 \neq 0$	$\bar{k}_1 = 0$	$\bar{q}_2 = 0$	Kelvin

where $T_{\text{reti}} = g_i/k_i$ and where in fact Eq. (C.12.3) represents $\mathfrak{I}$ Kelvin bodies coupled in series and *degenerated* so that $g_1 = 0$ and $k_2 = 0$, Fig. C.12.1, known as the *general Kelvin body* or *model*.

Similarly, Eq. (C.12.2) is a summation of viscoelastic Maxwell bodies. Since the total excitation is the sum of the excitations of the individual bodies, all these bodies are coupled in parallel. Here again, a more general form where the response is also a function of time is deduced from Eq. (C.7.3). It is known as the *general Maxwell body* or *model*, shown in Fig. C.12.2.

$$F(t) = \sum_{i=1}^{\mathfrak{I}'} k_i' \left(\exp \frac{-t}{T_{\text{reli}}} \right) \int_0^t \dot{E}(t') \exp \frac{t'}{T_{\text{reli}}} \, \mathrm{d}t' \qquad \text{(C.12.4)}$$

where $T_{\text{reli}} = g_i'/k_i'$.

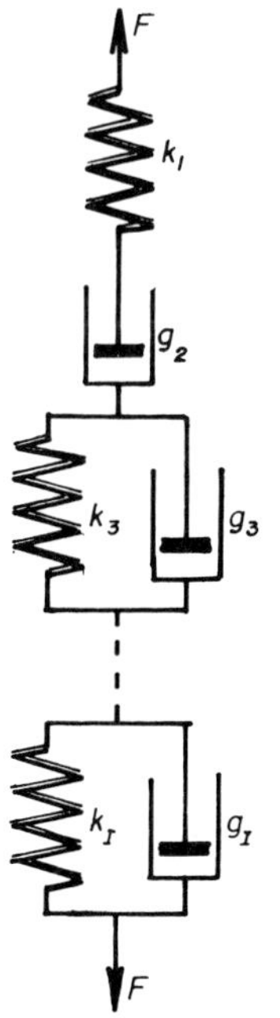

Fig. C.12.1. General Kelvin body.

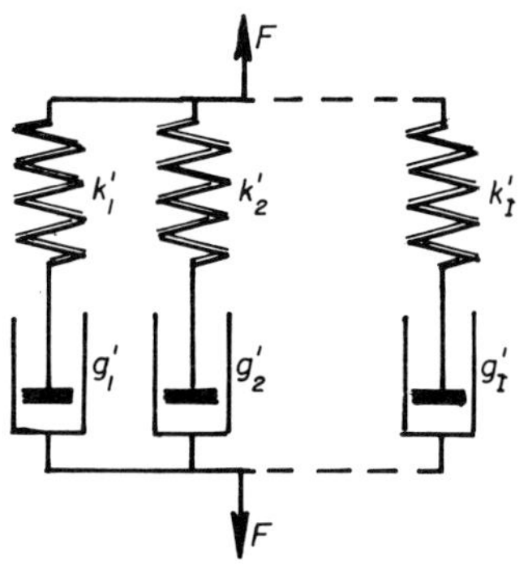

Fig. C.12.2. General Maxwell body.

There is an equivalence between the general Kelvin and general Maxwell models. Yet the coefficients k_i and q_i are different from those of k'_i and q'_i and certainly T_{reti} is different from T_{reli} and $\Im' = \Im - 1$. So, for instance, the four-element Kelvin model (the Burgers body) is equivalent to the four-element Maxwell model, Fig. C.12.3, and the coefficients relate

$$k_1 = k'_1 + k'_2; \quad k_3 = \frac{k'_1 k'_2 (q'_1 + q'_2)(k'^2_1 + k'^2_2)}{(q'_1 k'_2 + q'_2 k'_1)}$$

$$q_2 = q'_1 + q'_2; \quad q_3 = \frac{q'_1 q'_2 (k'_1 + k'_2)(q'^2_1 + q'^2_2)}{(q'_1 k'_2 + q'_2 k'_1)} \qquad \text{(C.12.5)}$$

We have thus two equivalent models, a model of $\Im + 2$ Kelvin bodies coupled in series with two bodies degenerated one into an elastic and one into a viscous element, and a model of $\Im' = \Im + 1$ Maxwell bodies coupled in parallel with no degenerated bodies. Each of the two models can appear in one of four particular forms, obtained by removing one, both or neither of the two degenerated elements, Fig. C.12.4. The four forms are summarized in Table C.12.1. The simple rheological models corresponding to the four forms for which $\Im = 3$ are also marked in Table C.12.1.

Many other, more complicated, linear models can be constructed; however, since they all contain either Kelvin or Maxwell bodies, they must be presented by sums of exponentials, and therefore they are reduced to Eqs. (C.12.3) or (C.12.4), respectively, with the distribution functions (C.11.11) and (C.11.6), respectively.

Although there is an equivalence between the Kelvin and the Maxwell models, the Kelvin model has certain advantages from an interpretative standpoint. Each of the terms in the Kelvin model has a direct physical significance, and is related to one particular *mechanism* of response to excitation. In the Maxwell model, on the other hand, the terms have only *indirect* physical significance. For this reason the Kelvin model seems preferable to the Maxwell model as a means of correlating macroscopic or phenomenological mechanical behavior. We will, consequently, apply more emphasis on the Kelvin model in our further treatment.

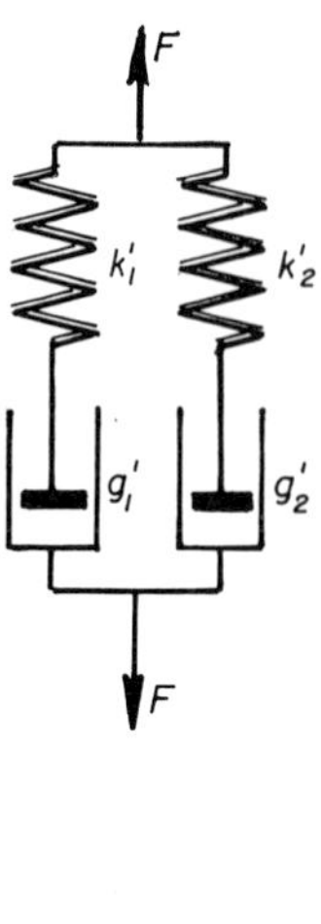

Fig. C.12.3. Four-element Maxwell model.

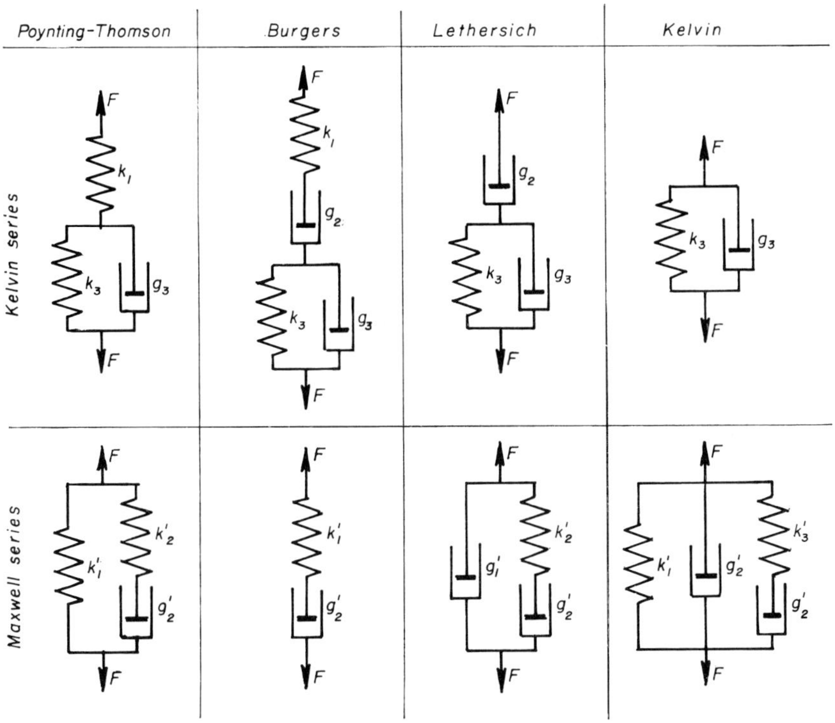

Fig. C.12.4. Equivalent canonic forms of models.

C.13 Elastic and Dissipative Excitations

Many viscoelastic phenomena are not only response dependent but also response-rate dependent. It is instructive, therefore, to investigate the strain-rate of the general Kelvin model which from Eq. (C.12.3) yields

$$\dot{E}(t) = \frac{\dot{F}(t)}{k_1} + \frac{F(t)}{g_2} - \sum_{i=3}^{\mathfrak{I}} \frac{1}{g_i T_{\text{ret}i}} \left(\exp \frac{-t}{T_{\text{ret}i}} \right) \int_0^t F(t') \exp \frac{t'}{T_{\text{ret}i}} \, \mathrm{d}t' + \frac{F(t)}{g} \tag{C.13.1}$$

where $\dot{E}(t)$ is the *response rate* of the $\mathfrak{I}$ Kelvin bodies representing the general Kelvin model.

From Eq. (C.8.2) it is seen that the excitation invested in each and every Kelvin body is made up of two parts, one related to its elastic spring and one related to its viscous dashpot. For the $\mathfrak{I}$ Kelvin bodies coupled in series and forming the general Kelvin model, we can define an *elastic* or *effective excitation* $F(t)_e$, as follows

$$F(t)_e = \frac{1}{\mathfrak{I}} \sum_{i=3}^{\mathfrak{I}} k_i E(t)_i = \frac{1}{\mathfrak{I}} \left\{ F(t) + \sum_{i=3}^{\mathfrak{I}} \frac{1}{T_{reti}} \left(\exp \frac{-t}{T_{reti}} \right) \int_0^t F(t') \exp \frac{t'}{T_{reti}} \, dt' \right\} \tag{C.13.2}$$

and a *dissipative* or *viscoelastic exitation* $F(t)_p$, as follows

$$\begin{aligned} F(t)_p &= \frac{1}{\mathfrak{I}} \sum_{i=3}^{\mathfrak{I}} g_i \dot{E}(t)_i \\ &= \frac{1}{\mathfrak{I}} \left\{ (\mathfrak{I} - 1) F(t) - \frac{1}{T_{reti}} \left(\exp \frac{-t}{T_{reti}} \right) \int_0 F(t') \exp \frac{t'}{T_{reti}} \, dt' \right\} \end{aligned} \tag{C.13.3}$$

It can be seen that the sum of the elastic and dissipative excitations is equal to the excitation $F(t)$

$$F(t)_e + F(t)_p = \frac{1}{\mathfrak{I}} \{ \mathfrak{I} F(t) \} = F(t) \tag{C.13.4}$$

which confirms Eq. (C.8.2) for a general Kelvin model as well. The corresponding equations for the four-element Kelvin model are

$$F(t)_e = \frac{1}{3} \left\{ F(t) + \frac{1}{T_{ret3}} \left(\exp \frac{-t}{T_{ret3}} \right) \int_0^t F(t') \exp \frac{t'}{T_{ret3}} \, dt' \right\} \tag{C.13.5}$$

$$F(t)_p = \frac{1}{3} \left\{ 2F(t) - \frac{1}{T_{ret3}} \left(\exp \frac{-t}{T_{ret3}} \right) \int_0^t F(t') \exp \frac{t'}{T_{ret3}} \, dt' \right\} \tag{C.13.6}$$

Eqs. (C.13.2) and (C.13.3), also (C.13.5) and (C.13.6), have far reaching importance in the mechanics of soils, by providing the basis for the effective pressure and pore pressure, respectively.

C.14 The Plastic Restraint

In the previous sections it was made clear that viscosity is a time-dependent function, and thus any energy dissipation due to viscosity is a function of time. There are, however, other dissipative energies, not directly dependent on time but on other variables, such as response or excitation. In a way, these dissipative energies indicate a *departure from equilibrium*, when, owing to an excitation only part of the expected response is attained, or when a response is obtained from a greater excitation than expected. In both cases part of the excitation dissipates, and thus one can say that a response due to an excitation is not fully recovered when the excitation is removed. A *hysteresis* loop in the excitation–response curve is evident.

In Sect. C.5 the St Venant element was introduced, and in Sect. C.6 it was shown that the St Venant element comes coupled in series with an elastic element, to form the Prandtl body. The St Venant element may also be coupled in parallel with the bodies discussed, including the general Maxwell and Kelvin rheological bodies.

While elasticity and viscosity are clearly defined within the well established linear theory of viscoelasticity, as discussed in Sects. C.9–C.12, the plastic restraint is still subject to speculations and revisions, and open to changes.

When a general Kelvin body is coupled in parallel with a St Venant body, Fig. C.14.1, the rules of parallel coupling outlined in Sect. C.4 are valid, and one can write

$$F = F_1 + F_2 + F_S = F_K + F_S; \quad F_1 + F_2 = F_K \tag{C.14.1}$$

$$E = E_1 = E_2 = E_K = E_S; \tag{C.14.2}$$

where F_1 is the excitation supported by the elastic spring element of the Kelvin body, F_2 is the excitation acting in the viscous dashpot of the body and F_S is the excitation acting in the St Venant element. From Eq. (C.14.1) we have

$$F = F_K + \mathfrak{F}_0\vartheta \tag{C.14.3}$$

where ϑ is the *coefficient of plastic restraint*, and where it is assumed that part of the excitation equal to $\mathfrak{F}_0\vartheta$ has been lost in the form of energy and therefore cannot produce work. Consequently, the active work produced by the *activating excitation* F_K is equal to $F_K = F - \mathfrak{F}_0\vartheta$, which is that part of the work that is free to activate the general Kelvin body.

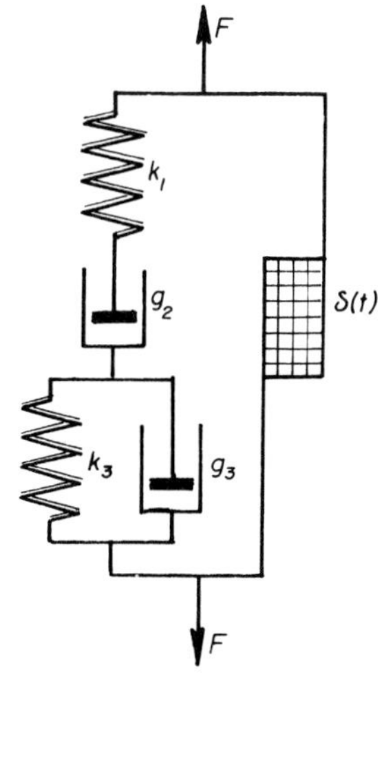

Fig. C.14.1. Degenerated general Kelvin body coupled in parallel with a St Venant element.

The plastic restraint is a function of the state of the material, and since this state changes with the response of the material, it turns out to be a function of the response, $\mathfrak{F}_0\vartheta(F) \equiv \mathfrak{F}_0\vartheta(t)$, and in turn a function of time.

The response of a general Kelvin body coupled in parallel with a St Venant element follows from Eq. (C.10.10)

$$\begin{aligned} E(t) &= \int_{-\infty}^{t} \frac{\mathrm{d}}{\mathrm{d}t'}\left[F(t')\left(\frac{1}{k_1} + \frac{t-t'}{g_2} + \psi(t-t')\right)\right]\mathrm{d}t' \\ &= \frac{1}{k_1}F(t) + \frac{1}{g_2}\int_{-\infty}^{t} F(t')\,\mathrm{d}t' + \int_{-\infty}^{t} \frac{\mathrm{d}F(t')}{\mathrm{d}t'}\psi(t-t')\,\mathrm{d}t' \\ &= \frac{1}{k_1}(F(t) - \mathfrak{F}_0\vartheta(t)) + \frac{1}{g_2}\int_{-\infty}^{t}(F(t') - \mathfrak{F}_0\vartheta(t))\,\mathrm{d}t' \\ &\quad + \int_{-\infty}^{t}\frac{\mathrm{d}}{\mathrm{d}t'}\left[[F(t') - \mathfrak{F}_0\vartheta(t)]\int_{-\infty}^{t}\alpha A(t')[1 - \exp(t'-t)]\,\mathrm{d}t'\right]\mathrm{d}t' \end{aligned} \tag{C.14.4}$$

and, in the discontinuous form, we obtain from Eqs. (C.10.10) and (C.11.12)

$$\begin{aligned} E(t) &= \frac{1}{k_1}[F(t) - \mathfrak{F}_0\vartheta(t)] + \frac{1}{g_2}\int_{-\infty}^{t}[F(t') - \mathfrak{F}_0\vartheta(t)]\,\mathrm{d}t' \\ &\quad + \int_{-\infty}^{t}\frac{\mathrm{d}}{\mathrm{d}t'}\left[[F(t') - \mathfrak{F}_0\vartheta(t)]\sum_{i=3}^{\mathfrak{Z}}\alpha A_i\left(1 - \exp\frac{-(t-t')}{T_{\mathrm{ret}i}}\right)\right]\mathrm{d}t' \end{aligned} \tag{C.14.5}$$

The corresponding response rates obtained by differentiating Eqs. (C.14.4) and (C.14.5) are

$$\begin{aligned} \dot{E}(t) &= \frac{\mathrm{d}}{\mathrm{d}t}\left\{\int_{-\infty}^{t}\frac{\mathrm{d}}{\mathrm{d}t'}\left[[F(t') - \mathfrak{F}\vartheta(t')]\left(\frac{1}{k_1} + \frac{t-t'}{g_2} + \psi(t-t')\right)\right]\mathrm{d}t'\right\} \\ &= \left(\frac{1}{g_2} + \frac{1}{g_3}\int_{0}^{t}\alpha A(t')\exp\frac{-(t-t')}{T_{\mathrm{ret}i}}\,\mathrm{d}t'\right)F(t) \\ &\quad + \left[\frac{1}{k_1} + \int_{0}^{\infty}\alpha A(t')\left(1 - \exp\frac{-(t-t')}{T_{\mathrm{ret}}}\right)\mathrm{d}t'\right][\dot{F}(t) - \mathfrak{F}_0\vartheta(t)] \end{aligned} \tag{C.14.6}$$

$$\begin{aligned} \dot{E}(t) &= \frac{\mathrm{d}}{\mathrm{d}t}\left\{\int_{-\infty}^{t}\frac{\mathrm{d}}{\mathrm{d}t'}\left[[F(t') - \mathfrak{F}\vartheta(t')]\left(\frac{1}{k_1} + \frac{t-t'}{g_2}\right.\right.\right. \\ &\quad \left.\left.\left. + \sum_{i=3}^{\mathfrak{Z}}\alpha A_i\left(1 - \exp\frac{t-t'}{t'}\right)\right)\mathrm{d}t'\right]\mathrm{d}t'\right\} \end{aligned}$$

$$= \left(\frac{1}{q_2} + \sum_{i=3}^{\mathfrak{I}} \frac{\alpha A_i}{T_{\text{reti}}} \exp \frac{-t}{T_{\text{reti}}}\right)[F(t) - \mathcal{F}_0 \vartheta(t)]$$

$$+ \left(\frac{1}{k_1} + \sum_{i=3}^{\mathfrak{I}} \alpha A_i\left(1 - \exp \frac{-t}{T_{\text{reti}}}\right)\right)[\dot{F}(t) - \mathfrak{F}_0 \dot{\vartheta}(t)] \tag{C.14.7}$$

where the activating excitation rate is equal to $\dot{F}(t) - \mathfrak{F}_0 \dot{\vartheta}(t)$.

The plastic restraint is a function of time and therefore it induces non-steady elements in the equations. Thus, what seems to be a steady-state problem, a creep experiment with constant excitation $F = \text{const} = \mathcal{F}$, turns out to be a non-steady problem with $\mathcal{F}(t)$ varying with time.

The medium thus defined was termed the *visco-elasto-plastic continuum* (Klausner 1961), although it was formulated differently at the time. Since it is derived from the visco-elastic continuum, it is a linear medium.

Accordingly, the many known functions that satisfy the mathematical requirements of $\vartheta(t)$ specified in Sect. C.5 must be equal to zero at the outset of the excitation, increase monotonically, and reach asymptotically the final value of unity as time approaches infinity

$$\vartheta(0) = 0; \quad \vartheta(\infty) = 1; \quad \dot{\vartheta}(\infty) = 0 \tag{C.14.8}$$

It would be desirable if $\vartheta(t)$, in addition to satisfying the mathematical requirements and providing a heuristic solution to the technical problem of excitation–response, could also be derived directly from the intricate physical interactions within the medium considered, either at the structural or the microscopic level. Presently, the links connecting these levels to the phenomenological level, in which $\vartheta(t)$ is defined, are few and sporadic.

An exponential function, which serves, in general, to determine energy dissipation, is perhaps the simplest expression to satisfy the mathematical conditions

$$\vartheta(t) = 1 - \exp(-Ct) = 1 - \exp(-t/T); \tag{C.14.9}$$

Another simple function satisfying the conditions is

$$\vartheta(t) = \tanh Bt = \tanh(1/T) \tag{C.14.10}$$

An elaborate form of the same is a Fourier series

$$\vartheta(t) = 1 - C \sum_{n=1}^{\infty} \frac{(-1)^n}{n} \exp(-n^2 B^2 t) \sin \frac{2n}{A}; \quad A, B, C = \text{const} \tag{C.14.11}$$

A more promising function, related to the standard distribution curve of particulate matter, is the error function

$$\vartheta(t) = \frac{2A}{\sqrt{\pi}} \int_0^t \exp -(A^2 t'^2)\, dt'; \quad A = \text{const} \tag{C.14.12}$$

where A, B and C are reciprocals of a *time factor* T, $[1/A] = [T]$, $[1/B] = [T]$ and $[1/C] = [T]$.

Further study is required on the topic of plastic restraint.

Rheological modeling can provide material for a voluminous manuscript on its own. Many significant topics, such as equivalence of models, comparison with other analogies, transient and oscillatory excitations, algebraic, Fourier, Laplace, Stieltjes inversions and transforms, energy and work considerations, etc. were not discussed here. Some of these topics are found in the literature mentioned in the text.

References

Adkins JE (1963a) Nonlinear diffusion. I. Diffusion and flow of mixtures of fluids. Phil Trans R Soc Lond A 255: 607–633

Adkins JE (1963b) Nonlinear diffusion. II. Constitutive equations for mixtures of isotropic fluids. Phil Trans R Soc Lond A 255: 634–648

Aitchison GD (1956) Some preliminary studies of saturated soils. The circumstance of unsaturation in soils with particular reference to Australian environment. In: Proceedings of the 2nd Australian New Zealand conference. SOMEFE, Christchurch, pp 173–191

Aitchison GD (1961a) Relationship of moisture stress and effective stress functions in unsaturated soils. In: Proceedings of the Brit Nat Soc SOMEFE conference on pore pressure and suction in soils. Butterworths, pp 47–52

Aitchison GD (1961b) Discussion to session on shallow foundations. In: Proceedings of the 5th INCOSOMEFE, Paris, vol 3, p 200

Akroyd TN (1961) Concrete under triaxial stress. Mag Conc Res 13: 111–118

Alfrey T Jr (1948) Mechanical behavior of high polymers. Interscience, 581 pp

Alfrey T Jr , Gurnee EF (1956) Dynamics of viscoelastic behavior. In: Eirich FR (ed.) Rheology, vol 1. Academic Press, pp 387–429

Alpan I (1959) A study of the principle of effective stress in partly saturated soils. PhD thesis, Imperial College, London

Alpan I (1961) The dissipation function for unsaturated soils. In: Proceedings of the 5th INCOSOMEFE, Paris, vol 1, pp 3–5

Alpan I (1962) An experimental method for determining the air permeability of partly saturated soils. Soil Sci 94/4: 263–269

Alpan I (1963) Effective stresses in partly saturated soils. Topics in applied mathematics, Memorial volume to the late Professor E Schwerin. Technion IIT

Arora HS, Coleman NT (1979) The influence of electrolyte concentration on flocculation of clay suspensions. Soil Sci 127: 134–139

Arthur JRF, Dunstan T, Al-Ami QAJL, Assadi A, Vaid YP, Finn WDL (1979) Static shear and liquefaction potential. ASCE J Geotech Div 105 GT10: 1249–1253

Aseltine (1958) Transform methods in linear system analysis. McGraw-Hill

Atkinson JH, Evans JS, Ho EWL (1985) Non uniformity of triaxial samples due to consolidation with radial drainage. Geotechnique 35/3: 353–355

Atterberg A (1911) Uber die physikalische Bodenuntersuchung, und uber die Plastizitat der Tone. Internationale Mitteilungen fur Bodenkunde. Verlag fur Fachliteratur, 1: pp 10–43

Babcock KL (1963) Theory of the chemical properties of soil colloid systems at equilibrium. Hilgardia 34: 417–542

Badiey M, Yamamoto T, Turgut A (1988) Laboratory and in situ measurements of selected geoacoustic properties of carbonate sediments. J Acoust Soc Am 84/2: 689–696

Baer J (1972) Dynamics of fluids in porous media. Elsevier, 756 pp

Balmer GG (1949) Shearing strength of concrete under high triaxial stress. US Bureau of Recl Str Res Lab Rep SP-23: 1–26

Barden L (1965) Consolidation of compacted and unsaturated clays. Geotechnique 15: 267–286

Barden L (1968) Primary and secondary consolidation of clay and peat. Geotechnique 18: 1–24

Bardet JP (1983) Application of plasticity theory to soil behavior: a new sand model. PhD thesis, Cal Inst Tech Pasadena

Bar-On P, Shainberg I (1970) Hydrolysis and decomposition of Na-montmorillonite in distilled water. Soil Sci 109: 241–246
Bar-On P, Shainberg I, Michaeli I (1970) Electrophoretic mobility of montmorillonite particles saturated with Na/Ca ions. J Colloid Interface Sci 33(3): 471–472
Barshad I (1964) Thermal analysis techniques for mineral identification amd mineralogical composition. Agronomy 9: 699–742
Basak P, Madhav MR (1976) Hydraulic conductivity in surface active soils. Austral J Soil Res 14: 121–127
Bassett RH (1967) Granular materials in simple shear apparatus. PhD thesis, University of Cambridge
Baver LD (1942) Retention and movement of soil moisture. In: Meinzer OE (ed) Hydrology, Dover, pp 364–384
Bazant ZP, Krizek RJ (1976) Endochronic constitutive law for liquefaction of sand. ASCE J Engng Mech Div 102 EM2: 225–238
Bedford A, Ingram JD (1971) A continuum theory of fluid saturated porous media. J Appl Mech 93: 1–7
Bedford A, Drumheller DS (1978) A variational theory of immiscible mixtures. Arch Rat Mech Anal 68 (1): 35–54
Bedford A, Drumheller DS (1983) Theories of immiscible and structural mixtures. Int J Engng Sci 21: 863–960
Been K, Sills GS (1981) Self weight consolidation of soft soils: an experimental and theoretical study. Geotechnique 31/4: 519–535
Beltrami E (1889) Sur la theorie de la deformation infiniment petite d'un milieu. Leçons de Cinematique C R 108: 502–504
Benedict M, Webb GB, Rubin LC (1940) Empirical equation for thermodynamic properties of light hydrocarbons and their mixtures. J Chem Phys 10: 747–758
Benyaako S, Ruth E, Kaplan IR (1974) Calcium carbonate saturation in northern Pacific. In situ determination and geomechanical implication. Deep-Sea Res 21 (3): 229
Bertrand JLF (1887) Thermodynamique. Gouthier–Villars, Paris
Besseling JF (1966) A thermodynamic approach to rheology. In: Parkin H, Sedov LI (eds) Proceedings IUTAM symposium on irreversible aspects of continuum mechanics, Vienna. Springer, pp 16–53
Biarez J, Bellier J, Bordes JL, Boucet B, Le Long L, Orliac M, Remy C (1969) Proprietes Mécaniques des sols divers sollicitations. In: Proceedings of the 7th INCOSOMEFE, Mexico, vol 1, pp 21–28
Biot MA (1939) Nonlinear theory of elasticity and the linearized case for a body under initial stress. Phil Mag 7 (27): 468–489
Biot MA (1941) General theory of three-dimensional consolidation. J Appl Phys 12: 155–164
Biot MA (1955) Theory of elasticity and consolidation for a porous anisotropic solid. J. Appl Phys 26 (2): 182–185
Biot MA (1956) General solutions of the equations of elasticity and consolidation for a porous material. Trans ASME E: J Appl Mech 78: 91–96
Biot MA (1963) Theory of stability and consolidation of a porous medium under initial stress. J Math Mech 12 (4): 194–198
Biot MA, Willis DG (1957) The elastic coefficient of the theory of consolidation. Trans ASME E: J Appl Mech 24: 1–8
Bishop AW (1959) The principle of effective stress. Lecture delivered in Oslo 1955. Printed Tek Ukeblad, vol 39, pp 859–863
Bishop AW (1960) The measurement of pore pressure in triaxial test. In: Proceedings of the Conference of the British Nat Soc of ISSMFE on pore pressures and suction in soils. Butterworths, pp 38–46
Bishop AW (1966) The strength of soils as engineering material. Sixth Rankine lecture. Geotechnique 16 (2): 91–130
Bishop AW, Eldin AKG (1950) Undrained triaxial tests and saturated sands and their significance in the general theory of shear strength. Geotechnique 2/1: 13–32
Bishop AW, Alpan I, Blight GE, Donald LB (1960) Factors controlling the strength of partly saturated cohesive soils. In: Proceedings of ASCE research conference on shear strength of cohesive soils, Boulder, pp 503–532
Bishop AW, Donald LB (1961) The experimental study of partly saturated soils in the triaxial apparatus. Proceedings of the 5th INCOSOMEFE Paris, vol 1, pp 13–21
Bishop AW, Henkel DJ (1962) The measurement of soil properties in the triaxial test. Edward Arnold, London, 228 pp
Bishop AW, Blight GE (1963) Some aspects of effective stress in saturated and partly saturated

soils. Geotechnique 2/1: 177–197
Bishop AW, Green GE (1965) The influence of end restraint on the compression strength of a cohesionless soil. Geotechnique 15 (3): 243–266
Bjerum L (1969) President's opening address. Proceedings 7th INCOSOMEFE, Mexico, vol 3, pp 100–104
Blackmore AV (1969) Water structure and viscosity in kaolinite and bentonite. Austral J Sci 31: 402–404
Blackmore AV, Miller RD (1961) Tactoid size and osmotic swelling in calcium montmorillonite. Proc Soil Sci Soc Am 25 (3): 169–173
Bland DR (1960) The theory of linear visco-elasticity. Pergamon, 125 pp
Blight GE (1966a) A study of effective stresses for volume change. Proceedings of the symposium on moisture equilibria and moisture change in soils. Butterworths, Sydney, pp 259–269
Blight GE (1966b) Strength characteristics of desiccated clays. ASCE J Soil Mech Div 92 (SM 6): 18–38
Blight GE (1967) Effective stress evaluation for unsaturated soils. ASCE J Soil Mech Div 93 (SM 2): 125–148
Blight GE (1983) Aspects of the capillary model for unsaturated soils. Proceedings of the 7th regional conference SOMEFE, Haifa, vol 1, pp 3–7
Boehler JP, Kabbaj M (1985) Etude de gonflement anisotrope de la bentonite. Proceedings of the 11th INCOSOMEFE, vol 2, pp 407–410
Böker R (1915) Die Mechanik der bleiben den Formanderungen in kristallinische aufgebauten Korpern. Mitt uber dem Vereines deutscher Ingenieure 175: 1–51
Bolt GH (1955) Analysis of the validity of the Gouy–Chapman theory of electric double layer. J Colloid Sci 10: 206–218
Bolt GH (1956) Physico-chemical analysis of the compressibility of pure clays. Geotechnique 6: 86–93
Bolt GH, Warkentin BP (1958) The negative adsorbtion of anions by clay suspensions. Koloid-Zeitschrift 156: 41–46
Bolton MD (1986) Strength and dilatancy of sand. Geotechnique 36/1: 65–78
Boltzmann L (1894) Zur Integrazion der Diffusionsgleichung bei Variablen Diffusionskoeffizienten. Ann Physik Chimie 53: 959–964
Borisenko AI, Tarapov IE (1968) Vector and tensor analysis with applications. Prentice-Hall
Borja RJ, Kavazanyan Jr E (1985) A constitutive model for the stress–strain time behavior of wet clay. Geotechnique 35/3: 283–298
Born M, Green HS (1946–47) A general kinetic theory of liquids, I–II. Proc R Soc Lond A 188: 10–18
Boussinesq J (1872) Sur une manière simple de determiner experimentalement la resistance au glissement maximum dans un solide ductile, homogene et isotrope. C R 75: 254–257
Bowden FP, Tabor D (1967) Friction and lubrication. Methuen
Bowen RM (1967) Toward a thermodynamics and mechanics of mixtures. Arch Rat Mech Anal 24: 370–403
Bowen RM (1973) Theory of mixtures. In: Eringen AC (ed) Continuum physics, vol 3. Academic Press
Bredthauer RO (1957) Strength characteristics of rock samples under hydrostatic pressure. Trans ASME 79: 695–708
Bridgman PW (1931) Physics of high pressure. Macmillan, New York
Bridgman PW (1939) Considerations on rupture under triaxial stress. Mech Engng 61: 107–111
Bridgman PW (1949) Volume changes in the plastic stages of simple compression. J Appl Phys 20: 1241–1251
Bridgman PW (1952) Studies in large plastic flow and fracture with special emphasis on the effects of hydrostatic pressure. McGraw-Hill, New York
British Standard Institution (1948) Methods for test of soil classification and compaction. BS 1377
Brønstedt JN (1955) Principles and problems in energetics. Interscience
Brown BM (1961) The mathematical theory of linear systems, Chapman and Hall
Bruce RR, Klute A (1956) The measurement of soil moisture diffusivity. Soil Sci Proc Am 20: 458–462
Bruce RR, Klute A (1962) Measurements of soil moisture diffusivity from tension plate outflow data. Soil Sci Proc Am 27: 18–21
Bruch Jr JL, Zyvoloski G (1974) Solution of equation for vertical unsaturated flow of soil water. Soil Sci 116/6: 417–422
Buckingham E (1907) Studies on the movement of soil moisture. USDA Bur of Soils Bull no 38, pp 29–61
Buisman ASK (1936) Results of long duration settlement tests. In: Proceedings of the INCO-

SOMEFE, Cambridge, vol 1, pp 103–106
Burland NB (1964) Effective stresses in partly saturated soils. Geotechnique (correspondence) 14/1: 64–68
Camp TR, Meserve RL (1974) Water and its impurities. Dowden, Hutchison and Ross, Stroudsburg, PA, 384 pp
Capper PL, Cassie WF (1953) The mechanics of engineering soils. McGraw-Hill, 315 pp
Caratheodory C (1909) Untersuchungen uber die Grundlagen der Thermodynamik. Mathematische Annalen 67: 355–386
Carillo N (1948) Influence of artesian wells in the sinking of Mexico City. In: Proceedings of the 2nd INCOSOMEFE, Rotterdam, vol 7, pp 28–30
Carman PC (1956) Flow of gases through porous media. Butterworths, London
Casagrande A (1936) The determination of the pre-consolidation load and its practical significance. Proceedings of the 1st INCOSOMEFE vol 3. Harvard University Press, pp 60–64
Casagrande A, Wilson SD (1950) Effect of rate of loading on the strength of clays and shales at constant water content. Geotechnique, 2: 251–263
Casagrande A, Hirschfeld RC (1960) Stress deformation and strength characteristics of a clay compacted to a constant dry unit weight. Proceedings of the ASCE conference on shear strength of cohesive soils. Boulder, pp. 359–417
Chapman DL (1913) A contribution to the theory of electrocapillarity. Phil Mag 25 (6): 475–481
Chapman S, Cowling TG (1939) The mathematical theory of non uniform theory of gases. Cambridge University Press
Chen WF, Baladi GY (1985) Soil plasticity: theory and implementation. Elsevier
Childs EC (1969) An introduction to the physical basis of soil water phenomena. J Wiley and Sons
Childs EC (1972) Concepts of soil–water phenomena. Soil Sci 113 (4): 246–253
Childs EC, Collis-George N (1950) Movement of moisture in unsaturated soils. In: Transactions 4th international congress soil science: Amsterdam, vol 1
Chinn J, Zimmermann R (1965) Behavior of plain concrete under various high compression loading conditions. US Air Force Weapons Lab, Tech Rep WL-TR 64–163
Christie IF (1965) Secondary compression effects during one-dimensional consolidation tests. In: Proceedings of the 6th INCOSOMEFE, vol 1, pp 198–202
Christoffel EB (1877) Untersuchungen uber die mit dem Fortbestehen linearer partieller Differentialgleichungen vertraglichen Unstetigkeiten. Ann Math 8: 81–113
Clausius R (1854) Uber der verenderte Form der Zweiten Hauptsatzes der mechanischen Warmetheorie. Ann Phys 93: 481–506
Clausius R (1865) Uber verschiedene fur die Anwendung bequeme Formen und Hauptgleichungen der mechanischen Warmetheorie. Ann Phys 125: 353–400
Cole ERL (1967) The behaviour of soils in simple shear apparatus. PhD thesis, University of Cambridge
Coleman JD (1961) Discussion to division 1 soil properties and their measurement. In: Proceedings of the 5th INCOSOMEFE, vol 3, pp 111–112
Coleman JM, Noll W (1961a) Foundations of linear viscoelasticity. Rev Mod Phys 33 (2): 239–249
Coleman JM, Noll W (1961b) Normal stress in second order viscoelasticity. Trans Soc Rheol 5: 205–220
Coleman JM, Prior DB (1978) Submarine landslides in the Mississippi River delta. In: 10th offshore technical conference, Houston, pp 1067–1074
Coulomb CA (1776) Essai sur une application des règles des maximis et minimis a quelques problèmes de statique relatifs à l'architecture. Mem Acad R Pres Divers Savants 5–7: 343
Coulomb CA (1821) Theorie des Machines Simples. Paris
Croney D, Coleman JD (1958) Soil moisture suction properties and their bearing on the moisture distribution in soils. In: Proceedings of the 3rd INCOSOMEFE, vol 1, pp 13–18
Croney D, Coleman JD (1961) Pore pressure and suction in soils. In: Proceedings of the Brit Nat Soc SOMEFE conference on pore pressure and suction in soils. Butterworths, pp 31–37
Croney D, Coleman JD, Black WPM (1958) Movement and distribution of water in soil in relation to highway design and performance. In: Winterkorn HF (ed) HRB special rep no 40. Symposium on water and its conduction in soils, pp 226–252
Crossland B, Dearden WH (1958) The plastic flow and fracture of a "brittle" material with particular references to the effect of fluid pressure. Disc Proc Inst Mech Engrs, 172: 805–820
Cuevas JA (1936a) Foundation conditions in Mexico City. In: Proceedings of the 1st INCOSOMEFE, Cambridge, vol 3, pp 233–237
Cuevas JA (1936b) The floating foundation of the new building for the national lottery of

Mexico. An actual size study of the deformation of a flocculent structured deep soil. In: Proceedings of the 1st INCOSOMEFE, Cambridge, vol 1, pp 294–301
Dafalias YE, Hermann LR (1981) Bounding surface formulation of soil plasticity. In: Pande G, Zienkievitz OC (eds) International symposium on soils under cyclic and transient loading. John Wiley and Sons, London, pp 253–282
Dallavalle JM (1943) Macromeritics: the technology of fine particles. Pitman, New York, Chicago, 428 pp
Dagan G (1964) The movement of the interface between two liquids in a porous medium with application to a coastal aquifer. PhD thesis. Technion, Haifa
Darcy H (1856) Les fontaines publiques de la ville de Dijon. Dalmont, Paris
Davidson DT, Sheeler JB (1953) Cation exchange capacity of loess and its relation to engineering properties. Symposium on exchange phenomena in soils. ASTM spec publ no 142, pp 354–361
Davis RO, Mullenger G (1978) A rate type constitutive model for soil with a critical state. Intl J Numer Anal Meth Geomech 20: 275–282
De Groot SR, Mazur P (1962) Non-equilibrium thermodynamics. North-Holland, 242 pp
De Haan FAM, Bolt GH (1963) Determination of anion adsorbtion of clays. Proc Soil Sci Soc Am 27: 636–640
De Jong E, Douglas JT, Gross MJ (1983) Gaseous diffusion in shrinking soils. Soil Sci 136 (1): 10–18
De Josselin De Jong G (1959) Statics and kinematics in the failable zone of a granular material. Delft
De Josselin De Jong G, Verruijt A (1965) Primary and secondary consolidation of a spherical clay sample. In: Proceedings of the 6th INCOSOMEFE, Toronto, vol 1, pp 254–258
Derjagin BV (1940) Repulsive forces between charged colloid particles and the theory of slow coagulation and stability of lyophobic sols. Trans Faraday Soc 36: 203
Derjagin BV, Krylov NA (1944) Works of the conference on the viscosity of liquids, vol 2. USSR Acad Sci Press
Derjagin BV, Melinkova NK (1958) Mechanism of moisture equilibrium and migration in soils. In: Proceedings of the international symposium on water and its conduction in soil. HRB Spec Rep, vol 40, pp 43–54
De Wet JA (1961) The use of the energy concept in soil mechanics. In: Proceedings of the 5th INCOSOMEFE, Paris, vol 1, pp 403–406
De Wet JA (1965) A formulation of three-dimensional moisture movement in a partially saturated soil. Butterworths, pp 33–38
De Wiest RJM (1969) Flow through porous media. Academic Press, 523 pp
Dobran F (1984) Constitutive equations for multiphase mixtures of fluids. Intl J Multiphase Flow 10 (3): 273–305
Dobran F (1985) Theory of multiphase mixtures. Intl J Multiphase Flow 11 (1): 1–30
Donald IB (1963a) Shearing strength of unsaturated soils. Dept CE Melbourne University Summer School in FE
Donald IB (1963b) Effective stress parameters in unsaturated soils. In: Proceedings 4th Australian–New Zealand COSOMEFE, Adelaide, pp 41–46
Dow RB (1956) Some rheological properties under high pressure. In: Reiner M (ed) Rheology, vol. 1, pp 243–319
Drescher A, Vardoulakis I (1982) Geometric softening in triaxial tests on granular material. Geotechnique 32/4: 291–304
Drew DA (1976) Two-phase flows: Constitutive equations for lift and Brownian motion and some basic flows. Arch Rat Mech Anal 62: 149–163
Drucker DC (1953) Limit analysis of two and three dimensional soil mechanics problems. J Mech Phys Solids 1: 217–226
Drucker DC (1959) A definition of stable inelastic material. Trans ASME E: J Appl Mech 26: 106–112
Drucker DC, Prager W (1952) Soil mechanics and plastic analysis or limit design. Q Appl Maths 10: 157–165
Drucker DC, Gibson RE, Henkel DJ (1955) Soil mechanics and work hardening theories of plasticity. Proc ASCE 81: paper 798
Drumheller DS (1978) The theoretical treatment of a porous solid, using a mixture theory. Intl J. Solids Struct 14: 441–456
Drumheller DS, Bedford A (1980) A thermomechanical theory of reacting immiscible mixtures. Arch Rat Mech Anal 73: 257–284
Duhem P (1903) Sur la viscosité en un millieu vitreux. C R Acad Sci Paris 136: 343–345, 592–595, 733–735, 858–860, 1032–1034

Duncan JM, Dunlop P (1969) Behavior of soils in simple shear tests. In: Proceedings of the 7th INCOSOMEFE, Mexico, vol. 1, pp 101–109
Eckhart C (1940) The thermodynamics of irreversible processes, I–II. Phys Rev 58: 269–275
Edlefsen NE, Anderson ABC (1943) Thermodynamics of soil moisture, vol 15. Hilgardia, pp 31–299
Edwards DG, Quirk JP (1962) Repulsion of chloride by montmorillonite. J Colloid Sci 17: 872–882
Edwards DG, Posner AM, Quirk JP (1965) Repulsion of chloride ions by negatively charged clay surfaces, I–III. Trans Faraday Soc. 61: 2808–2815
Eirich FR (ed) (1956) Rheology, vol 1–4. Academic Press
Emde F (1915) Zur Vectorrechnung. Archiv Math 24: 1–11
Enderby JA (1955) The domain model of hysteresis. Part I independent domains. Trans Faraday Soc 51: 835–848
Ericksen JL (1960a) Tensor fields. In: Flugge S (ed) Handbuch der Physik, vol. 3/1. Springer, pp 749–858
Ericksen JL (1960b) Anisotropic Fluids. Arch Rat Mech Anal 4: 231–237
Ericksen JL (1961) Poiseuille flow of certain anisotropic fluids. Arch Rat Mech Anal 8(1): 1–34
Ericksen JL, Rivlin RS (1954) Large elastic deformations of homogeneous anisotropic materials. J Rat Mech Anal 3(3): 281–301
Eringen AC (1962) Non-linear theory of continuous media. McGraw-Hill, 477 pp
Esrig MI, Kirby RC (1977) Implications of gas content for predicting the stability of submarine slopes. Mar Geotechnol 2: 81–100
Farouki OT (1963) Properties of granular systems. Res Rep Dept CE, Princeton University
Fatt I, Davis DH (1952) Reduction in permeability with overburden pressure. Tech Note J. Petroleum Technol Trans AIME, 195
Fellenius W (1918) Kaj-och Jordraset i Goteborg. (Swedish) Teknisk Tidskrift no 2
Fellenius W (1927) Erdstatische Berechnung mit Reibung und Kohasion. W Ernst u Sohn
Ferry JD (1961) Viscoelastic properties of polymers. John Wiley, New York
Feynman RP, Leighton RB, Sands M (1966) The Feynman lectures on physics, vol 1. Addison-Wesley
Fick A (1855) Uber Diffusion. Ann Physik 94: 59–86
Finger J (1892) Uber die gegenseitigen Beziehungen gewisser in der Mechanik mit Vorteil anwendbaren Flachen zweiter Ordnung nebst Anwendung auf Probleme der Astatik. Akad Wiss Wien Sitzungsber 101: 1105–1142
Finger J (1894) Uber die allgemeinste Beziehungen zwischen Deformazionen und den zugehoringen Spannungen in aeolotropen und isotropen Substanzen. Akad Wiss Wien Sitzungsber IIa 103: 1073–1100
Finn WDL, Wade NH, Lee KL (1967) Volume changes in triaxial and plane shear tests. ASCE J Soil Mech Div 93 (SM6): 297–308
Fitzgerald JE (1980) A tensorial Hencky measure of strain and strain rate for finite deformations. J Appl Phys 51 (10): 5111–5115
Florin VA (1959) Fundamentals of soil mechanics. Leningrad
Fourier J (1822) The analytical theory of heat. (Trans), Dover (1955)
Franzini JB (1949) The effect of porosity on permeability in the case of laminar flow through granular media. PhD thesis, Stanford University
Fredlund DG (1975) A diffused air volume indicator for unsaturated soils. Can Geotech J 12: 533–539
Fredlund DG (1979) Appropriate concepts and technology for unsaturated soils. Can Geotech J 16: 121–139
Fredlund DG (1985) Soil mechanics principles that embrace unsaturated soils. In: Proceedings of the 11th INCOSOMEFE, vol 2, pp 465–472
Fredlund DG, Morgenstern NR (1973) Pressure response below high entry discs. In: Proceedings of the 3rd International Conference on expansive soils, Haifa, vol 1, pp 97–208
Fredlund DG, Morgenstern NR (1976) Constitutive relations for volume change in unsaturated soils. Can Geotech J 13: 261–276
Fredlund DG, Morgenstern NR (1977) Stress state variables for unsaturated soils. ASCE J Geotech Div 103(GT5): 447–466
Fredlund DG, Hasan IU (1979) One dimensional consolidation theory: Unsaturated soils. Can Geotech J 16: 521–531
Frenkel J (1926) Ueber die Warmebewegung in festen und flussigen Koerpern. Zeit Physik 35: 652–669
Frenkel J (1946) Kinetic theory of liquids. Dover, 488 pp

Freudenthal AM (1950) The inelastic behavior of materials and structures. John Wiley and Sons, 587 pp
Freudenthal AM, Geiringer H (1958) The mathematical theories of the inelastic continuum. In: Flugge S (ed) Handbuch der Physik, vol 6. Springer, pp 229–433
Fukushima S, Tatsuoka F (1982) Deformation and strength of sand in torsional simple shear. IUTAM conference on deformation and failure of granular materal. Delft, pp 371–378
Gardner Willard (1920a) The capillary potential in relation to soil moisture constants. Soil Sci 10: 357–359
Gardner William (1920b) A capillary transmission constant and methods of determining it experimentally. Soil Sci 10: 103–126
Gardner WR (1956) Calculation of capillary conductivity from pressure plate outflow data. Proc Soil Sci Soc Am 20: 317–320
Gardner WR (1958) Mathematics of isothermal water conduction in unsaturated soil. International Symposium on water and its conduction in soils. HRB Spec Rep 40: 78–87
Gast RG (1969) Standard free energy for alkali metal cations on Wyoming bentonite. Proc Soil Sci Soc Am 33(1): 33–41
Gast RG, Van Bladel R, Deshpande KB (1969) Standard heats and entropies of exchange for alkali metal cations on Wyoming bentonite. Proc Soil Sci Soc Am 33/5: 661–664
Gauss CF (1813) Theoriae Attractionis corporum sphaerodicorum ellipticorum homogeneorum methodus nova tractata. Commun Soc Sci Gottingen, 2(5): 3–22
Gauss CF (1830) Principia generalia theoriae fluidorum in statu aequilibrii. Gottingen Commun Rec, vol 7
Gazetas G, Yegian MK (1979) Shear and Rayleigh waves in soil dynamics. ASCE J Geotech Div 105 (GT12): 1455–1470
Gersevanov NM (1934) The foundation of dynamics of soils (in Russian). Leningrad, Stroiizdat
Geuze ECWA (1953) General reporter introduction to discussions. Session 2 on laboratory investigation, including compaction tests, improvement of soil properties. Proceedings of the 3rd INCOSOMEFE, vol 3, pp 119–132
Geuze ECWA (1960) The effect of time on the shear strength of clays. ASCE convention, New Orleans
Geuze ECWA, Tan TK (1954) The mechanical behavior of clays. In: Harrison, VGW (ed) Proceedings of the 2nd international congress on rheology. Academic Press, pp 247–259
Gibbs JW (1873) A method of geometrical representation of the thermodynamic properties of substances by means of surfaces. Trans Connecticut Acad 2: 382–404
Gibbs JW (1875) On the equilibrium of heterogeneous substances. Trans Connecticut Acad 3: 108–248
Gibbs HJ, Hilf J, Holtz WJ, Walker FC (1960) Shear strength of cohesive soils. In: Proceedings of the ASCE conference on shear strength of cohesive soils, Boulder, pp 33–162
Gibbs HJ, Coffey CT (1969) Techniques for pore pressure measurements and shear testing of soils. In: Proceedings of the 7th INCOSOMEFE, Mexico, vol 1, pp 151–157
Gibson RE, Lo KY (1961) A theory of consolidation for soils exhibiting secondary compression. Norwegian Geotec Inst Publ no 41
Gibson RE, England GL, Hussey MGL (1967) The theory of one-dimensional consolidation of saturated clays. Geotechnique 17: 261–273
Gillott JE (1968) Clay in engineering geology. Elsevier
Glaeser R, Mering J (1958) Role of valence of exchangeable cations in hectorite. C R Acad Sci Paris 246: 1569–1572
Goldenblat II (1962) Some problems of the mechanics of deformable media. P Noordhoff, Groningen, 304 pp
Goldstein H (1953) Classical mechanics. Addison Wesley, 399 pp
Goodman MA, Cowin SC (1972) A continuum theory for granular materials. Arch Rat Mech Anal 44: 249–266
Gouy G (1910) Sur la constitution de la charge electrique à la surface d'un electrolyte. J Physique ser 4, 9: 457–467
Gouy G (1917) Sur la fonction electrocapillaire. J Physique ser 9, 7: 129–184
Govier GW, Aziz K (1972) The flow of complex mixtures in pipes. Van Nostrand-Reinhold 792 pp
Graf O (1930) Aufbau des Mortels und Betons, 3rd edn. Springer, Berlin
Gray DH, Mitchell JK (1967) Fundamental aspects of electro-osmosis in soils. ASCE J Soil Mech Div 93 (SM 6): 209–236
Green AE (1956) Hypoelasticity and plasticity. Proc R Soc Lond 234: 46–59
Green AE, Rivlin RS (1964) The mechanics of materials with structure. In: Kravtchenko J,

Sirieys PM (eds) IUTAM symposium on rheology and soil mechanics, Grenoble. Springer, pp 132–145
Green AE, Naghdi PM (1967) A theory of mixtures. Arch Rat Mech Anal 24: 243–263
Green AE, Naghdi PM (1969) On basic equations of mixtures. Q J Appl Maths 22: 427
Green AE, Naghdi PM (1971) Entropy inequalities for mixtures. Q J Appl Maths 24: 473–485
Green G (1828) An essay on the application of mathematical analysis to the theories of electricity and magnetism. Nottingham. (Repr. 1871 Mathematical papers of the late G. Green, London)
Green GE, Bishop AW (1969) A note on the drained strength of sand under generalized strain conditions. Geotechnique 19/1: 144–149
Green HS (1960) The structure of liquids. In: Flugge S (ed) Handbuch der Physik, vol 10. Springer, pp 1–133
Griggs DT (1936) Deformation of rocks under high confining pressures. J. Geol 44: 541–577
Grim RE (1968) Clay Mineralogy, 2nd edn. McGraw-Hill, 596 pp
Gross B (1947) On creep and relaxation. I. J Appl Phys 18: 212
Gross B (1948) On creep and relaxation. II. J Appl Phys 19: 257
Gross B (1952) On the inversion of the Voltera integral equation. Q Appl Maths 10: 74
Gross B (1953) Mathematical structure of the theories of visco-elasticity. Publ Inst Nat Tech Brazil. Herman, 74 pp
Gross B, Pelzer H (1951) On creep and relaxation. III. J Appl Phys 22: 1035–1039
Gudehus G (1973) Elasto-Plastische Stoffgleichungen fur trokenen Sand. Ingineur-Archiv 42: 151–169
Gudehus G, Kolymbas D (1979) A constitutive law of rate type for soils. In: White W (ed) Proceedings of the 3rd international congress on numerical methods in geomechanics, vol 1. Balkema, pp 319–320
Guest JJ (1900) On the strength of ductile materials under combined stress. Phil Mag 50: 69
Gurevich GB (1948) Foundations of the theory of algebraic invariants. Gosteknizdat
Gurtin ME (1972) The linear theory of elasticity. Truesdell C (ed) Handbuch der Physik, vol 6a/2. Springer pp 1–295
Habib, P (1953) Influence de la variation de la contrainte principale moyenne sur la resistance au cisaillement des sols. In: Proceedings 3rd INCOSOMEFE, Zurich, vol 1, pp 131–136
Habib P (1954) Etude experimental de la variation de la resistance et cisaillement des sols en fonction de la contrainte principale intermediaire. In: Harrison VG (ed) Proceedings of the 2nd international congress on Rheology, Oxford, pp 239–246
Hadamard J (1901) Sur la propagation des ondes. Bull Soc Math France 29: 50–60
Hadamard J (1903) Leçons sur la propagation des ondes et les equations de l'hydrodynamique (lectures of 1898–1900), Paris
Haefeli R, Amberg G (1948) Contribution to the theory of shrinking. In: Proceedings of the 2nd INCOSOMEFE, Rotterdam, vol 1, pp 13–17
Hagen G (1839) Ueber die Bewegung des Wassers in engen Zilindrischen Roeren. Prog Ann Phys Chem 46: 423–442
Haig BP (1920) The strain energy function and the elastic limit. Engineering 190/1: 158
Hammer MJ, Thompson DE (1966) Foundation clay shrinkage caused by large trees. ASCE J Soil Mech Div 92 (SM 6): 1–17
Handin J (1953) An application of high pressures in geophysics: experimental rock deformation. Trans ASME 75: 315–324
Handin J, Hager Jr RV (1957) Experimental deformation of rock under confining pressure: tests at room temperature on dry samples. Bull Am Assoc Petroleum Geologists 41: 1–50
Hanin M, Reiner M (1956) On isotropic tensor-functions and the measure of deformation. Z angew Math Phys 7: 377–393
Hannant DJ, Frederick CO (1968) Failure criteria for concrete in compression. Mag Conc Res 20: 137–144
Hansbo S (1960) Consolidation of clay with special reference to influence of vertical sand drains. Swedish Geotechnical Inst Proc no 18
Hansen JB (1961) A model law for simultaneous primary and secondary consolidation. In: Proceedings of the 5th INCOSOMEFE, Paris, vol 1, pp 133–136
Hansen B (1961) Shear box tests on sand. In: Proceedings of the 5th INCOSOMEFE, Paris, vol 1, pp 127–131
Harper RF (1904) The code of Hammurabi, King of Babylon, about 2550 BC. University of Chicago Press
Harr ME (1962) Groundwater and seepage. McGraw-Hill, 315 pp
Hayashi S, Ochiai H, Motoki M (1985) Yield criteria for anisotropic soil. In: Proceedings of the 11th INCOSOMEFE, vol 2, pp 503–506

Haythornthwaite RM (1960a) Mechanics of the triaxial cell for soils. ASCE J Soil Mech Div 86 (SM5): 35–62
Haythornthwaite RM (1960b) Stress and strain in soils. In: Proceedings of the 2nd symposium of naval structural mechanics. Pergamon, pp 185–193
Helm DC (1987) Three-dimensional consolidation theory in terms of the velocity of solids. Geotechnique 37(3): 369–392
Helmholtz H (1858) Uber Integrale der hydrodynamischen Gleichungen, welche der Wirbelbewegungen entsprechen. J Reine angew Math 55: 25–55
Helmholtz H (1882) Die Thermodynamik Chemisher Vorgange. Sitzgsber Akad Wiss Berlin 2: 958–978
Hengchaovanich D, Nelson JD (1970) Shear strength characteristics of the stiff Bangkok clay. Research Rep no 9, Asian Institute of Technology
Hencky H (1924) Zur Theorie Plastisch Deformablen und der hierdurch in Material Hervorgerufenen Nebenspanungen. In: Proceedings 1st international congress on applied mechanics, Delft, pp 312–317
Hencky H (1929) Das Superpositionsgesetz eines endlich deformierten relaxationsfehigen Elastischen Kontinuums und seine Bedeutung fur eine exakte Ableitung der Gleichungen fur die zahe Flussigkeit in der Eulerschen Form. Ann Physik 2: 617–630
Henkel DJ (1960) The relationship between the effective stress and water content in saturated clays. Geotechnique 10 (2): 41–53
Hertz H (1899) The principles of mechanics. Dover (1960), 271 pp
Hershey V (1952) A review of the definition of finite strain. In: Proceedings 1st US national congress applied mechanics, pp 473–478
Hight DW, Gens A, Symes MJ (1983) The development of a new hollow cylinder apparatus investigating the effects of principal stress rotation in soils. Geotechnique 33/4: 355–383
Hilf JW (1956) An investigation of pore water pressure in compacted cohesive soils. PhD thesis, University of Colorado, USDI Bur Recl Tech Memo 654
Hillel D (1980) Fundamentals of soil physics. Academic Press, p 413
Hirschfelder JO, Curtiss CF, Bird RB (1954) Molecular theory of gases and liquids. John Wiley and Son, New York
Ho KH (1962) A study of the strength parameters of partly saturated bentonite. MS thesis, McGill University
Hooke R (1678) Lectures de potentia restitutiva. Early science in Oxford. vol 8 (1931), pp 331–356. R T Gunther (Publ), London
Houlsby GT (1979) The work input to a granular material. Geotechnique 29/3: 354–358
Howe WL, Hudson CJ (1927) Studies in porosity and permeability characteristics of porous bodies. J Am Ceramic Soc 10: 443–448
Hu LW (1960) Plastic stress–strain relations and hydrostatic stress. In: Proceedings of the 2nd symposium of naval structural mechanics. Pergamon, New York, pp 194–201
Hu LW, Pae KD (1963) Inclusion of the hydrostatic stress components in formulation of the yield condition. J Franklin Inst 275/6: 491–502
Huber MT (1904) Wlasciwa Praca Odksztalcenia Iako Miara Wyteznia Materialu. Czasopismo Technizne, Lwow. vol 22
Hull HH (1961) The normal forces and their thermodynamic significance. Trans Soc Rheol 5: 115–131
Hvorslev MJ (1937) Ueber die Festigkeitseigenschaften Gestorter Bindiger Boden. Thesis. Danmarks Naturvidenskabelige Samfund Ingeniorvidenskabelige Skrifter, ser A, no 45, Copenhagen, 159 pp
Hvorslev MJ (1939) Torsion shear tests and their place in the determination of the shearing resistance of soils. Proc Am Soc Testing Materials 39: 999–1022
Hvorslev MJ (1960) Physical components of the shear strength of saturated clays. In: Proceedings ASCE research conference on shear strength of cohesive soils, Boulder, pp 169–273
Ilyushin AA, Ogybalov PM (1959) Effect of high pressure strengthening on a hollow cylinder. Izvestia AN USSR OTN MiM 6: 110–112. Transl Friedman MD
Irwin GR (1958) Fracture. In: Flugge S (ed) Handbuch der Physik, vol 6. Springer, pp 551–613
Iwan WD (1967) On a class of models for the yielding behavior of continuous and composite systems. Trans ASME E: J Appl Mech 34: 612–617
Jackson ML (1969) Soil chemical analysis – advanced course. Dept Soil Sci University of Wisconsin, 498 pp
Jáky J (1948a) Validity of Coulomb's law of stability. In: Proceedings of the 2nd INCOSOMEFE, Rotterdam, vol 1, pp 87–90
Jáky J (1948b) State of stress in great depth. In: Proceedings of the 2nd INCOSOMEFE,

Rotterdam, vol 1, pp 90–93
Jaumann G (1911) Geschlossenes System physikalischer und chemischer Differentialgesetze. Sitzgsber Akad Wiss Wien (II-a), 120: 385–530
Jennings JEB (1960) A revised effective stress law for use in the prediction of the behavior of unsaturated soils. In: Proceedings of the Brit Nat Soc SOMEFE conference on pore pressure and suction in soils. Butterworths, pp 26–30
Jennings JEB (1961) Discussion to session on shallow foundations. In: Proceedings of the 5th INCOSOMEFE, Paris, vol 3, p 300
Jennings JEB, Burland JB (1962) Limitations to the use of effective stresses in partly saturated soils. Geotechnique, 12(2): 125–144
Juarez-Badillo E (1985) General time volume change equations for soils. In: Proceedings of the 11th INCOSOMEFE, vol 2, pp 519–530
Jumikis AR (1962) Soil mechanics. D Van Nostrand, 791 pp
Kaplan IR (1966) Natural gases in marine sediments. Plenum Press
Kármán T von (1912) Festigkeitsfersuche unter Allseitigen Druck. Mitt Forschungsarb Gebiete Ingenierwesen, pp 1749–1757
Kármán T von (1913) Untersuchungen uber Knickfestigkeit. Mitt Forschungsarb Ver Deut Ing no 118
Kassif G, Ben-Shalom A (1971) Experimental relationship between swell pressure and suction. Geotechnique 21 (3): 245–255
Kavarajian E, Mitchell JK (1980) Time dependent deformation behavior of clays. ASCE J Geotech Div 106 (GT6): 611–630
Keinonen LS (1973) An energetic model of consolidation in cohesive soils. In: Proceedings of the 8th INCOSOMEFE, vol 1.1, pp 203–207
Kennedy HT, Bhagia NS (1969) Equation of state for condensate fluids. J Soc Petrol Engng 9(3): 279–286
Kenyon DE (1976) The theory of an incompressible solid–fluid mixture. Arch Rat Mech Anal 62: 131–147
Kirkpatrick WM (1957) The condition of failure for sands. In: Proceedings of the 4th INCOSOMEFE, vol 1, pp 172
Kirkpatrick WM, Belshaw DJ (1968) On the interpretation of the triaxial test. Geotechnique 18(3): 336–350
Kisiel J (1964) On experience acquired in the course of tests carried out with a shear box. Archwm Hydrotech 11: 415–442
Klausner Y (1960) Discussion to session 3. In: Proceedings of the ASCE research conference on shear strength of cohesive soils, Boulder, pp 1056–1062
Klausner Y (1961) Volume rheology of a two-phase system. Kolloid Zeitschrift 174(2): 109–113
Klausner Y (1962) Fundamentals of theoretical mechanics of soils. Wayne State University (mimeographed print)
Klausner Y (1964) Pure deviatoric loading of soils. Israel J Technol 2(3): 305–311
Klausner Y (1967) Certain properties of the pure deviatoric test. In: Proceedings of the 3rd Asian regional conference of SOMEFE, vol 1, pp 25–28
Klausner Y (1969a) The application of the Hencky measure of strain in simple extensions of soils. In: Proceedings of the international conference on materials in CE structure, solid mechanics and engineering design, Southampton, Wiley-Interscience (1971), vol 1, pp 613–620
Klausner Y (1969b) Pure deviatoric loading device for compressible materials. Improved pneumatic control. In: Proceedings of the international conference on materials in CE structure, solid mechanics and engineering design, Southampton, Wiley-Interscience (1971), vol 1, pp 157–167
Klausner Y (1970a) Volumetric behavior of clay soils. PhD dissertation, Princeton University
Klausner Y (1970b) Volumetric stress–strain relationships in soils. In: Proceedings of the 5th international congress on rheology, vol 2, University of Tokyo and University Park Press, pp 591–602
Klausner Y (1970c) Pure deviatoric shear test of soils. In: Proceedings of the 5th international congress on rheology, vol 2, University of Tokyo and University Park Press, pp 603–622
Klausner Y (1970d) Floors of ash treated soil at Tel Sheva. Report to Tel-Aviv University Dept of Archeology
Klausner Y (1987a) Soil test records, tests T1–T123. Period of 1952–1957. Technion, Israel Institute of Technology, Haifa
Klausner Y (1987b) Soil test records, tests P1–P138. Period of 1957–1960. Princeton University
Klausner Y (1987c) Soil test records, tests W1–W25. Period of 1960–1963. Wayne State University, Detroit

Klausner Y (1987d) Soil test records, tests N1–N258. Period of 1963–1970. Negev Institute Arid Zone Research, Beer-Sheva
Klausner Y (1989) Soil test records H1–H224. Period of 1986–1989. Practical Engineering College, Beer-Sheva
Klausner Y, Kraft SR (1965) Non-Poiseuille flow with axial forces. In: Proceedings of the 12th conference of Israel Soc Theor and Appl Mech Israel J Tech 3(2): 152–159
Klausner Y, Kraft SR (1966) A capillary model for non-Darcy flow through porous media. Trans Soc Rheol 10(2): 603–619
Klausner Y, Rooney JW (1967) Design and construction of a device for the pure deviatoric loading of soils. Re to NSF by the Dept CE, Wayne State University, Detroit
Klausner Y, Shainberg I (1967) Consolidation properties of arid-region soils. In: Proceedings of the 3rd Asian regional conference SOMEFE, Haifa. Jerusalem Academic Press, vol 1, pp 373–378
Klausner Y, Shainberg I (1971) Consolidation properties of Na^+ and Ca^{++} adsorbed montmorillonites. In: Proceedings of the 4th Asian regional conference SOMEFE, Bangkok, pp 373–378
Kleitz C (1873) Etudes sur les forces moleculaires dans les liquides en mouvement et application à l'hydrodynamique. Dunod, Paris
Klute A (1952) Some theoretical aspects of the flow of water in unsaturated soils. Proc Soil Sci Soc Am 16(2): 144–148
Klute A (1972) The determination of the hydraulic conductivity and diffusivity of unsaturated soils. Soil Sci 113(4), 264–276
Knight JH, Philip JR (1974) On solving the unsaturated flow equation. 2. Critique of Parlange's method. Soil Sci 116(6), 407–422
Kondo S (1956) Thermodynamical fundamental equation for spherical interface. J Chem Phys 25: 662
Koppula SD (1983) Pore-water pressure due to overburden removal. ASCE J Geotech Div 109(GT8): 1099–1112
Kotchine NE (1926) Sur la theorie des ondes de choc dans un fluide. Rend Circ Math Palermo 50: 305–344
Krynine DP (1947) Soil mechanics, 2nd edn. McGraw-Hill, 511 pp
Kunze RJ, Nielsen DR (1982) Finite difference solutions of the infiltration equation. Soil Sci 134(2), 81–88
Kunze RJ, Kar-Kuri HR (1983) Gravitational flow in infiltration. In: Proceedings of national conference on advances in infiltration, ASAE, pp 14–23
Kutilek M (1969) Non-Darcian flow of water in soils (laminar region). In: Proceedings of the 1st symposium on transport phenomena in porous media. Haifa, pp 327–340
Kutilek M, Vogel T, Nielsen DR (1985) Note on physical characteristics of linear soils. Soil Sci 139(6): 497–499
Ladd CC (1977) State of art report on soil properties. In: Proceedings of the 9th INCOSOMEFE, vol 3, pp 293–305
Lade PV (1972) The stress–strain characteristics of cohesionless soils. PhD thesis, University of Cambridge
Lade PV (1975) Elasto-plastic stress–strain theory for cohesionless soils. ASCE J Geotech Div 101 (GT10): 1037–1053
Lade PV (1978) Prediction of undrained behavior of sand. ASCE J Geotech Trans 104 (GT6): 721–735
Lade PV (1979) Elasto-plastic stress–strain theory of cohesionless soil with curved yield surfaces. Intl J Solids Structures 13: 1019–1035
Lade PV, Musante M (1978) Three dimensional behavior of remolded clay. ASCE J Geotech Div 104(GT2): 193–209
Lambe TW (1954) The permeability of fine grained soils. Symposium Proceedings Spec. Publ. 163
Lambe TW (1958) The structure of compacted clay. Proc ASCE J Soil Mech Div 84 (SM2): 1–34
Lambe TW (1960a) The engineering behaviour of compacted clay. Trans ASCE 125(1): 681–756
Lambe TW (1960b) A mechanistic picture of shear strength in clay. In: Proceedings of ASCE research conference on shear strength of cohesive soils, Boulder, Colorado, pp 555–580
Lambe TW, Whitman RV (1969) Soil mechanics. John Wiley and Sons, 553 pp
Langmuir L (1938) The role of attractive and repulsive forces in the formation of tactoids, thixotropic gels, protein crystals, and coacervates. J Chem Phys 6: 873–896
Lass H (1950) Vector and tensor analysis. McGraw-Hill, 347 pp
Lebedev AF (1927) The movement of water in soils and subsoils. Z Pflanzenernahr Dungung Bodenk 10A: 1–36
Legget RF (1962) Geology and engineering. McGraw-Hill, New York

Lee EH (1960) Stress analysis for visco-elastic bodies. In: Bergen JT (ed), Proceedings of symposium on visco-elasticity: phenomenological aspects. Academic Press, pp 1–26

Lee K, Sills GC (1979) A moving boundary approach to the large strain consolidation of a thin layer. Proceedings of the 3rd international conference on numerical methods in geomechanics, Aachen, pp 16–17

Lee KL (1978) End restraint effect on undrained static triaxial strength of sand. ASCE J Geotech Div 104 (GT6): 705–719

Leland TW Jr, Chappelear PS (1968) Corresponding states principle, review of current theory and practice. Ind Engng Chem 60 (7): 15–43

Leonards GA (1955) Strength characteristics of compacted clays. ASCE Trans 120: 1420–1455

Leonards GA, Girault P (1961) A study of the one dimensional consolidation test. Proceedings of the 5th INCOSOMEFE, Paris, vol 1, pp 213–218

Le Van Phuc, Morel-Seytoux HJ (1972) Effect of soil air movement and compressibility on infiltration rates. Proc Soil Sci Am 36 (2): 237–241

Levy M (1870) Memoire sur les equations generales des mouvements interieures des corps solides ductiles au dela des limites ou l'elasticite pourrait les ramener a leur premier etat. C R Acad Sci Paris J Math ser 2, 16 (1971): 396–372

Li Ping Seung (1960) On the initial gradient and the permeability of two clay soils. MSc thesis, Princeton University

Lloret A, Alonso EE (1980) Consolidation of unsaturated soils including swelling and collapse behaviour. Geotechnique 3 (4): 449–477

Lloret A, Alonso EE (1985) State surfaces for partly saturated soils. Proceedings of the 11th INCOSOMEFE vol 2, pp 557–562

Lo KY (1959) An investigation on the consolidation characteristics of clays exhibiting secondary effects. PhD thesis, Imperial College

Lodge QS (1951) On compatibility conditions for large strains. J Mech Appl Maths 4: 85–93

Lof P (1983) Minerals of the world. (Wall map)

Lohr E (1917) Entropieprinzip und geschlossenes Gleichungs-system. Denkschrift Akad Wiss 93: 339–421

Lomize GM, Kryzhanovsky AL (1967) On the strength of sand. Proceedings of the geotechnical conference on shear strength properties of natural soils and rocks, Oslo, vol 1, pp 215–219

Lomize GM, Kryzhanovsky AL, Vorontsov EI, Goldin AL (1969) Study on deformation and strength of soils under three dimensional state of stress. Proceedings of the 7th INCOSOMEFE, vol 1, pp 257–265

Lorenz H, Mewmeuer H, Gudehus G (1965) Tests concerning compaction and displacements performed on samples of sand in the state of plane deformation. Proceedings of the 6th INCOSOMEFE, Montreal, vol 1, pp 293–297

Love AEH (1892) A treatise on the mathematical theory of elasticity. Dover (1944), 738 pp

Low PF (1960) Viscosity of water in clay systems. Proceedings of the 8th national congress on clay and clay minerals, vol 8, pp 170–182

Low PF (1961) Physical chemistry of clay water interaction. In: Norman AG (ed) Advances in agronomy, vol 13, pp 269–327

Low PF (1968) Observations on activity and diffusion coefficients in Na-montmorillonite. Proceedings of the Mokady memorial symposium, pp 325–336

Lubahn JD (1948) Notch tensile testing-fracturing of metals, J Appl Mech 14 (A299): 90–132

Lutz JF, Kemper WD (1959) Intrinsic permeability of clay as affected by the clay–water interaction. Soil Sci 88: 83–90

Lynch WA, Truxal JG (1961) Introductory system analysis: signals and systems in electrical engineering. McGraw-Hill

Maksimovic M (1989) Nonlinear failure envelope for soils. ASCE J Geotech Div 115 (GT4): 581–586

Marsal RJ, Resines JS (1960) Pore pressure test. Proceedings of the ASCE research conference on shear strength of cohesive soils, Boulder, pp 965–983

Marshall CE (1964) The physical chemistry and mineralogy of soils. John Wiley and Sons, 388 pp

Matsuoka H, Nakai T (1974) Stress deformation and strength characteristics of soil under three different principal stresses. Proc J ASCE 232: 59–70

Matsuoka H, Koyama H (1983) A constitutive model for sands and cyclic shear stresses. ASCE J Geotech Div GT109

Matyas EL, Radhakrishna HS (1968) Volume change characteristics of partially saturated soils. Geotechnique 18(4): 432–448

Maxwell JC (1866) On the dynamical theory of gases. Phil Trans R Soc Lond 157: 49–88

Maxwell JC (1876) On the equilibrium of heterogeneous substances. Proc Camb Phil Soc 2: 427–430

Mayne PW (1985) Stress anisotropy effects on clay strength. ASCE J Geotech Div 111 (GT3): 356–366
McClintock FA, Walsh JB (1962) Friction on Griffith cracks in rocks under pressure. Proceedings of the 4th US national ASME congress of applied mechanics, vol 2, pp 1015–1022
McConnell AJ (1957) Application of tensor analysis. Dover, 318 pp
Meissner W (1938) Thermodynamische Behandlung stationarer Vorgange und Mehrphasensystemen. Ann Physik (5) 32: 115–127
Meixner J (1941) Zur Thermodynamik der Thermodiffusion. Ann Physik (5) 39: 333–356
Meixner J, Reik H (1959) Thermodynamik der irreversiblen Prozesse. In: Flugge S (ed) Handbuch der Physik, vol 3/2 Springer, pp 413–523
Mering J, Glaeser, R (1961) The X-ray identification and crystal structure of clay minerals. Mineralogical Society, London, vol 180, Brown G (ed)
Meyer OE (1874) Zur Theorie der Inneren Reibung. J Reine angew Mathematik 78: 130–135
Michaels AS, Lin CS (1954) The permeability of kaolinite. Ind Engng Chem 46: 1239–1246
Miller J, Low PF (1962) Threshold gradients for water flow in clay systems. Proc Am Soc Soil Sci 27(6): 605–609
Minkowski H (1898–1935) Capillarity. Entry in: Enzyklopadie Math Wiss vol 5.1, pp 558
Mises R Von (1913) Mechanik der fester Korper in plastischdeformablen Zustand Nachr Gottingen Math Physik Klasse, pp 582–592
Mitchell JK (1986) Practical problems from surprising soil behavior. ASCE J Geotech Div 112 (GT3): 255–289
Mohr O (1900) Welche Umstande bedingen die Elastizitaetgrenze und den Bruch eines Materials? Zeitschrift VDI 44: 1524–1530
Mohr O (1914) Abhandlungen aus den Gebiete der technischen Mechanik, 2nd edn. Wilhelm Ernst and Sohn, Berlin, pp 192–235
Mooney M (1948) The thermodynamics of a strained elastomer. I. General analysis. J Appl Phys 19: 434–444
Morel-Seytoux HJ (1961) Effect of boundary shape on channel seepage. Dept CE, Stanford University Tech Rep no 7
Morel-Seytoux HJ (1969) Flow of immiscible liquids in porous media. In: Flow through porous media De Wiest RJM (ed.) Academic Press, pp 456–516
Morgan JR, Gerrard CM (1973) Anisotropy and nonlinearity in sand properties. Proceedings of the 8th INCOSOMEFE, Moscow, vol 1.2, pp 287–292
Morgenstern NR, Tchalenko JS (1967a) Microstructural observation on shear zones from slips in natural clay. Proceedings of the geotechnical conference, Oslo, vol 1, pp 147–152
Morgenstern NR, Tchalenko JS (1967b) Microscopic structures in kaolin subjected to direct shear. Geotechnique 17(4): 309–328
Mroz Z (1967) On the description of anisotropic workhardening. J Mech Phys Solids 15: 163–175
Mroz Z, Piertruszczak S (1983) On hardening anisotropy of K_0-consolidated clays. Intl J Numer Anal Methods Geomech 7: 19–38
Mualem Y (1974) A conceptual model of hysteresis. Water Resources Res 10(3): 514–520
Mualem Y (1984a) A modified dependent-domain theory of hysteresis. Soil Sci 137(5): 283–291
Mualem Y (1984b) Prediction of the soil boundary wetting curve. Soil Sci 137(6): 379–390
Mualem Y, Dagan G (1975) A dependent domain model of capillary hysteresis. Water Resources Res 11(3): 452–460
Muller I (1968) A thermodynamic theory of mixtures of fluids. Arch Rat Mech Anal 28, 1–39
Murdoch AI (1985) A corpuscular approach to continuum mechanics: basic considerations. Arch Rat Mech Anal 88(4): 291–321
Murdoch AI, Morro A (1987) On the continuum theory of mixtures: motivation from discrete considerations. Intl J Multiphase Flow 25 (1): 9–25
Murnaghan FD (1941) The compressibility of solids under extreme pressures. Kármán Anniversary Volume, pp 121–136
Nadai A (1933) Theory of strength. J Appl Mech 1(3): 111–130
Nadai A (1937) Plastic behavior of metals in strain. Part I. J Appl Phys 8: 417
Nadai A (1950) Theory of flow and fracture of solids, vol 1. McGraw-Hill, New York, 572 pp
Nageswaran S (1983) Effect of gas bubbles on the sea bed behaviour. D Phil thesis, Oxford University
Naslin P (1965) The dynamics of linear and non-linear systems. Blackie, London, Glasgow, 586 pp
Natansohn L (1901) Ueber die Gesetze der Inneren Reibung. Z Physik Chem 38: 95–111
Neville AM (1981) Properties of concrete, 3rd edn. Longman, 532 pp
Newman K, Newman JB (1969) Failure theories and design criteria for plain concrete. Proceedings of the conference on structure, solid mechanics and engineering design, vol 2,

Southampton, pp 963–995
Newmark NM (1960) Failure hypotheses for soils. ASCE research conference on shear strength of soils, Boulder, pp 17–32
Newton I (1687) Philosophiae naturalis principia matematica. London, trans Motte A (1927)
Nicholls RL, Hebst Jr RL (1967) Consolidation under electrical pressure gradients. ASCE J Soil Mech Div 93 (SM5): 139–151
Nielsen DR, Biggar JW, Corey JC (1972) Application of flow theory to field situations. Soil Sci 113 4: 254–263
Nikolaevski VN (1968) Transfer phenomena in fluid saturated porous media. Proceedings of the IUTAM symposium on irreversible aspects of continuum mechanics. Springer, pp 250–258
Niwa Y, Kobayashi S, Koyanagi W (1967) Failure criterion of lightweight aggregate concrete subjected to triaxial compression. Mem Fac. Engng, Kyoto University, Japan 24: 119–131
Noll W (1958) A mathematical theory of the mechanical behaviour of continuous media. Arch Rat Mech Anal 2: 197–226
Norrish J (1954) The swelling of montmorillonite. Disc Faraday Soc 18: 120
Norrish J, Quirk JP (1954) Crystalline swelling of montmorillonite. Nature 173: 255
Novozhilov VV (1953) Foundation of the non-linear theory of elasticity. Graylock, 233 pp
Nunziato JW, Walsh EK (1980) On ideal multiphase mixtures with chemical reactions and diffusion. Arch Rat Mech Anal 73: 284–311
Nutting PG (1930) Physical analysis of oil sands. Bull Am Assoc Petr Geol 14: 1337–1349
Oda M (1981) Anisotropic strength of cohesionless sands. ASCE J Geotech Engng trans 107 (GT9): 1219–1231
Oldroyd JG (1950) Finite strains in an anisotropic elastic continuum. Proc R Soc Lond A200: 345–358
Olszak W, Perzyna P (1964) On thermodynamics of the differential type material. Proceedings of the IUTAM symposium on rheology and soil mechanics, Grenoble. Kravtchenko J, Sirieys PM (eds). Springer pp 279–291
Onsager L (1931) Reciprocal relations in irreversible processes. Phys Rev 37: 405–426
Overbeek, J. Th. G. (1952). The interaction between colloidal particles. In: Kruyt HR (ed) Colloid science, vol 1. Elsevier, pp 245–276
Painter S (1953) A history of the middle ages 1284–1500 AD. Alfred A Knopf
Palmer AC (1967) Stress–strain relations for clays. An energy theory. Geotechnique 17(4): 348–354
Parlange JY (1971a) Theory of water movement in soils. 1. One dimensional absorbtion. Soil Sci 111(3): 134–137
Parlange JY (1971b) Theory of water movement in soils. 2. One dimensional infiltration. Soil Sci 111(3): 170–174
Parlange JY (1971c) Theory of water movement in soils. 3. Two and three dimensional absorbtion. Soil Sci 112(5): 313–317
Parlange JY, Bradock RD, Simpson RW (1982) Optimization principle for air and water movement. Soil Sci 133(1): 4–9
Partom Y (1966) The dilatational and distortional partition of the energy function in finite strain. Israel J Tech 4(3) 218–223
Passman SL (1974) On the balance equations for a mixture of granular materials. Univ. Wisconsin Math Res Tech Rep no 1390
Passman SL (1977) Mixtures of granular materials. Intl J Engng Sci 15: 117–129
Peltier R (1953) Consideration geotechnique sur la portance des sols routiers. Proceedings of the 3rd INCOSOMEFE, Zurich, vol 3, pp 312–325
Perloff WH, Pombo LE (1969) End restraint effects in triaxial test. Proceedings of the 7th INCOSOMEFE, Mexico, vol 1, pp 327–333
Philip JR (1954) A infiltration equation with some physical significance. Soil Sci 77: 153–157
Philip JR (1955) The concept of diffusion applied to soil water. Proceedings Natl Acad Sci India, 24a: 93–104
Philip JR (1957a) The theory of infiltration. 1. The infiltration equation and its solution. Soil Sci 83: 345–357
Philip JR (1957b) The theory of infiltration. 2. The profile at infinity. Soil Sci 83: 435–448
Philip JR (1957c) The theory of infiltration. 3. Moisture profiles and relations to experiment. Soil Sci 84: 163–178
Philip JR (1957d) The theory of infiltration. 4. Sorbtivity and algebraic infiltration equations. Soil Sci 84: 257–264
Philip JR (1957e) The theory of infiltration. 5. The influence of the initial moisture content. Soil Sci 84: 329–339

Philip JR (1958a) The theory of infiltration. 6. Effect of water depth over soil. Soil Sci 85(5): 278–286
Philip JR (1958b) The theory of infiltration. 7. Soil Sci 85(6): 333–337
Philip JR (1958c) Physics of water movement in porous solids. Proceedings of the international symposium on water and its conduction in soils. HBR Spec. Rep. 40, pp 149–163
Philip JR (1966a) A linearization technique for the study of infiltration. UNESCO symposium on water in the unsaturated zone, Wageningen, vol 1, pp 524
Philip JR (1966b) Absorbtion and infiltration in two and three dimensional systems. UNESCO symposium on water in the unsaturated zone, Wageningen, vol 2, pp 503–525
Philip JR (1967) The second stage of drying. J Appl Met 6(3): 581–582
Philip JR (1968a) The theory of absorbtion in aggregated media. Austral J Soil Res 6: 1–19
Philip JR (1968b) Kinetics of sorbtion and volume change in clay colloid pastes. Austral J Soil Res 6: 249–267
Philip JR (1969a) Moisture equilibrium in the vertical in swelling soils: I. Basic theory. Austral J Soil Res 7: 90–120
Philip JR (1969b) Moisture equilibrium in the vertical in swelling soils: II. Applications. Austral J Soil Res 7: 121–141
Philip JR (1969c) Hydrostatics and hydrodynamics in swelling media. Proceedings of the IAHR 1st international symposium on fundamentals of transport phenomena in porous media, Haifa. Water Resources Res 5: 1070–1077
Philip JR (1970) Flow in porous media. Ann Rev Fluid Mech 177–204
Philip JR (1973) On solving the unsaturated flow equation: 1. The flux concentration relation. Soil Sci 115(5): 328–333
Philip JR, Smiles DE (1969) Kinetics of sorbtion and volume change in three-component systems. Austral J Soil Res 7: 1–19
Poncelet JV (1839) Introduction à la mécanique industrielle, physique et experimentale. Metz
Poulovassilis A (1962) Hysteresis of pore water, an application of the concept of independent domains. Soil Sci 93(6): 405–412
Poulovassilis A (1970) Hysteresis of pore water in granular porous bodies. Soil Sci 109(1): 5–12
Poulovassilis A, Tzimas E (1975) The hysteresis in the relationship between hydraulic conductivity and soil water content. Soil Sci 120(5): 327–333
Poulter TC (1932) Apparatus for optical studies of high pressure. Phys Rev 40: 860–876
Poynting JH (1905) Radiation pressure. Phil Mag ser 6, 9: 393–406
Poynting JH (1909) On pressure perpendicular to the shear planes in finite pure shears and on lengthening of loaded wires when twisted. Proc R Soc A 82: 546
Poynting JH (1912) On the changes in the dimensions of a steel wire when twisted. Proc R Soc A 86: 534
Pradhan TBS, Tatsuoka F, Horii N (1988) Simple shear testing on sand in a torsional shear apparatus. Soils foundat 28(2): 95–112
Prager W (1953) A geometrical discussion on the slip line field in plane plastic flow. Trans R Inst Tech no 65, Stockholm
Prager W (1959) Introduction to plasticity. Addison-Wesley, Reading, MA
Prevost JH (1978) Anisotropic undrained stress–strain behavior of clays. ASCE J Geotech Div 114 (GT8): 1075–1090
Prigogine J (1947) Etude thermodynamique des phenomenes irreversibles. Dunod-Desoer. English transl (1967) Wiley Interscience, 147 pp
Proctor RR (1948) Construction and operational details for a simple machine to test soils in double shear. Proceedings of the 2nd INCOSOMEFE vol 7, pp 61–64
Pyke R (1979) Discussion on "Definition of terms related to liquefaction" by the Committee on Soil Dynamics. ASCE J. Geotech. Div. 105 (GT10): 1260
Raats PAC (1965) Development of equations describing transport of mass and momentum in porous media with special reference to soils. PhD thesis, University of Illinois
Rabe WH (1941) Die Geschichte der Hochwassersicherung der Stadt Mexico. Die Bautechnik 17/18: 192
Radhakrishna HS (1967) Compressibility of partially saturated soils. PhD thesis, University of Waterloo
Radok JRM (1957) Viscoelastic stress analysis. Q Appl Maths 15: 198–202
Rankine WJM (1857) On the stability of loose earth. Phil Trans R Soc Lond 147: 9–27
Redlich O and Kwong JNS (1949) An equation of state. Fugacities of gaseous solutions. Chem Res 44: 233–244
Reik H (1953) Zur Theorie Irreversibler Vorgange. Ann Physik 11: 73–96 and 13: 407–428
Reiner M (1945) A mathematical theory of dilatancy. Am J Maths 67: 350–362

Reiner M (1948) Elasticity beyond the elastic limit. Am J Maths 70: 433–466
Reiner M (1949) Twelve lectures on rheology. North Holland, 162 pp
Reiner M (1954a) Rheology. In: Reiner M (ed) Building materials, their elasticity and inelasticity. North Holland, pp 3–37
Reiner M (1954b) Second order effects in elasticity and hydrodynamics. Bull Res Council Israel 3: 372–379
Reiner M (1958) Rheology. Springer Verlag, pp 434–550 (Handbuch der Physik, vol 6)
Reiner M (1960) Deformation strain and flow. Interscience
Reiner M (1966) The influence of dissipated stress work on the rupture of materials. In: Parkus H, Sedov LI (eds) IUTAM Symposium on irreversible aspects of continuum mechanics, Vienna. Springer pp 335–345
Rendulic L (1937) Ein Grundgesetz der Tonmechanik und sein experimenteller Beweis. Bauingenieur 18: 459–467
Reynolds O (1885) On the dilatancy of media composed of rigid particles in contact. With experimental illustrations. Phil. Mag. 20: 469–481
Reynolds O (1887) Experiments showing dilatancy, a property of granular material, possibly connected with gravitaton. Proc R Inst Gt. Britain 11: 354–363
Richards LA (1928) The usefulness of capillary potential to soil moisture and plant investigation. J Agric Res 37: 719–742
Richards LA (1950) Experimental demonstration of the hydraulic criterion for zero water flow in unsaturated soil. Trans Intl Congr Soil Sci 1: 67–68
Richart FE, Brandtzaeg A, Brown RL (1928) A study of the failure of concrete under combined compressive stresses. Univ. Illinois Engng Exp Stat Bull 185: 1–102
Richter H (1949) Verzerrungstensor, Verzerrungsdeviator, und Spannungstensor bei Endlichen Formenderungen. Z angew Math Mech 29: 65–75
Riemann B (1876) Commentatio mathematica, qua respondere tentatur quaestioni ab Illma Academia Parisiensi propositae. Werke 370–399
Rivlin RS (1948) Large elastic deformations of isotropic materials, I, II, III, IV. Phil Trans R Soc A 240: 459–490, 491–508, 509–525; 241: 379–397
Rivlin RS (1955) Further remarks on the stress–deformation relation for isotropic materials. J Rat Mech Anal 4: 681–701
Rivlin RS, Ericksen JL (1955) Stress–deformation relations for isotropic materials. J Rat Mech Anal 4: 323–425
Roscoe KH (1950) Mechanical models for the representation of visco-elastic properties. Brit J Appl Phys 1: 171
Roscoe KH (1953) An apparatus for the application of simple shear to soil samples. Proceedings of the 3rd INCOSOMEFE, vol 1, pp 186–191
Roscoe KH (1970) The influence of strains in soil mechanics. Geotechnique 20(2): 129–170
Roscoe KH, Schofield AN, Wroth CP (1958) On the yielding of soils. Geotechnique 9(1): 22–53
Roscoe KH, Basset RH, Coll ERL (1967) Principal axes observed during simple shear of a sand. Proceedings of the geotechnical conference, Oslo, vol 1, pp 231–238
Roscoe KH, Schofield AN, Thurairajah A (1963) Yielding of clays in states wetter than critical. Geotechnique 13(3): 111–133
Rosenqvist I Th (1959) Physico-chemical properties of soils. Proc ASCE Soil Mech Found Div 85(2): 31–53
Rosenqvist I Th (1961) Mechanical properties of a soil water system. Trans ASCE 126(1): 746–747
Rosenqvist I Th (1963) The influence of physico-chemical factors upon the mechanical properties of clays. Norwegian Geotech Inst Publ no 54
Rowe PW, Barden L, Lee IK (1964) Energy components during the triaxial cell and direct shear tests. Geotechnique 14(3): 247–261
Rutheford DE (1951) Classical mechanics. Oliver and Boyd, 200 pp
Rutledge PC (1940) Neutral and effective stresses in soils. Proceedings of the Purdue conference on soil mechanics, pp 124–129
Saada AS (1961) A rheological investigation into the shear behavior of saturated clay soils. PhD thesis, Princeton University
Saada AS (1967) A stress-controlled apparatus for triaxial testing. ASCE J Soil Mech Road Div 93 (SM6): 61–78
Saada AS (1968) A pneumatic computer for testing cross-anisotropic materials. ASTM Materials Research and Standards, vol 816/1, pp 17–23
Saffman PG (1959) A theory of dispersion in a porous medium. J Fluid Mech 6: 321–349
St Venant AJCB de (1855) De la torsion des prismes, avec des considerations sur leur flexion.

Mem Pres par Div Savant a l'Acad Sci Math Phys 14: 233–560
St Venant AJCB de (1864) Etablissement elementaire des formules et equations generales de la theorie de l'elasticite des corps solides. Appendix in: Resumé des leçons donnees a l'école de ponts et chaussées sur l'application de la méchanique , première partie, première section, de la resistance des corps solides par CLMH Navier 3rd edn, Paris
St Venant AJCB de (1870) Memoire sur l'etablissement des equations differentielles des mouvements interieures operes dans les corps solides ductiles au dela des limites ou l'elasticite pourrait les ramener a leur premiere etat. C R Acad Sci Paris 70: 473–480
Sampson E Jr (1950) Stabilization of granular materials with controlled precipitation of ground water type solutions. MSc thesis, Princeton University
Satake M (1981) On distortion of tensor and yield criteria of granular materials. Intl J Engng Sci 19(12): 1643–1650
Scheidegger AE (1961) General theory of dispersion in porous media. J Geophys Res 66: 3273–3278
Scheidegger AE (1963) The physics of flow through porous media. University of Toronto Press, 313 pp
Schiffman RL (1980) Finite and infinitesimal strain consolidation. ASCE J Geotech Div 106 (GT2): 203–207
Schiffman RL, Ladd CC, Chen WF (1964) The secondary consolidation of clay. IUTAM symposium on rheology and soil mechanics, Grenoble, pp 273–304
Schmertmann JH (1983) A simple question about consolidation. ASCE J Geotech Div 109 (GT1): 119–122
Schmertmann JH, Osterberg JO (1960) An experimental study of the development of cohesion and friction with axial strain in saturated cohesive soils. Proceedings of the ASCE conference on shear strength of cohesive soils, Boulder, pp 643–694
Schmid WE (1957) The permeability of soils and the concept of stationary boundary layer. Proc ASTM 57: 1195–1218
Schmid WE (1958) Water movement in soils under pressure potentials. Proceedings of the international symposium on water and its conduction. HRB Spec Rep 40: 164–177
Schmid WE (1962) New concepts of shearing strength for saturated clay soils. I–II. Sols–Soils 1: 31–42; 2: 19–26
Schmid WE, Klausner Y, Whitmore CF (1960) Rheological shear and consolidation behavior of soils. Prog Rep ONR Dept US Navy, Princeton University
Schofield RK (1935) The interpretation of the diffuse double layer surrounding soil particles. In: Transactions of the 3rd international congress on soil science, Oxford, vol 1, pp 30–33
Schofield RK (1947) Nature 160: 408–410
Schofield AN, Wroth CP (1968) Critical state soil mechanics. McGraw-Hill, 310 pp
Schouten JA (1965) Tensor calculus for physicists. Transl from Polish (1951)
Schuurman E (1966) The compressibility of an air/water mixture and a theoretical relation between the air and water pressure. Geotechnique 16(4): 269–281
Schwartz AE (1964) Failure of rock in the triaxial shear test. In: 6th symposium on rock mechanics, Rolla, MI
Scott RF (1965) Principles of soil mechanics. Addison-Wesley, 550 pp
Scott RF, Ko HY (1969) Stress-deformation and strength characteristics. In: Proceedings of the 7th INCOSOMEFE, Mexico. State of Art, vol 3, pp 1–47
Screwala FN (1988) Discussion on 'Plasticity and constitutive relations in soil mechanics by R.F. Scott from vol 111 GT5'. ASCE J Geotech Div 114 (GT5): 641
Seed HB (1987) Design problems in soil liquefaction. ASCE J Geotech Div 113 (GT8): 827–845
Seed HB, Chan CK (1959) Structure and strength characteristics of compacted clays. Proc ASCE 85: 87–128
Seed HB, Arango I, Chan CK, Gomez-Masso A, Ascoli RG (1981) Earthquake induced liquefaction near Lake Amatitlan – Guatemala. ASCE J Geotech Div 107 (GT4): 501–518
Seed HB, Idris JM, Arango I (1983) Evaluation of liquefaction potential using field performance data. ASCE J Geotech Div 109 (GT3): 458–482
Seifert R (1933) Untersuchungsmethoden um festzustellen ob sich ein gegebens Baumaterial fur den Bau eines Erddammes eignet. In: 1st congress on Grand Barrages, Stockholm, no 17
Selig HB, Chang CS (1981) Soil failure modes in undrained cycle loading. ASCE J Geotech Div 107 (GT5)
Serata S (1961) Transition from elastic to plastic states of rocks under triaxial compression. In: 4th symposium on rock mechanics, Penn State University, pp 73–82
Serdengecti S, Boozer GD (1961) The effect of strain rate and temperature on the behavior of rocks subjected to triaxial compression. In: 4th symposium on rock mechanics, Penn State

University, pp 83–98
Shainberg I (1968) Electrochemical properties of Na and Ca montmorillonite suspensions. In: Transactions of the 9th international congress on soil science, vol 1, p 577
Shainberg I (1970) Cation and anion exchange reactions. Internal report of the Negev Institute Arid Zone Research, Beer-Sheva and the Volcani Institute of agricultural research, Beit-Dagan, 60 pp
Shainberg I, Kemper WK (1966a) Electrostatic forces between clay and cations as calculated and inferred from electrical conductivity. Clays and Clay Minerals 26: 117
Shainberg I, Kemper WK (1966b) Conductance of adsorbed alkali cations in aqueous and alcoholic bentonite pastes. Proc Soil Sci Soc Am 30: 700–706
Shainberg I, Kemper WK (1966c) Hydration status of adsorbed cations. Proc Soil Sci Soc Am 30(6): 707–708
Shainberg I, Otoh H (1968) Size and shape of montmorillonite particle saturated with Na/Ca ions, inferred from viscosity and optical density measurements. Israel J Chem 6 (Mokady memorial issue): 251–259
Shainberg I, Bresler E, Klausner Y (1971) Studies of Na/Ca montmorillonite systems. 1. The swelling pressure. Soil Sci 111(4): 214–219
Shainberg I, Alperovitch NI, Keren R (1987) Charge density and Na–K–Ca exchange on smectites. Clays and Clay Minerals 35(1): 68–73
Shen ZJ (1985) A viscoelastic model of liquefaction of sands. In: Proceedings of the 11th INCOSOMEFE, vol 2
Shockley WG, Ahlvin RG (1960) Non uniform conditions in triaxial test specimens. In: ASCE research conference on shear strength of cohesive soils, Boulder, pp 341–359
Signorini A (1930) Sulle deformazioni finite dei sistemi continui. Rend Lincei (6) 12: 80–89
Sills GC (1987) The behavior of soil containing gas bubbles. Oxford University report no SM077/87
Sills GC, Austin G (1982) Pore pressure measurement in a seabed containing gas bubbles. Oxford University report no. SM028/SERC/82
Sills GC, Nageswaran S (1984) Compressibility of gassy soil. In: Oceanology international exhibition and conference, Brighton
Skempton AW (1961) President's opening address. In: Proceedings of the 5th INCOSOMEFE, Dunod Paris, vol 3, pp 39–42
Skempton AW, Bishop AW (1954) Soils. In: Reiner M (ed) Building materials, their elasticity and inelasticity, vol 3. North Holland, pp 417–482
Slichter CS (1897) Theoretical investigation of the motion of groundwater. US geological survey 19 annual report, part 2, pp 295–384
Smiles DE, Poulos HG (1969) The one-dimensional consolidation of columns of soil of finite length. Austral J Soil Res 7: 285–289
Smiles DE, Rosenthal MJ (1968) The movement of water in swelling materials. Austral J Soil Res 6: 237–248
Smith RM, Browning DR (1942) Persistent water-unsaturation of natural soils in relation to various soil and plant factors. Proc Soil Sci Soc Am 7: 114–119
Sokolnikov IS (1951) Tensor analysis, theory and applications. John Wiley and Sons
Sokolnikov IS (1956) Mathematical theory of elasticity. McGraw-Hill, 476 pp
Spain H (1953) Tensor calculus. Oliver and Boyd, 125 pp
Sparks ADW (1967) A theory of consolidation for partially saturated soils. CE S Africa 9 (July): 163–185
Spielrein J (1916) Lehrbuch der Vectorrechnung nach dem Bedurfwissen in der Technishe Mech und Electricitatslehre. Stuttgart
Standing MB, Katz DL (1942) Density of natural gases. Am Inst Mining Met Engng Tech Publ no 1323, 10 pp
Stefan J (1871) Uber das Gleichgewicht und die Bewegung, Insbesondere die Diffusion von Gasmengen. Sitz Akad Wiss Wien. 63/2: 63–124
Stern O (1924) Zur Theorie der Electrolytischen Doppelschicht. Z Electrochem 30: 508
Sternstein SS, Lederle GM (1968) Minimum free energy deformations. Rubber elasticity theory. Proceedings of the international congress on rheology, Kyoto, (Oral)
Stokes GG (1829) Memoirs of the general equation of equilibrium and motion of elastic solids J l'Ecole Politech. 13: 1
Stokes GG (1845) On the theories of the thermal friction of fluids in motion. Camb Phil Soc Trans 8: 287
Stokes GG (1854) Smith's prize examination paper, no 5
Stroud MA (1971) The behavior of sand at low stress levels in the simple shear apparatus. PhD

thesis, University of Cambridge
Suklje L (1967) A three dimensional consolidation analysis applied to non-linear viscous and anisotropic deformability of soils. Acta Geotech 10–11
Sullivan RR, Hertel KL (1942) The permeability methods for determining specific surface of fibers and powders. Adv Colloid Sci 1: 38–80
Swartzendruber D (1962) Non-Darcy behavior in liquid saturated porous media. J Geophys Res 67: 5205–5213
Synge JL, Schild A (1949) Tensor calculus. University of Toronto Press.
Talsma T (1970) Hysteresis in two sands and the independent domain model. Water Resources Res 6: 964–970
Tan Tjong Kie (1954) Investigations of the rheological properties of clays. PhD thesis, Technical University of Delft
Tan Tjong Kie (1957a) Structure mechanics of clays. Acad Sinica, Harbin, pp 58–75
Tan Tjong Kie (1957b) Secondary time effects of consolidation of clays. Acad Sinica, Harbin, pp 1–17
Tan Tjong Kie (1957c) Three-dimensional theory on the consolidation and flow of clay-layers. Sci Sinica 6(1): 203–215
Tatsuoka F, Maramatsu M, Sasaki T (1982) Cyclic undrained stress–strain behavior of dense sands by torsional simple shear test. Soils Foundat 20(2): 55–70
Tatsuoka F, Pradhan TBS, Yoshi-ie H (1989) A cyclic undrained simple shear testing method for soils. The Am Soc for testing and materials. Geotech Testing J 12(4): 269–280
Taylor DW (1944) 10th progress report to the US Engng Dept MIT, Soil Mechanics Lab
Taylor DW (1948) Soil mechanics. John Wiley and Sons, 700 pp
Taylor DW, Merchant W (1940) A theory of clay consolidation accounting for secondary compressions. J Maths Phys 19(3): 167–185
Taylor GI, Quinney H (1931) The plastic distortion of metals. Trans R Soc Lond A 230: 323–363
Terzaghi K (1923) Die Berechnung der Durchlassigkeitsziffer des Tones ausdem Verlauf der hydrodynamishen Spannungserscheinungem. Akad. Wissen. in Wien. Sitzungsberichte. Math. Naturwiss. Klasse Part IIa, 132 (3/4): 125–138
Terzaghi K (1925) Erdbaumechanik auf Bodenphysikalischer Grundlage. Fr Deuticke, Leipzig Wien
Terzaghi K (1934) Die Ursachen der Schiefstellung des Turmes von Pisa. Z Gesamte Bauwesen, 1/2: 1–4
Terzaghi K (1943) Theoretical soil mechanics. John Wiley and Sons, 510 pp
Terzaghi K (1944) End and means in soil mechanics. J Engng Canada 27(12): 608–615
Terzaghi K (1948) Opening address. In Proceedings of the 2nd INCOSOMEFE, Rotterdam, vol 6, pp 11–16
Terzaghi K (1953) Fifty years of subsurface exploration. In: Proceedings of the 3rd INCOSOMEFE, Zurich, vol 3, pp 227–237
Terzaghi K, Peck RB (1948) Soil mechanics in engineering practice. John Wiley and Sons, 566 pp
Thomas C (1889) The circular, square and octagonal earthworks of Ohio. Smithsonian Institution, Washington DC
Thomas SD (1987) D Phil thesis, Oxford University
Thomas TY (1957) Extended compatibility conditions for the study of surfaces of discontinuity in continuum mechanics. J Maths Mech 6: 311–322; 907–908
Thomson W (Lord Kelvin) (1858) Note on hydrodynamics. 2. On the equation of the bounding surface. Camb Publ Math J 3: 89–93
Thomson W (Lord Kelvin), Tait PG (1867) Treatise on natural philosophy, Cambridge University Press. Republished: Principles of mechanics and dynamics. Dover, vol 1, 508 pp; vol 2, 527 pp
Tiedemann B (1932) Die Bedeutung des Bodens im Bauwesen. Springer (Handbuch der Bodenlehre, vol 10)
Timoshenko SP (1953) History of strength of materials. McGraw-Hill
Topp CG (1969) Soil water hysteresis measured in a sandy loam compared with the hysteretic domain model. Proc Soil Sci Soc Am 33(6): 645–651
Topp CG (1971) Soil water hysteresis: the domain model theory extended to pore interaction conditions. Proc Soil Sci Soc Am 35(2): 219–225
Toupin RA (1956) The elastic dielectric. J. Rat Mech Anal 5: 849–915
Tresca H (1868) Memoire sur l'ecoulement des corps solides. Compte Rendu, Savants Etrangers, Memoires presente par Divers Savants, no 18, Paris (1886), pp 733–799
Tresca H (1869) Memoire sur le poinconnage des metaux et des matieres plastiques. C R Acad Sci Paris 70: 617–838
Trouton F (1905) On the viscosity of pitchlike substances. Proc Phys Soc 19: 47

Trouton F (1906) On the coefficient of viscous traction. Proc R Soc Lond A77: 326
Truesdell C (1952) The mechanical foundation of elasticity and fluid dynamics. J Rat Mech Anal 1: 125–300
Truesdell C (1954) The kinematics of vorticity. Bloomington Indiana University science series no 10
Truesdell C (1957) Sulle basi della termomeccanica. Reud Lincei 22: 33–38, 158–166
Truesdell C (1966) Six lectures on modern natural philosophy. Springer, 117 pp.
Truesdell C (1969) Rational thermodynamics. McGraw-Hill, 208 pp.
Truesdell C, Toupin R (1960) The classical field theories. Springer pp 226–793 (Handbuch der Physik, vol 3/1)
Tschebotarioff GP (1948) Report to the Bethlehem Steel Co (unpublished), Sparrows Point
Tschebotarioff GP (1952) Soil mechanics, foundation and earth structures. McGraw-Hill, 655 pp
Unwin WG (1888) The testing of materials of construction. London
US Engineers Corps (1938) Compaction tests and critical density investigations of cohesionless materials for Franklin Falls Dam. Boston District
Vaid YE, Chern JCC, Tumi H (1985) Confining pressure, grain angularity and liquefaction. ASCE J Geotech Div 111 (GT10): 1229–1231
Vaid YE, Finn WDL (1979) Static shear and liquefaction potential. ASCE J Geotech Div 105 (GT10): 1233–1246
Valanis KC, Read HE (1982) A new endochronic plasticity law for soils. In: Pande G, Zienkiewicz OC (eds) International symposium on soils under cyclic and transient loading. John Wiley and Sons, pp 375–417
Van der Waals, JD (1873) On the continuity of the liquid and gaseous states. Sighthoff, Leyden, trans from Dutch by Threlfall and Adair 1890, Physical Memoirs, The Physical Society, London
Van Olphen H (1957) Forces between suspended bentonite, clays and clay minerals. In: Proceedings of the 6th national conference on clay and clay minerals, Part 2, pp 204–224
Van Olphen H (1959) Clay and clay minerals, vol 6, pp 196–206
Van Olphen H (1963) An introduction of clay colloid chemistry. Interscience (2nd ed 1977), 301 pp
Verril AR (1942) Old civilisations of the new world. The New-York Home Library
Verwey EJ, Overbeek J Th G (1948) Theory of stability of lyophobic colloids. Elsevier
Viswanath DS, Su GJ (1965) Generalized thermodynamic properties of real gases. I, II. AIChE J 2: 202–206
Voigt W (1889) Uber die Innere Reibung der Festen Korper, Insbesondere der Krystalle. Abh Ges Wissen Gottingen 36: 743–759
Warburg EG (1879) Uber die Torsion. Freiburg, 499 pp
Warkentin BP (1961) Interpretation of the upper plastic limit of clays. Nature 190: 287–288
Warkentin BP, Bolt GH, Miller RD (1957) Swelling pressure of montmorillonite. Proc Soil Sci Soc Am 21: 495–497
Warkentin BP, Schofield RK (1962) Swelling pressure of Na-montmorillonite in NaCl solutions. J Soil Sci 13: 98–108
Warlam AA (1960) Recent progress in triaxial apparatus design. In: Proceedings of the ASCE research conference on shear strength of cohesive soils, Boulder pp 859–876
Watson WJ, Watson SR (1937) Bridges in history and legend. JW Jansen, Cleveland
Weidelich WC (1957) Physical-chemical factors influencing the consolidation of soils. MSc thesis, Princeton University
Weissenberg K (1947) A continuum theory of rheological phenomena. Nature 159: 310–311
Weissenberg K (1949) Specification of a rheological phenomena by means of a rheogoniometer. In: Proceedings of the international congress on rheology, vol. 1. North-Holland Publishing Company, Amsterdam, pp 29–46
Wheeler SJ (1986) The stress–strain behaviour of soils containing gas bubbles. DPhil thesis, Oxford University
Whelan III T, Lester GD (1980) Excess gas bubble pressures in modern Mississippi delta sediments. The Delta Project, vol 4, pp 1–9
White L (1940) Report on the collapse of silos
White WA (1949) Atterberg plastic limits of clay minerals. Am Mineral 34: 508–512
Winterkorn HF (1955) The science of soil stabilization. Symposium on soil and soil-aggregate stabilization. HRB Bull 108: 1–24
Winterkorn HF (1958) Mass transport phenomena in moist porous systems, as viewed from the thermodynamics of irreversible processes. Proceedings of international symposium on water and its conduction in soils. HRB Spec Rep Bull 40: 324–338
Winterkorn HF (1965) The bearing of soil–water interaction on water conduction under various

energy potentials. RILEM symposium on transfer of water in porous media. Paris Bull 27: 87–96
Winterkorn HF, Moorman RB (1941) A study of changes in physical properties of putnam soil induced by ionic substitution. Proc HRB 21: 415–434
Witt KJ, Brauns J (1983) Permeability anisotropy due to particle shape. ASCE J Geotech Div 109 (GT9): 1181–1187
Wolski W, Baranski T, Garbulewski K, Lechovics Z, Szymanski A (1985) Testing of anisotropic consolidation in organic soils. In: Proceedings of 11th INCOSOMEFE, vol 2, pp 699–702
Wong RKS, Arthur JRF (1985) Induced and inherent anisotropy in sand. Geotechnique 35(4): 471–481
Wood DM (1979) The behaviour of partly saturated soils: A review. University of Cambridge Dept Engng Rep CUED/D-Soils/TR 69
Wood DM (1984). On stress parameters. Geotechnique 34 (2): 282–287
Wooltorton FLD (1950) Movement in the desiccated alkaline soils of Burma. Proc ASCE 116
Wooltorton FLD (1954) The scientific basis for road design. Edward Arnold, 364 pp
Wroth CP, Houlsby GT (1985) Soil mechanics – property characterisation and analysis procedures. Theme lecture no 1 in: 11th INCOSOMEFE, San Francisco
Wu TH, El Rifai ANAA, Hsu JR, (1978) Creep deformation of clays. ASCE J Geotech Div 104 (GT1): 61–76
Wykoff RD, Botset HG, Muskat M, Reed DW (1933) The measurement of the permeability of porous media for homogeneous fluids. Rev Sci Instrum 4: 394–405
Wykoff RD, Botset HG (1936) The flow of gas–liquid mixture through unconsolidated sands. Physics 7: 325–345
Yamada K (1968) On the one dimensional consolidation by the three dimensional dehydration on the consolidation effect due to sand pile drainage works. Trans Japan Soc Civil Engng 149: 46–63
Yong RN, Warkentin BP (1960) Physico-chemical analysis of high swelling clays, subject to loading. In: Proceedings of the 1st Pan American conference SOMEFE, Mexico, pp 865–888
Yong RN, Taylor LO, Warkentin BP (1963) Swelling pressure of montmorillonite at depressed temperatures. In: Proceedings of the national conference on clay and clay minerals, vol 11 Pergamon Press, pp 268–281
Yong RN, Warkentin BP (1966) Introduction to soil behavior. Macmillan, 439 pp
Yoshimi Y, Osterberg JO (1963) Compression of partly saturated cohesive soils. ASCE J Soil Mech Div 89 (SM4): 1–24
Yoshimura Y (1953) On the natural shearing strain. In: Proceedings of the 2nd Japan mathematical congress on applied mechanics, pp 1–4
Zaremba S (1937) Sur une conception nouvelle des forces interieures dans un fluide en mouvement. Mem Sci Math no 82
Zaslavsky D, Mokady R, Ravina I (1967) Development of methods for measuring partial molar free energy of soil water. Annual Rep no 1, Technion
Zeevaert L (1953) Pore pressure measurements to investigate the main source of surface subsidence in Mexico City. In: Proceedings of the 3rd INCOSOMEFE, Zurich, vol 2, pp 299–304
Zeevaert L (1957) Foundation design and behavior of tower Latino Americana in Mexico City. Geotechnique 7/3: 115–133
Zur A (1969) A study of the collapse phenomenon in partially and fully saturated loess. DSc thesis, Technion, Haifa

Subject Index